HLA in Health and Disease

HLA in Health and Disease

Second edition

Edited by

Robert Lechler
Department of Immunology
Imperial College School of Medicine
Hammersmith Hospital
London, UK

and

Anthony Warrens
Departments of Immunology and Renal Medicine
Imperial College School of Medicine
Hammersmith Hospital
London, UK

ACADEMIC PRESS
A Harcourt Science and Technology Company

San Diego San Francisco New York Boston London Sydney Tokyo

The book is printed on acid-free paper ∞

Copyright 2000 Academic Press

All rights reserved.
No part of this publication may be reproduced or transmitted in any form or by any means electronic or mechanical, including photocopy, recording, or any information storage and retrieval system, without permission in writing from the publisher.

Academic Press Limited
A Harcourt Science and Technology Company
Harcourt Place, 32 Jamestown Road
London NW1 7BY
http://www.academicpress.com

Academic Press
A Harcourt Science and Technology Company
525 B Street, Suite 1900, San Diego, California 92101-4495, USA
http://www.academicpress.com

Library of Congress Catalog Card Number: 99-68335

A catalogue record for the book is available from the British Library

ISBN 0-12-440315-8

Produced and typeset by Gray Publishing, Tunbridge Wells, Kent
Printed in Great Britain, by the Bath Press, Bath, Somerset
00 01 02 03 04 BP 9 8 7 6 5 4 3 2 1

Contents

Foreword *Andrew J. McMichael* vii
Preface to the First Edition viii
Preface to the Second Edition x
Contributors xii
Abbreviations xv

SECTION I INTRODUCTION TO THE HLA SYSTEM

Chapter 1   The History of the Discovery of HLA   3
J.J. van Rood

Chapter 2   The Structure of the Major Histocompatibility Complex and its Molecular Interactions   23
Efrem Eren and Paul Travers

Chapter 3   Molecular Genetics of the Human Major Histocompatibility Complex   35
Caroline M. Milner, R. Duncan Campbell and John Trowsdale

Chapter 4   The Biology of Major Histocompatibility Complex Molecules – I   51
Richard J. Baker and Anthony N. Warrens

Chapter 5   The Biology of Major Histocompatibility Complex Molecules – II: Antigen Processing and Presentation   73
Chen Au Peh, Anthony W. Purcell and James McCluskey

Chapter 6   The Analysis of Genetic Susceptibility   107
Timothy J. Vyse and Bernard J. Morley

Appendix: Statistical Considerations in Analysing HLA and Disease Associations   129
Anthony N. Warrens

Chapter 7   Mechanisms of HLA and Disease Associations   139
Anthony N. Warrens and Robert I. Lechler

Chapter 8   HLA Nomenclature: The Name of the Rose   147
Julia G. Bodmer, Walter F. Bodmer and Steven G.E. Marsh

Chapter 9   The Evolution of the Major Histocompatibility Complex: Insights From Phylogeny   163
Ronald E. Bontrop

Chapter 10 The Generation and Selection of Major Histocompatibility Complex Polymorphisms   171
Jim Kaufman

SECTION II HLA AND DISEASE ASSOCIATIONS

Chapter 11	HLA and Rheumatoid Arthritis *Gerald T. Nepom*	181
Chapter 12	HLA and the Spondylarthropathies *Paul Bowness*	187
Chapter 13	HLA and Renal Disease *Richard G. Phelps and Andrew J. Rees*	195
Chapter 14	HLA and Neurological Diseases *Jan Hillert and Anna Fogdell-Hahn*	219
Chapter 15	HLA and Type I Diabetes *Gerald T. Nepom*	231
Chapter 16	HLA and Endocrine Disease *Anthony P. Weetman*	239
Chapter 17	HLA and Gastrointestinal Diseases *Ludvig M. Sollid, A. Spurkland and Erik Thorsby*	249
Chapter 18	HLA and Respiratory Disease *Cesare Saltini, Luca Richeldi and Massimo Amicosante*	263
Chapter 19	HLA and Eye Disease *Jose S. Pulido and J. Wayne Streilein*	279
Chapter 20	HLA and Infectious Diseases *Paul A. Glynne and Nicholas M. Price*	299
Chapter 21	HLA and Systemic Vasculitides, Systemic Lupus Erythematosus and Sjögren's Syndrome *Matthew C. Pickering, Mohini Perraudeau and Mark J. Walport*	327
Chapter 22	HLA and Dermatological Disease *Adam Friedmann*	365
Chapter 23	HLA and Transplantation – I: Allorecognition of HLA Molecules in Transplantation *Richard J. Baker*	387
Chapter 24	HLA and Transplantation – II: The role of HLA Matching in Clinical Transplantation *Gerhard Opelz*	405
Chapter 25	HLA and Psychiatric Disease *Padraig Wright, Vishwajit Nimgaonkar, Rohan Ganguli and Robin M. Murray*	419

SECTION III DEFINITION OF HLA POLYMORPHISM

Chapter 26	Serological Methods in HLA Typing *Philip A. Dyer, Susan Martin and Rachel E. Stanford*	429
Chapter 27	Cellular Methods in Testing Histocompatibility *Philip D. Mason and Robert I. Lechler*	441
Chapter 28	Polymerase Chain Reaction-based Methods of HLA Typing *Henry A. Erlich*	451

Index 463

A colour plate section appears between pages 14 and 15.

Foreword

The discovery of the HLA system in the 1960s was soon followed by the unravelling of some of its complexity and polymorphism. This early endeavour was largely driven by the then new surgical discipline of transplantation. At the same time, associations between HLA types and disease susceptibility or resistance were sought and found, the most striking being that between HLA B27 and ankylosing spondylitis. Once the HLA class II system was defined, a remarkable series of associations between particular alleles and the autoimmune diseases were shown. Whether these were primary associations or whether they were the result of linkage disequilibrium with unknown neighbouring genes was for a long time uncertain. Now that the full map and sequence of the HLA complex are known, it is clear that the HLA genes are the culprits. Indeed, now that there has been extensive mapping of DNA polymorphic sites throughout the genome, it is clear that the HLA genes show the strongest associations with the autoimmune diseases and contribute directly to susceptibility. However, the exact nature of these associations has defied detailed explanation, despite the explosive growth in our understanding of the functions of the major histocompatibility complex and its protein products. That the role of HLA class I and class II molecules in regulating immune responses is very complex is understood, and this creates plenty of scope for particular genes to influence the outcome of infections and the state of tolerance to autoantigens. The latter brings the subject back to transplantation where there is an ever-increasing demand from patients and, despite ever better drug-induced immunosuppression, a real need to improve outcomes by inducing states of tolerance.

This book assembles an impressive list of contributors who give a timely review of the HLA system, in relation to disease. The major autoimmune diseases, nearly all of which have strong HLA associations, are discussed. Transplantation and the influence of HLA matching on engraftment are dissected. The relationship between HLA polymorphism and resistance to infectious disease is explored. Finally, there is an update on the advances in HLA typing where advances in serology, cellular techniques and recombinant DNA methods have improved the characterization of the polymorphism and the practicalities of typing. This second edition has been extensively rewritten and updated with several new authors. This is a major addition to the field that will serve as a benchmark for further research in this area for many years to come.

Andrew J. McMichael
Institute of Molecular Medicine
University of Oxford
John Radcliffe Hospital
Oxford

Preface to the First Edition

The major histocompatibility complex (MHC), referred to in man as the HLA (human leucocyte antigen) region, has been one of the most intensively studied genetic regions during the past two decades. As a consequence of the application of modern cloning techniques, the HLA region is on the brink of full characterization and, together with recent advances in immunology, this creates the possibility of making significant progress in understanding the mechanisms responsible for the well-known associations of certain HLA types with human disease.

The MHC has been explored in several species, most notably in man and mouse, and is highly conserved. One of the conclusions that has emerged from recent cloning efforts is that the MHC is probably the area of the mammalian genome that is most densely populated with functional genes. Current estimates are that the region contains in the order of 200 genes. These include, of course, the genes encoding the MHC class I and class II molecules. The MHC genes are the most polymorphic of all mammalian genes. The functional significance of this sequence variation, and its possible relationship to disease susceptibility is one of the topics discussed in this book. In addition, there are a large number of other genes, some with known functions and some whose functions are as yet unidentified. Some of the well-characterized genes encode proteins with obvious relevance to the immune system. These include three complement genes, the tumour necrosis factor genes, and a cluster of genes that appear to be involved in the digestion of antigens inside antigen-presenting cells and the loading of antigen fragments on to MHC molecules for subsequent antigen presentation. Less is known of the polymorphism of these recently discovered genes but it may prove to be the case that sequence variation at one or more of these loci predisposes to autoimmune disease.

The first reports of an association between an HLA antigen and disease were those of Hodgkin's lymphoma with a cross-reactive group of HLA-B antigens, B5, B15, B18 and B35 (Amiel, 1967) and of acute lymphoblastic leukaemia with HLA-A2 (Walford et al., 1970). These particular associations have not proven to be consistently reproducible, but a short time later the remarkable association between the rheumatological disorder, ankylosing spondylitis, and HLA-B27 was described (Brewerton et al., 1973; Schlosstein et al., 1973). These observations provided the impetus for further studies of HLA and disease associations, and a large number of diseases have been shown to be associated with the possession of particular HLA types. The aim of this book is to provide a framework within which to interpret HLA and disease associations, and to design further studies.

The book is divided into two sections, A and B. Section A presents the current state of knowledge of the genetics and the structure of the HLA antigens. These topics are covered in Chapters 1 and 2. In Chapters 3 and 4 the biology of MHC molecules is discussed, with particular emphasis on the ways in which MHC antigens contribute to immune recognition and immune regulation. Section B begins with a consideration of the mechanisms underlying HLA and disease associations, in the light of the earlier discussions of the structure and functions of the MHC molecules (Chapter 5). Chapter 6 describes the principles of designing and interpreting an HLA and disease study, followed in Chapter 7 by a detailed description of some of the best characterized associations of HLA antigens with diseases that are thought to have an autoimmune pathogenesis. Chapter 8 focuses on a series of connective tissue diseases, some of which are more closely linked with the HLA class III region, an area that includes the genes for the complement proteins, C2, C4 and Factor B. Finally, in Chapter 9, the very exciting discoveries of the many additional genes within the MHC, some with important immune functions, are presented. Their possible relevance to disease pathogenesis is discussed.

The value of defining HLA and disease associations can be thought of in two ways. The first is that the definition of an association may help to define an 'at risk' population. This can be translated into practical benefit if a means of prevention becomes available. The second benefit of defining genetic susceptibility factors in any disease is that this can shed light on the mechanisms underlying the disease. This, in turn, can pave the way for more precisely targeted strategies for disease prevention or treatment.

There has never been a more exciting time for studies of this kind. The advent of new techniques for the precise definition of HLA polymorphisms, and the detailed molecular definition of how MHC molecules present antigens to T lymphocytes, and of how sequence variation influences this process, provides a conceptual context within which HLA and disease associations can be understood. As the autoantigens that are aberrantly recognized in autoimmune disease states are characterized, existing hypotheses can be tested directly. Furthermore, the discovery of the many additional genes within the HLA region creates new possibilities for identifying genetic susceptibility factors.

REFERENCES

Amiel, J.L. (1967). *Histocompatibility Testing 1967* (eds E.S. Curtoni, P.L. Mattiuz and R.M. Tosi), pp. 79–81, Munksgaard, Copenhagen.

Brewerton, D.A., Caffrey, M., Hart, F.D. et al. (1973). Ankylosing spondylitis and HL-A27. *Lancet* **i**: 904–907.

Schlosstein, L., Terasaki, P.I., Bluestone, J. et al. (1973). High association of the HL-A antigen, w27, with ankylosing spondylitis. *N. Engl. J. Med.* **288**: 704–706.

Walford, R.L., Finkelstein, S., Neerhout, R. et al. (1970). Acute childhood leukemia in relation to the HL-A human transplantation genes. *Nature* **225**: 461–462.

Preface to the Second Edition

In 1994, Academic Press published *HLA and Disease* edited by Professor Robert Lechler. The book reviewed the genetics, structure and function of human leucoctye antigen molecule and considered aspects of disease association.

All concerned were gratified with the positive response to this book and it seemed appropriate to consider a second edition. Anthony Warrens joined Robert Lechler as editor and decided that, given the warm reception of the previous edition, it would be appropriate to expand the scope of the book. In 1985, Tiwari and Terasaki published a comprehensive review of HLA and disease associations, which was for many years the authoritative word on the subject. However, our understanding of the extent of the polymorphism of the HLA system, resulting largely from the application of molecular biological techniques, as well as a significant increase in our understanding of the physiology of these molecules, means that much more can now be said about this subject. We therefore felt that it was timely to address this issue and this forms the largest section (Section II) of this book.

Finally, the methodology of tissue typing has become much more complex. No proper understanding of any clinical studies can be made without an appreciation of these techniques. For this reason we decided to add a third section to this book summarizing the history and the current practice of tissue typing.

All these "bright ideas" have their toll and we now find ourselves presenting a volume some four times the size of its immediate ancestor. Although substantially larger than the previous volume, we trust our readers feel the expansion is justified by the increased information provided.

We would have been unable to produce this book without the support and cajoling of Drs Lilian Leung and Tessa Picknett. In addition, the professionalism of Lesley Gray and her colleagues has been a major factor in the high technical quality of the finished product. To all these people and their associates, we offer our grateful thanks.

NOMENCLATURE USED IN THIS BOOK

In Chapter 1, van Rood, and in Chapter 8, Bodmer, Bodmer and Marsh describe how the nomenclature used to designate the genes and products of the HLA region has evolved with our increasing understanding of its complexity. Obviously, important data presented in this book were acquired at various stages during this process. For ease of reading and understanding, we have attempted, where possible, to use the same, most current nomenclature conventions, rather than those that were in use at the time the data were acquired. Sometimes, this has not been possible because of inadequate definition at the time of the study, which may lead to some inconsistencies within the book. We have tried to keep these to a minimum.

A second major problem arose from the common convention of presenting gene designation in italics (e.g. *HLA-DRB1*0101*) but proteins in

Roman style (e.g. DRβ1*0101). Unfortunately, this led to areas of ambiguity or complexity, for example reporting studies that used a mixture of serological and PCR-based techniques, or presenting haplotype data (clearly a genetic concept) that had been inferred from serologically defined data. Ultimately, we came to the conclusion that this convention served to confuse and distract more than to enlighten. For this reason, we have opted to follow the alternative convention and have not italicized genetic material in this book.

Robert Lechler
Anthony Warrens

Contributors

Massimo Amicosante
Department of Biology
University of Tor Vergata
Roma, Italy

Richard J. Baker
Department of Immunology
Imperial College School of Medicine
Hammersmith Hospital
London, UK

Julia G. Bodmer
ICRF Cancer and Immunogenetics Laboratory
Institute of Molecular Medicine
John Radcliffe Hospital
Headington, Oxford, UK

Walter F. Bodmer
ICRF Cancer and Immunogenetics Laboratory
Institute of Molecular Medicine
John Radcliffe Hospital
Headington, Oxford, UK

Ronald E. Bontrop
Department of Immunobiology
BPRC
Rijswijk, The Netherlands

Paul Bowness
Molecular Immunology Group
Institute of Molecular Medicine
John Radcliffe Hospital
University of Oxford
Headington, Oxford, UK

R. Duncan Campbell
UK HGMP Resource Centre
Hinxton, Cambridge, UK

Philip A. Dyer
Transplantation Laboratory
Manchester Royal Infirmary
Manchester, UK

Efrem Eren
Department of Immunology
Imperial College School of Medicine
Hammersmith Hospital
London, UK

Henry A. Erlich
Department of Human Genetics
Roche Molecular Systems Inc.
Alameda, California, USA

Anna Fogdell-Hahn
Karolinska Institute
Department of Biosciences at Novum
Huddinge, Sweden

Adam Friedmann
Department of Genetics
The Alex Silberman Institute for Life Sciences
The Hebrew University of Jerusalem
and the Unit for Development of Molecular
Biology and Genetic Engineering
Hadassah University Hospital
Jerusalem, Israel

Rohan Ganguli
Department of Psychiatry
University of Pittsburgh School of Medicine
Pittsburgh, Pennsylvania, USA

Paul A. Glynne
Department of Infectious Diseases
Imperial College School of Medicine
Hammersmith Hospital
London, UK

Jan Hillert
Karolinska Institute
Department of Neurology
Huddinge, Sweden

James Kaufman
Compton Laboratory
Institute for Animal Health
Compton, Newbury, UK

Robert I. Lechler
Department of Immunology
Imperial College School of Medicine
Hammersmith Hospital
London, UK

Steven G.E. Marsh
Anthony Nolan Research Institute
Royal Free Hospital
London, UK

Susan Martin
Transplantation Laboratory
Manchester Royal Infirmary
Manchester, UK

Philip Mason
Oxford Renal Unit
Churchill Hospital
Headington, Oxford, UK

Jim McCluskey
Department of Microbiology and Immunology
University of Melbourne
Parkville, Melbourne, Australia

Caroline M. Milner
MRC Immunochemistry Unit
Department of Biochemistry
Oxford, UK

Bernard J. Morley
Rheumatology Section
Division of Medicine
Imperial College School of Medicine
Hammersmith Hospital
London, UK

Robin M. Murray
Department of Psychological Medicine
Institute of Psychiatry
Denmark Hill, London, UK

Gerald T. Nepom
Virginia Mason Research Centre
Seattle, Washington, USA

Vishwajit Nimgaonkar
Department of Psychiatry
University of Pittsburgh School of Medicine
Pittsburgh, Pennsylvania, USA

Gerhard Opelz
Institute of Immunology
University of Heidelberg
Heidelberg, Germany

Chen Au Peh
Flinders Medical Centre and Flinders
University of South Australia
Bedford Park, South Australia, Australia

Mohini Perraudeau
Rheumatology Section
Imperial College School of Medicine
Hammersmith Hospital
London, UK

Richard G. Phelps
Department of Clinical and Surgical Sciences,
University of Edinburgh,
Edinburgh, UK

Matthew C. Pickering
Rheumatology Section
Imperial College School of Medicine
Hammersmith Hospital
London, UK

Nicholas M. Price
Department of Infectious Diseases
Imperial College School of Medicine
Hammersmith Hospital
London, UK

Jose S. Pulido
Department of Ophthalmology
UIC Eye Centre
Chicago, Illinois, USA

Anthony W. Purcell
Department of Microbiology and Immunology
University of Melbourne
Parkville, Melbourne, Australia

Andrew J. Rees
Department of Medicine and Therapeutics
University of Aberdeen
Institute of Medical Sciences Building
Foresterhill, Aberdeen, UK

Luca Richeldi
Department of Medical Sciences
Section of Respiratory Disease
Università degli Studi di Modena
Modena, Italy

Cesare Saltini
Department of Medical Sciences
Section of Respiratory Disease
Università degli Studi di Modena
Modena, Italy

Ludvig M. Sollid
The National Hospital, University of Oslo
Institute of Immunology
Oslo, Norway

A. Spurkland
The National Hospital, University of Oslo
Institute of Immunology
Oslo, Norway

Rachel E. Stanford
Tissue Typing Laboratory
Harefield Hospital,
Middlesex, UK

J. Wayne Streilein
Schepens Eye Research Institute
Boston, Massachusetts, USA

Erik Thorsby
The National Hospital, University of Oslo
Institute of Immunology
Oslo, Norway

Paul Travers
Anthony Nolan Bone Marrow Trust
Royal Free Hospital
London, UK

John Trowsdale
Division of Immunology
Department of Pathology
University of Cambridge
Cambridge, UK

Jon J. van Rood
Europdonar Foundation
Leiden University Hospital Medical Hospital
Leiden, The Netherlands

Timothy J. Vyse
Rheumatology Section
Division of Medicine
Imperial College School of Medicine
Hammersmith Hospital
London, UK

Mark J. Walport
Division of Medicine
Imperial College School of Medicine
Hammersmith Hospital
London, UK

Anthony N. Warrens
Renal Unit and Department of Immunology
Imperial College School of Medicine
Hammersmith Hospital
London, UK

Anthony P. Weetman
The University of Sheffield
Department of Medicine
Clinical Sciences Centre
Northern General Hospital
Sheffield, UK

Padraig Wright
Department of Psychological Medicine
Institute of Psychiatry
Denmark Hill, London, UK

Abbreviations

AAU	acute anterior uveitis	BHR	bronchial
Ab	antibody	BL	Burkitt's lymphoma
ABC	ATP-binding cassette	BMT	bone marrow transplantation
ACAID	anterior chamber associated immune deviation	BP	Behçet's disease/syndrome
		BSCR	birdshot retinochoroidopathy
AChR	acetylcholine receptor	CAH	congenital adrenal hyperplasia
AD	atopic dermatitis	CANCA	cytoplasmic anti-neutrophil cytoplasmic antibodies (ANCA)
Ag	antigen		
AGE	nonenzymatically glycosylated proteins	CD	coeliac disease
		CDC	complement dependent cytotoxicity
AIDP	acute inflammatory demyelinating polyneuropathy	CDR	complementary-determining region
		CFS	chronic fatigue syndrome
AIDS	acquired immune deficiency syndrome	CIDP	chronic inflammatory demyelinating polyneuropathy
AIF-1	allograft inflammatory factor-1	CIN	cervical intraepithelial neoplasia
AIH	autoimmune hepatitis	CLIP	class II-associated Ii-chain peptide
ANAP	agglutination negative absorption positive	CLP	cutaneous lichen planus
		CNS	central nervous system
ANCA	anti-neutrophil cytoplasmic antibodies	CREB	cyclic AMP response element binding protein
APCs	antigen presenting cells	CREG	cross-reactive groups
APMPPE	acute posterior multifunctional placoid pigment epitheliopathy	CTL(p)	cytotoxic T lymphocyte (precursor)
		CTS	collaborative transplant study
APN	affected pedigree number	CVID	common variable immunodeficiency
ARMS	amplification refractory mutation system		
		CYNAP	cytotoxicity ANAP
AS	ankylosing spondylitis	DC	dendritic cells
ASA	allele-specific amplification	DG3	desmoglein-3 glycoprotein
BACS	bacterial artificial chromosomes	DGGE	denaturing gradient gel electrophoresis
BCR	B cell receptor		
BE	beryllium	DH	dermatitis herpetiformis

DHF	dengue haemorrhagic fever	IMN	idiopathic membranous nephropathy
DM	diabetes mellitus	IM	infectious mononucleosis
DSCA	double-stranded conformation analysis	IPF	ideopathic pulmonary fibrosis
DTH	delayed type hypersensitivity	IR	immune response genes
EAAU	experimental anterior autoimmune uveitis	IRBP	interphotoreceptor binding protein
		ITAM	immunoreceptor tyrosine-based activating motif
EAU	experimental autoimmune uveitis		
EBV	Epstein–Barr virus	ITIN	immunoreceptor tyrosine-based inhibitory motif
EDTA	ethyl diamine tetra-acetic acid		
EF	etiological fraction	JME	juvenile myoclonic epilepsy
ELISA	enzyme-linked immunosorbent assay	JRA	juvenile rheumatoid arthritis
		KARS	killer activating receptors
ELP	erosive oral lichen planus	KIRS	killer (cell) inhibiting receptors
ER	endoplasmic reticulum	LA	Lyme arthritis
ESTS	expressed sequence tags	LCMV	lymphocyte choriomeningitis virus
FITC	fluorescein isothiocyanate	LCL	lymphoblastoid cell lines
FS	fogo selvagem	LDA	limiting dilution analysis
GBM	glomerular basement membrane	LL	lepromatous
GBS	Guillain–Barré syndrome	LMPs	latent membrane proteins
GCA	giant cell arthritis	LST-1	leucocyte specific transcript
GRR	genotype risk ratio	LT	lymphotoxin
GVHD	graft versus host disease	LTR	long terminal repeat
GVL	graft versus leukaemia	mAb	monoclonal antibody
HAR	hyperacute rejection	MAG	myelin-associated glycoprotein
HBV	hepatitis B virus	MBP	myelin basic protein
HCV	hepatitis C virus	MCNS	minimal change nephrotic syndrome
HEL	hen egg lysozyme	MG	myasthenia
HIV	human immunodeficiency virus	MHC	major histocompatibility complex
HLA(DR)	human leucocyte antigen-(D-related)	MICA	MHC class I chain-related gene A
HP	hypersensitivity pneumonitis	MLC/R	mixed lymphocyte culture/reaction
HPV	human papillomavirus	MMR	macrophage mannose receptor family
HRCT	high resolution computerized tomography		
		MOG	myelin oligodendrocyte glycoprotein
HSP70	heat shock protein 70	MS	multiple sclerosis
HTC	homozygous typing cells	NAIP	neuronal apoptosis inhibitory protein
HTL(P)	helper T lymphocyte (precursor)		
IBS	identity by state	NE	nephropathia epidemica
IDDM	insulin dependent diabetes mellitus	NHTR	non-haemolytic transfusion reactions
IFNγ	interferon gamma	NK	natural killer
Ig	immunoglobulin	NOD	non-obese diabetic
IHW	international histocompatibility workshop	NPC	nasopharyngeal carcinoma
		OLP	oral lichen planus

ON	optic neuritis	SMA	spinal muscular atrophy
ONTT	optic neuritis treatment trial	SMN	survival of motor neurone
OR	odds ratio	SNP	single nucleotide polymorphism
OspA	outer surface protein A	SnRNP	small nuclear ribonucleoprotein
PACS	page artificial chromosome	SO	sympathetic ophthalmia
PAN	polyarteritis nodosa	SOREM	sleep onset rapid eye movements
PANCA	perinuclear ANCA	SSCP	single-strand conformation polymorphisms
PBC	primary biliary cirrhosis	SSLP	simple sequence length polymorphisms
PBL	peripheral blood lymphocytes	SSO(P)	sequence specific oligonucleotide (probes)
PCC	pigeon cytochrome c	SSP	sequence specific primers
PCR	polymerase chain reaction	TAO	thyroid associated ophthalmopathy
PF	pemphigus foliaceus	TAP	transporter associated with antigen processing
PFGE	pulsed field gel electrophoresis	TB	tuberculosis
pfu	plaque forming units	TCCs	T cell clones
PG	pemphigoid gestationis	TCLs	T cell lines
PLL	poly-L-lysine	TCR	T cell receptor
PPD	purified protein derivative	TD	tardive dyskinesia
PPT-1	palmitoyl-protein thioesterase	TDT	transmission disequilibrium test
PMA	phorbal myrisate acetate	TGase	tissue transglutaminase
PMR	polymyalgia rheumatica	Th	T helper
PSC	primary sclerosing cholangitis	TM	trans-membrane
PTB	pulmonary tuberculosis	TMAC	tetramethylammonium chloride
PV(A)	pemphigus vulgaris (antigen)	TNF	tumour necrosis factor
QTL	quantitative trait loci	TSH	thyroid stimulating hormone
RA	rheumatoid arthritis	TSH-R	thyroid stimulating hormone receptor
RAGE	receptor for advanced glycosylation end products of proteins	UC	ulcerative colitis
ReA	reactive arthritis	UNOS	union network of organ sharing
RFLP	restriction fragment length polymorphism	VKH	Vogt–Koyanagi–Harada
Rh	rhesus	VNTR	variable number tandem repeat
RIP	rat insulin promoter	VUDs	volunteer unrelated donors
SAGE	serial analysis gene expression	YACs	yeast artificial chromosomes
SAGS	staphylococcal and streptococcal superantigens		
SALT	skin associated lymphoid tissue		
SBT	sequencing-based typing		
SCLE	subacute cutaneous lupus erythematosus		
SDS	sodium dodecyl sulphate		
SEB	staphylococcal enterotoxin B		
SIS	skin immune system		
SLE	systemic lupus erythematosus		

SECTION I

Introduction to the HLA System

CHAPTER 1

The History of the Discovery of HLA

J.J. van Rood

But those that were recovered had much compassion both on them that died and on them that lay sick, as having both known the misery themselves, and now no more subject to the danger. For this disease never took any man for the second time.
(Thucydides; in Major, 1965, p. 75)

INTRODUCTION

Each discovery goes through three phases. The first one, the discovery itself, is the realization that there is "something" which is or was unknown, not yet described, waiting to be discovered. In the second phase, the tools have to be selected with the help of which one can successfully proceed. This can require a discovery by itself. In the third and final phase, the real discovery begins; this can take years or even decades, as was and is the case for HLA. It is very similar to the discovery and charting of a new continent, such as the Americas. This chapter will therefore be divided into the above three phases.

PHASE ONE: HLA – A MAIDEN WAITING TO BE DISCOVERED

The quotation from Thucydides which was used to introduce this chapter could be regarded not only as an early description of acquired immunity ("... this disease never took any man for the second time"), but also as the recognition that all men are *not* equal and that some individuals are more prone to die than others. Because the plague hit the army, death could not be due only to age (the very young or old are more prone to die) or sex differences. In other words, humans are polymorphic in their resistance to disease.

The discovery by Landsteiner in 1901 that the sera of his colleagues agglutinated the red cells of some, but not of other, staff members was the beginning of the study of polymorphism *per se*, in this case by immunogenetics.

The first tool was simple: red cell agglutination in saline. Table 1.1 lists the blood group systems known in the early 1950s when the HLA story really started. The study of the genetics of these blood group systems was undertaken mainly by those working in blood transfusion and haematology, a new speciality in those days. The "take-home message" was that there were many different, unlinked inherited blood group systems, most of which were important in blood transfusion medicine.

TABLE 1.1 Listing of blood group systems known in the early 1950s, with their year of discovery

1.	ABO blood groups	1900
	A_1A_2 subdivisions	1911
2.	MN blood groups	1927
	MNSs subdivisions	1947
3.	P blood groups	1927
4.	Secretion of the ABO antigens in saliva	1932
5.	Rh blood groups	1940
	(many subsequent subdivisions)	
6.	Lutheran blood groups	1945
7.	Kell blood groups	1946
8.	Lewis blood groups	1946
9.	Duffy blood groups	1950
10.	Kidd blood groups	1951

Modified after Race and Sanger (1950, p. 7).

Another important line of research was the study of autoimmune disease: not only acquired haemolytic anaemia and idiopathic thrombocytopenia, but also thyroiditis and others. These could now be studied *in vitro* thanks to the discovery of anti-human gamma globulin antibodies by Coombs et al. (1945).

In the early 1950s, the work on red cells was extrapolated to leucocytes. Antibodies which could agglutinate leucocytes were described in patients with leukaemia (André et al., 1954), neutropenia of the newborn (Lehndorff, 1951), pyramidon-induced neutropenia (Moeschlin and Wagner, 1952a; Moeschlin and Moreno, 1954), agranulocytosis due to the occurrence of leucocyte agglutinins (Moeschlin and Wagner, 1952b) and haemolytic anaemia (Goudsmit and van Loghem, 1953). There was general uncertainty about the origin of these antibodies and their clinical importance. In 1954, two papers clearly implied the existence of leucocyte groups. The first was by Miescher and Fauconnet (1954), in which they described the presence of a leucocyte antibody which agglutinated only some of the donors of a panel. The article's title was "Mise en évidence de differents groupes leucocytaires chez l'homme": the first time that the term "leucocyte groups" was used. Another seminal article, "Leuco-agglutinins IV, leuco-agglutinins and blood transfusion", by Dausset (1954) demonstrated that blood transfusions induced leucocyte agglutinins which, however, did not agglutinate the patient's own leucocytes. In other words, they were alloimmune antibodies. The confusion and uncertainty which existed at that time is best illustrated by another publication, in 1954, in which Dausset described the presence of leucocyte agglutinins in patients suffering from pancytopenia and speculated that they might be autoimmune (Dausset et al., 1954). The confusion is understandable. Those who worked with these agglutinins often had a background in blood transfusion and/or haematology and in which both alloantibodies and autoantibodies against red cells were important; by extrapolation, one assumed that this would hold true also in the case of antibodies against leucocytes.

Studying these leucocyte agglutinins formed after blood transfusion turned out to be difficult. The antibodies were weak, "changed" their specificity and tended to disappear rather quickly. In the summer of 1957 Dausset met Ceppellini during a symposium in France. It was hot, the speaker was not very interesting, they were sitting on the last row of the hall and Ceppellini asked Dausset how his work on leucocyte agglutinins formed after blood transfusion was progressing. Dausset confessed that he was thinking of stopping it altogether, because the antibodies were so difficult to work with and because he was not sure whether he was looking at alloantibodies, antibodies recognizing determinants which differed between individuals or autoantibodies recognizing self-antigens. Ceppellini said:

> *well come on, you can solve that problem in an afternoon. All you have to do is to take the leucocytes of monozygotic and dizygotic twins and if the monozygotic twins even with your weak antisera show a similar pattern and the dizygotic twins differ, then you have proven your point. Your antibodies are recognizing genetically determined alloantigens and it is only a matter of technology.* (R. Ceppellini and J. Dausset, personal communication)

This turned out to be the case: these weak antibodies gave identical reactions in the monozygotic twins but quite disparate results in dizygotic twins (Table 1.2); and with that Dausset had proven that these antibodies, weak as they might be, indeed recognized alloantigens (Dausset and Brecy, 1958; Daussett, 1990).

This was the end of the first phase: the discovery itself. The publications of Miescher and Dausset had convinced most people working in the field that leucocyte groups existed, but there reigned widespread doubt that the leucocyte antibodies induced after blood transfusion would be suitable reagents for typing.

PHASE TWO: "GIVE US THE TOOLS AND WE WILL DO THE JOB" (WINSTON CHURCHILL)

The leucocyte antibodies might be poor typing reagents, but it was soon recognized that they were able to induce non-haemolytic transfusion reactions (NHTR). Van Loghem showed further-

TABLE 1.2 Agglutination of the leucocytes of twins with the aid of leuco-agglutinating sera

Monozygotic twins

	Leuco-agglutinating sera																												
	Au	Ch	Ce	Le	Di	Ma	So	Ls	Dr	Pa	Du	Lm	Bo	Te	Ro	De	Bn	Mg	Me	Bu	Ds	Rb	Be	Vc	Ba	Ca	Ve	Pu	Lx
P.Va	−	±	−	+	−	+	+		−		+		+	+	+	−	+	+	+	±		+			+	+	+	+	+
J.Va	−	±	−	+	−	+	+		−		+		+	+	+	−	+	+	+	±		+			+	+	+	+	+

Red cell groups: AB Cc D ee Fy(a−) K − M P

E.Ro	+	±		−	−			+	+	±	−	+	+	+		+	−			+	+	+	+	+		+	+	+	
H.Ro	+	±		−	−			+	+	±	−	+	+	+		+	−			+	+	+	+	+		+	+	+	

Red cell groups: A cc dd ee Fy(a−) K − M N P

D.Te	+	+		(+)+				+	+	−	+	+	+	+		−	−			+	+				+	+	+		
B.Te	+	+		(+)+				+	+	−	+	+	+	+		−	−			+	+				+	+	+		

Red cell groups: A CC D ee Fy(a−) K − N P

Dizygotic twins

L.Go	+	+		−	+			+	+	+	−	+	+	+	−	−	−			+	+				−	−	+	+	+	−	+
V.Go	+	+		−	+			+	+	+	−	+	+	+	+	+				+	+				+	+	+	+	+		+

Red cell groups: L. Go: A Cc D ee Fy(a−) K − N P V. Go: A cc D Ee Fy(a−) K − N P

Dausset and Brecy (1958).

more that if the buffy coat of the transfused blood was removed the NHTR did not occur (van Loghem et al., 1956). The present author was, at that time, director of the university hospital blood bank and a decision was made to start a consultancy. If a patient suffered a transfusion reaction (TR) an investigation was carried out into whether this was due to the presence of leucocyte antibodies or whether it had another cause.

Then serendipity stepped in. An NHTR occurred in a woman, Mrs H.-B., with multiple pregnancies. The patient had strong leucocyte antibodies, but when asked when and where she had received the blood transfusions, which supposedly had induced these antibodies, she sat up in her bed, looked at the doctors in horror and said:

> *This was the first transfusion I ever had.*
> *I have never been so sick in my life.*
> *Don't talk to me about transfusions.*

Only then did the idea form that, contrary to the dogma of these days, leucocyte antibodies might be formed not only after blood transfusion, but also during pregnancy. This turned out to be true. These leucocyte antibodies are best suited for HLA typing, because they are oligospecific, as they are only directed against the antigens that the child has inherited from the father and, for reasons which are still not understood today, continue to be produced many years after pregnancy. Sera were obtained which were excellent testing reagents obtained 20 years after the last pregnancy. In other words, the tools were available (van Rood et al., 1958). At the same time, Rose Payne came to a similar conclusion in a beautiful systematic study (Payne and Rolfs, 1958).

It was here that working in a blood bank that not only provided blood and blood products, but also screened for Rhesus (Rh) antibodies paid off. In the freezer were dozens of stored serum samples for women in whom such Rh and other antibodies had been found. In a few weeks time, they had been screened and it could be shown that in 10% of them leucocyte antibodies were indeed present. A first family study was done, in which it was shown that the antigens which were recognized by three sera showed Mendelian segregation.

These data were presented in Rome in 1958 during the haematology and blood transfusion meeting (van Rood et al., 1959).

Unknown to van Rood's group, Dausset had in the mean time performed an important experiment. Male anaemic patients were transfused repeatedly with blood from a single donor. The patients formed strong leucocyte antibodies reacting with a proportion of the leucocyte samples taken from a panel of randomly selected individuals. These antibodies did not react with the individuals who had formed these antibodies. The leucocyte antigen so identified was called MAC, not because the patients were Scottish, but to honour three volunteers participating in this experiment whose names began with an M, A and C, respectively. Mac turned out later to be similar to the cross-reacting antigens HLA-A2 and -A28. Unfortunately, the sera of the patients were never available to other investigators, because they soon lost their specificity (Dausset, 1958).

After Rome, efforts to improve leucocyte typing were continued. One such improvement was the finding that certain leucocyte samples, although they were not agglutinated by a given serum, sometimes were able to absorb the antibody out of the serum. This was called the ANAP phenomenon (agglutination negative, absorption positive), later also called CYNAP for the cytotoxicity assays (van Rood et al., 1961). This observation explained many of the discrepancies between the reactions of two or more sera with antibodies apparently recognizing the same antigen. The first serum so studied, no. 36, turned out to be the first of a cluster of sera which recognized the same antigen. It was named Antigen Four (now HLA-Bw4).

Next, the author set out to do what was, at that time, considered a very large experiment: 60 sera from women who had been pregnant and formed leucocyte agglutinins were tested against the leucocytes obtained from 100 panel donors selected at random. This is now a standard procedure which still forms an essential part of all workshops, but at that time (1960–1961) something like this had never been done. It took several weeks to conduct all of the tests (Figure 1.1). The only way to analyse the data was by comparing positive and negative reactions obtained by each serum against the positive and negative reactions of all other sera followed by 2×2 chi-squared tests.

A computer was needed, but the only computer that Leiden University owned at that time was located in the Department of Astronomy and was used for the calculations of distances of stars and their speeds. It was completely useless for the kind of analysis required here.

However, the Bureau of National Statistics was able to do the job, and together with the department that was responsible for calculating the salaries of the Dutch government's employees, the first really good analysis was achieved. The pattern that emerged is shown in Figure 1.2: of the 60 sera, eight recognized two very good clusters which were named 4a and 4b (now called HLA-Bw4 and -Bw6). All of this happened in 1961. The work was presented for the first time in Vienna during the meeting of the European Society of Haematology (van Rood et al., 1962). The studies were compiled in 1962 in the thesis "Leucocyte grouping, a method and its application" (van Rood, 1962) and in the *Journal of Clinical Investigation* (van Rood and van Leeuwen, 1963).

Rose Payne and Walter Bodmer were the first to introduce this computerized approach in their laboratory. A few months later the author received a letter from Rose Payne, saying:

> *Dear Jon,*
> *Could you send us another copy of your thesis. The copy you sent to me has travelled so often between Walter's lab and mine that it is completely falling to pieces and we would like to have a new one.*

It was sent immediately. In 1964 they described the LA1, LA2 and LA3 antigens (Payne et al., 1964; Payne, 1990). It was (and still is) of interest that using the defibrinated leucoagglutination technique, they identified the LA or HLA-A locus antigens, and later the HLA-B locus antigens while the author, using the EDTA leucoagglutination technique, identified the Group Four or HLA-B locus antigens only, with the exception of HLA-A2, which were also recognized.

During the same period two important technical improvements were realized. The first was the introduction of the complement-dependent cytotoxicity (CDC) assay by Batchelor (1990), Walford et al. (1964) (who showed that rabbit serum was the best complement source) and others, and the second was the development of the microcytotoxicity test by Terasaki and McClelland (1964), a truly revolutionary technical improvement. A very good researcher could carry out 20–25 agglutination tests with 1 ml of serum, but with Terasaki's microcytotoxicity test one could do 1000!

It is here that Bernard Amos should be mentioned. He has been important in the HLA effort in many ways, but especially because of two

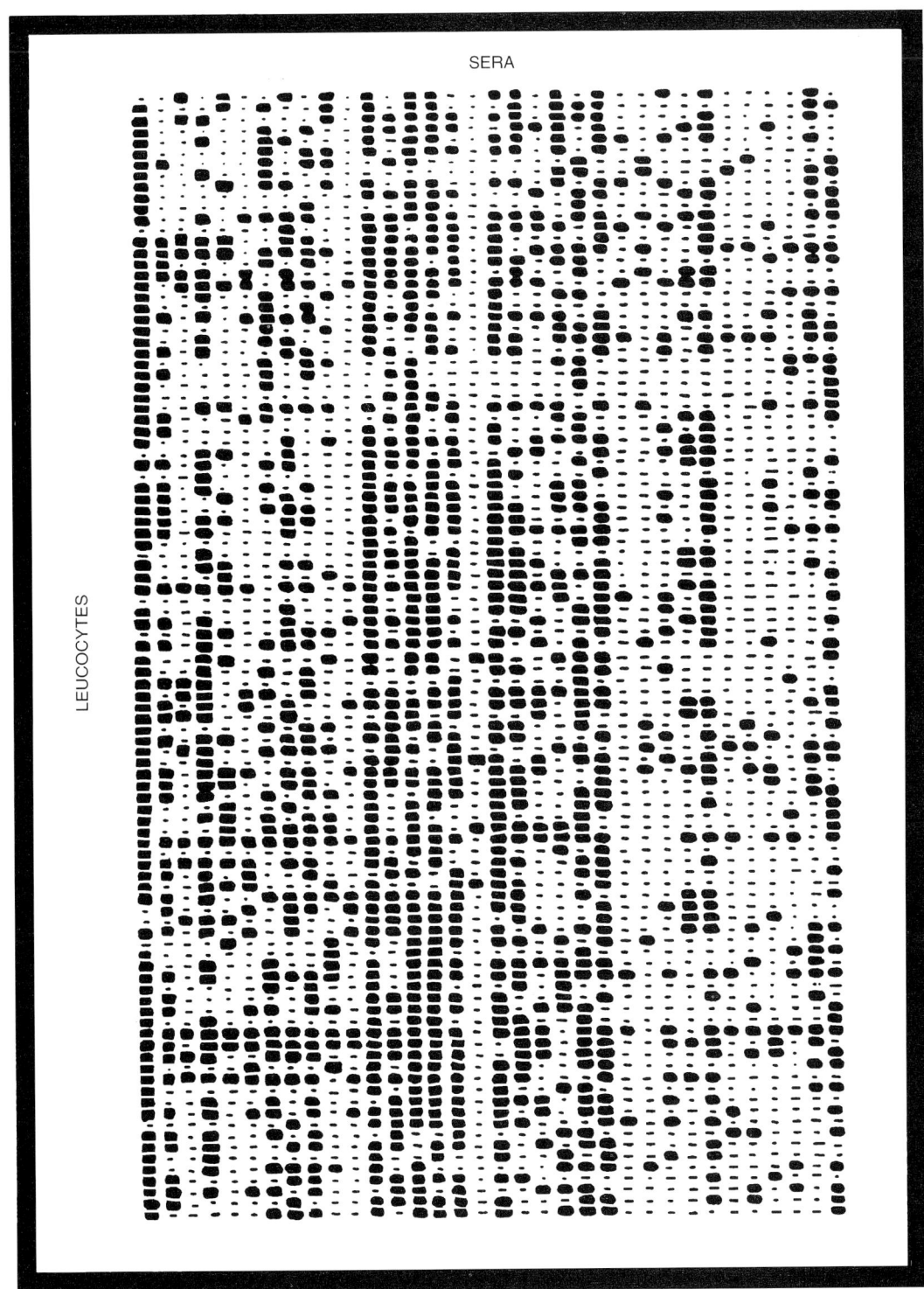

FIGURE 1.1 Agglutination pattern of 34 sera with leucocyte antibodies tested against 100 leucocyte samples. Each vertical column of squares and hyphens shows the results obtained with one serum against the panel; each horizontal row shows the results of the 34 sera with one leucocyte sample. ■: Agglutination positive; –: agglutination negative. Doubtful results were recorded as agglutination negative.

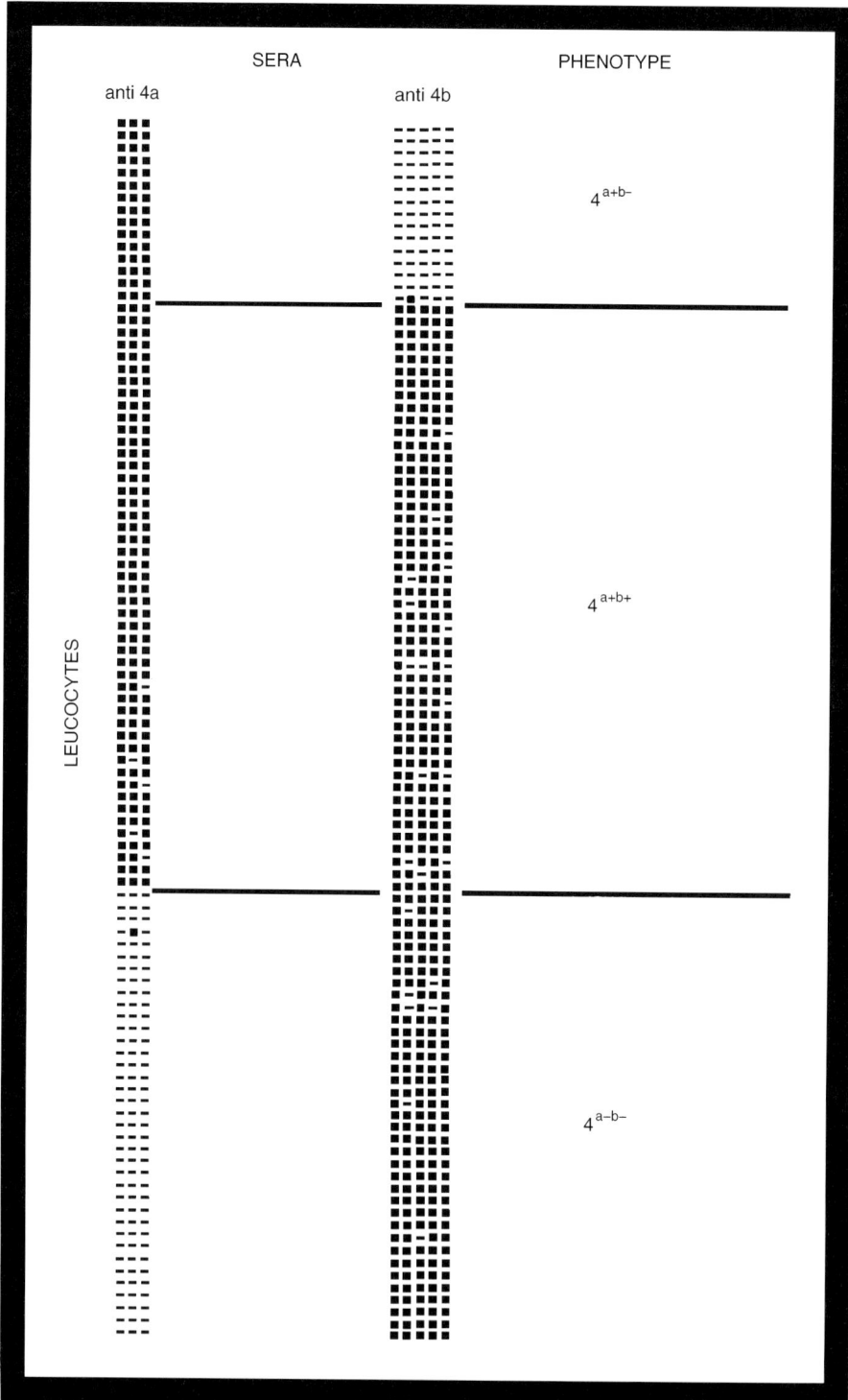

FIGURE 1.2 Agglutination pattern of eight sera with the leucocytes of the panel. Each vertical column of squares and hyphens shows the results obtained with one serum against the panel; each horizontal row shows the results of the eight sera with one leucocyte sample. ■: Agglutination positive; –: agglutination negative. Doubtful results were recorded as agglutination negative (van Rood et al., 1961).

major contributions. He was a student of Gorer, with whom he studied leucocyte antibodies in mice. Thereafter, he went to the USA. Amos was the one who persuaded the National Institutes of Health (NIH) to organize a workshop on HLA antigens, a field which at that time (1964) was represented by no more than 13 groups. That workshop set the stage for a series of workshops which is still continuing. The impact of the workshops cannot be overestimated and has been of major importance in developing our knowledge and understanding of the HLA system.

A second initiative of Amos of enormous importance was to persuade the NIH to set up a repository where every participant would deposit their best sera. These sera could then be requested by other workers in the field. This was in the 1960s, at a time in which patents were known about, but certainly not in the sense of today. Still, we were quite competitive and it took some persuasion before everybody joined in. These two initiatives of Amos were the start of an intensive world-wide exchange of reagents and standardized HLA typing.

With these acquisitions almost all of the tools were assembled: antibodies induced during pregnancy, many of which were monospecific, computer facilities to analyse the complexity of the results so obtained, the finding that some cell samples were negative in the agglutination or cytotoxicity technique, but could still absorb antibodies, a microcytotoxicity test which made the exchange of reagents much easier, workshops to discuss and compare results and a repository to facilitate serum exchange. However, a very important tool was still lacking: the cellular techniques and especially the mixed lymphocyte culture (MLC) test, developed at the same time by Fritz Bach and Kurt Hirschhorn in New York and by Barbara Bain and Löwenstein in Montreal. They have been and still are indispensable in order to explore the vastness of HLA (Bach and Hirschhorn, 1964; Bain et al., 1964).

PHASE THREE: THE DISCOVERY OF HLA

The third phase overlapped with the second. After describing the 4a and 4b antigens the author's group proceeded to analyse the other sera. This was done by looking for clusters of high chi-squared values obtained by comparing the positive and negative reactions of the individual sera. The first cluster that was identified and analysed turned out to identify a diallelic system, group 5, with the alleles 5a and 5b. In this case, the main work was performed by Aad van Leeuwen, the author's lifetime colleague, who together with George Eernisse steered the blood bank while he enjoyed a sabbatical in New York. The Group Five system was not linked to HLA; it was a "minor histocompatibility antigen" system (van Leeuwen et al., 1964). Although its role in transplantation has never been adequately studied, anti-5b antibodies can be the cause of such problems as NHTR (Table 1.3).

Further scrutiny of the chi-squared clusters identified three other sets of sera recognizing the antigens 6a, 6b and 6c (now HLA-B7). These data were presented at the first Histocompatibility Conference in Washington in 1964 organized by Amos, as already mentioned (van Rood and van Leeuwen, 1965). In the same meeting Rose Payne described the LA1 and LA2 (now HLA-A1 and -A2) antigens (Payne et al., 1964, 1965). There was no further mention of MAC.

Although the wet workshop achieved no more than a demonstration of the different techniques, all participants agreed that the effort had been worthwhile and should be continued. Leiden was chosen because of the presence of a

TABLE 1.3 Expected and observed frequencies of the phenotypes of group 5 in a group of 500 random donors

	Numbers	Fraction of total
Observed		
5(a+b−)	18	0.036
5(a+b+)	145	0.290
5(a−b+)	337	0.674
Total	500	
Gene freq. 5^a = 0.181		
Gene freq. 5^b = 0.819		
Expected		
5(a+b−) = $0.181^2 \times 500$	16.4	
5(a+b+) = $2 \times 0.181 \times 0.819 \times 500$	148.2	
5(a−b+) = $0.819^2 \times 500$	335.4	

$\chi^2 = 0.239 \quad 0.70 > p > 0.50$

Van Leeuwen et al. (1964).

panel already typed for 4a, 4b, 5a, 5b, 6a, 6b, 6c which would facilitate the comparison of the sera tested with techniques different from the EDTA-agglutination test which had been used previously. Here, the power of our tools was demonstrated clearly. During the last day of the conference the results of the analysis showed that most of the Leiden antigens had also been recognized by other groups (Histocompatibility Testing, 1965).

The 1965 workshop in Leiden was also a turning point because during this meeting family data from the author's group and population data from Dausset provided preliminary evidence that most of the antigens recognized belonged to one genetic system. At the time this was revolutionary. The idea that there could be one major integrated system was unthought of. After all, our thinking had been strongly influenced by the discovery of the many blood group systems which were mainly genetically independent, while most HLA workers of that time knew little or nothing of the beautiful work of Gorer (1947), Snell (1948) and others who studied H-2 in mice. It came as no surprise when Amos stated (in the discussion):

> *Probably quite a few people would like to take issue with you on the evidence that this is one system ... (Histocompatibility Testing, 1965).*

Amos was quite right. We had observed similar associations but had also shown that some of the antigen clusters that had significant associations between them included sera which recognized the Group Five system, a system independent of the other leucocyte antigens. We had called that one genetic system the Group Four system, Payne and Bodmer called it LA and Dausset, Hu-1. After an emotional meeting in Williamsburg Group Four was dropped, and HL-A, as it was then named, was born. Thirteen groups participated in the wet workshop organized by Amos in Duke in 1964, and 18 in 1965 during the Leiden workshop, of which 16 reconvened during the Torino workshop in 1967. They formed the hard core of HLA workers in the following years.

With so many people working on the different aspects of the discovery of HLA it becomes impossible to tell the story on a month-by-month or even a year-by-year basis. Therefore, the work after 1965 will be summarized in four strands: the genetics, the molecule and its function, its biological importance and its clinical importance.

CHARTING THE GENETIC MAP OF HLA

In both 1965 and 1967 workshops, the basic concept of the genetics of HLA was that it resembled a component system, such as the Rhesus blood group system (Bodmer and Payne, 1965).

It therefore came as a shock that shortly after the Torino meeting Ceppellini told the author during a telephone conversation that Flemming Kissmeyer-Nielsen in Aarhus had observed in a family study a cross-over between Group Four (now HLA-B) and LA (now HLA-A). I was flabbergasted, and at first could not believe it, but the data were beautiful and clear-cut (Kissmeyer-Nielsen et al., 1969). HLA was not a component system but a closely linked region with loci between which cross-overs could occur.

Flemming has been in many ways an important figure and contributor to the HLA field in general and to our group in particular. He helped to set up the microcytotoxicity technique, which was essential for the function of organ exchange organizations, such as Eurotransplant, which were started during that time. On the tenth anniversary of Eurotransplant, I wrote to thank him and Erik Thorsby for the technical help and the sera they provided while noticing that without that "many of our organ exchanges would have been 'blind dates'" (Kissmeyer-Nielsen et al., 1969). The discovery of HLA-C as a separate locus must be credited to the Aarhus group, together with Lena Sandberg from Sweden (Sandberg et al., 1970).

That HLA-D was a separate locus responsible for stimulation in the MLC test was described in a study by Edmond Yunis and Bernard Amos (1971). The dogma at that time was that the HLA-A and/or -B antigens were the stimulators and that for that reason HLA-A and -B identical siblings did not stimulate an MLC (Bach and Amos, 1967). However, they found an exception: an HLA-A and -B "identical" sibling pair that did stimulate an MLC. Thus, the HLA-D locus was born and soon confirmed (Eijsvoogel et al., 1972).

The MLC test made it possible to type for HLA-D. Bradley et al. (1972) were the first to introduce HLA-D typing in pigs, using homozygous typing cells (HTC). With these cells as stimulators in an MLC test, lack of stimulation implied that the responder cell shared an HLA-D antigen with the responder. Unaware of Bradley's paper, the author's group reinvented the wheel and looked for homozygous typing cells to be used for HLA-D typing. It was realized that they could best be found in the offspring of cousin marriages, but the problem was how to identify such people. Fortunately, the Roman Catholic church does not allow cousin marriages unless the Pope has given permission, and the church kept excellent archives on all matters including these cousin marriages. A letter was written to the Pope asking for his permission to use that information and after a few months a letter was received stating that we could contact the Bishop of Haarlem. We went there to discuss the problem, obtained his blessing and went ahead. It was a success and eight HLA-D determinants were identified. Independently, Mempel introduced the same approach (Mempel et al., 1973; van den Tweel et al., 1973).

Dausset heard the story and said:

> *what can be done in Holland should certainly be possible in France.*

He went to the Evêque de Paris and explained his case, but the Evêque de Paris said "Non" and that was the end of it! To solve the problem Degos and Colombani went to the Touareggs, who must marry their cousins. Unfortunately (and interestingly), they found far fewer homozygous children than expected. Was this due to chance (unlikely), lethal genes (W. Bodmer, personal communication) or diminished resistance against infections in people homozygous for HLA? All were possible, but none was proven.

Much of the pressure to search for methods to type for HLA-D and, later, DR came from those involved in clinical organ transplantation. It had already been recognized that low or negative MLC reactivity improved graft survival of both skin and kidneys in humans and monkeys (Cochrum et al., 1973; Koch et al., 1973). Although typing with homozygous typing cells made typing for HLA-D possible, it was too time consuming to be useful in clinical renal transplantation using cadaveric donors. A serological method would be the solution if such a method could be developed. At first, there was considerable scepticism that this would be feasible. Much of this scepticism was based on the fact that, in many other systems, T cell-dependent immunity differed in its specificity and antigenic requirements from B cell-dependent immunity.

A decision was made to study the problem of HLA-D-related serology on account of two observations, one of which was made by Grumet and Leventhal, and one by the author. Grumet had shown that sera with anti-HLA-A and -B antibodies were able to inhibit the MLC test (Grumet and Leventhal, 1970). Although MLC inhibition by HLA antibodies was thus a well-established phenomenon, the mechanism by which this occurred was not understood. Again serendipity opened our eyes. A woman, a mother of two children, had received an MLC-positive but HLA-A- and -B-identical experimental skin graft. On the basis of previous experience it had been expected to survive for 15 days at most, but at 23 days it was still not rejected. We wondered whether there might be something (perhaps an antibody induced by pregnancy) that neutralized the rejection-inducing effect of an HLA-D mismatch.

The MLC inhibition technique was used to screen for anti-HLA-D (now anti-DR) antibodies. It was clear from the beginning that the test system had to be rigorously controlled to be certain that irrelevant antibodies (i.e. other than anti-HLA-DR) did not cause MLC inhibition. To exclude interference with anti-HLA-A and -B antibodies, stimulator cells were used that were HLA-A- and -B identical with the responder cells. This approach was named the <u>M</u>LC <u>I</u>nhibition test using (<u>S</u>erological <u>D</u>efined) <u>SD</u> <u>I</u>dentical <u>S</u>timulator cells, or MISIS for short. The sera tested contained strong anti-HLA-A- and/or -B antibodies generally formed during pregnancy. The very first serum tested in the MISIS test was informative (van Leeuwen et al., 1973) (Table 1.4). Some responder–stimulator cell combinations were inhibited (P, N, T, U, V, W), while others were not (Q, R, S). This finding was reproducible and other sera that showed similar results were found easily.

TABLE 1.4 Typing for MLC LD: MLC inhibition and immunofluorescence

		Parous woman Sch: HL-A2,3,7,w10			
		AB serum (c.p.m.)	Serum Sch (c.p.m.)	AB serum/ serum Sch	Fluorescent (% positive cells)
Sch +	P	3300	400	*7.3*	17
	Q	14 000	8000	1.7	4
	N	9700	1100	*9.5*	17
	R	10 000	6400	1.6	9
	S	4200	2800	1.7	7
	T	5200	700	*7.4*	7
	U	13 500	1300	*10.0*	16
	V	26 500	1100	*24.1*	17
	W	1800	200	*10.0*	–

All stimulator cells were SD-identical with the responder cells Sch: HL-A2,3,7,w10. The column AB serum/serum Sch represents the inhibition index values obtained by dividing the c.p.m. obtained by culturing the cells in AB serum by the c.p.m. obtained when they were cultured in serum Sch. The significant inhibitions are in italics. Van Leeuwen et al. (1973).

More direct evidence that antibodies mediated the MLC inhibition was obtained when Aad van Leeuwen together with Riek Schuit developed a sandwich immunofluorescence test with anti-immunoglobulin G (IgG) which gave what was, at that time, an unusually low background with normal lymphocytes. While the background was 4–8%, some of the stimulator cells when tested with the serum used in the MISIS test showed bright staining of 16–17% of the cells (Table 1.4). These very important results were obtained because Aad van Leeuwen had, as usual, spent many long evenings counting each individual sample. The low background immunofluorescence was clearly essential to detect these small differences and was originally dependent on very rare and very pure rabbit anti-human-Ig sera which had been made by Jiri Ràdl in the Laboratory of Experimental Gerontology-TNO, in Rijswijk. When it was realized that the background staining was due to binding of fluorescein-labelled immunoglobulins by Fc-receptor molecules, commercially available sera were subjected to pepsin treatment to remove the Fc part of the anti-human Ig and the test became generally available. As illustrated in Table 1.4, the first serum to detect polymorphism in the MISIS test also showed significantly positive fluorescence. The results of the two tests correlated quite well, a finding substantiated in later studies. It was concluded that the MLC-inhibiting substance and the antibody that reacted in the immunofluorescence test were probably the same IgG molecule (van Leeuwen et al., 1973). Together with Bob Winchester, we showed that many of the sera contained complement-dependent cytotoxic antibodies which reacted with B cells and monocytes, but not with (most) T cells or with platelets (van Leeuwen et al., 1975). This made it possible to modify the test by enriching it for B cells and monocytes and using a standard CDC test. The paper was submitted first to *Nature* which, surprisingly, rejected the first clear-cut correlation of homozygous typing–cell typing and serology with HLA-D (Figure 1.3).

The test was further standardized and, with minor alterations, was used during the 6th and more extensively during the 7th Histocompatibility Workshop by several groups (van Rood et al., 1975a, b). A year later we modified the test as follows: after labelling with FITC-labelled anti-human gamma globulin, B cells could be recognized and differentiated from monocytes, which were labelled through their Fc receptors (van Rood et al., 1976). The test has been useful not only in HLA typing, but also for the study of monocyte antigens (Jager et al., 1988).

With this, the two requirements for the study of the HLA-DR antigens had been fulfilled: adequate techniques and reasonably pure antibodies were available. The genetics of the system also offered little problem, because it was assumed from the beginning that HLA-D

Donor	Typing cells Δ~HL-A7			Serum	Typing cells Δ~HL-A8			Serum		Typing cells	Serum
	V	XV	XIX	PO	VI	XIII	XI	BE	MO	XII	SI
1. 2, 3, 7, 12	■	■	■	–	–	–	–	–	–	–	–
2. 2, 3, W5, 8	■	■	■	–	–	–	–	–	–	–	–
3. 3, 7, T4	■	■	■	❐	–	–	–	–	–	–	–
4. 28, 30, 7, 15	■	■	■	❐	–	–	–	–	–	–	–
5. 2, 3, 7, W10	■	■	■	❐	–	–	–	–	–	–	–
6. 3, 11, 12, 16	■	■	–	❐	n.d.	–	–	–	–	–	–
7. 2, 7, W10	■	■	■	❐	–	–	–	–	–	–	–
8. 2, 5, 7	■	■	n.d.	❐	–	–	–	–	–	–	–
9. 1, 11, W5, 8	–	–	–	❐	■	■	■	❐	❐	–	–
10. 1, 9, 8, W10	–	–	–	–	■	■	■	❐	❐	–	–
11. 1, 8, 15	–	–	–	–	■	■	■	❐	❐	–	–
12. 1, 8, 28, 16	–	–	–	–	■	■	■	❐	❐	–	–
13. 1, 2, 8, W10	–	–	–	–	■	■	–	❐	❐	–	–
14. 1, 3, 8, W5	–	–	–	–	■	■	■	❐	❐	–	–
15. 1, 11, 8, 15	–	–	–	–	■	■	■	❐	❐	–	–
16. 1, 2, 5, 8	–	–	–	–	■	■	■	❐	❐	–	–
17. 2, 8, 15	–	–	–	–	■	■	■	❐	❐	–	–
18. 1, 2, 8, 15	–	–	–	–	■	■	■	❐	❐	–	–
19. 3, 9, W10, 16	–	–	–	–	■	■	■	❐	❐	–	–
20. 1, 28, 8, 16	–	–	–	–	■	■	■	❐	❐	–	–
21. 3, 10, 8, 13	–	–	–	–	■	■	■	❐	❐	■	❐
22. 1, 28, 8, 12	–	–	–	–	■	■	■	❐	❐	■	❐
23. 1, 9, W5, 13	–	–	–	–	–	–	–	–	–	■	❐
24. 2, 31, 7, 12	–	–	–	–	–	–	–	–	–	■	❐
25. 2, 13, 17	–	–	–	–	–	–	–	–	–	■	❐
26. 2, 30, 13, 15	–	–	–	–	–	–	–	–	–	■	❐
27. 2, 3, 12, 15	–	–	–	–	–	–	–	–	–	–	–
28. 2, 32, 15	–	–	–	–	–	–	–	–	–	–	–
29. 1, 3, 7, 8	–	–	–	–	–	–	–	–	–	–	–
30. 32, 8, 21	–	–	–	–	–	–	–	–	–	–	–
31. 1, 2, 8, 12	–	–	–	–	–	–	–	–	–	–	–

FIGURE 1.3 The first demonstration that determinants (or closely linked structures), so far only recognized by cellular techniques (HLA-D), can also be recognized by antibodies (HLA-DR). Lymphocytes were collected from donors who, with a few exceptions (nos 27–31), were known to carry the HLA-D determinants, LD V (now HLA-Dw2), LD VI (now HLA-Dw3) and LD XII (now HLA-Dw7). Sera exhaustively adsorbed with platelets were then tested in a cytotoxicity test, enriched for B lymphocytes. Note that the sera BE and MO and serum SI show a pattern of identity with LD VI and LD XII, respectively. ■: LD determinant present; ❐: cytotoxicity reaction was scored 2–5; —: LD determinant absent or serum scored as 0 and 1.

and -DR were closely related, if not identical to each other (DR stands for D-related, as decided after the 7th Workshop by Ruggero Ceppellini, Walter Bodmer and the present author in "Nomenclature for factors of the HLA system 1977").

The cell-mediated lympholysis test was also developed during these years. Several people contributed to this test, but among the most important were Ceppellini's group. Migiano presented the data and, when someone asked what CML meant answered, "Well, officially it is meant to be 'cell-mediated lympholysis', but in actual fact it is 'Ceppellini, Migiano and Lightbody', the three authors of the paper!" (Lightbody et al., 1971). We asked Els Goulmy, who had just arrived from Norway, to set up the CML test in our laboratory. Although she tried hard, she was unsuccessful. The test worked in Wisconsin, however, in Fritz Bach's laboratory, and he visited together with Barbara Alter and his wife Marilyn to Leiden for several months to set up the test. Unfortunately, even this attempt was unsuccessful. It took another 6 months to achieve success.

After Torino, the Los Angeles (1970) and Evian (1972) workshops took place, reinforcing the importance of HLA for organ transplantation and providing a first attempt at making an HLA map of the world. In Evian, Eijsvoogel presented his elegant data on a family with two cross-overs, one between HLA-D (class II) and HLA-B and another between HLA-A and HLA-B (class I). He could show that one needed both a class II and a class I disparity to obtain a positive CML test (Eijsvoogel et al., 1972).

At that time the dogma was that the CML test recognized class I antigens. A few years later Els Goulmy studied the cells of a woman, Mrs R, who had rejected a bone marrow transplant from her HLA-identical brother, and found that it recognized A2-positive cells, but not all of them. When Els presented the data during a work discussion, Alan Munro from Cambridge, who was on sabbatical in our department, said: "Now wait a moment, there has recently been a paper by Elizabeth Simpson on CTLs (Cytotoxic T Lymphocytes) recognizing the H-Y antigen in an MHC-restricted fashion (Gordon et al., 1975)." And indeed this turned out to be the case also for Mrs R (Goulmy et al., 1977).

The study of Mrs R provided another interesting observation. In her serum, Aad van Leeuwen could demonstrate an "MHC-restricted" antibody of the same specificity as the HLA-A2-restricted anti-H-Y CTLs (van Leeuwen et al., 1979). Unfortunately, we were not able to pursue this fascinating line of research further.

THE MOLECULE AND ITS FUNCTION

Although we and others had already attempted to define the molecular structure of the HLA (and Group Five) antigens in 1967 (all unsuccessfully), that fascinating line of research has not been described in detail. It will be mentioned only briefly and for the sake of completeness. The HLA antigens are encoded on the short arm of chromosome 6 (van Someren et al., 1974; Breuning et al., 1977). HLA class I antigens are composed of two subunits, of 44 kDa and 12 kDa. The larger one is encoded by the HLA genes, whereas the smaller subunit is characterized as β_2-microglobulin (Grey et al., 1973). The alloantisera detecting HLA-DR were shown to be similar to mouse Ia antigens (Giphart et al., 1977) and form a second class of transplantation antigens (Springer et al., 1977). It took until 1979 for Orr et al. to describe the complete amino acid sequence of B7 and for Ploegh et al. to describe cell-free mRNA translation. They both worked in Strominger's group (Orr et al., 1979; Ploegh et al., 1979).

The following decade showed immense progress in the unravelling of the genetics, not only of HLA, but also of the complete region of the major histocompatibility complex. Despite this progress, the function of the HLA molecules was not resolved until the crystal structure of HLA-A2 became available (Bjorkman et al., 1987a). HLA molecules bind peptides and interact with the T cell receptor (Bjorkman et al., 1987b). This discovery has revolutionized immunology.

In 1974 Zinkernagel and Doherty and also Shearer had already described the phenomenon of MHC restriction: a milestone in our understanding of the function of MHC on the molecular level. A more detailed discussion of the function of the HLA molecules is provided in Chapter 5 (Shearer, 1974; Zinkernagel and Doherty, 1974).

THE BIOLOGICAL IMPORTANCE OF HLA

By 1967 Ceppellini had already stated that:

> *Nature had certainly not maintained a polymorphism such as that of HLA only to frustrate transplant surgeons.*

True enough, while the HLA workers were discovering HLA and getting to grips with its clinical importance, those who were working with animals laid the foundation for an understanding of its function.

In 1963, during his sabbatical in New York, the author heard Benacerraf tell someone in an elevator that he had found that one strain of guinea pigs responded to one of Michael Sela's antigens and another strain did not: the discovery of the Ir-gene phenomenon (Levine et al., 1963). In 1964 Lilly, Boyse and Old showed that susceptibility to Gross Leukemia Virus was H-2 linked (Lilly et al.,

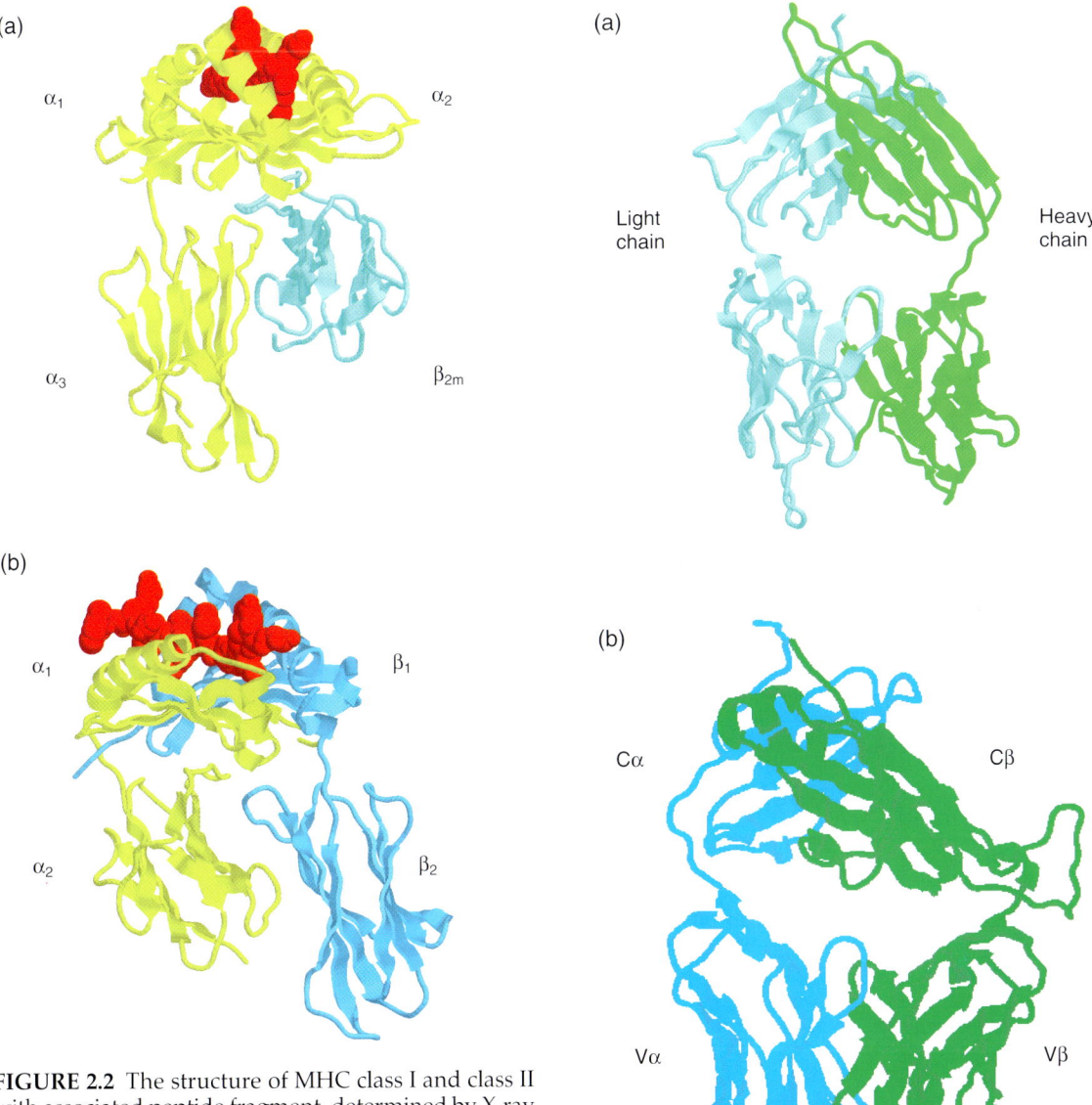

FIGURE 2.2 The structure of MHC class I and class II with associated peptide fragment, determined by X-ray crystallography. The crystal structure of MHC class I, (HLA B8), shows the molecule in association with a peptide fragment (HIV-1, Gag peptide) (Reid et al., 1996). The α_3 and β_2 microglobulin have a similar folded structure to immunoglobulin constant domains. The α_1 and α_2 domains are folded such as to produce two segmented α-helices lying on a sheet of eight antiparallel β-strands. This creates a binding site for the peptide fragment (red). The computer graphic representation of MHC class II I-Ak with its bound peptide (Hen Eggwhite Lysosyme) (Fremont et al., 1998), has a domain structure similar to that of the class I molecule. However, the domains forming the peptide-binding cleft are contributed by different chains. The peptide-binding groove is also open at both ends, as opposed to class I molecules, in which the ends of the peptide are buried within the molecule.

FIGURE 2.4 The TCR is similar to a membrane-bound antibody Fab fragment. Two ribbon diagrams, (a) an antibody F_{ab} fragment (anti-E-selectin monoclonal antibody 7A9 [Rodriguez-Romero et al., 1998]) and (b) a TCR αβ heterodimer (Garcia et al., 1996), are similar in comparison. The F_{ab} antibody fragment is a disulphide-linked heterodimer, with each chain (light or heavy chain) containing one immunoglobulin constant and one variable domain. The TCR is also a disulphide linked heterodimer, with each chain containing an immunoglobulin constant (Cα o + r Cβ) or variable-like (Vα or Vβ) domain. For both the TCR and F_{ab} molecules the antigen-binding site is formed by the juxtaposition of the variable domains.

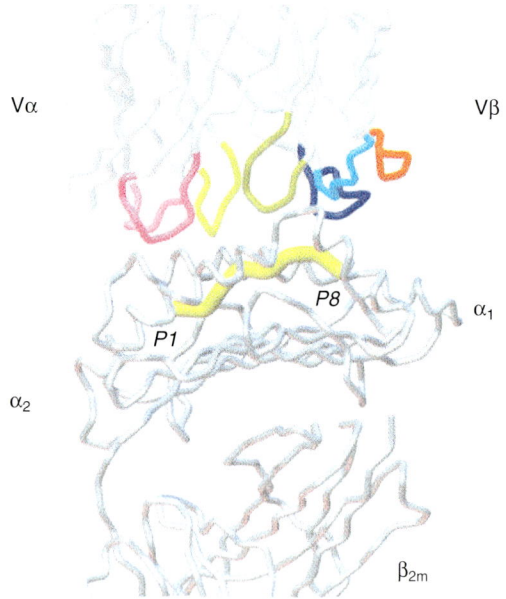

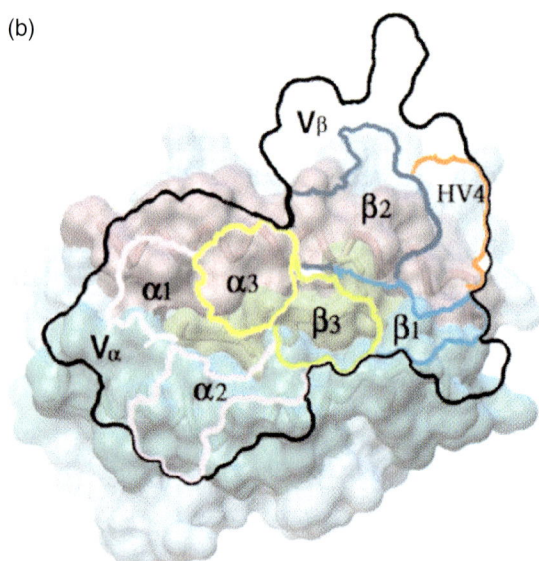

FIGURE 2.5 Interaction of the TCR complementary determining regions (CDRs) with a peptide-MHC complex. The TCR binding with a peptide-MHC complex is illustrated, showing the positions of the CDRs (coloured). (a) The key peptide anchor residues for MHC binding are indicated (P1 and P8). A more detailed map of the interaction of the CDRs with the peptide MHC surface is shown. (b) The TCR Vα and Vβ CDR regions are labelled as α_1 (for CDR1α) β_1 (for CDR1β), etc. (After Garcia et al., 1996.)

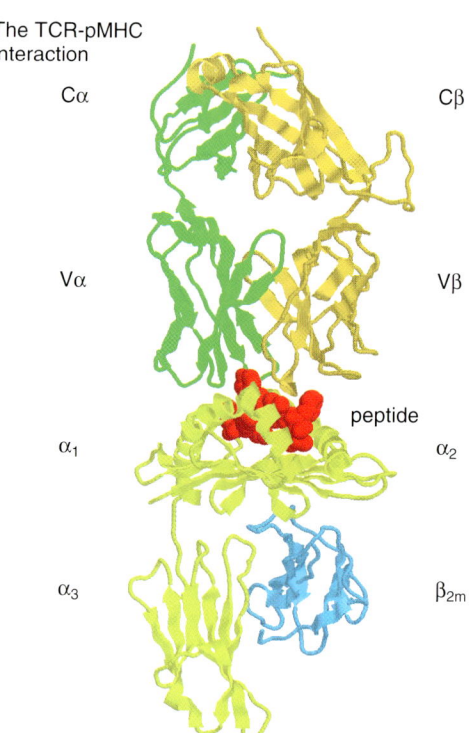

FIGURE 2.6 The TCR-peptide-MHC complex. The crystal structure illustrates the interaction between the TCR and a peptide (Tax peptide from human T lymphotropic virus type 1) and MHC class I (HLA-A2) (Garboczi et al., 1996).

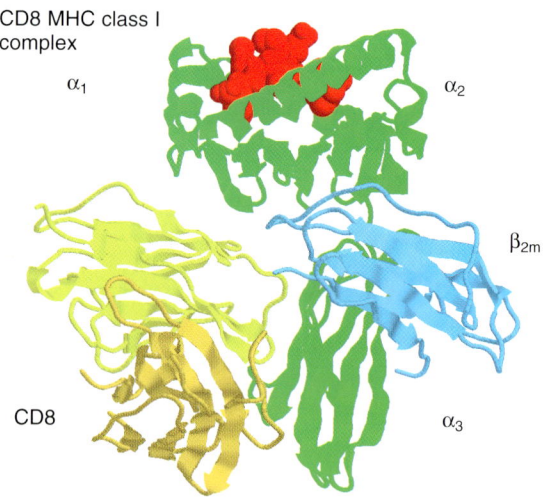

FIGURE 2.7 Crystal structure of the MHC class I-CD8 complex. The structure shown is that of a CD8 α-chain homodimer ($\alpha\alpha$), which both resembles and functions like the $\alpha\beta$ heterodimer. The CD8 binds to a site at the base of the α3 domain of the MHC class I (HLA-A2). The peptide fragment (red) is shown within the MHC-binding groove (Gao et al., 1997).

1964). A year later Hugh McDevitt and Michael Sela (1965) showed the same for the antibody response against artificial antigens.

Epidemiological studies on the importance of HLA in infections, probably its main biological function, are relatively rare. During the Evian Meeting in 1972 Piazza, Ceppellini and their colleagues presented their classic study on the association between endemic malaria and HLA in four Sardinian villages, a study which has recently been confirmed and extended by Hill and colleagues (Piazza et al., 1973; Hill et al., 1991). In 1976, René de Vries and colleagues did the same for leprosy, then much debated (De Vries et al., 1976). There are many examples of such endemics and epidemics. One of these concerned Dutch farmers who immigrated in the middle of the nineteenth century to Surinam in South America to start a "new life". A few weeks after their arrival, first an epidemic of typhoid fever and then one of yellow fever killed over half of the immigrants. They were very religious, married only between themselves and raised enormous families. René de Vries decided to compare the HLA gene frequencies of the descendants of the survivors of the epidemic with those of the present-day Dutch. Figure 1.4 depicts some of the results (De Vries et al., 1989). The black bars are those of the survivors of the epidemic and the white hatched bars those of the present-day Dutch. Although the frequencies of HLA-DR1, -DR3 and -DR7 are not different from the present-day Dutch, HLA-DR2 is absent, and HLA-DR4 and -DR13 are increased (De Vries et al., 1979). A founding effect due to recruitment from within a genetically similar pool seems to be an unlikely explanation, because the farmers had been recruited from all over The Netherlands. One possible explanation is that HLA-DR2-positive individuals could not overcome a typhoid and/or yellow fever attack, while those that were HLA-DR4 and -DR13 positive had a better than average chance of survival.

CLINICAL IMPORTANCE

Disease associations

Lilly's findings led Amiel to study the importance of the HLA antigens in Hodgkin's disease. During the third Histocompatibility Workshop in Torino in 1967 he reported a significant increase in HLA-B5 in Hodgkin's patients (Amiel, 1967). Amiel, who was extremely myopic, had to read the tests himself during the wet workshop because his technician had fallen ill. His results came out completely random and therefore nobody believed him. However, a few years later Morris confirmed his findings. Stokes et al. (1972) showed that HLA-B8 is associated with coeliac disease and in the same year Naito et al. (1972) from Terasaki's group reported the association of HLA-A3 with multiple sclerosis.

In the same year, Terasaki was contacted by Donald Kuba, who told him that four psoriasis patients all had the same HLA antigen, HLA-B13 (Russel et al., 1972), a finding that he confirmed (White et al., 1972). A year later, Terasaki again, now with the group of Brewerton from London, presented the linkage of HLA-B27 with ankylosing spondylitis (Brewerton et al., 1973; Scholsstein et al., 1973). The world went wild and it was predicted that internal medicine textbooks would have to be rewritten, the mechanism of immunopathogenetic disease could be understood, and so on. Although the clinical relevance was predicted to be very high, in practice so far it has been disappointingly low and it took another 25 years before insight into MHC restriction, its crystalline structure and peptide identification opened a new era in our understanding of disease associated with HLA.

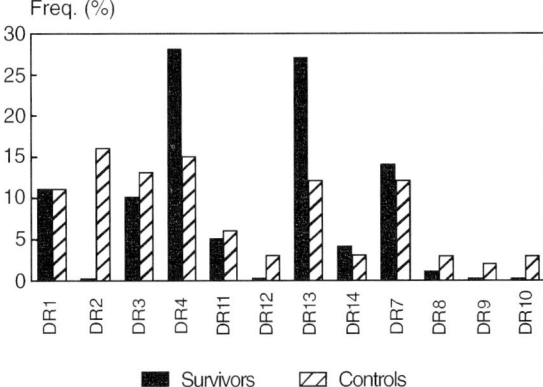

FIGURE 1.4 HLA-DR gene frequencies of survivors versus controls, after excluding six B58-DR7 haplotypes of the survivors, which may have been of Negroid origin.

Transplantation

The 1965 workshop was also memorable for the demonstration that the HLA antigens are transplantation antigens. The protocol was simple. Peter Medawar had shown in his classical experiment that if a piece of skin was exchanged between two rabbits, the transplant was rejected after 10 days, but when a second piece of skin from the same donor was transplanted to the same recipient, that skin was rejected in 5 days. Medawar concluded that the first skin had immunized the rabbit and that homograft sensitivity had been induced (Medawar, 1946). In the next experiment, Medawar did not exchange skin but instead injected the buffy coat cells of a blood sample from the donor intradermally in the recipient, and only after that transplanted skin from the buffy coat donor. Now this first transplant was rejected not after 10 days but after 5 days; in other words, an intradermal injection of leucocytes had induced the same homograft sensitivity as a skin transplant. It follows from this that these leucocytes must carry the same transplantation antigens as skin. With this the whole field of transplantation immunology began.

All that remained to be done was first to identify those transplantation antigens on leucocytes and secondly to show that matching for these antigens improved graft survival. The first had been done by several groups and the second was possible thanks to Felix Rapaport, who had developed a very good skin-grafting technique (Rapaport et al., 1962). Volunteers were needed. We approached many of the colleagues who had helped to build up the large panel by which 4a and 4b had been identified; they were very shocked when they were asked whether they would be willing to give and/or receive a skin graft. George Eernisse and I set the example, and then others agreed to collaborate. (This was long before the days of medical ethics committees.) Following a protocol worked out with Balner in monkeys (Balner et al., 1965), 2×10^8 leucocytes mismatched with the recipient for one antigen only (B7Y) were injected intradermally and 2 weeks later a skin graft was transplanted from two unrelated donors, one of whom was B7 negative and one B7 positive. Mean skin graft survival was 11 days for the B7-negative donors and 6 days for the B7-positive donors, proving that HLA antigens were transplantation antigens (Table 1.5) (van Rood et al., 1965).

TABLE 1.5 Survival time of 6^b6^c-compatible and incompatible skin grafts in recipients preimmunized against 6^b6^c antigens

Immunizing dose (i.d.)	Interval	Survival of skin graft (days)	
		$6^{b+}6^{c+}$	$6^{b-}6^{c-}$
12×10^7 leuc.	14	0	12
22×10^7	18	5	9
20×10^7	16	5.5	8.5
5×10^7	13	7	14
20×10^7	14	9	11
21×10^7	14	9.5	11.5
Mean		6	11

$6^{b+}6^{c+}$: now HLA-B7.
Van Rood et al. (1965).

After the Leiden workshop we travelled to Torino, where Ruggero Ceppellini and his group had exchanged skin transplants between family members to perform the HLA typing of these donors and recipients. It could indeed be shown that graft survival in non-immunized recipients also correlated with HLA matching. Matching for the independent Group Five system had no influence. Amos made similar observations at almost the same time (Amos et al., 1967).

In 1966, flying back from Zurich to London I discussed our findings with Sir Michael Woodruff, a surgeon, and one of the first to perform renal allograft transplantation from family donors. He also discovered anti-lymphocyte globulin (ALG). Woodruff became quite enthusiastic and persuaded us to come to Edinburgh to study his patients. After Edinburgh, we set out with Aad van Leeuwen, Ali Schippers and Hans Bruning on an odyssey around the transplant centres of the world. It was necessary to be on the spot to be able to do the leucocyte typing because an agglutination assay was used and the cells had to be fresh. In this way, data on over 100 patients and their donors were collected. When the sib transplants were analysed, there were far more HLA-identical siblings pairs and far fewer incompatible ones than expected. The 18 patients who could be typed came from a group of 40, of whom 22 had died in the meantime (Table 1.6) (van Rood et al., 1968). At that time

TABLE 1.6 Relation between kidney graft survival and leucocyte groups

	Sib to sib			
	Random families	Random expected	Patients found	Patients expected
	A	B	C	D
Identical	174	10.4	10	4.7
Compatible	137	8.3	2	3.7
Incompatible	358	21.3	6	9.6
Total alive				
Total operated		40		

In the various centres a total of 40 kidney transplants had been performed in which a sib had donated a kidney to another sib. Of these, 18 patients survived. If leucocyte groups were of no importance in early graft rejection it would be expected that the number of identical, compatible and incompatible donor and recipient pairs for the surviving patient material (C) would be the same as the number expected on the basis of the random family material (D). The expected patient frequencies (D) could be calculated on the basis of the data obtained from random families (A). If the distribution of identical, compatible and incompatible combinations had originally been about the same in the total operated patient material as in the random family material, the numbers shown under random expected (B) would have been found. It can be seen that significantly more identical than compatible or incompatible combinations have survived. Van Rood et al. (1968).

chronic dialysis was not readily available and if a patient rejected a transplanted kidney this meant certain death. The conclusion was clear: when donor and recipient are HLA identical the patient is less likely to reject the kidney. This study (and a similar study in which the donor was the parent) proved that the antigens that could be recognized by serology on leucocytes were transplantation antigens, which were important not only for skin graft but also for kidney graft survival (van Rood et al., 1968).

Shortly after the completion of this study the Third Histocompatibility Workshop took place in Turin, where Terasaki et al., Payne et al. and Dausset et al. presented similar data (1967). It was at that meeting that a proposal was made to start an international organization for organ exchange, Eurotransplant (van Rood, 1967). The name Eurotransplant had come to the author on passing a truck on which was written "Eurotransport". By changing only three letters, the name Eurotransplant came into existence.

Amazing as it may seem in hindsight, only a year before, in 1966, the Aarhus group had shown that a cross-match should be performed before kidney transplantation (Kissmeyer-Nielsen et al., 1966). It took several years, however, before the cross-match was implemented by all transplant teams.

In the beginning the proposal to start Eurotransplant received a rather mixed reception. With one or two exceptions, even close colleagues were rather sceptical. Nevertheless, the idea received the support of quite a few transplant centres, in the beginning especially in Belgium. The first kidney was flown in by helicopter from Louvain and transplanted in September 1967, much earlier than we had planned. In 1967 a total of 11 kidneys was transplanted under the auspices of Eurotransplant. In 1968 this number rose to 60 and then it steadily increased and now exceeds over 6000 kidneys, hearts, livers, lungs, pancreas and corneas annually. This was a good time. In a close co-operation between surgeons, nephrologists and immunologists, each patient was discussed before being selected for transplantation and the whole team was activated when a kidney became available. I remember well the direct involvement with the transplant procedure and how enjoyable this was.

The first analysis of the Eurotransplant data was presented at a meeting of the International Transplantation Society in The Hague in 1970 (van Rood et al., 1971). The meeting became quite famous, not so much for its scientific importance as for the very liberal availability of Dutch beer and gin and the excellent social programme. The intention was to give an overview of the activities of Eurotransplant at that meeting. However, when the data were analysed it was found that there was really very little to be said about the effect of matching. One thing that was noted, though, was that the results were much better than those reported by the International Registry for Kidney Transplantation. At the time we assumed that this was due to better than average matching, but no hard data were available to support this. With a longer follow-up, however, it became clear that in these early years long-term graft survival had indeed been improved (Figure 1.5) (van Hooff et al., 1985). The discussion on whether HLA matching improves organ graft (and patient) survival has remained until this day.

Another important contribution was made by Opelz and Terasaki, who showed that blood transfusions could not only immunize, but also downregulate allograft reactivity and improve kidney graft survival (Opelz et al., 1973). Persijn from Eurotransplant showed that even a single pretransplant blood transfusion could improve graft survival (Persijn et al., 1977). It has been a riddle for a very long time as to what downregulates and what upregulates allograft immunity. Malice Lagaaij showed convincingly that a pretransplant blood transfusion which shares a DR antigen with the blood transfusion recipient will cause improved graft survival, whereas if there is no sharing of DR antigens, survival will be poor (Lagaaij et al., 1989).

The conclusion was and still is that blood transfusions can downregulate the immune response if donor and recipient share some HLA antigens. These data have been confirmed, although not in all studies (van Hooff and van den Berg-Loonen, 1992; Middleton et al., 1994; Bayle et al., 1995; van der Mast and Balk, 1997) and remains a topic of ongoing investigations.

CLOSING REMARKS

Writing a history on any topic is an open-ended exercise. When one wants to stop, history goes on and it becomes an arbitrary decision as to what is history, the past, and what is the present, or better, present history. Whether it is medicine, art or the stock market one is always incomplete and makes arbitrary choices. Well, that is not quite true. The choices one makes are strongly influenced by the personal involvement and experiences of the writer in that field. As a result, the contributions of the group of the writer receive too much emphasis and those of others too little. The author has attempted to correct for this at least in part by including bits and pieces of unwritten history. These anecdotes are little spotlights on the individuals who participated in the amazing adventure that is the discovery of HLA.

That adventure had many ups and, of course, also its downs. It has, however, never lost its impetus because the HLA community as a whole had such open communication between all who participated in this exercise. Untroubled by patent rights reagents and technologies were exchanged long before they had been published. This has contributed in a major way to the speed and effectiveness by which HLA was placed on the map of biomedicine.

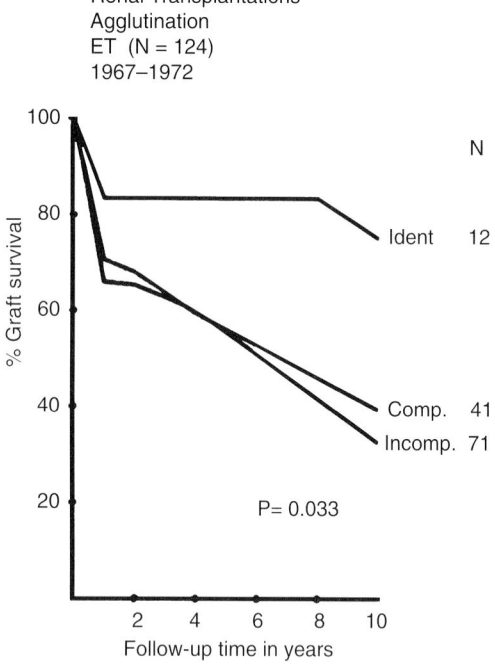

FIGURE 1.5 Long-term follow-up of the first 124 kidneys transplanted under the auspices of Eurotransplant. Transplants from donors which were identical with the recipient for the nine cross-reactive antigens then recognized did significantly better than those which were compatible or incompatible (van Rood et al., 1991).

ACKNOWLEDGEMENTS

I would like to thank Aad van Leeuwen for her assistance throughout and especially in the literature documentation, George Eernisse and Marius Giphart for proofreading and valuable suggestions and Hedy Richter and Wally van der Geest for secretarial help.

REFERENCES

Amiel, J.L. (1967). In *Histocompatibility Testing 1967, Report of a Conference and Workshop* (eds E.S. Curtoni, P.L. Mattiuz and R.M. Tosi), pp. 79–81, Munksgaard, Copenhagen, and Williams & Wilkins, Baltimore, MD.

Amos, D.B. (1965). Discussion. In *Histocompatibility Testing 1965, Report of a Conference and Workshop* (eds H. Balner, F.J. Cleton and J.G. Eernisse), p. 70, Munksgaard, Copenhagen.

Amos, D.B. and Cohen, I., with Nicks, J.P., MacQueen, M.M. and Mladick, E. (1967). In *Histocompatibility Testing 1967, Report of a Conference and Workshop* (eds E.S. Curtoni, P.L. Mattiuz and R.M. Tosi), pp. 129–138, Munksgaard, Copenhagen, and Williams & Wilkins, Baltimore, MD.

André, R., Dreyfus, B. and Bessis, M. (1954). Anticorps antileucocytaires dans un cas de leucémie lymhoide. *Rev. Hémat.* **9**: 50–67.

Bach, F.H. and Amos, D.B. (1967). Hu-1: major histocompatibility locus in man. *Science* **156**: 1506–1508.

Bach, F.H. and Hirschhorn, K. (1964). Lymphocyte interaction: a potential histocompatibility test *in vitro*. *Science* **143**: 813–814.

Bain, B., Vas, M.R. and Löwenstein, L. (1964). The development of large immature mononoclear cells in mixed leucocyte cultures. *Blood* **2**: 108–116.

Balner, H., Dersjant, H. and Rood, J.J. van (1965). A method to relate leukocyte antigens and transplantation antigens in monkeys. *Transplantation* **3**: 230–234.

Batchelor, J.R. (1990). Memories of HLA. In *History of HLA: Ten Recollections* (ed. P.I. Terasaki), p. 180, UCLA Tissue Typing Laboratory, Los Angeles, CA.

Bayle, F., Masson, D., Zaoui, P. et al. (1995). Beneficial effect of one HLA haplo- or semi-identical transfusion versus three untyped blood units on alloimmunization and acute rejection episodes in first renal allograft recipients. *Transplantation* **59**: 719–723.

Bjorkman, P.J., Saper, M., Samraoui, B. et al. (1987a). Structure of the human class I histocompatibility antigen HLA-A2. *Nature* **329**: 506–512.

Bjorkman, P.J., Saper, M., Samraoui, B. et al. (1987b). The foreign antigen binding site and T cell recognition regions of class I histocompatibility antigens. *Nature* **329**: 512–518.

Bodmer, W.F. and Payne, R. (1965). Cross-overs in component systems are unlikely to occur. In *Histocompatibility Testing 1965, Report of a Conference and Workshop* (eds H. Balner, F.J. Cleton and J.G. Eernisse), p. 141, Munksgaard, Copenhagen.

Bradley, B.A., Edwards, J.M., Dunn, D.C. and Caine, R.Y. (1972). Quantitation of mixed lymphocyte reaction by gene dosage phenomenon. *Nat. New Biol.* **240**: 54–56.

Breuning, M.H., Berg-Loonen, E.M. van den, Bernini, L.F. et al. (1977). Localization of HLA on the short arm of chromosome 6. *Human Genet.* **37**: 131–139.

Brewerton, D.A., Hart, F.D., Nicholls, A. et al. (1973). Ankylosing spondylitis and HL-A27. *Lancet* **i**: 904–907.

Cochrum, K.C, Perkins, H.A., Payne R. et al. (1973). The correlation of MLC with graft survival. *Transplant Proc.* **5**: 391–396.

Coombs, R.R.A., Mourant, A.E. and Race, R.R. (1945). Detection of weak and "incomplete" Rh agglutinins: a new test. *Lancet* **ii**: 15–16.

Dausset, J. (1954). Leuco-agglutinins IV, leuco-agglutinins and blood transfusion. *Vox Sang.* **4**: 190–198.

Dausset, J. (1958). Iso-leuco-anticorps. *Acta Haemat.* **20**: 156–166.

Dausset, J. (1990). The HLA adventure. In *History of HLA: Ten Recollections* (ed. P.I. Terasaki), p. 3, UCLA Tissue Typing Laboratory, Los Angeles, CA.

Dausset, J. and Brecy, H. (1958). Identical nature of the leukocyte antigens detectable in monozygotic twins by means of immune iso-1leuco-agglutinins. *Nature* **180**: 1430.

Dausset, J., Nenna, A. and Brecy, M. (1954). Leuko-agglutinins V. Leuko-agglutinins in chronic idiopathic or symptomatic pancytopenia and in paroxysmal nocturnal hemoglobinuria. *Blood* **9**: 696–720.

Dausset, J., Rapaport, F.T. and Legrand, L. (1967). Selection des donneurs par les groupes tissulaires. Advance in Transplantation. First International Congress of the Transplantation Society, Paris, Munksgaard, Copenhagen.

Eijsvoogel, V.P., Rood, J.J. van, Toit, E.D. Du and Schellekens, P.T. (1972). Position of a locus determining mixed lymphocyte reaction distinct from the known HL-A foci. *Eur. J. Immunol.* **2**: 413–418.

Giphart, M.J., Kaufman, J.F., Fuks, A. et al. (1977). HLA-D associated allo-antisera react with molecules similar to Ia antigens. *Proc. Natl Acad. Sci. USA* **74**: 3533–3536.

Gordon, R.D., Simpson, E. and Samelson, L.E. (1975). In vitro cell-mediated immune responses to the male specific (H-Y) antigen in mice. *J. Exp. Med.* **142**: 1108–1120.

Gorer, P.A. (1947). The antibody response to tumor inoculation in mice with secial reference to partial antibodies. *Cancer Res.* **7**: 634–641.

Goudsmit, R. and Loghem, J.J. van (1953). Studies on the occurrence of leucocyte-antibodies. *Vox Sang.* **3**: 58–67.

Goulmy, E., Termijtelen, A., Bradley, B.A. and Rood J.J. van (1977). Y-antigen killing by women is restricted by HLA. *Nature* **266**: 544–545.

Grey, H.M., Kubo, R.T., Colon, S.M. et al. (1973). The small subunit of HL-A antigens is b2-microglobulin. *J. Exp. Med.* **138**: 1608–1612.

Grumet, F.C. and Leventhal, B.G. (1970). Inhibition of the response in mixed leukocyte cultures by alloimmune plasma. *Transplantation* **9**: 405–409.

Hill, A.V.S., Allsopp, C.E.M., Kwiatkowski, D. et al. (1991). Common West African HLA antigens are associated with protection from severe malaria. *Nature* **352**: 595–600.

Hooff, J.P. van and Berg-Loonen, P.M. van den (1992). The influence of DR match of blood donor and recipient on the formation of T- and B-cell antibodies and on renal allograft outcome. *Transplant. Int.* **5** (Suppl. 1): S599–S600.

Hooff, J.P. van, Leeuwen, A. van, Paul, L. et al. (1985). The influence of matching for broadly reacting antigens on long-term kidney graft survival. *Transplant. Proc.* **XVII**: 2205–2206.

Jager M.J., Claas, F.H.J., Schot, J.D.L. et al. (1988). Genetics of two human monocyte antigens. *Human Immunol.* **22**: 163–170.

Kissmeyer-Nielsen, F., Olsen, S., Petersen, V.P. and Fjeldborg, O. (1966). Hyperacute rejection of kidney allografts, associated with pre-existing humoral antibodies against donor cells. *Lancet* **ii**: 662–665.

Kissmeyer-Nielsen, F., Svejgaard, A., Ahrons, S. and Staub-Nielsen, L.S. (1969). Crossing-over within the HL-A system. *Nature* **224**: 75–76.

Koch, C.T., Hooff, J.P. van, Leeuwen, A. van et al (1973). The relative importance of matching for the MLC versus the HL-A loci in organ transplantation. In *Histocompatibility Testing 1972* (eds J. Dausset and J. Colombani), pp. 521–524, Munksgaard, Copenhagen.

Lagaaij, E.L., Henneman, I.Ph.H., Ruigrok, M. et al. (1989). Effect of one HLA-DR antigen matched and completely HLA-DR mismatched blood transfusions on survival of heart and kidney allografts. *N. Engl. J. Med.* **321**: 701–705.

Landsteiner, K. (1901). Uber Agglutinationserscheinungen normalen Menschlichen Blutes. *Wien. Klin. Wschschr.* **14**: 1132–1134.

Leeuwen, A. van, Eernisse J.G. and Rood J.J. van (1964). A new leucocyte group with two alleles: leucocyte group Five. *Vox Sang.* **9**: 431–446.

Leeuwen, A. van, Schuit, H.R. and Rood, J.J. van (1973). Typing for MLC (LD). II. The selection of nonstimulator cells by MLC inhibition tests using SD-identical stimulator cells (MISIS) and fluorescence antibody studies. *Transplant. Proc.* **5**: 1539–1542.

Leeuwen, A. van, Winchester, R.J. and Rood, J.J. van (1975). Serotyping for MLC. II. Technical aspects. *Ann. N.Y. Acad. Sci.* **254**: 289–295.

Leeuwen, A. van, Goulmy, E. and Rood, J.J. van (1979). MHC restricted antibody reactivity mainly, but not exclusively, directed against cells from male donors. *J. Exp. Med.* **150**: 1075–1083.

Lehndorff, H. (1951). Transitorische Granulocytopenie beim Neugeborenen. *Helvet. Paediat. Acta* **6**: 173–183.

Levine, B.B., Ojeda, A. and Benacerraf, B. (1963). Studies on artificial antigens. III. The genetic control of the immune response to hapten-poly-L-lysine conjugates in guinea pigs. *J. Exp. Med.* **118**: 953–957.

Lightbody, J.J., Bernoco, D., Miggiano, V.C. and Ceppellini, R. (1971). Cell mediated lympholysis in man after sensitization of effector lymphocytes through mixed leukocyte cultures. *J. Bact. Virol. Immunol.* **64**: 243–254.

Lilly, F., Boyse, E.A. and Old, L.J. (1964). Genetic basis for susceptibility to viral leukaemogenesis. *Lancet* **ii**: 1207–1209.

Loghem, J.J. van, Sauer, A.J., Hart, M. van der et al. (1956). Zeldzame immunologische afwijkingen als oorzaak van bloedtransfusiereacties bij een lijder aan verworven haemolytische anaemie. *Ned. Tijdschr. Geneeskd.* **100**: 314–323.

Major, R.H. (1945). *Thucydides History of the Peleponnesian War, Classic Description of Disease*, p. 75, Charles C. Thomas, Springfield, IL.

Mast, B.J. van der and Balk, A.H.M.M. (1997). The effect of HLA-DR shared blood transfusion on the clinical outcome of heart transplantation. *Transplantation* **63**: 1514–1519.

McDevitt, H.O. and Sela, M. (1965). Genetic control of the antibody response. I. Demonstration of determinant-specific differences in response to synthetic polypeptide antigens in two strains of inbred mice. *J. Exp. Med.* **122**: 517–531.

Medawar, P.B. (1946). Immunity to homologous grafted skin. II. The relation between the antigens of blood and skin. *Br. J. Exp. Pathol.* **27**: 15–24.

Mempel, W., Grosse-Wilde, H., Baumann, P. et al. (1973). Population genetics of the MLC response: typing for MLC determinants using homozygous and heterozygous reference cells. *Transpl. Proc.* **5**: 1529–1534.

Middleton, D., Martin, J., Douglas, J.F. and McClelland, M. (1994). Transfusion of one HLA-DR antigen-matched blood to potential recipients of a renal allograft. *Transplantation* **58**: 845–848.

Miescher, P. and Fauconnet, M. (1954). Mise en évidence de différents groupes leucocytaires chez l'homme. *Schweiz. Med. Wschschr.* **84**: 597–599.

Moeschlin, S. and Wagner, K. (1952a). Leukocytenagglutinine als Ursache von Agranulocytosen (Pyramidon usw.) *Schweiz. Med. Wschschr.* **82**: 1104–1107.

Moeschlin, S. and Wagner, K. (1952b). Agranulocytosis due to the occurrence of leukocyte-agglutinins (pyramidon and cold agglutinins). *Acta Haemat.* **8**: 29–41.

Moeschlin, S. and Moreno, R. (1954). Pyramidon agranulocytose mit Agglutininen für Arteigene und Artfremde Leukocyten. *Klin. Wschschr.* **32**: 799–803.

Naito, S., Namerow, N., Mickey, M.R. and Terasaki, P.I. (1972). Multiple sclerosis: association with HL-A3. *Tissue Antigens* **2**: 1–4.

Nomenclature for Factors of the HLA System (1977). In *Histocompatibility Testing 1977* (eds W.F. Bodmer, J.J. Batchelor, J.G. Bodmer, H. Festenstein and P.J. Morris), p. 14, Munksgaard, Copenhagen.

Opelz, G., Sengar, D.P.S., Mickey, M.R. and Terasaki, P.I. (1973). Effect of blood transfusions on subsequent kidney transplantation. *Transplant. Proc.* **4**: 253–259.

Orr, H.T., Lopez de Castro, J.A., Lancet D. and Strominger, J.L. (1979). Complete amino acid sequence of a papain-solubilized human histocompatibility antigen, HLA-B7. 2. Sequence determination and search for homologies. *Biochemistry* **18**: 5711–5720.

Payne, R. (1965). Discussion. In *Histocompatibility Testing 1964, Report of a Conference and Worksho*p (eds P.S. Russell, H.J. Winn and D.B. Amos), pp. 37–40, National Academy of Sciences, Washington, DC.

Payne, R. (1990). Early history of HLA. In *History of HLA: Ten Recollections* (ed. P.I. Terasaki), p. 23, UCLA Tissue Typing Laboratory, Los Angeles, CA.

Payne, R. and Rolfs, M.R. (1958). Fetomaternal leukocyte incompatibility. *J. Clin. Invest.* **37**: 1756–1763.

Payne, R., Tripp, M., Weigle, J. et al. (1964). A new leukocyte isoantigenic system in man. *Cold Spring Harbor Symp. Quant. Biol.* **29**: 285–295.

Payne, R., Perkins, H.A. and Najarian, J.S. (1967). Compatibility for specific leukocyte antigens in kidney transplants using the agglutination test. *Advances in Transplantation*. First International Congress of the Transplantation Society, Paris, Munksgaard, Copenhagen.

Persijn, G.G., Hooff, J.P. van, Kalff, M.W. et al. (1977). Effect of blood transfusion and HLA matching on renal transplantation in the Netherlands. *Transplant. Proc.* **IX**: 503–505.

Piazza, A., Belvedere, M.C., Bernoco, D. et al. (1973). HL-A variation in four Sardinian villages under differential selective pressure by malaria. In *Histocompatibility Testing 1972* (eds J. Dausset and J. Colombani), pp. 73–84, Munksgaard, Copenhagen.

Ploegh, H.L., Cannon, L.E. and Strominger, J.L. (1979). Cell-free translation of the mRNAs for the heavy and light chains of HLA-A and HLA-B antigens. *Proc. Natl Acad. Sci. USA* **76**: 2273–2277.

Race, R.R. and Sanger, R. (1950) *Blood Groups in Man*, Blackwell Scientific Publications, Oxford.

Rapaport, F.T., Lawrence, H.S., Thomas, L. et al. (1962). Cross-reactions to skin homografts in man. *J. Clin. Invest.* **41**: 2166–2172.

Rood, J.J. van (1962). *Leucocyte grouping. A method and its application*, Thesis, Rijksuniversiteit Leiden, Den Haag, J.H. Pasmans, pp. 1–58.

Rood, J.J. van (1967). A proposal for international cooperation in organ transplantation: Eurotransplant. In *Histocompatibility Testing 1967* (eds E.S. Curtoni, P.L. Mattiuz and R.M. Tosi), p. 451, Munksgaard, Copenhagen, and Williams & Wilkins, Baltimore, MD.

Rood, J.J. van and Leeuwen, A. van (1965). Defined leucocyte antigenic groups in man. In *Histocompatibility Testing 1964, Report of a Conference and Workshop* (eds P.S. Russell, H.J. Winn and D.B. Amos), pp. 21–37, National Academy of Sciences, Washington, DC.

Rood, J.J. van, Eernisse, J.G. and Leeuwen, A. van (1958). Leucocyte antibodies in sera from pregnant women. *Nature* **181**: 1735–1736.

Rood, J.J. van, Leeuwen, A. van and Fernisse, J.G. (1959). Leucocyte antibodies in sera of pregnant women. *Vox Sang.* **4**: 427–444.

Rood, J.J. van, Leeuwen, A. van and Fernisse, J.G. (1961). Antibodies against leucocytes in sera of pregnant women. *Vox Sang.* **6**: 240–241.

Rood, J.J. van, Leeuwen, A. van and Bosch, L.J. (1962). Leucocyte antigens and transplantation immunity. *Proc. 8th. Congr. Eur. Soc. Hemat.*, Vienna 1961, 199 pp., Basel, Karger.

Rood, J.J. van and Leeuwen, A. van (1963). Leucocyte grouping. A method and its application. *J. Clin. Invest.* **42**: 1382–1390.

Rood, J.J. van, Leeuwen, A. van, Schippers, A.M.J. et al. (1965). Leukocyte groups, the normal lymphocyte transfer test and homograft sensitivity. In *Histocompatibility Testing 1965, Report of a Conference and Workshop* (eds H. Balner, F.J. Cleton and J.G. Fernisse), pp. 37–50, Munksgaard, Copenhagen.

Rood, J.J. van, Leeuwen, A. van, Bruning, J.W. and Porter, K.A. (1968). The importance of leukocyte antigens in renal transplantation. A study of patients of: G.R.J. Alexandre, J. van Geertruyden, W.D. Kelly, J.P. Merrill, J. Morelle, J.F. Mowbray, J. Murray, P.S. Russell, T.E. Starzl, Ch. Toussaint and M.F.A. Woodruff. In *Advances in Transplantation* (eds J. Dausset, J. Hamburger and G. Mathé), pp. 213–219, Munksgaard, Copenhagen.

Rood, J.J. van, Freudenberg, J., Leeuwen, A. van et al. (1971). Eurotransplant. *Transpl. Proc.* **III**: 933–941.

Rood, J.J. van, Leeuwen, A. van, Parlevliet, A et al. (1975a). LD typing by serology. VI. Description of a new locus with three alleles. In *Histocompatibility Testing 1975* (ed. F. Kissmeyer-Nielsen), pp. 629–636, Munksgaard, Copenhagen.

Rood, J.J. van, Leeuwen, A. van, Keuning, J.J. and Oud Alblas, A.B. van (1975b). The serological recognition of the human MLC determinants using a modified cytotoxicity technique. *Tissue Antigens* **5**: 73–79.

Rood, J.J. van, Leeuwen, A. van and Ploem, J.S. (1976). Simultaneous detection of two cell populations by two-colour fluorescence and application to the recognition of B-cell determinants. *Nature* **262**: 795–797.

Rood, J.J. van, Leeuwen, A. van and Fernisse, J.G. (1991). The Eurotransplant story. In *History of Transplantation: Thirty-Five Recollections* (ed. P.I. Terasaki), pp. 497–510, UCLA Tissue Laboratory, Los Angeles, CA.

Russel, T.J., Schultes, L.M. and Kuban, D.J. (1972). Histocompatibility (HLA-A) antigens associated with psoriasis. *N. Engl. J. Med.* **287**: 738–740.

Sandberg, L., Thorsby, E., Kissmeyer-Nielsen, F. and Lindblom, A. (1970). Evidence of a third sublocus within the HL-A chromosomal region. In *Histocompatibility Testing 1970* (ed. P.I. Terasaki), pp. 165–169, Munksgaard, Copenhagen.

Scholsstein, L., Terasaki, P.I., Bluestone, R. and Pearson, C.M. (1973). High association of an HLA antigen, W27, with ankylosing spondylitis. *N. Engl. J. Med.* **288**: 704–705.

Shearer, G.M. (1974). Cell-mediated cyto-toxicity to trinitophenyl-modified syngeneic lymphocytes. *Eur. J. Imm.* **4**: 527–533.

Snell, G.D. (1948). Methods for the study of histocompatibility genes. *J. Genet.* **49**: 87–108.

Someren, H. van, Westerveld, A., Hagemeyer, A. et al. (1974). Human antigen and enzyme markers in man–Chinese hamster somatic cell hybrids: evidence for synteny between the HL-A, PGM3, ME1 and IPO-B foci. *Proc. Natl Acad. Sci. USA* **71**: 962–965.

Springer, T.A., Kaufman, J.F., Siddoway, L.A. et al. (1977). Chemical and immunological characterization of HL-A linked B-lymphocyte alloantigens. *Cold Spring Harbor Symp. Quant. Biol.* **41**: 387–396.

Stokes, P.L., Asquith, P., Holmes, G.K.T. et al. (1972). Histocompatibility antigens associated with adult coeliac disease. *Lancet* **ii**: 162–164.

Terasaki, P.I. and McClelland, J.D. (1964). Microdoplet assay of human serum cytotoxins. *Nature* **204**: 998–1000.

Terasaki, P.I., Mickey, M.R. and McClelland, J.D. (1967). *Serotyping for homotransplantation XII. Evaluation of 200 antisera for their ability to detect transplantation antigens.* Advances in Transplantation. First International Congress of the Transplantation Society, Paris, Munksgaard, Copenhagen.

Tweel, J.G. van den, Oud Alblas, A.B. van, Keuning, J.J. et al. (1973). Typing for MLC (LD). I. Lymphocytes from cousin-marriage offspring as typing cells. *Transplant. Proc.* **5**: 1535–1538.

Vries, R.R.P. de, Lai A Fat, R.F.M., Nijenhuis, L.F. and Rood, J.J. van (1976). HLA-linked genetic control of host response to *Mycobacterium leprae. Lancet* **ii**: 1328–1330.

Vries, R.R.P. de, Meera Khan, P., Bernini, L.F., Loghem, E. van and Rood, J.J. van (1979). Genetic control of survival in epidemics. *J. Immunogenet.* **6**: 271–287.

Vries, R.R.P. de, Schreuder, G.M.Th., Naipal, A., D'Amaro, J. and Rood, J.J. van (1989). Selection by typhoid and yellow fever epidemics witnessed by the DR locus. In *Immunobiology of HLA: Immunogenetics and Histocompatibility* (ed. B. Dupont), pp. 461–466, Springer, New York.

Walford, R.L., Gallagher, R. and Sjaarda, J.R. (1964). Serologic typing of human lymphocytes with immune serum obtained after homografting. *Science* **144**: 868–870.

White, S., Newcomer, V.D., Mickey, M.R. and Terasaki, P.I. (1972). Disturbance of HLA antigen frequency in psoriasis. *N. Engl. J. Med.* **287**: 740–743.

Yunis, F.J. and Amos, D.B. (1971). Three closely linked genetic systems relevant to transplantation. *Proc. Natl Acad. Sci. USA* **68**: 3031–3035.

Zinkernagel, R.M. and Doherty, P.C. (1974). Activity of sensitized thymus-derived lymphocytes in lymphocyte chorionmeningitis reflects immunological surveillance against altered self components. *Nature* **251**: 547–548.

CHAPTER 2

The Structure of the Major Histocompatibility Complex and its Molecular Interactions

Efrem Eren and Paul Travers

INTRODUCTION

The recognition task faced by the immune system is complicated by the diversity of microorganisms (and some macroorganisms, such as parasitic worms) against which we must be protected. Not only must the immune system deal with the diversity of antigens that it encounters, but also the variety of locations in which they are found. Pathogens live in different sites of the body and invade different cell types. Some bacteria, fungi and parasites live either on the external surfaces of cells or free in the extracellular fluids. Other pathogens are phagocytosed, but remain alive within vesicles as they are able to withstand the enzymatic machinery of the phagosome. Yet others can enter into the cytosol of the cell from these vesicles and, of course, all viruses are obligate intracellular parasites, relying on the synthetic apparatus of the host cell for their replication. Different effector mechanisms must be applied in each of these cases to eliminate the infection: the production of neutralizing and opsonizing antibody for extracellular pathogens, the activation of macrophages to eliminate phagocytosed pathogens, and cytotoxic effectors to eliminate virally infected cells or those harbouring intracytoplasmic pathogens. The immune system must therefore perform another recognition task in addition to identifying the presence of pathogens within the body: it must identify the compartment in which the pathogen is found to ensure that the appropriate effector mechanisms are engaged.

It is this dual requirement for the recognition both of the presence and location of a pathogen that differentiates recognition by T cells from that by B cells and by antibody. The solution to these twin problems is to be found in the major histocompatibility complex (MHC), whose products are specialized to carry information about both the presence and the location of pathogens.

The MHC was defined originally on the basis of mapping loci which were the most significant determinants of graft rejection, hence the terms "major" and "histocompatibility". A second set of genes, the immune response (Ir) genes controlling antibody and delayed type hypersensitivity responses to protein antigens, was also mapped to the same region of the genome. With this observation came the first major division of the MHC, into what subsequently became known as class I and class II genes, initially thought to be those genes controlling graft rejection and immune responsiveness, respectively. As we shall see, the MHC class I and class II genes, while evolutionarily related, and generally very similar to each other, yet have subtle differences in both structure and function.

Many genes have been identified within the genetic boundaries of the MHC, encoding complement components, cytokines and other molecules with a role in immune responses. These other genes are sometimes referred to as MHC class III genes, although they are unrelated to the class I and class II genes. Yet others have no immunological function (although the function of many is as yet undetermined) and their presence in this particular part of the genome appears to be an accident of fate.

What we are concerned with here, however, is recognition by the immune system, and further discussion will be focused on the MHC class I and class II genes, and a subset of other MHC genes whose products act in concert with those of the class I and class II genes to facilitate antigen recognition.

FUNCTION OF MHC CLASS I AND CLASS II MOLECULES

The role of the MHC molecules in the immune system is, as has been alluded to above, in the presentation of antigens to T cells. The antigen receptors of T cells are unable to recognize antigen directly; they can only recognize foreign antigens in the form of short segments of peptide bound to MHC molecules. This role of the MHC molecules in the presentation of antigen was first suggested by the phenomenon of MHC-restricted recognition of antigens (Zinkernagel and Doherty, 1975), but now demonstrated directly by using purified MHC molecules and purified T cell receptors (TCR) (Matsui et al., 1994; Cox et al., 1995; Margulies et al., 1996; Allan et al., 1999) and has even been visualized by X-ray crystallography (Garboczi et al., 1996a). Neither peptide antigens nor MHC molecules alone can stimulate T cell responses, which require the formation of a peptide–MHC complex. However, in some specialized cases T cell receptors are able to recognize non-peptide antigens. For example, some T cells responding to mycobacterial infections have been found to recognize a lipid, mycolic acid, bound to the non-classical MHC molecule, CD1 (Beckman et al., 1994).

STRUCTURE OF MHC CLASS I AND CLASS II MOLECULES

The two distinct classes of MHC molecules that can be detected on the surface of cells differ biochemically on the basis of subunit structure. MHC class I molecules consist of two chains, an α or heavy chain associated non-covalently on the cell

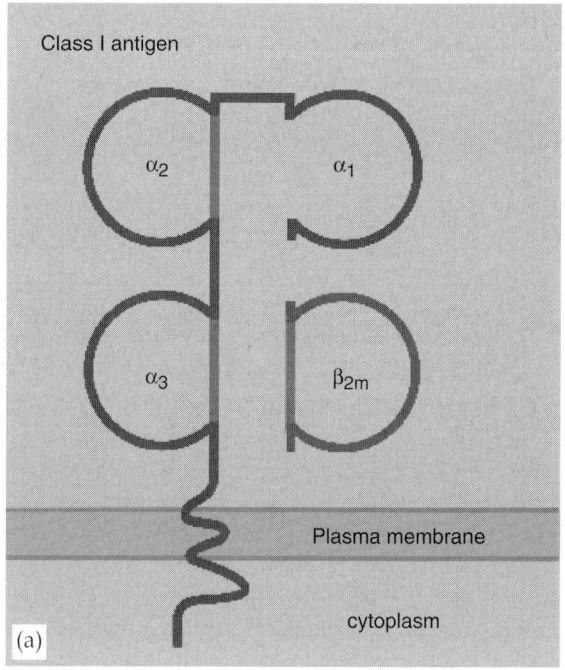

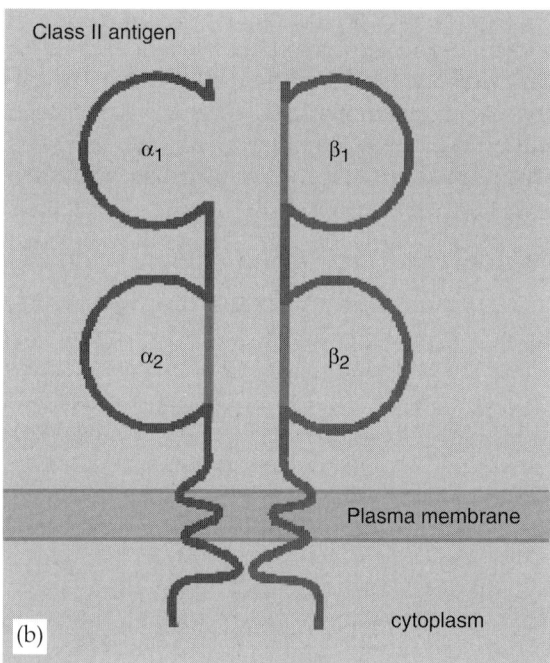

FIGURE 2.1 Schematic diagram of (a) class I and (b) class II MHC molecules. Class I molecules contain a large α-chain (43 kDa) associated non-covalently with the smaller β_2 microglobulin (β_2m) (12 kDa). The α-chain is composed of three external domains α_1, α_2 and α_3. Class II molecules consist of two transmembrane glycoprotein chains α (34 kDa) and β (29 kDa) which associate by non-covalent interactions.

surface with β_2-microglobulin. Only the class I α-chain spans the membrane. MHC class II molecules consist of a complex of two chains, α and β, both of which span the membrane and which are associated non-covalently on the cell surface (Figure 2.1).

The detailed structures of MHC class I and class II molecules have been determined by X-ray crystallography (Bjorkman et al., 1987a, b; Garboczi et al., 1992; Madden et al., 1992; Garboczi et al., 1994; Stern et al., 1994; Ghosh et al., 1995; Madden, 1995; Saper et al., 1991; Dessen et al., 1997; Smith et al., 1998) and are remarkably similar (Figure 2.2, in the colour plate section). Each molecule contains an extracellular portion composed of four domains. In the case of MHC class I molecules, three domains (α_1, α_2 and α_3) are contained within the α, or heavy, chain transmembrane glycoprotein while the fourth domain is contributed by β_2 microglobulin. In the case of the class II αβ heterodimers, each chain contributes two domains. In their general features, the structures of the class I and class II molecules are very similar. In both, the membrane distal domains, the α_1 and α_2 domains of class I and the α_1 and β_1 domains of class II, fold together to form a long groove. The remaining extracellular domains adopt an immunoglobulin-domain-like structure. Comparing the sequences of the two classes of MHC molecule, the class I α_3 domain corresponds to the class II β_2 domain and β_2 microglobulin corresponds in position with the class II α_2 domain.

While the overall structures of class I and class II molecules are similar, there are a number of small differences that have an important effect on the way that these molecules interact with peptide antigens.

ANTIGEN BINDING BY MHC CLASS I AND CLASS II MOLECULES

The form of antigen that is bound by both class I and class II MHC molecules exists as relatively short peptide fragments. In both cases, the same general principle applies, although the fine details of how the peptides are bound differ between the two classes of molecule (Madden, 1995). MHC molecules bind peptides in the first instance through conserved interactions between invariant residues on the MHC molecule and the peptide backbone. Superimposed on these interactions, which are independent of the sequence of the bound peptide, side chains of the peptide bind into specific pockets distributed along the peptide-binding cleft. It is these interactions that allow the MHC molecules to discriminate between peptides, binding some and not others.

MHC class I

In the case of MHC class I molecules, the peptides that are bound are predominantly short, usually octamers and nonamers, although longer peptides can be bound. The peptide is bound by both ends, with interactions between the MHC molecule and the *N*- and *C*-termini of the peptide (Matsumura et al., 1992). These interactions are critical for the ability of the class I molecule to bind peptide; peptide analogues with modified amino and carboxyl terminal groups fail to form stable complexes with MHC class I molecules (Bouvier and Wiley, 1994; Collins et al., 1994; Bouvier et al., 1998). In addition to these terminal and backbone interactions, which will be the same no matter what peptide is bound, MHC molecules also interact with side chains of the peptide and, as mentioned above, it is these interactions that give the different MHC molecules their differing specificities (Garrett et al., 1989; Matsumura et al., 1992; Young et al., 1994; Strominger and Wiley, 1995). For most class I molecules, the most important MHC interactions are those with the side chains of the second amino acid of the peptide (referred to as P2) and the final amino acid (depending on the length of the peptide this could be P8, P9, etc.; we will designate this residue as PC). For some class I molecules, notably the mouse H-2K^b and H-2D^b molecules, the residue at PC-3 (i.e. P5 for an octamer peptide and P6 for a nonamer) is a dominant contact residue and the contribution of the P2 residue is diminished (Fremont et al., 1992, 1995; Zhang et al., 1992; Young et al., 1994).

MHC class II

Peptide binding to class II molecules differs in a number of important respects from binding to class I. The most obvious difference is in the length of the peptides that are bound (Chicz et al., 1992). The majority of the peptides bound to class II molecules are greater than 13 amino acids in length and there is, in principle, no upper limit to the length of peptide that can be bound, with the constraint that the segment of the peptide that interacts with the class II molecule is both unfolded and accessible. This feature results from two important differences in the structure of the class II peptide-binding groove compared to that of class I molecules (Stern et al., 1994; Strominger and Wiley, 1995). Instead of the amino-terminal portion of the peptide being buried within the groove, in class II molecules an extended region of the polypeptide chain forms a short piece of parallel β sheet with a section of the peptide backbone of class II. This interaction has the effect of pulling up the peptide and allows the amino-terminal end of the peptide to extend beyond the end of the peptide-binding groove. At the carboxy-terminal end of the peptide, rather than the peptide being pulled up, the end of the α_1 domain is pulled down by the membrane proximal α_2 domain (remember that in class I molecules the end of the α_1 domain remains within the membrane-distal portion of the complex and becomes the class I α_2 domain, the homologue of the class II β_1 domain). This again allows the end of the peptide to protrude beyond the end of the peptide-binding groove.

Given that the ends of the peptide do not bind to the MHC class II molecule, how then are peptides bound? Instead of interactions focused at each end, peptides interact with MHC class II molecules along their length (Stern et al., 1994). As was mentioned above, a short section of parallel β sheet is formed between the class II α-chain and the peptide. In addition, the class II α-chain contributes a number of hydrogen bonds between conserved side chains, principally asparagine residues at 64 and 71, and the peptide backbone. The class II β-chain contributes a hydrogen bond from the conserved residues Trp61 (this interaction is identical to that between Trp147 and the PC-1 backbone carbonyl in class I–peptide complexes) and Asn 82. In some MHC molecules, the residues at positions 71, 78 and 81 are also able to interact with the peptide backbone.

As with the class I molecules, side chains of the peptide interact with specific pockets in the class II molecules. Since the peptides that bind to class II molecules can extend by variable lengths at both ends of the groove, it is less simple to provide a uniform nomenclature for the peptide side chains that interact with the class II molecule. The first side chain that interacts with the class II molecule is designated P1, although this is, in fact, the third amino acid which interacts with the MHC molecule; it is the backbone of the preceding two amino acids which interacts. The other side chains that are bound by the class II molecule are the P4, P6, P7 and P9 residues.

One distinctive feature of the class II molecules is the way in which a series of double hydrogen bonds fix the peptide backbone in place and straddle the upward-pointing residues, the potential TCR contacts. In contrast, for the peptides bound by class I molecules, the central portion of the peptide is relatively unconstrained by the MHC molecule and can adopt a number of conformations (Madden et al., 1993). It would appear that the T cells that discriminate class II–peptide complexes do so mainly on the basis of the sequence of the peptide and less on its conformation.

POLYMORPHISM

Because there are three genes encoding classical MHC class I molecules and three or four sets of classical MHC class II genes on each chromosome, every individual will express at least three different MHC class I proteins and three MHC class II proteins on his or her cells. Of course, the chances of an individual being homozygous at all these loci are very small. Hence, most individuals express twice these numbers (or slightly less), allowing for some homozygosity. Some MHC genes have several hundred alleles underscoring the low likelihood of homozygosity. The large number of distinct MHC molecules expressed by each cell in an individual thereby increases the diversity already available on account of the existence of multiple functionally equivalent genes (polygeny). For the MHC class II genes the number of different products may be increased still further by the combination of α- and β-chains encoded by different chromosomes (so that two α-chains and two β-chains can give rise to four different products). In mice, it has been shown that not all combinations of α- and β-chains can pair to form stable dimers and so, in practice, the exact number of different MHC class II molecules expressed will depend on which alleles are present on each chromosome.

Pairing between a class II α-chain of one isotype and a β-chain from another can occur (for example, DRβ pairing with DQα), but it is unlikely that such combinations play any significant role *in vivo*.

All MHC products are polymorphic to a greater or lesser extent, with the exception of the DRα chain. This chain does not vary in sequence between different individuals and is said to be monomorphic. All other MHC class I and class II genes are polymorphic. This might indicate a functional constraint that prevents variation in the DRα protein, but no such special function has yet been found.

INFLUENCE OF POLYMORPHISM ON PEPTIDE BINDING

We have discussed above the important side chains on antigenic peptides that bind to MHC molecules. Clearly, it is the sequence of the MHC molecule itself which determines the properties of the "pockets" into which these side chains bind and thus determines peptide specificity.

Individual MHC alleles can differ from one another by up to 20 amino acids, making each allele quite distinct. Most of these differences are localized to the exposed surfaces of the membrane-distal domains of the molecule, and to the peptide-binding groove in particular. The polymorphic residues that line the peptide-binding groove determine the peptide-binding properties of the different MHC molecules and the "pockets" in particular. The set of anchor residues that allow binding to a given MHC class I molecule is called a sequence motif. Different allelic variants of MHC class II molecules also bind different peptides, but the more open structure of the MHC class II peptide-binding groove, and the greater length of the peptides bound in it, allow greater flexibility in peptide binding, so that it is more difficult to predict which peptides will bind to MHC class II molecules and more difficult to define motifs in the peptides which do bind.

How the sequence variation in MHC molecules is able to change the specificity for peptide binding can be illustrated by comparing the effects of sequence changes in the first side chain pocket of MHC class I and class II molecules. In MHC class I molecules, this pocket binds the second, or P2, side chain. The residue in the MHC molecule that appears to have the greatest influence on the nature of the residue bound is residue 45 of the class I heavy chain, which, in HLA-B27 for example, is glutamic acid, a negatively charged residue. The side chain that is preferentially bound by the P2 pocket of HLA-B27 is arginine, a positively charged residue. Conversely, position 45 of HLA-B44 is lysine, a positively charged residue and in this case the preferred side chain is glutamic acid. In the P1 pocket of MHC class II molecules, an important variable residue is position 86 of the β-chain, which in HLA-DR molecules is present either as a glycine or as a valine. Those P1 pockets containing glycine are deep pockets that can accommodate both large and small residues, but have a preference for large hydrophobic amino acids such as tyrosine, phenylalanine or tryptophan. When a valine is present at position 86, the side chain of the valine fills up part of the pocket, and large residues such as tyrosine or tryptophan can no longer bind; instead, smaller hydrophobic residues (leucine, valine or methionine) bind this pocket.

The sites in MHC molecules that interact with peptide side chains all have some contribution from polymorphic residues; for example, in HLA-DR molecules, the P1, P4, P6, P7 and P9 pockets are all formed in part by variable amino acids. Moreover, all (bar one) of the most polymorphic residues in class II molecules are to be found amongst the residues forming these pockets, suggesting that the role of the polymorphism is indeed to vary the nature of the side-chain-binding pockets and thus alter the specificity of the class II molecule for peptide.

Since the role of the immune system is to protect against infection, it is here that we must look for the nature of the selective advantage conferred by the extensive polymorphism of the MHC proteins. Pathogens have a number of possible strategies for avoiding an immune response either by evading detection or by suppressing the ensuing response. The requirement for presentation by a MHC molecule provides at least one possible means of evasion. Although MHC molecules have a broad specificity of peptide binding, only certain peptides will bind to a given MHC molecule, and those that do not bind are not immunogenic. A pathogen could therefore escape detection by undergoing mutations that eliminated from its proteins all peptides able to bind MHC molecules. Failures in responsiveness to protein antigens by such a mechanism were reported in inbred animals, where they were called immune response (Ir) gene defects, long before the function of MHC molecules was understood. These defects could be shown genetically to map to genes within the MHC, and were the first clue to the antigen-presenting function of MHC molecules. Ir gene effects are common in inbred strains of mice because the mice are homozygous for all their MHC genes and thus express only one allelic variant from each gene locus. This process of evasion of the immune response is plainly much more difficult for a potential pathogen if there are many different MHC molecules, and the presence of different loci encoding functionally related proteins may have been an evolutionary adaptation to this strategy. As noted above, polymorphism at each locus can potentially double the number of different MHC molecules expressed by an individual, since most individuals will be heterozygotes. Polymorphism has the additional advantage that different individuals in a population will differ in the combinations of MHC molecules they express and will thus present different sets of peptides from each pathogen. This makes it unlikely that all individuals in a population will be equally susceptible to any given pathogen, thereby limiting its spread. Ir genes are discussed in greater detail in Chapter 4.

These arguments raise the question that if having three MHC class I molecules is advantageous, and six even more so, why are there not far more MHC class I loci? A possible explanation is that each time a new MHC molecule is added, all T cells that can recognize self-peptides bound to that molecule must be removed to maintain self-tolerance. It appears that the numbers of loci present in most species is roughly optimal to balance between the advantages of presenting an increased range of foreign peptides and the disadvantages of increased presentation of self-peptides and the consequent loss of T cells that accompanies it.

MHC polymorphism appears to have been strongly selected by evolutionary pressures. However, for selection to work efficiently in slowly reproducing organisms like humans, there must also be powerful mechanisms to generate the variability in MHC alleles on which the selective processes act. While the generation of polymorphism in MHC molecules is not yet fully understood since it is an evolutionary problem not readily analysed in the laboratory, it is clear that several genetic mechanisms contribute to the generation of new alleles. Some new alleles arise by point mutations, but many arise from combining the sequences of other alleles either through genetic recombination or by gene conversion, in which one sequence is replaced in part by another from a homologous gene.

Evidence for gene conversion comes from studies of the sequences of different alleles of MHC proteins, which have revealed that some changes involve clusters of several amino acids and require multiple nucleotide changes in a contiguous stretch of the gene. More significantly, the same sequences are found within other MHC genes on the same chromosome, a prerequisite for gene conversion.

Recombination between allelic variants at a single locus may, however, have been more important than gene conversion for generating MHC polymorphism. By comparing sequences of MHC allelic variants, it can be seen that different alleles represent recombination events between ancestral alleles. If one postulates a small number of ancestral alleles, most contemporary alleles can be generated by one or more recombination events occurring within polymorphic exons of MHC genes.

The effects of selective pressure in favour of polymorphism can be seen clearly in the pattern of point mutations in the MHC genes. Point mutations can be classified as replacement substitutions, which change an amino acid, or silent substitutions, which simply change the codon, but leave the amino acid the same. Replacement substitutions are seen within the MHC at a higher frequency relative to silent substitutions than would be expected, providing evidence that polymorphism is actively selected in the evolution of the MHC. The evolution of the MHC is discussed in greater detail in Chapter 9.

THE TRIMOLECULAR COMPLEX OF TCR, MHC MOLECULE AND BOUND PEPTIDE

The specialization of MHC molecules to present antigens from cytoplasmic versus extracellular sources is reflected in the specialization of effector functions in the T cells that recognize class I and class II molecules, respectively. Class I molecules present antigen to cytolytic T cells. Class II molecules are recognized by T helper cells that direct the immune response to develop a particular effector mechanism such as by activating macrophages to become more cytotoxic or B cells to secrete antibody. The effector functions of the T cells that recognize antigen in the context of class I and class II MHC molecules are therefore quite distinct, and are tailored to eliminate pathogens from the two major compartments (intracytoplasmic versus extracellular and vesicular) in which they occur. It is important for the T cell, therefore, that it is able to recognize not just the antigen, but also which class of MHC molecule is presenting the antigen in order for the correct effector mechanisms to be activated. However, it is not obviously the case that the TCR *per se* can discriminate class I from class II molecules. Instead, this task is carried out by the coreceptor molecules, CD4 and CD8.

Knowledge of the structure of the TCR, its coreceptors and the MHC–peptide complex (pMHC), has broadened our understanding of the complex interactions made by these molecules during an immune response (Davis and Bjorkman, 1988; Bjorkman, 1997; Garcia and Teyton, 1998). Most structural information has come from analysis of their amino acid sequence and more recently from use of X-ray crystallography. The structure of pMHC has already been discussed in detail. We now move to the other components of this crucial interaction.

The TCR is a glycosylated protein consisting of two disulfide-linked chains ($\alpha\beta$ or $\gamma\delta$) (Figure 2.3). Of these, the structure of the $\alpha\beta$ TCR has been most studied, although the $\gamma\delta$ TCR is similar in many respects (Allison et al., 1984; Hamilos and Wedner, 1984; Hannum et al., 1984; Meuer et al., 1984; Acuto et al., 1985; Hood et al., 1985; Yague et al., 1985; Marrack and Kappler, 1986, 1987; Lanier et al., 1987; Livingstone and Fathman, 1987).

More detailed information concerning the TCR structure has come from its X-ray crystal structure (Young et al., 1994; Bentley et al., 1995; Fields et al., 1995; Garboczi et al., 1996b; Garcia et al., 1996a, 1998; Dessen et al., 1997; Housset et al., 1997). Complete heterodimer TCR structures have now been analysed by crystallography, both alone and together with the peptide–MHC complex, in humans and mouse. The overall architecture of the TCR structure generally resembles

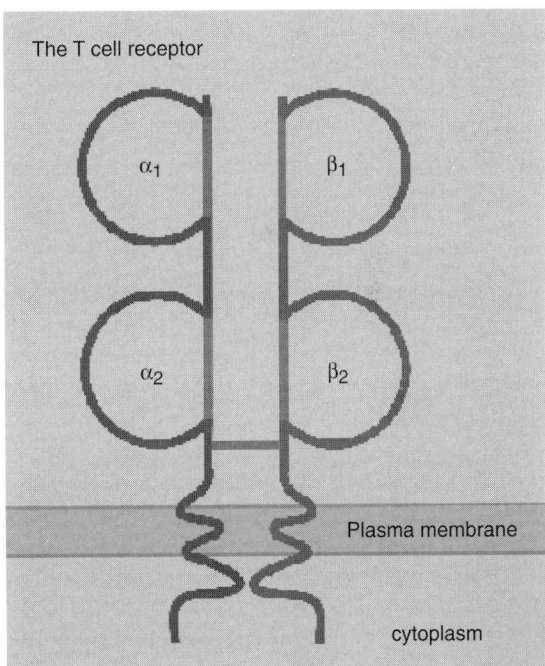

FIGURE 2.3 Schematic diagram of the T-cell receptor. The TCR is a heterodimer composed of α- and β-chains. The extracellular portion of each chain consists of two domains, linked by a membrane-proximal disulfide bond. The most distal (α_1 and β_1) of these resemble immunoglobulin variable domains and the most proximal (α_2 and β_2) immunoglobulin constant domains.

the F_{ab} fragment of antibody (Claverie et al., 1989) (Figure 2.4, in the colour plate section).

Just as immunoglobulins have regions of amino acid diversity, so too does the TCR. As with antibodies the variability resides in the N-terminal (V) domains of the α and β polypeptides which, again like antibodies, are the result of gene segment rearrangements (by convention named, V, D and J gene segments for the β-chain and V and J for α). Within each V domain, there are areas of greater variability known as complementary determining regions (CDRs) (Amzel and Poljak, 1979). These correspond to immunoglobulin hypervariable regions and exhibit the classical β-barrel structure of immunoglobulins. In each V region, at the end of the β-barrel there are four loops, three of which form the CDRs described above. The fourth loop (HV4) has been implicated in superantigen binding (Kline and Collins, 1997). The six CDRs (1, 2 and 3) from each of the α- and β-chains, plus HV4 make an antigen-combining site which is quite flat and consistent with its function in binding to the generally flat, undulating surface of the pMHC (Figure 2.5, in the colour plate section).

Examples of complexes studied include the murine TCR 2C in combination with its MHC–peptide complex, $H-2K^b$-dEV8 and the human TCRs A6 and B7 as complexes with the same MHC–peptide combination (HLA-A2 with the HTLV1-derived Tax peptide) (Fremont et al., 1992; Garboczi et al., 1996b; Garcia et al., 1996b; Ding et al., 1998; Garcia and Teyton, 1998).

The first murine TCR-pMHC complex to be reported was that of the 2CTCR with $H-2K^b$ and a mouse self-peptide dEV8 (Ding et al., 1998). Two human T cell receptors bind in a similar diagonal mode to the HLA-A2/Tax peptide complex using different TCR amino acids). The X-ray crystal structure revealed that the TCR was placed in an approximate diagonal orientation over the pMHC composite surface, with the α chain over the N-terminal half of the peptide and the β chain over the C-terminal half. The CDR3α and β both lie over the central positions (P4–P6) of the peptide, with the latter (CD3β) having minimal peptide contact. The CDR1s of both chains are in contact with both peptide and MHC, straddling the centre of the N- and C-terminal ends of the peptide-binding groove. The CDR2s predominantly contact the α helices of the MHC heavy chain.

To further our knowledge a detailed energy map of the antigen recognition site of the 2C TCR was produced. This was done by alanine scanning in which numerous single-site alanine substitutions within the 2C TCR were generated and the interactions with $H-2L^d$-QL9 (another alloligand) and anti-TCR antibodies were studied (Manning et al., 1998). As a result of this experiment, it was suggested that during a TCR–MHC–peptide interaction, there exists enough binding energy between the MHC heavy chain and the TCR to enable the TCR to dock on the MHC–peptide complex and analyse the peptide contents. If there are peptides that provide sufficient energetically favourable contacts with the TCR there may be sufficient kinetic stabilization for signalling to occur.

The A6 TCR complex with the HTLV-derived Tax peptide was the first human TCR complex whose structure was reported (Garboczi et al., 1996b) (Figure 2.6, in the colour plate section). In comparison with the 2C TCR described above, the Vα chain contacts the MHC helices in a similar manner to 2C. The β-chain, however, tilts off the peptide–MHC surface, with only the CDR3 region having extensive contact with the peptide and MHC molecule. The contact between the A6 TCR and the Tax peptide is closer than that of 2C and its pMHC. This is consistent with higher affinity of the human complex.

The B7 TCR/HLA-A2-Tax crystal structure has also been resolved (Ding et al., 1998). This consists of another human TCR (B7), in complex with the same pMHC. The B7 TCR differs in its Vα region and although 13 TCR and 18 MHC contact positions are identical between the two complexes, the amino acids used by the two TCRs are completely different. This illustrates how one pMHC can be recognized by different TCRs.

Analysis of all the TCR-pMHC class I crystal structures studied to date has shown that the TCR makes a diagonal footprint over the pMHC (diagonal docking). The areas of the TCR domain that interact with the peptide-MHC complex vary, particularly in the Vβ region. In contrast, the Vα regions appear to have an identical relationship with the pMHC. This region may be important in steering the TCR-pMHC orientation.

With the recent publication of the X-ray crystal structure of the murine self-MHC class II (I-A^k) in complex with peptide (conalbumin) and TCR (murine D10) (Reinherz et al., 1999), comparisons can now be made between the TCR pMHC class I and class II interaction. The most striking and unexpected finding was that of the docking mode, with the TCR sitting on top of the pMHC class II with its longer dimension crossing the bound peptide in an orthogonal (right-angled) manner, as opposed to the diagonal docking seen with the TCR-pMHC class I interaction. The size of the TCR footprint on the MHC covers nine peptide residues (considerably less than the size of the potentially bound peptide) and hence only a subset of the bound peptides is "read out" by the bound TCR. The importance of the P5 residue of the MHC-bound peptide also indicated. This has a central position and is very important in the TCR binding process. Minor changes to this residue could therefore destroy binding or lead to altered peptide ligands with very weak agonist or antagonistic activity.

CORECEPTORS IN THE TCR COMPLEX

The TCR binds to the pMHC as a complex of coreceptors which also includes CD8 and CD4. These are not polymorphic and therefore do not affect the specificity of the TCR, but play an important role in its interaction with pMHC. CD8 (either as α_2 homodimer or $\alpha\beta$ heterodimer) and CD4 have been shown to augment this interaction both *in vitro* and *in vivo* (Garcia et al., 1996). The dynamics of this process are unclear. However, elucidation of the X-ray structure of CD8 α_2 bound to HLA-A2 has clarified matters (Kern et al., 1998) (Figure 2.7, in the colour plate section). In this complex the CD8 α_2 subunits (which resemble an antibody Fv fragment) bind via their CDR-like loops with the MHC α3 domain in a similar way to antibody binding to antigen. This complex, when compared with previously reported isolated HLA molecules, reveals a shift in the position of the HLA-A2 α3 domain. It is also interesting to note that the HLA-A2 orientation in the CD8 complex is similar to that adopted by the TCR/Tax/HLA-A2 complex. The biological significance of this is uncertain.

No structures are available for MHC class II TCR/pMHC in association with their coreceptors. Site-directed mutagenesis studies have shown, however, that an area on the membrane-proximal β2 domain of class II molecules is necessary for binding to CD4 and for eliciting CD4 coreceptor activity (Nag et al., 1993). Interestingly, an additional binding site on the α2 domain was later identified (Konig et al., 1995). The positioning of these two sites is such that the same CD4 molecule cannot bind simultaneously to both α2 and β2 regions. A possible explanation would be that two CD4 molecules may associate with a single TCR-MHC class II complex, one binding to the α2 site, the other to the β2 site. This would form a clamp, tethering the class II ligand to the TCR. With the eventual emergence of X-ray crystallographic data, improved knowledge of these coreceptor interactions will be possible

The past 15 years has seen a considerable amount of structural data being published on the MHC molecules and its ligands. This has provided a physical basis to the several decades of functional data and has allowed us to address previously insurmountable questions. More than anything else, this elucidation of MHC structures has created new insights concerning thymic selection, Ir gene effects and the diverse outcomes of antigen recognition by T cells.

REFERENCES

Acuto, O., Fabbi, M., Bensussan, A. et al. (1985). The human T-cell receptor. *J. Clin. Immunol.* **5**: 141–157.

Allan, D.S., Colonna, M., Lanier, L.L. et al. (1999). Tetrameric complexes of human histocompatibility leukocyte antigen (HLA)-G bind to peripheral blood myelomonocytic cells. *J. Exp. Med.* **189**: 1149–1156.

Allison, J.P., Ridge, L., Lund, J., Gross-Pelose, J., Lanier, L., and McIntyre, B.W. (1984). The murine T cell antigen receptor and associated structures. *Immunol. Rev.* **81**: 145–160.

Amzel, L.M. and Poljak, R.J. (1979). Three-dimensional structure of immunoglobulins. *Annu. Rev. Biochem.* **48**: 961–997.

Beckman, E.M., Porcelli, S.A., Morita, C.T., Behar, S.M., Furlong, S.T. and Brenner, M.B. (1994). Recognition of a lipid antigen by CD1-restricted alpha beta+ T cells [see comments]. *Nature* **372**: 691–694.

Bentley, G.A., Boulot, G., Karjalainen, K. and Mariuzza, R.A. (1995). Crystal structure of the beta chain of a T cell antigen receptor [see comments]. *Science* **267**: 1984–1987.

Bjorkman, P.J. (1997). MHC restriction in three dimensions: a view of T cell receptor/ligand interactions. *Cell* **89**: 167–70.

Bjorkman, P.J., Saper, M.A., Samraoui, B., Bennett, W.S., Strominger, J.L. and Wiley, D.C. (1987a). The foreign antigen binding site and T cell recognition regions of class I histocompatibility antigens. *Nature* **329**: 512–518.

Bjorkman, P.J., Saper, M.A., Samraoui, B., Bennett, W.S., Strominger, J.L. and Wiley, D.C. (1987b). Structure of the human class I histocompatibility antigen, HLA-A2. *Nature* **329**: 506–12.

Bouvier, M. and Wiley, D.C. (1994). Importance of peptide amino and carboxyl termini to the stability of MHC class I molecules. *Science* **265**: 398–402.

Bouvier, M., Guo, H.C., Smith, K.J. and Wiley, D.C. (1998). Crystal structures of HLA-A*0201 complexed with antigenic peptides with either the amino- or carboxyl-terminal group substituted by a methyl group. *Proteins* **33**: 97–106.

Claverie, J.M., Prochnicka-Chalufour, A. and Bougueleret, L. (1989). Implications of a Fab-like structure for the T-cell receptor. *Immunol. Today* **10**: 10–14.

Collins, E.J., Garboczi, D.N. and Wiley, D.C. (1994). Three-dimensional structure of a peptide extending from one end of a class I MHC binding site. *Nature* **371**: 626–629.

Cox, J.H., Galardy, P., Bennink, J.R. and Yewdell, J.W. (1995). Presentation of endogenous and exogenous antigens is not affected by inactivation of E1 ubiquitin-activating enzyme in temperature-sensitive cell lines. *J. Immunol.* **154**: 511–519.

Davis, M.M. and Bjorkman, P.J. (1988). T-cell antigen receptor genes and T-cell recognition [published erratum appears in *Nature* (1988) **335**: 744]. *Nature* **334**: 395–402.

Dessen, A., Lawrence, C.M., Cupo, S., Zaller, D.M. and Wiley, D.C. (1997). X-ray crystal structure of HLA-DR4 (DRA*0101, DRB1*0401) complexed with a peptide from human collagen II. *Immunity* **7**: 473–481.

Ding, Y.H., Smith, K.J., Garboczi, D.N., Utz, U., Biddison, W.E. and Wiley, D.C. (1998). Two human T cell receptors bind in a similar diagonal mode to the HLA-A2/Tax peptide complex using different TCR amino acids. *Immunity* **8**: 403–411.

Fields, B.A., Ober, B., Malchiodi, E.L. et al. (1995). Crystal structure of the V alpha domain of a T cell antigen receptor. *Science* **270**: 1821–1824.

Fremont, D.H., Matsumura, M., Stura, E.A., Peterson, P.A. and Wilson, I.A. (1992). Crystal structures of two viral peptides in complex with murine MHC class I H-2Kb [see comments]. *Science* **257**: 919–927.

Fremont, D.H., Stura, E.A., Matsumura, M., Peterson, P.A. and Wilson, I.A. (1995). Crystal structure of an H-2Kb–ovalbumin peptide complex reveals the interplay of primary and secondary anchor positions in the major histocompatibility complex binding groove. *Proc. Natl Acad. Sci. USA* **92**: 2479–2483.

Fremont, D.H., Monnaie, D., Nelson, C.A. Hendrickson, W.A. and Unanue, E.R. (1998). Crystal structure of I-Ak in complex with a dominant epitope of lysozyme. *Immunity* **8**: 305–317.

Gao, G.F., Tormo, J., Gerth, U.C. et al. (1997). Crystal structure of the complex between human CD8alpha(alpha) and HLA- A2. *Nature* **387**: 630–634.

Garboczi, D.N., Hung, D.T. and Wiley, D.C. (1992). HLA-A2–peptide complexes: refolding and crystallization of molecules expressed in *Escherichia coli* and complexed with single antigenic peptides. *Proc. Natl Acad. Sci. USA* **89**: 3429–3433.

Garboczi, D.N., Madden, D.R. and Wiley, D.C. (1994). Five viral peptide–HLA-A2 co-crystals. Simultaneous space group determination and X-ray data collection. *J. Mol. Biol.* **239**: 581–587.

Garboczi, D.N., Ghosh, P., Utz, U., Fan, Q.R., Biddison, W.E. and Wiley, D.C. (1996a). Structure of the complex between human T-cell receptor, viral peptide and HLA-A2 [comment]. *Nature* **384**: 134–141.

Garboczi, D.N., Utz, U., Ghosh, P. et al. (1996b). Assembly, specific binding, and crystallization of a human TCR-alpha-beta with an antigenic Tax peptide from human T lymphotropic virus type 1 and the class I MHC molecule HLA-A2. *J. Immunol.* **157**: 5403–5410.

Garcia, K.C. and Teyton, L. (1998). T-cell receptor peptide–MHC interactions: biological lessons from structural studies. *Curr. Opin. Biotechnol.* **9**: 338–343.

Garcia, K.C., Degano, M., Stanfield, R.L. et al. (1996a). An alphabeta T cell receptor structure at 2.5 Å and its orientation in the TCR–MHC complex [see comments]. *Science* **274**: 209–219.

Garcia, K.C., Scott, C.A., Brunmark, A. et al. (1996b). CD8 enhances formation of stable T-cell receptor/MHC class I molecule complexes [see comments] [published erratum appears in *Nature* (1997) **387**: 634]. *Nature* **384**: 577–581.

Garcia, K.C., Degano, M., Pease, L.R. et al. (1998). Structural basis of plasticity in T cell receptor recognition of a self peptide–MHC antigen. *Science* **279**: 1166–1172.

Garrett, T.P., Saper, M.A., Bjorkman, P.J., Strominger, J.L. and Wiley, D.C. (1989). Specificity pockets for the side chains of peptide antigens in HLA-Aw68 [see comments]. *Nature* **342**: 692–696.

Ghosh, P., Amaya, M., Mellins, E., and Wiley, D.C. (1995). The structure of an intermediate in class II MHC maturation: CLIP bound to HLA-DR3. *Nature* **378**: 457–462.

Hamilos, D. and Wedner, H.J. (1984). The T lymphocyte antigen receptor. A critical review of recent experimental literature. *Surv. Synth. Pathol. Res.* **3**: 292–310.

Hannum, C.H., Kappler, J.W., Trowbridge, I.S., Marrack, P. and Freed, J.H. (1984). Immunoglobulin-like nature of the alpha-chain of a human T-cell antigen/MHC receptor. *Nature* **312**: 65–67.

Hood, L., Kronenberg, M. and Hunkapiller, T. (1985). T cell antigen receptors and the immunoglobulin supergene family. *Cell* **40**: 225–229.

Housset, D., Mazza, G., Gregoire, C., Piras, C., Malissen, B. and Fontecilla-Camps, J.C. (1997). The three-dimensional structure of a T-cell antigen receptor V alpha V beta heterodimer reveals a novel arrangement of the V beta domain. *EMBO J.* **16**: 4205–4216.

Kern, P.S., Teng, M.K., Smolyar, A. et al. (1998). Structural basis of CD8 coreceptor function revealed by crystallographic analysis of a murine CD8alphaalpha ectodomain fragment in complex with H-2Kb. *Immunity* **9**: 519–530.

Kline, J.B. and Collins, C.M. (1997). Analysis of the interaction between the bacterial superantigen streptococcal pyrogenic exotoxin A (SpeA) and the human T-cell receptor. *Mol. Microbiol.* **24**: 191–202.

Konig, R., Shen, X. and Germain, R.N. (1995). Involvement of both major histocompatibility complex class II alpha and beta chains in CD4 function indicates a role for ordered oligomerization in T cell activation. *J. Exp. Med.* **182**: 779–787.

Lanier, L.L., Serafini, A.T., Ruitenberg, J.J. et al. (1987). The gamma T-cell antigen receptor. *J. Clin. Immunol.* **7**: 429–440.

Livingstone, A.M. and Fathman, C.G. (1987). The structure of T-cell epitopes. *Annu. Rev. Immunol.* **5**: 477–501.

Madden, D.R. (1995). The three-dimensional structure of peptide–MHC complexes. *Annu. Rev. Immunol.* **13**: 587–622.

Madden, D.R., Gorga, J.C., Strominger, J.L. and Wiley, D.C. (1992). The three-dimensional structure of HLA-B27 at 2.1 Å resolution suggests a general mechanism for tight peptide binding to MHC. *Cell* **70**: 1035–1048.

Madden, D.R., Garboczi, D.N. and Wiley, D.C. (1993). The antigenic identity of peptide–MHC complexes: a comparison of the conformations of five viral peptides presented by HLA-A2 [published erratum appears in *Cell* (1994) **76**: following 410]. *Cell* **75**: 693–708.

Manning, T.C., Schlueter, C.J., Brodnicki, T.C. et al. (1998). Alanine scanning mutagenesis of an alphabeta T cell receptor: mapping the energy of antigen recognition. *Immunity* **8**: 413–425.

Margulies, D.H., Plaksin, D., Khilko, S.N. and Jelonek, M.T. (1996). Studying interactions involving the T-cell antigen receptor by surface plasmon resonance. *Curr. Opin. Immunol.* **8**: 262–70.

Marrack, P. and Kappler, J. (1986). The T cell and its receptor. *Sci. Am.* **254**: 36–45.

Marrack, P., and Kappler, J. (1987). The T cell receptor. *Science* **238**: 1073–1079.

Matsui, K., Boniface, J.J., Steffner, P., Reay, P.A. and Davis, M.M. (1994). Kinetics of T-cell receptor binding to peptide/I-Ek complexes: correlation of the dissociation rate with T-cell responsiveness. *Proc. Natl Acad. Sci. USA* **91**: 12862–12866.

Matsumura, M., Fremont, D.H., Peterson, P.A. and Wilson, I.A. (1992). Emerging principles for the recognition of peptide antigens by MHC class I molecules [see comments]. *Science* **257**: 927–934.

Meuer, S.C., Acuto, O., Hercend, T., Schlossman, S.F. and Reinherz, E.L. (1984). The human T-cell receptor. *Annu. Rev. Immunol.* **2**: 23–50.

Nag, B., Wada, H.G., Passmore, D., Clark, B.R., Sharma, S.D. and McConnell, H.M. (1993). Purified beta-chain of MHC class II binds to CD4 molecules on transfected HeLa cells. *J. Immunol.* **150**: 1358–1364.

Reid, S.W., McAdam, S., Smith, K.J. et al. (1996). Antagonist HIV-1 Gag peptides induce structural changes in HLA B8. *J. Exp. Med.* **184**: 2279–2286.

Reinherz, E.L., Tan, K., Tang, L. et al. (1999). The crystal structure of a T cell receptor in complex with peptide and MHC class II. *Science* **286**: 1913–1921.

Rodriguez-Romero, A., Almog, O., Tordova, M., Randhawa, Z. and Gilliland, G.L. (1998). Primary and tertiary structures of the Fab fragment of a monoclonal anti-E-selectin 7A9 antibody that inhibits neutrophil attachment to endothelial cells. *J. Biol. Chem.* **273**: 11770–11775.

Saper, M.A., Bjorkman, P.J. and Wiley, D.C. (1991). Refined structure of the human histocompatibility antigen HLA-A2 at 2.6 Å resolution. *J. Mol. Biol.* **219**: 277–319.

Smith, K.J., Pyrdol, J., Gauthier, L., Wiley, D.C. and Wucherpfennig, K.W. (1998). Crystal structure of HLA-DR2 (DRA*0101, DRB1*1501) complexed with a peptide from human myelin basic protein. *J. Exp. Med.* **188**: 1511–1520.

Stern, L.J., Brown, J.H., Jardetzky, T.S. et al. (1994). Crystal structure of the human class II MHC protein HLA-DR1 complexed with an influenza virus peptide. *Nature* **368**: 215–221.

Strominger, J.L. and Wiley, D.C. (1995). The 1995 Albert Lasker Medical Research Award. The class I and class II proteins of the human major histocompatibility complex. *J. Am. Med. Assoc.* **274**: 1074–1076.

Yague, J., White, J., Coleclough, C., Kappler, J., Palmer, E. and Marrack, P. (1985). The T cell receptor: the alpha and beta chains define idiotype, and antigen and MHC specificity. *Cell* **42**: 81–87.

Young, A.C., Zhang, W., Sacchettini, J.C. and Nathenson, S.G. (1994). The three-dimensional structure of H-2Db at 2.4 Å resolution: implications for antigen-determinant selection. *Cell* **76**: 39–50.

Zhang, W., Young, A.C., Imarai, M., Nathenson, S.G. and Sacchettini, J.C. (1992). Crystal structure of the major histocompatibility complex class I H-2Kb molecule containing a single viral peptide: implications for peptide binding and T-cell receptor recognition. *Proc. Natl Acad. Sci. USA* **89**: 8403–8407.

Zinkernagel, R.M. and Doherty, P.C. (1975). H-2 compatability requirement for T-cell-mediated lysis of target cells infected with lymphocytic choriomeningitis virus. Different cytotoxic T-cell specificities are associated with structures coded for in H-2K or H-2D. *J. Exp. Med.* **141**: 1427–1436.

CHAPTER 3

Molecular Genetics of the Human Major Histocompatibility Complex

Caroline M. Milner, R. Duncan Campbell and John Trowsdale

INTRODUCTION

The human major histocompatibility complex (MHC) or human leucocyte antigen (HLA) region encompasses over 4 Mb of DNA (~0.1% of the genome) on the short arm of chromosome 6 at 6p21.3 (Campbell and Trowsdale, 1993, 1997). Over 220 genes have been located in the MHC and at least 10% of these have functions related to the immune system. The MHC is traditionally divided into the class I, class II and class III regions, in the order: centromere – class II, class III, class I – telomere (see Figure 3.1).

Genetic studies have indicated that genes within the MHC contribute to a large number of immune-related disorders, e.g. insulin-dependent diabetes mellitus (IDDM), rheumatoid arthritis (RA), ankylosing spondylitis (AS), common variable immunodeficiency (CVID) and immunoglobulin A deficiency (IgAD) (Davies et al., 1994; Tiwari and Terasaki, 1985; Sinha et al., 1990; Lechler, 1994). Strong associations have been found between many diseases and alleles of genes in the MHC class II region. Other disease susceptibility genes may also lie within the class I and III regions. For example, a recent study has localized the susceptibility locus for IgAD and CVID to the class III region between G1 and HLA-B (Schroeder et al., 1998). Genes in or near the MHC have also been implicated in a number of non-immune-related diseases such as haemochromotosis, congenital adrenal hyperplasia (CAH) and sialidosis (Morel and Miller, 1991; Bonten et al., 1996; Feder et al., 1996). The MHC is one of the most extensively characterized regions of the human genome and has now been entirely cloned in cosmids, YACs or PACS. This has allowed detailed molecular mapping of the region and has led to the identification of a large number of new genes.

THE CLASS I REGION

The class I region spans 2 Mb of DNA. This is the least well-characterized region of the MHC but the results of recent studies involving cDNA selection and genomic DNA sequencing suggest that many transcripts are encoded in this region (Gruen and Weissman, 1997; Janer and Geraghty, 1998; Shiina et al., 1998).

The class I region contains three main functional class I loci, HLA-A, HLA-B and HLA-C, all of which are highly polymorphic (Lawlor et al., 1990). These genes are expressed by most somatic tissues at varying levels. In addition to their roles in the presentation of antigens to cytotoxic T lymphocytes (CTLs) the expression of class I molecules at the cell surface can have other important effects. For example, class I molecules appear to be expressed selectively on electrically silent neurons which are then targeted for destruction, thus directing immunosurveillance by CTLs on to functionally impaired neurons (Neumann et al., 1995). Lack of class I expression at the maternofoetal interface (villous and extravillous cytotrophoblast and syncytiotrophoblast) may facilitate the survival of the foetal tissue as an allograft in the maternal host. In addition, there is increasing evidence for a role of HLA-C in target recognition

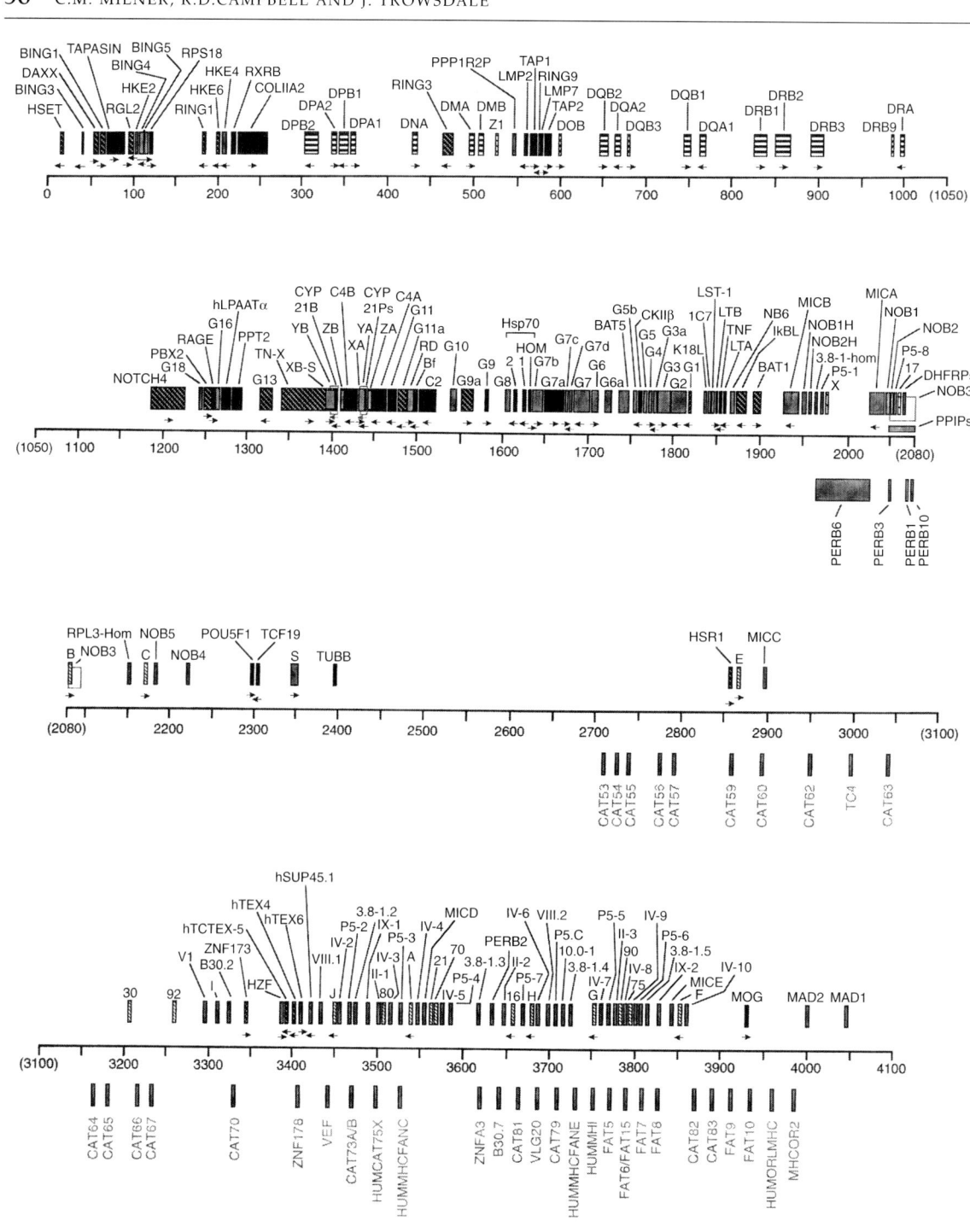

FIGURE 3.1 Map of the human MHC. The map is taken from Campbell and Trowsdale (1997) with modifications (Lepourcelet et al., 1996, 1998; Herberg et al., 1998; Shiina et al., 1998).

by natural killer (NK) cells (Colonna et al., 1993).

The non-classical class I loci include HLA-E, HLA-F and HLA-G (Geraghty, 1993). These genes, which are generally much less polymorphic than HLA-A -B or -C, lead to the production of class I-related molecules of restricted tissue distribution. HLA-E is expressed on a number of tissues at low surface levels and is retained in the endoplasmic reticulum (ER) unless it receives a peptide from the extreme N-terminus of another class I molecule, when it is involved in instructing NK cell receptors (Ulbrecht et al., 1992; Braud et al., 1997, 1998). HLA-G is expressed on foetal trophoblast cells and is thought to play a role in the maternal tolerance of the foetus (Loke and King, 1991; Parham, 1996). In addition to these loci, the class I region contains at least 12 pseudogenes or gene fragments that are closely related to the class I heavy chain genes: HLA-17, -X, -30, -L/92, -J/59, -80, -21, -K/70, -16, -H/54, -90 and -75 (Geraghty, 1993; Le Bouteiller, 1994).

Yet another class I-related gene has been found at a considerable distance telomeric of the MHC (>4 Mb telomeric of HLA-A), namely the HFE locus. A mutation in this gene has been implicated in the development of hereditary haemochromotosis (Feder et al., 1996).

Other genes in the class I region

The list of genes in the extended class I region, in addition to those related to the classical class I genes, is expanding rapidly (Gruen et al., 1996). The genes fall into a number of families, some of which have a potential role in the immune system. One of the largest contains a number of members of the butyrophilin family (Henry et al., 1997). In addition, several members of the olfactory receptor family (Fan et al., 1995), as well as a number of ubiquitin-like proteins and zinc-finger-containing proteins, are encoded within the class I region.

THE CLASS II REGION

The class II region spans over 800 kb of DNA and contains one gene every 40 kb on average. These include all of the known class II α- and β-chain genes. In humans HLA-DP, -DQ and -DR are expressed on the surfaces of antigen-presenting cells (APCs), where they present peptides to T helper cells. The α- and β-chain loci are all arranged as matched pairs (i.e. DRA and DRB, DQA and DQB, DPA and DPB) but the number of DRB genes and pseudogenes has been shown to vary between haplotypes (Rollini et al., 1985; Andersson et al., 1987; Gorski et al., 1987; Kawai et al., 1989). Both the DQ and DP regions include paired pseudogenes (DQA2 and DQB2, DPA2 and DPB2), although the coding regions of the pair of DQ sequences do not contain any obvious deleterious mutations (Trowsdale et al., 1984; Gustafsson et al., 1987; Jonsson et al., 1987).

Also in the class II region are two other pairs of class II-like genes. The products of HLA-DNA and -DOB form a heterodimeric protein, HLA-DO, although they are not adjacent to each other on the chromosome (Trowsdale et al., 1984; Tonnelle et al., 1985; Young and Trowsdale, 1990; Karlsson et al., 1992). HLA-DO is closely related to the classical class II molecules, but shows only limited polymorphism. The tissue distribution of HLA-DO is more restricted than that of the classical class II molecules and HLA-DM and its physiological role is not entirely clear. Recently, HLA-DO has been shown to regulate HLA-DM activity and thus may modulate the repertoire of class II–peptide complexes presented at the cell surface (Denzin et al., 1997; van Ham et al., 1997; Jensen, 1998). This activity in B cells may be important in the induction of peripheral T cell tolerance.

HLA-DMA and -DMB are only distantly related to the other class II sequences and show relatively little polymorphism (Cho et al., 1991; Sanderson and Trowsdale, 1995). The putative structure of the DM molecule is similar to that of classical class II molecules, such as DR, DQ or DP, but the α_1 and β_1 domains are each predicted to contain an additional disulfide bridge which is likely to preclude peptide binding. The membrane-proximal β_2 and α_2 domains are almost as similar to class I α_3 as they are to β_2 and α_2 class II sequences. This relationship suggests that DM sequences arose at around the same time that class I and class II sequences diverged from each other, several hundred million years ago. The products of the DM genes form a heterodimer which is not expressed on the cell surface and does not contain a recognizable CD4-binding site. DM is mostly found in intracellular vesicles, thought to be the sites of loading of peptides on to class II molecules (Sanderson et al., 1994). The DM molecule participates in the formation of the class II–peptide complexes (Fling et al., 1994; Morris et al., 1994), probably by inducing a conformational change or stabilizing a transition state that favours release of the invariant chain-derived CLIP peptide (Karlsson et al., 1994; Vogt et al., 1996). DM genes are expressed in similar circumstances to other class II molecules, consistent with a complementary function in peptide loading.

The TAP and LMP genes

The processing of antigens to generate peptides for presentation by MHC class I molecules is a highly regulated process (Lehner and Trowsdale, 1998). Some of the proteins involved in this pathway are encoded by a tight cluster of genes that is inserted into the class II region. The products of the two TAP genes, TAP1 and TAP2, are members of the ABC (ATP-binding cassette) transporter superfamily (Townsend and Trowsdale, 1993). Members of this family of transmembrane molecules are involved in transport of a wide selection of different substances across membranes, including oligopeptides, proteins and ions (Higgins et al., 1990). The products of the two TAP genes form a complex in the ER membrane which translocates peptides from the cytoplasm into the lumen of the ER (Momburg et al., 1994; Androlewicz and Cresswell, 1994). Here, the transported peptides can participate in the assembly of class I molecules and there is strong evidence for an association between the TAPs and class I molecules since they may be co-immunoprecipitated (Ortmann et al., 1994). Cells deficient in either TAP1 or TAP2 have a reduction in the level of cell surface class I expression and are unable to present the usual intracellular antigens to CTLs (Spies and DeMars, 1991; Kelly et al., 1992).

The LMP2 and LMP7 genes are located very close to the TAP genes and encode components of the proteasome complex (Glynne et al., 1991; Kelly et al., 1991; Goldberg, 1995). The seven different β-subunits of this complex form a ring-like structure. A free threonine at the N-terminus of each of the mature subunits provides part of the active site of each subunit (Glynne et al., 1993). LMP2 and LMP7 can replace two other constitutive proteasome β-subunits, δ and ε (MB1) (Belich et al., 1994) and this may alter the proteolytic activity of the proteasome to favour the production of peptides appropriate for binding to the grooves of class I molecules (Gaczynska et al., 1994). Consistent with these ideas is the finding that expression of LMP2 and LMP7, like that of TAP and the MHC class I molecules, is inducible by interferon-γ (IFNγ).

Other sequences in the class II region

RING3 is the only gene in the class II region of the MHC that encodes a protein with no obvious function in the immune system. The ubiquitously expressed product of this locus shares homology with the *Drosophila* gene, female sterile homeotic, *fsh* (Beck et al., 1992).

Two pseudogenes, located between the DMB and LMP2 loci, are of interest. This region is thought to be the result of an insertion of genetic material into the class II region from elsewhere, as there is no trace of a similar region in the mouse MHC between the Ob and LMP2 genes. One pseudogene, lying about 40 kb centromeric of LMP2, is a copy of a phosphatase inhibitor gene, IPP-2 (Sanseau et al., 1994). The other pseudogene, lying between the IPP-2 gene and DMB, is a short fragment of a class I-like gene. Its presence in the class II region suggests that this segment of DNA originated in the class I region and became inserted into the class II region, carrying IPP-2.

Additional immune system-related loci centromeric of the class II region

The tapasin molecule is a member of the Ig superfamily required for correct association between TAP and MHC molecules in the ER (Ortmann et al., 1997). Tapasin contains an Ig domain, which lies adjacent to the ER membrane and is encoded by a separate exon (Herberg et al., 1998). The rest of the molecule exhibits only limited identity with Ig sequences and the introns are in atypical positions, suggesting that the tapasin sequence has a distinct phylogeny. The tapasin locus was mapped recently to a position centromeric of HLA-DP, between the HSET and HKE1.5 genes, about 500 kb from the TAP loci. The gene is located in an equivalent position in mice and rats. The localization of this gene, the product of which is intimately associated with the antigen-processing machinery for producing peptides that bind to class I molecules, may be of regulatory or functional significance.

Evidence that yet another molecule involved in antigen processing is encoded at the centromeric end of the MHC came from the study of the Cim2 gene (Simmons et al., 1997). This locus was identified by studying the response of HLA-B27-restricted CTL to the H-Y minor histocompatibility antigen in mice transgeneic for HLA-B27 and human β_2-microglobulin. A polymorphism was identified near the H-2 complex that resulted in two distinct, overlapping sets of B27-presented peptides. Subsequently, the Cim2 locus was mapped between the H-2K and Pb loci, equivalent to the set of genes in humans spanning RING1 to HLA-DP. The locus, which remains unidentified to date, appears to exert an influence on the class I antigen-presentation pathway.

THE CLASS III REGION

The 1100 kb segment of DNA that separates the class I and II regions is generally termed the class III region and contains at least 70 genes (Figure 3.1; reviewed in Aguado et al., 1996). The gene density in the class III region is approximately one gene every 10 kb on average and many of the genes are very closely packed: for example G11 and C4 are separated by 611 bp, while RAGE and PBX2 overlap slightly. The 120 kb segment of DNA containing the duplicated C4/P450c21/X/Y gene cluster is one of the most complex in the human genome and contains overlapping genes and genes within genes (Bristow et al., 1993a; Tee et al., 1995). Whilst the class I and II regions are typified by a high level of gene duplication and extensive polymorphism, these features are less apparent in the class III region. However, this does contain several clusters of related genes. These include those encoding the complement components C2, C4 and factor B (Bf), two members of the 70 kDa heat shock protein (hsp70) family, and the cytokines tumour necrosis factor (TNF), lymphotoxin (LT)α and LTβ. The MHC class I chain-related (MIC) gene family has two members (MICA and MICB) between HLA-B and BAT1 and a further three related gene fragments within the class I region (Bahram et al., 1994). The PERB (Perth MHC β-block) sequences also include class I-like pseudogenes and gene fragments (Marshall et al., 1993).

GENES ENCODING PROTEINS WITH KNOWN OR PUTATIVE ROLES IN THE IMMUNE SYSTEM

It is becoming apparent that many of the genes in the class III region encode proteins with immune-related functions. Of particular interest is a cluster of genes at the telomeric end of the region, which includes the genes encoding TNF, LTα and LTβ and at least three other proteins with potential roles in inflammation (Holzinger et al., 1995; Nalabolu et al., 1996; Utans et al., 1996). These genes and their products will now be discussed in order, moving from centromere to telomere.

G17 (PBX2)

The G17 gene encodes PBX2, a homeodomain-containing protein related to PBX1 (Aguado and Campbell, 1995). It has been suggested that the PBX proteins can form heterodimers with homeobox proteins and thereby modulate their DNA specificities. Since the homeobox proteins show different expression patterns in various developmental stages and lineages of the haematopoietic system, the PBX proteins may be important in regulating the expansion of haematopoietic precursors (Lawrence et al., 1996).

RAGE

RAGE (a receptor for advanced glycosylation end-products of proteins) is a member of the Ig superfamily of cell surface molecules (Neeper et al., 1992; Sugaya et al., 1994). AGEs are non-enzymatically glycosylated proteins that accumulate in vascular tissue as a result of ageing, and at an accelerated rate in diabetes, owing to the prolonged exposure of proteins to aldoses such as glucose and ribose. RAGE mediates monocyte migration and activation in response to AGEs, whilst the binding of AGEs to endothelial RAGE induces cellular oxidative stress resulting in increased vascular permeability and activation of nuclear factor (NF)-κB. RAGE could contribute to diabetic complications by modulating cellular function following interaction with AGEs and expression of RAGE is upregulated in the vasculature in a range of vasculopathies, in artherosclerotic vascular lesions and in immune vasculitides (Schmidt et al., 1995).

G15 hLPAATα

The G15 gene encodes a protein that shares significant sequence identity with lysophosphatidic acid acetyltransferase (LPAAT) from yeast, plant and bacterial species (Aguado and Campbell, 1998). The recombinant protein has been shown to have LPAAT activity: specifically converting LPA to PA, which is the precursor of all glycerolipids. The MHC-linked LPAAT has been designated hLPAATα and shares 48% identity with hLPAATβ, which is encoded on chromosome 9. hLPAATα is localized in the ER and is predicted to be an integral membrane protein with a potential active site in the cytosol (Aguado and Campbell, 1998).

hLPAATα has been shown to have an affinity for fatty acids with acyl chains of 12–18 carbons in length, with a slight dependence on the degree of saturation of the fatty acid (Aguado and Campbell, 1998). Of particular interest is the observation that hLPAATα can incorporate arachidonoyl-coenzyme A (CoA) into LPA to form PA. Arachidonic acid is the most important prostaglandin precursor in humans and can be released from PA by the action of phospholipase A_2. Prostaglandins play important roles in the inflammatory response, the production of pain and fever, the regulation of blood pressure, induction of blood clotting, control of several reproductive functions and regulation of the sleep–wake cycle (Voet and Voet, 1995).

A major function of phospholipids is the formation of biological membranes. However, some phospholipids and their metabolites have been implicated in signalling pathways (Moolenaar, 1995). Interleukin-1β (IL-1β), TNFα and platelet activating factor are among species that may activate and signal, at least in part, through a common lipid intracellular signalling pathway following the activation of LPAAT (see references in Aguado and Campbell, 1998). Furthermore, inhibitors of LPAAT have been shown to block the inflammatory response mediated by an increase in PA.

Complement components C2 and C4 and factor B

Three components of the complement system, the principal effector mechanism of humoral immunity (Law and Reid, 1995), are encoded in the class III region (Reid and Campbell, 1993). C2 and C4 participate in the classical pathway, which is activated by the binding of C1 to complexes of IgG or IgM, with antigen, or following the interaction of mannan-binding protein with carbohydrate ligands on the surfaces of bacteria or viruses. Bf is a component of the alternative pathway, which is activated by a diverse set of substances including components of yeast and bacterial cell walls. The C2 and Bf genes are separated by only 421 bp, while the two C4 genes, C4A and C4B, lie ~30 kb from Bf and are separated from each other by ~10 kb. The close linkage and similar organization of the C2 and Bf genes, together with the fact that they encode functionally related proteins, suggests that these genes arose by duplication of an ancestral locus. This is also thought to be the case for C4A and C4B, which encode proteins that differ by only four amino acids, but nevertheless have profoundly different covalent binding activities. Both isotypes of C4 are highly polymorphic, whilst C2 and Bf show more limited polymorphism. Duplication and deletion of the C4 genes is relatively common, such that the number of expressed C4 genes on an individual chromosome can range from none to four. Complete deficiency of C4 is rare and correlates with severe immune complex disease, whereas partial deficiency of C4 is more common and is thought to be associated with an increased susceptibility to systemic lupus erythematosus (SLE), scleroderma and primary biliary cirrhosis (Hauptmann et al., 1988).

G9

The G9 gene encodes a protein with significant sequence identity to bacterial sialidases (Bonten et al., 1996; Milner et al., 1997; Pshezhetsky et al., 1997). Recombinant G9 protein has sialidase activity with an acidic pH optimum and has been shown to be localized in lysosomes, where it forms a complex with protective protein/cathepsin A (PPCA) and β-galactosidase (van der Spoel et al., 1998). Sialidases are important in modulating cellular functions by regulating the sialic acid content of glycoproteins and glycolipids. Deficiency in lysosomal sialidase activity, caused by point mutations in the sialidase itself or in PPCA, can result in various forms of sialidosis or in galactosialidosis, lysosomal storage disorders characterized by various developmental and neurological abnormalities (Bonten et al., 1996; Pshezhetsky et al., 1997; Rudenko et al., 1998). The murine equivalent of G9, known as Neu1, was originally localized to the H2-S region by genetic linkage studies in back-cross mice (Figueroa et al., 1982). Three allelic variants of Neu1 have been described, of which Neu1^a confers only ~17% of normal sialidase activity (Klein and Klein, 1982). Mice carrying this allele are small at birth, and show a high incidence of amyloidosis and low lymphocyte response to the mitogen phytohaemagglutinin. Comparison of mice expressing the different Neu1 allotypes has revealed that the activity of the MHC-linked sialidase is upregulated in activated T cells with a concomitant reduction in cell surface sialic acid content (Landolfi et al., 1985). The latter is probably due to hyposialylation of cell surface proteins and is a requirement for T cell responsiveness to alloreactive B cells. Furthermore, Neu1 has been shown to be required for the regulation of IL-4 production during T cell activation, which has important implications for the initiation of Th2- versus Th1-type responses (Chen et al., 1997a). Imbalance in T cell responses may lead to autoimmune or allergic responses.

Heat shock proteins

The HSP70–1 and -2 genes both encode the major heat-inducible hsp70 and their expression is upregulated in response to heat shock (Milner and Campbell, 1990). HSP70-HOM encodes a protein 90% identical to HSP70–1/-2, but its expression is not heat inducible (Milner and Campbell, 1990). All hsp70s can bind unfolded proteins and peptides and act as chaperones in the synthesis, folding, assembly, translocation and degradation of proteins during normal cellular processes and following stress. The cytosolic hsp70s (including HSP70 and HSC70), HSP90 and the ER-resident gp96, are thought to form a relay line chaperoning antigenic peptides generated in the cytosol until they bind MHC class I molecules in the ER (Srivastava et al., 1994). In support of this, hsp preparations from tumour cells, virally infected cells and also normal cells and tissues have been shown to contain complexes of hsp and peptide. Furthermore, hsp70 and gp96 purified from murine MethA sarcomas have been found to give rise to tumour-specific immunity mediated by T cells when injected into syngeneic mice (Srivastava et al., 1994). This indicates that an immune response is being initiated by tumour-specific peptides carried by hsps. In the case of gp96, this has been shown to arise following uptake of the hsp–peptide complexes by a subset of macrophages (Suto and Srivastava, 1995). These professional APCs prime antigen-specific CTLs, presumably as a result of intracellular hsp–peptide dissociation and channelling of the peptides into the MHC class I antigen-presenting pathway.

The interactions of the hsp70s with proteins and peptides are mediated by a C-terminal peptide binding domain of ~28 kDa. Investigation of the peptide-binding properties of HSC70 and GRP78 has indicated that sequence variation in the peptide-binding domain is associated with differential binding specificities (Fourie et al., 1994). Two polymorphic residues have been identified in the peptide-binding domain of HSP70-HOM: Met493Thr (Milner and Campbell, 1992) and Glu602Lys (S.C. Jenkins, R.D. Campbell and C.M. Milner, manuscript in preparation). This may influence the spectrum of proteins/peptides bound by different allotypes of HSP70-HOM.

Cytosolic members of the hsp70 family have been detected on the surface of some tumour cells, where they mediate a strong anti-tumour response by rendering cells sensitive to NK cell-mediated lysis (Multhoff et al., 1997). The mechanisms of transport to the cell surface and association with the cell membrane are not known. However, there is evidence to suggest that the C-terminal region of the hsp70(s) is recognized by receptors that control the proliferative and cytolytic responses of NK cells (Multhoff et al., 1999).

Increased expression of hsp70s has been observed in autoimmune diseases such as scleroderma and SLE, but it is not clear whether this is a contributory factor or a consequence of disease. However, exposure of keratinocytes to ultraviolet light results in accumulation of hsp70 in the nucleoli and a concomitant increase in binding sites for autoantibodies to nuclear antigens, which are a feature of erythematosis lesions (Furukawa et al., 1993).

G1

cDNAs corresponding to two transcripts, G1 and allograft inflammatory factor-1 (AIF-1), which appear to be the result of alternative splicing of the G1 gene, have been described (Olavesen et al., 1993; Utans et al., 1996). These show distinct patterns of expression: AIF-1 is expressed at high levels only in cells of the macrophage lineage, whilst G1 is also expressed at significant levels in T cells and to a lesser extent B cells. The G1 and AIF-1 transcripts encode proteins of 93 and 147 amino acids, respectively, which both contain a potential "EF-hand"-type calcium-binding domain, although this is truncated at the N-terminus in G1. The expression of AIF-1 is upregulated in response to IFNγ and its elevated expression in experimental cardiac transplants has led to the suggestion that AIF-1 plays a role in the processes leading to chronic allograft rejection (Utans et al., 1996). An involvement in the progression of IDDM has also been suggested since AIF-1 can modulate insulin secretion and has been detected at high levels in macrophages isolated from pancreatic extracts of prediabetic rats, but not control rats (Chen et al., 1997b). The precise function of AIF-1, and that of G1, remains to be determined.

1C7

A novel gene, 1C7, has been described that appears to be expressed only in spleen (Nalabolu et al., 1996). The function of the encoded protein is unknown, but this restricted pattern of expression suggests a possible role in the immune system.

LST-1

Leucocyte specific transcript (LST-1) is the human equivalent of the murine B144 gene (Holzinger et al., 1995). It is expressed predominantly in monocytes, where it is inducible by IFNγ, and also, at lower levels, in some T cells. This has led to the suggestion that LST-1 plays a role in the immune response. Evidence for a complex pattern of alternative splicing suggests that multiple protein isoforms of LST-1 may exist (Neville and Campbell, 1997; de Baey et al., 1997). Furthermore, it appears that both membrane-bound and soluble forms of LST-1 are expressed, depending on the usage of two possible open reading frames. Treatment of monocytes with IFNγ has been shown to result in a shift from expression of both soluble and membrane-bound LST-1 to expression of the soluble form alone (de Baey et al., 1997). However, there is no indication as to the function of this protein.

Cytokines tumour necrosis factor, lymphotoxin-α and lymphotoxin-β

The cytokines TNF, LTα and LTβ all belong to the TNF ligand superfamily (Guss et al., 1995). TNF occurs as membrane-bound and secreted homotrimers, both of which are biologically active. LTα also forms secreted homotrimers, but can be retained at the cell surface as heterotrimers with LTβ. TNF is produced by a variety of cells and exhibits numerous inflammatory and immunomodulatory activities as well as being involved in tumour cachexia. LTα is expressed exclusively by lymphocytes. It shows similar activity *in vitro*, but is often less potent than TNF. Recent studies suggest a unique role for secreted LTα in the generation of germinal centres, and membrane-bound LTα has been implicated in the development of the spleen, lymph nodes and Peyer's patches (Matsumoto et al., 1996a, b; Neumann et al., 1996; Pasparakis et al., 1996). Polymorphisms in the promoter of the TNF gene are thought to play a role in increased susceptibility to fatal cerebral malaria (McGuire et al., 1994) and inflammatory bowel disease (Louis et al., 1996), and may also influence the onset of autoimmune diseases such as RA.

IKBL

The IKBL protein contains two copies of the 33 amino acid ANK repeat followed by a third poorly conserved ANK repeat and a short region rich in acidic amino acids (Albertella and Campbell, 1994). A similar pattern of sequence elements is found in the IκB family of transcription factor inhibitors. The Rel family of transcription factors, which includes NF-κB, regulates the expression of a wide variety of genes including cytokines, cytokine receptors and stress proteins. In unstimulated cells these transcription factors are maintained in the cytosol owing to their association with inhibitors belonging to the IκB family. Stimulation of cells by agents such as phorbol 12-myrisate 13-acetate (PMA), IL-1β, TNF and lipopolysaccharide results in the phosphorylation and subsequent degradation of the IκB protein, allowing translocation of the free Rel proteins to the nucleus.

MICA and MICB

MICA and MICB encode proteins that share 83% identity and have the same domain structure as the class I heavy chains (Bahram et al., 1994; Bahram and Spies, 1996). However, they are highly divergent from the classical and non-classical class I genes and differ in their patterns of expression. The MIC proteins lack the residues of the class I heavy chains involved in CD8 binding. Furthermore, MICA does not associate with β$_2$-microglobulin and is stable in the absence of peptide ligands. MICA is expressed predominantly in the gastrointestinal epithelium and is upregulated in response to cell stress, suggesting a possible role in the gut mucosal immune system (Groh et al., 1996). MICA and MICB are both polymorphic: more than 20 alleles of MICA and 11 alleles of MICB have been described (Fodil et al., 1996; Ando et al., 1997a). Three related pseudogenes MICC, MICD and MICE, are localized in the class I region (Bahram and Spies, 1996).

GENES ENCODING PROTEINS WITH NO OBVIOUS INVOLVEMENT IN THE IMMUNE SYSTEM

NOTCH-4

NOTCH-4, the human counterpart of mouse mammary tumour gene, int-3, encodes a member of the Notch family of transmembrane proteins (Sugaya et al., 1994). Notch was first identified as a *Drosophila* neurogenic locus required for cell lineage determination in embryogenesis. However, Notch and its homologues are widely expressed during embryonic and adult development and participate in cell–cell interactions, mediating a variety of regulatory events (Fortini and Artavanis-Tsakonas, 1993).

PPT1

Palmitoyl-protein thioesterase (PPT1) is a lysosomal hydrolase that removes long-chain fatty acyl groups from modified cysteines of target proteins. Fatty acid modification may contribute to the intracellular localization of proteins or may be of structural importance. Mutations in PPT1 have been shown to underlie the hereditary neurodegenerative disorder infantile neuronal ceroid lipofuscinosis. Recently, PPT2, which shares 18% amino acid identity with PPT1, was described (Soyombo and Hofmann, 1997) and has been shown to correspond to the G14 gene product (Aguado, 1998). PPT2 is a lysosmal protein with palmitoyl-CoA hydrolase activity, but has a substrate specificity distinct from that of PPT1. The physiological function of this enzyme remains to be determined.

Creb-rp

The G13 (Creb-rp) protein product contains a leucine zipper consensus sequence preceded by two basic clusters, separated by an alanine spacer (bZip domain) (Min et al., 1995; Khanna and Campbell, 1996). This domain is conserved in many transcription factors including the cyclic AMP response element binding protein (CREB) family, which comprises proteins that show sequence-specific binding to DNA (via a region rich in basic amino acids) and uses dimerization to control function (via a leucine zipper domain).

Steroid 21-hydroxylase

The P450c21B and P450c21A genes share 97% overall nucleotide sequence identity (Higashi et al., 1986). P450c21B encodes the adrenal enzyme steroid 21-hydroxylase, which is involved in the conversion of progesterone and 17-hydroxyprogesterone to 11-deoxycorticosterone and 11-deoxycortisol, the intermediate steps in mineralocorticoid and glucocorticoid biosynthesis, respectively. P450c21A contains deleterious mutations and so cannot encode functional protein. Deficiency of P450c21B is the most frequent cause (95%) of CAH, which is characterized by accumulation of 17-hydroxyprogesterone and production of excess androgen (Morel and Miller, 1991). Classical CAH occurs as the milder simple virilizing form and the more severe, and life-threatening, salt-wasting form. In addition, there is a non-classical late-onset form where individuals remain asymptomatic or develop the symptom of an excess of androgens in childhood or at puberty.

XB and XA

The XB (TN-X) and XA genes overlap the 3' ends of the P450c21B/A genes by 481 bp, respectively, and are transcribed in the opposite orientation to the C4, P450c21 and Y genes. The protein product of XB has been termed tenascin-X (TN-X), owing to its structural similarity to tenascin (TN-C) (Bristow et al., 1993b). The tenascins are a family of extracellular matrix proteins, the functions of which are unclear, but TN-C can facilitate cell movement by counteracting the cell adhesion and spreading activity of fibronectin *in vitro* and may also be important in embryogenesis. A short form of XB (XB-S) is transcribed from a promoter within intron 26 of XB (Tee et al., 1995). The nucleotide sequence of the XA gene is >99% identical to XB-S, but XA contains premature stop codons, so it is unlikely to encode a protein. The YB and YA genes utilize the P450c21B/A promoters, and are transcribed in the same orientation as the latter, but use a different array of exons and introns. It has not been established whether the YA/YB transcripts are translated into protein.

G11

Genomic sequence analysis of G11 has shown that its 3' end was involved in the duplication event giving rise to the two C4/P450c21/X/Y gene clusters (Sargent et al., 1994; Shen et al., 1994). Long and short forms of G11 have been identified, where G11-Y lacks the N-terminal 110 amino acids of G11-Z. Both of these protein isoforms exist in two alternative forms owing to the differential use of two splice sites at the end of exon 5. The G11 proteins have been shown to possess Ser/Thr protein kinase activity that phosphorylates the non-specific kinase substrates α-casein and histone (Gomez-Escobar et al., 1998). In addition, G11 has been shown to be localized in the nucleus. Kinases participate in events including the regulation of differentiation, cell division, transcription and DNA repair. However, the specific role of G11 remains to be determined.

G11a

The G11a (or Ski2W) protein contains an RNA helicase domain and a leucine zipper motif, and shows extensive similarity to the yeast antiviral protein Ski2p and the human protein KIAA0052 (Dangel et al., 1995; Albertella et al., 1996). Helicases are essential enzymes, being required to unwind DNA during replication, repair, recombination and transcription, and RNA during splicing and translation. Many RNA viruses do not contain the poly(A) tail and/or this is not capped. Ski2p exerts its antiviral action by translational inhibition of poly(A)$^-$ RNA.

G9a

G9a contains six contiguous copies of the 33 amino acid ANK repeat (Milner and Campbell, 1993). This repeat is thought to mediate protein–protein interactions and is found in proteins of diverse function, including *Drosophila* Notch and NF-κB. In addition, the C-terminal region of G9a (which binds zinc *in vitro*) comprises a SET domain. This 150 amino acid motif is also found in the C-terminal regions of *Drosophila* trithorax (TRX) and its human homologue ALL-1 (or HRX). The SET domain appears to play an important role in the activity of these proteins as regulators of transcription during development (Rozenblatt-Rosen et al., 1998).

G7a

The G7a protein shows 48.3% identity with the valyl-tRNA synthetase (Trs^{Val}) of *Saccharomyces cerevisiae* and functional studies on recombinant G7a have confirmed that it is the human Trs^{Val} (Hsieh and Campbell, 1991; Vilata et al., 1993). G7a has a unique N-terminal region which may be involved in the interaction of human Trs^{Val} with elongation factor 1H.

Casein kinase II β-subunit

The G5a gene encodes the casein kinase II (ckII) β-subunit (Albertella et al., 1996). CkII is a highly conserved serine protein kinase, consisting of a heterotetramer of α and α′ catalytic subunits and two regulatory β-subunits. The β-subunit is required for high-activity phosphorylation of the α-subunit. CkII has been linked to the regulation of cell growth and metabolism, as many of its substrates (such as insulin, epidermal growth factor and a variety of cellular and viral oncoproteins) play key roles in cell growth. CkII activity appears to be upregulated in the neurofibrillary tangles in the brains of Alzheimer's disease (AD) patients, and associations of HLA-A2 and DR3 with AD have been reported.

BAT1

BAT1 encodes a nuclear protein with significant sequence similarity to the DEAD-box protein family of ATP-dependent RNA helicases (Peelman et al., 1995). These proteins participate in diverse cellular functions such as initiation of translation, RNA splicing, ribosome assembly, spermatogenesis, oogenesis, cell growth and division.

POLYMORPHISM AND THE MAJOR HISTOCOMPATIBILITY COMPLEX AS A GENE CLUSTER

There is a case for the coevolution of combinations of alleles at different MHC loci, in the form of haplotypes. Indeed, it has been argued that the positioning of genes such as the TAPs and LMPs, as well as some other immune system genes, within the MHC may not have occurred by chance and may serve to maintain successful combinations of alleles in *cis* (Trowsdale, 1993). Studies of the rat TAP/class I allelic combinations are particularly informative on this issue (Howard, 1993; Joly et al., 1994). In zebrafish the class I and class II genes are on different chromosomes and it has recently been shown that the LMP and TAP genes are linked to the class I genes in this species (Takami et al., 1997). Furthermore, in *Xenopus*, the single class I gene lies between the class II and class III genes, presumably in close linkage with the TAP and LMP loci (Flajnik et al., 1993; Sato et al., 1993; Salter-Cid et al., 1994; Nonaka et al., 1997). The close linkage of these genes would facilitate the perpetuation of favourable combinations of alleles. There are other explanations for keeping combinations of loci together on extended haplotypes. One is coregulation of expression of genes. Another is gene conversion: keeping class I or class II genes together may be advantageous in permitting microsequence exchange.

RECOMBINATION AND LINKAGE DISEQUILIBRIUM

Recombination mapping places HLA-B 0.2 cM centromeric of HLA-C and HLA-A 0.8 cM telomeric of HLA-C, in reasonable agreement with the physical map. In other regions of the MHC the concordance between physical distance and recombination distance is less good. This can be attributed to linkage disequilibrium, i.e. the non-random association of alleles at linked loci. An example from the Caucasian population is HLA-A1 and HLA-B8, which occur at individual gene frequencies of about 27.5% and 15.7%, respectively. The expected frequency of the two alleles being together on the same chromosome is, therefore, around 4.3%, but this haplotype is present on 9.8% of chromosomes 6, indicating a low frequency of recombination between HLA-A and HLA-B. Recombination is not observed between HLA-DQ and -DR. There is only weak linkage disequilibrium between DP and DQ, suggesting hotspots of recombination, and indeed several cross-overs have been mapped between DP and DQ, in the second intron of the TAP1 gene, for example. An example of extreme distortion between the physical map and recombination data is to be found telomeric of the class I region, where studies of the iron-uptake disease haemochromatosis and HLA class I markers reveal a region of over 4 Mbp in which recombination is extremely rare (Feder et al., 1996).

Explanations for linkage disequilibrium range from models of selection for apt combinations of alleles at linked loci to population bottlenecks, which have not had time to become randomized and the occurrence of non-random recombination sites.

Recent measurements of recombination rates have been made using microsatellite probes (Cullen et al., 1997). The recombination rate within the class II region (DRB1-DPB1) was shown to be 0.74%, while in the class III region (HLA-B-DRB1) the rate was 0.94%. Both of these rates are within the expected range, given the standard of 1% recombination per Mbp of DNA per meiosis. However, the recombination rate between HLA-A and HLA-B was found to be 0.31%, which is smaller than expected for a 1.4 Mbp segment of DNA.

PARALOGOUS CHROMOSOMES

Constellations of loci similar to some of those near the MHC but on human chromosomes 1, 9 and 19 have been identified. For example, NOTCH4, PBX2, TN-X and hLPAATα on chromosome 6p21.3 are related to NOTCH1, PBX3, TN-C and hLPAATβ on chromosome 9q34 (Kasahara et al., 1996; Katsanis et al., 1996; Ando et al., 1997b; Endo et al., 1997; Aguado and Campbell, 1998). This is consistent with the view that the duplication of a primordial chromosome with a group of genes already in place gave rise to the four chromosomes by a tetraploidization event. The explanation for the current arrangement compatible with most of the data is that chromosome duplication took place, but it is puzzling that sequence comparisons indicate that different members of the set of duplicated genes diverged at wildly different times (Hughes and Yeager, 1997). Perhaps the duplications were followed by subsequent recruitment of a subset of genes on some of the paralogous chromosomes.

CONCLUSION

The complexities of this fascinating section of the genome have only begun to become clear during the 1990s. It is likely that the next decade will see further elucidation of the class II and III regions and an opening up of the class I region. It is likely that this will bring with it deeper insights into immunology as well as the diverse regions of biology in which this unique stretch of DNA has already become involved, and others besides.

REFERENCES

Aguado, B. (1998). *Sequence analysis and functional characterization of three novel genes located in the human MHC.* D.Phil. Thesis, University of Oxford, Oxford.

Aguado, B. and Campbell, R.D. (1995) The novel gene G17, located in the human major histocompatibility complex, encodes PBX2 a homeodomain-containing protein. *Genomics* **25**: 650–659.

Aguado, B. and Campbell, R.D. (1998). Characterization of a human lysophosphatidic acid acyltransferase that is encoded by a gene located in the class III region of the human major histocompatibility complex. *J. Biol. Chem.* **273**: 4096–4105.

Aguado, B., Milner, C.M. and Campbell, R.D. (1996). Genes of the MHC class III region and the functions of the proteins they encode. In *HLA and MHC: Genes, Molecules and Function* (eds M. Browning and A. J. McMichael), pp. 34–45, Bios Scientific Publishers Ltd, Oxford, UK.

Albertella, M.R. and Campbell, R.D. (1994). Characterisation of a novel gene in the human major histocompatibility complex that encodes a potential new member of the I kappa B family of proteins. *Hum. Mol. Genet.* **3**: 793–799.

Albertella, M.R., Jones, H., Thomson, W. et al. (1996). Localisation of eight additional genes in the human major histocompatibility complex, including the gene encoding the casein kinase II beta subunit. *Genomics* **36**: 240–251.

Andersson, G., Larhammar, D., Widmark, E. et al. (1987). Class II genes of the human major histocompatibility complex. Organization and evolutionary relationship of the DR-beta genes. *J. Biol. Chem.* **262**: 8748–8758.

Ando, H., Mizuki, N., Ota, M. et al. (1997a). Allelic variants of the human MHC class I chain-related B gene (MICB). *Immunogenetics* **319**: 81–89.

Ando, A., Inoko, H. and Endo, T. (1997b). Gross similarity of genes in the HLA region on chromosome 6 and 9q33–34. *Tanpakushitsu Kakusan Koso* **42**: 2704–2713.

Androlewicz, M.J. and Cresswell, P. (1994). Human transporters associated with antigen processing possess a promiscuous peptide-binding site. *Immunity* **1**: 7–14.

Baey, A. de, Fellerhoff, B., Maier, S. et al. (1997). Complex expression pattern of the TNF region gene LST1 through differential regulation, initiation, and alternative splicing. *Genomics* **45**: 591–600.

Bahram, S. and Spies, T. (1996). The MIC gene family. *Res. Immunol.* **147**: 328–333.

Bahram, S., Bresnahan, M., Geraghty, D.E. and Spies, T. (1994). A second lineage of mammalian major histocompatibility complex class I genes. *Proc. Natl Acad. Sci. USA* **91**: 6259–6263.

Beck, S., Hanson, I., Kelly, A. et al. (1992). A homologue of the *Drosophila fsh* gene in the human MHC class II region. *DNA Sequence* **2**: 203–210.

Belich, M.P., Glynne, R.J., Senger, G. et al. (1994). Proteasome components with reciprocal expression to that of the MHC-encoded LMP proteins. *Curr. Biol.* **4**: 769–776.

Bonten, E., Spoel, A. van der, Fornerod, M. et al. (1996) Characterization of human lysosomal neuraminidase defines the molecular basis of the metabolic storage disorder sialidosis. *Genes Dev.* **10**: 3156–3169.

Braud, V.M., Allan, D.S.J., Wilson, D. and McMichael, A.J. (1997). TAP- and tapasin-dependent HLA-E surface expression correlates with the binding of an MHC class I leader peptide. *Curr. Biol.* **8**: 1–10.

Braud, V.M., Allan, D.S.J., O'Callaghan, C.A. et al. (1998) HLA-E binds to natural killer cell receptors. *Nature* **391**: 795–797.

Bristow, J., Tee, M.K., Gitelmann, S.E. et al. (1993a). Tenascin-X: a novel extracellular matrix protein encoded by the human XB gene overlapping P450c21B. *J. Cell Biol.* **122**: 265–278.

Bristow, J., Gitelman, S.E., Tee, M.K. et al. (1993b). Abundant adrenal-specific transcription of the human P450c21A "pseudogene". *J. Biol. Chem.* **268**: 12 919–12 924.

Campbell, R.D. and Trowsdale, J. (1993). Map of the human MHC. *Immunol. Today* **14**: 349–352.

Campbell, R.D. and Trowsdale, J. (1997). Map of the human MHC. *Immunol. Today* **18**: centrefold.

Chen, X.-P., Enioutina, E.Y. and Daynes, R.A. (1997a). The control of IL-4 gene expression in activated murine T lymphocytes. *J. Immunol.* **158**: 3070–3080.

Chen, Z.-W., Ahren, B., Ostenson, C.-G. et al. (1997b). Identification, isolation, and characterization of daintain (allograft inflammatory factor 1), a macrophage polypeptide with effects on insulin secretion and abundantly present in the pancreas of prediabetic BB rats. *Proc. Natl Acad. Sci. USA* **94**: 13 879–13 884.

Cho, S., Attaya, M. and Monaco, J.J. (1991). New class II-like genes in the murine MHC. *Nature* **353**: 573–576.

Colonna, M., Brooks, E.G., Falco, M. et al. (1993). Generation of allospecific natural killer cells by stimulation across a polymorphism of HLA-C. *Science* **260**: 1121–1124.

Cullen, M., Beck, S., Erlich, H. et al. (1997). Characterisation of recombination hotspots in the HLA class II region. *Am. J. Hum. Genet.* **60**: 392–407.

Dangel, A.W., Shen, L.M., Mendoza, A.R. et al. (1995). Human helicase gene Ski2W in the HLA class III region exhibits striking structural similarities to the yeast antiviral gene Ski2 and to the human gene KIAA0052 – emergence of a new gene family. *Nucl. Acids Res.* **23**: 2120–2126.

Davies, J.L., Kawaguchi, Y., Bennett, S.T. et al. (1994). A genome-wide search for human type 1 diabetes susceptibility genes. *Nature* **371**: 130–136.

Denzin, L.K., Sant'Angelo, D.B., Hammond, C. et al. (1997). Negative regulation by HLA-DO of MHC Class II-restricted antigen processing. *Science* **278**: 106–110.

Endo, T., Imanishi, T., Gojobori, T. and Inoko, H. (1997). Evolutionary significance of intra-genome duplications on human chromosomes. *Gene* **205**: 19–27.

Fan, W., Liu, Y.-C., Parimoo, S. and Weissman, S.M. (1995). Olfactory receptor-like genes are located in the human major histocompatibility complex. *Genomics* **27**: 119–123.

Feder, J.N., Gnirke, A., Thomas, W. et al. (1996). A novel MHC class I-like gene is mutated in patients with heriditary haemochromatosis. *Nature Genetics* **13**: 399–403.

Figueroa, F., Klein, D., Tewarson, S. and Klein, J. (1982). Evidence for placing the Neu-1 locus within the mouse H-2 complex. *J. Immunol.* **129**: 2089–2093.

Flajnik, M.F., Kasahara, M., Shum, B.P. et al. (1993). A novel type of class I gene organisation in vertebrates: a large family of non-MHC-linked class I genes is expressed at the RNA level in the amphibian *Xenopus*. *EMBO* **12**: 4385–4396.

Fling, S.P., Arp, B. and Pious, D. (1994). HLA-DMA and -DMB genes are both required for MHC class II/peptide complex formation in antigen-presenting cells. *Nature* **368**: 554–558.

Fodil, N., Laloux, L., Wanner, V. et al. (1996). Allelic repertoire of the human MHC class I MICA gene. *Immunogenetics* **44**: 351–357.

Fortini, M.E. and Artavanis-Tsakonas, S. (1993). Notch: neurogenesis is only part of the picture. *Cell* **75**: 1245–1247.

Fourie, A.M., Sambrook, J.F. and Gething, M.J. (1994). Common and divergent peptide binding specificities of hsp70 molecular chaperones. *J. Biol. Chem.* **269**: 30 470–30 478.

Furukawa, F., Ikai, K., Matsuyoshi, N. et al. (1993). Relationship between heat shock protein induction and the binding antibodies to the extractable nuclear antigens on cultured human keratinocytes. *J. Invest. Dermatol.* **101**: 191–195.

Gaczynska, M., Rock, K.L., Spies, T. and Goldberg, A.L. (1994). Peptidase activities of proteasomes are differentially regulated by the major histocompatibility complex-encoded genes for LMP2 and LMP7. *Proc. Natl Acad. Sci. USA* **91**: 9213–9217.

Geraghty, D.E. (1993). Structure of the HLA class I region and expression of its resident genes. *Curr. Opin. Immunol.* **5**: 3–7.

Glynne, R., Powis, S.H., Beck, S. et al. (1991). A proteasome-related gene between the two ABC transporter loci in the class II region of the human MHC. *Nature* **353**: 357–360.

Glynne, R., Kerr, L.A., Mockridge, I. et al. (1993). The MHC-encoded proteasome component LMP7: alternative first exons and post-translational processing. *Eur. J. Immunol.* **23**: 860–866.

Goldberg, A.L. (1995). Functions of the proteasome: the lysis at the end of the tunnel. *Science* **268**: 522–523.

Gomez-Escobar, N., Chou, C.-F., Lin, W.-W. et al. (1998). The G11 gene located in the major histocompatibility complex encodes a novel nuclear serine/threonine protein kinase. *J. Biol. Chem.* **273**: 30 954–30 960.

Gorski, J., Rollini, P. and Mach, B. (1987). Structural comparisons of the genes of two HLA-DR supertypic groups: the loci encoding DRw52 and DRw53 are not truly allelic. *Immunogenetics* **25**: 397–402.

Groh, V., Bahram, S., Bauer, S. et al. (1996). Cell stress-regulated human major histocompatibility complex class I gene expressed in gastrointestinal epithelium. *Proc. Natl Acad. Sci. USA* **93**: 12 445–12 450.

Gruen, J.R. and Weissman, S.M. (1997). Evolving views of the MHC. *Blood* **90**: 4252–4265.

Gruen, J.R., Nalabolu, S.R., Chu, T.W. et al. (1996). A transcription map of the major histocompatibility complex (MHC) class I region. *Genomics* **36**: 70–85.

Guss, H.-J. and Dower, S. (1995). Tumor necrosis factor ligand superfamily: involvement in the pathology of malignant lymphomas. *Blood* **85**: 3378–3404.

Gustafsson, K., Widmark, E., Jonsson, A.-K. et al. (1987). Class II genes of the human major histocompatibility complex. Evolution of the DP region as deduced from nucleotide sequences of the four genes. *J. Biol. Chem.* **262**: 8778–8786.

Ham, S.M. van, Tjin, E.P.M., Lillemeier, B.F. et al. (1997). HLA-DO is a negative modulator of HLA-DM-mediated MHC class II peptide loading. *Curr. Biol.* **7**: 950–957.

Hauptmann, G. Tappeiner, G. and Schifferli, J.A. (1988). Inherited deficiency of the fourth component of human complement. *Immunodefic. Rev.* **1**: 3–32.

Henry, J., Ribouchon, M.T., Depetris, D. et al. (1997). Cloning, structural analysis and mapping of the B30 and B7 multigenic families to the MHC and other chromosomal regions. *Immunogenetics* **46**: 383–395.

Herberg, J.A., Sgouros, J., Jones, T. et al. (1998). Genomic analysis of the Tapasin gene, located close to the TAP loci in the MHC. *Eur. J. Immunol.* **28**: 459–467.

Higashi, Y., Yoshioka, H., Yamane, M. et al. (1986). Complete nucleotide sequence of two steroid 21-hydroxylase genes tandemly arranged in human chromosome: a pseudogene and a genuine gene. *Proc. Natl Acad. Sci. USA* **83**: 2841–2845.

Higgins, C.F., Hyde, S.C., Mimmack, M.M. et al. (1990). Binding protein-dependent transport systems. *J. Bioerg. Biomem.* **22**: 571–592.

Holzinger, I., Baey, A. de, Messer, G. et al. (1995). Cloning and genomic characterisation of LST1: a new gene in the human TNF region. *Immunogenetics* **42**: 315–322.

Howard, J.C. (1993). Restrictions on the use of antigenic peptides by the immune system. *Proc. Natl Acad. Sci. USA* **90**: 3777–3779.

Hsieh, S.L. and Campbell, R.D. (1991). Evidence that gene G7a in the human major histocompatibility complex is valyl-tRNA synthetase. *Biochem. J.* **278**: 809–816.

Hughes, A.L. and Yeager, M. (1997). Molecular evolution of the vertebrate immune system. *Bioassays* **19**: 777–786.

Janer, M. and Geraghty, D.E. (1998). The human major histocompatibility complex: 42,221 bp of genomic sequence, high-density sequence-tagged site map, evolution, and polymorphism for HLA class 1. *Genomics* **51**: 35–44.

Jensen, P.E. (1998). Antigen processing: HLA-DO – a hitchhiking inhibitor of HLA-DM. *Curr. Biol.* **8**: 128–131.

Joly, E., Deverson, E.V., Coadwell, J.W. et al. (1994). The distribution of Tap2 alleles among laboratory rat RT1 haplotypes. *Immunogenetics* **40**: 45–53.

Jonsson, A.-K., Hyldig-Nielsen, J.-J., Servenius, B. et al. (1987). Class II genes of the human major histocompatibility complex. Comparisons of the DQ and Dx alpha and beta genes. *J. Biol. Chem.* **262**: 8767–8777.

Karlsson, L., Surh, C.D., Sprent, J. and Peterson, P.A. (1992). An unusual class II molecule. *Immunol. Today* **13**: 469–470.

Karlsson, L., Peleraux, A., Lindstedt, R. et al. (1994). Reconstitution of an operational MHC class II compartment in nonantigen-presenting cells. *Science* **266**: 1569–1573.

Kasahara, M., Hayashi, M., Tanaka, K. et al. (1996). Chromosomal localisation of the proteasome Z subunit gene reveals an ancient chromosomal duplication involving the major histocompatibility complex. *Proc. Natl Acad. Sci. USA* **93**: 9096–9101.

Katsanis, N., Fitzgibbon, J. and Fisher, E.M.C. (1996). Paralogy mapping: identification of a region in the human MHC triplicated onto human chromosomes 1 and 9 allows the prediction and isolation of novel PBX and NOTCH loci. *Genomics* **35**: 101–108.

Kawai, J., Ando, A., Sato, T. et al. (1989). Analysis of gene structure and antigen determinants of DR2 antigens using DR gene transfer into mouse L cells. *J. Immunol.* **142**: 312–317.

Kelly, A., Powis, S.H., Glynne, R. et al. (1991). Second proteasome-related gene in the human MHC class II region. *Nature* **353**: 667–668.

Kelly, A., Powis, S.H., Kerr, L.A. et al. (1992). Assembly and function of the two ABC transporter proteins encoded in the human major histocompatibility complex. *Nature* **355**: 641–644.

Khanna, A. and Campbell, R.D. (1996). The G13 gene in the class III region of the human MHC encodes a potential DNA binding protein. *Biochem. J.* **319**: 81–89.

Klein, D. and Klein, J. (1982). Polymorphism of the Apl (Neu-1) locus in the mouse. *Immunogenetics* **16**: 181–184.

Landolfi, N.F., Leone, J., Womack, J.E. and Cook, R. (1985). Activation of T lymphocytes results in an increase of H-2-encoded neuraminidase. *Immunogenetics* **22**: 159–167.

Law, S.K.A. and Reid, K.B.M. (1995). *Complement*, 2nd edn, IRL Press, Oxford.

Lawlor, D.A., Zemmour, J., Ennis, P.D. and Parham, P. (1990). Evolution of class-I MHC genes and proteins: from natural selection to thymic selection. *Annu. Rev. Immunol.* **8**: 23–63.

Lawrence, H.J., Sauvageau, G., Humphries, R.K. and Largman, C. (1996). The role of HOX homeobox genes in normal and leukemic hematopoiesis. *Stem Cells* **14**: 281–291.

Le Bouteiller, P. (1994). HLA class I chromosomal region, genes and products: facts and questions. *Crit. Rev. Immunol.* **14**: 89–129.

Lechler, R. (1994). *HLA and Disease*, Academic Press, London.

Lehner, P.J. and Trowsdale, J. (1998). Antigen processing: coming out gracefully. *Curr. Biol.* **8**: R605–R608.

Lepourcelet, M., Andrieux, N., Giffon, T. et al. (1996). Systematic sequencing of the human HLA-A/HLA-F region: establishment of a cosmid contig and identification of a new gene cluster within 37 kb of sequence. *Genomics* **37**: 316–326.

Lepourcelet, M., Coriton, O., Hampe, A. et al. (1998). HTEX4, a new human gene in the MHC class I region, undergoes alternative splicing and polyadenylation processes in testis. *Immunogenetics* **47**: 491–496.

Loke, Y.W. and King, A. (1991). Recent developments in the human maternal–fetal immune interaction. *Curr. Opin. Immunol.* **3**: 762–766.

Louis, E., Satsangi, J., Roussomoustakaki, M. et al. (1996). Cytokine gene polymorphisms in inflammatory bowel disease. *Gut* **39**: 705–710.

Marshall, B., Leelayuwat, C., Degli-Esposti, M.A. et al. (1993). New major histocompatibility complex genes. *Hum. Immunol.* **38**: 24–29.

Matsumoto, M., Lo, S.F., Carruthers, C.J. et al. (1996a). Affinity maturation without germinal centres in lymphotoxin-alpha-deficient mice. *Nature* **382**: 462–466.

Matsumoto, M., Mariathasan, S., Nahm, M.H. et al. (1996b). Role of lymphotoxin and the type I TNF receptor in the formation of germinal centers. *Science* **271**: 1289–1291.

McGuire, W., Hill, A., Allsopp, C.E. et al. (1994). Variation in the TNF-alpha promoter region associated with susceptibility to cerebral malaria. *Nature* **371**: 508–510.

Milner, C.M. and Campbell, R.D. (1990). Structure and expression of the three MHC-linked HSP70 genes. *Immunogenetics* **32**: 242–251.

Milner, C.M. and Campbell, R.D. (1992). Polymorphic analysis of the three MHC-linked HSP70 genes. *Immunogenetics* **36**: 357–362.

Milner, C.M. and Campbell, R.D. (1993). The G9a gene in the human major histocompatibility complex encodes a novel protein containing ankyrin-like repeats. *Biochem. J.* **290**: 811–818.

Milner, C.M., Smith, S.V., Carrillo, M.B. et al. (1997). Identification of a sialidase encoded in the human major histocompatibility complex. *J. Biol. Chem.* **272**: 4549–4558.

Min, J., Shukla, H., Kozono, H. et al. (1995). A novel creb family gene telomeric of HLA-DRA in the HLA complex. *Genomics* **30**: 149–156.

Momburg, F., Roelse, J., Hammerling, G.J. and Neefjes, J.J. (1994). Peptide size selection by the major histocompatibility complex-encoded peptide transporter. *J. Exp. Med.* **179**: 1613–1623.

Moolenaar, W.H. (1995). Lysophosphatidic acid, a multifunctional phospholipid messenger. *J. Biol. Chem.* **270**: 12 949–12 952.

Morel, Y. and Miller, W.L. (1991). Clinical and molecular genetics of congenital adrenal hyperplasia due to 21-hydroxylase deficiency. *Adv. Human Genet.* **20**: 1–68.

Morris, P., Shaman, J., Attaya, M. et al. (1994). An essential role for HLA-DM in antigen presentation by class II major histocompatibility molecules. *Nature* **368**: 551–554.

Multhoff, G., Botzler, C., Jennen, L. et al. (1997). Heat shock protein 72 on tumor cells: a recognition structure for natural killer cells. *J. Immunol.* **158**: 4341–4350.

Multhoff, G., Mizzen, L., Winchester, C.C. et al. (1999). Heat shock protein 70 (Hsp70) stimulates proliferation and cytolytic activity of natural killer cells. *Exp. Hematol.* **27**: 1627–1636.

Nalabolu, S.R., Shukla, H., Nallur, G. et al. (1996). Genes in a 220-kb region spanning the TNF cluster in human MHC. *Genomics* **31**: 215–222.

Neeper, M., Schmidt, A.M., Brett, J. et al. (1992). Cloning and expression of a cell surface receptor for advanced glycosylation end products of proteins. *J. Biol. Chem.* **267**: 14 998–15 004.

Neumann, B., Luz, A., Pfeffer, K. and Holzmann, B. (1996). Defective Peyer's patch organogenesis in mice lacking the 55-kd receptor for tumor necrosis factor. *J. Exp. Med.* **184**: 259–264.

Neumann, H., Cavalie, A., Jenne, D.E. and Wekerle, H. (1995). Induction of MHC class I genes in neurons. *Science* **269**: 549–552.

Neville, M.J. and Campbell, R.D. (1997). Alternative splicing of the LST-1 gene located in the major histocompatibility complex on human chromosome 6. *DNA Sequence* **8**: 155–160.

Nonaka, M., Namikawa, C., Kato, Y. et al. (1997). Major histocompatibility complex gene mapping in the amphibian *Xenopus* implies a primordial organization. *Proc. Natl Acad. Sci. USA* **94**: 5789–5791.

Olavesen, M.G., Thomson, W., Cheng, J. and Campbell, R.D. (1993). Characterisation of a novel gene (G1) in the class III region of the human MHC. In *Proceedings of the 11th International Histocompatibility Workshop*, Vol. 2 (eds K Tsuji, M. Aizawa and T. Sasazuki), pp. 190–193, Oxford University Press, Oxford.

Ortmann, B., Androlewicz, M.J. and Cresswell, P. (1994). MHC class I/beta 2m associate with TAP transporters before peptide binding. *Nature* **369**: 864–867.

Ortmann, B., Copeman, J., Lehner, P.J. et al. (1997). A critical role for tapasin in the assembly and function of MHC class I/TAP complexes. *Sciences* **277**: 1306–1309.

Parham, P. (1996). Immunology: keeping mother at bay. *Curr. Biol.* **6**: 638–641.

Pasparakis, M., Alexopoulou, L., Episkopou, V. and Kollias, G. (1996). Immune and inflammatory responses in TNF-alpha-deficient mice – a critical requirement for TNF-alpha in the formation of primary B cell follicles, follicular dendritic cell networks and germinal centers, and in the maturation of the humoral immune response. *J. Exp. Med.* **184**: 1397–1411.

Peelman, L.J., Chardon, P., Nunes, M. et al. (1995). The BAT1 gene in the MHC encodes an evolutionarily conserved putative nuclear RNA helicase of the DEAD family. *Genomics* **26**: 210–218.

Pshezhetsky, A.V., Richard, C., Michaud, L. et al. (1997). Cloning, expression and chromosomal mapping of human lysosomal sialidase and characterization of mutations in sialidosis. *Nat. Genet.* **15**: 316–320.

Reid, K.B.M. and Campbell, R.D. (1993). Structure and organization of complement genes. In *Complement in Health and Disease*, 2nd edn (eds K. Whaley, M. Loos and J.M. Weiler), pp. 94–98, Kluwer Academic Publishers, Dordrecht.

Rollini, P., Mach, B. and Gorski, J. (1985). Linkage map of three HLA-DRbeta chain genes: evidence for a recent duplication event. *Proc. Natl Acad. Sci. USA* **82**: 7197–7201.

Rozenblatt-Rosen, O., Rozovskaia, T., Burakov, D. et al. (1998). The C-terminal SET domains of ALL-1 and TRITHORAX interact with the INI1 and SNR1 proteins, components of the SW1/SNF complex. *Proc. Natl Acad. Sci. USA* **95**: 4152–4157.

Rudenko, G., Bonten, E., Hol, W.G.J. and D'Azzo, A. (1998). The atomic model of the human protective protein/cathepsin

A suggests a structural basis for galactosialidosis. *Proc. Natl Acad. Sci. USA* **95**: 621–625.

Salter-Cid, L., Kasahara, M. and Flajnik, M.F. (1994). Hsp70 genes are linked to the *Xenopus* major histocompatibility complex. *Immunogenetics* **39**: 1–7.

Sanderson, F., Kleijmeer, M.J., Kelly, A.P. et al. (1994). Accumulation of HLA-DM, a regulator of antigen presentation, in MHC class II compartments. *Science* **266**: 1566–1569.

Sanderson, F. and Trowsdale, J. (1995). Kissing cousins exchange CLIP. *Curr. Biol.* **5**: 1372–1376.

Sanseau, P., Jackson, A., Alderton, R.P. et al. (1994). Cloning and characterization of human phosphatase inhibitor-2 (IPP-2) sequences. *Mamm. Genome* **5**: 490–496.

Sargent, C.A., Anderson, M.J., Hsieh, S.L et al. (1994). Characterisation of the novel gene G11 lying adjacent to the complement C4A gene in the human major histocompatibility complex. *Hum. Mol. Genet.* **3**: 481–488.

Sato, K., Flajnik, M.F., DuPasquier, L. et al. (1993). Evolution of the MHC: isolation of class II beta-chain cDNA clones from the amphibian *Xenopus laevis*. *J. Immunol.* **150**: 2831–2843.

Schmidt, A.M., Yan, S.D. and Stern, D.M. (1995). The dark side of glucose. *Nature Med.* **1**: 1002–1004.

Schroeder, H.W.J., Zhu, Z.B., March, R.E. et al. (1998). Susceptibility locus for IgA deficiency and common variable immunodeficiency in the HLA-DR3, -B8, -A1 haplotypes. *Mol. Med.* **4**: 72–86.

Shen, L., Wu, L.C., Sanlioglu, S. et al. (1994). Structure and genetics of the partially duplicated gene RP located immediately upstream of the complement C4A and the C4B genes in the HLA class III region. Molecular cloning, exon–intron structure composite retroposon, and breakpoint of gene duplication. *J. Biol. Chem.* **269**: 8466–8476.

Shiina, T., Tamiya, G., Oka, A. et al. (1998). Nucleotide sequencing analysis of the 146-kilobase segment around the IκBL and MICA genes at the centromeric end of the HLA Class I region. *Genomics* **47**: 372–382.

Simmons, W.A., Roopenian, D.C., Summerfield, S.G. et al. (1997). A new MHC locus that influences class I peptide presentation. *Immunity* **7**: 641–651.

Sinha, A.A., Lopez, M.T. and McDevitt, H.O. (1990). Autoimmune diseases: the failure of self tolerance. *Science* **248**: 1380–1388.

Soyombo, A.A. and Hofmann, S.L. (1997). Molecular cloning and expression of palmitoyl-protein thioesterase 2 (PPT2), a homolog of lysosomal palmitoyl-protein thioesterase with a distinct substrate specificity. *J. Biol. Chem.* **272**: 27 456–27 463.

Spies, T. and DeMars, R. (1991). Restored expression of major histocompatibility class I molecules by gene transfer of a putative peptide transporter. *Nature* **351**: 323–324.

Spoel, A. van der, Bonten, E. and d'Azzo, A. (1998). Transport of human lysosomal neuraminidase to mature lysosomes requires protective protein/cathepsin A. *EMBO J.* **17**: 1588–1597.

Srivastava, P.K., Udno, H., Blanchere, N.E. and Li, Z. (1994). Heat shock proteins transfer peptides during antigen processing and CTL priming. *Immunogenetics* **39**: 93–98.

Sugaya, K., Fukagawa, T., Matsumoto, K.I. et al. (1994). Three genes in the human MHC class III region near the junction with the class II: gene for receptor of advanced glycosylation end products, PBX2 homeobox gene and a notch homolog, human counterpart of mouse mammary tumor gene int-3. *Genomics* **23**: 408–419.

Suto, R. and Srivastava, P.K. (1995). A mechanism for the specific immunogenicity of heat shock protein-chaperoned peptides. *Science* **269**: 1585–1588.

Takami, K., Zaleska-Rutczynska, Z., Figueroa, F. and Klein, J. (1997). Linkage of LMP, TAP, and RING3 with MHC class I rather than class II genes in the zebrafish. *J. Immunol.* **159**: 6052–6060.

Tee, M.K., Thomson, A.A., Bristow, J. and Miller, W. (1995). Sequences promoting the transcription of the human XA gene overlapping P450c21A correctly predicts the presence of a novel, adrenal-specific truncated form of Tenascin-X. *Genomics* **28**: 171–178.

Tiwari, J.L. and Terasaki, P.I. (1985). *HLA and Disease*, Springer, New York.

Tonnelle, C., DeMars, R. and Long, E.O. (1985). DO beta: a new beta chain gene in HLA-D with a distinct regulation of expression. *EMBO J.* **4**: 2839–2847.

Townsend, A. and Trowsdale, J. (1993). The transporters associated with antigen processing. *Semin. Cell Biol.* **4**: 53–61.

Trowsdale, J. (1993). Genomic structure and function in the MHC. *Trends Genet* **9**: 117–122.

Trowsdale, J., Kelly, A., Lee, J. et al. (1984). Linkage map of two HLA-SBbeta-related genes: an intron in one of the SBbeta-related genes contains a processed pseudogene. *Cell* **38**: 241–249.

Ulbrecht, M., Kellermann, J., Johnson, J.P. and Weiss, E.H. (1992). Impaired intracellular transport and cell surface expression of nonpolymorphic HLA-E: evidence for inefficient peptide binding. *J. Exp. Med.* **176**: 1083–1090.

Utans, U., Quist, W.C., McManus, B.M. et al. (1996). Allograft inflammatory factory-1. A cytokine-responsive macrophage molecule expressed in transplanted human hearts. *Transplantation* **61**: 1387–1392.

Vilata, A., Donovan, D., Wood, L. et al. (1993). Cloning, sequencing and expression of a cDNA encoding mammalian valyl-tRNA synthetase. *Gene* **123**: 181–186.

Voet, D. and Voet, J.G. (1995). *Biochemistry*, 2nd edn, John Wiley & Sons, New York.

Vogt, A.B., Kropshofer, H., Moldenhauer, G. and Hammerling, G.J. (1996). Kinetic analysis of peptide loading onto HLA-DR molecules mediated by HLA-DM. *Proc. Natl Acad. Sci. USA* **93**: 9724–9729.

Young, J.A. and Trowsdale, J. (1990). The HLA-DNA (DZA) gene is correctly expressed as a 1.1 kb mature mRNA transcript. *Immunogenetics* **31**: 386–388.

CHAPTER 4

The Biology of Major Histocompatibility Complex Molecules – I

Richard J. Baker and Anthony N. Warrens

INTRODUCTION

An organism's requirement to defend itself from internal attack led to the development of an immune system. In general, the higher up the phylogenetic tree a species is to be found, the more sophisticated is the immune system that it has evolved. Almost all vertebrates boast two limbs to this defence mechanism. The first is an innate immune system, which recognizes its prey through some motif that is common in microorganisms but is not a feature of its own cells, an example being the recognition of bacterial cell walls by phagocytes. The second limb is the adaptive immune response, which adapts according to the organism's experience of exposure.

The cells which display this adaptivity are the T and B lymphocytes (usually referred to simply as T and B cells). It has become clear that each of these cells can be divided into interacting subgroups. At the centre of an immune response is a group of $CD4^+$ T cells, also known as helper or helper/inducer cells. These cells determine the nature of an immune response. Although part of the adaptive system, and thus antigen specific themselves, helper T cells may direct a local inflammatory response by recruiting non-antigen-specific cells. Another possible outcome is that they may generate an effector limb that is itself antigen specific, either by inducing B cells to produce antigen-specific antibodies or by inducing the differentiation of $CD8^+$ cytotoxic T lymphocytes (CTLs). In addition to these responses, there are numerous checks and balances which limit immune responsiveness. Characteristic of the learning process of the adaptive immune response is immunological memory.

One of the remarkable features of the immune system is the fact that each lymphocyte has a single surface receptor that is probably unique, making available an enormous range of specificities. [Note that, because each chromosome may rearrange to form a different T cell receptor gene, a minority of cells may have a second receptor (Heath and Miller, 1993; Padovan et al., 1993).] The ability to respond and learn results from the selection of cells with specificities appropriate for a given situation. However, during its development, the system must be educated to distinguish self from non-self. [The absolute centrality of the self–non-self discrimination paradigm for immune recognition has recently been challenged and a model based on the recognition of "danger" suggested as an alternative (Matzinger, 1994).] The recognition event for both developing and mature T cells (to which is attributed the phenomenon of antigen specificity and self–non-self discrimination) involves the interaction of the T cell receptor (TCR) with a complex comprising of the major histocompatibility complex (MHC) molecule and a short antigen-derived peptide. This chapter will review our current knowledge of the role of MHC in the immune response.

MAJOR HISTOCOMPATIBILITY COMPLEX RESTRICTION

Early studies demonstrated that the interaction of lymphocytes and macrophages was associated with lymphocyte activation, and that the macrophages took up antigen (Cline and Swett, 1968). The ability of alloantisera to block this activation suggested that a product of MHC was involved in this activation. The concept that antigen-reactive T cells have specificity for both antigen and an MHC molecule arose from experiments carried out in guinea-pigs (Rosenthal and Shevach, 1973). T cells from strain 2 guinea-pigs immunized *in vivo* against purified protein derivative (PPD) would only proliferate *in vitro* in response to PPD presented by peritoneal antigen-presenting cells (APCs) from strain 2, and not from MHC-incompatible strain 13 animals. In contrast, immune T cells from strain $(2 \times 13)F_1$ animals would proliferate in response to APCs from either parental strain. They further noted that, for the F_1 T cells, alloantisera directed against the MHC type of the APCs caused complete inhibition of the response. These results demonstrated that MHC products were acting as recognition elements for T cells, and that T cells exhibited allelic specificity for the co-recognized MHC molecule. This phenomenon became known as MHC restriction: the antigen is recognized only if presented in the context of native "self" MHC products.

A similar phenomenon, for which the authors were awarded a Nobel prize, was described for CTLs (Zinkernagel and Doherty, 1974a). $H-2^k$ and $H-2^d$ mice were both infected with lymphocytic choriomeningitis virus (LCMV) and the reactivity of their splenocytes was subsequently tested *in vitro*. It was shown that lymphocytes from an $H-2^k$ mouse could lyse fibroblasts taken from a syngeneic mouse and infected with LCMV. However, those $H-2^k$ lymphocytes could not lyse LCMV-infected fibroblasts from a $H-2^d$ mouse.

These experiments were part of the evolving understanding of the function of the MHC and tied it into the process of T cell recognition. Simultaneously, data were accumulating that suggested that the MHC appeared to influence the range and magnitude of the immune response, known as Ir gene effects.

IR GENE EFFECTS

It was recognized during the nineteenth century that there was a genetic component to immune responsiveness when, for example, a familial pattern of susceptibility to *Corynebacterium diphtheriae* was identified. This observation was extended with the availability of inbred strains of animals. Guinea-pig strain 2 has a "responder" phenotype to poly-L-lysine (PLL) and the random copolymer of L-glutamine and L-alanine (GA), but strain 13 guinea-pigs are non-responders to both. Conversely, strain 13 are responders to the random copolymer of L-glutamine and L-tyrosine (GT). These genes are linked and segregate with each other but cross-overs between genes determining responsiveness to PLL and GA were identified in the outbred Hartley guinea-pigs (Benacerraf and McDevitt, 1972). The location of Ir genes within the MHC had previously been established following the observations that inbred mice were able to make a high antibody response to a series of branched multichain synthetic copolymers and that this ability was inherited in an autosomal dominant fashion from a locus linked to H-2 (McDevitt and Chinitz, 1969).

MHC incompatibility has a quantitative, gene dosage effect, in that fully MHC-incompatible allografts usually induce much stronger responses than semi-allogeneic tissues (Lechler and Batchelor, 1982). This is reflected in clinical human transplantation where kidney graft survival correlates with the number of matched loci (Opelz, 1987). However, this does not explain the many idiosyncratic Ir gene effects. These may be manifest in responses to single antigens, such as the responses of different guinea-pig strains to PPD in the studies referred to above (Rosenthal and Shevach, 1973). Another example, demonstrated in a series of papers by Sercatz's group, is that of the immune response to different fragments of hen egg lysozyme (HEL). Preimmunization of B10 mice with HEL induced suppression of the anti-HEL response to HEL-coated erythrocytes in spleen cell cultures, whereas if B10.A mice were used, preimmunization resulted in helper activity (Adorini et al., 1979). In a third example, Jensen et al. demonstrated that immunization of $H-2^b$ mice with porcine insulin gave rise to a population of T cells, identified by their radiosensitivity, which could suppress a secondary antibody response to porcine insulin, but not bovine insulin. Since these two types of insulin differ at only a small region of the molecule, and since the region is identical in both murine and porcine insulin, these authors postulated that this Ir gene effect was mediated by the induction of antigen-specific suppressor cells, and that these might be important in the maintenance of self-tolerance to murine insulin (Jensen et al., 1984).

Alternatively, Ir gene effects may be manifest in responses to much more complex antigens: for example, rejection of (DA × Lewis) F_1 rat kidneys by Lewis recipients is very difficult to suppress, but the response of DA recipients to kidneys from the same F_1 strain is weak and easily suppressed. Using a series of congenic rat strains that differed only in the MHC, clear variation amongst MHC-disparate strains was detected in response to the same erythrocyte alloantigen (Butcher et al., 1982).

There is no unifying mechanism to explain Ir gene effects. Although there are likely to be others, three established mechanisms will now be reviewed: (i) determinant selection: some MHC molecules, but not others, will select a given antigen; (ii) "holes" in the T cell repertoire: in the presence of a particular haplotype, the organism does not possess T cells that can respond to a given antigen; and (iii) suppression: a particular MHC type induces active inhibition of a response.

The phenomenon of determinant selection can be understood in terms of MHC structure, as elucidated in Chapter 1, namely, whether or not a class II molecule is able to bind a particular peptide. Lin et al. studied the Ir gene effect of non-responsiveness of H-2^{bm12b} mice to beef insulin (Lin et al., 1981). They found that immune (B6 × B6.C-H-2^{bm12b}) F_1 mice would proliferate in response to antigen presented by F_1 or B6 macrophages, but not B6.C-H-2^{bm12b} cells. Ronchese et al. mapped an Ir gene effect to a single residue on a class II molecule (Ronchese et al., 1987). Their study was based on the observation that H-2^k is a non-responder to pigeon cytochrome c (PCC), but H-2^b is a responder. L cells transfected with H2-E^k molecules, mutated to express the H2-E^b residue at position 29 of the β-chain (β29), were able to present PCC peptide to PCC-specific T cell clones. The most likely explanation of these data is that the non-responders did not express a class II molecule which could bind the specific peptide and that β29 played a crucial role in determining binding. Babbitt et al. demonstrated that the immunogenicity of HEL peptide 46–61 in H-2^k, but not H-2^d, correlated with the ability of purified detergent-solubilized A^k, but not A^d, molecules to physically bind the peptide (Babbitt et al., 1985).

A second possible mechanism of Ir gene effects is a "hole" in the organism's repertoire: that there is no T cell with a receptor which can recognize a given antigen. Given current understanding, it may be suggested that such a TCR would have been deleted as a result of being coincidentally self-reactive (the process of repertoire selection is discussed in greater detail below). Ishii et al. (1981) demonstrated that the p, q, r, s, u and v haplotypes were low responders to the random copolymer of glutamic acid and alanine, GA. However, macrophages from all of these haplotypes (except for v) were able to present GA to T cells of responder haplotypes. This implied that the defect lay at the level of the T cell, not in the APC's ability to bind peptide. Müllbacher et al. (1981) addressed the issue of haplotype preference in the H-Y (male antigen) response in (b × s)F_1 mice. They found that T cells from (b × s)F_1 mice could not be primed to lyse b haplotype male cells, unlike T cells from parental type b haplotype or (b × d)F_1 mice. They demonstrated that, in the presence of the s haplotype, the cells bearing TCRs that would recognize H-Y in the context of b were deleted on account of being autoreactive to the s haplotype. Indeed, they showed that H-2^b-restricted, H-Y-specific T cells were also alloreactive on H-2^s.

A third possible explanation for Ir gene effects is the induction of suppression. Jensen et al. performed cell-mixing experiments to investigate the lack of response of H-2^b mice to porcine insulin (Jensen et al., 1984). They found that primed T cells would not support an antibody response to porcine insulin. However, if they were irradiated before being added to the B cells, a response could indeed be obtained. This suggested the destruction of a population of suppressor cells by irradiation and was confirmed when the suppressive effect was found to be dominant: irradiated primed T cells were mixed with non-irradiated cells that were otherwise identical, and the mixture failed to support an antibody response. Adorini et al. examined the Ir gene effect in which H-2^a, but not H-2^b, mice produce an antibody response to HEL (Adorini et al., 1979). They showed that it was possible to induce an anti-HEL antibody response in H-2^b mice if the HEL was coupled to an immunogenic carrier (red blood cells) and complete Freund's adjuvant was used. However, this response was abolished if cells from this animal were cocultured in a 9:1 ratio with H-2^b cells that had been primed with HEL alone. This implied that in these latter mice, non-responsiveness could be accounted for by the induction of suppressor cells which exert a dominant effect, even if greatly outnumbered.

Given the absence of inbred strains in humans, there are few such clear-cut Ir gene effects as those described in mice. However, numerous immunological disease associations with MHC have been described. Two examples will suffice. Hill et al. (1992) sequenced peptides eluted from the HLA-B53 molecule known to be protective in severe malaria, and identified a plasmodium antigen with compatible sequence. They confirmed their hypothesis, that the B53 effect was due to the binding of a protective peptide, by demonstrating that prepulsing with this peptide rendered B53$^+$ B cells susceptible to specific lysis by T cells from B53$^+$ malaria-immune individuals. Another recent example focused on an association between DPB1 alleles and berylliosis (Richeldi et al., 1993). DPB1*0201 and DPB1*0401 were found in 11% and 52%, respectively, of a control population and 10% and 48% of an unaffected beryllium-exposed population. However, their prevalence within patients with the disease was 30% and 14%, respectively. Comparing the polymorphic residues of DP$_\beta$, 32 of 33 patients had a glutamic acid at position 69 (as do DPB1*0201 and several other less common alleles, but not DPB1*0401), compared with only 25% of the unaffected beryllium-exposed population. Whether specific associations of autoimmune disease with HLA alleles are due to Ir gene effects will only be clarified when autoantigenic peptides have been further characterized.

Non-MHC genes affecting immune responsiveness (and thus Ir genes) have now also been described. For example, the ity locus on chromosome 1 confers resistance to *Salmonella typhimurium* by modulating interferon-γ (IFN-γ) production and is probably responsible for similar effects in the response to other pathogens (Ramarathinam et al., 1993). Polymorphism in peptide transporters in the rat have been shown to have an effect on the array of MHC-bound peptides (Powis et al., 1992), although no effect of peptide transporter polymorphism has been demonstrated in humans.

THE T CELL RECEPTOR

An adaptive immune response based on the selection of relevant clones requires a large pool of TCRs from which to choose. Like the B cell receptor, the immunoglobulin molecule, TCRs have two distinct regions, coded by separate gene segments, one constant and the other highly variable, with the maximum variability at the regions in which their gene segments join. The functional characteristics of the T cell recognition mechanism, i.e. antigen specificity and MHC restriction, were defined long before its structure and gave rise to a debate centring on whether the T cell expressed one receptor (assuming that antigen and MHC interacted) (Matzinger, 1980), two receptors (assuming that they were independently recognized) (Blanden and Ada, 1978), or even three receptors (specific for antigen, self-MHC and allo-MHC, respectively) (Williamson, 1980).

Several experiments established that a T cell had dual specificity for antigen and the MHC restriction element through a single receptor. In one study two murine T cell hybridomas with specificity for MHC1 ± antigen x, and MHC2 ± antigen y were fused. Fusion products that carried both of the parental reaction patterns were obtained, but no cells that had a "crossed" specificity for MHC1 ± antigen y or for MHC2 ± antigen x were detected (Kappler et al., 1981). These data implied that T cells recognized a complex composed of an MHC molecule and an antigen, and that recognition occurred through a single receptor, although alternative explanations, for example based on conformational loss of specificity, remained a possibility. Further evidence came from studies of the response to the copolymer GAT (L-glutamic acid^{60}-L-alanine^{30}-L-tyrosine^{10}), which was under the control of H2-A, i.e. another Ir gene effect. Rock and Benacerraf demonstrated that GAT was able to block the response of a percentage of H2-A-, but not H2-E, allospecific clones (Rock and Benacerraf, 1984). This implied that GAT and class II were recognized together. The first direct demonstration of an antigenic peptide binding to purified MHC molecules by equilibrium dialysis (Babbitt et al., 1985) was followed by many other descriptions of the same observation using a series of different approaches. Furthermore, there were multiple instances in which a close correlation was seen between the efficiency of binding of a particular peptide to a particular MHC allelic product and the use of that MHC molecule as a restriction element for the peptide in question (Buus et al., 1987). Perhaps most convincingly, it was demonstrated that the transfer of TCR α- and β-chain genes from one cytotoxic T cell clone to a second transferred the antigen specificity and MHC restriction of the donor cell to the recipient (Dembić et al., 1986). These results established that the ligand specifically recognized by T cells is a binary complex of MHC molecule and peptide.

At approximately the same time, the TCR was characterized biochemically, and the genes encoding it were cloned from human (Yanagi et al., 1984) and murine (Hedrick et al., 1984a, b) T cells. Genetic analysis revealed that the TCR genes, like their immunoglobulin counterparts, are rearranging genes, and generate diversity by an apparently random process of joining variable (V), diversity (D) and joining (J) segments to give rise to a functional, expressible gene. The predicted three-dimensional structure of the TCR αβ dimer closely resembles that of an Fab fragment of an immunoglobulin molecule with a single major binding site formed by the interaction of the variable domains of the two chains (Davis and Bjorkman, 1988). A second dimeric TCR, γδ, was defined subsequently (Brenner et al., 1986) and is found on a small subpopulation of T cells in the thymus and in the peripheral immune system, particularly in epithelial tissues.

The TCR polypeptides are encoded by members of the immunoglobulin gene superfamily. Like antibodies, the most membrane-distal domains are coded for by variable genes that are spliced to constant region genes by a process of somatic gene rearrangement. The variable domains form regions analogous to the complementarity-determining regions (CDRs) of immunoglobulin, and it is predicted that these interact with the peptide–MHC antigen complex. This has recently been confirmed by the cocrystallization of TCR with MHC (Chang et al., 1997).

THE BIOLOGY OF T CELL RECOGNITION

Self–non-self-discrimination

The idea that the immune system allowed an organism to discriminate between self and non-self has a long and honourable history, dating back to Paul Ehrlich's concept of *horror autotoxicus* (Ehrlich, 1900). Burnet first proposed that this was achieved by clonal deletion (Burnet, 1957) and Lederberg suggested that a cell passes from a tolerizable to an immunizable state (Lederberg, 1959). In large measure, these pioneers were all correct, although the absolute centrality of the self–non-self discrimination paradigm for recognition within the immune system has recently been challenged and a model based on the recognition of "danger" suggested as an alternative (Matzinger, 1994). It is now believed that the thymus selects potentially useful T cell clones (positive selection) and deletes potentially autoreactive ones (negative selection) (Von Boehmer et al., 1989). In addition, the system is always amenable to peripheral modulation by various mechanisms that induce tolerance. In the following sections, current ideas on positive and negative thymic selection and peripheral tolerance will be reviewed.

The thymus: positive selection of T cells

The concept of positive selection implies that only T cells with receptors that can usefully interact with self-MHC antigens are selected in the thymus for maturation and export. This is the mechanism underlying the phenomenon of MHC restriction (discussed above). Early evidence in favour of positive selection came from experiments in which (A × B)F₁ stem cells were used to reconstitute a strain A mouse whose lymphocytes had been eliminated by irradiation. Several groups found that T cell responses in such animals were restricted by strain A MHC only. For example, in one study (BALB/c × A)F₁ cells were used to reconstitute either parental type strain which was then infected with vaccinia virus. Only target cells infected with vaccinia and expressing MHC antigens of that parental strain could be lysed by cells from the reconstituted animal (Zinkernagel et al., 1978).

Subsequent work located the site of positive selection within the thymus. In a study in which radiation chimeras, similar to those described above, were generated, F₁ animals were thymectomized and grafted with parental type thymuses prior to reconstitution. It was found that pretreatment with deoxyguanosine (which affects bone marrow-derived cells, but not thymic epithelium) had no effect on the selection of thymus MHC antigens as restriction elements for antigen responses (Lo and Sprent, 1986). Later work used a murine system in which different promoter region deletions caused differential expression of class II within the thymus (Benoist and Mathis, 1989). Taking advantage of the observation that expression of H-2E leads to the positive selection of $V_\beta 6$ T cells, this group demonstrated that mice with no H-2E expression in the thymic medulla still underwent positive selection of $V_\beta 6$ T cells, but those lacking H-2E expression in the epithelial cells of the thymic cortex did not. This implies that the principal cell responsible for positive selection is the thymic cortical epithelial cell.

The process of maturation of αβ TCR T cells requires the binding of TCR to the complex of MHC. This was established in a murine system in which animals were rendered transgenic for the α- and β-chains of a TCR specific for H-Y and restricted by D^b. Cells expressing the transgenic TCR were selected in a female D^{b+}, but not a D^{b-}, mouse (Kisielow et al., 1988a). It is now clear that peptides also play an important role in this process. Mutation of residues lying within the groove, and thus inaccessible to T cells, still affected positive selection (Nikolić-Zugić and Bevan, 1990), as could varying the available peptides (Ashton-Rickardt et al., 1993). In an *in vitro* foetal thymic lobe culture system, using cells from a mouse transgenic for a single TCR, it was possible to alter the development of mature T cells. For example, one peptide was able to induce positive selection if the class I was present at low levels, but actually induced negative selection if class I was present at high levels (Hogquist et al., 1994). Apart from indicating the sensitivity of the thymic selective processes, it also implied strong peptide specificity in these processes. However, the recent observation that peptides may act as partial agonists rather complicates the interpretation of these experiments. Disruption of the genes encoding TAP proteins, essential for the transport of peptides to the compartment in which they bind class I, led to a dramatic decrease in the production of mature $CD8^+$ cells. This could not simply be explained by the lack of peptide destabilizing the class I molecule, since in a TCR transgenic system, only the use of the peptide for which the TCR was specific would allow its positive selection *in vitro*.

Although these data made it clear that peptides were involved in the process of positive selection, they raised the issue of how closely the peptides available in the thymus needed to mirror the array of peptides which the mature T cell repertoire might be expected to deal. It was suggested that the thymus is able to produce a special set of peptides, possibly through mistakes in translation and thus produce almost all possible peptides (Kourilsky and Claverie, 1986). In this scenario, the T cell undergoes exactly the same interaction in the process of positive selection as could trigger it to activation in later life. This does not appear to be the case. Recent work has begun to elucidate which peptides actually participate in positive selection within the thymus. Two groups have extracted self-peptides from class I molecules which were able to restore positive selection of TCR transgenic thymocytes in a TAP-deficient thymus organ culture (Hogquist et al., 1997; Hu et al., 1997). The relative ease with which naturally presented self-peptides that can participate in positive selection have been found implies that no TCR is dependent on a single peptide for the process of positive selection, but rather that a mixture of self-peptides is responsible (Bevan, 1997).

By contrast, the MHC allele specificity of the process of positive selection seems to be profound, selecting TCR repertoires with limited overlap (Fink and Bevan, 1995). What is perhaps most remarkable is the extent to which the coevolution of MHC and TCR genes (on different chromosomes) has resulted in a very high degree of compatibility. It has been shown that about 20% of randomly assembled TCRs can recognize the small subset of MHC molecules that are present in a given individual (Merkenschlager et al., 1997; Zerrahn et al., 1997).

One final issue to be discussed in connection with positive selection is the association of CD4 single positive (SP) status ($CD4^+CD8^-$) with class II restriction and of CD8 SP status ($CD4^-CD8^+$) with class I restriction. This long-established observation was confirmed by the demonstration that class I (Zijlstra et al., 1990) and class II (Cosgrove et al., 1991) deficient mice lack $CD8^+$ and $CD4^+$ SP cells, respectively. However, it remains unclear how an individual cell develops one of these phenotypes. Two hypotheses formed the basis of the initial stages of this debate. The "instructive" hypothesis suggests that the binding of TCR and CD8 to class I generates different signals from the binding of TCR and CD4 to class II. Thus, the ability to bind an MHC molecule of a given class dictates the signal generated which, in turn, gives rise to a particular phenotype (Robey et al., 1991). The alternative, "stochastic/selective" hypothesis suggests that a double-positive (DP) ($CD4^+CD8^+$) cell randomly downregulates one coreceptor and only those with a fitting TCR–coreceptor combination would thrive. Thus, a T cell with a class I binding-TCR and CD8 would be positively selected, but one with the same TCR and CD4 would die (Borgulya et al., 1991). The data aimed at distinguishing between these hypotheses are conflicting. More recently, evidence has suggested that neither is correct, but rather that the lineage of the cell is dictated by the nature of the signal it receives, and that this is not necessarily a function of the class of MHC being recognized. Volkmann et al. (1998) demonstrated that an antagonist peptide is able to alter a lineage decision *in vitro* from CD4 to CD8. They postulate that this may be due to the level of $p56^{lck}$ recruited, which varies between agonist and antagonist peptides.

The thymus: negative selection of T cells

As a result of the randomness of the mutation process in the generation of both T and B cell receptors, there must exist a mechanism whereby any cell expressing a potentially self-reactive receptor may be prevented from generating an autoreactive immune response. The first-line mechanism for preventing this is the moulding of the T cell repertoire by the thymus to silence such cells. This is the process of negative selection.

Some of experimental approaches that led to the confirmation of negative selection in the thymus have already been alluded to. MacDonald et al. (1988) took advantage of the observation that $V_\beta 6^+$ T cells interact with the superantigen Mls-1^a allele, and studied the elimination of such T cells in Mls-1^{a+} mice, for whom that antigen constituted "self" (MacDonald et al., 1988). They noted that thymocytes expressing $V_\beta 6$ were present at the stage of low TCR expression, but were eliminated before achieving high levels of TCR expression. Kappler et al. (1987) showed the same phenomenon for the H2-E-reactive $V_\beta 17a$ gene segment. In the transgenic system described above, Kisielow et al. (1988a) demonstrated that cells expressing a transgenic TCR with H-Y (male antigen) specificity and D^b restriction would be expanded to form the majority of T cells in a female D^b mouse, but deleted in a male D^b mouse. Indeed, in the male mouse, the thymus was only 5% of its normal weight. It has been shown in this system that deletion occurs at the immature DP stage (Kisielow et al., 1988b).

It is clear that the cell delivering the lethal hit is in many, if not all cases, bone marrow derived, principally the dendritic cell. Data suggesting that it may also be possible for thymic epithelial cells to induce negative selection have been criticized for their lack of rigour in excluding an effect of other cell lineages. The quantitative importance of dendritic cells has been emphasized by a study which shows that the addition of DC to a thymic culture to constitute 1% of cells results in a reduction in 80% of $CD4^+$ cell numbers. The role of peptides in negative selection has also been established: it has been shown that an antagonist peptide may rescue thymocytes from deletion in a foetal thymic organ culture system (Williams et al., 1998).

The whole basis of the process of repertoire selection by the thymus is predicated on the assumption of a differential threshold for positive and negative selection (Von Boehmer et al., 1989). This has now been demonstrated experimentally. In the experiments described above (Hogquist et al., 1994), in which the level of expression of class I + peptide is varied, as well as in work involving the direct measurement of affinity using surface plasmon resonance (Alam et al., 1996), positive selection is associated with low-affinity interactions with MHC + peptide and negative selection with high-affinity interactions. Those MHC + peptide complexes which had undetectably low affinities had no functional effects on the T cell. These two data sets can be brought together by further *in vitro* thymic culture studies, which suggest that it is avidity, i.e. the combination of ligand density and affinity, which is important in determining whether or not positive selection occurs (Wang, et al., 1998).

Peripheral tolerance

Negative selection by the thymus does not alone fully explain self-tolerance. It is now clear that it is also possible to silence T cells through peripheral mechanisms of tolerance. A particularly rigorous recent study demonstrates how T cells may be tolerized peripherally without cells being deleted (Alferink et al., 1995). In this work, mice expressing two transgenes, one for K^b under a keratinocyte promoter and the other an anti-K^b TCR, were used. These mice were tolerant to K^b skin grafts but the anti-K^b TCR remained detectable in undiminished numbers. However, if tolerant anti-K^b TCR-expressing T cells were transferred from these mice into K^b-negative mice, these new hosts were able to reject K^b skin grafts after a brief period. This implies that widespread expression of K^b on non-professional APCs actively induces and maintains tolerance in the anti-K^b T cells. In the absence of antigen this tolerance is soon broken.

The cognate recognition of the MHC–peptide ligand by TCR does not automatically lead to T cell activation. Bretscher and Cohn (1970) originally proposed a two-signal model of activation which postulated that, in the absence of a second costimulatory signal from the APC, cognate recognition results in tolerance. Several experimental models have been established in which tolerance is determined by the nature of the APC involved. Lamb et al. (1983) induced tolerance in a human T cell clone using high doses of its specific antigen in the absence of added APC. Given that human T cells can express cell-surface MHC class II molecules, the inference drawn from these data was that the T cells themselves were acting as APCs, and that recognition of the MHC and its bound peptide on another T cell leads to a state of non-responsiveness. Similarly, the presentation of antigen by keratinocytes which had been induced to express class II molecules using IFN-γ has been shown to induce T cell tolerance in murine (Gaspari et al., 1988) and human (Bal et al., 1990) systems. Comparable experiments have been performed using other "non-professional" APCs, such as renal tubular cells (Frasca et al., 1998), thyroid epithelium (Lombardi et al., 1997) and muscle cells (Warrens et al., 1994). Tolerance in mouse T cell clones was also induced by splenocytes which had been pretreated with the cross-linking agent ECDI [1-ethyl-3-(3-dimethylaminopropyl)-carbodiimide] (Jenkins and Schwartz, 1987). Markmann et al. (1988) generated transgenic mice that only expressed H-2E molecules on the pancreatic islet beta cells. Not only did these animals fail to reject the pancreas, but when the H2-E-expressing islet cells were used to present specific peptide to an H2-E-restricted T cell clone *in vitro*, tolerance was induced. In yet another model, APCs were replaced altogether by an inert artificial planar membrane impregnated with the mouse class II MHC molecule, H2-A^d (Quill and Schwartz, 1987). Culture of an H2-A^d-restricted mouse T cell clone in the presence of specific peptide and the H2-A^d-containing planar membranes also induced tolerance. The molecular interactions involved in delivering the second signal will be discussed in a later section.

Even amongst those APCs which do induce responsiveness, it is clear that there is a hierarchy, ranging from the highly efficient professional APC, through to cells which can be induced to act in that role, but which cannot do so in a resting state. Most antigen presentation *in vivo* utilizes cells derived from the bone marrow which constitutively express MHC class II molecules, although not all such cells are able to act as efficient APCs, as judged by their ability to stimulate primary alloresponses. For example, MHC-incompatible resting mouse B cells fail to stimulate proliferation in an MLR (Steinman and Inaba, 1986), while activated B cells are capable of primary *in vitro* allostimulation. This cannot be accounted for by differing levels of MHC expression. The cell type with the greatest potency as an MLR stimulator is the bone marrow-derived dendritic cell (Steinman and Inaba, 1986).

In addition to the absence of a second signal, it is now clear that an altered first signal, transduced through the TCR–CD3 complex, may induce tolerance. This is likely to be the mechanism by which tolerance is induced by altered peptide ligands. Using the model of the T cell response to HA 307-329, it was shown that analogues of this peptide, in which single bases had been changed, bound class II and inhibited proliferation in an antigen-specific fashion, even in the presence of otherwise stimulatory amounts of unaltered antigenic peptide (De Magistris et al., 1992).

Factors other than the nature of the interactions with the APC may also determine the outcome of the T cell–ligand interaction. The T cell's level of maturity may have a profound effect on its responsiveness. For example, Geiger et al. (1992) generated transgenic mice which began expressing simian virus 40 large tumour antigen in the pancreas only 4–25 days after birth, and crossed them with specific TCR-transgenic mice. They found that the transgenic TCR-bearing T cells were not tolerized but rather infiltrated and destroyed the expressing tissue. This suggests that an antigen appearing later may encounter more mature T cells which are less susceptible to tolerization.

As well as functional silencing, there is evidence that clonal deletion may be a mechanism of peripheral tolerance in certain circumstances. For example, using the $V_\beta 8$ marker of response to staphylococcal enterotoxin B (SEB), following SEB injection the number of $V_\beta 8^+$ cells increases initially but is then followed by a rapid decline in $V_\beta 8$ cell numbers owing to their selective death and a concomitant functional state of tolerance (MacDonald et al., 1991). Another group introduced T cells from a female H-Y-specific TCR transgenic mouse and put them into a syngeneic male nude mouse (Rocha and von Boehmer, 1991). The transgene-positive cells proliferated initially, then most disappeared. Those that remained appeared to have been rendered nonresponsive. It is not clear why the end result in these mice is anergy and not memory, but other models have produced similar findings. It is also not clear that this pattern – a rise and then a rapid fall in cell numbers – actually represents tolerance due to deletion, rather than the normal pattern in an immune response.

One postulated mechanism of peripheral deletional tolerance may be the overwhelming and exhausting of a specific immune response. For example, mice infected with 10^2 plaque-forming units (pfu) of the LCMV strain DOCILE (LCMV-D) completely cleared the virus with a brisk CTL response. However, infection with 10^{7-} pfu resulted in a weak CTL response, prolonged tolerance and viral persistence. This was the same in thymectomized mice and appeared to result from peripheral deletion (Moskophidis et al., 1993).

An alternative to all of these mechanisms is that the lack of a response may not be the result of any active event: the antigen may just be ignored. This has been inferred from situations in which tolerance has been broken. It is possible to imagine a TCR with affinity for self that falls below the threshold for negative selection. However, in certain circumstances, for example if antigen concentration were to increase, stimulation could result. One group has demonstrated how it is possible to manipulate a situation in such a way that stimulation may result (Miller and Heath, 1993). They developed three sets of transgenic mice and crossed them. The transgenic molecules were K^b, an anti-K^b TCR and interleukin-2 (IL-2), all under the rat insulin promoter (RIP). In the TCR/K^b double transgenic, the TCR^{hi} T cells were deleted, but there was neither insulitis nor tolerance to K^b skin grafts. A population of TCR^{lo} T cells was present in the periphery. In the triple transgenic system, the mice became diabetic. The IL-2 appeared to have circumvented the ignorance. Two other possible mechanisms for clonal ignorance, holes in the T cell repertoire and failure of peptide binding, were discussed in the section on Ir gene effects.

However, clonal ignorance may occur when there is ample antigen available and the immunological potential to respond. In another transgenic model the LCMV glycoprotein was introduced under the control of the RIP and was shown to be present in detectable amounts (Ohashi et al., 1991). However, these animals showed neither activation of T cells nor induction of tolerance: the antigen was not noticed. This ignorance could be abolished by infecting the animals with LCMV.

Another reason for clonal ignorance might be the sequestration of intracellular antigenic peptides that never enter the class II presentation pathway. One postulated mechanism of the breaking of tolerance and subsequent autoimmunity is the leakage of such self-peptides into the class II presentation pathway. Alternatively, antigens may be sequestered in tissues with poor lymphatic supply and therefore anatomically segregated.

In summary, there are several mechanisms that circumvent inappropriate immune responsiveness. Not only does the thymus negatively select potentially autoreactive T cells, but in the periphery there are numerous checks and balances which together protect the individual from autoimmune attack.

T cell costimulatory interactions

In the discussion of peripheral tolerance above, the evidence supporting a two-signal hypothesis of T cell activation was reviewed. Arguably the most important molecule on the surface of the APC which results in the transduction of a second signal is B7 (previously called BB1). B7 occurs in at least two forms, B7-1 (Freeman et al., 1989) and B7-2 (Freeman et al., 1993), now designated as CD80 and CD86. Its T cell ligands are CD28 and CTLA4 (CD152). Immobilized anti-CD28 antibody alone can provide costimulatory function (Harding et al., 1992) and, in the presence of a soluble chimeric ligand for B7 (CTLA4Ig), APCs that would otherwise activate T cells actually tolerize them (Tan et al., 1993). CTLA4, although not present on naive T cells, has greater affinity for B7 than CD28 but its ligation is inhibitory of T cell responses. CTLA4's inhibitory function is not merely one of competition with CD28, since it inhibits activation in the absence of CD28 (Fallarino et al., 1998). The importance of costimulatory potential *in vivo* is underlined by the observation that otherwise non-immunogenic tissue can be rendered immunogenic by the transfection of the B7 gene (Chen et al., 1992).

B7 costimulation, however, is not the whole story. The interactions of CD40 and CD40 ligand (CD40L; CD154) have also been demonstrated to be important in costimulation. Initially thought to be important only in the regulation of B cell responses with respect to proliferation, isotype switching, antibody production and memory formation, it has become clear that that CD40 can be expressed on several other APC types such as macrophages, dendritic cells and fibroblasts. Indeed, in the inhibition of an alloresponse, the blocking of the interaction of CD40 and CD40L has a greater effect than the blocking of the interaction of CD28 with B7 (Bumgardner et al., 1998). One mechanism underlying this is that the ligation of CD40 on APCs by CD40L is thought to induce expression of CD80 (B7-1) and CD86 (B7-2). The blockade of both the B7/CD28 and CD40/CD40L interactions can have a profound effect on lymphocyte stimulation.

Numerous other molecules are involved in the cellular interactions of lymphocytes. Some may only have adhesive properties: the affinity of the TCR for the MHC–peptide complex is low and additional adhesion is highly desirable. However, these interactions may have additional signalling functions. For example, ligation of either CD4 or CD8 alone inhibits subsequent activation (Tite et al., 1986), and the concurrent binding of CD4 and TCR with MHC augments signalling (Rojo and Janeway, 1988). The transmembrane tyrosine phosphatase, CD45, is also required for effective signalling, and its different isoforms may be important in effecting different signals.

All data on signalling requirements must be treated with some caution. For example, it is important to note that signalling by the TCR appears not to be a homogeneous phenomenon. The responses to superantigen and peptide antigen are different: superantigens result in deletion and unresponsiveness but peptides cause activation and memory, yet each binds the same TCR (Webb et al., 1990). Similarly, modified peptide ligands have been described that bind to the same

MHC but induce tolerance rather than activation in the T cell, implying that a qualitatively different signal is transduced.

Allorecognition

Thus far this chapter has discussed the current understanding of the physiological role of the MHC products. However, their very name – major histocompatibility complex – emphasizes the fact that they were discovered in the highly non-physiological situation of tissue transplantation between genetically disparate individuals. The rest of this chapter will discuss how the phenomenon of allorecognition fits into our current understanding of MHC physiology, as well as some of the mechanisms by which this process brings about the vigorous activation of T cells.

It is the allogeneic HLA antigens that determine the foreignness of transplanted tissue. Although minor (non-HLA) histocompatibility antigens play a role, especially in bone marrow transplantation, it is predominantly the immune response to foreign HLA molecules that define the antigenicity of organ grafts. The foreign HLA molecules on transplanted tissue are unique as antigens in that they can be recognized by the host's T cells in two different fashions (see Figure 4.1).

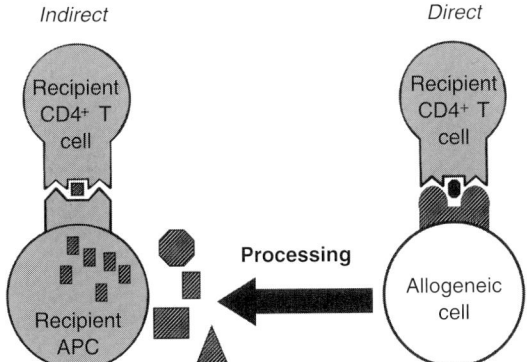

FIGURE 4.1 Two pathways of allorecognition of CD4$^+$ T cells. Indirect allorecognition mimics conventional self-restricted responses. The antigen processed through the class II pathway consists of a disproportionately high representation of HLA-derived peptides. In the other pathway the TCR recognizes the allogeneic HLA molecule as an intact protein directly on the allogeneic cell.

In the case of solid organ grafts, the foreign HLA molecules may be recognized as an intact protein on the cell surface by the recipient's T cells. This may occur on the graft parenchymal cells or on the passenger leucocytes that are contained within the graft. This is called the direct pathway of allorecognition and forms the basis of the *in vitro* mixed lymphocyte reaction. Secondly, HLA molecules may be shed from donor cells, digested by recipient specialist APCs and presented as peptide fragments of HLA molecules in association with recipient HLA, as in a conventional self-MHC-restricted response. This is called the indirect pathway of allorecognition.

Direct allorecognition

This immune reaction is extremely vigorous. In fact, when measured *in vitro* by limiting dilution analysis, up to 2% of an individual's T cells will proliferate in response to allogeneic stimulator cells (Lindahl and Wilson, 1977a, b). This constitutes a response that is up to 1000 times stronger than responses to conventional self-restricted antigens. Such vigour is perhaps surprising given the tuning that the T cell repertoire undergoes during positive and negative selection in the thymus. To explain this phenomenon it is worth considering the trimolecular interaction between the TCR and allogeneic MHC–peptide complex. This is presumed to be similar to the interaction discussed above in the context of self-restricted responses. So, why are there so many T cell clones in any individual's repertoire that attain a sufficient energy threshold for activation?

There are two main theories to explain the widespread activation of T cells, and they are not necessarily mutually exclusive. The first is called the high determinant density theory and it postulates that the allogeneic HLA molecules are intrinsically antigenic, irrespective of any bound peptide (Bevan, 1984). If this were true then the density of antigen on the surface of an MHC-incompatible cell would be huge, with upwards of 10^5 allogeneic MHC molecules, all capable of stimulating T cells. This would represent an unprecedented number of stimulatory MHC molecules when contrasted with a conventional self-restricted response, where tiny numbers of peptide–HLA complexes have been calculated to be sufficient for T cell activation (Sykulev et al., 1996). The outcome of this

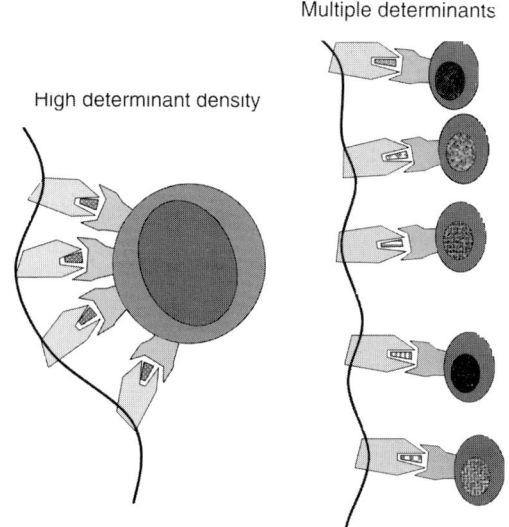

FIGURE 4.2 Schematic representation of the two theories to account for the high precursor frequency of allogeneic T cells.

form of stimulation would be widespread activation of T cell clones involving the activation of many T cells with low-affinity TCRs. This is facilitated by the increased avidity when so many TCRs can be cross-linked by the high ligand density on the target cell (see Figure 4.2). Experimental evidence for this hypothesis derives from a number of sources. First, allogeneic responses can be demonstrated, *in vitro*, against reconstituted HLA molecules mounted on plastic, supposedly without peptides, in a cell-free system (Elliott and Eisen, 1990). Secondly, mutant cell lines, such as T2, which are unable to load normal cellular peptides on to nascent HLA molecules in the endoplasmic reticulum, owing to the absence of the TAP transporter, are able to provide an allogeneic stimulus. This implies that the immunogenicity of allogeneic HLA molecules on the stimulator cells is not dependent on the normally diverse peptide repertoire (Heath et al., 1989, 1991). These results have been confirmed in experiments with mutated MHC molecules. Mutations in the upward-facing residues of the α helix can affect allorecognition in a number of experimental systems (Ajitkumar et al., 1988; Hogan et al., 1989; Clayberger et al., 1990). Indeed, in one system it was clearly shown that a normal cellular peptide, eluted from an autologous MHC molecule, became immunogenic when presented by an allogeneic MHC molecule, implying that the molecular determinants on the allogeneic MHC molecule brought about an alloresponse (Grandea and Bevan, 1992).

If indeed foreign HLA molecules are intrinsically antigenic it would appear to contradict one of the fundamental tenets of immunology, that of self-MHC restriction (Zinkernagel and Doherty, 1974b; Von Boehmer, 1994; Zil and Doherty, 1997). Although recent research has shown that TCRs are inherently alloreactive prior to thymic selection (Merkenschlager et al., 1997; Zerrahn et al., 1997), positive selection should not select those TCRs with reactivity to non-self HLA molecules. However, it is possible that TCRs positively selected on self-MHC molecules for intermediate affinity (Alam et al., 1996) could by chance cross-react with allogeneic HLA molecules, because they could actually attain a better fit with the upward-facing residues of an allogeneic HLA molecule, i.e. form a higher affinity interaction. Of course, such a high-affinity interaction during the thymic selection process with self-MHC would have resulted in negative selection and subsequent apoptosis. Such an example is shown for DR17 and DR11 in Figure 4.3. This phenomenon would only have to occur in 1 in 1000 T cells to explain the measured alloreactive frequency (Lechler and Lombardi, 1991).

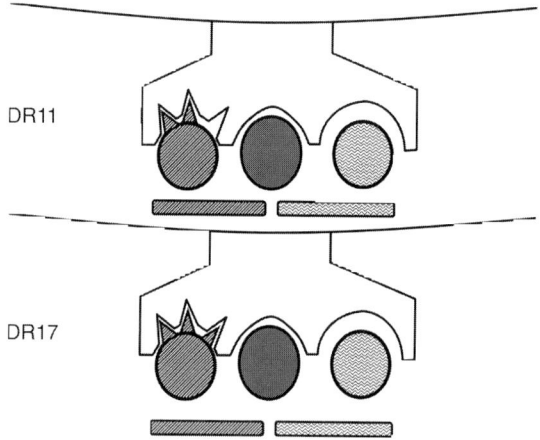

FIGURE 4.3 The better fit model to explain why positively selected T lymphocytes are capable of activation by allogeneic HLA molecules.

A second theory to explain the vigorous nature of the alloresponse makes use of the fact that the bulk of HLA polymorphism lies in the β-pleated sheet on the floor of the peptide-binding groove. This means that for many HLA molecules the upward-facing residues in the α-helices are in fact the same. For T cells approaching the HLA molecule the structure of the HLA molecule itself will be similar. This means that the T cell interaction mimics self-restricted antigen recognition. However, the peptide-binding motifs will be altered owing to the different residues lining the floor of the peptide binding groove (Rammensee et al., 1995). The peptide repertoire would then be largely different, with every HLA molecule on the cell surface binding a distinct array of peptides and thus presenting neoepitopes, not previously encountered during thymic selection. This would stimulate thousands of different clones of responding T cells. This model has been called the multiple determinant theory (Matzinger and Bevan, 1977). This form of allorecognition could use the existing self-restricted T cell repertoire and is heavily dependent on novel peptides. It has been estimated that at least 2000 different peptides may be bound to a single type of HLA molecule at any one time (Engelhard, 1994). This mechanism has been demonstrated for the alloresponse between an HLA DR1 responder and certain HLA-DR4 subtypes (see Figure 4.4) (Lombardi et al., 1989). Further experimental evidence for this model comes from peptide-loading mutant cell lines where allorecognition could only be restored upon addition of peptides to the cells (Heath et al., 1989, 1991). Incubation of cells expressing MHC class II with exogenous peptide capable of binding class II can in some cases abrogate allorecognition, implying that these peptides competitively displace the allogeneic peptides (Rock and Benacerraf, 1984; Eckels et al., 1988). In addition, there are many examples of alloreactive T cell clones that discriminate between MHC molecules differing at residues that lie within the peptide-binding groove (Lombardi et al., 1989; Mattson et al., 1989; Hunt et al., 1990; Dellabona et al., 1991; Matsui et al., 1993; Frelinger and McMillan, 1996). It may also explain why some alloreactive T cells only recognize MHC molecules when they are on certain cell types, indicating that the peptide repertoire has changed (Heath

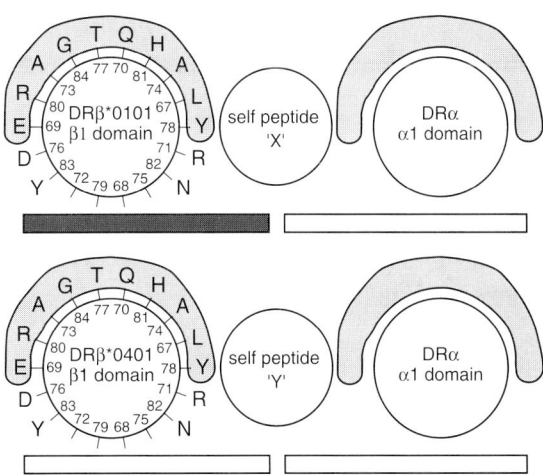

FIGURE 4.4 Activation of T lymphocytes by allogeneic HLA molecules may resemble conventional self-restricted responses at a molecular level in certain circumstances permitting cross-reactivity.

and Sherman, 1991). An interesting example of this phenomenon was an allogeneic clone that only recognized HLA-A3 when it was expressed on renal cells, but no other human cell lines (Poindexter et al., 1995, 1997), the implicit assumption being that the clone was specific for a peptide expressed in renal tissue alone.

Both theories involve a certain amount of cross-reactivity with self-restricted responses, although the latter relies more heavily on peptide interactions. If this were indeed the case then it would be true that:

$$\text{Autologous HLA} + \text{peptide X} = \text{Allogeneic HLA and peptide Y.}$$

It has certainly been shown many times that self-restricted clones show some alloreactivity (Lombardi et al., 1989). More recently, it was demonstrated that a well-characterized alloreactive clone could indeed be activated by autologous MHC with a different peptide (Brock et al., 1996). So, there is good evidence to suggest that alloreactivity occurs due to cross-reactivity by self-restricted T cells, rather than to a distinct component of the T cell repertoire.

A recent study compared the interaction, at a molecular level, between a single TCR and, on the one hand, self-MHC loaded with agonist peptide and, on the other hand, allogeneic MHC and stimulating peptides (Daniel et al., 1998).

Interestingly, while the autologous activation was critically dependent on the peptide sequence, allorecognition was far less exacting, tolerating a wide diversity of peptides in the groove. The molecular mechanism for this degenerate recognition was contingent upon six residues in the α helix of the MHC class II β-chain. Substitution of these amino acids in the autologous class II molecule produced a higher affinity interaction with the TCR β-chain (probably the CDR1 and CDR2 regions), which meant that activation became less dependent on the CDR3–peptide interaction. Thus, peptide degeneracy was observed. The authors contended that this mechanism might be widespread in allorecognition and, indeed, in a class I system, similar peptide degeneracy has been observed (Wang et al., 1998).

In summary, it would appear that allorecognition depends on a complex sum of interactions between the TCR and MHC–peptide complexes. In different situations the contribution of peptide–TCR (CDR3) interactions and allo-MHC-TCR (CDR1 and 2) interactions may vary.

Indirect allorecognition by T lymphocytes

Just like any other protein antigen, HLA molecules on grafted tissue can be endocytosed by antigen-processing cells, degraded within the endoplasmic reticulum and loaded on to class II molecules to be presented as peptide fragments in a self-restricted manner (Shoskes and Wood, 1994; Auchincloss and Sultan, 1996; Chitilian and Auchincloss, 1997). This is called indirect allorecognition (see Figure 4.1), but it is important to recognize that this is the normal physiological mechanism of antigen processing. It is a mechanism that has been overshadowed to some extent by the direct pathway. This is probably because the direct response is so impressively vigorous *in vitro*. Experimentally, the dominance of the direct pathway *in vitro* has made it difficult to design experiments showing the influence of indirect presentation. However, it has been shown that immunization of graft recipients with peptide fragments of donor MHC antigens can cause accelerated graft rejection (Fangmann et al., 1992; Benham et al., 1995). By using peptide fragments only indirect presentation was possible in this system. It has also been possible to demonstrate rejection of skin grafts from class II knock-out mice, transplanted to normal allogeneic mice (Auchincloss et al., 1993). This process was shown to be dependent on $CD4^+$ T cells and so only the indirect pathway of presentation was possible. These experiments show that T cell activation by the indirect pathway is certainly possible and, in animal models, can contribute to graft rejection. In addition, in models of skin transplantation, the T cell help required for the production of IgG alloantibodies against MHC molecules is dependent on the indirect T cell response (Steele et al., 1996).

Therefore, it is established that the indirect pathway may be of some importance in allograft rejection, but what are the important antigenic peptides? Probably encouraged by the fact that HLA matching is extremely influential to graft acceptance, it has been speculated that fragments of HLA peptides are the immunodominant peptides in graft rejection (Gould and Auchincloss, 1999). Analysis of peptides eluted from HLA molecules reveals a surprisingly high representation of HLA peptide fragments (Rudensky et al., 1991; Chicz et al., 1992, 1993). In animal models, indirectly primed clones specific for allogeneic MHC–peptide fragments have been generated from rats undergoing acute rejection (Waaga et al., 1998). In addition, in animal models of late rejection, indirectly primed recipient T cells have been implicated in the generation of allospecific DTH responses, the provision of B cells helps to make alloantibodies and the development of allospecific $CD8^+$ cytotoxic lymphocytes (Vella et al., 1997).

The derivation of indirectly primed $CD4^+$ T cells might not be causal, and indeed could just be an epiphenomenon related to parenchymal tissue damage with consequent antigen release. However, a recent paper showed that rejection of skin grafts could be transferred by a self MHC-restricted T cell line that was reactive for a single immunodominant allopeptide (Valujskikh et al., 1998). Fortunately for therapeutic purposes, there seems to be some evidence for immunodominance amongst allopeptide epitopes (Benichou et al., 1994). Indeed, it has proved possible to block *in vivo* alloresponses with limited numbers of synthetic peptides (MacDonald et al., 1997).

It is widely thought that this mechanism is important in the rejection of human allografts, in particular chronic rejection. These data are reviewed in a later chapter on transplantation.

B cell recognition

So far the discussion has only dealt with T cell activation, while recognizing the part that $CD4^+$ T cells play in helping B cells to differentiate and class switch. IgM antibodies against allogeneic MHC molecules can be detected within 3 days of a blood transfusion or an allogeneic organ transplantation in non-immunosuppressed individuals (Wasowska et al., 1992, 1993; Wray et al., 1992). This response peaks at 7 days and then declines. IgG antibodies are apparent at 3–7 days, after which the quantity and affinity increase. The response declines in the weeks following transfusion but remains elevated for many months after transplant rejection (Norman et al., 1985; Wasowska et al., 1992; Wray et al., 1992). B-cell activation is a significant clinical problem, with about one-quarter of patients on renal transplant waiting lists having circulating IgG alloantibodies directed against foreign HLA determinants.

Little is known about the intrinsic alloreactivity of the B cell receptor; however, alloantibodies directed against HLA molecules are extremely rare unless sensitization has occurred (Scornik et al., 1988). The isotype switch to IgG is T cell dependent and can be abrogated in the absence of T cells (Tilney et al., 1981). Thus, the B cell alloresponse is heavily dependent on the $CD4^+$ T cell response described above. The production of alloantibodies depends on costimulation of T cells, in particular those mediated via the CD28/B7 family (Akalin et al., 1996; Borriello et al., 1997). In animal models the production of alloantibodies directed against allogeneic MHC molecules in response to both blood transfusions and allografts can be markedly attenuated by CTLA4-Ig, which blocks such interactions (Akalin et al., 1996; Ibrahim et al., 1996). Another crucial interaction between B cells and T cells is the binding of CD40Ligand (CD154) on the T cell and CD40 on the B cell. This interaction is crucial to achieve isotype switching and affinity maturation of antibodies. Interference with this interaction is an extremely potent way to stop alloantibody production (Foy et al., 1993; Banchereau et al., 1994). Interference with numerous other costimulatory, coreceptor and adhesion molecule ligand pairs can also affect alloantibody production (Kupiec-Weglinski et al., 1993; Talento et al., 1993). The types of APC that can potentially initiate antibody responses are unclear but the role of B cells themselves deserves special consideration. Although most B cells are not equipped to endocytose and process allogeneic cells, the 10^5 IgM molecules on the surface of allospecific B cells render them particularly efficient at binding very low concentrations of soluble alloantigens. These can then be internalized and presented, in association with class II molecules (indirect allorecognition), to T cells in the germinal centres of lymphoid organs (Buelow et al., 1995).

Many variables determine the effects of alloantibody production on organ grafts, including the specificity, concentration, class and subclass of antibody produced. As might be predicted, the most injurious type of antibody to grafts are therefore high-affinity, complement-activating, IgG antibodies to HLA class I antigens (see below). IgG1 and IgG3 subclasses fix complement most efficiently and are thus the most efficient at causing graft damage. Although IgM antibodies to class I HLA antigens fix complement very efficiently, they are not as harmful (Taylor et al., 1989). This is probably because they are of lower affinity and also have a limited circulation as a result of their larger size.

Natural killer cell recognition

Natural killer (NK) cells represent the third member of the lymphocyte family present in peripheral blood. They are distinguished morphologically by being large and granular. They are also relatively radioresistant and differ from T and B cells in that their receptors do not undergo rearrangement and their effector function exhibits no memory response. NK cells are certainly present in biopsies of rejecting organ transplants but their significance is uncertain. HLA molecules are fundamental in determining the behaviour of NK cells which are able to distinguish autologous cells from allogeneic cells. This characteristic is embodied in the missing-self hypothesis, whereby NK cells are inhibited from cytolysis whenever they encounter self-HLA molecules on a cell

(Karre et al., 1986; Bix et al., 1991; Liao et al., 1991; Moretta et al., 1992). This ability accounts for the phenomenon of hybrid resistance whereby hybrid F_1 (A × B) strains of mice are resistant to bone marrow engraftment from either parental strain (A or B) (Cudkowicz and Stimpfling, 1964; Murphy et al., 1987). Inevitably, the bone marrow graft from the parental strain lacks one of the recipient's inhibitory HLA antigens and is therefore destroyed. Antigen recognition occurs via both stimulatory killer activating receptors (KARs) and killer inhibitory receptors (KIRs) (Lanier et al., 1997; Lanier, 1998a). More is known about the KIRs, which are members of the immunoglobulin superfamily and are clonally distributed on groups of NK cells. This means that any given individual has discernible subsets of cells whose function is fashioned by the collective group of KIRs and KARs expressed. As would be expected from an evolutionary perspective, each person has the matching KIRs for self-HLA molecules, thus inhibiting the destruction of autologous cells. Note that the converse is not true, in that an individual may express KIRs specific for an HLA class I molecule that is not necessarily expressed on autologous cells (Gumperz et al., 1996). An illuminating insight into the functional behaviour of the NK repertoire was recently gained by Peter Parham's group, in California, who made more than 100 NK clones from two individuals (Valiante et al., 1997). All of the NK clones generated lysed a cell line that expressed no HLA class I molecules and none lysed autologous B cell lines. The lysis of six transfectants was then analysed, each target expressing only one of the donor HLA alleles. Significantly, they found that almost all (>98%) of the clones were inhibited by one or other of the self-HLA-class I molecules, thus confirming the missing self hypothesis. However, the clones themselves could express between two and nine KIRs and while the repertoire, as a whole, will distinguish self from non-self, the individual component NK cells are themselves complex and heterogeneous. It was suggested that persons expressing different class I HLA molecules select a repertoire during their development so that NK cells without a KIR for self-HLA molecules do not appear in the peripheral circulation.

In humans at least 50 KIRs have been discovered. These can be divided into the p140, p70 and p58 families, corresponding to the inhibitory ligands HLA-A, HLA-B and HLA-C, respectively (Wagtmann et al., 1995; Dohring et al., 1996). Human KIRs were formerly thought to be distinguishable from their murine counterparts by being members of the immunoglobulin superfamily, as opposed to lectin-type molecules. This distinction has now become blurred with the identification of a human counterpart, CD94, also from the lectin family which, when associated with a novel protein NKG2, forms a KIR that can bind both classical and non-classical HLA class I molecules (Phillips et al., 1996; Soderstrom et al., 1997). It is fascinating that these KIRs recognize the atypical class I molecules, HLA-G and HLA-E, and that these interactions may have a role in the tolerance to the foetal allograft (Soderstrom et al., 1997; Braud et al., 1998; Lanier, 1998b). Intriguingly, the stable surface expression of HLA-E molecules further requires the binding of peptides derived from the signal sequences of HLA-A, B, C or G molecules. While the presence of HLA class I molecules is clearly central to the inhibition of NK cytolytic function, it is worth pointing out that certain non-HLA ligands may also be instrumental in defining NK function, through alternative inhibitory receptors (Meyaard et al., 1997).

Less is known about KARs but their extracellular structure is thought to be similar to KIRs and indeed they may bind the same HLA class I molecules, as well as carbohydrates. Intracellularly, though, they lack the immunoreceptor tyrosinebased inhibitory motif (ITIM). Instead, they seem to associate intracellularly with an activating protein, DAP12, which contains an activating motif (ITAM) (Lanier et al., 1998a, b). Overall, NK function is clearly a complex integration of sometimes conflicting signalling, but the foremost antigenic influences on outcome are the HLA class I molecules.

Despite their ability to distinguish self from non-self, NK cells have not been shown to have a prominent role in solid organ transplantation (Manilay and Sykes, 1998). At the present time any role remains speculative, despite the presence of NK cells in biopsy specimens of rejecting organs (Young et al., 1997). They may have a greater role in bone marrow transplantation, potentially being involved in the graft-versus-host, graft rejection and graft-versus-leukaemia effects (Lanier, 1995). NK cells seem to play a prominent role in the rejection of xenogeneic tissues and this represents an area of considerable interest (Seebach and Waneck, 1987; Manilay and Sykes, 1998).

CONCLUSION

The development of our understanding of the biology of the MHC molecules has followed a rather unusual path. Originally defined because they stood in the way of tissue transplantation, it has now become clear that they function as the central players in normal immune recognition by T cells. This made the original role as the agent of tissue rejection somewhat paradoxical. Only in recent years has this circle been squared. The next chapter will deal in greater detail with how MHC molecules subserve their function as antigen-presenting molecules to the immune system.

REFERENCES

Adorini, L., Harvey, M.A., Miller, A. and Sercarz, E.E. (1979). Fine specificity of regulatory T cells. II. Suppressor and helper T cells are induced by different regions of hen egg-white lysozyme in a genetically nonresponder mouse strain. *J. Exp. Med.* **150**: 293–306.

Ajitkumar, P., Geier, S.S., Kesari, K.V. et al. (1988). Evidence that multiple residues on both the alpha-helices of the class I MHC molecule are simultaneously recognized by the T cell receptor. *Cell* **54**: 47–56.

Akalin, E., Chandraker, A., Russell, M.E. et al. (1996). CD28-B7 T cell costimulatory blockade by CTLA4Ig in the rat renal allograft model: inhibition of cell-mediated and humoral immune responses in vivo. *Transplantation* **62**: 1942–1945.

Alam, S.M., Travers, P.J., Wung, J.L. et al. (1996). T-cell-receptor affinity and thymocyte positive selection. *Nature* **381**: 616–620.

Alferink, J., Schitteck, B., Schonrich, G. et al. (1995). Long life span of tolerant T cells and the role of antigen in maintenance of peripheral tolerance. *Int. Immunol.* **7**: 331–336.

Ashton-Rickardt, P., Kaer, L., Schumacher, T. et al. (1993). Peptide contributes to the specificity of positive selection of CD8$^+$ T cells in the thymus. *Cell* **73**: 1041–1049.

Auchincloss, H. and Sultan. H. (1996). Antigen processing and presentation in transplantation. *Curr. Opin. Immunol.* **8**: 681–687.

Auchincloss, H., Jr, Lee, R., Shea, S. et al. (1993). The role of "indirect" recognition in initiating rejection of skin grafts from major histocompatibility complex class II-deficient mice. *Proc. Natl Acad. Sci. USA* **90**: 3373–3377.

Babbitt, B.P., Allen, P.M., Matsueda, G. et al. (1985). Binding of immunogenic peptides to Ia histocompatibility molecules. *Nature* **317**: 359–361.

Bal, V., McIndoe, A., Denton, G. et al. (1990). Antigen presentation by keratinocytes induces tolerance in human T cells. *Eur. J. Immunol.* **20**: 1893–1897.

Banchereau, J., Bazan, F., Blanchard, D. et al. (1994). The CD40 antigen and its ligand. *Annu. Rev. Immunol.* **12**: 881–922.

Benacerraf, B. and McDevitt, H. (1972). Histocompatibility-linked immune response genes. *Science* **175**: 273–279.

Benham, A., Sawyer, G. and Fabre, J. (1995). Indirect T cell allorecognition of donor antigens contributes to the rejection of vascularised kidney allografts. *Transplantation* **59**: 1028–1032.

Benichou, G., Fedoseyeva, E., Lehmann, P. et al. (1994). Limited T cell response to donor MHC peptides during allograft rejection. *J. Immunol.* **153**: 938–945.

Benoist, C. and Mathis, D. (1989). Positive selection of the T cell repertoire: where and when does it occur? *Cell* **58**: 1027–1033.

Bevan, M. (1984). High determinant density may explain the phenomenon of alloreactivity. *Immunol. Today* **4**: 128–130.

Bevan, M. (1997). In thymic selection, peptide diversity gives and takes away. *Immunity* **7**: 175–178.

Bix, M., Liao, N.S., Zijlstra, M. et al. (1991). Rejection of class I MHC-deficient haemopoietic cells by irradiated MHC-matched mice. *Nature* **349**: 329–331.

Blanden, R. and Ada, G. (1978). A dual recognition model for cytotoxic T cells based on thymic selection of precursors with low affinity for self H-2 antigens. *Scand. J. Immunol.* **7**: 181–190.

Borgulya, P., Kishi, H., Muller, U. et al. (1991). Development of the CD4 and CD8 lineage of T cells: instruction versus selection. *EMBO J.* **10**: 913–918.

Borriello, F., Sethna, M.P., Boyd, S.D. et al. (1997). B7-1 and B7-2 have overlapping, critical roles in immunoglobulin class switching and germinal center formation. *Immunity* **6**: 303–313.

Braud, V.M., Allan, D.S., CA, O'Callaghan, A. et al. (1998). HLA-E binds to natural killer cell receptors CD94/NKG2A, B and C. *Nature* **391**: 795–799.

Brenner, M., McLean, J., Dialynas, D. et al. (1986). Identification of a putative second T cell receptor. *Nature* **322**: 145–149.

Bretscher, P. and Cohn, M. (1970). A theory of self–nonself discrimination. *Science* **169**: 1042–1049.

Brock, R., Wiesmuller, K.H., Jung, G. and Walden, P. (1996). Molecular basis for the recognition of two structurally different major histocompatibility complex/peptide complexes by a single T-cell receptor. *Proc. Natl Acad. Sci. USA* **93**: 13 108–13 113.

Buelow, R., Burlingham, W.J. and Clayberger, C. (1995). Immunomodulation by soluble HLA class I. *Transplantation* **59**: 649–654.

Bumgardner, G.L., Li, J., Heininger, M. and Orosz, C.G. (1998). Costimulation pathways in host immune responses to allogeneic hepatocytes. *Transplantation* **66**: 1841–1845.

Burnet, F. (1957). A modification of Jerne's theory of antibody production using the concept of clonal deletion. *Aust. J. Sci.* **20**: 67–69.

Butcher, G.W., Corvalan, J.R., Licence, D.R. and Howard, J.C. (1982). Immune response genes controlling responsiveness to major transplantation antigens: specific major histocompatibility complex-linked defect for antibody responses to class I alloantigens. *J. Exp. Med.* **155**: 303–320.

Buus, S., Sette, A., Colon, S.M. et al. (1987). The relation between major histocompatibility complex (MHC) restriction and the capacity of Ia to bind immunogenic peptides. *Science* **235**: 1353–1358.

Chang, H.C., Smolyar, A., Spoerl, R. et al. (1997). Topology of T cell receptor-peptide/class I MHC interaction defined by charge reversal complementation and functional analysis. *J. Mol. Biol.* **271**: 278–293.

Chen, L., Ashe, S., Brady, W. et al. (1992). Costimulation of anti-tumor activity by the B7 counterreceptor for the T lymphocyte molecules CD28 and CTLA-4. *Cell* **71**: 1093–1102.

Chicz, R., Urban, R., Gorga, J. et al. (1993). Specificity and promiscuity among naturally processed peptides bound to HLA-DR alleles. *J. Exp. Med.* **178**: 27–47.

Chicz, R.M., Urban, R.G., Lane, W.S. et al. (1992). Predominant naturally processed peptides bound to HLA-DR1 are derived from MHC-related molecules and are heterogeneous in size. *Nature* **358**: 764–768.

Chitilian, H.V. and Auchincloss, H., Jr (1997). The indirect pathway in graft rejection. *Curr. Opin. Organ Transpl.* **2**: 3–7.

Clayberger, C., Rosen, M., Parham, P. and Krensky, A.M. (1990). Recognition of an HLA public determinant (Bw4) by human allogeneic cytotoxic T lymphocytes. *J. Immunol.* **144**: 4172–4176.

Cline, M. and Swett, V. (1968). The interaction of human monocytes and lymphocytes. *J. Exp. Med.* **128**: 1309–1325.

Cosgrove, D., Gray, D., Dierich, A. et al. (1991). Mice lacking MHC class II molecules. *Cell* **66**: 1051–1066.

Cudkowicz, G. and Stimpfling, J. (1964). Hybrid resistance to parental marrow graft: association with the K region of H-2. *Science* **144**: 1339–1341.

Daniel, C., Horvath, S. and Allen, P. (1998). A basis for alloreactivity: MHC helical residues broaden peptide recognition by the TCR. *Immunity* **8**: 543–552.

Davis, M.M. and Bjorkman, P.J. (1988). T-cell antigen receptor genes and T-cell recognition. *Nature* **334**: 395–402.

De Magistris, M., Alexander, J., Coggeshall, M. et al. (1992). Antigen analog–major histocompatibility complexes act as antagonists of the T cell receptor. *Cell* **68**: 625–634.

Dellabona, P., Wei, B.Y., Gervois, N. et al. (1991). A single amino acid substitution in the Ak molecule fortuitously provokes an alloresponse. *Eur. J. Immunol.* **21**: 209–213.

Dembić, Z., Haas, W., Weiss, S. et al. (1986). Transfer of specificity by murine a and b T-cell receptor genes. *Nature* **320**: 232–238.

Dohring, C., Scheidegger, D., Samaridis, J. et al. (1996). A human killer inhibitory receptor specific for HLA-A1, 2. *J. Immunol.* **156**: 3098–3101.

Eckels, D.D., Gorski, J., Rothbard, J. and Lamb, J.R. (1988). Peptide-mediated modulation of T-cell allorecognition. *Proc. Natl Acad. Sci. USA* **85**: 8191–8195.

Ehrlich, P. (1900). On immunity with special reference to cell life. *Proc. R. Soc. Lond.* **66**: 424–448.

Elliott, T.J. and Eisen, H.N. (1990). Cytotoxic T lymphocytes recognize a reconstituted class I histocompatibility antigen (HLA-A2) as an allogeneic target molecule. *Proc. Natl Acad. Sci. USA* **87**: 5213–5217.

Engelhard, V.H. (1994). Structure of peptides associated with class I and class II MHC molecules. *Annu. Rev. Immunol.* **12**: 181–207.

Fallarino, F., Fields, P.E. and Gajewski, T.F. (1998). B7-1 engagement of cytotoxic T lymphocyte antigen 4 inhibits T cell activation in the absence of CD28. *J. Exp. Med.* **188**: 205–210.

Fangmann, J., Dalchau, R. and Fabre, J. (1992). Rejection of skin allografts by indirect allorecognition of donor class I major histocompatibility peptides. *J. Exp. Med.* **175**: 1521–1529.

Fink, P.J. and Bevan, M.J. (1995). Positive selection of thymocytes. *Adv. Immunol.* **59**: 99–133.

Fischer-Lindahl, K.F. and Wilson, D.B. (1977a). Histocompatibility antigen-activated cytotoxic T lymphocytes. I. Estimates of the absolute frequency of killer cells generated in vitro. *J. Exp. Med.* **145**: 500–507.

Fischer-Lindahl, K.F. and Wilson, D.B. (1977b). Histocompatibility antigen-activated cytotoxic T lymphocytes. II. Estimates of the frequency and specificity of precursors. *J. Exp. Med.* **145**: 508–522.

Foy, T.M., Shepherd, D.M., Durie, F.H. et al. (1993). In vivo CD40-gp39 interactions are essential for thymus-dependent humoral immunity. II. Prolonged suppression of the humoral immune response by an antibody to the ligand for CD40, gp39. *J. Exp. Med.* **178**: 1567–1575.

Frasca, L., Marelli Berg, F., Imami, N. et al. (1998). Interferon-gamma-treated renal tubular epithelial cells induce allospecific tolerance. *Kidney Int.* **53**: 679–689.

Freeman, G.J., Freedman, A.S., Segil, J.M. et al. (1989). B7, a new member of the Ig superfamily with unique expression on activated and neoplastic B cells. *J. Immunol.* **143**: 2714–2722.

Freeman, G., Gribben, J., Boussiotis, V. et al. (1993). Cloning of B7-2: a CTLA-4 counter-receptor that costimulates human T cell proliferation. *Science* **262**: 909–911.

Frelinger, J.A. and McMillan, M. (1996). The role of peptide specificity in MHC class I-restricted allogeneic responses. *Immunol. Rev.* **154**: 45–58.

Gaspari, A., Jenkins, M. and Katz, S. (1988). Class II MHC-bearing keratinocytes induce antigen-specific unresponsiveness in hapten-specific Th1 clones. *J. Immunol.* **141**: 2216–2220.

Geiger, T., Gooding, L. and Flavell, R. (1992). T-cell responsiveness to an oncogenic peripheral protein and spontaneous autoimmunity in transgenic mice. *Proc. Natl Acad. Sci. USA* **89**: 2985–2989.

Gould, D. and Auchincloss, H. (1999). Direct and indirect recognition: the role of MHC antigens in graft rejection. *Immunol. Today* **20**: 77–82.

Grandea, A.G.d. and Bevan, M.J. (1992). Single-residue changes in class I major histocompatibility complex molecules stimulate responses to self peptides. *Proc. Natl Acad. Sci. USA* **89**: 2794–2798.

Gumperz, J.E., Valiante, N.M., Parham, P. et al. (1996). Heterogeneous phenotypes of expression of the NKB1 natural killer cell class I receptor among individuals of different human histocompatibility leukocyte antigens types appear genetically regulated, but not linked to major histocompatibililty complex haplotype. *J. Exp. Med.* **183**: 1817–1827.

Harding, F., McArthur, J., Gross, J. et al. (1992). CD28-mediated signalling costimulates murine T cells and prevents the induction of anergy in T cell clones. *Nature* **356**: 607–609.

Heath, W. and Miller, J. (1993). Expression of two a chains on the surface of T cells in T cell receptor transgenic mice. *J. Exp. Med.* **178**: 1807–1811.

Heath, W.R. and Sherman, L.A. (1991). Cell-type-specific recognition of allogeneic cells by alloreactive cytotoxic T cells: a consequence of peptide-dependent allorecognition. *Eur. J. Immunol.* **21**: 153–159.

Heath, W.R., Hurd, M.E., Carbone, F.R. and Sherman, L.A. (1989). Peptide-dependent recognition of H-2Kb by alloreactive cytotoxic T lymphocytes. *Nature* **341**: 749–752.

Heath, W.R., Kane, K.P., Mescher, M.F. and Sherman, L.A. (1991). Alloreactive T cells discriminate among a diverse set of endogenous peptides. *Proc. Natl Acad. Sci. USA* **88**: 5101–5105.

Hedrick, S., Cohen, D., Nielsen, E. and Davis, M. (1984a). The isolation of cDNA clones encoding T cell-specific membrane-associated proteins. *Nature* **308**: 149–153.

Hedrick, S., Nielsen, E., Kavaler, J. et al. (1984b). Sequence relationships between putative T-cell receptor polypeptides and immunoglobulins. *Nature* **308**: 153–158.

Hill, A., Elvin, J., Willis, A. et al. (1992). Molecular analysis of the association of HLA-B53 and resistance to severe malaria. *Nature* **360**: 434–439.

Hogan, K.T., Clayberger, C., Bernhard, E.J. et al. (1989). A panel of unique HLA-A2 mutant molecules define epitopes recognized by HLA-A2-specific antibodies and cytotoxic T lymphocytes. *J. Immunol.* **142**: 2097–2104.

Hogquist, K., Jameson, S., Heath, W. et al. (1994). T cell receptor antagonist peptides induce positive selection. *Cell* **76**: 17–27.

Hogquist, K.A., Tomlinson, A.J., Kieper, W.C. et al. (1997). Identification of a naturally occurring ligand for thymic positive selection. *Immunity* **6**: 389–399.

Hu, Q., Bazemore Walker, C.R., Girao, C. et al. (1997). Specific recognition of thymic self-peptides induces the positive selection of cytotoxic T lymphocytes. *Immunity* **7**: 221–231.

Hunt, H.D., Pullen, J.K., Dick, R.F. et al. (1990). Structural basis of Kbm8 alloreactivity. Amino acid substitutions on the beta-pleated floor of the antigen recognition site. *J. Immunol.* **145**: 1456–1462.

Ibrahim, S., Jakobs, F., Kittur, D. et al. (1996). CTLA4Ig inhibits alloantibody responses to repeated blood transfusions. *Blood* **88**: 4594–4600.

Ishii, N., Baxevanis, C., Nagy, Z. and Klein, J. (1981). Responder T cells depleted of alloreactive cells react to antigen presented on allogeneic macrophages from nonresponder strains. *J. Exp. Med.* **154**: 978–982.

Jenkins, M.K. and Schwartz, R.H. (1987). Antigen presentation by chemically modified splenocytes induces antigen specific T cell unresponsivenesss in vitro and in vivo. *J. Exp. Med.* **165**: 302–319.

Jensen, P.E., Pierce, C.W. and Kapp, J.A. (1984). Regulatory mechanisms in immune responses to heterologous insulins. II. Suppressor T cell activation associated with non-responsiveness in H-2b mice. *J. Exp. Med.* **160**: 1012–1026.

Kappler, J., Roehm, N. and Marrack, P. (1987). T cell tolerance by clonal elimination in the thymus. *Cell* **49**: 273–280.

Kappler, J.W., Skidmore, B., White, J. and Marrack, P. (1981). Antigen-inducible, *H-2* restricted, interleukin-2-producing T cell hybridomas. Lack of independent antigen and *H-2* recognition. *J. Exp. Med.* **153**: 1198–1214.

Karre, K., Ljunggren, H.G., Piontek, G. and Kiessling, R. (1986). Selective rejection of H-2-deficient lymphoma variants suggests alternative immune defence strategy. *Nature* **319**: 675–678.

Kisielow, P., Teh, H.S., Bluthmann, H. and Von Boehmer, H. (1988a). Positive selection of antigen-specific T cells in thymus by restricting MHC molecules. *Nature* **335**: 730–733.

Kisielow, P., Blüthmann, H., Staerz, U. et al. (1988b). Tolerance in the T-cell receptor transgenic mice involves deletion of nonmature $CD4^+8^+$ thymocytes. *Nature* **333**: 742–746.

Kourilsky, P. and Claverie, J.M. (1986). The peptidic self model: a hypothesis on the molecular nature of the immunological self. *Ann. Inst. Pasteur/Immunol.* **137D**: 3–21.

Kupiec-Weglinski, J.W., Wasowska, B., Papp, I. et al. (1993). CD4 mAb therapy modulates alloantibody production and intracardiac graft deposition in association with selective inhibition of Th1 lymphokines. *J. Immunol.* **151**: 5053–5061.

Lamb, J., Skidmore, B., Green, N. et al. (1983). Induction of tolerance in influenza virus-immune T lymphocyte clones with synthetic peptides of influenza haemagglutinin. *J. Exp. Med.* **157**: 1434–1447.

Lanier, L.L. (1995). The role of natural killer cells in transplantation. *Curr. Opin. Immunol.* **7**: 626–631.

Lanier, L.L. (1998a). NK cell receptors. *Annu. Rev. Immunol.* **16**: 359–393.

Lanier, L.L. (1998b). Follow the leader: NK cell receptors for classical and nonclassical MHC class I. *Cell* **92**: 705–707.

Lanier, L.L., Corliss, B. and Phillips, J.H. (1997). Arousal and inhibition of human NK cells. *Immunol. Rev.* **155**: 145–154.

Lanier, L.L., Corliss, B., Wu, J. and Phillips, J.H. (1998a). Association of DAP12 with activating CD94/NKG2C NK cell receptors. *Immunity* **8**: 693–701.

Lanier, L.L., Corliss, B.C., Wu, J. et al. (1998b). Immunoreceptor DAP12 bearing a tyrosine-based activation motif is involved in activating NK cells. *Nature* **391**: 703–707.

Lechler, R. and Lombardi, G. (1991). Structural aspects of allorecognition. *Curr. Opin. Immunol.* **3**: 715–721.

Lechler, R.I. and Batchelor, J.R. (1982). Restoration of immunogenicity to passenger cell-depleted kidney allografts by the addition of donor strain dendritic cells. *J. Exp. Med.* **155**: 31–41.

Lederberg, J. (1959). Genes and antibodies: do antigens bear instructions for antibody specificity or do they select lines that arise by mutation? *Science* **129**: 1649–1653.

Liao, N.S., Bix, M., Zijlstra, M. et al. (1991). MHC class I deficiency: susceptibility to natural killer (NK) cells and impaired NK activity. *Science* **253**: 199–202.

Lin, C.-C., Rosenthal, A., Passmore, H. and Hensen, T. (1981). Selective loss of antigen-specific *Ir* gene function in *IA* mutant B6.C-*H-2*^{bm12b} is an antigen presenting cell defect. *Proc. Natl Acad. Sci. USA* **78**: 6406–6410.

Lo, D. and Sprent, J.S. (1986). Identity of cells that imprint H-2-restricted T-cell specificity in the thymus. *Nature* **318**: 672–675.

Lombardi, G., Sidhu, S., Batchelor, J.R. and Lechler, R.I. (1989). Allorecognition of DR1 by T cells from a DR4/DRw13 responder mimics self-restricted recognition of endogenous peptides [published erratum appears in *Proc. Natl Acad. Sci. USA* (1989) **86**: 10 074]. *Proc. Natl Acad. Sci. USA* **86**: 4190–4194.

Lombardi, G., Arnold, K., Uren, J. et al. (1997). Antigen presentation by interferon-gamma-treated thyroid follicular cells inhibits interleukin-2 (IL-2) and supports IL-4 production by B7-dependent human T cells. *Eur. J. Immunol.* **27**: 62–71.

MacDonald, C.M., Bolton, E.M., Jaques, B.C. et al. (1997). Reduction of alloantibody response to class I major histocompatibility complex by targeting synthetic allopeptides for presentation by B cells. *Transplantation* **63**: 926–932.

MacDonald, H., Schneider, R., Lees, R. et al. (1988). T-cell V_β use predicts reactivity and tolerance to *Mls*^{a} encoded antigens. *Nature* **332**: 40–45.

MacDonald, H., Baschieri, S. and Lees, R. (1991). Clonal expansion precedes anergy and death of $V_\beta 8^+$ peripheral T cells responding to staphylococcal enterotoxin B *in vivo*. *Eur. J. Immunol.* **21**: 1963–1966.

Manilay, J. and Sykes, M. (1998). Natural killer cells and their role in graft rejection. *Curr. Opin. Immunol.* **10**: 532–538.

Markmann, J., Lo, D., Naji, A. et al. (1988). Antigen presenting function of class II MHC expressing pancreatic beta islet cells. *Nature* **336**: 476–479.

Matsui, M., Hioe, C.E. and Frelinger, J.A. (1993). Roles of the six peptide-binding pockets of the HLA-A2 molecule in allorecognition by human cytotoxic T-cell *clones*. *Proc. Natl Acad. Sci. USA* **90**: 674–678.

Mattson, D.H., Shimojo, N., Cowan, E.P. et al. (1989). Differential effects of amino acid substitutions in the beta-sheet floor and alpha-2 helix of HLA-A2 on recognition by alloreactive viral peptide-specific cytotoxic T lymphocytes. *J. Immunol.* **143**: 1101–1107.

Matzinger, P. (1980). A one-receptor view of T-cell behaviour. *Nature* **292**: 497–501.

Matzinger, P. (1994). Tolerance, danger, and the extended family. *Annu. Rev. Immunol.* **12**: 991–1045.

Matzinger, P. and Bevan, M.J. (1977). Hypothesis: why do so many lymphocytes respond to major histocompatibility antigens? *Cell Immunol.* **29**: 1–5.

McDevitt, H. and Chinitz, A. (1969). Genetic control of the antibody response: relationship between immune response and histocompatibility (H-2) type. *Science* **163**: 1207–1208.

Merkenschlager, M., Graf, D., Lovatt, M. et al. (1997). How many thymocytes audition for selection? *J. Exp. Med.* **186**: 1149–1158.

Meyaard, L., Adema, G.J., Chang, C. et al. (1997). LAIR-1, a novel inhibitory receptor expressed on human mononuclear leukocytes. *Immunity* **7**: 283–290.

Miller, J. and Heath, W. (1993). Self ignorance in the peripheral T cell pool. *Immunol. Rev.* **133**: 131–150.

Moretta, L., Ciccone, E., Moretta, A. et al. (1992). Allorecognition by NK cells: nonself or no self? *Immunol. Today* **13**: 300–306.

Moskophidis, D., Lechner, F., Pircher, H. and Zinkernagel, R. (1993). Virus persistence in acutely infected immunocompetent mice by exhaustion of antiviral cytotoxic effector T cells. *Nature* **362**: 758–761.

Müllbacher, A., Sheena, J.H., Fierz, W. and Brenan, M. (1981). Specific haplotype preference in congenic F1 hybrid mice in the cytotoxic T cell response to the male specific antigen H-Y. *J. Immunol.* **127**: 686–689.

Murphy, W.J., Kumar, V. and Bennett, M. (1987). Acute rejection of murine bone marrow allografts by natural killer cells and T cells. Differences in kinetics and target antigens recognized. *J. Exp. Med.* **166**: 1499–1509.

Nikolić-Zugić, J. and Bevan, M. (1990). Role of self-peptides in positively selecting the T-cell repertoire. *Nature* **344**: 65–67.

Norman, D.J., Barry, J.M. and Wetzsteon, P.J. (1985). Successful cadaver kidney transplantation in patients highly sensitized by blood transfusions. Unimportance of the most reactive serum in the pretransplant crossmatch. *Transplantation* **39**: 253–255.

Ohashi, P., Oehen, S., Bueki, K. et al. (1991). Ablation of "tolerance" and induction of diabetes by virus infection in viral antigen transgenic mice. *Cell* **65**: 305–317.

Opelz, G. (1987). Effect of HLA matching in 10 000 cyclosporin treated cadaver kidney transplants. *Transplant. Proc.* **19**: 641–649.

Padovan, E., Casorati, G., Dellabona, P. et al. (1993). Expression of two T cell receptor a chains: dual receptor T cells. *Science* **262**: 422–424.

Phillips, J.H., Chang, C., Mattson, J. et al. (1996). CD94 and a novel associated protein (94AP) form a NK cell receptor involved in the recognition of HLA-A, HLA-B, and HLA-C allotypes. *Immunity* **5**: 163–172.

Poindexter, N.J., Naziruddin, B., McCourt, D.W. and Mohanakumar, T. (1995). Isolation of a kidney-specific peptide recognized by alloreactive HLA-A3-restricted human CTL. *J. Immunol.* **154**: 3880–3887.

Poindexter, N.J., Steward, N.S., Shenoy, S. et al. (1997). Renal allograft infiltrating lymphocytes: frequency of tissue specific lymphocytes. *Hum. Immunol.* **55**: 140–147.

Powis, S., Deverson, E., Coadwell, W. et al. (1992). Effect of polymorphism of an MHC-linked transporter on the peptides assembled in a class I molecule. *Nature* **357**: 211–215.

Quill, H. and Schwartz, R.H. (1987). Stimulation of normal inducer T cell clones with antigen presented by purified Ia molecules in planar lipid membranes: specific induction of a long-lived state of proliferative nonresponsiveness. *J. Immunol.* **138**: 3704–3712.

Ramarathinam, L., Niesel, D. and Klimpel, G. (1993). Ity influences the production of IFN-γ by murine splenocytes stimulated in vitro with *Salmonella typhimurium*. *J. Immunol.* **150**: 3965–3972.

Rammensee, H.G., Friede, T. and Stevanoviic, S. (1995). MHC ligands and peptide motifs: first listing. *Immunogenetics* **41**: 178–228.

Richeldi, L., Sorrentino, R. and Saltini, C. (1993). HLA-DPB1 glutamate 69: a genetic marker of beryllium disease. *Science* **262**: 242–244.

Robey, E.A., Fowlkes, B.J., Gordon, J.W. et al. (1991). Thymic selection in CD8 transgenic mice supports an instructive model for commitment to a CD4 or CD8 lineage. *Cell* **64**: 99–107.

Rocha, B. and Boehmer, H. von (1991). Peripheral selection of the T cell repertoire. *Science* **251**: 1225–1228.

Rock, K.L. and Benacerraf, B. (1984). Selective modification of a private I-A allo-stimulating determinant(s) upon association of antigen with an antigen-presenting cell. *J. Exp. Med.* **159**: 1238–1252.

Rojo, J. and Janeway, C., Jr (1988). The biological activity of anti-T cell receptor variable region monoclonal antibodies is determined by the epitope recognized. *J. Immunol.* **140**: 1081–1088.

Ronchese, F., Schwartz, R.H. and Germain, R.N. (1987). Functionally distinct subsites on a class II major histocompatibility complex molecule. *Nature* **329**: 254–256.

Rosenthal, A.S. and Shevach, E.M. (1973). Function of macrophages in antigen recognition by guinea pig T lymphocytes. I. Requirement for histocompatible macrophages and lymphocytes. *J. Exp. Med.* **138**: 1194–1212.

Rudensky, A., Preston-Hurlburt, P., Hong, S.C. et al. (1991). Sequence analysis of peptides bound to MHC class II molecules . *Nature* **353**: 622–627.

Scornik, J.C., Salomon, D.R., Howard, R.J. and Pfaff, W.W. (1988). Evaluation of antibody synthesis in broadly sensitized patients. *Transplantation* **45**: 95–100.

Seebach, J. and Waneck, G. (1987). Natural killer cells in xenotransplantation. *Xenotransplantation* **4**: 201–211.

Shoskes, D.A. and Wood, K.J. (1994). Indirect presentation of MHC antigens in transplantation. *Immunol. Today* **15**: 32–38.

Soderstrom, K., Corliss, B., Lanier, L.L. and Phillips, J.H. (1997). CD94/NKG2 is the predominant inhibitory receptor involved in recognition of HLA-G by decidual and peripheral blood NK cells. *J. Immunol.* **159**: 1072–1075.

Steele, D., Laufer, T., Smiley, S. et al. (1996). Two levels of help for B cell alloantibody production. *J. Exp. Med.* **183**: 699–703.

Steinman, R.M. and Inaba, K. (1986). Stimulation of the primary mixed lymphocyte reaction. *CRC Crit. Rev. Immunol.* **5**: 331–333.

Sykulev, Y., Joo, M., Vturina, I. et al. (1996). Evidence that a single peptide–MHC complex on a target cell can elicit a cytolytic T cell response. *Immunity* **4**: 565–571.

Talento, A., Nguyen, M., Blake, T. et al. (1993). A single administration of LFA-1 antibody confers prolonged allograft survival. *Transplantation* **55**: 418–422.

Tan, P., Anasetti, C., Hansen, J. et al. (1993). Induction of alloantigen-specific hyporesponsiveness in human T lymphocytes by blocking interaction of CD28 with its natural ligand B7/BB1. *J. Exp. Med.* **177**: 165–173.

Taylor, C.J., Chapman, J.R., Ting, A. and Morris, P.J. (1989). Characterization of lymphocytotoxic antibodies causing a positive crossmatch in renal transplantation. Relationship to primary and regraft outcome. *Transplantation* **48**: 953–958.

Tilney, N.L., MacLennon, I., Strom, T.B. and Baldwin, W.M.d. (1981). Effects of thymectomy, thoracic duct drainage, and radiation of T and B lymphocyte distribution and cardiac allograft survival in rats. *Transplantation* **31**: 26–30.

Tite, J., Sloan, A. and Janeway, C., Jr (1986). The role of L3T4 in T cell activation: L3T4 may be both an Ia binding protein and a receptor that transduces a negative signal. *J. Mol. Cell. Immunol.* **2**: 179–190.

Valiante, N.M., Uhrberg, M., Shilling, H.G. et al. (1997). Functionally and structurally distinct NK cell receptor repertoires in the peripheral blood of two human donors. *Immunity* **7**: 739–751.

Valujskikh, A., Matesic, D., Gilliam, A. et al. (1998). T cells reactive to a single immunodominant self-restricted allopeptide induce skin graft rejection in mice. *J. Clin. Invest.* **101**: 1398–1407.

Vella, J.P., Vos, L., Carpenter, C.B. and Sayegh, M.H. (1997). Role of indirect allorecognition in experimental late acute rejection. *Transplantation* **64**: 1823–1828.

Volkmann, A., Barthlott, T., Weiss, S. et al. (1998). Antagonist peptide selects thymocytes expressing a class II major histocompatibility complex-restricted T cell receptor into the CD8 lineage. *J. Exp. Med.* **188**: 1083–1089.

Von Boehmer, H. (1994). Positive selection of lymphocytes. *Cell* **76**: 219–228.

Von Boehmer, H., Teh, H.S. and Kisielow, P. (1989). The thymus selects the useful, neglects the useless and destroys the harmful. *Immunol. Today* **10**: 57–61.

Waaga, A.M., Chandraker, A., Spadafora Ferreira, M. et al. (1998). Mechanisms of indirect allorecognition: characterization of MHC class II allopeptide-specific T helper cell clones from animals undergoing acute allograft rejection. *Transplantation* **65**: 876–883.

Wagtmann, N., Rajagopalan, S., Winter, C.C. et al. (1995). Killer cell inhibitory receptors specific for HLA-C and HLA-B identified by direct binding and by functional transfer. *Immunity* **3**: 801–809.

Wang, R., Nelson, A., Kimachi, K. et al. (1998). The role of peptides in thymic positive selection of class II major histocompatibility complex-restricted T cells. *Proc. Natl Acad. Sci. USA* **95**: 3804–3809.

Wang, W., Man, S., Gulden, P.H. et al. (1998). Class I-restricted alloreactive cytotoxic T lymphocytes recognize a complex array of specific MHC-associated peptides. *J. Immunol.* **160**: 1091–1097.

Warrens, A., Zhang, J., Sidhu, S. et al. (1994). Myoblasts fail to stimulate T cells but induce tolerance. *Intl Immunol.* **6**: 847–853.

Wasowska, B., Baldwin, W.M.d. and Sanfilippo, F. (1992). IgG alloantibody responses to donor-specific blood transfusion in different rat strain combinations as a predictor of renal allograft survival. *Transplantation* **53**: 175–180.

Wasowska, B., Baldwin, W.M.d., Howell, D.N. and Sanfilippo, F. (1993). The association of enhancement of renal allograft survival by donor-specific blood transfusion with host MHC-linked inhibition of IgG anti-donor class I alloantibody responses. *Transplantation* **56**: 672–680.

Webb, S., Morris, C. and Sprent, J. (1990). Extrathymic tolerance of mature T cells: clonal elimination as a consequence of immunity. *Cell* **63**: 1249–1256.

Williams, O., Tarazona, R., Wack, A. et al. (1998). Interactions with multiple peptide ligands determine the fate of developing thymocytes. *Proc. Natl Acad. Sci. USA* **95**: 5706–5711.

Williamson, A. (1980). Three-receptor, clonal expansion model for selection of self-recognition in the thymus. *Nature* **283**: 527–532.

Wray, D.W., Baldwin, W.M.d. and Sanfilippo, F. (1992). IgM and IgG alloantibody responses to MHC class I and II following rat renal allograft rejection. Effects of transplantectomy and posttransplantation blood transfusion. *Transplantation* **53**: 167–174.

Yanagi, Y., Yoshikai, Y., Leggett, K. et al. (1984). A human T cell-specific cDNA clone encodes a protein having extensive homology to immunoglobulin chains. *Nature* **308**: 145–149.

Young, N.T., Bunce, M., Morris, P.J. and Welsh, K.I. (1997). Killer cell inhibitory receptor interactions with HLA class I molecules: implications for alloreactivity and transplantation. *Hum. Immunol.* **52**: 1–11.

Zerrahn, J., Held, W. and Raulet, D.H. (1997). The MHC reactivity of the T cell repertoire prior to positive and negative selection. *Cell* **88**: 627–636.

Zijlstra, M., Bix, M., Simister, N. et al. (1990). b2-microglobulin deficient mice lack $CD4^-8^+$ cytolytic cells. *Nature* **344**: 742–746.

Zinkernagel, R., Callahan, G., Klein, J. and Dennert, G. (1978). Cytotoxic T cells learn specificity for self H-2 during differentiation in the thymus. *Nature* **271**: 251–253.

Zinkernagel, R.M. and Doherty, P.C. (1974a). Activity of sensitized thymus-derived lymphocytes in lymphocytic choriomeningitis reflects immunological surveillance against altered self-components. *Nature* **251**: 547–548.

Zinkernagel, R.M. and Doherty, P.C. (1974b). Restriction of in vitro T cell-mediated cytotoxicity in lymphocytic choriomeningitis within a syngeneic or semiallegeneic system. *Nature* **248**: 701–702.

Zinkernagel, R.M. and Doherty, P.C. (1997). The discovery of MHC restriction. *Immunol. Today* **18**: 14–17.

CHAPTER 5

The Biology of Major Histocompatibility Complex Molecules – II: Antigen Processing and Presentation

Chen Au Peh, Anthony W. Purcell and James McCluskey

HISTORY AND OVERVIEW

The historical background leading to the concept of antigen (Ag) processing has already been reviewed elsewhere (Unanue, 1984; Unanue and Allen, 1987). The biochemical definition of Ag processing began with studies of T cell responses to Ags from the bacterium *Listeria monocytogenes* (Ziegler and Unanue, 1981). In these experiments a time delay was observed before macrophages incubated with bacteria acquired the ability to bind to Ag-specific T cells (Ziegler and Unanue, 1981; Unanue, 1984; Unanue and Allen, 1987). Bacteria were internalized by the macrophages during this lag period. Macrophages coincubated with bacteria and then lightly fixed with paraformaldehyde retained the ability to react with T cells. However, fixation of macrophages before the addition of bacteria destroyed the T cell–macrophage interaction, suggesting that antigen-presenting cells (APCs) needed to be metabolically active in order to present foreign Ag. This finding helped to explain the observed requirement for accessory cells in order to evoke immune responses from T-dependent Ags (Unanue and Askonas, 1968; Unanue and Cerottini, 1970). These data were also reconcilable with the finding that cellular immune responses appeared to be specific for unfolded, sequential antigenic determinants (Gell and Benacerraf, 1959).

The metabolic events necessary for Ag presentation by APCs were termed "Ag processing". Processing of Ag was found to be inhibited by lysosomotropic agents such as the weak bases, chloroquine and ammonium chloride (Ziegler and Unanue, 1982; Unanue and Allen, 1987). In these experiments APCs were pulsed with Ag (or bacteria) for a variable period in the presence and absence of the lysosomotropic agent, then fixed and tested for their ability to react with helper T cells. Because weak bases inhibit the acidification of lysosomes and endosomes, this finding implicated an acidic vesicular compartment as the likely site of Ag processing. Similar findings were obtained in other systems (Chesnut et al., 1982a; Lee et al., 1982; Streicher et al., 1984a), including studies of the soluble Ag ovalbumin, for which a chloroquine-sensitive lag period was also observed prior to Ag presentation (Chesnut et al., 1982a).

The biochemistry of Ag processing and presentation is now much better understood and its critical role in eliciting T cell immunity is clear. These discoveries have opened up the new field of Ag processing and presentation, which transcends immunology and embraces biochemistry, intracellular pathways and structural biology. Understanding the cell biology of Ag processing and presentation may hold the key to unravelling the mechanisms underlying many human leucocyte antigen (HLA) associations with disease.

MAJOR HISTOCOMPATIBILITY COMPLEX CLASS II ANTIGEN PROCESSING AND PRESENTATION

Capture and processing of exogenous antigens

Both class I and class II molecules capture peptide Ags for display on the cell surface, where they are scrutinized by T cells. The pathways by which peptide Ags are created (Ag processing) and then captured (Ag presentation) differ significantly for class I and class II molecules (reviewed in Castellino et al., 1997; Pieters, 1997; Watts, 1997). Thus, class I molecules mainly specialize in the capture of endogenous Ags expressed or localized in the cytoplasm of APC. A subset of highly specialized APCs can capture exogenous Ag for presentation by class I molecules but this is a minor pathway in most cell types (Bevan, 1995; Jondal et al., 1996; Rock, 1996; Watts, 1997). By contrast, class II molecules are well designed for the capture of peptides derived from exogenous Ag because they sample peptides created in the endosomal/lysosomal compartment of the APC. This property of class II molecules means that the peptides constitutively presented by these cells are a mixture derived from self-Ag, which intersect the vacuolar pathway and Ag taken up from the extracellular environment. The efficiency of presentation of exogenous Ag by class II molecules therefore depends on the efficiency of their uptake.

Uptake mechanisms

The mechanisms of uptake of Ag for class II-restricted presentation include receptor-mediated endocytosis, phagocytosis and macropinocytosis (Lanzavecchia, 1990; Watts, 1997) (Figure 5.1). The relative contribution of these different uptake mechanisms varies between cell types. Cells with receptors specific for Ag have the advantage of being able to capture Ag under limiting conditions and potentially deliver this to the correct compartment of the cell for processing. These include: (i) Ag-specific B cells bearing surface immunoglobulin with reactivity for the particular Ag; (ii) Fc-receptor positive APC loaded with Fc-bound specific immunoglobulins (Ig) or immune complexes of Ag and Ig; (iii) C3d-coated Ag taken up by complement receptors; and (iv) APC-expressing members of the macrophage mannose receptor family (MMR), including DEC 205, which augments uptake of mannosylated and fucosylated microbial Ag (Jiang et al., 1995).

In the absence of specific receptor-mediated uptake of Ag, phagocytic cells can engulf particulate microbial Ag and all cell types can take Ag up non-specifically by constitutive endocytosis. Accordingly, the endocytic and phagocytic activity of an APC will determine the level of Ag uptake by these mechanisms. In general, soluble exogenous Ag are sampled by endocytosis and microbial debris is taken up by phagocytosis. These pathways of Ag uptake are partly regulated by inflammatory stimuli which can activate APC and boost Ag uptake rates (Cella et al., 1997a).

Receptor-mediated antigen uptake by B cells

Resting B cells are notoriously poor at Ag presentation but activated B cells acquire many of the properties necessary for Ag presentation and stimulation of specific T cells (Rock et al., 1984). The ability of specific B cells selectively to take up Ag under limiting conditions gives them an important edge in achieving the crucial requirement of cognate T–B interaction (Rock et al., 1984; Lanzavecchia, 1985). Selective uptake of foreign Ag is depicted schematically in Figure 5.1. Specific B cells can present Ag to T cells at concentrations of Ag three to four orders of magnitude less than those necessary for Ag presentation by non-specific B cells (Rock et al., 1984; Lanzavecchia, 1985; Kanost and McCluskey, 1994). Moreover, Ag selectively taken up by surface Ig is endocytosed into clathrin-coated pits and delivered to an appropriate compartment for further processing and loading into class II molecules. The internalization of Ag through specific membrane-bound Ig (mIg) apparently requires mIg to associate with the Igα and Igβ disulfide heterodimer which together form the B cell receptor (BCR) (Venkitaraman et al., 1991; van Noesel et al., 1992). Signals for internalization of mIg may depend not on cross-linking of the BCR but on signalling events controlled by the cytoplasmic tail of the Igβ-subunit of the BCR (Patel and Neuberger, 1993). However, the exact roles of

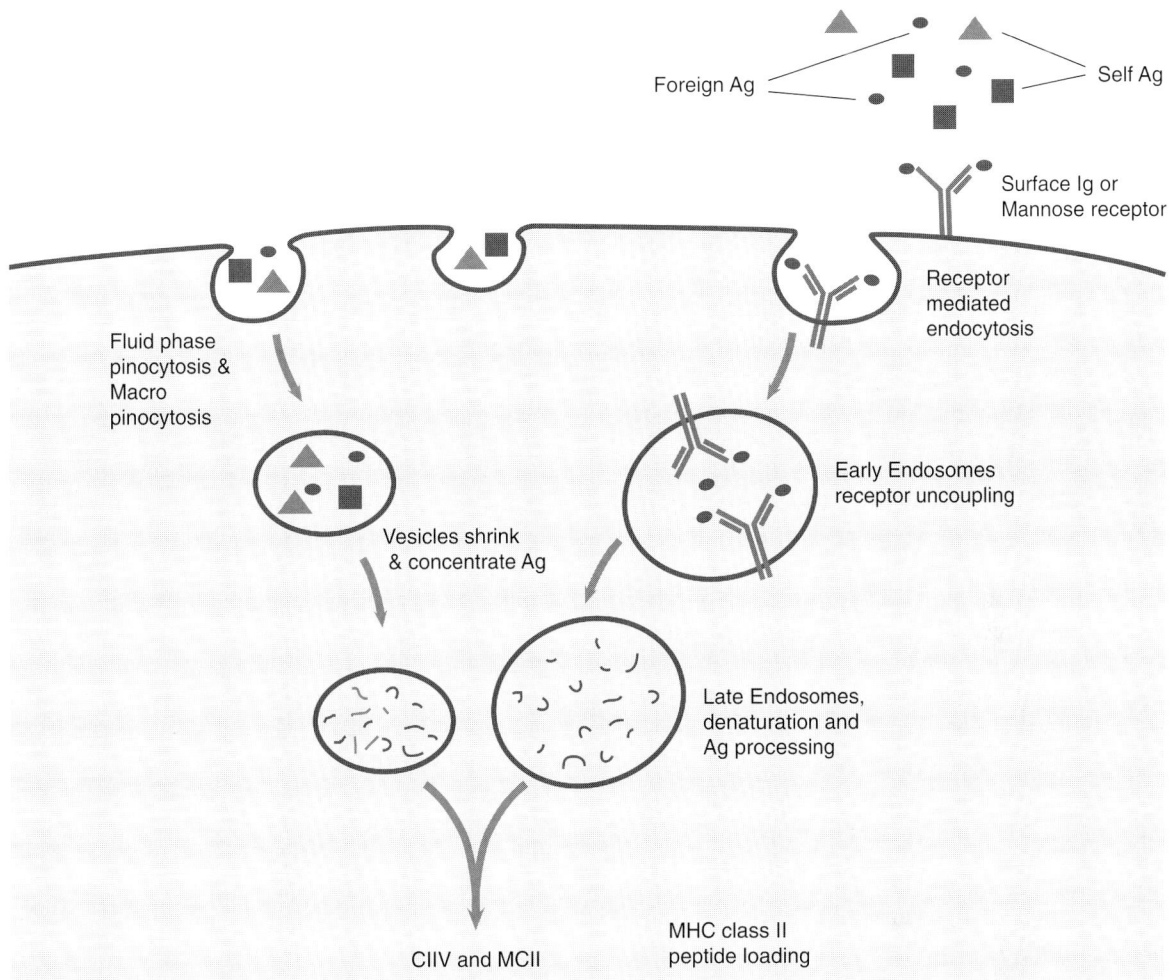

FIGURE 5.1 Capture of antigens by APC. Self and foreign antigens are depicted as symbols which enter the cell via fluid phase endocytosis (left) or receptor-mediated endocytosis into clathrin-coated pits (right). In the diagram, receptor-mediated endocytosis is mediated by antigen-specific surface immunoglobulin or through binding of mannosylated or fucosylated bacterial Ag via a member of the mannose receptor family (e.g. DEC 205). Selective uptake allows foreign antigens to be concentrated in endosomes where they are routed to late endosomes and a lysosomal-like compartment rich in class II molecules and HLA–DM (CIIV or MIIC). Processing of Ag by denaturation and proteolysis begins as the endosome becomes acidified. Fluid phase endocytosis captures self and foreign antigens simultaneously and so is much less efficient than receptor-mediated uptake.

different structural elements of the BCR in controlling intracellular signalling, internalization and Ag delivery have not yet been determined in detail (Watts, 1997). For instance, the transmembrane domain of IgM may also play a role in the intracellular sorting of the BCR (Shaw et al., 1990; Parikh et al., 1992). None the less, the properties of selective uptake of Ag and enhanced Ag presentation reside in all mIg isotypes and serve to give specific B cells a distinct advantage under conditions of limiting Ag concentration (Venkitaraman et al., 1991; van Noesel et al., 1992).

The enhancement of Ag presentation observed in Ag-specific B cells varies for different T cell epitopes (Kanost and McCluskey, 1994). This might reflect the different processing needs among T cell epitopes, some of which require processing in late endosomes/lysosomes with reduction of disulfide bonds, while others can be generated in less catabolic endosomal compartments. Delivery of internalized Ag to different compartments may depend on the level of oligimerization of surface Ig and so might be influenced by Ag valency (Watts, 1997). As discussed later, the masking of certain topographical regions of native Ag by Ig binding might also alter subsequent proteolysis of Ag by protection or exposure of Ag epitopes.

Fc receptors and B cells

Many Ag are present in extracellular environments as immune complexes with and without attached complement proteins. The attachment of C3d to hen egg lysozyme (HEL) increases the efficiency of Ag presentation up to 1000-fold by boosting BCR signalling and B cell activation through coligation of the CR2 complement receptor and CD19 (Fearon and Carter, 1995; Dempsey et al., 1996). The exact mechanism by which B cell activation improves Ag presentation under these circumstances is not resolved but one possibility is the enhanced targeting of Ag to a compartment containing newly synthesized class II molecules. Otherwise, Ag presentation by B cells is not helped by Fc receptor (FcR)-mediated uptake because engagement of the FcγRII expressed by B cells produces a negative regulatory influence which tends to dampen B cell responses (D'Ambrosio et al., 1995). This results in part from a distinct signalling by this FcR and partly because of failure of the FcγRII to internalize through coated pits (Amigorena et al., 1992).

Phagocytosis and macropinocytosis: B cells and dendritic cells

Phagocytically active cells such as activated macrophages are capable of taking up quite large volumes of fluid and bigger subcellular particles than those taken into coated pits (~0.1 μm diameter) by either B cells or dendritic cells (DCs) (Watts, 1997). Phagocytosis can be activated through FcR engagement, which signals membrane engulfment of surface material through phosphorylation of sites within the γ-subunit of the FcR (reviewed in Greenberg, 1995). Alternatively, phagocytosis can be independent of opsonization of Ag. Once internalized, the phagosome undergoes successive remodelling and exchange of contents with lysosomes. This exchange process allows progressive denaturation, reduction and proteolysis of protein Ag to create potential ligands for class II molecules. Many microorganisms, such as *Mycobacterium tuberculosis*, have evolved mechanisms of disrupting this process and thereby impairing Ag presentation to T cells. Ultimately, newly synthesized class II molecules are charged with peptides either by transporting them into a compartment containing lysosomal degradation products or by cycling through phagolysosomes en route to the cell surface.

The phagocytic and endocytic potential of DCs appears to depend on their state of maturation and activation (Cella et al., 1997b). DCs in tissues are initially activated by innate inflammatory stimuli to increase their rates of Ag uptake but they remain poor at T cell activation until they have returned to local lymphoid tissue. Thus, DCs expanded by culture in growth factors [granulocyte–macrophage colony-stimulating factor and interleukin-4 (IL-4)] are immature and demonstrate a high level of fluid phase uptake by macropinocytosis, using a form of endosome (macropinosomes are approximately 2 μm in diameter) with a favourable surface area-to-volume ratio for pinocytosis (Steinman and Swanson, 1995; Sallusto et al., 1995; Lutz et al., 1996; Norbury et al., 1996). Macropinocytosis does not involve coated pits and is associated with a membrane ruffling inducible by phorbol esters (Swanson and Watts, 1995). How this is normally regulated is unclear but this pathway of Ag uptake is very active in immature or newly activated DCs and helps to make the presentation of Ag by these cells upon maturation as efficient as that of specific B cells in which pinocytosis is much less active (20–30-fold) (Sallusto et al., 1995; Lutz et al., 1996). Phagocytosis may also be more active in growth factor-induced DCs; however, the most important auxiliary mechanisms for boosting Ag uptake operate through FcRs (FcγRII) and members of the MMR family such as DEC205 (Sallusto et al., 1995). Since many microbial Ag contain mannosylated and fucosylated molecules these can be selected from host Ag by specific receptors recognizing these sugars. The mannose receptors release their Ag at low pH and then recycle to the cell surface where they can re-engage new ligands for iterative rounds of capture and release (Sallusto et al., 1995).

These multiple mechanisms for Ag uptake are most active in newly activated DCs located at sites of injury and mucosal surfaces. Tissue DCs normally recycle class II molecules rapidly, with a half-life of around 10 h (Cella et al., 1997b). Once they are loaded with exogenous Ag acquired in an inflammatory context they begin their migration to regional lymph nodes. On the way they start to downregulate Ag-uptake activities and acquire attributes which promote loading of processed determinants and presentation to T cells (Sallusto and Lanzavecchia, 1994; Sallusto et al., 1995; Lutz et al., 1996). The differentiation towards an Ag presentation phenotype includes an increase in surface expression of costimulatory molecules and adhesion molecules. This change in phenotype can be mimicked *in vitro* by the exposure of DCs to tumour necrosis factor-α, IL-1, CD40L and lipopolysaccharide factors (Sallusto and Lanzavecchia, 1994; Sallusto et al., 1995), suggesting that the maturation and migration of DCs *in vitro* is regulated by similar mechanisms (Roake et al., 1995). During maturation of DCs there is a transient boost in class II synthesis and a change in the half-life of the class II molecules to > 100 h, maximizing the generation and loading of peptides derived during the preceding period of fluid phase and receptor-mediated Ag uptake (Cella et al., 1997b). In this way DC function is cleverly regulated to sample Ags constitutively at mucosal and other surfaces. Inflammatory stimuli lead to a sustained burst of Ag uptake at these surfaces, coupled with migration to regional lymph nodes, where DCs acquire the ability to prime naive T cells. This latter property has led DCs to be called professional APCs.

Non-professional antigen-presenting cells

The ability of a variety of different major histocompatibility complex (MHC) class II-positive cell types to process and present Ag to primed T cells became apparent when clonal populations of APCs were first studied for Ag presentation function. Several independent B lymphoma cell lines which expressed class II molecules were shown to present Ag to T hybridomas (Chesnut et al., 1982b; Glimcher et al., 1982; Walker et al., 1982). Perhaps more surprising was the observation that mouse L cells (a presumed fibroblast line which does not express class II proteins) could process and present Ag when transfected with the genes encoding the appropriate MHC restricting element (Germain and Malissen, 1986). Some class II-transfected L cells fail to present foreign Ag, presumably reflecting a specific processing defect in these clonally variable cells (Shastri et al., 1985).

Many MHC class II-negative cell types reveal the capacity to process and present Ag once they are rendered class II positive by cytokine induction or gene transfection (Sunshine et al., 1980; Chesnut et al., 1982a, b; Glimcher et al., 1982; Walker et al., 1982; Shastri et al., 1985; Germain and Malissen, 1986; Metlay et al., 1989). The generality of these findings suggests that the intracellular events occurring during Ag processing and class II–peptide assembly involve properties of most cell types rather than being unique features of differentiated bone-marrow-derived cells. Nevertheless, the role of these potential APCs in the initial expansion of naive T cells (priming) is not clear (Metlay et al., 1989).

Antigen processing

Prior to loading into the cleft of class II molecules, protein Ag generally need to be cleaved to yield peptides of average length around 12–18 amino acids (i.e. to be "processed") (Rudensky et al., 1991; Chicz et al., 1992, 1993, 1994; Hunt et al., 1992; Rudensky et al., 1992). However, Ag processing involves more than simple proteolysis of polypeptides to create shorter peptides. A significant amount of peptide trimming probably occurs after binding to class II molecules, giving rise to ragged ends of class II-bound peptides which vary in length at their termini. Accordingly, Ag processing should be considered as the sum of all Ag modifications necessary to liberate MHC class II ligands. For some protein Ags intracellular processing by APC is unnecessary. For instance, some T cells can recognize fibrinogen without the need for any processing by binding a flexible free helical region of the Ag present in the native polypeptide (Lee et al., 1988). Unfolded but intact HEL (Allen and Unanue, 1984), ribonuclease (Lorenz et al., 1988), myoglobin (Streicher et al., 1984b), influenza haemagglutinin (Mills et al., 1986) and perhaps even free influenza nucleoprotein (Wraith and Vessey, 1986) can also be presented to some T cells without the need for further processing. Similarly, some Ags do not require endogenous proteolytic processing when delivered to APC in liposomes (Walden et al., 1985). These observations imply that processing of Ag may sometimes involve its denaturation or unfolding without necessarily requiring its proteolysis. Proteolysis may only be required for presentation of determinants encrypted within complex Ags (internally disulfide-bonded, multimeric or globular proteins with multiple hidden hydrophobic regions; Bersofsky et al., 1988). Accordingly, flexible or easily denatured molecules may reveal sites suitable for presentation to T cells without the need for proteolysis. In practice, however, most proteins need to be denatured and to undergo some degree of proteolysis in order to generate suitable peptide ligands for class II binding. Reduction of disulfide bonds is also essential for processing of some Ag such as HEL. Thus, the complexity and structure of a given Ag will determine the extent to which it is required to undergo processing for Ag presentation. The flip side of intracellular proteolysis is the destruction of potentially antigenic peptides by endosomal proteases. Thus, the peptide Ag available for binding to class II molecules is a balance between generation and destruction of potential peptides. Not surprisingly, many peptides present within a given Ag can bind to class II molecules but do not form natural determinants in an immune response to the native Ag. Presumably, some of these peptides are poorly created by processing events while others are generated but rapidly catabolized within the APC.

Proteases and antigen processing

Four classes of endoprotease have been described according to the functional group involved in the active site of the enzyme. These include metalloproteases, thiol proteases, serine proteases and the acid or carboxyl proteases. These proteases can be selectively inhibited using a variety of commercially available protease inhibitors. A number of studies indicates that specific protease inhibitors dramatically impair Ag processing and the specific proteases involved in Ag processing may differ for different Ags. For example, inhibition of the proteolytic processing of some Ags has been described when disrupting the normal function of the thiol (cysteine) proteases (cathepsin B and D) (Bersofsky et al., 1988; Wederlin et al., 1988), cysteine or serine proteases (Streicher et al., 1984b; Puri et al., 1986; Yoshikawa et al., 1987) and the acid proteases (Buus and Wederlin, 1986; Chain et al., 1988), either in combination or alone. Although some of these proteolytic enzymes (e.g. cathepsin A and B) are located in endocytic vesicles, the inferred involvement of certain proteases in Ag processing does not localize the exact intracellular site where this occurs. In electron-microscopic studies of Ag-pulsed B lymphoblastoid cells, class II molecules, invariant chain, soluble Ag and cathepsin B and D colocalized in early endocytic vesicles within minutes following the addition of Ag to cells (Guagliardi et al., 1990). Ultrastructural data also demonstrate that soluble Ag is internalized into endosomes but then reappears on the cell membrane (Lin et al., 1988). These findings provide anatomical support for the assembly of peptide Ag with MHC molecules in early endosomes prior to their presentation on the cell surface. Such complexes may include fragments of Ag much larger than the peptides eluted from purified class II molecules (Lee et al., 1988). The major site of Ag proteolysis and class II loading is likely to be later in the endosomal pathway.

Proteolytic enzymes likely to be responsible for creating most Ag peptides are members of the aspartate protease (cathepsin D and E) family, while the cysteine protease family (cathepsin B, L, S and H) contains some of the enzymes more likely to be destructive of class II epitopes (Watts, 1997). The specificity of the different proteases depends in part on the pH of the endosomal compartment. Late endosomes and lysosomes achieve a lower pH (pH ~4.5) than the early endosomal compartment (pH ~5.5). The contribution of individual proteases in Ag processing has been studied using protease inhibitors; however, these tend to be multispecific and results are therefore difficult to interpret. Inhibitors of endosomal acidification such as the weak base chloroquine or the H^+/K^+ ion pump inhibitors, e.g. bafilomycin, frequently abolish Ag processing and have been used as markers of the class II pathway of Ag processing and presentation (Brooks et al., 1993).

Sequences which flank antigenic peptides can play a profound role in their release from the parent polypeptide. However, apart from the negative effect of flanking proline residues, there do not appear to be general rules which allow confident prediction of these influences (Falk et al., 1994). Moreover, in vaccine studies where a series of class II determinants was strung together as a single contiguous "polytope" encoded within a minigene, preliminary data indicate that all of the Ag peptides within the polytope were liberated and presented to T cells by class II molecules (Suhrbier, 1997). However, this finding is unlikely to be universal since co-linearity of two determinants each presented by different class II molecules can lead to interference with presentation of one epitope associated with successful presentation of the other epitope (Sercarz et al., 1993). This kind of antigenic competition might presumably operate to confer protection from some autoimmune diseases by certain class II molecules. For instance, HLA DR2–DQ1 is strongly protective of insulin-dependent diabetes (IDDM), which is positively associated with DR3–DQ2 and DR4–DQ3 (Nepom and Ehrlich, 1991). This protection is dominant, as would be expected if the mechanism involved Ag competition impairing the presentation of critical autoantigenic determinants.

Proteolysis of Ag is also altered by its binding to immunoglobulin molecules (Davidson and Watts, 1989). Binding of tetanus toxoid to certain monoclonal antibodies (mAbs) results in protection of particular protein domains and class II determinants (Simitsek et al., 1995). In contrast to steric interference of epitope release by Ig binding, some antibodies can catalyse the liberation of peptide epitopes, perhaps by stabilizing crucial proteolytic intermediates in the processing pathway (Simitsek et al., 1995). Thus, B cells of defined Ig specificity might be fastidious in the specificity of T cells required to provide T–B helper functions necessary for their proliferation and differentiation in an immune response.

Major histocompatibility complex class II synthesis, maturation and transport

MHC class II molecules are highly polymorphic $\alpha\beta$ heterodimers which assemble cotranslationally with a monomorphic protein called invariant chain (Ii) in the lumen of the endoplasmic reticulum (ER) (Cresswell, 1994, 1996) (Figure 5.2). The stoichiometry of the $\alpha\beta$Ii complex is nonameric $(\alpha\beta Ii)_3$ and the binding of Ii to class II molecules prevents endogenous peptides from loading in the ER (Cresswell, 1996). In human cells there are four different isoforms of Ii; IiP33, IiP41, IiP35 and IiP43 (reviewed in Pieters, 1997). In the mouse, two isoforms are created by alternate splicing of pre-mRNA transcripts, giving rise to IiP31 and IiP41. The human IiP35 and IiP43 isoforms result from alternate initiation of translation and contain a 16-residue sequence which signals ER retention. The IiP41 and IiP43 both contain an extra 64-residue domain which inhibits cysteine proteases involved in Ag processing and Ii proteolysis. Ii is a type 2 membrane protein, so the amino-terminus is orientated into the cytoplasm and the carboxyl terminus forms the ectodomain. Trimerization of Ii is controlled by the carboxyl-terminus and perhaps the transmembrane domain. The binding of class II molecules by Ii is mediated by a segment comprising residues 81–104 known as CLIP (class II-associated invariant-chain peptide).

The $(\alpha\beta Ii)_3$ complex is transported to the trans-Golgi and then to the endocytic pathway through dileucine-based targeting signals in the cytoplasmic tail of the Ii molecule. The molecules involved in diverting $\alpha\beta$Ii complexes to the endocytic compartment are not identified. In the absence of the targeting sequences, Ii molecules are expressed on the cell surface and even some wild-type Ii molecules are found on the surface membrane of normal APC. During transport to the endocytic compartment, the luminal domain of Ii is progressively degraded, creating various Ii fragments and producing free $\alpha\beta$–CLIP complexes. In mice, the IiP41 form of Ii contains a cysteine-rich domain (also found in thyroglobulin) which appears to result in enhanced Ag presentation (see review by Pieters, 1997). This is thought to occur because the cysteine-rich domain inhibits cathepsin L and either rescues antigenic epitopes or saves class II–CLIP complexes from further proteolysis. Interestingly, inhibition studies suggest that cathepsin S, which is strongly related to cathepsin L, might be the key protease responsible for Ii processing, implying some redundancy in this function (Riese et al., 1996).

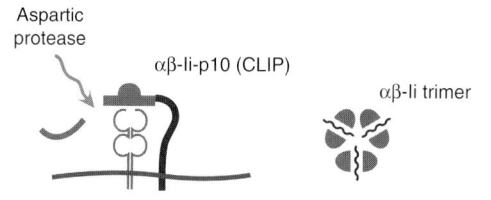

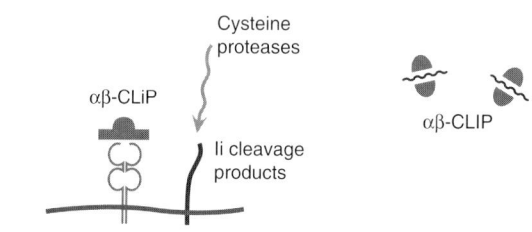

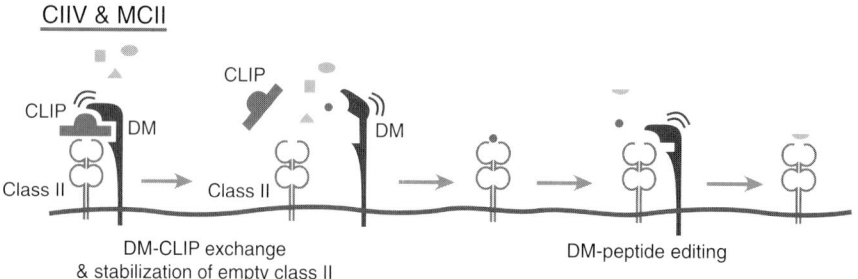

FIGURE 5.2 Invariant chain assembly and processing. The class II αβ dimer associates cotranslationally with Ii in the lumen of the endoplasmic reticulum (ER). The αβIi complex forms a nonamer (a trimer of trimers) in which the CLIP region (residues 81–104) of Ii occupies the antigen-binding cleft of the αβ dimer. CLIP is depicted as a bulldog clip to reflect the obstruction which its binding presents to the loading of new class II molecules with endogenous peptides. The Ii molecule undergoes progressive proteolytic processing as the αβIi complex travels through the early and late endosomal compartment to create a cohort of SDS-sensitive αβCLIP complexes (αβ-Ii-p10). However, most αβ-Ii-p10 complexes are probably created in a specialized lysosomal compartment known in different studies as the CIIV or MIIC compartment. Subtle differences may exist between these compartments but for the purposes of simplicity they are amalgamated here. The exchange of CLIP for antigenic peptides probably occurs mostly in this CIIV or MIIC compartment. However, some exchange might also occur in earlier compartments since some DM and stable class II–peptide complexes can be detected throughout the endosomal pathway. In this diagram, DM is shown as actively evicting CLIP, but in reality it probably stabilizes a peptide-receptive intermediate in the dissociation of class II molecules from CLIP or other peptides, thus facilitating peptide editing.

A special compartment for antigen loading of class II molecules

At a late stage in their transport, many class II–CLIP complexes appear to be localized in multilamellar, lysosome-like structures which ultrastructurally are highly enriched in class II molecules (Pieters et al., 1991a, b; Amigorena et al., 1994; Qui et al., 1994; Tulp et al., 1994; West et al., 1994) (Figure 5.3). These compartments have been named MIIC or CIIV by different investigators and even though they may be subtly different in their exact endocytic lineage they are likely to be closely related functionally (reviewed in Castellino et al., 1997; Watts, 1997). Regardless of this, they both possess high levels of MHC class II and little Ii is detectable, presumably because of its processing en route to, and within, the MIIC/CIIV compartment. The exact trafficking pathway of newly synthesized class II molecules to the MIIC/CIIV compartment is still a matter of some debate. However, in mouse B cells the evidence favours the transit of new class II molecules through conventional early and late endosomes as judged by biochemical markers. Inhibition of class II processing by leupeptin causes an accumulation of αβ–CLIP and prevents the appearance of peptide-loaded class II molecules on the cell surface (Neefjes and Ploegh, 1992). Therefore, the MIIC/CIIV compartment is likely to be the major site of CLIP removal and Ag peptide loading of newly synthesized molecules, but peptide loading also occurs in multiple endocytic sites along the pathway to the cell surface. The process of CLIP removal and peptide loading is enhanced at low pH but this alone is insufficient to prevent the accumulation of large amounts of αβ–CLIP complexes on the cell surface. These complexes are easily dissociated in sodium dodecyl sulphate (SDS), unlike the more stable αβ complexes containing antigenic peptides. The major molecular mechanism by which CLIP is removed from αβ–CLIP complexes involves the action of HLA–DM, a MHC-linked class II-like dimer which catalyses the removal and exchange of peptides from class II molecules (Figure 5.2) (Denzin and Cresswell, 1995; Sherman et al., 1995; Sloan et al., 1995). HLA–DM is concentrated in the MIIC/CIIV compartment but is also present throughout the entire endocytic pathway (Pierre et al., 1996). In HLA–DM-mutant cell lines large amounts of class II complexes loaded with CLIP are expressed on the cell surface. CLIP binds class II molecules in a fashion which is structurally almost identical to that of antigenic peptides and so, effectively, the cleft is totally obstructed by the presence of CLIP (Ghosh et al., 1995). The mechanism of action of HLA–DM involves direct interaction with class II molecules which are thought to be stabilized in the process of undergoing peptide or CLIP dissociation. This action presumably allows HLA–DM to facilitate peptide exchange (including CLIP exchange) by preserving class II molecules in a transient peptide-receptive conformation (Denzin et al., 1995). Some class II alleles do not have an absolute requirement for DM-facilitated CLIP exchange because of a lower affinity of CLIP binding (Brooks et al., 1994; Sette et al., 1995). Allelic variation in affinity for CLIP might conceivably play a role in HLA class II-restricted susceptibility to certain autoimmune diseases, especially if autoantigenic peptides are loaded by APC in the absence of DM function.

Although the vast majority of peptides complexed to class II molecules is derived from the endosomal compartment, some peptides are also derived from resident ER proteins (Brooks et al., 1991) and cytoplasmic Ags (Brooks and McCluskey, 1993). The exact mechanism of loading of these Ag is unclear. Cytoplasmic Ags might load by the process of autophagy whereby cytosolic contents are sampled through invaginated vacuoles which then reorganize their membranes to introduce the cytosolic contents into the vacuolar lumen. Once cytosolic Ag is contained in vacuoles it is presumably processed and captured in a manner similar to other endosomal Ag. ER loading of endogenous peptides might occur by the displacement of Ii from newly synthesized class II molecules. Given the high affinity of Ii for class II molecules, non-endosomal Ag represent a very minor pathway and probably involve only very stable high-affinity peptide Ags.

Recycling of class II molecules and antigen presentation

The majority of class II–peptide complexes appear to be very stable on the cell surface, where

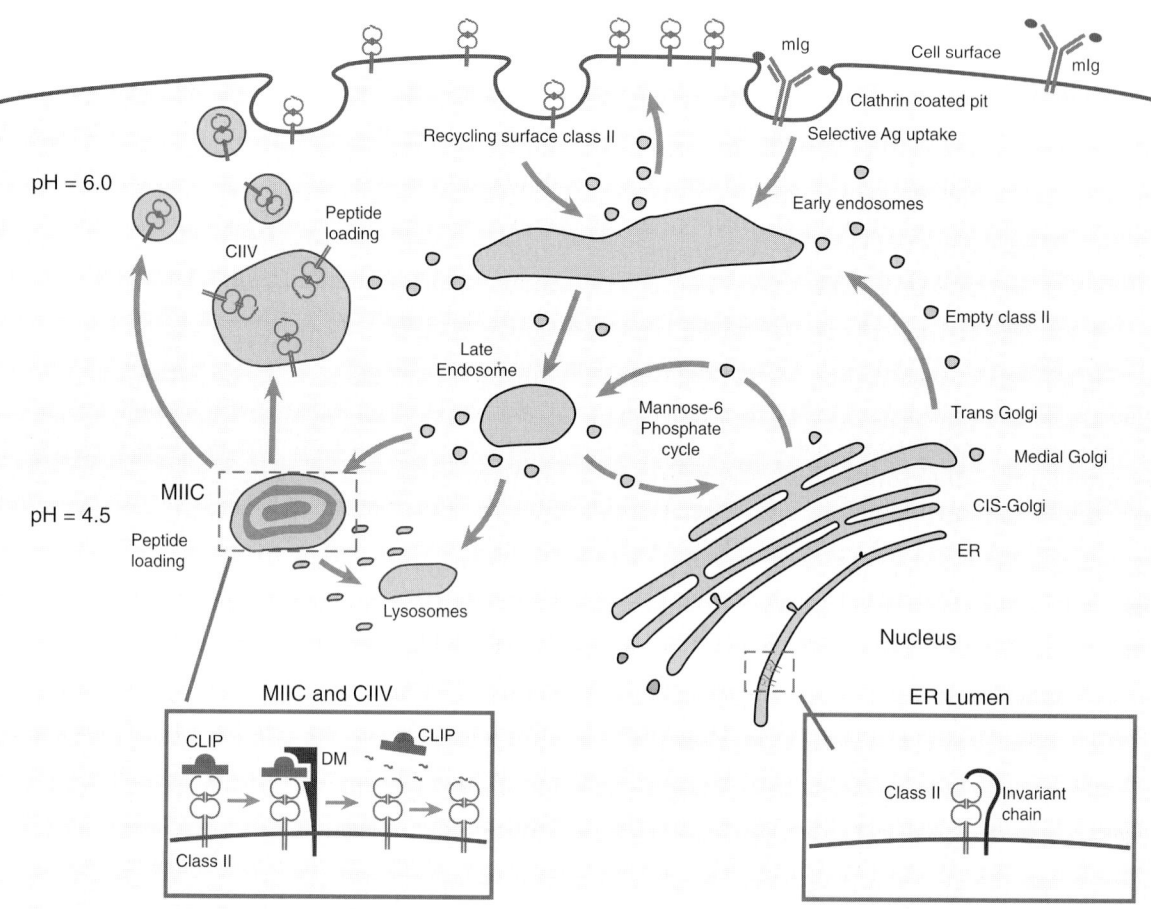

FIGURE 5.3 Relationships of the vacuolar compartments and class II transport. Class II molecules are synthesized in the ER and transit the cis, medial and trans Golgi associated with Ii. One school of thought believes that αβIi complexes transit directly to the specialized CIIV or MCII compartment where Ii processing is completed and CLIP is exchanged for antigenic peptides through the editorial action of HLA–DM. A second school of thought is that newly synthesized class II molecules travel through multiple distinct endosomal vesicles (early and late) where they experience some Ii processing and even some peptide loading as Ag are denatured and degraded to a form suitable for capture by MHC class II molecules. In this model there is a progressive decrease in pH and corresponding increase in efficiency of CLIP displacement and peptide loading as molecules transit the endosomal pathway. In both models the most efficient site of peptide loading is the CIIV or MIIC compartment from where peptide loaded class II molecules proceed direct to the cell-surface membrane for scrutiny by T cells. Recycling of a small number of surface class II–peptide complexes permits some reloading of poorly formed complexes with peptide Ag generated in early endosomes.

they remain for 1–2 days (Lanzavecchia et al., 1992). However, there is considerable indirect evidence that some class II–peptide complexes may recycle through peripheral acidic endosomes where occasional peptide exchange may occur, particularly in the case of poorly bound complexes (reviewed in Castellino et al., 1997; Watts, 1997). The sorts of epitopes created by recycling might be less stringent in their processing requirements and are likely to be Ii independent. There may also be allelic variation in the activity of this pathway, which could be related to the function of class II alleles in susceptibility to autoimmune disorders. The major features of the class II-restricted pathway of Ag presentation are summarized in Figure 5.3.

MAJOR HISTOCOMPATIBILITY COMPLEX CLASS I ANTIGEN PROCESSING AND PRESENTATION

Class I MHC molecules bind and present peptides of eight to 11 amino acids in length. Class I MHC molecules bound to antigenic peptides are recognized by $CD8^+$ cytotoxic T lymphocytes (CTLs) (Rammensee et al., 1993). The presentation of peptides derived from endogenous cytoplasmic proteins by class I MHC molecules thus allows immune surveillance of the cellular interior. Peptides derived from self-proteins can be presented by class I MHC molecules but are generally ignored by the immune system. Conversely, presentation of peptides derived from foreign proteins may stimulate an immune response if the class I MHC complex encounters Ag-specific CTL. As discussed in the first part of this chapter, class II MHC molecules present slightly longer peptides (eight to 25 amino acids) mainly derived from exogenous sources and are recognized by cells of the $CD4^+$ T helper phenotype. The differences in the nature and origin of presented peptides are largely maintained through the segregation of class I and II molecules into distinct pathways of assembly and export.

The polymorphism in class I MHC alleles is mostly found in the Ag binding cleft and consequently alters the peptide ligand-binding repertoire of different class I allotypes (Falk et al., 1991; Schumacher et al., 1991). Sequencing of peptides which naturally associate with class I MHC molecules has revealed that the majority of bound peptides possesses conserved sequence motifs (Rammensee et al., 1995). Analysis of the crystal structures of several peptide–class I MHC complexes demonstrated that the side chains of these conserved amino acids (termed anchor residues) make crucial intermolecular contacts with the binding groove of the class I MHC–β_2m complex (Madden 1995). In addition, sequencing of mixtures of pooled peptides eluted from particular class I MHC molecules has further delineated preferred and subdominant anchor residues. Thus, biochemical analysis of the peptide species associated with specific class I MHC molecules provides a set of global binding rules. These rules or motifs highlight obligatory amino acids along the sequence of the antigenic peptide responsible for binding specificity and preferred amino acids at other positions in the sequence of the determinant, and also provide information on non-tolerated amino acids (Rammensee et al., 1995).

The identification of consensus binding motifs for the majority of common murine and human class I MHC molecules provides a powerful tool for predicting T cell determinants from the full amino acid sequence of candidate Ags. Some caution must be used, however, as many high-affinity class I MHC ligands lack consensus binding motifs. In addition, synthesis of peptides based on binding motifs may produce ligands which fail to bind target class I MHC molecules with significant affinity *in vitro*. Further complicating the use of predicted motifs is the observation that post-translational modification of peptides may alter the immunogenicity of a peptide Ag. For example, melanoma tyrosinase-specific T cells were found to recognize a peptide that had undergone post-translational conversion of its asparagine residue to aspartic acid (Skipper et al., 1996). The asparagine residue was at a site for N-linked glycosylation. Recycling of glycosylated tyrosinase from the ER into the cytosol could have exposed the glycan side chain to degradation by peptide N-glycanase, leading to deamidation of the asparagine residue. In another example, HLA-A*0201-restricted CTLs specific for a peptide determinant from the H-Y Ag only recognized the post-translationally altered form of this peptide in which a second cysteine residue had been covalently linked to a cysteine in the peptide sequence (Meadows et al., 1997). These examples indicate that the peptide structures presented by MHC molecules are not always predictable from the amino acid sequence of a given protein.

There are numerous checkpoints in the MHC class I Ag processing and presentation pathway which may affect determinant generation and the focus of the CTL response *in vivo*. Polymorphism in the MHC may influence the rate of peptide generation, the specificity of peptide transport from the cytosol into the ER, the stability of the peptide species, the specificity and affinity of peptide binding by the class I molecule, and the structural assembly of the class I MHC complex. Some of these and other pertinent aspects involved in the generation and presentation of peptide Ags by class I MHC will now be addressed.

Generation of peptide antigens in the cytosol

The majority of class I-restricted peptide Ags are thought to result from the degradation of cytosolic polypeptides. The proteasome is the structure primarily responsible for the intracellular, non-lysosomal, degradation of short-lived proteins and the generation of many oligopeptides associated with the Ag presentation pathway (Goldberg, 1995). The catalytic core of the proteasome, also termed the 20S proteasome, is a 700 kDa multisubunit complex comprising four stacked seven-member rings. The solution of the crystal structure (Lowe et al., 1995) of the 20S proteasome from the archaebacterium *Thermoplasma acidophilum* has provided critical insight into the structure and function of this complex. The bacterial molecule contains two different subunits, α and β. The two outer rings of the four-ringed structure are each composed of seven α-subunits, while both the inner rings are formed from seven β-subunits (i.e. the molecule adopts an $\alpha_7\beta_7\beta_7\alpha_7$ structure). Each of the 14 β-subunits has catalytic sites on the internal surface of this cylindrical structure. Although the composition of the ancestrally related mammalian 20S proteasome is more complex, with seven different α-subunits and ten different β-subunits identified so far, the overall structure of the complex is similar to that of the bacterial protein (Groll et al., 1997).

The 20S proteasome cleaves proteins independently of ubiquitination and *in vitro* studies have revealed that it may only cleave denatured substrates. *In vivo*, the catalytic core of the proteasome must associate with a 19S regulator in order to degrade native proteins. This complex is referred to as the 26S proteasome. Proteins are tagged for proteolysis by this structure via conjugation of multiple ubiquitin molecules in an ATP-dependent process (Goldberg and Rock, 1992; Michalek et al., 1993). Another protein complex known to associate with the 20S proteasome is PA28 (the 11S regulator) (Dubiel et al., 1992; Ma et al., 1992). PA28 is induced by interferon-γ (IFNγ) and has been demonstrated to alter the quantity and quality of peptide cleavage products (Groettrup et al., 1995). The influence of PA28 upon peptide generation by the 20S proteasome was further defined using peptide precursors 19–25 residues long (Dick et al., 1996). In the absence of PA28, the 20S proteasome generated peptides resulting from a single cleavage of the precursor, while the presence of PA28 resulted in products generated by two cleavages. Functionally, it was shown that induction of PA28 resulted in marked enhancement of CTL recognition (Groettrup et al., 1996a).

The inflammatory cytokine IFNγ also induces expression of three proteasome subunits, LMP2, LMP7 and MECL-1. Incorporation of these subunits into the proteasome has been reported to change proteasomal substrate specificity (Driscoll et al., 1993; Gaczynska et al., 1993; Groettrup et al., 1996b). Whilst the proteolytic specificity of LMP2 and LMP7 may favour the generation of peptides suited to binding class I MHC molecules (Driscoll et al., 1993; Gaczynska et al., 1993; Cerundolo et al., 1995), these subunits are not essential for stable class I molecule surface expression and presentation of some viral epitopes (Arnold et al., 1992; Momburg et al., 1992; Yewdell et al., 1994). In LMP2$^{-/-}$ mice, cell-surface class I MHC expression was the same as in wild-type mice, but the number of CD8$^+$ T cells was reduced by 30–40% (Van Kaer et al., 1994). In contrast, there was a 25–40% reduction in expression of class I MHC in LMP7$^{-/-}$ mice but no difference in CD8$^+$ T cell numbers (Fehling et al., 1994). Analysis of 20S proteasomes isolated from tissues of LMP7$^{-/-}$ mice revealed altered proteolytic activities compared with non-transgenic littermates (Stowasser, 1996). Cells from both mice were limited in their ability to present certain Ags to CD8$^+$ T cells, although there is no evidence for dramatic alterations in the T cell repertoire. Therefore, it is likely that LMP2 and LMP7 are important for the generation of some determinants but not others.

There is much biochemical evidence highlighting the importance of the proteasome in the generation of antigenic oligopeptides. The most compelling of these data is the ability of different protease inhibitors (particularly peptidyl aldehydes and the antibiotic lactacystin) to inhibit specifically the function of the 20S and 26S proteasomes. These inhibitors irreversibly occupy the catalytic sites of the targeted β-subunit, subsequently blocking presentation of antigenic determinants derived from protein precursors. Despite a large body of evidence implicating the proteasome as the principal cytosolic protease for the generation of antigenic peptides, a recent study which employed a panel of protease inhibitors with overlapping ranges of inhibitory activities suggested that non-proteasomal cytosolic proteases may also be involved in the generation of a significant portion of class I MHC binding peptides (Vinitsky, 1997).

Transport of peptide antigens into the endoplasmic reticulum by TAP

Peptides generated in the cytosol must pass into the lumen of the ER before they can assemble with class I heavy chain and β2m to form stable class I MHC complexes. The structure responsible for this translocation is called the transporter associated with antigen processing (TAP) and is made up of two subunits, TAP 1 and TAP 2. TAP belongs to the ABC (ATP-binding cassette) superfamily of transporters which includes the human multidrug-resistance protein, and a series of transporters from bacteria and eukaryotic cells capable of transporting a range of substrates, including peptides. The genes encoding TAP1 and TAP2 have been mapped to the MHC class II region, and their role in Ag presentation was confirmed when transfection of these genes corrected Ag presentation defects in mutant cell lines (Deverson et al., 1990; Spies et al., 1990; Trowsdale et al., 1990; Spies and De Mars, 1991; Kelly et al., 1992). TAP1 and TAP2 proteins are approximately 76 kDa and 70 kDa, respectively. Each subunit contains an N-terminal transmembrane domain which spans the ER membrane six times and a C-terminal cytoplasmic domain with an ATP binding site (Nijenhuis and Hammerling, 1996) (see Figure 5.4). The expression and non-covalent association of both molecules is required to form a stable, functional TAP complex (Androlewicz et al., 1994). In the absence of TAP, most peptides cannot be transported into the ER and the formation of stable class I complexes is impeded. This results in the reduced expression of cell-surface class I molecules which requires exogenous loading of peptides in order to stimulate Ag-specific CTL. The addition of exogenous peptides also markedly increases the surface expression of the restricting class I allele (Townsend et al., 1990).

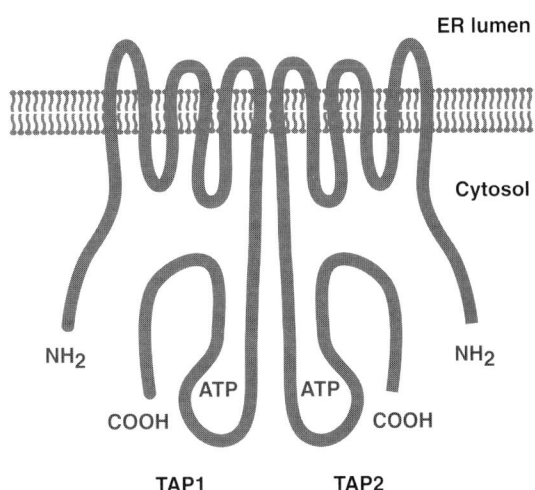

FIGURE 5.4 Model of the TAP (transporter associated with Ag presentation). TAP comprises two subunits, TAP 1 and TAP 2, and belongs to the ABC (ATP-binding cassette) superfamily of transporters which includes the human multidrug-resistance transporters. Each subunit contains an N-terminal transmembrane domain which spans the ER membrane six times and a C-terminal cytoplasmic domain with an ATP binding site. TAP-binding peptides show broad sequence specificity but peptides with a length of 8–13 residues are preferentially translocated.

Translocation is thought to involve a closely coupled two-step process, commencing with the ATP-independent binding of peptides to TAP followed by the transport into the ER via an ATP-dependent mechanism (Androlewicz and Cresswell, 1994, 1996; van Endert et al., 1995). Recently, four regions (two each on TAP1 and TAP2) on the cytosolic aspects of the transmembrane segments and close to the ATP-binding sites have been identified that make major contributions to peptide binding (Momburg et al., 1996; Nijenhuis and Hammerling, 1996). These findings are consistent with a model in which the transmembrane segments form a pore in the ER membrane with peptides binding to the cytosolic mouth of the pore prior to translocation.

TAP molecules exhibit selection for both peptide length and sequence. Studies of peptide translocation either in streptolysin O-permeabilized cell systems or in isolated microsomes have shown that peptides with a length of eight to 13 residues are preferentially translocated by TAP (Heemels and Ploegh, 1994; Momburg et al., 1994a; Schumacher et al., 1994). This is consistent with the binding constraints of class I MHC molecules. By studying a range of peptides of varying lengths all containing a consensus sequence for ER-specific N-linked glycosylation at opposite ends and a radiolabelled tyrosine elsewhere, Koopmann et al. (1996) were able to ensure that only non-degraded products were monitored in peptide translocation and competition assays. This study reaffirmed that peptide translocation was most efficient for peptides eight to 12 residues in length. Whilst longer peptides (up to 40 residues) may be transported by TAP, typically these peptides are partially degraded by proteases prior to translocation (Momburg et al., 1994a). This lends support to earlier observations that TAP-dependent peptides up to 33 residues could be isolated from a subset of cell-surface HLA-B27 molecules (Urban et al., 1994) and that H-2K^b may bind to peptides 15 residues long (Joyce et al., 1995). It also raises the possibility that longer peptides transported into the ER may require trimming either before or after loading on to peptide-receptive class I MHC molecules.

While the specificity of peptides translocated by TAP is necessarily broad, it appears to vary between species. The human TAP does not possess a preference for any particular C-terminal residue, whereas mouse TAP shows a clear preference for peptides with hydrophobic C-termini (Momburg et al., 1994b; Neefjes et al., 1995). The rat TAP is atypical as the two alleles of TAP2, TAP2^a and TAP2^u, determine the type of peptides translocated into the ER. Thus, the TAP1/TAP2^u complex transports peptides ending with hydrophobic C-termini, whereas TAP1/TAP2^a does not have such a preference. This polymorphism results in a phenomenon termed the cim (class I modification) effect (Livingstone et al., 1991), whereby intracellular assembly of the rat class I RT1.A^a molecule has been shown to be much slower in cells homozygous for the TAP2^u than the TAP2^a allele. It was later shown that the set of peptides translocated by TAP1/TAP2^a met the peptide structural preference of RT1.A^a (Powis et al., 1996).

The preference for length and in some instances the C-terminal residue of peptides translocated by TAP does not appear to play a major role in determining the hierarchy of CTL responses. For instance, Neisig et al. (1995) examined the translocation efficiencies of 16 immunogenic peptides and found that the four least efficiently translocated peptides were immunodominant. Certain peptide sequences influence the efficiency of TAP-mediated translocation. For example, substitution of proline in position 3 of poorly translocated peptides improved the rate of translocation of immunodominant viral determinants (Neisig et al., 1995). The addition of naturally occurring flanking sequences may improve the transport of otherwise poorly translocated peptides by TAP. This again suggests that some translocated peptides require further proteolytic processing in the ER prior to or during association with class I MHC molecules. Studies of peptide translocation rates using isolated microsomes have shown that murine TAP can transport cytosolic peptides at high rates (K_m in the micromolar range) in a concentration-dependent manner (Hahn et al., 1996). Therefore, high level expression of a foreign peptide can potentially overcome poor translocation efficiencies and still result in significant levels of presentation.

The expression of the TAP molecules, like that of LMP2, LMP7 and MHC class I molecules, is inducible by IFNγ. It is interesting to note that the LMP2 and LMP7 molecules are encoded in the class II region in close proximity to the genes encoding TAP1 and TAP2, with shared transcriptional control (Brown et al., 1991; Glynne et al., 1991; Martinez and Monaco, 1991; Ortiz-Navarrete and Hammerling, 1991). Kinetic studies have shown that upon induction by IFNγ, the transcription of TAP1 and TAP2 mRNA occurs much more quickly than that of HLA class I mRNA. The same kinetic findings applied to the expression of the proteins that they encoded (Ma et al., 1997). The increased expression of TAP molecules and corresponding peptide translocation into the ER may facilitate the increased expression of cell-surface class I molecules after induction with IFNγ.

The biological importance of the TAP molecule in Ag presentation has been emphasized by a recent clinical report of a family in which two of five siblings suffered from recurrent respiratory infections (de la Salle et al., 1994). Both parents expressed one normal and one truncated, non-functional allele of TAP2. The two index cases had inherited the defective TAP2 genes and hence had non-functional TAP molecules and very low levels of cell-surface class I. In another report, a patient who carried one functional and one null TAP1 allele developed a small cell lung carcinoma which failed to express TAP1 protein due to the loss of the functional allele (Chen et al., 1996). Again, these cells expressed low levels of surface class I molecules. It is possible that the loss of class I surface expression and Ag presentation function contributed to the escape of tumour cells from immune surveillance. The majority of clinical studies has not shown any association between TAP polymorphism and susceptibility to autoimmune diseases. These findings are supported by *in vitro* assays which demonstrated no significant influence of human TAP polymorphism on peptide selection (Obst et al., 1995).

TAP-independent pathways of antigen presentation

In the absence of TAP, antigenic peptides that are preceded by hydrophobic signal sequences can be delivered efficiently into the ER. The signal sequence is released by the action of signal peptidase, and the resultant peptides may subsequently be presented by class I MHC molecules (Anderson et al., 1992). Furthermore, it has been reported that peptides derived from signal sequences associate with HLA-A2 molecules expressed in a TAP-deficient cell line (Henderson et al., 1992; Wei and Cresswell, 1992). The length of the signal sequence far exceeds the length of peptides presented by HLA-A2, raising the possibility of proteolytic activity other than signal peptidase within the ER. In addition, these observations raise the possibility that longer peptides may be processed further in the ER to generate shorter and better fitting peptides, as suggested by Falk et al. (1990).

The possibility of peptide trimming in the ER was addressed systematically by studying the liberation of two distinct peptide determinants synthesized in tandem and targeted into the ER of TAP deficient cells by a N-terminal signal sequence (Snyder et al., 1994). Irrespective of the order of the peptides, the C-terminal peptide was always presented much more efficiently than the N-terminal peptide. The results implied that aminopeptidases were responsible for the major trimming activity in the ER, whereas carboxypeptidases and endopeptidases contributed only marginally. The finding in another study that carboxypeptidase found within an intracellular secretory compartment was required for the presentation of a peptide extended by two residues from its natural C-terminus also supported the notion that carboxypeptidase may be deficient or absent in the ER (Eisenlohr et al., 1992). Thus, as suggested by Falk and colleagues, the C-terminal end of a peptide may fit into the class I MHC binding groove, with trimming activities being limited to the N-terminal end. This arrangement may reflect the critical nature of the interaction of the C-terminal residues of antigenic peptides with class I molecules.

These findings raise the possibility that antigenic peptides may be generated from secreted and membrane-bound proteins in a signal sequence-dependent and TAP-independent manner. This has been investigated by targeting various full-length viral proteins to the ER of TAP-deficient cells by N-terminal signal sequences. Presentation of known peptide epitopes derived from these proteins was assayed by peptide-specific CTL. None of full-length influenza nucleoprotein, haemagglutinin or OVA was processed this way (Snyder et al., 1994; Elliott et al., 1995; Bacik et al., 1997). However, it has been shown that a subset of epitopes could be processed from HIV-1 gp120 in a signal sequence-dependent and TAP-independent manner, possibly through the action of proteases located in an early secretory compartment (Hammond et al., 1995).

Hombach and colleagues studied the TAP-dependent presentation of an immunodominant epitope (gp33) from the lymphocytic choriomeningitis virus (LCMV). This epitope is located at positions 33–41 within the signal sequence of the LCMV glycoprotein. TAP$^{-/-}$ mice failed to process and present the gp33 LCMV epitope despite normal surface expression of the glycoprotein (Hombach et al., 1995). Generation of this epitope would have required a combination of carboxypeptidase or endopeptidase in addition to aminopeptidase activities. These requirements may not be met unless the antigenic peptide epitope was sited at the C-terminal end of the signal sequence. Two of the three peptides isolated from the HLA-A2 molecule from T2 cells were thought to have been derived from the C-terminus of the hydrophobic signal sequence of IP30 (Wei and Cresswell, 1992). Given these constraints, proteolytic activity in the ER may be limited to the trimming of longer than usual peptide precursors, whilst native proteins are relatively resistant to proteolysis.

Contrary to expectation, the studies which examined the processing of full-length proteins tagged with signal sequences demonstrated strict TAP dependence (Hammond et al., 1995; Hombach et al., 1995; Bacik et al., 1997). TAP-deficient T2-K^d cells were unable to present a peptide from the influenza nucleoprotein preceded by a signal sequence unless they had been restored with TAP (Bacik et al., 1997). This implied that only peptides derived from proteins degraded in the cytosol and transported into the ER were available for MHC loading. These antigenic peptides could be derived from proteins that aberrantly had not been transported into the ER (Yewdell and Benninck, 1992). Alternatively, such proteins could have been recycled from the ER back into the cytosol to be processed by cytosolic proteases. Roelse et al. (1994) demonstrated that the majority of translocated peptides which failed to bind assembling class I MHC molecules were recycled from the ER back into the cytosol via an energy-dependent process. Likewise, introduction of an N-linked glycosylation site was found to reduce selectively the presentation of the NP 147–155 determinant secreted (i.e. tagged with an ER insertion sequence) but not from cytosolic nucleoprotein (Bacik et al., 1997). In this case the presentation of NP 147–155 was strictly TAP dependent, regardless of whether the nucleoprotein was cytosolic or in the signal sequence form. It was argued that glycosylation of the signal sequence form of the substituted nucleoprotein occurred in the ER prior to its relocation into the cytosol. Thus, glycosylation may have interfered with TAP-mediated transport of the peptide determinant, binding with the relevant MHC class I molecule or recognition by CTL. Alternatively, removal of the oligosaccharide from the asparagine by cytosolic peptide N-glycanase could have resulted in conversion to an aspartate residue at this position, disrupting recognition by the NP 147–155-specific CTL. In the cytosol these peptides are rapidly degraded, with a small proportion of trimmed peptides returning to the ER via TAP-mediated translocation. Thus, peptides which do not fit class I MHC binding requirements may be trimmed in the ER (a relatively slow process) or recycled into the cytosol.

The immunodominance of peptide determinants reflects differences in antigen processing

The number of peptides in a protein that can potentially bind a class I MHC molecule is generally larger than the number of epitopes recognized in a CTL response to that protein. For example, the H-2K^b-restricted CTL response in C57BL/6 mice immunized with chicken OVA is dominated by the epitope OVA 257–264. Although the protein contains six peptides that satisfy the binding requirements of the H-2K^b molecule, the OVA 257–264 peptide is immunodominant (Chen et al., 1994). In contrast, OVA 55–62 is poorly immunogenic. Although it contains the necessary binding motif, the binding affinity of K^b to OVA 55–62 is 50-fold lower than that to OVA 257–264 (Chen et al., 1994). Niedermann et al. (1995, 1996) also addressed this issue by examining the generation of these two epitopes by purified 20S proteasomes. They found that the generation of the OVA 52–62 epitope was limited by inefficient proteolytic cleavage in the sequences flanking the epitope and proteolytic destruction of the epitope. This suggests that proteasomes may also play an important role in determining immunodominance.

The ability of flanking sequences to affect the generation of CTL determinants was first suggested by observations that presentation of a viral epitope inserted into a different protein was significantly altered (Del Val et al., 1992). These results implied that the position of a peptide determinant in an Ag and the composition of the neighbouring residues could determine the efficiency with which an epitope was generated and subsequently presented to CTL. Similarly, flanking sequence has been shown to affect the presentation of the dominant OVA 257–264 determinant during *in vitro* biochemical studies using synthetic peptides (Niedermann et al., 1995). Mutations of the residues flanking an epitope may also affect the presentation of that epitope. An alanine-to-proline mutation N-terminal to the influenza nucleoprotein 146–155 epitope severely reduced the generation of this epitope, not by blocking the cleavage of the flanking residues, but by enhancing cleavage within the epitope (Yellen-Shaw et al., 1997). Mutant strains of viruses may adopt this strategy of evading immune detection by targeting immunodominant epitopes for proteolytic destruction (Ossendorp et al., 1996).

Structural assembly of major histocompatibility complex class I molecules in the endoplasmic reticulum

The formation of stable MHC class I complexes requires the assembly of class I heavy chains with β2m and peptides, the lack of which results in poor cell-surface expression and impaired Ag presentation (Townsend et al., 1989). In the absence of β2m, most class I heavy chains cannot form stable MHC complexes and are not expressed on the cell surface (Hyman and Stallings, 1976). Furthermore, these heavy chains fail to traffic out of the ER, as indicated by the immature status of their carbohydrate side chains (Owen et al., 1980). Misfolded heavy chains that have been retained in the ER are eventually translocated into the cytoplasm and degraded by proteasomes (Hughes et al., 1997). Failure to translocate peptides into the ER results in weak heavy chain–β2m interaction, failure of heavy chains to fold properly and attain defined serological epitopes, and prolonged retention of immature complexes in the ER (Townsend et al., 1989, 1990). These deficiencies can be corrected by incubating the immature heavy chain–β2m complexes *in vitro* with allele specific peptides (Townsend et al., 1990). Class I instability is also seen when proteasome inhibitors attenuate the supply of antigenic peptides (Hughes et al., 1996; Suh et al., 1996). Heavy chain–β2m complexes formed in the absence of peptides are thermolabile but can be stabilized by incubation of these cells at 26°C, leading to increased levels of surface class I MHC (Ljunggren et al., 1990).

Central to the process of loading class I–β2m complexes with peptides is the physical association of the peptide receptive class I complex with the TAP molecule (Ortmann et al., 1994; Suh et al., 1994). Although the precise mechanism of peptide loading remains unknown, physical colocalization of the peptide-importing machinery with a peptide-receptive complex at the site of maximal peptide concentration may be sufficient to facilitate peptide loading. Thus, up to four class I–β2m complexes may cluster around one TAP molecule (Ortmann et al., 1997). The physical interaction between class I–β2m complexes with TAP seems to be mediated via the TAP1 subunit, even though functional translocation of peptides from the cytosolic compartment into the ER requires heterodimerization of the TAP1 and the TAP2 subunits (Androlewicz et al., 1994). The addition of allele-specific peptides leads to the release of the class I–β2m–peptide trimeric complex from TAP (Ortmann et al., 1994; Suh et al., 1994). Conversely, shutting down peptide supply with proteasome inhibitors leads to prolongation of association between class I–β2m and the TAP molecule (Hughes et al., 1996; Suh et al., 1996). Presumably, the acquisition of suitable peptide induces a conformational change in the structure of the class I complex, allowing its release from the TAP molecule.

Recently, a 48 kDa ER transmembrane glycoprotein called tapasin (for TAP-associated glycoprotein) has been found to play a pivotal role in the physical association between empty class I–β2m complexes and TAP (Sadasivan et al., 1996). The gene encoding tapasin has been mapped to a site within approximately 10^6 base pairs of the MHC (Ortmann et al., 1997). Thus, in the human cell line 721.220 which lacks tapasin and both copies of its HLA-A and HLA-B alleles (Greenwood et al., 1994), transfected class I molecules are poorly expressed on the surface (Grandea et al., 1995) and fail to present viral epitopes to peptide specific CTL (Ortmann et al., 1997). These biochemical and functional defects are fully restored by coexpression of tapasin, suggesting that class I–β2m complexes cannot associate with TAP in the absence of tapasin (Ortmann et al., 1997 and personal observations). Although the precise mechanism by which tapasin carries out its function is unclear, it may stabilize the empty class I–β2m complex sufficiently to allow interaction with TAP. Alternatively, it may act as a binding intermediary between the class I–β2m complex and TAP. These possibilities are not mutually exclusive. Soluble tapasin has recently been shown to restore surface expression and Ag presentation of HLA class I molecules in the 721.220 cell line without restoring the association of class I molecules with TAP (Lehner et al., 1998). This finding lends support to a role for tapasin in stabilizing empty class I complexes in a peptide-receptive conformation (Lehner et al., 1998). Tapasin also seems to increase the expression of TAP, presumably enhancing the supply of peptides for class I assembly (Lehner et al., 1998).

Besides tapasin, other ER-resident chaperone molecules such as calnexin and calreticulin are known to participate in the structural assembly of class I MHC complexes. Both are ubiquitous Ca^{2+}-binding ER proteins whose role in relation to the class I pathway is to promote maintenance of partially folded class I heavy chains and their subsequent intermediaries in the peptide loading pathway, preventing premature degradation. They may also play a housekeeping role in retaining misfolded complexes that have failed to capture peptides within the ER for subsequent degradation. Both of these chaperones are lectins which interact specifically with partially trimmed, monoglycosylated, N-linked oligosaccharides (Krause and Michalak, 1997).

Calnexin is an 88 kDa protein anchored to the ER lumen via a transmembrane domain. In mouse cells, calnexin associates with free class I heavy chains and class I–β2m complexes (Degen et al., 1992). In mouse cells, the binding of calnexin to heavy chain may be maintained during and beyond the interaction of the class I–β2m complex with TAP (Suh et al., 1996). In contrast, calnexin has been found to associate only with free HLA class I heavy chains in human cells (Ortmann et al., 1994; Sugita and Brenner, 1994; Nöbner and Parham, 1995). In human cells, calnexin may not be essential for the assembly of HLA class I molecules as they are fully functional in a calnexin-negative cell line (Sadasivan et al., 1995; Scott and Dawson, 1995).

As shown in Figure 5.5A, upon binding of β2m to HLA class I heavy chains calnexin is replaced by calreticulin, an ER-resident 46 kDa chaperone (Sadasivan et al., 1996). Unlike calnexin, calreticulin is not anchored to the ER membrane, but instead contains an ER-retention signal. In the absence of β2m, not only do class I heavy chains fail to associate with TAP (Ortmann et al., 1994; Suh et al., 1994), but heavy chains also fail to associate with calreticulin (Sadasivan et al., 1996). Studies by Cresswell and colleagues (Sadasivan et al., 1996) have also shown that calreticulin operated concomitantly with tapasin to facilitate the association of empty class I–β2m complexes with TAP. Thus, in the presence of tapasin, calreticulin and TAP can be coprecipitated with class I–β2m complexes. In the absence of tapasin, class I–β2m complexes and calreticulin fail to associate with TAP. The coexistence of calreticulin and TAP in such complexes is unlikely to be the result of the direct binding of calreticulin to TAP because the association is dependent on coexpression of tapasin and β2m (Sadasivan et al., 1996; Solheim et al., 1997). Significantly, in the absence of β2m alone, TAP but not calreticulin can continue to associate with tapasin (Sadasivan et al., 1996), arguing that calreticulin does not associate directly with tapasin. Furthermore, in the absence of TAP alone, complexes consisting of class I–β2m, calreticulin and tapasin can still be found (Sadasivan et al., 1996). Lastly, it is interesting to note that β2m was found to associate with TAP as well as calreticulin in the 721.221 cell line which lacks HLA-A, B and C alleles but expresses tapasin (Solheim et al., 1997). The association of β2m with TAP and calreticulin persisted even after extensive preclearance of trace amounts of class I heavy chain protein molecules expressed by this cell line. Therefore, a model is proposed in Figure 5.6 which incorporates the above observations.

Knowledge of the precise sites of these intermolecular interactions awaits further study. Several sites have been identified on the class I heavy chain which may be involved in the interaction with TAP. A threonine-to-lysine substitution at amino acid residue 134 on the α2 domain of the HLA-A*0201 molecules was found to affect TAP association and peptide loading (Lewis et al., 1996; Peace-Brewer et al., 1996). The

(A)

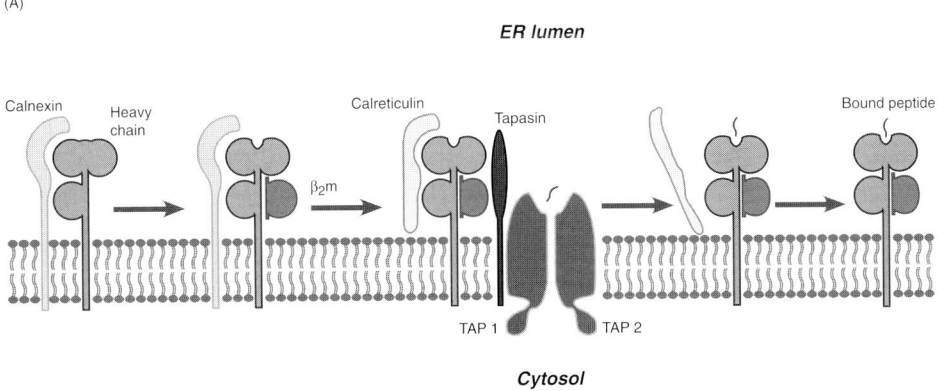

(B)

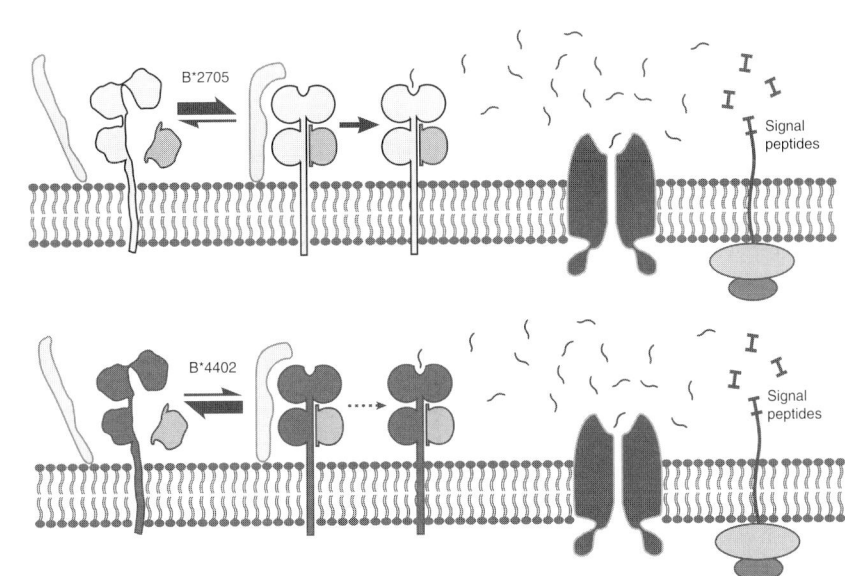

FIGURE 5.5 A model for peptide loading of class I molecules in the lumen of the endoplasmic reticulum. As shown in the upper panel (A), newly synthesized free human class I heavy chains assemble with calnexin but upon binding of β2m they become associated with calreticulin. Tapasin and calreticulin–class I–β2m complexes associate with each other, allowing the colocalization of the complex with TAP, which also binds tapasin. This association of empty class I–β2m complexes may also act to stabilize them in a peptide-receptive conformation. The potential stabilization of peptide-receptive class I molecules results in efficient sampling of newly transported peptides which are custom delivered by the close proximity of the TAP lumen. As shown in the lower panels (B), in the absence of tapasin, class I molecules do not colocalize with TAP and peptide sampling is constrained by the distribution, concentration and $t_{1/2}$ of potential peptide ligands. Allelic variation in the efficiency of peptide loading in the absence of tapasin might depend on the structural stability of the peptide-receptive class I–β2M complex at 37°C, which differs between class I alleles. Thermostability could be an intrinsic property of the complex or might rely on chaperone interactions. For thermolabile molecules such as B*4402 the equilibrium between loaded and denatured class I molecules is easily shifted by culture of cells at 26°C, which renders the molecules stable for sufficient time to capture specific peptides. Thermostable molecules such as B*2705 continue to load with peptides even in the absence of tapasin. It remains to be determined whether tapasin is required for loading of signal-sequence derived peptides which are TAP independent in their translocation into the ER lumen.

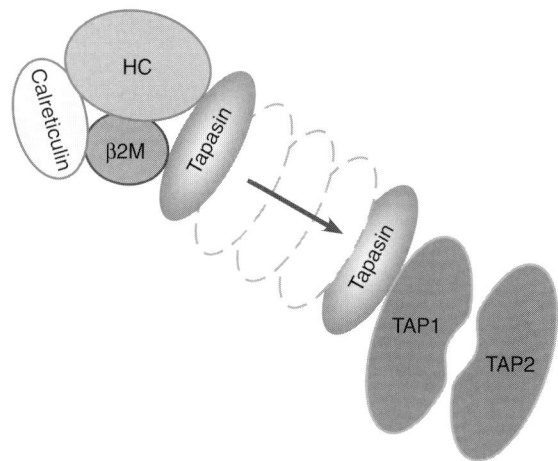

FIGURE 5.6 Model depicting tapasin bridging calreticulin–class I–β2m complexes to the TAP through simultaneous physical association of tapasin with molecular surfaces on the class I heavy chain, β2m and TAP.

α3 domain is also a likely site of interaction with TAP based on the inability of α3 domain-specific mAbs to precipitate TAP–class I complexes and the fact that a point mutation of residue 222 on the α3 domain of the H-2D^d molecule rendered this mutant class I molecule unable to bind to TAP (Suh et al., 1996). In view of the complexity of the intermolecular interactions, as shown by Figures 5.5 and 5.6, interpretation of these findings warrants careful consideration. These putative sites may affect interaction of the class I–β2m complex with TAP indirectly by affecting the binding of the complex to other molecules such as calreticulin and/or tapasin.

Although the tapasin-mediated association of class I–β2m complexes with TAP may be important in the promotion of peptide loading, certain MHC alleles can function in its absence. Thus, HLA-A2 can capture and present peptides derived from signal sequences in the absence of functional TAP molecules in T2 cells (Henderson et al., 1992; Wei and Cresswell, 1992). However, the peptide repertoire presented by HLA-A2 under these conditions is extremely limited. Our group, like De Mars and colleagues (Greenwood et al., 1994), has observed allelic variation in the dependence on tapasin for the surface expression of various HLA-B class I molecules (Peh et al., 1998). Thus, HLA-B*4402 is highly dependent on tapasin for its surface expression, whereas HLA-B*2705 is relatively independent and HLA-B8 demonstrates an intermediate dependence. Furthermore, compared with HLA-B*4402, HLA-B*2705 was found to be relatively independent of tapasin in its ability to present viral-derived epitopes to peptide-specific CTL. Importantly, HLA-B*2705, like HLA-B*4402, required tapasin to mediate its association with TAP. In view of this finding, the high-level surface expression of HLA-B*2705 and its persistent capability to present Ags in the absence of tapasin imply that the HLA-B*2705 molecule is capable of acquiring peptides without associating with TAP. Unlike HLA-A2, HLA-B*2705 is dependent on TAP as a source of peptides. This was demonstrated by the poor surface expression of HLA-B*2705 when transfected into T2 cells. In addition, signal sequence-derived peptides are not known to associate with HLA-B*2705 molecules expressed on normal cells (Jardetzky et al., 1991).

To reconcile these findings, a model is proposed whereby the ability of HLA-B*2705 compared with HLA-B*4402 to present peptides in the absence of tapasin reflects the inherent relative stability of empty HLA-B*2705–β2m complexes (Figure 5.5B). This model proposes that in the absence of tapasin, empty B*2705–β2m complexes remain stable for sufficient time to acquire ER peptides without the need to interact directly with TAP. By comparison, empty B*4402–β2m complexes appear to be unstable in the absence of tapasin. This property of HLA-B*2705 may reflect either stronger binding of the HLA-B*2705 heavy chain to the β2m molecule or stronger interaction of the HLA-B*2705–β2m complex with other chaperone molecules, for example calreticulin, or both. However, "empty" B*2705 molecules still require peptides for their ultimate stabilization and transport to the cell surface, explaining the low expression of these molecules on the surface of the TAP-negative T2 cell line. We predict that, in the absence of tapasin, the HLA-B*2705 molecule may bind to highly abundant peptides and long-lived peptides. Thus, HLA-B*2705 molecules that are expressed in the absence of tapasin may present a modified repertoire of peptides compared with those presented by HLA-B*2705 molecules in the presence of tapasin. It can be further speculated that the ability of HLA-B*2705 to present a skewed repertoire of either self- or pathogen-derived peptides, such as observed in the absence of tapasin, may contribute to the pathogenesis of HLA-B27-associated diseases such as ankylosing spondylitis.

Major histocompatibility complex class I presentation of exogenous antigens

The class I MHC pathway generally involves the processing and presentation of peptide Ags derived from endogenously synthesized proteins, be they self or foreign. However, it is known that under certain conditions CTL can be primed *in vitro* and *in vivo* by professional APCs pulsed with exogenous Ag. This topic has been extensively reviewed by several authors (Bevan, 1995; Jondal et al., 1996; Rock, 1996; Watts, 1997). Conventionally, it is thought that priming of $CD8^+$ CTL takes place most efficiently in the lymphoid regions. Endowed with the necessary costimulatory and adhesion molecules, professional APCs may phagocytose exogenous cellular debris derived from infected or mutated somatic cells and generate peptides for presentation by MHC class I molecules to prime naive CTLs. Furthermore, concomitant Ag presentation by MHC class II molecules on these cells may solicit $CD4^+$ T cell help.

Conceptually, MHC class I presentation of exogenous Ags by healthy APCs seems paradoxical since it would expose them to lysis by CTL. Furthermore, the mechanisms by which peptides can be generated from exogenous proteins and yet intersect the class I MHC pathway in order to load class I molecules for presentation have not been established. Nevertheless, this process can explain the phenomenon of cross-priming (Bevan, 1976), a process whereby a host animal that has been transplanted with an MHC-mismatched graft can be primed to reject a subsequent MHC-matched graft if the priming and rejected grafts share the same minor histocompatibility Ag, different from the host.

In an elegant study (Kurts et al., 1996), when OVA-specific $CD8^+$ T cells from a T cell receptor-transgenic line were injected into unirradiated transgenic mice expressing OVA only in their renal tubular cells and pancreatic β-islet cells, the T cells homed into the draining lymph nodes of the kidneys and pancreas only. After unilateral nephrectomy, OVA-specific CTL injected after an interval of 4 h continued to accumulate in the draining lymph nodes on both sides, but CTL injected after an interval of 7–13 days accumulated only in the draining lymph nodes of the remaining kidney. These observations suggest that the T cells recognized OVA Ags that had drained from the kidneys for subsequent presentation in the lymph nodes. Furthermore, replacement of bone marrow of the OVA expressing mice by MHC class I mismatched marrow led to failure of the OVA-specific T cells to home into the draining lymph nodes, which suggested that the APCs responsible for the processing and presentation of exogenous Ag were of bone marrow origin.

The efficiency with which soluble Ags are presented via the exogenous pathway of phagocytes is greatly increased when the Ag is bound to micrometre-sized particles. These particles may access phagosomes and endocytic vacuoles via the processes of phagocytosis and macropinocytosis. Many different types of particle can target Ags to the class I pathway of phagocytes, ranging from latex bead-bound OVA (Kovacsovics-Bankowski et al., 1993), liposome-associated Ags (Zhou et al., 1992), viral subparticles such as hepatitis B surface Ags (Schirmbeck et al., 1995), cell-associated recombinant Ags (Carbone and Bevan, 1990), to heat-inactivated whole Sendai viruses (Liu et al., 1997). Although some of these means of CTL priming may not be entirely physiological, they are highly relevant to the development of $CD8^+$ CTL responses to vaccine immunization. Apoptotic bodies that have been phagocytosed by macrophages may also serve as a source of exogenous Ags (Bellone et al., 1997). Finally, this pathway may offer a line of defence against pathogens that can survive in phagosomes, such as mycobacteria.

The routes by which peptides derived from exogenous proteins enter the class I MHC pathway for presentation are unknown. The context in which an exogenous protein is introduced may influence the processing pathway (Wick and Pfeifer, 1996). Presentation of OVA conjugated to latex beads by macrophages was insensitive to leupeptin and the weak base chloroquine, which elevates the pH of endocytic compartments and thus inactivates pH-sensitive proteinases. However, its presentation was sensitive to peptide aldehyde inhibitors of the proteasome and also brefeldin-A, which inhibits the egress of peptides out of the ER and Golgi apparatus (Kovacsovics-Bankowski and Rock, 1995). Consistent with this, presentation of exogenous OVA was dependent on an operational TAP peptide transport system. Therefore, these findings support a route where Ags gain access to the cytosol via phagosomes.

In contrast to the TAP-dependent pathway of presentation, exogenous Ags may also be processed in the vacuoles of phagocytes for presentation by class I MHC molecules. For example, exogenous heat-killed Sendai virus can prime Sendai virus-specific $CD8^+$ T cells *in vivo* via a choloroquine-sensitive but TAP-independent route (Liu et al., 1997). Equally, 22 nm size subviral particles of hepatitis B surface Ag can sensitize cells for CTL lysis via an endosomal, TAP-independent processing pathway that can be blocked by chloroquine and leupeptin but not by brefeldin A (Schirmbeck et al., 1995). It has been proposed that peptides generated from this pathway may have been regurgitated on to the cell surface to load class I MHC molecules (Pfeifer et al., 1993) or, alternatively, that these class I molecules may have been recycled into phagocytic vacuoles where they could be loaded with peptides (Schirmbeck and Reimann, 1996). Invariant chain may also direct newly synthesized class I molecules into the endocytic compartment (Sugita and Brenner, 1995). However, it is unclear how specific peptides with suitable class I MHC-binding characteristics can be generated by vacuolar processing.

Another potential mechanism of introducing exogenous proteins into the class I pathway is the heat shock protein-mediated delivery of Ag. In seminal studies by Srivastava and co-workers tumour-specific immunity was generated by immunization with preparations from MHC-mismatched tumour cells (Tamura et al., 1997). In these studies, heat shock proteins such as hsp70 and hsp96 from tumour cells but not heat shock proteins from normal cells (even though the heat shock proteins from both sources were identical) could prime animals for tumour-specific immunity (Udono and Srivastava, 1994), in a TAP-dependent process (Huang et al., 1996). It was proposed that heat shock proteins released from tumour cells may act as chaperones of antigenic peptides (Udono and Srivastava, 1994). Immunization with hsp96 isolated from β-galactosidase-expressing cells induced CTLs specific for β-galactosidase peptide, whether the host animals were MHC class I-matched or mismatched compared with the cell source of the hsp96 (Arnold et al., 1995). It has been proposed that they may bind to unknown receptors on the surface of macrophages (Srivastava et al., 1994), prior to being routed to the class I processing pathway. The way in which exogenous heat shock protein associated peptides gain access to the class I pathway, however, is unclear.

Viral evasion of T cell-mediated immunity

The dependence of viral replication and transmission on the integrity of host cell biosynthetic pathways has led to the coevolution of host immune defences and viral mechanisms to avoid detection. In order to prevent clearance by the host cellular immune responses, viruses have evolved a diverse array of evasion mechanisms. Paradoxically, viral evasion of immune surveillance has revealed several key host proteins involved in the processing, assembly and trafficking of immunogenic cell-surface MHC molecules. Viruses evade CTL recognition by affecting nearly every stage of Ag processing (Figure 5.7). The herpes group of viruses is perhaps the most successful at evading immune surveillance. Both human and murine cytomegalovirus (CMV) express an impressive array of proteins intended to dysregulate viral Ag processing and presentation. These and other examples will be discussed for each of several important checkpoints which occur during antigenic peptide generation, translocation, assembly with class I and trafficking of the complex to the cell surface.

Downregulation of class I expression

A common mechanism of immune evasion by viruses is the downregulation of class I MHC expression on the surface of cells. For example, adenovirus accomplishes downregulation of class I expression by inhibiting transcription of class I

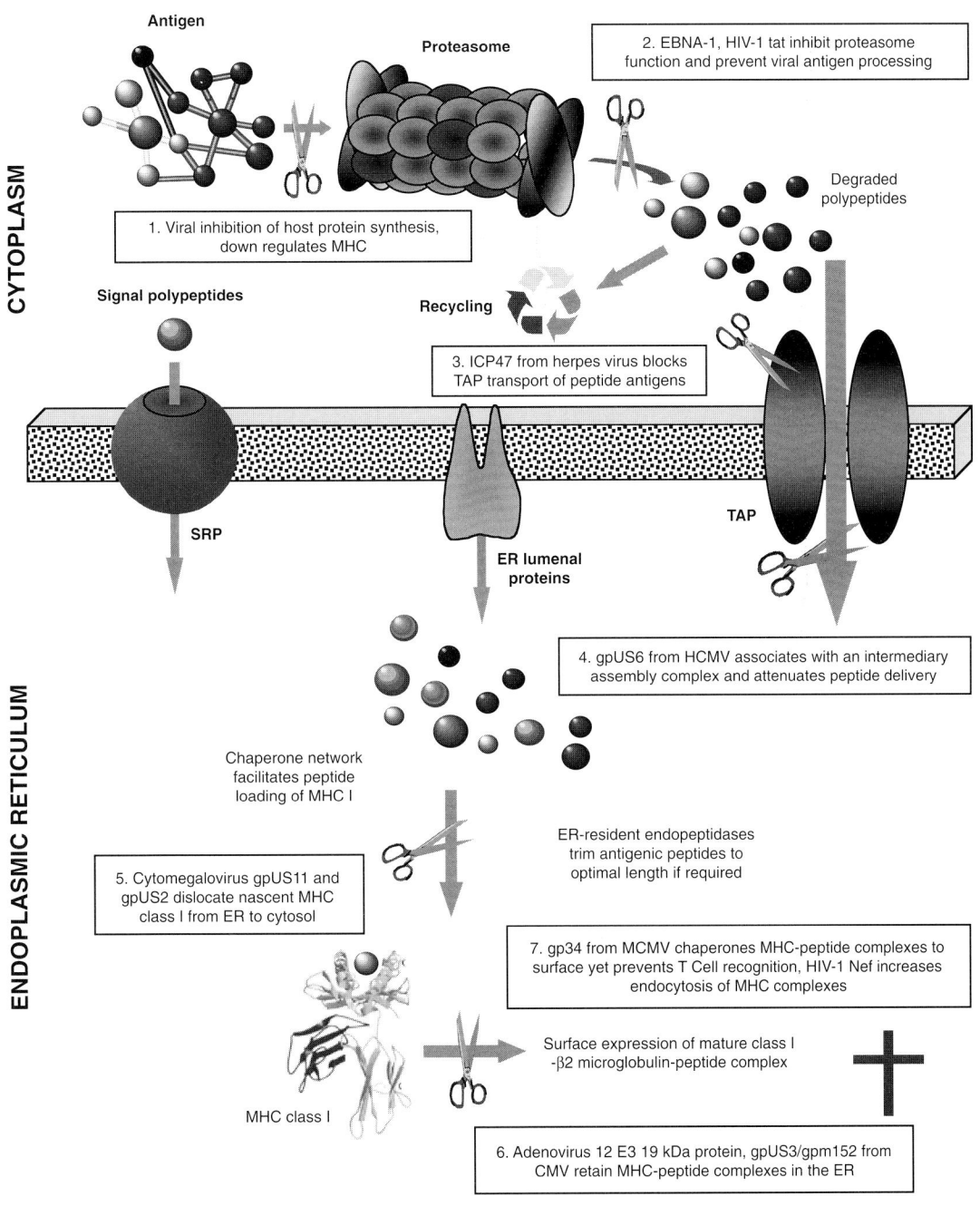

FIGURE 5.7 Mechanisms by which viral gene products interfere with class I-restricted Ag presentation.

mRNA. Specifically, the E1A protein inhibits the processing of precursors of the transcription factor NFκB and consequently interferes in the activity of this transcriptional activator of class I (Schouten et al., 1995). Although downregulation of cell-surface class I MHC molecules is frequently accomplished at the transcriptional level by suppression of host cell protein synthesis (Maudsley and Pound 1991), blockade or dysregulation of numerous molecules involved in processing and presentation of viral peptides can have the same outcome.

Prevention of viral antigen proteolysis

The most direct way to prevent recognition of infected cells is for a period of latency where no viral polypeptides are translated (Wiertz et al., 1997). Thus, many viruses undergo cyclical periods of latency to avoid CTL recognition of infected cells. Downregulation of surface class I expressing viral Ags may also result from attenuation of their supply. For example, the Epstein–Barr virus nuclear Ag, EBNA 1, contains a Gly-Ala-rich repeat sequence which is presumed to impair proteasome function and results in poor presentation of *cis*-linked peptide determinants during EBV infection (Levitskaya et al., 1995). The human immunodeficiency virus type 1 (HIV-1) Tat protein also modulates proteasomal function. Interestingly, this protein seems to enhance 26S proteasome activity yet inhibits the peptidase activity of the 20S proteasome and interferes with the formation of the 20S proteasome-PA28 (11S regulator) complex (Seeger et al., 1997). This effect may alter the type of peptides generated in the cytosol for subsequent translocation into the ER. It has been proposed that the 20S proteasome–PA 28 complex (sometimes referred to as the immunoproteasome) is principally involved in the generation of viral determinants recognized by antiviral CTL involved in eradicating infection (Griffin et al., 1998). These observations are supported by studies which have argued against earlier suggestions of a transcriptional block of class I MHC molecule expression by Tat (Matsui et al., 1996).

A different mechanism of attenuating peptide recognition by CTL is utilized in a peptide-specific manner by human CMV. During human CMV infection CTL recognition of IE-1 derived peptides (a viral protein expressed at very high levels during the intermediate–early phase of viral replication) is specifically blocked by ppUL83, another human CMV encoded peptide (Gilbert et al., 1996). The exact mechanism is unclear; however, the blocking appears to occur prior to loading of IE-1 peptides. Therefore, viruses may reduce the level of viral Ag recognized by CTL by either not expressing viral polypeptides during a period of latency, protecting epitopes from proteolysis or inhibiting specific recognition of peptide–MHC complexes.

Blockade of TAP translocation of viral oligopeptides

Among the repertoire of evasion factors produced by herpes simplex virus (HSV) is a 9 kDa protein (ICP47) which is a potent inhibitor of peptide Ag translocation from the cytosol into the ER. ICP47 coimmunoprecipitates with the TAP heterodimer and appears to interact with the cytoplasmic aspect of the translocation pore (York et al., 1994; Fruh et al., 1995; Hill et al., 1995). This blockade of the TAP molecule leads to a rapid decrease in cell-surface class I MHC molecules and effectively eliminates recognition by antiviral CTL. Human CMV expresses a 21 kDa glycoprotein encoded by the US6 gene (gpUS6). Unlike ICP47, gpUS6 immunoprecipitates with a transient assembly complex comprised of the TAP heterodimer, class I heavy chain, β2m, tapasin and calreticulin. Thus, rather than physically blocking the translocation pore, gpUS6 seems to inhibit peptide translocation at a point beyond peptide association at the cytosolic mouth of the TAP molecule. Other viruses also affect the TAP molecule. For instance, EBNA-1-positive cells have reduced levels of TAP (Rowe et al., 1995), although the exact mechanism for this reduction in TAP expression is not known. Human adenovirus 12-transformed cells demonstrate dramatically decreased levels of class I heavy chain, TAP1 and LMP2 mRNA, reflecting a transcriptional elimination of these species and downregulation of cell-surface class I MHC molecules (Rotem-Yehuder et al., 1996; Proffitt and Blair, 1997).

Dislocation of class I molecules from the endoplasmic reticulum to the cytosol

Human and murine CMV cause a downregulation of class I MHC molecules on the surface of infected cells, partly by destabilizing class I molecules in the ER. This phenomenon occurs despite normal class I mRNA production and stability. In particular, human CMV produces two glycoproteins, gpUS2 and gpUS11, which act to destabilize class I by different mechanisms. Both polypeptides dislocate class I heavy chains from the ER to the cytosol, where they are rapidly degraded by the proteasome (Jones et al., 1995; Wiertz et al., 1996a, b). It appears that gpUS11 interferes with translocation of the nascent class I heavy chain into the lumen of the ER either by interfering with the molecular-ratchet mechanism of translocation pore or by dysregulating the stop-transfer signal. gpUS2 seems to act at a later stage, since normal translocation of the nascent class I heavy chain can occur in the presence of this viral glycoprotein. It therefore seems that gpUS2 recruits translocated class I heavy chain and dislocates it back through the translocation pore where both polypeptides are rapidly degraded.

Alteration of class I trafficking

The HIV-1 *vpu* gene product is a phosphoprotein with two established functions: degradation of the viral coreceptor CD4 in the ER and a role in viral particle release from the infected cell. Recently, it has been shown to decrease expression of class I MHC molecules (Kerkau et al., 1997). This effect is restricted to rapid degradation of nascent class I MHC molecules in the ER and is not due to attenuation of the peptide supply. It is possible vpu-induced decrease of class I MHC molecules may contribute to the inability of $CD8^+$ T cells to eradicate HIV-1 from infected individuals. Likewise, the HIV-1 Nef protein has also recently been demonstrated to downregulate cell-surface class I molecules (Le Gall et al., 1997). This may be caused by stimulation of class I endocytosis in a manner similar to that reported for CD4, since class I MHC molecules matured normally and trafficked through the Golgi, but were rapidly internalized, accumulated in endosomal vesicles and ultimately degraded in Nef^+ infected cells (Schwartz et al., 1996).

Murine CMV differs from human CMV in that it promotes the accumulation of relatively stable class I–peptide complexes. gpm152 is a 40 kDa glycoprotein expressed by murine CMV responsible for the retention of such complexes in an intermediate compartment between the ER and the Golgi (Thäle et al., 1995; Ziegler et al., 1997). It appears that gpm152 transiently associates with class I complexes and renders them unable to undergo normal egress through the ER–Golgi network to the cell surface. A human CMV protein, gpUS3, also appears to retain class I in a similar ER compartment (Ahn et al., 1996; Jones et al., 1996). It is worthwhile noting that this process is relatively independent of the influences exerted by the gpUS11 and gpUS2 molecules. Similarly, adenovirus expresses a 19 kDa protein, E3, which also retains MHC–peptide complexes in the ER and Golgi apparatus (Burgert and Kvist, 1985). In contrast to these molecules which retain or dislocate class I complexes during maturation in the ER, the murine CMV gp34 molecule escorts class I MHC molecules to the cell surface, yet blocks T cell recognition of these complexes (Kleijnen et al., 1997).

Repression of natural killer cell activity and expression of class I homologues

Downregulation of MHC I surface complexes renders virally infected cells sensitive to killing by natural killer (NK) cells. Thus, human and murine CMV also synthesize class I analogues (gpUL18 and gpm144, respectively), which are proposed to help virally infected cells to evade NK cell killing (Fahnestock et al., 1995; Farrell et al., 1997). The expression of these analogues on the cell surface allows the NK cell receptor to be engaged and renders the infected host cells safe from lysis. These analogues bind to β2m and the gpUL18 appears to bind self-peptides in a manner similar to conventional class I molecules. It has been reported recently that peptide profiles of species associated with gpUL18 and normal class I are similar, whereas no peptides were detected associated with the murine CMV gpm144 molecule (Fahnestock et al., 1995; Chapman and Bjorkman, 1998).

Unlike the other viruses, flavivirus upregulates class I MHC expression, owing to a TAP-independent translocation of peptides into the ER, leading to a pathological loss of self-tolerance (Müllbacher and Lobigs, 1995). This translocation of peptides is not restricted to flavivirus peptides and is independent of IFNγ expression. Finally, there is the well-described ability of viruses to escape immune recognition by mutation of immunodominant epitopes or by expressing natural variants of cytotoxic epitopes with antagonist activity for antiviral CTL (Bertoletti et al., 1994; Klenerman et al., 1994).

CONCLUSION

It is interesting that each group of viruses has developed multiple mechanisms to evade class I restricted immunity. This presumably reflects the viruses' response to the polymorphism of the class I pathway. For example, some class I allotypes are more resistant to gpUS11 dislocation than others. Other alleles seem more stable and bind peptides delivered via TAP-independent pathways more efficiently than others. It seems only a matter of time before viral proteins will be identified which are responsible for inhibiting other parts of the Ag processing pathway. In addition, molecules may be identified as potential inhibitors of class II processing and, like the class I pathway, several key participants in the pathway may be targeted (HLA–DM, Ii, proteins associated with lysosomal acidification, etc.). This is especially apt since many viruses are susceptible to surveillance by CD4$^+$ T cells and class II expression is significantly decreased in several types of infection. The identification of these viral polypeptides provides a new emphasis for antiviral drug design but, perhaps more interestingly, these species may prove to be of more use as immunotherapeutics for certain infectious and autoimmune diseases.

ACKNOWLEDGEMENTS

This work was supported by the NHMRC, Australia, Australian Research Council and the Arthritis Foundation of Australia. C.A.P. is an NHMRC postgraduate scholar. We are grateful to Andrew Hillam for help with the illustrations.

REFERENCES

Ahn, K., Angulo, A., Ghazal, P. et al. (1996). Human cytomegalovirus inhibits antigen presentation by a sequential multistep process. *Proc. Natl Acad. Sci. USA* **93**: 10 990–10 995.

Allen, P.M. and Unanue, E.R. (1984). Differential requirements for antigen processing by macrophages for lysozyme-specific T cell hybridomas. *J. Immunol.* **132**: 1077.

Amigorena, S., Bonnerot, C., Drake, J.R. et al. (1992). Cytoplasmic domain heterogeneity and functions of IgG Fc receptors in B lymphocytes. *Science* **256**: 1808–1812.

Amigorena, S., Drake, J.R., Webster, P. et al. (1994). Transient accumulation of new MHC molecules in a novel endocyic compartment in B lymphocytes. *Nature* **369**: 113–120.

Anderson, K., Cresswell, P., Gammon, M. et al. (1992). Endogenously synthesised peptide with an endoplasmic reticulum signal sequence sensitises antigen processing mutant cells to class I-restricted cell-mediated lysis. *J. Exp. Med.* **174**: 489–492.

Androlewicz, M.J. and Cresswell, P. (1994). Human transporters associated with antigen processing possess a promiscuous peptide-binding site. *Immunity* **1**: 7–14.

Androlewicz, M.J. and Cresswell, P. (1996). How selective is the transporter associated with antigen processing? *Immunity* **5**: 1–5.

Androlewicz, M.J., Ortmann, B., Endert, P.M. van et al. (1994). Characteristics of peptide and major histocompatiblity complex class I/β2m-microglobulin binding to the transporters associated with antigen processing (TAP1 and TAP2). *Proc. Natl Acad. Sci. USA* **91**: 12 716–12 720.

Arnold, D., Driscoll, J., Androlewicz, M. et al. (1992). Proteosome subunits encoded in the MHC are not generally required for the processing of peptides bound by MHC class I molecules. *Nature* **360**: 171–173.

Arnold, D., Faath, S., Rammensee, H.-G. et al. (1995). Cross-priming of minor histocompatibility antigen-specific cytotoxic T cells upon immunisation with the heat shock protein gp96. *J. Exp. Med.* **182**: 885–889.

Bacik, I., Snyder, H.L., Anton, L.C. et al. (1997). Introduction of a glycosylation site into a secreted protein provides evidence for an alternative antigen processing pathway: transport of precursors of major histocompatibility complex class I-restricted peptides from the endoplasmic reticulum to the cytosol. *J. Exp. Med.* **186**: 479–487.

Bellone, M., Iezzi, G., Rovere, P. et al. (1997). Processing of engulfed apoptotic bodies yields T cell epitopes. *J. Immunol.* **159**: 5391–5399.

Bersofsky, J.A., Brett, S.J., Streicher, H.Z. et al. (1988). Antigen processing for presentation to T lymphocytes: Function, mechanisms and implications for the T-cell repertoire. *Immunol. Rev.* **106**: 5.

Bertoletti, A., Sette, A., Chisari, F.V. et al. (1994). Natural variants of cytotoxic epitopes are T cell receptor antagonist for antiviral cytotoxic T cells. *Nature* **369**: 407–410.

Bevan, M.J. (1976). Cross-priming for a secondary cytotoxic response to minor H antigens with *H-2* congenic cells which do not cross-react in the cytotoxic assay. *J. Exp. Med.* **143**: 1283–1288.

Bevan, M.J. (1995). Antigen presentation to cytotoxic T lymphocytes in vivo. *J. Exp. Med.* **182**: 639–641.

Brooks, A. and McCluskey, J. (1993). Class II-restricted presentation of a hen egg lysozyme determinant derived from endogenous antigen sequestered in the cytoplasm or endoplasmic reticulum of the antigen presenting cells. *J. Immunol.* **150**: 3690–3697.

Brooks, A., Hartley, S., Kjer-Nielsen, L. et al. (1991). Class II-restricted presentation of an endogenously-derived immunodominant T cell determinant of hen egg lysozyme. *Proc. Natl Acad. Sci. USA* **88**: 3290–3294.

Brooks, A.G., Campbell, P.L., Reynolds, P. et al. (1994). Antigen presentation and assembly by mouse I-Ak class II molecules in human APC containing deleted or mutated HLA DM genes. *J. Immunol.* **153**: 5382–5392.

Brown, M.G., Driscoll, J. and Monaco, J.J. (1991). Structural and serological similarity of MHC-linked LMP and proteosome (multicatalytic proteinase) complexes. *Nature* **353**: 355–357.

Burgert, H.-G. and Kvist, SD. (1985). An adenovirus type 2 glycoprotein blocks cell surface expression of histocompatibility class I antigens. *Cell* **41**: 987–997.

Buus, S. and Wederlin, O. (1986). A group-specific inhibitor of lysosmal proteinases selectively inhibits both proteolytic degradation and presentation of the antigen DNP-poly-L-lysine by guinea pig accessory cells to T cells. *J. Immunol.* **136**: 459.

Carbone, F.R. and Bevan, M.J. (1990). Class I-restricted processing and presentation of exogenous cell-associated antigen in vivo. *J. Exp. Med.* **171**: 377–387.

Castellino, F., Zhong, G. and Germain, R.N. (1997). Antigen presentation by MHC class II molecules: invariant chain function, protein trafficking and the molecular basis of diverse determinant capture. *Hum. Immunol.* **54**: 159–169.

Cella, M., Engering, A., Pivet, V. et al. (1997a). Inflammatory stimuli induce accumulation of MHC class II complexes on dendritic cells. *Nature* **388**: 782–787.

Cella, M., Sallusto, F., and Lanzavecchia, A. (1997b). Origins, maturation and antigen presenting function of dendritic cells. *Curr. Opin. Immunol.* **9**: 10–16.

Cerundolo, V., Kelly, A., Elliott, T. et al. (1995). Genes encoded in the major histocompatibility complex affecting the generation of peptides for TAP transport. *Eur. J. Immunol.* **25**: 554–562.

Chain, B.M., Kaye, P.M. and Shaw, M-A. (1988). The biochemistry and cell biology of antigen processing. *Immunol. Rev.* **106**: 33.

Chapman, T.L. and Bjorkman, P.J. (1998). Characterization of a murine cytomegalovirus class I major histocompatibility complex (MHC) homolog: comparison to MHC molecules and to the human cytomegalovirus MHC homolog. *J. Virol.* **72**: 460–466.

Chen, H.L., Gabrilovich, D., Tampe, R. et al. (1996). A functionally defective allele of TAP1 results in loss of MHC class I antigen presentation in a human lung cancer. *Nature Genet.* **13**: 210–213.

Chen, W., Khilko, S., Fecondo, J. et al. (1994). Determinant selection of major histocompatiblity complex class I-restricted antigenic peptides is explained by class I-peptide affinity and is strongly influenced by nondominant anchor residues. *J. Exp. Med.* **180**: 1471–1483.

Chesnut, R.W., Colon, S.M. and Grey, H.M. (1982a). Requirements for the processing of antigens by antigen-presenting B cells. I. Functional comparison of B cell tumors and macrophages. *J. Immunol.* **129**: 2382.

Chesnut, R.W., Colon, S.M. and Grey, H.M. (1982b). Antigen presentation by normal B cells, B cell tumors, and macrophages: functional and biochemical comparison. *J. Immunol.* **128**, 1764.

Chicz, R.M., Urban, R.G., Lane, W.S. et al. (1992). Predominant naturally processed peptides bound to HLA-DR1 are derived from MHC-related molecules and are heterogenous in size. *Nature* **358**: 764.

Chicz, R.M., Urban, R.G., Gorga, J.C. et al. (1993). Specificity and promiscuity among naturally processed peptides bound to HLA-DR alleles. *J. Exp. Med.* **178**: 27.

Chicz, R.M., Lane, W.S., Robinson, R.A. et al. (1994). Self-peptides bound to the type I diabetes associated class II MHC molecules HLA-DQ1 and HLA-DQ8. *Int. Immunol.* **6**: 1639.

Cresswell, P. (1994). Assembly, transport, and function of MHC class II molecules. *Annu. Rev. Immunol.* **12**: 259.

Cresswell, P. (1996). Invariant chain structure and MHC class II function. *Cell* **84**: 505–507.

D'Ambrosio, D., Hippen, K.L., Minskoff, S.A. et al. (1995). Recruitment and activation of PTP1C in negative regulation of antigen receptor signalling by Fc gamma RIIB1. *Science* **268**: 293–297.

Davidson, H.W. and Watts, C. (1989). Epitope-directed processing of specific antigen by B lymphocytes. *J. Cell. Biol.* **109**: 85–92.

De la Salle, H., Hanau, D., Fricker, D. et al. (1994). Homozygous human TAP peptide transporter mutation in HLA class I deficiency. *Science* **265**: 237–241.

Degen, E., Cohen-Doyle, M.F. and Williams, D.B. (1992). Efficient dissociation of the p88 chaperone from major histocompatibility complex class I molecules requires both b2-microglobulin and peptide. *J. Exp. Med.* **175**: 1653–1661.

Del Val, M., H. Hengel, H., Macher, U. et al. (1992). Cytomegalovirus prevents antigen presentation by blocking the transport of peptide loaded major histocompatibility complex class I molecules into the medial-golgi compartment. *J. Exp. Med.* **176**: 729–738.

Dempsey, P.W., Allison, M.E.D., Akkaraiu, S. et al. (1996). C3d of complement as a molecular adjuvant: bridging innate and acquired immunity. *Science* **271**: 348–350.

Denzin, L.K. and Cresswell, P. (1995). HLA-DM induces CLIP dissociation from MHC class II dimers and facilitates peptide loading. *Cell* **82**: 155–165.

Denzin, L.K., Hammond, C. and Cresswell, P. (1996). HLA-DM interactions with intermediates in HLA-DR maturation and a role for HLA-DM in stabilizing empty HLA-DR molecules. *J. Exp. Med.* **184**: 2153–2165.

Deverson, E.V., Gow, I.R., Coadwell, W.J. et al. (1990). MHC class II region encoding proteins related to the multidrug resistance family of transmembrane transporters. *Nature* **348**: 739–741.

Dick, T.P., Ruppert, T., Groettrup, M. et al. (1996). Co-ordinated dual cleavages induced by the protesome regulator PA28 lead to dominant MHC ligands. *Cell* **86**: 253–262.

Driscoll, J., Brown, M.G., Finley, D. et al. (1993). MHC-linked LMP gene products specifically alter peptidase acivities of the proteosome. *Nature* **365**: 262–264.

Dubiel, W., Pratt, G., Ferrell, K. et al. (1992). Purification of an 11S regulator of the multicatalytic protease. *J. Biol. Chem.* **267**: 22 369–22 377.

Eisenlohr, L.C., Bacik, I., Bennink, J.R. et al. (1992). Expression of a membrane protease enhances presentation of endogenous antigens to MHC class I-restricted T lymphocytes. *Cell* **71**: 963–972.

Elliott, T., Willis, A., Cerundolo, V. et al. (1995). Processing of major histocompatibility class I-restricted antigens in the endoplasmic reticulum. *J. Exp. Med.* **181**: 1481–1491.

Fahnestock, M.L., Johnson, J.L., Feldman, R.M.R. et al. (1995). The MHC class I homolog encoded by human cytomegalovirus binds endogenous peptides. *Immunity* **3**: 583–590.

Falk, K., Rotschke, O. and Rammensee, H.-G. (1990). Cellular peptide composition governed by major histocompatibility complex class I molecules. *Nature* **348**: 248–251.

Falk, K., Rotzschke, O., Stevanovic, S. et al. (1991). Allele-specific motifs revealed by sequencing of self peptides eluted from MHC molecules. *Nature* **351**: 290–296.

Falk, K., Rotzschke, O., Stevanovic, S. et al. (1994). Pool sequencing of natural HLA DR, DQ and DP ligands reveals detailed peptide motifs, constraints of processing, and general rules. *Immunogenetics* **39**: 230.

Farrell, H.E., Vally, H., Lynch, D.M. et al. (1997). Inhibition of natural killer cells by a cytomegalovirus MHC class I homologue *in vivo*. *Nature* **386**: 510–514.

Fearon, D.T. and Carter, R.H. (1995). The CD19/CR2/TAPA-1 complex of B lymphocytes: linking natural to acquired immunity. *Annu. Rev. Immunol.* **13**: 127–149.

Fehling, H.J., Swat, W., Laplace, C. et al. (1994). MHC class I expression in mice lacking the proteasome subunit LMP-7. *Science* **265**: 1234–1237.

Fruh, K., Ahn, K., Djaballah, H. et al. (1995). A viral inhibitor of peptide transporters for antigen presentation. *Nature* **375**: 415–418.

Gaczynska, M., Rock, K.L. and Goldberg, A.L. (1993). Gamma-interferon and expression of MHC genes regulate peptide hydrolysis by proteasomes. *Nature* **365**: 264–267.

Gell, P.G.H. and Benacerraf, B. (1959). Studies on hypersensitivity. II. Delayed hypersensitivity to denatured proteins in guinea pigs. *Immunology* **2**: 64.

Germain, R.N. and Malissen, B. (1986). Analysis of the expression and function of class-II major histocompatibility complex-encoded molecules by DNA-mediated gene transfer. *Annu. Rev. Immunol.* **4**: 281.

Ghosh, P., Amaya, M., Mellins, E. and Wiley, D.C. (1995). The structure of an intermediate in class II MHC maturation: CLIP bound to HLA-DR3. *Nature* **378**: 457–462.

Gilbert, M.J., Riddell, S.R., Plachler, B. et al. (1996). Cytomegalovirus selectively blocks antigen presentation of its intermediate-early gene product. *Nature* **383**: 720–722.

Glimcher, L., Kim, H., Green, K.J. et al. (1982). Ia antigen-bearing B cell tumour lines can present protein antigen and alloantigen in a major histocompatibility complex-restricted fashion to antigen-reactive T cells. *J. Exp. Med.* **155**: 445.

Glynne, R., Powis, S.H., Beck, S. et al. (1991). A proteosome-related gene between the two ABC transporter loci in the class II region of the human MHC. *Nature* **353**: 357–360.

Goldberg, A.L. and Rock, K.L. (1992). Proteolysis, proteosomes and antigen presentation. *Nature* **357**, 375–379.

Grandea III, A.G., Androlewicz, M.J., Athwal, R.S. et al. (1995). Dependence of peptide binding by MHC class I molecules on their interaction with TAP. *Science* **270**: 105–108.

Greenberg, S. (1995). Signal transduction in phagocytosis. *Trends Cell Biol.* **5**: 93–99.

Greenwood, R., Shimizu, Y., Sekhon, G.S. et al. (1994). Novel allele-specific, post-translational reduction in HLA class I surface expression in a mutant human B cell line. *J. Immunol.* **153**: 5525–5536.

Griffin, T.A., Nandi, D., Cruz, M. et al. (1998). Immunoproteasome assembly: cooperative incorporation of interferon gamma (IFN-gamma)-inducible subunits. *J. Exp. Med.* **187**: 97–104.

Groettrup, M., Ruppert, T., Kuehn, L. et al. (1995). The interferon-γ-inducible 11S regulator (PA28) and the LMP2/LMP7 subunits govern the peptide production by the 20S proteasome in vitro. *J. Biol. Chem.* **270**: 23 808–23 815.

Groettrup, M., Soza, A., Eggers, M. et al. (1996a). A role for the proteasome regulator PA28α in antigen presentation. *Nature* **381**: 166–168.

Groettrup, M., Kraft, R., Kostka, S. et al. (1996b). A third interferon-γ-induced subunit exchange in the 20S proteasome. *Eur. J. Immunol.* **26**: 863–869.

Groll, M., Ditzel, L., Lowe, L. et al. (1997). Structure of 20S proteosome from yeast at 2.4 Å resolution. *Nature* **386**: 463–471.

Guagliardi, L.E., Koppelman, B., Blum, J.S. et al. (1990). Co-localization of molecules involved in antigenprocessing and presentation in an early endocytic compartment. *Nature* **343**: 133.

Hahn, Y.S., Yang, B. and Braciale, T.J. (1996). Regulation of antigen processing and presentation to class I MHC restricted CD8[+] T lymphocytes. *Immunol. Rev.* **151**: 31–49.

Hammond, S.A., Johnson, R.P., Kalams, S.A. et al. (1995). An epitope-selective, transporter associated with antigen presentation (TAP)-1/2-independent pathway and a more general (TAP)-1/2-dependent antigen-processing pathway allow recognition of the HIV-1 envelope glycoprotein CD8[+] CTL. *J. Immunol.* **154**: 6147–1775.

Heemels, M.-T. and Ploegh, H.L. (1994). Substrate specificity of allelic variants of the TAP pepnsporter. *Immunity* **1**: 775–784.

Henderson, R.A., Michel, H., Sakaguchi, K. et al. (1992). HLA-A2.1-associated peptides from a mutant cell line: a second pathway of antigen presentation. *Science* **255**: 1263–1266.

Hill, A., Jugovich, P., York, I. et al. (1995). Herpes simplex virus turns off the TAP to evade host immunity. *Nature* **375**: 411–415.

Hombach, J., Pircher, H., Tonegawa, S. et al. (1995). Strictly transporter of antigen presentation (TAP)-dependent presentation of an immunodominant cytotoxic T lymphocyte epitope in the signal sequence of a virus protein. *J. Exp. Med.* **182**: 1615–1619.

Huang, A.Y.C., Bruce, A.T., Pardoll, D.M. et al. (1996). In vivo cross-priming of MHC class I-restricted antigens requires the TAP transporter. *Immunity* **4**: 349–355.

Hughes, E.A., Ortmann, B., Surman, M. et al. (1996). The protease inhibitor, N-acetyl-L-Leucyl-L-Leucyl-L-Norleucinal, decreases the pool of major histocompatibility complex class I-binding peptides and inhibits peptide trimming in the endoplasmic reticulum. *J. Exp. Med.* **183**: 1569–1578.

Hughes, E.A., Hammond, C. and Cresswell, P. (1997). Misfolded major histocompatibility complex class I heavy chains are translocated into the cytoplasm and degraded by the proteosome. *Proc. Natl Acad. Sci. USA* **94**: 1896–1901.

Hunt, D.F., Michel, H., Dickinson, T.A. et al. (1992). Peptides presented to the immune system by the murine class II major histocompatibility complex molecule I-A^d. *Science* **256**: 1817.

Hyman, R. and Stallings, V. (1976). Characterisation of a TL$^-$ variant of a homozygous TL$^+$ mouse lymphoma. *Immunogenetics* **3**: 75–84.

Jardetzky, T.S., Lane, W.S., Robinson, R.A. et al. (1991). Identification of self peptides bound to purified HLA-B27. *Nature* **353**: 326–329.

Jiang, W., Swiggard, W.J., Heufler, C. et al. (1995). The receptor DEC-205 expressed by dendritic cells and thymic epithelial cells is involved in antigen processing. *Nature* **375**: 151–155.

Jondal, M., Schirmbeck, R. and Reimann, J. (1996). MHC class I-restricted CTL responses to exogenous antigens. *Immunity* **5**: 295–302.

Jones, T.R., Hanson, L.K., Sun, L. et al. (1995). Multiple independent loci within the human cytomegalovirus unique short region downregulate expression of major histocompatibility complex class I heavy chain. *J. Virol.* **69**: 4830–4841.

Jones, T.R., Wiertz, E.J.H.J., Sun, L. et al. (1996). Human cytomegalovirus US3 impars transport and maturation of major histocompatibility complex class I heavy cahins. *Proc. Natl Acad. Sci. USA* **93**: 11 327–11 333.

Joyce, S., Kuzushima, K., Kepecs, G. et al. (1995). Characterisation of an imcompletely assembled major histocompatibility class I molecule (H-2Kb) associated with unusually long peptides: implications for antigen processing and presentation. *Proc. Natl Acad. Sci.* **91**: 4145–4159.

Kanost, D. and McCluskey, J. (1994). Anergic B cells constitutively present self-antigen: enhanced immunoglobulin-receptor mediated presentation of antigenic determinants by B cells is hierarchical. *Eur. J. Immunol.* **25**: 1186–1193.

Kelly, A., Powis, S.H., Kerr, L.-A. et al. (1992). Assembly and function of the two ABC transporter proteins encoded in the human major histocompatibility complex. *Nature* **355**: 641–644.

Kerkau, T., Bacik, I., Bennink, J.R. et al. (1997). The human immunodeficiency virus type 1 (HIV-1) Vpu protein interferes with an early step in the biosynthesis of major histocompatibility complex (MHC) class I molecules. *J. Exp. Med.* **185**: 1295–1305.

Kleijnen, M.F., Huppa, J.B., Lucin, P. et al. (1997). A mouse cytomegalovirus glycoprotein, gp34, forms a complex with folded class I MHC molecules in the ER which is not retained but is transported to the cell surface. *EMBO J.* **16**: 685–694.

Klenerman, P., Rowland-Jones, S., McAdam, S. et al. (1994). Cytotoxic T cell activity antagonized by naturally occurring HIV-1 Gag variants. *Nature* **369**: 403–407.

Koopmann, J.-O., Post, M., Neefjes, J.J. et al. (1996). Translocation of long peptides by transporters associated with antigen processing (TAP). *Eur. J. Immunol.* **26**: 1720–1728.

Kovacsovics-Bankowski, M. and Rock, K.L. (1995). A phagosome-to-cytosol pathway for exogenous antigens presented on MHC class I molecules. *Science* **267**: 243–246.

Kovacsovics-Bankowski, M., Clark, K., Benacerraf, B. et al. (1993). Efficient major histocompatibility complex class I presentation of exogenous antigen upon phagocytosis by macrophages. *Proc. Natl Acad. Sci. USA* **90**: 4942–4946.

Krause, K.-H. and Michalak, M. (1997). Calreticulin. *Cell* **88**: 439–443.

Kurts, C., Heath, W.R., Carbone, F.R. et al. (1996). Constitutive class I-restricted exogenous presentation of self antigens in vivo. *J. Exp. Med.* **184**: 923–930.

Lanzavecchia, A. (1985). Antigen-specific interaction between T and B cells. *Nature* **314**: 537–539.

Lanzavecchia, A. (1990). Receptor-mediated antigen uptake and its effect on antigen presentation to class II-restricted T lymphocytes. *Annu. Rev. Immunol.* **8**: 773–793.

Lanzavecchia, A., Reid, P.A. and Watts, C. (1992). Irreversible association of peptides with class II MHC molecules in living cells. *Nature* **357**: 249–252.

Lee, K.C., Wong, M. and Spitzer, D. (1982). Chloroquine as a probe of antigen processing by accessory cells. *Transplantation* **34**: 150.

Lee, P., Matsueda, G.R. and Allen, P.M. (1988). T cell recognition of fibrinogen. A determinant on the A$_\alpha$-chain does not require processing. *J. Immunol.* **140**: 1063.

Le Gall, S., Prevost, M.C., Heard, J.M. and Schwartz, O. (1997). Human immunodeficiency virus type I Nef independently affects virion incorporation of major histocompatibility class I molecules and virus infectivity. *Virology* **229**: 295–301.

Lehner, P.J., Surman, M.J. and Cresswell, P. (1998). Soluble tapasin restores MHC class I expression and function in the tapasin-negative cell line .220. *Immunity* **8**: 221–231.

Levitskaya, J., Coram, M., Levitsky, V. et al. (1995). Inhibition of antigen processing by the internal repeat region of the Epstein–Barr virus nuclear antigen-1. *Nature* **375**: 685–688.

Lewis, J.W., Neisig, A. and Neefjes, J. (1996). Point mutations in the alpha 2 domain of HLA-A2.1 define a functionally relevant interaction with TAP. *Current Biology* **6**: 873–883.

Lin, J., Bersofsky, J.A. and Delovitch, T.I. (1988). Ultrastructural study of internalization and recycling of antigen by antigen presenting cells. *J. Mol. Cell. Immunol.* **3**: 321.

Liu, T., Chambers, B., Diehl, A.D. et al. (1997). TAP peptide transporter-independent presentation of heat-killed Sendai virus antigen on MHC class I molecules by splenic antigen-presenting cells. *J. Immunol.* **159**: 5364–5371.

Livingstone, A.M., Powis, S.J., Diamond, A.G. et al. (1991). A trans-acting major histocompatability complex-linked gene whose alleles determine gain and loss changes in the antitgenic structure of a classical class I molecule. *J. Exp. Med.* **170**: 777–795.

Ljunggren, H., Stam, N.J., Ohlen, C. et al. (1990). Empty MHC class I molecules come out in the cold. *Nature* **346**: 476–480.

Lorenz, R.G., Tyler, A.N. and Allen, P.M. (1988). T cell recognition of bovine ribonuclease: self/non-self discrimination at the level of binding to the I-Ak molecule. *J. Immunol.* **141**: 4124.

Lowe, J., Stock, D., Jap, B. et al. (1995). Crystal structure of the 20S proteasome from the archaeon T.acidophilum at 3.4 Å resolution. *Science* **268**: 533–539.

Lutz, M.B., Assman, C.U., Gi, G. et al. (1996). Different cytokines regulate antigen uptake and presentation of a precursor dendritic cell line. *Eur. J. Immunol.* **26**: 586–594.

Ma, C.P., Slaughter, C.A. and DeMartino, G.N. (1992). Identification, purification, and characterisation of a protein activator (PA28) of the 20S proteasome (macropain). *J. Biol. Chem.* **267**: 10 515–10 523.

Ma, W., Lehner, P.J., Cresswell, P. et al. (1997). Interferon-γ rapidly increases peptide transporter (TAP) subunit expression and peptide transport capacity in endothelial cells. *J. Biol. Chem.* **272**: 16 585–16 590.

Madden, D.R. (1995). The three-dimensional structure of peptide–MHC complexes. *Annu. Rev. Immunol.* **15**: 587–622.

Martinez, C.K. and Monaco, J.J. (1991). Homology of proteosome subunits to a major histocompatibility complex-linked LMP gene. *Nature* **353**: 664–667.

Matsui, M., Warburton, R.J., Cogswell, P.C. et al. (1996). Effects of HIV-1 Tat on expression of HLA class I molecules. *J. Aquir. Immune Defic. Syndr. Hum. Retrovirol.* **11**: 233–240.

Maudsley, D.J. and Pound, J.D. (1991). Modulation of MHC antigen expression by viruses and oncogenes. *Immunol. Today* **12**: 429–431.

Meadows, L., Wang, W., Haan, J. den et al. (1997). The HLA-A*0201-restricted H-Y antigen contains a posttranslationally modified cysteine that significantly affects T cell recognition. *Immunity* **6**: 273–281.

Metlay, J.P., Pure, E. and Steinman, R.M. (1989). Control of the immune response at the level of antigen-presenting cells: a comparison of the function of dendritic cells and B lymphocytes. *Adv. Immunol.* **47**: 45.

Michalek, M.T., Grant, E.P., Gramm, C. et al. (1993). A role for the ubiquitin-dependent proteolytic pathway in MHC class I-restricted antigen presentation. *Nature* **363**: 552–554.

Mills, K.H.G., Skehel, J.J. and Thomas, D.B. (1986). Conformational-dependent recognition of influenza virus hemagglutinin by murine T helper clones. *Eur. J. Immunol.* **16**: 276.

Momburg, F., Ortiz-Navarrete, V., Neefjes, J. et al. (1992). Proteosome subunits encoded by the major histocompatibility complex are not essential for antigen presentation. *Nature* **360**: 174–177.

Momburg, F., Roelse, J., Hammerling, G.J. et al. (1994a). Peptide size selection by the major histocompatibility complex-encoded peptide transporter. *J. Exp. Med.* **179**: 1613–1623.

Momburg, F., Roelse, J., Howard, J. et al. (1994b). Selectivity of MHC-encoded peptide transporters from human, mouse and rat. *Nature* **367**: 648–651.

Momburg, F., Armandola, E.A., Post, M. et al. (1996). Residues in TAP2 peptide transporters controlling substrate specificity. *J. Immunol.* **156**: 1756–1763.

Müllbacher, A. and Lobigs, M. (1995). Up-regulation of MHC class I by flavivirus-induced peptide translocation into the endoplasmic reticulum. *Immunity* **3**: 207–214.

Neefjes, J.J. and Ploegh, H.L. (1992). Inhibition of endosomal proteolytic activity by leupeptin blocks surface expression of MHC class II molecules and their conversion to SDS resistance alpha beta heterodimers in endosomes. *EMBO J.* **11**: 411–416.

Neefjes, J., Gottfried, E., Roelse, J. et al. (1995). Analysis of the fine specificity of rat, mouse and human TAP peptide transporters. *Eur. J. Immunol.* **25**: 1133–1136.

Neisig, A., Roelse, J., Sijts, A.J.A.M. et al. (1995). Major differences in transporter associated with antigen presentation (TAP)-dependent translocation of MHC class I-presentable peptides and the effect of flanking sequences. *J. Immunol.* **154**: 1273–1279.

Nepom, G.T. and Ehrlich, H. (1991). MHC class-II molecules and autoimmunity. *Annu. Rev. Immunol.* **9**: 493–525.

Niedermann, G., Butz, S., Ihlenfeldt, H.G. et al. (1995). Contribution of proteosome-mediated proteolysis to the hierarchy of epitopes presented by major histocompatibility complex class I molecules. *Immunity* **2**: 289–299.

Niedermann, G., King, G., Butz, S. et al. (1996). The proteolytic fragments generated by vertebrate proteosomes: structural relationshops to major histocompatibility complex class I binding peptides. *Proc. Natl Acad. Sci. USA* **93**: 8572–8577.

Nijenhuis, M. and Hammerling, G.J. (1996). Multiple regions of the transporter associated with antigen processing (TAP) contribute to its peptide binding site. *J. Immunol.* **157**: 5467–5477.

Nöbner, E. and Parham, P. (1995). Species-specific differences in chaperone interaction of human and mouse major histocompatibility complex class I molecules. *J. Exp. Med.* **181**: 327–337.

Noesel, C.J.M. van, Brouns, G.S., van Schijndel, G.M.W. et al. (1992). Comparison of human B cell antigen receptor complexes: membrane-expressed forms of immunoglobulin (Ig)M, IgD, and IgG are associated with structurally related heterodimers. *J. Exp. Med.* **175**: 1511–1519.

Norbury, C.C., Chambers, B.J., Prescott, A.R. et al. (1996). Constitutive macropinocytosis allows TAP-dependent presentation of exogenous antigen on class I MHC molecules by bone marrow derived dendritic cells. *Eur. J. Immunol.* **27**: 280–288.

Obst, R., Armandola, E.A., Nijenhuis, M. et al. (1995). TAP polymorphism does not influence transport of peptide variants in mice and humans. *Eur. J. Immunol.* **25**: 2170–2176.

Ortiz-Navarrete, V. and Hammerling, G.J. (1991). Surface appearance and instability of empty H-2 class I molecules under physiological conditions. *Proc. Natl Acad. Sci. USA* **88**: 3594–3497.

Ortmann, B., Androlewicz, M.J. and Cresswell, P. (1994). MHC class I/β2-microglobulin complexes associate with TAP transporters before peptide binding. *Nature* **368**: 864–867.

Ortmann, B., Copeman, J., Lehner, P.J. et al. (1997). A critical role for tapasin in the assembly and function of multimeric MHC class I-TAP complexes. *Science* **277**: 1306–1309.

Ossendorp, F., Eggers, M., Neisig, A. et al. (1996). A single residue exchange within a viral CTL epitope alters proteosome-mediated degradation resulting in lack of antigen presentation. *Immunity* **5**: 115–124.

Owen, M.J., Kissonerghis, A.-M. and Lodish, H.F. (1980). Biosynthesis of HLA-A and HLA-B antigens in vivo. *J. Biol. Chem.* **255**: 9678–9684.

Parikh, V.S., Bishop, G.A., Liu, K.-J. et al. (1992). Differential transmembranal structure–function requirements of the transmembranal domain of the B cell antigen receptor. *J. Exp. Med.* **176**: 1025–1031.

Patel, K.J. and Neuberger, M.S. (1993). Antigen presentation by the B cell antigen receptor is driven by the alpha-beta sheath and occurs independently of its cytoplasmic tyrosines. *Cell* **74**: 939–946.

Peace-Brewer, A.L., Tussey, L.G., Matsui, M. et al. (1996). A point mutation in HLA-A*0201 results in failure to bind the TAP complex and to present virus-derived peptides to CTL. *Immunity* **4**: 505–514.

Peh, C.A., Burrows, S.R., Barnden, M. et al. (1998). HLA-B27 restricted antigen presentation in the absence of taparin reveals polymorphism in mechanisms of HLA class I peptide loading. *Immunity* **8**: 531–542.

Pierre, P., Denzin, L.K., Hammond, C. et al. (1996). HLA-DM is localized to conventional and unconventional MHC class II-containing endocytic compartments. *Immunity* **4**: 229–239.

Pieters, J. (1997). MHC class II restricted antigen presentation. *Curr. Opin. Immunol.* **9**: 89–96

Pieters, J.P., Neefjes, J.J., Oorschot, V. et al. (1991a). Segregation of MHC class II molecules from MHC class I molecules in the Golgi complex for transport to lysosomal compartments. *Nature* **349**: 669–676.

Pieters, J., Horstmann, H., Bakke, O. et al. (1991b). Intracellular transport and localization of major histocompatibility complex class II molecules and associated invariant chain. *J. Cell Biol.* **115**: 1213–1223.

Pfeifer, J.D., Wick, M.J., Roberts, R.L. et al. (1993). Phagocytic processing of bacterial antigens for class I MHC presentation to T cells. *Nature* **361**: 359–362.

Powis, S.J., Young, L.L., Joly, E. et al. (1996). The rat cim effect: TAP allele-dependent changes in a class I MHC anchor motif and evidence against C-terminal trimming of peptides in the ER. *Immunity* **4**: 159–165.

Proffitt, J.A. and Blair, G.E. (1997). The MHC-encoded TAP1/LMP2 bidirectional promotor is down-regulated in highly oncogenic adenovirus type 12 transformed cells. *FEBS Lett.* **400**: 141–144.

Puri, J., Lonai, P. and Friedman, V. (1986). Antigen/Ia interaction and the proteolytic processing of antigen: the structure of the antigen determines its restriction to the A or E molecules of the MHC. *Eur. J. Immunol.* **16**: 1093.

Qui, Y., Xu, X., Wandinger-Ness, A. et al. (1994). Separation of subcellular compartments containing distinct functional forms of MHC class II. *J. Cell Biol.* **125**: 595–605.

Rammensee, H.-G., Falk, K. and Rotzschke, O. (1993). Peptides naturally presented by MHC class I molecules. *Annu. Rev. Immunol.* **11**: 213–244.

Rammensee, H.-G., Friede T. and Stevanovic, S. (1995). MHC ligands and peptide motifs: first listing. *Immunogenetics* **41**: 178–228.

Riese, R.J., Wolf, P.R., Bromme, D. et al. (1996). Essential role for cathepsin S in MHC class II-associated invariant chain processing and peptide loading. *Immunity* **4**: 357–366.

Roake, J.A., Rao, A.S., Morris, P.J. et al. (1995). Dendritic cell loss from nonlymphoid tissues after systemic administration of lipopolysaccharide, tumor necrosis factor, and interleukin 1. *J. Exp. Med.* **181**: 2237–2247.

Rock, K.L., Benacerraf, B. and Abbas, A.K. (1984). Antigen presentation by hapten-specific B lymphocytes I: Role of surface immunoglobulin receptors. *J. Exp. Med.* **160**: 1102–1125.

Rock, K.L. (1996). A new foreign policy: MHC class I molecules monitor the outside world. *Immunol. Today* **17**: 131–137.

Roelse, J., Gromme, M., Momburg, F. et al. (1994). Trimming of TAP-translocated peptides in the endoplasmic reticulum and in the cytosol during recycling. *J. Exp. Med.* **180**: 1591–1597.

Rotem-Yehuder, R., Groettrup, M., Soza, A. et al. (1996). LMP-associated proteolytic activities and TAP-dependent peptide transport for class 1 MHC molecules are suppressed in cell lines transformed by the highly oncogenic adenovirus 12. *J. Exp. Med.* **183**: 499–514.

Rowe, M., Khanna, R., Jacob, C.A. et al. (1995). Restoration of endogenous antigen processing in Burkitt's lymphoma cells by Epstein–Barr virus latent memebrane protein-1: coordinate up-regulation of peptide transporters and HLA-class I antigen expression. *Eur. J. Immunol.* **25**: 1374–1384.

Rudensky, A., Preston-Hurlburt, P., Hong, S.C. et al. (1991). Sequence analysis of peptides bound to MHC class II molecules. *Nature* **353**: 622.

Rudensky, A., Preston-Hurlburt, P., al-Ramadi, B.K. et al. (1992). Truncation variants of peptides isolated from MHC class II molecules suggest sequence motifs. *Nature* **359**: 429.

Sadasivan, B.K., Cariappa, A., Waneck, G.L. et al. (1995). Assembly, peptide loading, and transport of MHC class I molecules in a calnexin-negative cell line. *Cold Spring Harbor Symp. Quant. Biol.* **LX**: 267–275.

Sadasivan, B., Lehner, P.J., Ortmann, B. et al. (1996). Roles for calreticulin and a novel glycoprotein, tapasin, in the interaction of MHC class I molecules with TAP. *Immunity* **5**: 103–114.

Sallusto, F. and Lanzavecchia, A. (1994). Efficient presentation of soluble antigen by cultured human dendritic cells is maintained by granulocyte/macrophage colony-stimulating factor plus interleukin 4 and downregulated by tumor necrosis factor alpha. *J. Exp. Med.* **179**: 1109–1118.

Sallusto, F., Cella, M., Danieli, C. et al. (1995). Dendritic cells use macropinocytosis and the mannose receptor to concentrate macromolecules in the major histocompatibility complex class II compartment: down regulation by cytokines and bacterial products. *J. Exp. Med.* **182**: 389–400.

Schirmbeck, R. and Reimann, J. (1996). Empty L^d molecules capture peptides from endocytosed hepatitis B surface antigen particles for major histocompatibility complex class I-restricted presentation. *Eur. J. Immunol.* **26**: 2812–2822.

Schirmbeck, R., Melber, K. and Reimann, J. (1995). Hepatitis B virus small surface antigen particles are processed in a novel endosomal pathway for major histocompatibility complex class I-restricted epitope presentation. *Eur. J. Immunol.* **25**: 1063–1070.

Schouten, G.J., Eb, A.J. van der and Zantema, A. (1995). Downregulation of MHC class I expression due to interference with p105-NFkB1 processing by Ad12E1A. *EMBO J.* **14**: 1498–1507.

Schumacher, T.N.M., De Bruijn, M.L.H., Vernie, L.N. et al. (1991). Peptide selection by MHC class I molecules. *Nature* **350**: 703–706.

Schumacher, T.N.M., Kantesaria, D.V., Heemels, M.-T. et al. (1994). Peptide length and sequence specificity of the mouse TAP1/TAP2 translocator. *J. Exp. Med.* **179**: 533–540.

Schwartz, O., Marechal, V., Le Gall, S. et al. (1996). Endocytosis of major histocompatibility complex class I molecules is induced by the HIV-1 Nef protein. *Nature Med.* **2**: 338–342.

Scott, J.E. and Dawson, J.R. (1995). MHC class I expression and transport in a calnexin-deficient cell line. *J. Immunol.* **155**: 143–148.

Seeger, M., Ferrell, K., Frank, R. et al. (1997). HIV-1 tat inhibits the 20 S proteasome and its 11 S regulator-mediated activation. *J. Biol. Chem.* **272**: 8145–8148.

Sercarz, E.E., Lehmann, P.V., Ametani, A. et al. (1993). Dominance and crypticity of T cell antigenic determinants. *Annu. Rev. Immunol.* **11**: 729–766.

Sette, A., Southwood, S., Miller, J. et al. (1995). Binding of major histocompatibility complex class II to the invariant chain-derived peptide, CLIP, is regulated by allelic polymorphism in class II. *J. Exp. Med.* **181**: 677–683.

Shastri, N., Malissen, B. and Hood, L. (1985). Ia-transfected L-cell fibroblasts present a lysozyme peptide but not the native protein to lysozyme-specific T cells. *Proc. Natl Acad. Sci. USA* **82**: 5885.

Shaw, A.C., Mitchell, R.N., Weaver, Y.K. et al. (1990). Mutations of immunoglobulin transmembrane and cytoplasmic domains: effects on intracellular signalling and antigen presentation. *Cell* **63**: 381–392.

Sherman, M.A., Weber, D.A. and Jensen, P.E. (1995). DM enhances peptide binding to class II MHC by release of invariant chain-derived peptide. *Immunity* **3**: 197–205.

Simitsek, P.D., Campbell, D.G., Lanzavecchia, A. et al. (1995). Modulation of antigen processing by bound antibodies can boost or suppress class II major histocompatibility complex presentation of different T cell determinants. *J. Exp. Med.* **181**: 1957–1963.

Skipper, J.C.A., Hendrickson, R.C., Gulden, P.H. et al. (1996). An HLA-A2-restricted tyrosinase antigen on melanoma cells results from posttranslational modification and suggests a novel pathway for processing of membrane proteins. *J. Exp. Med.* **183**: 527–534.

Sloan, V.S., Cameron, P., Porter, G. et al. (1995). Mediation by HLA-DM of dissociation of peptides from HLA-DR. *Nature* **375**: 802–806.

Snyder, H.L., Yewdell, J.W. and Bennink, J.R. (1994). Trimming of antigenic peptides in an early secretory compartment. *J. Exp. Med.* **180**: 2389–2394.

Solheim, J.C., Harris, M.R., Kindle, C.S. et al. (1997). Prominence of β2-microglobulin, class I heavy chain conformation, and tapasin in the interactions of class I heavy chain with calreticulin and the transporter associated with antigen processing. *J. Immunol.* **158**: 2236–2241.

Spies, T. and DeMars, R. (1991). Restored expression of major histocompatability class I molecules by gene transfer of a putative peptide transporter. *Nature* **351**: 323–324.

Spies, T., Bresnahan, M., Bahram, S. et al. (1990). A gene in the human major histocompatibility complex class II region controlling the class I antigen presentation pathway. *Nature* **348**: 744–747.

Srivastava, P.K., Udono, H., Blachere, N.E. et al. (1994). Heat-shock proteins transfer peptides during antigen processing and CTL priming. *Immunogenetics* **39**: 93–98.

Steinman, R.M. and Swanson, J. (1995). The endocytic activity of dendritic cells. *J. Exp. Med.* **182**: 283–288.

Stowasser, R., Kuckelkorn, U., Kraft, R. et al. (1996). 20S proteosomes from LMP7 knock out mice reveals altered proteolytic activities and cleavage site preferences. *FEBS Lett.* **383**: 109–113.

Streicher, H.Z., Berkower, I., Busch, M. et al. (1984a). The role of antigen conformation in determining requirements for antigen processing T cell activation. In *Regulation of The Immune Response*, (eds E. Sercarz, H. Cantor and L. Chess), p. 163, Liss, New York.

Streicher, H.Z., Berkower, I.J., Busch, M. et al. (1984b). Antigen conformation determines processing requirements for T-cell activation. *Proc. Natl Acad. Sci. USA* **81**: 6831.

Sugita, M. and Brenner, M.B. (1994). An unstable b2-microglobulin: major histocompatibility complex class I heavy chain intermediate dissociates from calnexin and then is stabilized by binding peptide. *J. Exp. Med.* **180**: 2163–2171.

Sugita, M. and Brenner, M.B. (1995). Association of the invariant chain with major histocompatibility complex class I molecules directs trafficking to endocytic compartments. *J. Biol. Chem.* **270**: 1443–1448.

Suh, W.-K., Cohen-Doyle, M.F., Fruh, K. et al. (1994). Interaction of MHC class I molecules with the transporter associated with antigen processing. *Science* **264**: 1322–1326.

Suh, W.-K., Mitchell, E.K., Yang, Y. et al. (1996). MHC class I molecules form ternary complexes with calnexin and TAP and undergo peptide-regulated interaction with TAP via their extracellular domains. *J. Exp. Med.* **184**: 337–348.

Suhrbier, A. (1997). Multi-epitope DNA vaccines. *Immunol. Cell Biol.* **75**: 402–408.

Sunshine, G.H., Katz, D.R. and Feldmann, M. (1980). Dendritic cells induce T cell proliferation to synthetic antigens under Ir gene control. *J. Exp. Med.* **152**: 1817.

Swanson, J.A. and Watts, C. (1995). Macropinocytosis. *Trends Cell Biol.* **5**: 424–428.

Tamura, Y., Peng, P., Liu, K. et al. (1997). Immunotherapy of tumors with autologous tumor-derived heat shock protein preparations. *Science* **278**: 117–120.

Thäle, R., Szepan, U., Hengel, H. et al. (1995). Identification of the mouse cytomegalovirus genomic region affecting major histocompatibility complex class I molecule transport. *J. Virol.* **69**: 6098–6105.

Townsend, A., Ohlen, C., Bastin, J. et al. (1989). Association of class I major histocompatiblity heavy and light chains induced by viral peptides. *Nature* **340**: 443–448.

Townsend, A., Elliott, T., Cerundolo, V. et al. (1990). Assembly of MHC class I molecules analysed in vitro. *Cell* **62**: 285–295.

Trowsdale, J., Hanson, I., Mockridge, I. et al. (1990). Sequence encoded in the class II region of the MHC related to the 'ABC' superfamily of transporters. *Nature* **348**: 741–744.

Tulp, A., Verwoerd, D., Dobberstein, B. et al. (1994). Isolation and characterization of the intracellular MHC class II compartment. *Nature* **369**: 120–126.

Udono, H. and Srivastava, P.K. (1994). Comparison of tumour-specific immunogenicities of stress-induced proteins Gp96, Hsp90, and Hsp70. *J. Immunol.* **152**: 5398–5403.

Unanue, E.R. (1984). Antigen presenting function of the macrophage, *Annu. Rev. Immunol.* **2**: 395.

Unanue, E.R. and Allen, P.M. (1987). The basis for the immunoregulatory role of macrophages and other accessory cells. *Science* **236**: 551.

Unanue, E.R. and Askonas, B.A. (1968). Persistence of immunogenicity of antigen after uptake by macrophages. *J. Exp. Med.* **127**: 915.

Unanue, E.R. and Cerottini, J.C. (1970). The immunogenicity of antigen bound to the plasma membrane of macrophages. *J. Exp. Med.* **131**: 711.

Urban, R.G., Chicz, R.M., Lane, W.S. et al. (1994). A subset of HLA-B27 molecules contains peptides much longer than nonamers. *Proc. Natl Acad. Sci. USA* **91**: 1534–1538.

Van Endert, P.M., Riganelli, D., Greco, G. et al. (1995). The peptide-binding motif for the human transporter associated with antigen processing. *J. Exp. Med.* **182**: 1883–1895.

Van Kaer, L., Ashton-Rickardt, P.G., Eichelberger, M. et al. (1994). Altered peptidase and viral-specific T cell response in LMP2 mutant mice. *Immunity* **1**: 533–541.

Venkitaraman, A.R., Williams, G.T., Dariavach, P. et al. (1991). The B-cell antigen receptor of the five immunoglobulin classes. *Nature* **352**: 777–781.

Vinitsky, A., Anton, L.C., Snyder, H.L. et al. (1997). The generation of MHC class I-associated peptides is only partially inhibited by proteosome inhibitors. Involvement of nonproteosomal cytosolic proteases in antigen processing? *J. Immunol.* **159**: 554–564.

Walden, P., Nagy, Z.A. and Klein, J. (1985). Induction of regulatory T lymphocyte responses by liposomes carrying major histocompatibility complex molecules and foreign antigens. *Nature* **315**: 327.

Walker, E., Warner, E.L., Chesnut, R.W. et al. (1982). Antigen specific, *I*-region restricted interactions *in vitro* between tumor cell lines and T cell hybridomas. *J. Immunol.* **128**: 2164.

Watts, C. (1997). Capture and processing of exogenous antigens for presentation on MHC molecules. *Annu. Rev. Immunol.* **15**: 821–850.

Werderlin, O., Mouritsen, S., Laub Petersen, B. et al. (1988). Facts on the fragmentation of antigens in presenting cells, on the association of antigen fragments with MHC molecules in cell free systems, and speculation on the cell biology of antigen processing. *Immunol. Rev.* **106**: 181–193.

Wei, M.L. and Cresswell, P. (1992). HLA-A2 molecules in an antigen-processing mutant cell contain signal sequence-derived peptides. *Nature* **356**: 443–446.

West, M.A., Lucocq, J.M. and Watts, C. (1994). Antigen processing and class II MHC peptide-loading compartments in human B-lymphoblastoid cells. *Nature* **369**: 147–151.

Wick, M.J. and Pfeifer, J.D. (1996). Major histocompatibility complex class I presentation of ovalbumin peptide 257–264 from exogenous sources: proteins context influences the degree of TAP-independent presentation. *Eur. J. Immunol.* **26**: 2790–2799.

Wiertz, E.J.H.J., Tortorella, D., Bogyo, M. et al. (1996a). Sec-61 mediated transfer of a membrane protein from the endoplasmic reticulum to the proteasome for destruction. *Nature* **384**: 432–438.

Wiertz, E.J.H.J., Jones, T.R., Sun, L. et al. (1996b). The human cytomegalovirus US11 gene product dislocates MHC class I heavy chain from the endoplasmic reticulum to the cytosol. *Cell* **84**: 769–779.

Wiertz, E.J.H.J., Mukherjee, S. and Ploegh, H.L. (1997). Viruses use stealth technology to escape from the host immune system. *Molec. Med. Today* **3**: 116–123.

Wraith, D.C. and Vessey, A.E. (1986). Influenza virus-specific cytotoxic T cell recognition: stimulation of nucleoproteinspecific clones with intact antigen. *Immunology*, **59**: 173.

Yellen-Shaw, A.J., Wherry, E.J., Dubois, G.C. et al. (1997). Point mutation flanking a CTL epitope ablates in vitro and in vivo recognition of a full-length viral protein. *J. Immunol.* **158**: 3227–3234.

Yewdell, J. and Benninck, J. (1992). Cell biology of antigen processing and presentation to major histocompatibility complex class I molecule-restricted T lymphocytes. *Adv. Immunol.* **52**: 1–123.

Yewdell, J., Lapham, C., Bacik, I. et al. (1994). MHC-encoded proteosome subunits LMP2 and LMP7 are not required for efficient antigen presentation. *J. Immunol.* **152**: 1163–1170.

York, I.A., Roop, C. Andrews, D.C. et al. (1994). A cytosolic herpes simplex virus protein inhibits antigen presentation to CD8+ T lymphocytes. *Cell* **77**: 523–535.

Yoshikawa, M., Watanabe, M. and Hozumi, N. (1987). Analysis of proteolytic processing during specific antigen presentation. *Cell. Immunol.* **110**: 431.

Ziegler, H.K. and Unanue, E. R. (1981). Identification of a macrophage antigen-processing event required for I-region-restricted antigen presentation to T lymphocytes. *J. Immunol.* **127**: 1869.

Ziegler, H.K. and Unanue, E.R. (1982). Decrease in macrophage antigen catabolism caused by ammonia and chloroquine is associated with inhibition of antigen presentation to T cells. *Proc. Natl Acad. Sci. USA* **79**: 175.

Ziegler, H. Thäle, R., Lucin, P. et al. (1997). A mouse cytomegalovirus glycoprotein retains MHC class I complexes in the ERGIC/cis-Golgi compartments. *Immunity* **6**: 57–66.

Zhou, F., Rouse, B.T. and Huang, L. (1992). Induction of cytotoxic T lymphocytes in vivo with protein entrapped in membranous vesicles. *J. Immunol.* **149**: 1599–1604.

CHAPTER 6

The Analysis of Genetic Susceptibility

Timothy J. Vyse and Bernard J. Morley

INTRODUCTION

The majority of diseases that commonly affect humans is not directed by single gene events. In major histocompatibility complex (MHC)-associated disease, this is underlined by the fact that no single MHC allele is 100% linked to disease; the overwhelming majority of ankylosing spondylitis patients is B27, but the majority of B27 individuals does not have ankylosing spondylitis (Tiwari and Terasaki, 1988). Common diseases are multigenic and potentially involve multiple environmental insults. Consequently, they are described as multifactorial or complex traits. Even for those diseases thought to be caused solely by a mutation in a single gene, such as sickle cell anaemia, there is now increasing evidence that the broad spectrum of pathology frequently presented may be controlled by a host of modifying genes (Craig et al., 1996). An additional layer of complexity is added by gene interaction or epistasis. Demonstration of epistasis implies that genes act in functionally related pathways with respect to the disease phenotype. The temptation is to give up at this point, bowing to this inevitably overwhelming complexity. However, the rapid evolution of techniques in genetic analysis, coupled with the developing discipline of bioinformatics, has enabled an informative and logical approach to be adopted to understanding these complex diseases.

The first step in the identification of disease susceptibility genes is to ensure that there is indeed a major genetic component in the disease. This is usually estimated by studying concordance rates in identical twins compared with non-identical twins and next by estimating the risk ratio (λ). The most common value is that of λ_S or sibling risk, which is calculated from:

$$\frac{\text{risk to sibs}}{\text{population prevalence}} = \lambda_S$$

For type I diabetes $\lambda_S = 15$, and for haemophilia (a disease associated with a defect in a single gene) $\lambda_S = 50\,000$ (Vyse and Todd, 1996). If the λ_S is much less than 10, then the genetic contribution is relatively small and if more than one gene is involved, individual gene effects may be very difficult to identify.

When a significant genetic component is established, the next goal is to identify a chromosomal region or interval that may play a role in disease aetiology. The next step is to narrow down that interval as much as possible before finally identifying a known or novel gene within that region which is the disease susceptibility gene. This last stage of gene identification is the rate-limiting step in the procedure. The whole process is based on two critical sources of information: (i) the ability to distinguish individuals from one another; and (ii) a large resource of physical information about DNA sequences. This can range from genetic and physical maps that place genes in particular regions of chromosomes, through genomic and cDNA libraries [either full-length or containing short marker sequences known as expressed sequence tags (ESTs); Hillier et al., 1996], to the total human sequence such as that generated by the Human

Genome Project. Both the methodology to facilitate these processes and the fundamental information itself are growing exponentially.

Once disease susceptibility genes can be identified for these complex traits, they will indicate pathways of disease progression that merit further direct study. When physiological pathways are understood, this will implicate environmental features which can affect these pathways. Ultimately, a better understanding of disease will facilitate the development of better treatment and even indicate novel avenues of therapeutic intervention.

IDENTIFICATION OF DISEASE GENES

Tests for the presence of disease genes on chromosomes can basically be divided into two types: those dependent on family data (linkage studies and whole genome searches) and those performed on populations (association studies). The vast majority of data for the role of MHC in disease is based on association studies (for which reason this approach and the statistics underlying it are discussed in much greater detail in the following chapter). To perform either sort of analysis, it is necessary to establish a system of "marking" DNA so that individuals can be distinguished.

Selecting markers

In its simplest form a marker can be a protein polymorphism, e.g. PA and PB. This polymorphism could then be followed in families to determine whether PA nearly always follows the disease trait in one family, while PB may track disease in another. Alternatively, two populations of matched individuals with and without disease can be compared to demonstrate whether PA is present with disease more often than it should be by chance. The drawback with PA and PB is that there are only two variants. An ideal marker should be highly variable, yet stable through generations; randomly distributed in the genome and easily measurable, to facilitate rapid analysis. Early markers used included phenotypes in early *Drosophila* work, restriction fragment length polymorphisms and minisatellites, but each of these had its own drawbacks. However, since 1989 marker technology has been revolutionized by the use of simple tandem repeat sequences or microsatellites (Weber and May, 1989). These repeated sequences, for example, repeats of the same dinucleotide, are scattered randomly throughout eukaryotic genomes at a relatively high frequency, occurring approximately every 10^5 base pairs (bp). Among human individuals or inbred rodent strains, microsatellites exhibit marked allelic variation in the length of the repeat sequence, which is referred to as simple sequence length polymorphism (SSLP). Since each microsatellite is flanked by unique sequences, specific oligonucleotides can be designed to match these flanking sequences and used to amplify the region surrounding a particular microsatellite using the polymerase chain reaction (PCR). The allelic form of a microsatellite is determined by size separation of PCR products. Because of their frequency, high allelic polymorphism and technological ease, microsatellites have become the most useful markers for genetic mapping studies (Love et al., 1990; Dietrich et al., 1994, 1996; Dib et al., 1996). Thousands of SSLPs have now been mapped in the human and mouse genomes which can be used to analyse genetic linkage in human families or experimental crosses of mouse strains. A whole genome scan consists of about 100 evenly spaced markers in a mouse cross, or about 300 markers in a study of human families in which there are affected and unaffected members.

The next generation of probe will probably be the single-nucleotide polymorphisms (SNPs) (Kruglyak, 1997). Although these are not as polymorphic as microsatellites, with only four possible variants at any position, they occur more frequently and analysis can be fully automated using oligonucleotide probes fixed to silicon chips, or probe arrays for detection (Fodor et al., 1991, 1993). This will facilitate the very large studies which will be necessary to identify disease susceptibility genes with small effects (Wang et al., 1998).

Approaches to analysis[1]

Association test

In an association study, the frequency of an allele in a sample of affected individuals is compared

---

[1]For a more detailed discussion of the statistics involved, see the appendix to this chapter.

with its frequency in a case–control population. The major problem with this type of study lies in the control population, which is the most common source of false-association data. For this reason, this population must be carefully selected for ethnicity, age, gender and environment. Indeed, it is now generally accepted that such studies should use internal controls that comprise unaffected family members, who are, by definition, matched for ethnic ancestry (Falk and Rubinstein, 1987; Terwilliger and Ott, 1992). In general, association studies are performed on candidate genes. These are genes where the known or suspected biological functions would indicate a likely role in the pathophysiological process of the disease under investigation; hence the study of human leucocyte antigen (HLA) in immune disorders, especially autoimmunity.

The statistical analysis of an association study is relatively straightforward: the frequency of a marker allele in patients is compared with its frequency in controls. The data obtained can be expressed in a 2 × 2 contingency table and statistical significance quantified by the chi-square (χ^2) test. Such data are often expressed as an odds ratio with 95% confidence intervals. Association, however, is not causation. Although it is formally possible that an allele is associated with a disease because it has an aetiological role, a positive association can arise because the marker used is in linkage disequilibrium (see below) with a disease susceptibility gene or, more likely, is a spurious association because of inadequate control data (Lander and Schork, 1994). A very good example of linkage disequilibrium explaining an HLA association is provided by the haemochromatosis disease gene. This disease was originally associated with HLA-A3 and HLA-B14 in population studies (Simon et al., 1976). Later, a novel MHC class I-like molecule was identified (HLA-H) in linkage disequilibrium with HLA-A3. This gene, later called HFE, was demonstrated to be the disease gene (reviewed in Worwood, 1998). In type I diabetes, many studies showed an association with HLA-DR3 and HLA-DR4. More extensive analysis indicated that stronger associations were present with certain DQ alleles that were in linkage disequilibrium with the DR alleles (Thomson et al., 1988).

Linkage analysis

Linkage describes the tendency of two markers or genes to be inherited together. In general, this is due to the two markers lying on the same chromosome, i.e. they are physically linked. However, cosegregation of two markers within a family can also occur as a result of epistasis and this is referred to as genetic linkage. In a linkage study, one of the markers is the disease phenotype and the analysis proceeds based on an assumed hypothesis, i.e. a dominant or recessive mode of inheritance and a particular level of penetrance. The analysis involves studying the inheritance of disease in pedigrees, and ascertaining the probability that a marker is inherited with disease and the probability that it is *not* inherited with disease (Haldane and Smith, 1947). The ratio of these two probabilities is termed the likelihood ratio (LR) and is usually expressed after $\log_{10}$ transformation. This value is termed the lod score (Z):

$$Z(\theta) = \log_{10}\left[\frac{L(\theta)}{L(0.5)}\right]$$

where θ is the recombination frequency between the marker and the putative disease gene (0 for no recombination, i.e. the marker and the gene under study are the same locus; and 0.5 if they are unlinked). The term $L(\theta)$ represents the likelihood of the marker being a distance q from the disease gene; that is the cumulative probability in the pedigrees analysed. $L(0.5)$ represents the likelihood of there being no linkage, i.e. when $\theta = 0.5$. The lod score can then be differentiated with respect to θ in order to determine the maximum lod score (MLS) in the families studied (Edwards, 1992). The evident complexity of the analysis means that most linkage studies employ computer programs to determine MLS values. A detailed explanation of the mathematics involved is provided in Cavalli-Sforza and Bodmer (1971).

As with all statistical tests, it is somewhat arbitrary where one draws the line between significance and non-significance. Tiwari and Terasaki (1985) suggest significance levels of 1000 to 1 ($Z = +3.000$) and 100 to 1 ($Z = -2.00$) for positive and negative linkages, respectively. However, the threshold for significance that should be applied to linkage analyses is currently a contentious issue. It has recently been proposed that the significance level be raised from $Z = 3$ to $Z = 3.3$ (equivalent to a single point $p = 1 \times 10^{-4}$), which theoretically gives a false-positive probability of 0.05 in a genome-wide scan (Lander and Kruglyak, 1995). However, the most powerful means of eliminating false-positive findings may be replication of linkage in an independent data set.

Because of low penetrance and genetic epistasis, traditional linkage methods usually have limited applicability in complex polygenic disorders (Ott, 1992). It has been suggested recently by Risch and Merikangas (1996) that association studies are the best method to identify disease susceptibility genes in complex disorders. They argue that many of the genes involved in common diseases have fairly limited effects. Thus, the numbers of individuals required to establish significant linkage will be prohibitively large. However, the numbers for association studies will be acceptable, given high-throughput analysis techniques. For a genotypic relative risk (GRR, see section on epistasis below) of 2.0 and a disease allele frequency of 0.1, 5382 families are required for linkage analysis compared with 695 singletons for an association test. If the disease gene frequency drops to 0.01, then 296 710 families would be required compared with 5823 singletons. Obviously, it would not be possible to identify 296 710 families – 5382 would prove difficult enough! Thus, they argue that only genes with significant effects will be identified by linkage studies. From a physiological point of view, these are likely to be the most important and therefore linkage analysis is not yet redundant. In addition, although the number of individuals will be smaller, every gene in the genome is a candidate for a disease of unknown aetiology. Thus, once the entire genome is sequenced and the full complement of genes known, each can be studied individually in association tests for different diseases.

Linkage analysis can be applied if a single gene has high penetrance with respect to a complex trait. An example of using traditional linkage analysis in autoimmune disease is the mapping of an autosomal recessive gene linked with autoimmune polyglandular disease type I (APECED) to the long arm of chromosome 21. This was accomplished in a Finnish population which has a relatively limited genetic diversity (Aaltonen et al., 1994). Traditional linkage methods have been modified to allow the examination of two loci concurrently. Such methods have been applied to multiple sclerosis in Finnish pedigrees, in which linkage was found with the MHC (identified previously by several groups) and with the myelin basic protein (MBP) gene on chromosome 18 (Tienari et al., 1994). Linkage with MBP has not been reported in studies outside Finland (Wood et al., 1994).

Allele-sharing methods

Traditional linkage methods, as described above, have limited applicability in studies of the genetics of complex polygenic diseases, and alternative approaches such as allele-sharing methods can be used to examine the inheritance of marker alleles in multiplex families (Day and Simons, 1976; Fishman et al., 1978; Blackwelder and Elston, 1985; Weeks and Lange, 1988). An advantage of such methods is that they are non-parametric, that is they are not model based and therefore do not specify a particular model of inheritance for any disease gene. Markers covering the entire genome are used in what is termed a genome-wide linkage analysis. Linkage studies in multiplex families (i.e. families with more than one affected member) operate on the principle that family members can be expected, under Mendelian inheritance, to share a given proportion of their alleles. The fraction depends on their family relationship. Allele-sharing methods can be applied to any form of multiplex family, but have been used most frequently for sets of affected sib pairs. Sib pairs have the advantage of relative ease of case ascertainment and sampling in comparison with larger pedigrees.

For sib pairs, the inheritance of alleles can be determined by two means: identity-by-descent (IBD) or identity-by-state (IBS). To determine IBD, parental genotypes must be identified, although occasionally, IBD may be inferred. From Mendelian principles, sib pairs would be expected to share zero, one or two alleles with probabilities of 0.25, 0.50 and 0.25, respectively (Suarez et al., 1978). A departure from this expected distribution, such that the affected sib pairs exhibit an increased likelihood of sharing two alleles, and a decreased likelihood of sharing zero alleles, suggests that the marker used may be in linkage with a disease. Several tests have been developed to quantify any observed deviation from the expected distribution (Blackwelder and Elston, 1985). The most powerful methods employ a test statistic derived from the mean number of alleles that are shared IBD by sib pairs (Lange, 1986). The results can be expressed as a maximum lod-score analogous to that used in traditional linkage studies (Risch, 1990a). IBD analysis of sib pairs has been used successfully to map non-MHC loci in type I diabetes.

In comparison to IBD methods, the IBS approach has the advantage that it does not require knowledge of parental genotypes. For example, siblings with genotypes A_1A_2 and A_1A_3 at locus A may not have inherited A_1 from the same parent. Thus, they share allele A_1 by state. This is illustrated in Figure 6.1. IBS methods are especially useful in situations where disease onset occurs later in life and parental material may not be available. Although IBS-based methods are less powerful than those based on IBD, if highly polymorphic markers are screened, genotyping by IBS may achieve the power of IBD methods. It is critical to include data specifying allele frequencies, since common allelic variants would be expected to be shared more commonly than rarer variants simply because of their frequency in the population. IBS methods can be made more powerful by extending the family members studied to include all affected individuals in a pedigree. This modification is termed the affected-pedigree-member (APM) method (Weeks and Lange, 1988). To date, IBS methods (with APM) have been utilized successfully in primary hypertension (Caulfield et al., 1994) and schizophrenia (Wang et al., 1995).

The statistical threshold at which linkage is declared using allele-sharing methods has been the subject of constant review. Recent recommendations by Kruglyak and Lander propose a lod score (or MLS) of 2.2 ($p = 7.4 \times 10^{-4}$) to conclude suggestive linkage and a lod score of 3.6 ($p = 2.2 \times 10^{-5}$) to conclude significant linkage when studying sib pairs (Lander and Kruglyak, 1995). These calculations are based on the probability of false-positive linkage in the context of a genome-wide scan. However, in polygenic traits, because of the number of disease-susceptibility loci that operates, an individual locus may make a relatively small contribution to the total genetic risk (Risch, 1990a, b). Thus, it may also be reasonable to employ a lower stringency threshold to avoid missing an actual locus and to rely on confirmation in a separate data set (Weeks and Lathrop, 1995).

Related to the issue of statistical thresholds is that of statistical power. Power is influenced by the number of sib pairs studied and the polymorphism of the marker used. For example, for a locus with $\lambda = 1.6$, 200 sib pairs would have a power of 0.9 to detect such a locus at a threshold of MLS = 1.0 (Risch, 1990a). If the threshold were increased to 3.0, then the same number of sib pairs would have the same power to detect a locus with $\lambda = 2.3$, which would reflect a major contribution from a single locus to a polygenic trait. As this

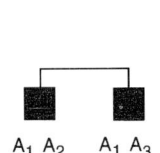 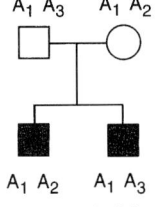 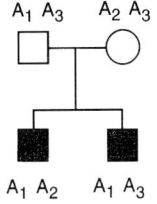

The affected siblings share A_1 allele. A_1 may have been inherited from one parent only, or from each parent (see adjacent panels). **Identity-by-state.**	The affected siblings share A_1 but each sibling may have inherited the allele from a different parent.	The affected siblings share A_1 and each inherited the allele from the father. **Identity-by-descent.**

FIGURE 6.1

example shows, a disadvantage in sib-pair analyses is that a large number of families is required to detect loci with relatively weak linkage (Kruglyak and Lander, 1994).

NARROWING DOWN THE INTERVAL

The mapping techniques described above are used to identify loci in linkage with disease or disease-related traits. In traditional linkage studies for single genes, initially a large interval was identified. Often this was in the order of 10 centiMorgans (cM), where a centiMorgan is a genetic distance based on a 1% recombination frequency and which approximates to 10^6 bp (1 Mb) of DNA (Fields et al., 1994). In complex polygenic diseases, there is an inherent inaccuracy in the process and the genetic intervals identified are usually large. Studies in mice have generally mapped disease-linked loci within intervals ranging from 10 to 20 cM. Since 1 Mb of DNA (1 cM) may encode as many as 25 genes, the intervals identified by mapping may contain up to 500 genes. Moreover, only a small fraction, albeit an ever-increasing one, of the genes within such an interval has been identified to date. These limitations make selection of the appropriate candidate gene difficult. Therefore, to identify the gene that is responsible for the observed linkage, it is usually necessary to reduce the size of the genetic interval in which the disease gene is located.

Linkage disequilibrium

One technique for fine mapping human loci depends on linkage disequilibrium for loci that are closely spaced together on a chromosome. All loci in an infinitely sized population of infinite age should be in complete random assortment, i.e. linkage equilibrium. Two closely linked loci are said to be in linkage equilibrium when the population frequency of a combination of alleles at these loci equals the product of the population allele frequencies at the two loci. For example, if the frequency of allele Y_1 is 0.3 and the frequency of allele Z_1 is 0.2 in a population, then Y_1 and Z_1 would be expected to occur together with frequency of 0.06 (0.2 × 0.3), given random segregation. A frequency greater than 0.06 implies linkage disequilibrium. Since the human race is not of infinite age or size, linkage disequilibrium is common over short distances on chromosomes. In a population, linkage disequilibrium will decline with the number of meiotic events (generations) in proportion to the recombination distance between genes. It can be shown, given random recombination over 2000 years (~80 generations) within a human population, that disequilibrium has been maintained over a maximum of about 1–2 cM (Charmley and Concannon, 1995; Peterson et al., 1995). Thus, if a marker lies within 1 cM of a disease mutation, the marker and mutation will remain sufficiently linked to show non-random segregation in a population study. At genetic distances of greater than 1–2 cM the probability of recombination between a marker and disease gene is such that linkage disequilibrium is unlikely to be detectable. Although disequilibrium is a good indication for the proximity of two loci, greater than expected allele frequencies may also arise because of non-random assortment of alleles related to selection, genetic drift within a population, introduction of a new mutation in a population or incomplete admixture of genetically diverse populations.

If an interval is found to be linked with disease by linkage analysis of affected sib pairs, the transmission disequilibrium test (TDT) can be used to determine the position of the disease gene with much greater accuracy (Spielman et al., 1993). This process requires that polymorphic markers are identified at intervals of approximately 1 cM within the larger interval linked to disease in a sib-pair study. This test requires not affected sib pairs, but rather a single affected family member. The test ascertains the frequency of transmission of an allele, previously shown to be associated with disease, from a heterozygous parent to an affected child. Disequilibrium in inheritance of a particular parental allele strongly suggests linkage. This form of investigation is most powerful when limited to relatively homogeneous genetic populations, for which the same aetiological mutation is likely to play a role and be detected.

Linkage disequilibrium mapping has been used to confirm that the human insulin gene is linked with type I diabetes (designated IDDM2).

Moreover, the maximal chromosomal interval containing the aetiological mutation has been reduced to 4 kilobases (kb) of DNA upstream of this gene (Spielman et al., 1993; Bennett et al., 1995). The linkage disequilibrium method is usually more powerful than linkage analyses in multiplex families (i.e. families in which several members carry the disease phenotype). For example, the statistical support for IDDM7 being linked to disease from sib-pair analysis was marginal ($p = 0.005$ from a meta-analysis of over 550 sib pairs), whereas the evidence supporting the linkage of IDMM7 to disease IDDM7 from linkage disequilibrium testing was striking ($p = 7 \times 10^{-14}$; Copeman et al., 1995). Because of the limited genetic regions over which disequilibrium may act, it is not practical to screen a genome with this method. However, it is suitable for narrowing intervals of ~20 cM as identified by linkage analyses. An additional advantage that linkage disequilibrium methods offer is that although they are most informative when carried out in families (i.e. transmission disequilibrium testing), multiplex families are not required. The use of different single affected children greatly increases the number of available subjects for genetic studies compared with allele sharing linkage studies in multiplex families.

Physical mapping

At some stage in the identification of novel susceptibility genes, it will be necessary to clone and map the interval of interest. A number of different vectors exists for cloning large pieces of genomic DNA which vary in the amount of insert that they can accommodate. These are yeast (0.5–1 Mb) bacterial (0.2 Mb) and P1 phage (0.1 Mb) artificial chromosomes, or YACs, BACs and PACs (Sclessinger, 1990; Larin et al., 1991; Haldi et al., 1996; and for review see Schalkwyk et al., 1995). The much larger pieces of YACs are very useful for mapping purposes; however, the larger the insert, the more unstable it is. In addition, large YACs often contain non-contiguous regions of genomic DNA. Consequently, they must be mapped and compared with genomic DNA. This is usually performed using pulsed-field gel electrophoresis (PFGE), a technique of agarose gel electrophoresis which relies on alternating current direction to separate large DNA fragments (Burmeister and Ulanovsky, 1992).

Once a gross map has been obtained for a region by YAC mapping and PFGE, high-resolution mapping can be performed using cosmid subclones. Entire YACs or BACs can be fragmented by partial restriction endonuclease digestion ("shot-gunning") and cloned into cosmid (40 kb) or plasmid (<10 kb) vectors. These are then ordered along the YAC map using restriction mapping, PCR and cosmid fingerprinting (Vollrath et al., 1988; Ledbetter et al., 1990; Lovett et al., 1991; Parimoo et al., 1991; Bohlander et al., 1992; Heng et al., 1992; Valdes et al., 1994). Sequence data can now be obtained and ESTs and sequence-tagged sites positioned in the cosmid. Novel genes can be identified using four main approaches: (i) the use of "zoo" or "ark" blots to identify regions of sequence conservation across different species imposed by the presence of expressed sequences (Hanson et al., 1991); (ii) exon-trapping techniques to locate intron–exon boundaries (Buckler et al., 1991; Church et al., 1994); (iii) screening cDNA libraries with genomic DNA from the candidate region (Lovett et al., 1991; Parimoo et al., 1991); and (iv) searching for regions rich in unmethylated CG dinucleotides, CpG islands, which are often associated with transcribed DNA (reviewed in Bird, 1992). However, the use of banks of ESTs which can be mapped back to cloned regions is superseding most of these approaches (Hillier et al., 1996). Once target genes have been identified for analysis, variation can be determined either by direct sequencing or by using one of the many techniques available to identify sequence polymorphism, such as single-strand conformation polymorphism analysis (SSCP) (Orita et al., 1989a, b; Sheffield et al., 1993), heteroduplex cleavage analysis by either RNase (RNase protection; Myers et al., 1985) or chemicals (chemical cleavage; Cotton et al., 1988), denaturing gradient gel electrophoresis (DGGE) (Lerman and Silverstein, 1987; Sheffield et al., 1989) and many others (reviewed in Grompe, 1993; Spanakis and Day, 1997). Any stage in the procedure, except for the polymorphic sequence screen of the gene itself, can be skipped by the use of a candidate gene analysis. Without linkage data, this would involve an association test. However, once an interval has been determined from linkage data, then candidate genes can be identified and screened.

This entire approach from linkage to physical mapping to gene identification was initially named "reverse genetics" following the identification of the chronic granulomatous disease on the X chromosome by Orkin and co-workers (Royer-Pokora et al., 1986). It is now more aptly described as positional cloning. Approximately 50 disease genes, almost exclusively from single-gene disorders, have been identified using positional cloning methods (Collins, 1995). In single-gene disorders it is a relatively straightforward matter to prove that the identified gene is indeed the disease gene. More patients are identified, preferably from unrelated families. The implicated disease gene is sequenced and mutations are identified. In a single-gene disorder, all patients must have a mutation in the one gene. The disease-causing mutation need not necessarily be the same, e.g. β-thalassaemia, but the same gene (β-globin) must contain a mutation. If it does not, then identification of the "disease" gene must be brought into question.

For complex traits, however, a further problem awaits. Where no single gene is involved and mutations that confer susceptibility will, in all probability, be "normal" polymorphisms with slight functional differences, the problem of proof is very real. Even in single-gene disorders it can prove problematic. Spinal muscular atrophy (SMA) offers an example. This disease was mapped to 5q11.2–13.3. YAC cloning resulted in the identification of two disease genes in this region (Lewin, 1995), both showing mutations in patients with SMA. The two genes, survival of motor neuron (SMN) and neuronal apoptosis inhibitory protein (NAIP), were studied in several other families before resolution in favour of the SMN gene (reviewed in Lefebvre et al., 1998). Inevitably, some form of functional screening process is necessary to prove finally that an identified polymorphism confers increased risk (see below).

The process of positional cloning is currently undergoing revolutionary change on account of the rapid accumulation of data generated by the Human Genome Project. Sequence data are currently being accrued for both genome sequences and expressed (which includes cDNA) sequences (Boguski and Schuler, 1995). Over 400 000 partial human cDNA sequences (ESTs), representing approximately 45 000 independent sequences, have been published. These sequences are available at World Wide Web sites (Boguski and Schuler, 1995) and many more are stored in private databases held by a number of independent companies. It is possible that all potential transcripts have been cloned and are present in a library of ESTs somewhere. Anchoring expressed sequences to the physical genome map is one of the next goals of the Human Genome Project, which should be largely completed by the year 2000. Similar work is underway with respect to the mouse genome.

FUNCTIONAL SCREENING

An alternative approach to physical mapping is to perform a functional screen. This involves using phenotypic expression to determine regions that are important and narrow down the interval by defining the smallest region which still confers disease susceptibility. The most obvious application of this approach is the use of congenic mice (see below). However, congenic mice are bred specifically to look at segments of the mouse genome and are not applicable to other species, including humans. However, the use of transgenic animals to express regions of human genomic DNA and hence produce an artificial congenic is distinctly possible. Transgenic animals, unlike congenics, carry exogenous fragments of DNA introduced by microinjection into single egg cells (reviewed in Hogan et al., 1994). Copy number can be controlled by injecting variable amounts of target DNA and so adult animals can be produced which have a "normal" copy number (two copies) or which might be expected to overexpress the transgene. For this approach to be useful, large regions of genomic DNA must be introduced in one transgenic line. The most obvious source of DNA would therefore be YAC clones. An inherent problem in the use of YACs, however, is that these large clones tend to fragment when manipulated. The physically violent method of microinjection for transgenesis exacerbates this. However, recently, Rubin's group in the Lawrence Berkeley National Labs in California (Smith et al., 1997) used this fact to their advantage. They used lines produced from accidentally fragmented YACs in addition to the full-length YACs to strongly implicate a 180 kb region in human chromosome 21 in learning defects associated with Down's syndrome. This functional approach (reviewed in Rubin and Smith, 1997), resulting in the construction of *in vivo* transgenic mouse libraries, which can be screened for disease directly or interbred with other mouse lines to produce a phenotype, may well represent the future for susceptibility gene mapping.

DIFFERENTIAL ANALYSIS TECHNIQUES

These techniques have been designed to increase the speed and capacity of the initial screening process. At present, they are in their infancy; however, given time, these approaches will form valuable tools in the geneticists' armoury. Essentially, they fall into two groups; those associated with fundamental differences in the genomic DNA and those designed to look at variations in gene expression.

All genomic screening techniques are based on an initial hybridization between the two genomes that are undergoing comparison. Essentially, one source of genomic DNA is screened against the other (Lander, 1993; Sagerstrom et al., 1997). In representational difference analysis, heterologous fragments are preferentially amplified by PCR after removing homologous fragments using a highly efficient subtractive hybridization protocol (Lisitsyn et al., 1994; reviewed in Lisitsyn, 1995). The alternative approach is to select the sequences that show mismatches and identify those mismatches using *Escherichia coli* repair enzymes (Nelson et al., 1993). These mismatch sequences can then be removed leaving only regions of identity behind. This technique of genomic mismatch screening has been used to perform an IBD linkage analysis without the use of markers (Cheung et al., 1998). Regions that were identified as similar between individuals were used to probe a microarray (Schena et al., 1995) of genomic fragments from regions of interest. Using this technique, Cheung and colleagues correctly identified a known disease gene. However, the procedure needs significant optimization before an entire genome-wide search could be performed (reviewed in Kruglyak and McAllister, 1998).

As a result of developments in cancer genetics aimed at identifying altered expression patterns in tumours, a large number of techniques has been developed to investigate differential gene expression. All of them are highly effective and in their most expansive (and expensive) forms may well be capable of recognizing differences in mRNA levels as low as two-fold (Adams, 1996). The drawback for complex traits is that it is not always obvious what tissue should be screened. Ideally, several different cell types would be studied, but the problem then arises that the time involved may well be greater than the time necessary for a complicated genome search. All of these techniques are also critically dependent on the quality of mRNA isolated. It must be representative of the message level in the cell and therefore it is vital to devote a considerable amount of time to this first step. Generally, cDNA is synthesized, followed in most cases by a random amplification using degenerate PCR primers. This random amplification uses a variety of conditions and different primer pairs in order to produce an accurate representation of mRNA species present. The amplicons are then analysed by a variety of different methods.

In differential display (Liang and Pardee, 1992, 1997) the degenerate PCR reactions are performed side by side in tissue from diseased and non-diseased subjects. Pattern differences are studied and confirmed with the aim of identifying PCR fragments present in one cell and not the other. The background to this procedure can be very significant and figures for its efficiency vary from 10% to 80% (Bertioli et al., 1995; Sompayrac et al., 1995). Optimization of the PCR and primers can help to reduce background, but reliable results are difficult to obtain. The major advantage is in its simplicity.

Suppression subtractive hybridization techniques (Diatchenko et al., 1996; Jin et al., 1997; Lukyanov, 1997), often known by the tradename PCR select™ (Clontech Laboratories, Palo Alto, CA, USA), are probably more robust and reliable, but harder to perform. These techniques rely once more on a subtractive hybridization to remove similar sequences. After repeated subtractions, the remaining amplicons are cloned and sequenced. Again, one cell type is compared with another.

In serial analysis of gene expression (SAGE), a copy number proportional to the level of each mRNA species is generated for each cell type (Velculescu et al., 1995; Polyak et al., 1997; Bertelsen and Velculescu, 1998). Consequently, correlations can be performed between three and more cell types without the necessity for pairwise comparison. Essentially, specialized restriction enzymes are used to generate small (c. 14 bp) sequence tags from specific positions within transcripts. Tags are ligated together in pairs and amplified by PCR, and multiple tags cloned together in a single fragment to facilitate sequencing. The number of times that a tag appears is then counted. For medium copy mRNAs, approximately 20 000–40 000 tags need to be analysed to distinguish differences in expression of about five-fold. This represents about 3 months of sequencing for a single individual, a significantly greater initial input than the other methods. However, the tag numbers generated are fixed and can be used in any future comparisons.

The final method is potentially the most exciting and also by far the most rapid and expensive. This involves preparing microchips which carry oligonucleotide probes (Fodor et al., 1991, 1993) that are representative of all the messages under investigation. Thus, it would be possible to have an immune regulation microchip with cytokines, adhesion molecules, etc., or a chip for intermediary metabolic components. This chip is then screened with mRNA from the cell types of interest generating a signal of magnitude dependent on the amount of mRNA present in the probe sample (Lockhart et al., 1996; Wodicka et al., 1997; reviewed in Fodor, 1997). This technique is essentially the same as a reverse Northern, since probe DNA is screened with labelled mRNA to determine levels. The screening and analysis of the chip takes very little time, less than a week, and a chip can carry a huge amount of information (500–200 000 sequences). In addition, the information is accurate to a two-fold increase, whereas the majority of other techniques relies on five-fold or greater differences. One problem is which cDNA sequences to use, but with the vast number of ESTs now available, the answer would be to use them all. The only problem at present is that the prohibitive cost of chip production and analysis means that these studies are only being undertaken in the commercial sector.

This summary is not intended to be a comprehensive guide to specific techniques but an outline indicating the underlying principles. All of the techniques described are adjuncts to the initial linkage analysis, facilitating gene identification. Even the chip technology will not answer all questions since at this juncture it only analyses differential gene expression. If the disease susceptibility gene is functionally different or secondary modification varies, then these techniques will not identify it. However, recently, the techniques of *in vivo* expression of green fluorescent protein fusions, two-dimensional gel electrophoresis, mass spectrometry and EST database searches have been linked to identify the protein components of the human spliceosome (Neubauer et al., 1998). Applications of multiple state-of-the-art techniques in this way will doubtless render old-fashioned linkage analysis redundant in the not too distant future.

ANIMAL MODELS

There are obvious advantages to using animal models in genetic analysis. For example, it is easy to conduct directed breeding programmes which generate the large numbers of affected and non-affected progeny needed for a linkage analysis. In the case of inbred strains, all mice are isogenic (homozygous and identical at every locus). The environmental conditions can be kept fairly constant for all animals. The parental strain usually shows a high frequency of disease which is often of rapid onset (i.e. less than 12 months). Furthermore, especially in the case of a disease such as systemic lupus erythematosus, in which there is heterogeneity in human patients, lupus-prone mice offer the opportunity to study a single phenotype (e.g. lupus nephritis, anti-dsDNA autoantibody production).

However, there are limitations to studying animal models. From a genetic perspective, studying one mouse strain is comparable to studying one single human subject with disease. Furthermore, because relatively large genetic regions (containing many linked genes) are inherited from one parent in a genetic cross, it is difficult to narrow down intervals containing the disease-linked gene. Possibly most important of all, a large-scale study of disease in an animal model will generate data about disease in that model, but it is not clear that this will translate into humans. Genes may be different, pathways different and unless animal husbandry is the goal, it may prove to be medically useless. This may well be the case and it is entirely possible that different genes will be found to be important in animal models and humans. However, it is highly likely that pathways of disease progression will be similar, provided the model chosen is an accurate reflection of the disease under investigation. In mice, the non-obese diabetic (NOD) strain provides such a model, as do the New Zealand strains for lupus. Indeed, there is already one example of a disease gene identified in mice which has since proved to play a role in human disease, the Fas antigen. The lymphoproliferative (lpr) mutation in mice was mapped to a region on mouse chromosome 19. Separate studies mapped the gene encoding the Fas antigen, a molecule known to be involved in lymphocyte apoptosis (Itoh et al., 1991), to the same chromosomal region (Watanabe et al., 1991). Mutations in Fas were subsequently shown to account for the lpr phenotype (Watanabe-Fukunaga et al., 1992). Mice homozygous for this mutation typically develop massive lymphoproliferation, mostly related to the accumulation of an unusual subset of T cells. lpr is also an accelerator of lupus-like autoimmunity. Mutations in the human homologue of Fas have since been identified in patients with lymphoproliferative disease (Rieux-Laucat, 1995) and also in a number of lupus patients (Wu et al., 1996).

Considerable linkage data have been acquired with studies into rodents, especially mice, and it is these on which this section will concentrate. However, the basic ideas apply to any similar studies. To determine linkage with disease traits in mice, it is necessary to conduct breeding experiments to create a genetically heterogeneous sample. The usual approach is to first cross a disease-prone strain with a disease-resistant strain – the progeny are F_1 mice. These F_1 mice are then crossed to one of the parental strains. This type of cross is known as a backcross. To identify which cross will be more informative, the incidence and severity of disease are first investigated in the F_1 animals. This provides an indication of the overall recessive nature of the process. If the trait has a very low incidence in F_1 animals then an informative study would be to backcross to the disease-prone parental strain. If the majority of the F_1 animals exhibits the trait (indicating an overall dominant contribution) then the appropriate backcross is to the disease-resistant parent. As an alternative to a backcross, the F_1 animals are crossed together to produce F_2 progeny. Such an F_1 intercross may be particularly informative because there are twice as many meioses than in a backcross study and homozygosity for either parental alleles can be examined in the same cross. However, more animals may be required to produce significance (see Chi-square test below). Whether a backcross or intercross is used, the principle of identifying linkage is the same. Experimental animals are scored for disease or for a disease-related trait, and the mice are genotyped. The segregation of marker alleles with disease is determined. If a marker is located near a disease susceptibility locus, then its allele frequency will vary from the expected Mendelian ratios in affected and unaffected animals. The expected distribution is 1:1 for a backcross and 1:2:1 for an intercross. Statistical analysis is readily achieved by employing the χ^2 statistic when the trait is a qualitative yes–no (dichotomous) disease phenotype such as presence or absence of disease. An example of such data is given in Table 6.1, with respect to a study on lupus-prone BXSB mice using a (BXSB × [B10 × BXSB]F_1) backcross (Hogarth et al., 1998). Non-dichotomous, complex phenotypes (quantitative traits), such as degree of insulitis or autoantibody levels, can also be examined in this way using an extreme phenotype analysis. Thus, only animals at the extreme ends of the trait's distribution (i.e. very low and very high) are studied. Such an analysis of extremes can be a powerful approach to analyse continuous traits (Lander and Botstein, 1989). In mice, such quantitative trait loci (QTL) can also be mapped using computer programs, such as MAPMAKER/QTL, which was developed by

TABLE 6.1 Distribution of genotypes in BXSB × (B10 × BXSB)F_1 backcross mice on chromosome 1

Marker	Position (cM)	χ^2	Nephritis		No nephritis	
			BXSB	BXSB/B10	BXSB	BXSB/B10
D1Mit3	11.0	13.6*	44	35	11	38
D1Mit5	32.8	15.7**	48	35	13	42
D1Nds2	53.5	12.7*	49	34	14	37
D1Mit12	63.1	20.2**	47	36	10	45
D1Mit102	73.0	9.2*	43	42	14	42
D1Mit33	81.6	4.5	40	45	16	39
D1Mit107	85.0	5.1	41	44	16	39
D1Mit15	87.9	5.7	40	43	15	39
D1Mit36	92.3	4.6	38	46	15	40
D1Mit403	100.0	6.9	40	44	14	41
D1Mit223	106.3	3.6	40	45	17	38

*$p < 0.0034$, indicating suggestive linkage; **$p < 1 \times 10^{-4}$.

Eric Lander's group (Beutler and Cerami, 1987; Paterson et al., 1988; Lincoln et al., 1992). In disease models, QTL mapping has been utilized to investigate apoptosis defects in NOD mice (Garchon et al., 1994; Gill et al., 1995; Penha-Gonçalves et al., 1995) and autoantibody production in lupus mice (Vyse et al., 1996a, b). Mapping QTLs in humans is mathematically more complex than in the mouse; however, extreme phenotype analysis (Carey and Williamson, 1991) and regression-based techniques (Bonney et al., 1988; Fulker and Cardon, 1994) can be employed. The latter approach can be modified so that it is applicable in allele-sharing based studies. Some of the methods used to detect human QTLs have been reviewed recently (Ghosh and Schork, 1996).

An inherent part of conducting a genome-wide scan is multiple hypothesis testing related to the large number of loci that is screened for linkage. To avoid false-positive results, it is essential to study a sufficiently large number of mice (usually at least 150) and to take a low probability (p) value as a statistical threshold for linkage. Recently, rigorously derived criteria have been proposed based on the probability of a false-positive result in the context of whole genome scan (Lander and Kruglyak, 1995). For example, in an analysis of backcross mice, the proposed threshold is $p < 0.0034$ ($\chi^2 > 8.58$; lod > 1.9) to conclude suggestive linkage, and $p < 1 \times 10^{-4}$ ($\chi^2 > 15.1$; lod > 3.3) to conclude probable linkage. For p-values between 1×10^{-4} and 0.01, further breeding experiments should be performed for confirmation. Replication of statistically significant intervals (at $p < 0.01$) in separate crosses has been emphasized as an important criterion in the confirmation of true disease linkage. It has been recommended that new names be provided for only those disease-linked loci that meet criteria for probable linkage or confirmed linkage.

An additional means by which the genetic basis of disease can be investigated in mice is by the use of recombinant inbred (RI) strains. Such mice are generated by F_2 intercrossing followed by repeated brother–sister mating of progeny lines. The result is a mix of parental genes, and with continued breeding the RI strains approach homozygosity at each locus. Historically, RI strains have been very useful for identifying the positions of genes involved in single-gene traits, which can then be confirmed in a backcross analysis in which the genotyping is directed to that region. However, RI strains have not been useful in the analysis of complex genetic traits such as autoimmune disease (Drake et al., 1995). This usually relates to the relatively limited number of RI strains within a set and the need for many more strains when the genetic basis involves multiple susceptibility alleles, incomplete penetrance, genetic heterogeneity and complex genetic interactions. Furthermore, with the use of microsatellites and the feasibility of genome-wide linkage analysis, the use of RI strains for mapping has been obviated. However, the use of an RI strain in a backcross for a genome-wide analysis may be advantageous. Since the RI strain is homozygous for some regions which will be shared by the parental strain chosen for the backcross, these regions will not vary in the cross and are hence fixed. The number of markers necessary for total coverage of the genome will be reduced but any potential contribution from the fixed intervals cannot be investigated. If the full spectrum of disease is maintained in the RI strain, then these probably have little effect and the results will be valid.

Narrowing the interval for mice: congenics

It is important to emphasize that, especially in murine studies, adding more mice to a linkage analysis is not likely to narrow the interval appreciably. This is mostly related to the imprecise nature of disease expression in relation to the genotype (at any particular linked locus) and the relatively low frequency of recombination in the region around a linked locus. In murine studies, narrowing an interval containing an aetiological allele can best be achieved by breeding congenic mice. As an example, if strain A contains a disease-linked locus that is penetrant when strain A is crossed with strain B, then the interval surrounding this strain A locus can be bred on to the B background. To do this, strains A and B are crossed and the progeny are mated to strain B. The backcross mice are screened with genetic markers for the interval of interest from strain A. The selected mice are then repeatedly backcrossed to strain B for a minimum of six generations, with selection for the required strain A interval. After n generations of backcrossing, the congenic strain will contain the interval around the disease-linked locus, but only $2^{-(n+1)}$ of the genome from strain A. Once the strain A interval is fixed on a relatively homogeneous genetic background and shown to increase expression of disease, additional crossing and recombination within the interval allow more accurate localization of the disease gene. If the interval is reduced to 1–2 cM, positional cloning techniques and more directed candidate gene choices can be employed. More recently, a modification of the method for production of congenic strains, termed "speed congenic", has been employed. This approach uses a marker-assisted method to screen for contamination with the unwanted genome (strain A). Speed congenics reduce the time required to generate a congenic mouse strain from around 4 years to around 2 years (Markel et al., 1997; Wakeland et al., 1997).

EPISTASIS

This analysis of gene effects is complicated by epistatic interaction. Genes may act either independently or together. The precise nature of an interaction cannot be determined by linkage analyses. However, it is possible, given sufficient numbers, to determine whether the data best fit an epistatic or a non-interactive (also termed additive or heterogeneity) model (Risch, 1990b; Risch et al., 1993). The easiest form of epistasis to understand, and the most extreme form of interaction, is a multiplicative model. In this model, the total risk of disease is the product of the individual risks associated with each locus. In humans, λ_s would be the product of the λ of each locus. In mice, a function termed the genotype risk ratio (GRR) is defined with the same properties as λ_s with respect to modelling interactions. The GRR in a backcross is defined as follows: GRR = $A/2A_2$, in which A represents the total number of affected animals and A_2 is the number of affected animals that is homozygous at a susceptibility locus. Thus, in a fully recessive single-gene disorder, GRR = 0.5 and 50% of the animals would be affected. The use of the GRR function is discussed in greater depth by Risch et al. (1993).

MAPPING STUDIES IN TYPE I DIABETES

The greater part of the techniques described has been used in the study of one or more complex trait. However, there are more data from both human and animal studies for insulin-dependent diabetes than for the majority of other diseases. There are currently about 14 corroborated non-MHC mouse Idd loci in studies of NOD mice and in studies of type I diabetes patients, more than 10 IDDM loci have been reported, but some of these are likely to be false positive. These data will be used to illustrate the power of the techniques discussed.

Analysis of human type I diabetes mellitus

Until the advent of genome-wide linkage studies, genetic predisposition to autoimmune disease in humans was studied by examining the relationship between disease and candidate genes, especially MHC, selected because of their biological relevance to the disease (see below). For example, in type I diabetes, association studies had demonstrated the predisposing influence of various MHC haplotypes (designated IDDM1), as well as the insulin gene (INS), located on chromosome 11p15 and designated IDDM2 (Julier et al., 1991). The contribution from IDDM2 has also been suggested in linkage analyses and confirmed by transmission disequilibrium testing.

Genome-wide linkage studies of sib pairs with type I diabetes have been conducted by several groups (Davies et al., 1994; Field et al., 1994; Hashimoto et al., 1994; Copeman et al., 1995). The loci that are in probable and suggestive linkage with disease are reviewed in Todd and Farrall (1996). The MHC was identified to be in linkage with type I diabetes using genome-wide scans. The relative risk, λ_s, for IDDM1 was estimated to be 2.6, compared with a total risk equal to 15 (Risch, 1987). An estimate of the contribution of the MHC to familial clustering is influenced by the error in the risk assessment of the locus itself and whether there is epistasis between the MHC and other susceptibility loci. Thus, the contribution of the MHC to the total genetic risk is between 20% and 53%, with the remainder being accounted for by the combined effect of non-MHC genes (Todd, 1995). None of the non-MHC loci has a disproportionate contribution to disease. The locus-specific λ_s values for all of the loci lie between 1.07 and 1.45, a relatively low value, indicating that each confers a low risk. As mentioned above, the MHC (IDDM1) and the insulin genes (IDDM2) have been corroborated as genetic contributors to disease. There are four additional loci for which there is currently sufficient evidence from linkage and association studies to justify fine mapping experiments: IDDM4 (FGF/11q13), IDDM5 (ESR/6q22), IDDM8 (D6S281/6q27) and IDDM12 (CTLA4/2q33). IDDM4, 5 and 8 were detected by genome-wide scanning, whereas IDDM12 was initially noted by a candidate gene approach (Nisticó et al., 1996). So far, two genome-wide scans using at least 250 markers have been completed in 61 (Hashimoto et al., 1994) and 96 (Davies et al., 1994) sib-pair families. Larger numbers of families (up to at least 600) have been used in combined analyses to confirm IDDM2, 4, 5 and 8 (Davies et al., 1996; Luo et al., 1996). It has been estimated, however, that over 1000 affected sib-pair families will be required to be confident that all genes with effects equivalent to that of the insulin gene locus, IDDM2 ($\lambda_s = 1.25$), have been detected.

Several named type I diabetes loci, including IDDM2, IDDM6, IDDM7, IDDM10 and IDDM12, have been analysed by linkage disequilibrium mapping. The power of this technique has been amply illustrated by the localization of IDDM2 to polymorphism within a variable number tandem repeat (VNTR) at the 5' end of the insulin gene (Bell et al., 1984; Julier et al., 1991). The localization of IDDM2 to within a ~4 kb portion of the genome, in the insulin gene VNTR, was achieved because of recombinants at the 5' end of the gene (Bennett et al., 1995). This enabled the dissociation of polymorphic markers that were associated with type I diabetes from those that were not associated with disease. Such recombinants are essential to determine which polymorphisms were associated with IDDM directly from those that were in association because of linkage disequilibrium. The VNTR has been shown to influence the transcription of the insulin gene within the thymus, such that protective VNTR alleles are associated with higher levels of insulin mRNA (Vafiadis et al., 1997; Pugliese et al., 1997). Since insulin is a target autoantigen in type I diabetes, thymic expression might alter central tolerance. However, the role of VNTR is further complicated by the observation that this region of the genome is subject to genomic imprinting (Polychronakos et al., 1995). Thus, expression from this region is dependent on the parental origin, varying with inheritance of maternal or paternal alleles. Such a parent-of-origin effect was observed in a transmission analysis in over 1000 families (Bennett et al., 1997). In this study, the function of the VNTR was influenced by the type of the untransmitted paternal VNTR allele at IDDM2.

Analysis in Sardinian and Italian families has narrowed the interval containing IDDM7 to within 2 cM (~2 Mb of DNA) on chromosome 2q31 (Copeman et al., 1995). It is worth noting that the MLS of this locus by the initial sib-pair (96 pairs) analysis was only 1.2. An interval of 2 Mb is sufficiently small to allow a positional cloning strategy to be employed to define the aetiological allele at this locus. Human chromosome 2 has been the focus of mapping because it contains the homologue of the NOD susceptibility gene, Idd5, on mouse chromosome 1 (Cornall et al., 1991). In addition, there are several strong candidate genes in this region, including the related CD28 and CTLA4 genes. CD28 is a cell-surface receptor that interacts with B7 molecules on antigen-presenting cells and is involved in costimulation of T cells. CTLA-4 is a related cell-surface receptor but it is a powerful negative regulator of T cell activation and proliferation. Localization of IDDM12 to CTLA4 was reported in Italian families, and this result was extended in a collection of Italian and Spanish families by transmission disequilibrium testing for an adenine-to-guanine polymorphism in the first exon of the gene (Nisticó et al., 1996). Strong additional support for the association of the CTLA4 region has been provided by a case–control study in Belgium (Nisticó et al., 1996). The CTLA4 gene is also of interest because it has been associated with susceptibility to Graves' disease in separate studies (Todd and Farrall, 1996).

Modelling epistasis in humans with type I diabetes is very challenging. It is far easier in animal models and hence there is much more data in NOD mice (see below). There is, however, evidence for interaction between the MHC (IDDM1) and the insulin gene (IDDM2) (Dizier et al., 1994; Knapp et al., 1994). Likewise, it has been shown that in the UK population, there is little evidence for epistasis between IDDM1 and IDDM3 and also between IDDM1 and IDDM4 (Cordell et al., 1995). In the latter instances the data best fit an additive or a heterogeneity model. Such conclusions require verification and may be subject to population bias.

Analysis of insulin-dependent diabetes in NOD mice

NOD mice are genetically predisposed to develop insulitis and diabetes. Full expression of the disease is more prevalent in female mice, and 70% of females are affected by 6 months of age. From 3 months of age there is a mononuclear infiltrate in the pancreatic islets. These cells are predominantly macrophages and $CD4^+$ T cells; later, $CD8^+$ T cells are recruited. Evidence from transfer and blocking experiments indicates that all three cell types are involved in disease pathogenesis (Hanafusa et al., 1988; Shizuru et al., 1988; Katz et al., 1993). Transfer of disease with antigen-specific T-cell clones suggests that T cells with a Th1 cytokine profile are pathogenic in this model of diabetes (Haskins and McDuffie, 1990; Shimizu et al., 1993).

Several studies have shown linkage of NOD diabetes with the $H2^{g7}$ (MHC) locus (Wicker et al., 1995). This locus is designated Idd1, and the remaining named NOD loci, Idd2 to Idd15, Idd17 and Idd18, are non-MHC loci. The identity of Idd1 was thought to be solely the I-Ag7 molecule (there is no I-Eg7 product) (Todd et al., 1987). However, although the I-Ag7 molecule is certainly responsible for much of the Idd1 linkage, recent studies of mice with recombination within the MHC region suggest that other MHC genes may also contribute to disease susceptibility, and these have been termed Idd16 (Ikegami et al., 1995).

Evidence that non-MHC genes play a vital role in murine diabetes was most clearly derived from the study of various strains congenic for the NOD MHC, $H2^{g7}$. When $H2^{g7}$ was bred on to other non-autoimmune strains, $B10.H2^{g7}$ and $B6.H2^{g7}$, neither congenic mouse strain developed diabetes (Ikegami and Makino, 1993; Wicker et al., 1993). The variable effect of the non-MHC background is evidenced by another congenic strain, $NON.H2g^7$. The NON mouse is genetically much more similar to NOD than B6 or B10. Hence, it is not surprising that $NON.H2^{g7}$ mice have a 50% incidence of insulitis (Ikegami and Makino, 1993).

The first non-MHC locus to be identified, Idd2, was found in (NOD × NON)F_1 × NOD backcross mice (Prochazka et al., 1987). Idd2 has been shown to be closely linked to the Thy1 gene on chromosome 9 (Wicker et al., 1995). Subsequently, Idd2 to Idd10 were mapped in (NOD × $B10.H2^{g7}$)F_1 × NOD backcross mice (Cornall et al., 1991; Todd et al., 1991; Ghosh et al., 1993; Wicker et al., 1994). In contrast to the first cross, this cross with B10 congenic mice is matched with NOD mice at the MHC, thus only non-MHC genes were mapped. Of the nine non-MHC Idd genes mapped, all but two (Idd7 and Idd8) showed an increased incidence of disease in NOD homozygous mice. For Idd7 and Idd8, disease was increased in heterozygous mice, indicating that B10 genes at these positions contributed to disease susceptibility (Ghosh et al., 1993). The contribution of a disease gene from the low-susceptibility strain, or a protective allele from the high-susceptibility strain, has been reported in other studies (Hogarth et al., 1998) and these alleles are termed "transgressive" (Frankel, 1995). Other Idd loci, Idd11 and Idd12, were mapped by meta-analysis of backcross studies with non-autoimmune B6 and SJL mice (Morahan et al., 1994). A dominant NOD allele on chromosome 13, Idd14, was mapped in a (NOD × NON.H2^{g7})F$_2$ intercross (McAleer et al., 1995).

Is there evidence of epistasis in NOD mice? In backcross studies, less than 50% of animals are affected, suggesting an epistatic model for disease. In one study, 13% of (NOD × B6.H2^{g7})F$_1$ × NOD mice developed diabetes (Todd et al., 1991). When the GRRs of the individual linked loci were determined, the range of values for NOD susceptibility alleles was 0.51–0.85 (Risch et al., 1993). This indicates the variability in penetrance of the individual loci and that none was fully recessive. When various models were employed to fit all of the individual locus GRRs to the observed disease incidence in backcross mice, two conclusions were apparent: (i) no model fitted the data exactly, although a multiplicative model fitted the data with much greater confidence than did an additive one; and (ii) not all of the observed risk could be accounted for by the observed loci, suggesting that there were additional weaker Idd loci to be mapped (Wicker et al., 1995). The interbreeding of congenic mice will enable conclusive demonstration of epistasis. Until small intervals or actual loci are mapped for the majority of susceptibility loci, proof will remain elusive.

A clue as to the function of some of the above Idd genes has been given by examining (NOD × B10.H2^{g7})F$_1$ × NOD backcross mice for linkage with respect to insulitis rather than overt diabetes. Of the Idd loci, only Idd3 and Idd10 which are linked on chromosome 3, and Idd5 on chromosome 1, were found to be in linkage with histological evidence of islet inflammation (Wicker et al., 1995).

Three loci, Nod1, Nod2 and Nod3, have been mapped as quantitative traits. Nod1 and Nod3 are linked to T cell resistance to apoptosis induction (Garchon et al., 1994; Penha-Gonçalves et al., 1995). Nod1 does not colocalize with any Idd loci, although it does map close to the Bcl2 gene (Garchon et al., 1994), which is involved in apoptosis resistance. Nod3 colocalizes with Idd6 on chromosome 6 (Penha-Gonçalves et al., 1995). Nod2 maps an NOD subphenotype that involves proliferative unresponsiveness to TCR/CD3 complex cross-linking. This gene maps to chromosome 11, close to Idd4, which is at the site of a cluster of b-chemokine genes (Gill et al., 1995).

Although the identity of any of the mapped non-MHC Idd genes has yet to be established, the position of some Idd genes raise some interesting possibilities regarding candidate genes. For example, Idd3 on chromosome 3 is close to the interleukin-2 (IL-2) gene. Recently, a congenic analysis of the Idd3 interval has located this disease gene within a 0.35 cM interval (Denny et al., 1997). Moreover, the establishment of YAC contigs across this interval and the analysis of novel polymorphic markers have revealed that a point mutation in exon 1 of the Il2 gene consistently segregates with Idd3. Whether this mutation alters function and how this IL-2 allele would predispose to disease remain to be established. The value of congenics is further demonstrated in the analysis of Idd10 on chromosome 3. Subdivision of the Idd10 interval revealed that there were two distinct genes, Idd10 and the newly designated Idd17, some 10 cM apart. Furthermore, recombination within the congenic mice showed that neither Idd10 nor Idd17 cosegregated with Fcgr1, a previously favoured candidate gene in this area (Podolin et al., 1997). Additional congenics have now subdivided the Idd10 effect further, yielding the new locus, Idd18 (Podolin et al., 1998). This complex situation, resulting in the identification of four loci (Idd3, Idd10, Idd17 and Idd18) from the initially defined single locus on chromosome 3 (Idd3), underlines the problems in mapping multifactorial disease. Idd2 on chromosome 9 is close to the genes encoding the CD3 components of the TCR complex. Idd5 maps on chromosome 1 to the region of the Ctla4 and Cd28 genes, both of which encode costimulatory molecules involved in T cell activation. Finally, Idd13 maps to the site of the IL-1 gene, Il1a. It must be remembered, however, that the list of candidate genes at many Idd loci is large and that none has yet been identified as an aetiological mutation.

What is the relation between the loci mapped in the NOD mouse and those mapped in human IDDM? The human loci can be positioned on the mouse map when there is known to be an area of the human genome that contains homologous genes to the mouse, regions termed syntenic. When the human IDDM loci are placed in syntenic mouse regions it can be seen that the only non-MHC loci that appear to colocalize are Idd5 with IDDM7 and IDDM12. This process almost certainly underestimates colocalization because of the incomplete data regarding synteny. A striking reflection of the mouse and human data exists at the MHC, where some of the alleles contributing to disease have been identified. Thus, the HLA-DQβ variable sequences that influence disease in humans are similar to differences in the murine H2-A locus that affects NOD diabetes (Acha-Orbea and McDevitt, 1987; Todd et al., 1987; Tisch and McDevitt, 1996). The NOD mouse MHC encodes the I-Ag7 molecule, which is strongly linked with diabetes, and both the human and mouse class II alleles that enhance disease expression (or protect from disease) have been shown to have similar sequence variations. In humans IDDM2 has been mapped to INS VNTR, the mouse equivalent being Ins2 on distal chromosome 7 (Bennett et al., 1995). This locus has yet to be mapped as being linked with murine diabetes.

Many of the techniques discussed have been brought to bear in the study of diabetes both in the human and in an animal model, the NOD mouse. A large number of intervals has been implicated in the disease process as a whole and, indeed, in specific parts of the disease. However, identification of a disease gene remains elusive. It is inconceivable that much more time will elapse before at least one of the loci, particularly in the mouse, is known and then a number of as yet unanswered questions can be broached. What is the nature of disease susceptibility in a gene? Will disease genes fall into a particular category such as low expressors or low responders? Can information be translated from mouse to human? And, potentially the most elusive and important question of all: will increased understanding of the genetics underlying a disease aid in the treatment of that disease? It is debatable whether or not understanding the genetic basis of sickle cell anaemia, for example, has affected treatment. Is genetics the ultimate panacea that it is often held out to be? The answers may soon be known.

REFERENCES

Aaltonen, J., Björses, P., Sandkuijl, L. et al. (1994). An autosomal locus causing using autoimmune disease: autoimmune polyglandular disease type I assigned to chromosome 21. *Nature Genet.* **8**: 83–87.

Acha-Orbea, H. and McDevitt, H.O. (1987). The first external domain of the non-obese diabetic mouse class II I-Ab chain is unique. *Proc. Natl Acad. Sci. USA* **84**: 2435–2439.

Adams, M.P. (1996). Serial analysis of gene expression: ESTs get smaller. *Bioessays* **18**: 261–262.

Bell, G.I., Horita, S. and Karam, J.H. (1984). A polymorphic locus near the human insulin gene is associated with insulin-dependent diabetes mellitus. *Diabetes* **33**: 176–183.

Bennet Lucassen, A.M., Gough, S.C., Powell, E.E. et al. (1995). Susceptibility to human type 1 diabetes at *IDDM2* is determined by tandem repeat variation at the insulin gene minisatellite locus. *Nature Genet.* **9**: 284–292.

Bennett, S.T., Wilson, A.J., Esposito, L. et al. (1997). Insulin VNTR allele-specific effect in type 1 diabetes depends on the identity of the untransmitted paternal allele. *Nature Genet.* **17**: 350–352.

Bertelsen, A.H. and Velculescu, V.E. (1998). High-throughput gene expression analysis using SAGE™. *Drug Discovery Today* **3**: 152–159.

Bertioli, D.J., Schlichter, U.H.A., Adams, M.J. et al. (1995). An analysis of differential display shows a strong bias towards high copy number mRNAs. *Nucl. Acids Res.* **23**: 4520–4523.

Beutler, B. and Cerami, A. (1987). Cachectin: more than a tumor necrosis factor. *N. Engl. J. Med.* **316**: 379–385.

Bird, A. (1992). The essentials of DNA methylation. *Cell* **70**: 5–8.

Blackwelder, W.C. and Elston, R.C. (1985). A comparison of sib-pair linkage tests for disease susceptibility loci. *Genet. Epidemiol.* **2**: 85–97.

Boguski, M.S. and Schuler, G.D. (1995). Establishing a human transcript map. *Nature Genet.* **10**: 369–371.

Bohlander, S.K., Espinosa III, R., Le Beau, M.M. et al. (1992). A method for the rapid sequence-independent amplification of microdissected chromosomal material. *Genomics* **13**: 1322–1324.

Bonney, G.E., Lathrop, G.M. and Lalouel, J.-M. (1988). Combined linkage and segregation analysis using regressive models. *Am. J. Hum. Genet.* **43**: 29–37.

Buckler, A.J., Chang, D.D., Graw, S.L. et al. (1991). Exon amplification: a strategy to isolate mammalian genes based on RNA splicing. *Proc. Natl Acad. Sci. USA* **88**: 4005–4009.

Burmeister, M. and Ulanovsky, L., eds (1992). *Methods in Molecular Biology*, Vol. 12, *Pulsed-Field Gel Electrophoresis*, Humana Press, Totowa, NJ.

Carey, G. and Williamson, J. (1991). Linkage analysis of quantitative traits: increased power by using selected samples. *Am. J. Hum. Genet.* **49**: 786–796.

Caulfield, M., Lavender, P., Farrell, M. et al. (1994). Linkage of the angiotensin gene to essential hypertension. *N. Engl. J. Med.* **330**: 1629–1633.

Charmley, P. and Concannon, P. (1995). PCR-based genotyping and haplotype analysis of human TCRBV gene segment polymorphisms. *Immunogenetics* **42**: 254–261.

Cheung. V.G., Gregg, J.P., Gogolin-Ewens, K.J. et al. (1998). Linkage-disequilibrium mapping without genotyping. *Nature Genet.* **18**: 225–230.

Church, D.M., Stotler, C.J., Rutter, J.L. et al. (1994). Isolation of genes from complex sources of mammalian genomic DNA using exon amplification. *Nature Genet.* **6**: 98–105.

Collins, F. (1995). Positional cloning. *Nature Genet.* **9**: 347–350.

Copeman, J.B., Cucca, F., Hearne, C.M. et al. (1995). Linkage disequilibrium mapping of a type I diabetes susceptibility gene (*IDDM7*) to chromosome 2q31–q33. *Nature Genet.* **9**: 80–85.

Cordell, H.J., Todd, J.A., Bennett, S.T. et al. (1995). Two-locus maximum lod score analysis of a multifactorial trait: joint consideration of *IDDM2* and *IDDM4* with *IDDM1* in type I diabetes. *Am. J. Hum. Genet.* **57**: 920–934.

Cornall, R.J., Prins, J.B., Todd, J.A. et al. (1991). Type 1 diabetes in mice is linked to the interleukin-1 receptor and *Lsh/Ity/Bcg* genes on chromosome 1. *Nature (London)* **353**: 262–265.

Cotton, R.G., Rodrigues, N.R. and Campbell, R.D. (1988). Reactivity of cytosine and thymine in single-base-pair mismatches with hydroxylamine and osmium tetroxide and its application to the study of mutations. *Proc. Natl Acad. Sci. USA* **8**: 4397–4401.

Craig, J.E., Rochette, J., Fisher, C.A. et al. (1996). Dissecting the loci controlling fetal haemoglobin production on chromosomes 11p and 6q by the regressive approach. *Nature Genet.* **12**: 58–64.

Davies, J.L., Kawaguchi, Y., Bennett, S.T. et al. (1994). A genome-wide search for human type I diabetes susceptibility genes. *Nature* **371**: 130–136.

Davies, J.L., Cucca, F., Goy, J.V. et al. (1996). Saturation multipoint linkage mapping of chromosome 6q in type 1 diabetes. *Hum. Mol. Genet.* **5**: 1071–1074.

Day, N.E. and Simons, M.J. (1976). Disease susceptibility genes – their identification by multiple case family studies. *Tissue Antigens* **8**: 109–119.

Denny, P., Lord, C.J., Hill, N.J. et al. (1997). Mapping of the insulin-dependent diabetes mellitus locus, *Idd3*, to a 0.35 cM interval containing the *Interleukin-2* gene. *Diabetes* **46**: 695–700.

Diatchenko, L., Lau, Y.F., Campbell, A.P. et al. (1996). Suppression subtractive hybridization: a method for generating differentially regulated or tissue-specific cDNA probes and libraries. *Proc. Natl Acad. Sci. USA* **93**: 6025–6030.

Dib, C., Fauré, S., Fizames, C. et al. (1996). A comprehensive genetic map of the human genome based on 5,264 microsatellites. *Nature* **380**: 152–154.

Dietrich, W.F., Miller, J.C., Steen, R.G. et al. (1994). A genetic map of the mouse with 4,006 simple sequence length polymorphisms. *Nature Genet.* **7**: 220–245.

Dietrich, W.F., Miller, J., Steen, R. et al. (1996). A comprehensive map of the mouse genome. *Nature* **380**: 149–151.

Dizier, M.-H., Babron, M.-C. and Clerget-Darpoux, F. (1994). Interactive effect of two candidate genes in a disease: extension of marker-association-segregation chi^2 method. *Am. J. Hum. Genet.* **55**: 1042–1049.

Drake, C.G., Rozzo, S.J., Hirschfeld, H.F. et al. (1995). Analysis of the New Zealand Black contribution to lupus-like renal disease. Multiple genes that operate in a threshold manner. *J. Immunol.* **154**: 2441–2447.

Edwards, A.W.F. (1992). *Likelihood*. Johns Hopkins University Press, Baltimore, MD.

Falk, C.T. and Rubinstein, P. (1987). Haplotype relative risk: an easy reliable way to construct a proper control sample for risk calculations. *Ann. Hum. Genet.* **51**: 227–233.

Field, L.L., Tobias, R. and Magnus, T. (1994). A locus on chromosome 15q26 (*IDDM3*) produces susceptibility to insulin-dependent diabetes mellitus. *Nature Genet.* **8**: 189–194.

Fields, C., Adams, M.D., White, O. et al. (1994). How many genes in the human genome? *Nature Genet.* **7**: 345–346.

Fishman, P.M., Suarez, B., Hodge, S.E. et al. (1978). A robust method for the detection of linkage in familial diseases. *Am. J. Hum. Genet.* **30**: 308–321.

Fodor, S.P.A. (1997). Massively parallel genomics. *Science* **277**: 393–395.

Fodor, S.P., Rava, R.P., Huang. X.C. et al. (1993). Multiplexed biochemical assays with biological chips. *Nature* **364**: 555–556.

Fodor, S.P., Read, J.L., Pirrung, M.C. et al. (1991). Light-directed, spatially addressable parallel chemical synthesis. *Science* **251**: 767–773.

Frankel, W.N. (1995). Taking stock of complex trait genetics in mice. *Trends Genet.* **11**: 471–477.

Fulker, D.W. and Cardon, L.R. (1994). A sib-pair approach to interval mapping of quantitative trait loci. *Am. J. Hum. Genet.* **54**: 1092–1103.

Garchon, H.-J., Luan, J.-J., Eloy, L. et al. (1994). Genetic analysis of immune dysfunction in non-obese diabetic (NOD) mice: mapping of a susceptibility locus close to the *Bcl-2* gene correlates with increased resistance of NOD T cells to apoptosis induction. *Eur. J. Immunol.* **24**: 380–384.

Ghosh, S. and Schork, N.J. (1996). Genetic analysis of NIDDM: the study of quantitative traits. *Diabetes* **45**: 1–14.

Ghosh, S., Palmer, S.M., Rodrigues, N.R. et al. (1993). Polygenic control of autoimmune diabetes in nonobese diabetic mice. *Nature Genet.* **4**: 404–409.

Gill, B.M., Jaramillo, A., Ma, L. et al. (1995). Genetic linkage of thymic T cell proliferative unresponsiveness to mouse chromosome 11 in NOD mice. *Diabetes* **44**: 614–619.

Grompe, M. (1993). The rapid detection of unknown mutations in nucleic acids. *Nature Genet.* **5**: 111–117.

Haldane, J.B.S. and Smith, C.A.B. (1947). New estimate of linkage between genes for colour-blindness and haemophilia in man. *Ann. Eugenics* **14**: 10–31.

Haldi, M.L., Strickland, C., Lim, P. et al. (1996). A comprehensive large-insert yeast artificial chromosome library for physical mapping of the mouse genome. *Mamm. Genome.* **7**: 767–769.

Hanafusa, T., Sugihara, S., Fujino-Kurihara, H. et al. (1988). Induction of insulitis by adoptive transfer with L3T4^+Lyt2^- T-lymphocytes in T-lymphocyte-depleted NOD mice. *Diabetes* **37**: 204–208.

Hanson, A.U., Poustka, A. and Trowsdale, J. (1991). New genes in the class II region of the human major histocompatibility complex. *Genomics* **10**: 417–424.

Hashimoto, L., Habita, C., Beressl, J.P. et al. (1994). Genetic mapping of a susceptibility locus for insulin-dependent diabetes mellitus on chromosome 11q. *Nature* **371**: 161–164.

Haskins, K. and McDuffie, M. (1990). Acceleration of diabetes in young NOD mice with a $CD4^+$ islet-specific T cell clone. *Science* **249**: 1433–1436.

Heng, H.H.Q., Squire, J. and Tsui, L.C. (1992). High-resolution mapping of mammalian genes by *in situ* hybridization to free chromatin. *Proc. Natl Acad. Sci. USA* **89**: 9509–9513.

Hillier, L.D., Lennon, G., Becker, M. et al. (1996). Generation and analysis of 280,000 human expressed sequence tags. *Genome Res.* **6**: 807–828.

Hogan, B., Beddington, R., Costantini, F. et al. (1994). *Manipulating the Mouse Embryo, A Laboratory Manual,* 2nd edn, Cold Spring Harbor Laboratory Press, Cold Spring Harbor.

Hogarth, M.B., Slingsby, J.H., Allen, P.J. et al. (1998). Multiple lupus susceptibility loci map to chromosome 1 in BXSB mice. *J. Immunol.* **161**: 2753–2761.

Ikegami, H. and Makino, S. (1993). Genetic susceptibility to insulin-dependent diabetes mellitus: from NOD mice to humans. In *Lessons from Animal Diabetes* (ed. E. Shafrir), pp. 39–50, Smith-Gordon, London.

Ikegami, H., Makino, S., Yamoto, E. et al. (1995). Idenitification of a new susceptibility locus for insulin-dependent diabetes mellitus by ancestral haplotype congenic mapping. *J. Clin. Invest.* **96**: 1936–1942.

Itoh, N., Yonehara, S., Ishii, A. et al. (1991). The polypeptide encoded by the cDNA for human cell surface antigen Fas can mediate apoptosis. *Cell* **66**: 233–243.

Jin, H., Cheng, X., Diatchenko, L. et al. (1997). Differential screening of a subtracted cDNA library: a method to search for genes preferentially expressed in multiple tissues. *Biotechniques* **23**: 1084–1086.

Julier, C., Hyer, R.N., Davies, J. et al. (1991). *Insulin-IGF2* region on chromosome 11p encodes a gene implicated in *HLA-DR4*-dependent diabetes susceptibility. *Nature* **354**: 155–159.

Katz, J., Benoist, C. and Mathis, D. (1993). Major histocompatibility complex class I molecules are required for the development of insulitis in nonobese diabetic mice. *Eur. J. Immunol.* **23**: 3358–3360.

Knapp, M., Seuchter, M.A. and Baur, M.P. (1994). Two-locus disease models with two marker loci: the power of affected-sib-pair tests. *Am. J. Hum. Genet.* **55**: 1030–1041.

Kruglyak, L. (1997). The use of a genetic map of biallelic markers in linkage studies. *Nature Genet.* **17**: 21–24.

Kruglyak, L. and Lander, E.S. (1994). High-resolution genetic mapping of complex traits. *Am. J. Hum. Genet.* **56**: 1212–1223.

Kruglyak, L. and McAllister, L. (1998). Who needs genetic markers? *Nature Genet.* **18**: 200–202.

Lander, E. and Kruglyak, L. (1995). Genetic dissection of complex traits: guidelines for interpreting and reporting linkage results. *Nature Genet*, **11**: 241–247.

Lander, E.S. (1993). Finding similarities and differences among genomes. *Nature Genet.* **4**: 5–6.

Lander, E.S. and Botstein, D. (1989). Mapping Mendelian factors underlying quantitative traits using RFLP linkage maps. *Genetics* **121**: 185–199.

Lander, E.S. and Schork, N.J. (1994). Genetic dissection of complex traits. *Science* **265**: 2037–2048.

Lange, K. (1986). A test statistic for the affected-sib-set method. *Ann. Hum. Genet.* **50**: 283–290.

Larin, Z., Monaco, A.P. and Lehrach, H. (1991). Yeast artificial chromosome libraries containing large inserts from mouse and human DNA. *Proc. Natl Acad. Sci. USA* **88**: 4123–4127.

Ledbetter, S.A., Nelson, D.L., Warren, S.T. et al. (1990). Rapid isolation of DNA probes within specific chromosome regions by interspersed repetitive sequence polymerase chain reaction. *Genomics* **6**: 475–481.

Lefebvre, S., Burglen, L., Frzal, J. et al. (1998). The role of the *SMN* gene in proximal spinal muscular atrophy. *Hum. Mol. Genet.* **7**: 1531–1536.

Lerman, L.S. and Silverstein, K. (1987). Computational simulation of DNA melting and its application to denaturing gradient gel electrophoresis. *Methods Enzymol.* **155**: 482–501.

Lewin, B. (1995). Genes for SMA: multum in parvo. *Cell* **80**: 1–5.

Liang, P. and Pardee, A.B. (1992). Differential display of eukaryotic messenger RNA by means of the polymerase chain reaction. *Science* **257**: 967–971.

Liang, P. and Pardee, A.B. (1997). Differential display. A general protocol. *Methods Mol. Biol.* **85**: 3–11.

Lincoln, S.E., Daly, M. and Lander, E.S. (1992). *Mapping Genes Controlling Quantitative Traits with MAPMAKER/QTL 1.1*, 2nd edn, Whitehead Institute Technical Report, Cambridge, MA.

Lisitsyn, N.A. (1995). Representational difference analysis: finding the differences between genomes. *Trends Genet.* **11**: 303–306.

Lisitsyn, N.A., Segre, J.A., Kusumi, K. et al. (1994). Direct isolation of polymorphic markers linked to a trait by genetically directed representational difference analysis. *Nature Genet.* **6**: 57–63.

Lockhart, D.J., Dong, H., Byrne, M.C. et al. (1996). Expression monitoring by hybridization to high-density oligonucleotide arrays. *Nature Biotechnol.* **14**: 1675–1689.

Love, J.M., Knight, A.M., McAleer, M.A. et al. (1990). Towards construction of a high resolution map of the mouse genome using PCR-analysed microsatellites. *Nucl. Acids Res.* **18**: 4123–4130.

Lovett, M., Kere, J. and Hinton, L.M. (1991). Direct selection: a method for the isolation of cDNAs encoded by large genomic regions. *Proc. Natl Acad. Sci. USA* **88**: 9628–9632.

Lukyanov, K., Diatchenko, L., Chenchik, A. et al. (1997). Construction of cDNA libraries from small amounts of total RNA using the suppression PCR effect. *Biochem. Biophys. Res. Commun.* **230**: 285–288.

Luo, D.-F., Buzzetti, R., Rotter, J.I. et al. (1996). Confirmation of three susceptibility genes to insulin-dependent diabetes mellitus: *IDDM4*, *IDDM5* and *IDDM8*. *Hum. Mol. Genet.* **5**: 693–698.

Markel, P., Shu, P., Ebeling, C. et al. (1997). Theoretical and empirical issues for marker-assisted breeding of congenic mouse strains. *Nature Genet.* **17**: 280–284.

McAleer, M.A., Reifsnyder, P., Palmer, S.M. et al. (1995). Cross of NOD mice with the related NON strain: a polygenic model for IDDM. *Diabetes* **44**: 1180–1195.

Morahan, G., McClive, P.J., Huang, D. et al. (1994). Genetic and physiological association of diabetes susceptibility with raised Na^+/H^+ exchange activity. *Proc. Natl Acad. Sci. USA* **91**: 5898–5902.

Myers, R.M., Larin, Z. and Maniatis, T. (1985). Detection of single base substitutions by ribonuclease cleavage at mismatches in RNA: DNA duplexes. *Science* **230**: 1242–1246.

Nelson, S.F., McCusker, J.H., Sander, M.A. et al. (1993). Genomic mismatch scanning: a new approach to genetic linkage mapping. *Nature Genet.* **4**: 11–17.

Neubauer, G., King, A., Rappsilber, J. et al. (1998). Mass spectrometry and EST-database searching allows characterization of the multi-protein spliceosome complex. *Nature Genet.* **20**: 46–50.

Nisticó, L., Buzzetti, R., Pritchard, L.E. et al. (1996). The *CTLA-4* gene region of chromosome 2q33 is linked to, and associated with, type 1 diabetes. Belgian Diabetes Registry. *Hum. Mol. Genet.* **5**: 1075–1080.

Orita, M., Iwahana, H., Kanazawa, H. et al. (1989a). Detection of polymorphisms of human DNA by gel electrophoresis as single-strand conformation polymorphisms. *Proc. Natl Acad. Sci. USA* **86**: 2766–2770.

Orita, M., Suzuki, Y., Sekiya, T. et al. (1989b). Rapid and sensitive detection of point mutations and DNA polymorphisms using the polymerase chain reaction. *Genomics* **5**: 874–879.

Ott, J. (1992). The future of multilocus linkage analysis. *Ann. Med.* **24**: 401–403.

Parimoo, S,. Patanjali, S.R., Shukla, H. et al. (1991). cDNA selection: efficient PCR approach for the selection of cDNAs encoded in large chromosomal DNA fragments. *Proc. Natl Acad. Sci. USA* **88**: 9623–9627.

Paterson, A.H., Lander, E.S., Hewitt, J.D. et al. (1988). Resolution of quantitative traits into Mendelian factors by using a complete RFLP linkage map. *Nature* **335**: 721–726.

Penha-Gonçalves, C., Leijon, K., Persson, L. et al. (1995). Type I diabetes and the control of dexamethasone-induced apoptosis in mice maps to the same region on chromosome 6. *Genomics* **28**: 398–404.

Peterson, A.C., Di Renzio, A., Lehesjoki, A.E. et al. (1995). The distribution of linkage disequilibrium over anonymous genome regions. *Hum. Mol. Genet.* **4**: 887–894.

Podolin, P.L., Denny, P., Lord, C.J. et al. (1997). Congenic mapping of the insulin dependent diabetes (*Idd*) gene, *Idd10*, localizes two genes mediating the *Idd10* effect, and eliminates the candidate *Fcgr1*. *J. Immunol.* **159**: 1835–1843.

Podolin, P.L., Denny, P., Armitage, N. et al. (1998). Localization of two insulin-dependent diabetes (*Idd*) genes to the *Idd10* region on mouse chromosome 3. *Mamm. Genome* **9**: 283–286.

Polyak, K., Xla, Y., Zweler, J. et al. (1997). A model for p53-induced apoptosis. *Nature* **389**: 300–305.

Polychronakos, C., Kukuvitis, A., Giannoukakis, N. et al. (1995). Parental imprinting effect at the *INS-IGF2* diabetes susceptibility locus. *Diabetologia* **38**: 715–719.

Prochazka, M., Leiter, E.H., Serreze, D.V. et al. (1987). Three recessive loci required for insulin-dependent diabetes in nonobese diabetic mice [published erratum appears in *Science* (1988) **242**: 945]. *Science* **237**: 286–289.

Pugliese, A., Zeller, M., Fernandez, A.J. Jr et al. (1997). The insulin gene is transcribed in the human thymus and transcription levels correlate with allelic variation at the *INS VNTR-IDDM2* susceptibility locus for type 1 diabetes. *Nature Genet.* **15**: 293–297.

Rieux-Laucat, F., Le Deist, F., Hivroz, C. et al. (1995). Mutations in Fas associated with human lymphoproliferative syndrome and autoimmunity. *Science* **268**: 1347–1349.

Risch, N. (1987). Assessing the role of HLA-linked and unlinked determinants of disease. *Am. J. Hum. Genet.* **40**: 1–14.

Risch, N. (1990a). Linkage strategies for genetically complex traits. I. Multilocus models. *Am. J. Hum. Genet.* **46**: 222–228.

Risch, N. (1990b). Linkage strategies for genetically complex traits. II. The power of affected relative pairs. *Am. J. Hum. Genet.* **46**: 229–241.

Risch, N. and Merikangas, K. (1996). The future of genetic studies of complex human disease. *Science* **273**: 1516–1517.

Risch, N., Ghosh, S. and Todd, J.A. (1993). Statistical evaluation of multiple-locus linkage data in experimental species and relevance to human studies: application to nonobese diabetic (NOD) mouse and human insulin-dependent diabetes mellitus (IDDM). *Am. J. Hum. Genet.* **53**: 702–714.

Royer-Pokora, B., Kunkel, L., Monaco, A.P. et al. (1986). Cloning the gene for an inherited human disorder – chronic granulomatous disease – on the basis of its chromosomal location. *Nature* **322**: 32–38.

Rubin, E.M. and Smith, D.J. (1997). Optimizing the mouse to sift sequence for function. *Trends Genet.* **13**: 423–426.

Sagerstrom, C.G., Sun, B.I. and Sive, H.L. (1997). Subtractive cloning: past, present and future. *Annu. Rev. Biochem.* **66**: 751–783.

Schalkwyk, L.C., Francis, F. and Lehrach, H. (1995). Techniques in mammalian genome mapping. *Curr. Opin. Biotech.* **6**: 37–43.

Schena, M., Shalon, D., Davis, R.W. et al. (1995). Quantitative monitoring of gene expression patterns with a complementary DNA microarray. *Science* **270**: 467–470.

Schlessinger, D. (1990). Yeast artificial chromosomes: tools for mapping and analysis of complex genomes. *Trends Genet.* **6**: 248–258.

Sheffield, V.C., Cox, D.R., Lerman, L.S. et al. (1989). Attachment of a 40 base-pair G + C-rich sequence (GC-clamp) to genomic DNA fragments by the polymerase chain reaction results in improved detection of single-base changes. *Proc. Natl Acad. Sci. USA* **86**: 232–236.

Sheffield, V.C., Beck, J.S., Kwitek, A.E. et al. (1993). The sensitivity of single-strand conformation polymorphism analysis for the detection of single base substitutions. *Genomics* **16**: 325–332.

Shimizu, J., Kanagawa, O. and Unanue, E.R. (1993). Presentation of b-cell antigens to CD4$^-$ and CD8$^-$ T cells in nonobese diabetic mice. *J. Immunol.* **151**: 1723–1730.

Shizuru, J.A., Taylor-Edwards, C., Banks, B.A. et al. (1988). Immunotherapy of the nonobese diabetic mouse: treatment with an antibody to T-helper lymphocytes. *Science* **240**: 659–662.

Simon, M., Bourel, M., Fauchet, R. et al. (1976). Association of HLA-A3 and HLA-B14 antigens with idiopathic haemochromatosis. *Gut* **17**: 332–334.

Smith, D.J., Stevens, M.E., Sudanagunta, P. et al. (1997). Functional screening of 2Mb of human chromosome 21q22.2 in transgenic mice implicates *minibrain* in learning defects associated with Down Syndrome. *Nature Genet.* **16**: 28–36.

Sompayrac, L., Jane, S., Burn, T.C. et al. (1995). Overcoming limitations of the mRNA differential display technique. *Nucl. Acids Res.* **23**: 4738–4739.

Spanakis, E. and Day, I.N.M. (1997). In *Genetics of Common Diseases* (eds I.N.M. Day and S.E Humphries), pp. 33–63, BIOS Scientific Publishers, Oxford.

Spielman, R.S., McGinnis, R.E. and Ewens, W.J. (1993). Transmission test for linkage disequilibrium: the insulin gene region and insulin-dependent diabetes mellitus (IDDM). *Am. J. Hum. Genet.* **52**: 506–516.

Suarez, B.K., Rice, J. and Reich, T. (1978). The generalized sib pair IBD distribution: its use in the detection of linkage. *Ann. Hum. Genet.* **42**: 87–94.

Terwilliger, J.D. and Ott, J. (1992). A haplotype-based "haplotype relative risk" approach to detecting allelic associations. *Hum. Hered.* **42**: 337–346.

Thomson, G., Robinson, W.P., Kuhner, M.K. et al. (1988). Genetic heterogeneity, modes of inheritance and risk estimates for a joint study of Caucasians with insulin-dependent diabetes mellitus. *Am. J. Hum. Genet.* **43**: 799–816.

Tienari, P.J., Terwilliger, J.D., Ott, J. et al. (1994). Two-locus linkage analysis in multiple sclerosis. *Genomics* **19**: 320–325.

Tisch, R. and McDevitt, H.O. (1996). Insulin-dependent diabetes mellitus. *Cell* **85**: 291–297.

Tiwari, J.L. and Terasaki, P. (1988). *HLA and Disease Associations*, Springer, New York.

Todd, J.A. (1995). Genetic analysis of type I diabetes using whole genome approaches. *Proc. Natl Acad. Sci. USA* **92**: 8560–8565.

Todd, J.A. and Farrall, M. (1996). Panning for gold: genome-wide scanning in type 1 diabetes. *Hum. Mol. Genet.* **5**: 1443–1448.

Todd, J.A., Bell, J.I. and McDevitt, H.O. (1987). *HLA-DQb* gene contributes to susceptibility and resistance to insulin-dependent diabetes. *Nature* **329**: 599–604.

Todd, J.A., Aitman, T.J., Cornall, R.J. et al. (1991). Genetic analysis of autoimmune type 1 diabetes mellitus in mice. *Nature* **351**: 542–547.

Vafiadis, P., Bennett, S.T., Todd, J.A. et al. (1997). Insulin expression in the human thymus is modulated by *INS* VNTR alleles at the *IDDM2* locus. *Nature Genet.* **15**: 289–292.

Valdes, J.M., Tagle, D.A. and Collins, F.S. (1994). Island rescue PCR: a rapid and efficient method for isolating transcribed sequences from yeast artificial chromosomes and cosmids. *Proc. Natl Acad. Sci. USA* **91**: 5377–5381.

Velculescu, V.E., Zhang, L., Vogelstein, B. et al. (1995). Serial analysis of gene expression. *Science* **270**: 484–487.

Vollrath, D., Davis, R.W., Connelly, C. et al. (1988). Physical mapping of large DNA by chromosome fragmentation. *Proc. Natl Acad. Sci. USA* **85**: 6027–6031.

Vyse, T.J. and Todd, J.A. (1996). Genetic analysis of autoimmune disease. *Cell* **85**: 311–318.

Vyse, T.J., Drake, C.G., Rozzo, S.J. et al. (1996a). Genetic evidence for the importance of anti-gp70 versus antinuclear autoantibodies in murine lupus nephritis. *J. Clin. Invest.* **98**: 1762–1772.

Vyse, T.J., Morel, L., Tanner, F.J. et al. (1996b). Backcross analysis of genes linked to autoantibody production in New Zealand white mice. *J. Immunol.* **157**: 2719–2727.

Wakeland, E.K., Morel, L., Achey, K. et al. (1997). Speed congenics: a classic technique in the fast lane (relatively speaking). *Immunol. Today* **18**: 472–477.

Wang, D.G., Fan, J.-B. Siao, C.-J. et al. (1998). Large-scale identification, mapping and genotyping of single-nucleotide polymorphisms in the human genome. *Science* **280**: 1077–1082.

Wang, S., Sun, C.E., Walczak, C.A. et al. (1995). Evidence for a susceptibility locus for schizophrenia on chromosome 6pter-p22. *Nature Genet.* **10**: 41–46.

Watanabe, T., Sakai, Y., Miyawaki, S. et al. (1991). A molecular genetic linkage map of mouse chromosome 19, including the *lpr*, *Ly-44*, and *Tdt* genes. *Biochem. Genet.* **29**: 325–335.

Watanabe-Fukunaga, R., Brannan, C.I., Copeland, N.G. et al. (1992). Lymphoproliferation disorder in mice explained by defects in Fas antigen that mediates apoptosis. *Nature* **356**: 314–317.

Weber, J.L. and May, P.E. (1989). Abundant class of human DNA polymorphisms which can be typed using the polymerase chain reaction. *Am. J. Hum. Genet.* **44**: 388–396.

Weeks, D.E. and Lange, K. (1988). The affected-pedigree-member method of linkage analysis. *Am. J. Hum. Genet.* **42**: 315–326.

Weeks, D.E. and Lathrop, G.M. (1995). Polygenic disease: methods for mapping complex disease traits. *Trends Genet.* **11**: 513–519.

Wicker, L.S., DeLarato, N.H., Pressey, A. et al. (1993). Genetic control of diabetes and insulitis in the nonobese diabetic mouse: analysis of the NOD.H-2^b and B10.H-2^b strains. In *Molecular Mechanisms of Immunological Self-Recognition* (eds F.W. Alt and H.J. Vogel), pp. 173–181, Academic, New York.

Wicker, L.S., Todd, J.A., Prins, J.-B. et al. (1994). Resistance alleles at two non-major histocompatibility complex-linked insulin-dependent diabetes loci on chromosome 3, *Idd3* and *Idd10*, protect nonobese diabetic mice from diabetes. *J. Exp. Med.* **180**: 1705–1713.

Wicker, L.S., Todd, J.A. and Peterson, L.B. (1995). Genetic control of autoimmune diabetes in the NOD mouse. *Annu. Rev. Immunol.* **13**: 179–200.

Wodicka, L., Dong, H., Mittman, M. et al. (1997). Genome-wide expression monitoring in *Saccharomyces cerevisiae*. *Nature Biotechnol.* **15**: 1359–1367.

Wood, N.W., Holmans, P., Clayton, D. et al. (1994). No linkage or association between multiple sclerosis and the myelin basic protein gene in affected sibling pairs. *J. Neurol. Neurosurg. Psychiatry* **57**: 1191–1194.

Worwood, M. (1998). Haemochromatosis. *Clin. Lab. Haematol.* **20**: 65–75.

Wu, J., Wilson, J., He, J. et al. (1996). Fas ligand mutation in a patient with systemic lupus erythematosus and lymphoproliferative disease. *J. Clin. Invest.* **98**: 1107–1113.

Appendix: Statistical Considerations in Analysing HLA and Disease Associations

Anthony N. Warrens

ASSOCIATION STUDIES

Basic statistical analysis

If individuals with different HLA antigens have varying risks of disease, then a difference should be seen in the distribution of HLA phenotypes between patients and healthy controls. The objective of association (population) studies is to compare the frequencies of specific HLA antigens in a series of patients with those seen in a series of controls. As is discussed in the main body of Chapter 6, the most important issue is selecting appropriate control groups. In addition, it is important to remember that the finding of a statistically significant association has several biological interpretations. Again, this is discussed in more detail above.

The chi-square test

The statistical analysis of an association study rests on the chi-square (χ^2) test. The application of this test will be illustrated using some sample data. Table 6.A1 shows a fictitious study in which 96 patients with a defined condition were typed for the products of the DR locus and 504 individuals from the same racial group were typed as controls. Each line of the table shows the number of patients or controls that typed positively for each of the antigens. Although the absolute numbers in the table are not easily interpreted, examination of the percentages shows that DR2 is increased in frequency among the cases, while all of the other antigens (except for DR1) are reduced in frequency compared with the controls. (The frequency of DR1 is almost the same in both groups.) The main question arising from these data is whether or not there is a statistically significant association between this disease and the presence of DR2. The difference between 29.8% and 75.0% is striking. However, until statistically analysed, it remains possible that this apparent difference arose by chance. Not all changes in frequency are meaningful. If DR2 is positively

TABLE 6.A1 Example of data from a population study of HLA and disease association

	Controls ($n = 504$)		Patients ($n = 96$)	
Antigen	Number positive	Percentage positive	Number positive	Percentage positive
DR1	105	20.8	21	21.9
DR2	150	29.8	74	75.0
DR3	150	29.8	11	11.5
DR4	172	34.2	29	30.2
DR6	107	21.2	17	18.0
DR7	122	24.2	12	12.5
DR8	16	3.4	1	1.0
DR9	8	1.7	1	1.0
DR10	5	1.0	0	0
DR11	46	9.1	6	6.3
DR12	5	1.0	0	0

TABLE 6.A2 2 × 2 Contingency table summarizing the DR2 data in Table 6.A1

	$DR2^+$ individuals	$DR2^-$ individuals	Totals
Patients	74	22	96
Controls	150	354	504
Totals	224	376	600

associated with disease and there are no other associations (positive or negative), there will be a compensatory decrease in the frequency of other DR allotypes.

To decide whether the discrepancy in the frequency of DR2 in the two groups is statistically significant, the summary given in Table 6.A2 is constructed. This table is termed a "2 × 2 contingency table". Cases and controls are documented in separate rows of the table, the number positive for DR2 is shown in the first column and the number negative for DR2 is shown in the second column. The row sum for the first row of the table must equal the number of cases studied, while the second row must summate to the number of controls in the series.

How does one estimate the probability of obtaining these data by chance? If it is assumed that there is no association (which is the "null hypothesis"), being a case or a control is independent of the presence or absence of the antigen. How far do the observed data (O) differ from what would be expected (E)? Of the total 600 individuals examined, 224 were $DR2^+$. In the absence of an association, of the 96 patients, 35.84 [i.e. (224/600) × 96] would be expected to be $DR2^+$ and the remaining 60.16 would be expected to be $DR2^-$. The frequencies expected if the null hypothesis is correct can be calculated in a similar way for the controls. These observed (O) and expected (E) data are used to calculate a test statistic (approximating to the χ^2 statistic) to estimate the significance of this apparent association. This statistic is calculated using the formula:

$$\Sigma(O - E)^2/E.$$

Large values of the test statistic imply that the observed data have not arisen by chance. To determine whether a test statistic is "large", one consults tables of the χ^2 distribution. To do so, it is necessary to know the number of "degrees of freedom" and the level of statistical significance required (usually $p < 0.05$) since, for each number of degrees of freedom, the χ^2 distribution is different. If the test statistic is larger than the relevant value found in the tables of the χ^2 distribution, the null hypothesis falls and it may be concluded, with statistical justification, that there is an association between the HLA antigen and this disease. For the data illustrated in Table 6.A1, the test statistic is 77.19. This is a far greater value than the threshold of the χ^2 distribution for one degree of freedom at a significance level of 0.05 (3.84) or a significance level of 0.01 (6.64). Hence, these data show a statistically highly significant association between the antigen and the disease.

The χ^2 distribution is only an approximation to the true distribution appropriate for the test statistic, $\Sigma(O - E)^2/E$. There are two sources of error if it is applied inappropriately. Firstly, it cannot be applied to small samples. A rule of thumb which is widely accepted is that 80% of *expected* frequencies (E) must exceed 5 and none must be less than 1 (Bland, 1987). However, the exact limits of size for the χ^2 test remain an area of debate among statisticians. For a 2 × 2 contingency table, this means, in practice, that all four values of E must exceed 5.

The second problem arises from the use of the continuous χ^2 distribution to deal with a discrete distribution. This discrete distribution can be made to fit more closely to the continuous χ^2 distribution by introducing Yates' continuity correction, which is how the statistic is often tabulated. The Yates' continuity correction helps to make the discrete data generated by the test statistic, $\Sigma(O - E)^2/E$, approximate more closely to the χ^2 distribution. With the correction, the statistic calculated is $\Sigma(|O-E| - 1/2)^2/E^2$ rather than $\Sigma(O - E)^2/E$. (Flanking a number or expression with vertical lines indicates "modulus". This means the number is represented without a sign. Positive numbers are unchanged, and negative numbers are altered to their positive equivalent.) This may be explained by imagining a discrete distribution over the range 0–20. There will be 20 lines drawn, the first at 1 and the twentieth at 20, whereas a continuous distribution extending over that range will begin at 0. If one were to draw the axis of symmetry for each of these distributions, it would be at 10 for the continuous distribution, but at 10.5 for the discrete distribution. The two are brought more closely into line by subtracting 0.5 of an interval from each of the data items. In essence, one is assuming that the data at, say, point 10, actually represent the data in the range $9 < x \leq 10$, and plotting it at 9.5.

However, Yates' correction may result in an unreasonably conservative test and, consequently, a significant association may be missed. Statisticians are divided on whether or not it should be used (Bland, 1987; Daniel, 1987). The correction, strictly, is not necessary for large sample sizes, but it only meaningfully affects the result when the sample size is small.

Fisher's exact test

If the expected frequencies are too small to use the χ^2 test, an alternative statistic, generated by Fisher's exact test, is used. To explain the principle of the test an example of a fictitious study that is clearly too small to be of value in practice is summarized in Table 6.A3. The null hypothesis states that there is no association between this disease and DR2.

Fisher's exact test asks this question: given the four totals (in bold letters in Table 6.A3), what is the likelihood that this combination of figures arose by chance? In fact, there are only four possible ways of filling in the gaps in this table, since there are only four possible figures that can be inserted into the DR2$^-$/disease box: 0, 1, 2 and 3, and all other figures follow from that. The four possibilities are outlined in Table 6.A4.

The eight subjects may be labelled a–h, with the DR2$^-$ subjects called a–c and the DR2$^+$ individuals d–h. What are the possible combinations in each of these tables? In panels A and D of Table 6.A4, subjects a, b and c are committed to a single box; the single individual in the DR2$^+$/control

TABLE 6.A3 Fictitious data set for demonstration of Fisher's exact test

	DR2$^-$ individuals	DR2$^+$ individuals	Totals
Disease	1	3	4
Controls	2	2	4
Totals	**3**	**5**	**8**

TABLE 6.A4 Worked examples of Fisher's exact test

	DR2$^-$ individuals	DR2$^+$ individuals	Totals
Disease			4
Controls			4
Totals	3	5	8

The four possible ways of filling the empty spaces in this table are:

A	DR2$^-$	DR2$^+$	Totals
Disease	0	4	4
Controls	3	1	4
Totals	3	5	8

B	DR2$^-$	DR2$^+$	Totals
Disease	1	3	4
Controls	2	2	4
Totals	3	5	8

C	DR2$^-$	DR2$^+$	Totals
Disease	2	2	4
Controls	1	3	4
Totals	3	5	8

D	DR2$^-$	DR2$^+$	Totals
Disease	3	1	4
Controls	0	4	4
Totals	3	5	8

box is one and only one of the others; thus, there are five possible permutations of each of panels A and D. Panels B and C similarly have identical numbers of permutations: the numbers in the two rows have merely been swapped. In panel B, the DR2$^-$/disease subject could be a, b or c (three possibilities). There are 10 possible combinations of d, e, f, g and h to make up the trio in the DR2$^+$/disease box: def, deg, deh, dfg, dfh, dgh, efg, efh, egh and fgh. The total number of permutations for each of panels B and C is thus $3 \times 10 = 30$. Add to this the 5 for each of panels A and D, and the total number of permutations for these tables is

TABLE 6.A5 Worked examples of Fisher's exact test: probabilities of each of the four panels in Table 6.A4 occurring

Panel	Probability
A	5/70 = 0.071
B	30/70 = 0.429
C	30/70 = 0.429
D	5/70 = 0.071

70. From this it is possible to work out the possibility of each table arising (Table 6.A5).

The observed data conformed to panel B of Table 6.A4, which has a probability of occurring by chance of 0.429. Hence, the null hypothesis stands and it is not possible to infer an association between DR2 and this disease from these data. In fact, none of the possible tables for these data had a probability of occurring less than 0.05, so no combination of figures with these row and column totals could have given a statistically significant association using Fisher's exact test. However, it is clear that almost any more complex table or one with larger numbers is likely to have one or more combinations that would occur by chance only rarely. This illustrates the potential of this approach. In practice, Fisher's exact test is applied by entering the data into an appropriate computer program.

Additional complications of HLA studies

Some additional features complicate the analysis of disease associations with HLA antigens. Owing to the extreme polymorphism of the HLA system, most of the phenotypes observed will be seen in only a small proportion of cases and controls. It is possible to include multiple alleles in the same comparison. However, since there is a risk of generating false-positive results, a correction factor for the number of tests must be included.

An additional complication arises because of the presence of two or more copies of all loci in each individual. A case or control who is typed positive for only a single antigen by serology may carry either two copies of the same allele, or one copy of the allele in question and a second allele encoding an antigen for which no antisera are available. The lack of serological definition can only be removed if family information is available. Such issues are not likely to arise in future studies since DNA analysis of the HLA system, rather than serological analysis, is now the gold standard.

Measuring the strength of the association

If there is evidence of a significant association, then a method for indicating the strength of the association is required. The usual measure of the strength of an association is the risk of disease

TABLE 6.A6 Symbolic representation of a 2×2 contingency table to illustrate the calculation of relative risk

	Number of individuals	
	Positive for the antigen	Negative for the antigen
Patients	a	b
Controls	c	d

among those positive for the antigen compared with those without the antigen, as suggested initially by Woolf (1955). A strong association is one in which the risk among carriers is either much larger or much smaller than the risk to non-carriers. Using the nomenclature of Table 6.A6, if the disease is rare, then the ratio of these risks can be estimated by ad/bc.

This ratio for the estimation of relative risk (RR) can be derived as follows. The prevalence of the disease in individuals positive for the antigen can be expressed as $(a/c) \times k$ and the prevalence of disease in antigen-negative individuals as $(b/d) \times k$. An unknown constant (k) has been introduced because no information is available about the absolute prevalence of the disease. However, the same constant is applied to both a/c and b/d. The relative incidence or relative risk can be calculated using the formula:

$$RR = \frac{(a/c) \times k}{(b/d) \times k} = \frac{(a/c)}{(b/d)} = \frac{ad}{bc}$$

A RR close to 1.0 implies that those carrying the antigen have no increased risk of disease, i.e. there is no association between the antigen and disease expression. A RR greater than 1.0 implies that those carrying the antigen are at increased risk of disease (a positive association), while a RR less than 1.0 implies those carrying the antigen are at lower risk of expressing the disease than those not carrying the antigen (a negative association). The further the number deviates from 1, the stronger the association.

Statistical interpretation of an association between an antigen and the disease

How does one evaluate the results of a particular study? Clearly, a RR of 3.0 would be much more exciting if it were obtained with a sample of

1000 patients and controls than if achieved with a sample size of 20 in each group. The RR obtained from the smaller sample size could simply be due to chance because an unusual sample could produce this inflated estimate even if there were really no association; such an explanation would be much less plausible for the larger sample size. The interpretation then depends on the sample size and is measured by the statistical significance.

With the 2×2 contingency table, the χ^2 test was used to measure how far each of the observed frequencies deviates from an expected value (expected if all values were taken from the same population). The statistical significance of a RR can similarly be calculated using the χ^2 distribution, by making an estimate of the deviation of each of the figures represented by a, b, c, and d in the above formula.

Using the data in Table 6.A7 as an example, the estimated RR for the whole population for the association between DR3 and the disease is 4.7 $[(65 \times 440)/(151 \times 40)]$. This means that is there is evidence to suggest a strong positive association between the antigen and the disease.

The statistical significance of a RR can be calculated using the χ^2 distribution. Using the nomenclature of Table 6.A6, the variance is first calculated using the formula:

$$V = 1/a + 1/b + 1/c + 1/d.$$

Next, the test statistic is calculated using the expression:

$$(1/V) \times (\log_e RR)^2.$$

The value obtained is then compared with the χ^2 distribution for 1 degree of freedom. As before, to obtain statistical significance ($p < 0.05$) the test statistic must be >3.84. This is discussed further in Tiwari and Terasaki (1985). These data gener-ate a test statistic for the χ^2 test of 48.60, much greater than 3.84, which makes this RR highly statistically significant.

Analysing more complex data

The application of statistical tests to the simple 2×2 table and the evaluation of RR have been discussed. The application of statistical analysis becomes much more involved with more complex data, as is illustrated in the following example.

Let us postulate a disease in whose aetiology HLA is believed to be involved. An association study is conducted in the standard way. In particular, the association between the disease and each of the alleles of this particular HLA locus is examined. On performing a χ^2 test, it is found that several antigens show evidence of an association. Before becoming too enthusiastic about the results, one should re-examine the meaning of the phrase "statistical significance". The significance level calculated for each antigen is the probability that a result as extreme as this would occur if there truly were no association. A significance level of 0.05 means that one out of 20 analyses conducted when there is really no association would produce a result as extreme as this by chance, i.e. would erroneously suggest an association. Put another way, even if there is no relationship between HLA and this disease, were one to examine 20 antigens using a significance level of 0.05, one of the antigens would be expected to show evidence of an association by having a χ^2 test statistic with a value of at least 3.84.

How then does one decide on the importance of the results of a particular study? Since the analysis of each antigen that is typed for can spuriously produce evidence of an association, the usual approach is to calculate the significance level of each test and to multiply that result by the number of statistical tests performed. This "corrected" significance level is then reported as the statistical significance of the test. For instance, if one particular HLA locus is examined, then one statistical test is performed for each antigen that was typed, so the number of statistical tests is equal to the number of antigens. If 20 HLA-DR types are defined in a particular study and the significance level for the association with one of the DR types is 0.001, then the reported value for that analysis

TABLE 6.A7 Example of a 2×2 contingency table to illustrate the calculation of relative risk

	Number of individuals	
	DR3$^+$	DR3$^-$
Patients ($n = 105$)	65	40
Controls ($n = 591$)	151	440

should be modified to 0.02 (= 20 × 0.001). This approach is called the Bonferoni correction (see Dunn, 1961, for a detailed discussion, or a statistics textbook for a more general presentation). If the corrected significance level is still less than 0.05 then this is taken as an indication of a genuine association between that antigen and the disease. Any antigen for which the corrected value is greater than 0.05 would not be considered to be associated with the disease.

Table 6.A8 reproduces one of the original studies that described the association between ankylosing spondylitis (AS) and HLA-B27 (Schlosstein et al., 1973). Initial inspection suggests a strong association, with a much larger incidence of B27 in the AS group than in the controls. Application of the Bonferoni correction factor confirms the positive association with B27 and a much weaker negative one with B7.

Table 6.A8 Data from original study (Schlosstein et al., 1973) demonstrating the association of HLA-B27 with ankylosing spondylitis (used here to illustrate the application of the Bonferoni correction factor)

Antigen	Cases of ankylosing spondylitis (%) ($n = 40$)	Controls (%) ($n = 906$)
A1	10	27
A2	60	48
A3	18	24
A9	30	22
A10	8	11
A11	10	12
A28	18	12
A29	10	7
A30	15	10
A32	13	8
B51	5	11
B7	3*	25
B8	15	21
B44	18	24
B13	3	4
B35	18	22
B40	18	14
B14	5	7
B15	8	8
B17	5	8
B18	5	9
B21	3	5
B22	3	5
B27	88**	8

*$\chi^2 = 9.32$ ($p < 0.05$ when multiplied by 24);
**$\chi^2 = 236.41$ ($p < 0.0001$ when multiplied by 24, number of specificities tested).

One way to circumvent problems caused by applying Bonferoni's correction factor is to focus the study design on a very specific question. Suppose a colleague has published a study reporting evidence of an association of the disease in question with a specific HLA antigen. In order to confirm this association, one may study a group of patients with the same disease and examine in particular the association with that antigen, using the statistical method previously described. The advantage of this kind of confirmatory study is that, because of information from the previous study, attention can be focused on a single HLA antigen, thus greatly reducing the correction factor. The earlier study should also provide an estimate of the strength of the association. On the basis of that estimate, and allowing for a margin of error, it should be possible to estimate the size of sample required to perform a study with adequate statistical power to detect an association of similar magnitude. This is preferable to finding after the study is completed that the statistical analysis does not achieve significance and then attempting to achieve significance by collecting more samples.

LINKAGE STUDIES

The principles involved in linkage analysis are outlined in the main text of this chapter. The following example is provided to elucidate those principles. Figure 6.A1 represents the pedigree of a family in which a number of individuals have hereditary elliptocytosis (filled rather than empty symbols). In this figure, each individual's rhesus blood group is presented to allow the possibility of linkage between rhesus phenotype and elliptocytosis to be studied. The three rhesus genotypes within the family are designated A, B and C. It can be predicted that, if there is no link-

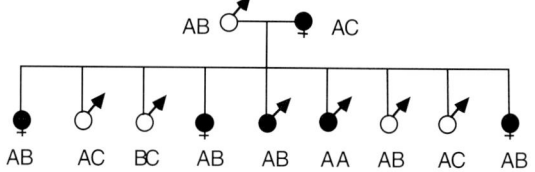

FIGURE 6.A1 Sample pedigree for method of lod scores.

age between elliptocytosis and rhesus phenotype, 50% of the affected progeny would have maternal rhesus genotype A and 50%, C. Conversely, if the two phenotypes are so closely linked that there are no recombination events between them, then all affected progeny would have the same rhesus phenotype. The probability of the observed pedigree arising in the case of either non-linkage or non-recombination, as well as all intermediate levels of linkage, may be calculated. It is therefore possible to obtain the value for linkage (q) (percentage recombination between the two loci) that gives the maximum probability of obtaining the observed pedigree.

In this example, the value which gives the maximum probability of obtaining the observed pedigree is 0.11, or an 11% recombination frequency. The log odds ratio (or z, the lod score) is a ratio of this probability divided by the probability of obtaining this pedigree if there were no linkage (see main text of this chapter). In this example, with a value of q of 0.11, the formula gives a z-value of 1.044, which is the logarithm of the ratio of odds 11:1. In other words, there is an 11 to 1 odds in favour of linkage between rhesus blood groups and elliptocytosis.

The question of whether or not this constitutes statistical significance is discussed further in the main text of this chapter.

ALLELE-SHARING METHODS

If a disease-causing gene is not itself an HLA gene, but is located close to such a gene, then a family study is usually the most convenient and often the only way to identify the association. The method relies on linkage between genes, i.e. the tendency for both genes to segregate together in families (see Cavalli-Sforza and Bodmer, 1971, for more details of linkage). If the linkage is very tight so that the disease gene essentially always segregates with HLA, and the disease-causing mutation has occurred recently enough, then it may be possible to detect the association using a population-based method (as described above). The conditions under which such an association could be identified are relatively restrictive and, in their absence, a family study will usually be required to identify the linkage.

Family studies are based on the following observations. Suppose a family has two children. Both parents are typed: the father's haplotypes are A and B and the mother's are C and D. There are four possible haplotype pairings for each child (AC, AD, BC and BD), making 16 possible combinations of HLA types for the two children (see Table 6.A9). A, B, C and D could represent the typing results from either a single or multiple HLA loci. In the latter case, they are termed "haplotypes". The 16 possibilities can be classified usefully to one of three categories depending on the number of haplotypes that the two sibs share. If the two sibs are identical at both HLA loci (combinations 1, 6, 11 and 16 in Table 6.A9), then they are said to share two haplotypes (i.e. the same haplotype from each parent), while two sibs who are entirely discordant for the HLA alleles inherited from the parents (combinations 4, 7, 10 and 13) are said to share zero haplotypes. The other possibility is that the sibs share exactly one set of alleles, in which case the sibs have inherited the same haplotype from one parent and different haplotypes from the other parent. The haplotypes A, B, C and D will be different in different families but for each family the typing results are reduced to the number of shared haplotypes among the children.

The method of identifying disease-linked genes using family studies was originally suggested by Penrose (1935), who realized that the sharing of haplotypes would be modified if the typed locus were genetically linked to a disease-causing gene. In particular, sibs affected with the same disease should show more sharing of parental haplotypes than would be predicted by chance. To identify a linkage, the observed pattern of haplotype sharing among sibs must be compared with the distribution of haplotypes that would occur if the disease gene and the typed gene were not linked to each other. In the latter case, the 16 possibilities of Table 6.A6 would be equally likely, so that the probability that the sibs share two haplotypes is 0.25, one haplotype is 0.5 and zero haplotypes is 0.25. Any evidence of excess sharing over this distribution is consistent with linkage. For instance, the result of a study by Ebers et al. (1982) of sibs with multiple sclerosis (MS) is shown in Table 6.A10. In 32 families, the haplotype sharing among two affected siblings was

TABLE 6.A9 Possible combinations of parental haplotypes in two siblings

Combination	Child 1	Child 2	Combination	Child 1	Child 2
1	AC	AC	5	AC	AD
2	AD	AC	6	AD	AD
3	BC	AC	7	BC	AD
4	BD	AC	8	BD	AD

Combination	Child 1	Child 2	Combination	Child 1	Child 2
9	AC	BC	13	AC	BD
10	AD	BC	14	AD	BD
11	BC	BC	15	BC	BD
12	BD	BC	16	BD	BD

Table 6.A10 Results of sibling study of Ebers et al. (1982)

Haplotypes shared	Expected	Observed
2	10	12
1	20	18
0	10	10

identified: 12 shared two haplotypes, 15 shared one haplotype and five shared no haplotypes. This can be compared with the expected result of eight pairs sharing two haplotypes, 16 pairs sharing one haplotype and eight sharing no haplotypes. There is therefore some modest increase in HLA sharing among the MS siblings. Again, the observed and expected sharing can be compared using a χ^2 test. In this case, the test statistic is not significant, showing that there is no evidence of excess HLA sharing among the MS siblings.

The statistical method described above is simplistic in that it does not take into account any features of the disease transmission in families. Such information is often not well known, in which case this method is appropriate. However, if there is further information, such as an indication as to whether the disease is inherited as an autosomal dominant or a recessive trait, then this can be incorporated and utilized to produce a more powerful test of linkage.

REFERENCES

Bertrams, J. and Baur, M.P. (1979). HLA-A, B, Bf three point associations of 1072 haplotypes in a German population. *Tissue Antigens* **14**: 317–324.

Bidwell, J.L. and Bignon, J.D. (1991). DNA-RFLP methods and interpretation scheme for HLA-DR and DQ typing. *Eur. J. Immunogenet.* **18**: 5–22.

Bland, M. (1987). *An Introduction to Medical Statistics,* Oxford University Press, Oxford.

Brewerton, D.A., Caffrey, M., Hart, F.D. et al. (1973). Ankylosing spondylitis and HL-A27. *Lancet* **i**: 904–907.

Cavalli-Sforza, L.L. and Bodmer, W.F. (1971). *The Genetics of Human Populations,* W.H. Freeman, San Francisco.

Collier, S., Sinnott, P.J., Dyer, P.A. et al. (1989). Pulsed field gel electrophoresis identifies a high degree of variability in the number of tandem 21-hydroxylase and complement C4 gene repeats in 21-hydroxylase deficiency haplotypes. *EMBO J.* **8**: 1393–1402.

Cross, S.J., Tonks, S., Trowsdale, J. et al. (1991). Novel detection of restriction fragment length polymorphisms in the human major histocompatibility complex. *Immunogenetics* **34**: 376–384.

Davidson, J.A., Kippax, R.L. and Dyer, P.A. (1988). A study of HLA-A,B,DR and Bf bearing haplotypes derived from 304 families resident in the north west of England. *J. Immunogenetics* **15**: 227–237.

Daniel, W.W. (1987). *Biostatistics: A Foundation for Analysis in the Health Sciences,* pp. 552–553, John Wiley, New York.

Day N.E. and Simons M.J. (1976). Disease susceptibility genes – their identification by multiple case family studies. *Tissue Antigens* **8**: 109–119.

Dunn, O.J. 1961. Multiple comparisons among means. *Am. J. Statist. Assoc.* **56**: 52–64.

Dyer, P.A. and Martin, S. (1991). Techniques used to define human MHC antigens: serology. *Immunol. Letts* **29**: 15–22.

Ebers, G.C., Paty, D.W., Stiller, C.R. et al. (1982). HLA typing in multiple sclerosis sibling pairs. *Lancet* **ii**: 88–90.

Fernandez-Vina, M.A., Gao, X., Moraes, M.E. et al. (1991). Alleles at four HLA class II loci determined by oligonucleotide hybridization and their associations in five ethnic groups. *Immunogenetics* **34**: 299–312.

Festenstein, H. and Demant, P. (1978). *HLA and H-2 Basic Immunogenetics, Biology and Clinical Relevance* (ed. J. Turk Edward), pp. 26–29, Arnold, London.

Fishman, P.M., Suarez, B., Hodge, S.E. et al. (1978). A robust method for the detection of linkage in familial disease. *Am. J. Hum. Genet.* **30**: 308–321.

Green, J.R. and Woodrow, J.C. (1977). Sibling method for detecting HLA-linked genes in disease. *Tissue Antigens* **9**: 31–35.

Gyllensten, U.B. and Erlich, H.A. (1988). Generation of single stranded DNA by the polymerase chain reaction and its application to direct sequencing of the HLA-DQA locus. *Proc. Natl Acad. Sci. USA* **85**: 7652–7656.

Haldane, J.B.S. (1956). The estimation and significance of the logarithm of a ratio of frequencies. *Ann. Hum. Genet.* **20**: 309–311.

Jongeneel, C.V., Briant, L., Udalova, I.A. et al. (1991). Extensive genetic polymorphism in the human tumour necrosis factor region and relation to extended HLA haplotypes. *Proc. Natl Acad. Sci. USA* **88**: 9717–9721.

Kendall, E., Todd, J.A. and Campbell, R.D. (1991). Molecular analysis of the MHC class II region in DR4, DR7 and DR9 haplotypes. *Immunogenetics* **34**: 349–357.

Knowles, R.W. (1989). Structural polymorphism of the HLA class II α and β chains: summary of the 10th Workshop 2-D gel analysis. In *Immunobiology of HLA* (ed. B. Dupont), Vol. 1, pp. 365–380, Springer, New York.

Mathews, J.D. (1984). Statistical and genetic aspects of specificity. In *Detection of Immune Associated Genetic Markers of Human Disease* (eds M.J. Simons and B.D. Tait), Churchill Livingstone, Edinburgh.

Neitzel, H. (1986). A routine method for the establishment of permanent growing lymphoblastoid cell lines. *Hum. Genet.* **73**: 320–326.

Nevinny-Stickel, G., Hinzpeter, M., Andreas, A. et al. (1991). Non-radioactive oligotyping for HLA-DR1-DRw10 using polymerase chain reaction, digoxigenin-labelled oligonucleotides and chemiluminescence detection. *Eur. J. Immunogenet.* **18**: 323–332.

Olerup, O. and Zetterquist, H. (1992). HLA-DR typing by PCR amplification with sequence-specific primers (PCR-SSP) in 2 hours: an alternative to serological DR typing in clinical practice including donor-recipient matching in cadaveric transplantation. *Tissue Antigens* **39**: 225–235.

Ott, J. (1974). Estimation of the recombination fraction in human pedigrees: efficient computation of the likelihood for human linkage studies. *Am. J. Hum. Genet.* **26**: 588–597.

Ott, J. (1991). *Analysis of Human Genetic Linkage*, Johns Hopkins University Press, Baltimore, MD.

Penrose, L.S. (1935). The detection of autosomal linkage in data which consists of pairs of brothers and sisters of unspecified parentage. *Ann. Eugenics* **6**: 133–138.

Read, A.P., Kerzin-Storrar, L., Dyer, P.A. et al. (1986). Possible maternal effects in genetic susceptibility to myasthenia gravis. *Lancet* **i**: 167–168.

Rodey, G.E. (1991). *HLA Beyond Tears*, De Novo, Atlanta, GA.

Schlosstein, L., Terasaki, P.I., Bluestone, R. et al. (1973). High association of an HL-A antigen, w27, with ankylosing spondylitis. *N. Engl. J. Med.*, **288**: 704–706.

Sinnott, P.J., Collier, S., Costigan, C. et al. (1990). Genesis by meiotic unequal crossover of a de novo deletion that contributes to steroid 21-hydroxylase deficiency. *Proc. Natl Acad. Sci. USA* **87**: 2107–2111.

Thomson, G. and Bodmer, W. (1977). The genetic analysis of HLA and disease associations. In *HLA and Disease* (eds J. Dausset, A. Svejgaard), pp. 84–93, Munksgaard, Copenhagen.

Tiwari, J.L. and Terasaki, P.I. (1985). *HLA and Disease Associations*, New York. Springer.

Trowsdale, J., Ragoussis, J. and Campbell, R.D. (1991). Map of the human MHC. *Immunol. Today* **12**: 443–446.

Uryu, N., Maeda, M., Ota, M. et al. (1990). A simple and rapid method for HLA-DRB and -DQB typing by digestion of PCR-amplified DNA with allele specific restriction endonucleases. *Tissue Antigens* **35**: 20–31.

Vries, R.R.P. de, Nijenhuis, L.E., Lai-A-Fat, R.F.M. et al. (1976). HLA linked control of host response to *M. leprae. Lancet* **ii**: 1328–1330.

Woolf, B. (1955). One estimating the relation between blood group and disease. *Ann. Hum. Genet.* **19**: 251–253.

Yang, S.Y. (1989). Population analysis of class I HLA antigens by one-dimensional isoelectric focussing gel electrophoresis: workshop summary report. In *Immunobiology of HLA* (ed. B. Dupont), Vol. 1, pp. 309–331, Springer, New York.

CHAPTER 7

Mechanisms of HLA and Disease Associations

Anthony N. Warrens and Robert I. Lechler

INTRODUCTION

In Chapter 4, the basic principles underlying major histocompatibility (MHC)-restricted recognition of antigens by T cells were outlined. This included discussion of the phenomenon of MHC-determined variation in immune responsiveness. Armed with these concepts, in this chapter the mechanisms underlying the association between the inheritance of particular MHC alleles and disease susceptibility will be discussed. To illustrate these points, reference will be made to some of the well-studied animal models of autoimmune disease and less closely defined examples from human disease associations. The latter will be discussed in much greater detail in Section II of this book.

Before considering the mechanisms responsible for an individual disease association, it is important to distinguish autoimmune, immune complex-mediated and non-immune diseases. These are clearly important distinctions because they will help in the interpretation of human leucocyte antigen (HLA) associations. For these purposes, autoimmune diseases are defined as those diseases whose pathogenesis results from the direct effects of a specific immune response (either an autoantibody or an autoreactive T cell response) against host tissues. A classical example of an autoimmune disease that fits this definition is anti-glomerular basement membrane (anti-GBM) disease, otherwise known as Goodpasture's disease. Tissue damage in this condition appears to be caused by the circulating anti-GBM antibody, which can be detected with a characteristic pattern of linear deposition on the renal glomerular and the pulmonary alveolar basement membranes.

This contrasts with an immune complex disease such as systemic lupus erythematosus (SLE). Autoantibodies, in this case against double-stranded DNA, are also a feature of SLE (although the pathogenicity of the anti-DNA antibodies in this disease is a point of continuing debate). SLE is an archetypal immune complex-mediated disease, and the systemic vasculitis that can affect multiple sites in this condition appears to be the result of the deposition, or the *in situ* formation, of insoluble immune complexes. Some of these may be DNA–anti-DNA complexes.

The third category of HLA-associated diseases does not have any known immune component in their pathogenesis. Narcolepsy is a classical example of this kind of disease, in that a strong association has been documented with HLA-DR15 (Juji et al., 1983), but there is little to suggest an autoimmune or immune complex-mediated pathogenesis in this condition.

Ir GENE EFFECTS AND DISEASE SUSCEPTIBILITY

As described in Chapter 4, many experimental examples have been documented of genetically determined variation in immune responsiveness to individual antigenic determinants that map to the MHC. The three major mechanisms underlying this MHC-linked variation were also reviewed in Chapter 4, and were: (i) determinant selection; (ii) holes in the T cell repertoire; and (iii) T cell-mediated suppression. Experimental evidence implicating each of these mechanisms has been assembled. The following discussion describes two animal models of autoimmune disease which illustrate these principles, after which some data on human immune response (Ir) gene effects are summarized.

Experimental allergic encephalomyelitis

This is a demyelinating disease which can be induced in several animal strains and causes varying degrees of paralysis or even death. It is characterized by infiltration of the white matter of the central nervous system (CNS) by mononuclear cells, and is clearly immunologically mediated. This disease is induced by the injection of the CNS autoantigen, myelin basic protein (MBP). The route of immunization in most of these experimental models is intradermal, and the induction of immunity requires the use of a potent adjuvant such as complete Freund's adjuvant. Models of the disease have been described in rabbits (Stuart and Krikorian, 1928), mice (Olitsky and Yager, 1949), rats and guinea pigs (Paterson, 1976). The reason that so much research effort has been channelled into the study of experimental allergic encephalomyelitis (EAE) is that it is one of the best characterized experimental models of the human demyelinating disease multiple sclerosis (MS).

In the mouse model of EAE, it is clear that genetic factors control susceptibility to developing an autoimmune response. The best defined of these genetic factors is the possession of certain MHC class II H-2A alleles. Two of these, namely H-2A^s and H-2A^u, have been studied in detail, and the precise specificity of autoimmune T cells has been determined (Zamvil et al., 1987). As befits this MHC class II association, CD4$^+$ T cells are the major disease-inducing population in EAE. Thus, anti-CD4 antibodies were able to prevent disease in mice and rats (Brostoff and Mason, 1984; Waldor et al., 1985). Furthermore, CD4$^+$, MBP-specific T cell clones were able to induce EAE following transfer to naïve recipient mice (Zamvil et al., 1985). In addition, the injection of anti-MHC class II antibodies was able to confer protection from EAE (Steinman et al., 1983).

There are three sets of observations regarding EAE that relate to the issue of MHC and disease associations. The first is that, for both of the H-2A types which cause disease susceptibility, a distinct peptide specificity has been defined in the disease-mediating MBP-specific T cells. For the H-2A^s-expressing mice, the T cell response appears to be almost entirely directed against an MBP sequence comprising amino acid residues 89–101 (Sakai et al., 1988). In contrast, the specificity of T cells from H-2A^u-expressing mice is for the N-terminal region of MBP between residues 1 and 11 (Zamvil et al., 1988). This provides a clear example of the influence of MHC polymorphism on T cell specificity.

These findings have been further explored by examining the efficiency of binding of these peptides of MBP to the H-2A^s and H-2A^u molecules. As might be expected, the 1–11 peptide bound efficiently to H-2A^u, and this binding could not be significantly inhibited by addition of a large molar excess of the H-2A^s-restricted 89–101 peptide (Wraith et al., 1989). Although these results illustrate the influence of H-2A polymorphism on the specificity of an autoimmune response, it remains unclear whether the fact that other H-2A types are not associated with susceptibility to EAE is because they do not bind and present available peptides of MBP efficiently. This would be the case if "determinant selection" were the sole explanation for the MHC association. Alternatively, H-2A^s- and H-2A^u-associated disease susceptibility may reflect the influence of these MHC products on the selection of the T cell repertoire or on the induction of suppressor T cells.

A second series of important observations relating to the pathogenesis of EAE have focused attention on T cell receptor usage in MBP-specific T cells in autoimmune animals. The remarkable finding was that the large majority of MBP-reactive T cell clones isolated from autoimmune mice and rats used the same Vβ gene segment, and one of two Vα segments (Vβ8.2 and Vα2 or 4) (Zamvil et al., 1988; Acha-Orbea et al., 1989). Given the proportion of mouse peripheral T cells which express these V gene segments, the frequency of T cells expressing these particular combinations of Vβ8.2 and Vα2 or 4 should be in the region of 1 in 1000, assuming a random αβ-chain association. It is particularly surprising that the same receptor should appear in MBP-immune T cells from the rat as well as the mouse (Heber-Katz and Acha-Orbea, 1989). The MBP peptide specificity of the T cells from the two species is quite different, and the degree of sequence homology between the rat and mouse MHC class II restriction elements is only in the order of 80%. Although these observations are difficult to explain, they do suggest that T cell repertoire selection is influential in the aetiology of this disease. It is interesting to note that preferential usage of the Vα11 gene segment has been reported in T cells isolated from the cerebrospinal fluid of MS patients (Oksenberg et al., 1990).

The third set of data relating to the autoimmune response mediating EAE suggests that suppressor T cells can play an important role in regulating anti-MBP immunity. It was noted several years ago that if guinea pigs were injected intravenously with MBP prior to intradermal immunization with Freund's adjuvant, the animals were protected from the development of disease. Furthermore, T cells from these pretreated animals conferred protection from EAE when transferred to naive, syngeneic recipients (Arnon, 1982). This illustrates that autoimmunity of this kind is susceptible to T cell regulation. As yet there is no evidence that regulation of this sort plays a role in the protection of non-susceptible animals from developing EAE.

The conclusion that can be drawn about the MHC association with EAE from these studies is that all three of the mechanisms proposed in Chapter 4, and outlined at the start of this chapter, could contribute to MHC-linked susceptibility to anti-MBP autoimmunity. Further elucidation of the underlying mechanisms may be furnished by carefully designed transgenic mouse experiments.

The non-obese diabetic mouse

In the same way that EAE serves as a model for the human disease MS, the non-obese diabetic (NOD) mouse serves as an experimental model for human insulin-dependent diabetes (IDD). Diabetes in these mice is characterized by autoimmune insulitis and the destruction of pancreatic β-cells. Unlike EAE, diabetes in this mouse strain occurs spontaneously; in common with EAE, it is clear that more than one genetic locus is responsible for disease susceptibility. One of the key susceptibility genes for the development of disease in NOD mice is a unique allele of the H-2Ab gene, known as H-2Ab^{NOD} (Acha-Orbea and McDevitt, 1987). This Ab allele is accompanied by an entirely normal Aa^d gene. It is of particular interest that the polypeptide encoded by the Ab^{NOD} gene lacks the negatively charged aspartic acid at position 57. Lack of aspartic acid at this position in the human homologue of Aβ, DQβ, appears to be associated with susceptibility to human IDD (Todd et al., 1987). The other feature of the class II region of NOD mice is a defect in the H-2E genes that encode the second mouse class II product, so that no H-2E molecule is expressed.

Investigation of the mechanisms responsible for the MHC and disease association in this model is hampered by the fact that the pancreatic autoantigen (or autoantigens) has not been defined. It is not possible, therefore, to grow disease-mediating T cells and explore their specificity or T cell receptor gene usage. Nonetheless, a series of genetic and immunological interventions has led to protection from the onset of disease, and these results provide clues about the contribution of the MHC haplotype to this autoimmune reaction. There is evidence to suggest that both the possession of the Ab^{NOD} gene and the lack of H-2E expression may contribute to the development of diabetes in these mice. The most informative experiments in this area have involved the production of transgenic mice into which have been transferred genes encoding "normal" or mutated H-2A and wild-type H-2E molecules. Nishimoto et al. (1987) observed that the introduction of an H-2E transgene into NOD mice prevented the development of insulitis. This result argues strongly that the lack of H-2E expression is an important factor in the tendency of these mice to autoimmune insulitis. The data of Nishimoto et al. (1990), Slattery et al. (1990), and Lund et al. (1990) provide evidence that disease susceptibility in NOD mice can be attenuated or abolished by introduction of H-2A transgenes. Protection was achieved by expression of wild-type H-2A^k (Slattery), H-2A^k with serine rather than the charged aspartic acid at position 57 of the H-2Aβ-chain (Nishimoto et al., 1987; as discussed above, it has been suggested that a charged residue in DQβ and H-2Aβ-chains at this position confers protection from disease), and expression of H-2A with proline rather than histidine at position 56 in the β-chain (Lund et al., 1990). In addition to having an uncharged residue at position 57, the Aβ^{NOD}-chain is unique in having histidine at position 56. Despite the fact that the Aβ$^{NOD-56pro}$-chain retained the uncharged serine at position 57, this modified Aβ-chain still conferred protection, further emphasizing that the contribution of MHC class II genes to susceptibility to diabetes cannot be entirely accounted for by polymorphism at position 57 in DQβ and H-2Aβ-chains.

The protective effects of introducing a wild-type H-2E gene or mutated H-2Ab genes into NOD mice shed some light on the probable mechanisms of susceptibility to autoimmune destruction of pancreatic β cells. The most likely explanation for these effects is that the introduction of additional class II genes alters the expressed T cell repertoire in the transgenic mice. This could be the result of intrathymic deletion of potentially autoreactive T cells due to reactivity of such cells with the additional expressed class II products, or the result of selection of regulatory suppressor T cells that prevent the development of disease in the transgenic animals. Thus, it is possible that all three of the major mechanisms of Ir gene effects contribute to susceptibility to diabetes in NOD mice and susceptibility to demyelination in mice that develop EAE.

Human studies

It has not been possible to study human diseases in as much detail, for obvious reasons. No inbred or congenic strains are available and in very few cases have the relevant antigens been defined.

In type I diabetes mellitus, the protective effect of having a charged residue in position 57 of the DQβ-chain has already been alluded to. In addition, it has been noted that those HLA alleles that are most closely linked with disease (DQA1*0201 and *0302) code for products which share a common peptide-binding motif (Kwok et al., 1995) which distinguishes them from other DQ molecules. These seem to affect the binding of peptide residues 4 and 9. This implies a role in determinant selection, although how this affects the predisposition to diabetes (e.g. through binding an immunogenic peptide or through failing to delete a series of particular potentially self-reactive T cell clones) remains unclear. This is discussed in much greater detail in Chapter 15.

The most compelling example implicating an Ir gene effect in human disease has arisen from studies of severe falciparum malarium. HLA-B*5301 was found to be very closely linked with protection from severe malaria (defined as cerebral malaria or severe anaemia). It is thought that HLA-B53 binds a nonapeptide derived from a liver stage-specific plasmodium antigen, since cytotoxic lymphocytes were found in the malaria-immune population which were B53-restricted and specific for this peptide (Hill et al., 1992). Given the functional assay used to identify this peptide, it seems likely that this HLA association operates by a given peptide and HLA allele combining to stimulate a protective immune response, a phenomenon of determinant selection, rather than eliminating or suppressing a pathogenic one. This subject is dealt with in greater detail in Chapter 20.

Similar hypotheses to explain associations between certain HLA alleles and other infectious diseases have been suggested. For example, a peptide composed of amino acid residues 18–27 of the hepatitis B virus core antigen is widely recognized by cytotoxic T cells of HLA-A2-positive patients (Vitiello et al., 1995).

Goodpasture's disease has also been particularly useful in teasing out the mechanisms of autoimmune disease. Not only does it have well-defined HLA associations but, unlike most other human autoimmune disease, the target auto-antigen is known. The class II alleles known to be positively associated with Goodpasture's disease, DRB1*1501 and DRB1*0401, were found to share structural features in an important pocket (pocket 4) within the peptide binding groove. Importantly, this was shown to differ in those alleles found to be protective. These data also support the hypothesis that it is the peptide-binding characteristics of an HLA molecule which determine its ability to promote or protect from autoimmunity to a particular antigen. Again, no comment can be made as to how this works functionally.

A final example may be drawn from the chronic lung disease, berylliosis. As discussed in Chapter 18, susceptibility may be linked to the sequence of the HLA-DPβ-chain, in particular to the presence of glutamic acid at position 69. The importance of this residue is borne out by the observation that beryllium-specific clones from patients with berylliosis are restricted by DP alleles with a glutamic acid at position 69 on the β-chain.

MOLECULAR MIMICRY

It has long been known that cross-reactions between microbial antigens and autoantigens can arise and can lead to an autoimmune response. Two of the earliest examples of this kind of molecular mimicry to be defined were between *Trypanosoma cruzi* and streptococcal antigens and the myocardium, giving rise to the manifestations of Chagas' disease and rheumatic fever, respectively (Williams, 1983).

A more contentious hypothesis was put forward several years ago to account for the pathogenesis of the rheumatological disease ankylosing spondylitis. Two groups reported that antibodies raised against the gut pathogen *Klebsiella* reacted with lymphocytes from HLA-B27-positive patients with ankylosing spondylitis (Ebringer, 1979; Seager et al., 1979). They also observed that anti-B27 antibodies cross-reacted with various enterobacteria, including *Klebsiella*. This was followed up by the finding that monoclonal antibodies raised against B27 also recognized antigens expressed by *Klebsiella*, *Salmonella*, *Shigella* and *Yersinia*. The significance of these cross-reactions is that these organisms are the very ones that can provoke reactive arthritis in B27-positive individuals.

There are, however, several problems with this hypothesis that are responsible for its failure to achieve general acceptance. At a conceptual level it is unclear why this kind of cross-reaction should give rise to a disease that is confined, in most cases, to the joints. Furthermore, other groups have failed to reproduce these original observations in other groups of patients (Archer, 1981). Indeed, in one recent study, antibodies specific for a shared determinant carried by HLA-B27 and by a strain of *Klebsiella* were found with equal frequency in patients and controls (Tsuchiya et al., 1989). This is discussed in more detail in Chapter 12.

Another set of molecules that has received considerable attention in recent years, because of the possibility of cross-reactivity between pathogen and host homologues, is the heat shock proteins. These molecules are highly conserved across a remarkably wide range of organisms, from bacteria to humans. It has been postulated that the induction of T cell immunity against a bacterial or a mycobacterial heat shock protein could lead to cross-reactive recognition of autologous heat shock proteins (Kaufman, 1991). There is every reason to think that either the initial response to the foreign antigen or the cross-reaction on the autoantigen will be under HLA-linked immune response gene control.

Rather contentiously, molecular mimicry has been proposed as the underlying mechanism of another infectious disease, namely human immunodeficiency virus (HIV) infection. Considerable homology has been noted between part of the HIV-1 envelope V3 loop and HLA-DR5 and DR6. Certain DR5 and DR6 alleles have been associated with accelerated disease. These observations have given rise to the suggestion that acquired immunodeficiency syndrome (AIDS) is an autoimmune disease. This is discussed in greater detail in Chapter 20.

MAJOR HISTOCOMPATIBILITY COMPLEX CLASS I AND II MOLECULES AS RECEPTORS FOR MICROBES AND DRUGS

Another possible mechanism underlying HLA and disease associations derives from the observation that HLA molecules may contribute to the uptake of certain pathogens by cells. The mechanism by which many viruses enter cells involves attachment to cell-surface glycoproteins, followed by fusion of their membranes with the cell's plasma membrane, either at the cell surface or after endocytosis. Such cellular receptor proteins have to be glycoproteins that recycle or undergo endocytosis after virus-induced cross-linking. HLA antigens fit these criteria. It has even been suggested that HLA molecules act as cell-surface receptors for bacteria (e.g. HLA-B27 and *Salmonella*, thus explaining reactive arthritis). Some of the documented molecules that are known to be involved in viral entry into cells are the C3d receptor for the Epstein–Barr virus (Jonsson et al., 1982) and the CD4 molecule for HIV-1 (Dalgleish et al., 1984). Three viruses have been shown to adhere specifically to mouse or human MHC molecules. The Semliki forest virus and adenovirus type 2 bind to MHC class I (Helenius et al., 1978; Signas et al., 1982) and the lactate dehydrogenase virus to class II molecules (Inada and Mims, 1984). It is possible that some viruses enter cells by forming specific attachments to HLA antigens. This would create the possibility of some HLA types conferring resistance to particular viruses. This would lead to a strong selective advantage for mutant viruses that "escaped" from this allelic restriction; however, given that there is an enormous difference in the generation times of viruses and humans it is difficult to see how viral glycoproteins and HLA molecules could exist in a stable state of balanced polymorphism.

ROLE OF NON-HLA GENES WITHIN THE HUMAN MAJOR HISTOCOMPATIBILITY COMPLEX

One of the important implications of the strong linkage disequilibrium that characterizes the MHC region is that caution must always be exercised when interpreting a correlation between a raised frequency of an HLA allelic marker and a human disease. As discussed in detail in Chapter 3, many genes are present within the MHC in addition to those that encode the HLA antigens. Some of these have already been characterized and include genes that encode proteins with obvious relevance to the immune system. These include genes in the class III region that encode the complement components C2, C4 and factor B, tumour necrosis factor (TNF) and the heat shock protein, hsp70. More recently, genes encoding peptide transporters and proteasome units have been identified within the class II region.

One example from human disease occurs in the archetypal immune complex autoimmune disease systemic lupus erythematosus (SLE). Hereditary total deficiency of C2 or C4 increases susceptibility to SLE (Walport and Lachmann, 1990). It is now clear that the presence of a null allele of the C4A locus is also a risk factor for the disease. This occurs independently of an association with class I or class II alleles with which C4A0 might be in linkage disequilibrium, since it is found in many different populations. Having said that, it is clear that there are certain HLA alleles which also contribute to disease susceptibility, potentially through any or all of the mechanisms of Ir gene effects outlined above. This is discussed in much greater detail in Chapter 21.

The genes encoding TNFα and β, and the hsp70 genes, have tantalizing possible links with the immune response. TNFβ is one of the soluble factors which can induce and/or enhance the expression of MHC class II genes in some cell types (Arenzana-Seisdedos et al., 1988). It has already been established that there is restriction fragment length polymorphism at this locus (Partanen and Koskimies, 1988; Jacob et al., 1990); it is not yet known whether this reflects coding region or regulatory sequence variation which could, in turn, lead to functional differences in the induction of the TNFB gene or in its efficiency in inducing class II expression. Given the key role that class II molecules play in T cell immune responses, variation in the effects of TNF could exert a strong influence on autoimmunity. Of particular relevance to this suggestion is the observation that quantitative variation in the expression of MHC class II molecules can have profound effects on T cell responses. This has been demonstrated using class II-expressing transfectants (Lechler et al., 1985). Whether this class II induction has an amplifying or a dampening effect on a local autoimmune reaction remains a point of debate. Nonetheless, individual variation in the inducibility of class II molecules in tissues, related to allelic variation in TNF genes, could very well act as an important contributor to disease susceptibility.

In a similar manner, allelic variation in the hsp70 gene could affect T cell immunity. It is suggested that this stress protein plays a role in the intracellular handling of internalized antigens (Parham, 1990), and polymorphisms have been defined (Milner and Campbell, 1982).

The genes that have been discovered to date may represent only the tip of the iceberg: there is a large number of other genes within the MHC which remain to be characterized during the next few years. It remains to be seen whether these additional genes encode products which influence or are directly involved in the immune system and whether they exhibit significant polymorphism. Nonetheless, the possibility exists that the true susceptibility genes underlying some HLA-associated diseases may be one of these non-HLA genes. The apparent HLA association may reflect the fact that tight linkage disequilibrium exists between allelic variants of the non-HLA and the HLA genes.

OTHER MECHANISMS

Several decades of observing HLA–disease associations have led to a number of other hypotheses, supported by very variable, and often very limited, bodies of evidence. One has suggested that the arthritogenic potential of HLA-B27 derives from an unusual (but importantly not unique) unpaired cysteine residue at position 67. It has been suggested that this is susceptible to forming covalent thiol bonds with other molecules, thus generating an "altered (immunogenic) self" (Benjamin and Parham, 1990). There is little evidence to support this although, as Bowness points out in Chapter 12, rats transgenic for a mutant HLA-B27, which do not have cysteine at 67, develop little arthritis.

Another suggestion is that certain HLA molecules have very different biology and thus may circumvent normal pathways which prevent the effective presentation of self-antigens. For example, HLA-B27 is unlike other class I molecules in that it appears more stable than most in the absence of peptide (Benjamin et al., 1991). It may also be expressed on the cell surface in the absence of tapasin. The implications of this follow from the discussion of antigen processing in Chapter 5. Such a class I molecule might represent an immunological "loose cannon", able to bind exogenous antigens and stimulate autoimmunity.

Yet another hypothesis suggests that autoimmunity may result from the random activation of self-reactive T cells by bacterial superantigens and that susceptibility to autoimmunity follows from the presence of an HLA allele which has the ability to bind such a superantigen. This is very different to the determinant selection described above in which the peptide binds within the peptide-binding groove of the HLA molecule. In this case, the binding of a superantigen may be similarly dependent on allele-specific sequence, but use a completely difference part of the molecule.

CONCLUSION

Several different mechanisms are clearly important in HLA associations with disease. Mechanisms originally suggested from first principles were confirmed as being of biological significance in animal models. An exciting advance in recent years has been the identification of such mechanisms in human disease and this is the subject of much of Section II of this book. It is likely that more and more such clinically important mechanisms will be defined over the next few years.

REFERENCES

Acha-Orbea, H., Steinman, L. and McDevitt, H. (1989). T cell receptors in murine autoimmune diseases. *Annu. Rev. Immunol.* **7**: 371–405.

Benjamin, R., Madrigal, J.A. and Parham, P. (1991). Peptide binding to empty HLA-B27 molecules on viable human cells. *Nature* **351**: 74–77.

Benjamin, R. and Parham, P. (1990). Guilt by association: HLA B27 and ankylosing spondylitis. *Immunol. Today* **11**: 137.

Bottazzo, G., Pujol-Borrell, R., Hanafusa, T. (1983). Role of aberrant HLA-DR expression and antigen presentation in induction of endocrine autoimmunity. *Lancet* **ii**: 1115–1119.

Bottazzo, G., Dean, B., McNally, J. et al. (1985). In situ characterization of autoimmune phenomena and expression of HLA molecules in the pancreas in diabetes mellitus. *N. Engl. J. Med.* **313**: 353–360.

Brostoff, S. and Mason, D. (1984). Experimental allergic encephalomyeitis: successful treatment in vivo with a monoclonal antibody that recognizes T helper cells. *J. Immunol.* **133**: 1938–1942.

Castano, L. and Eisenbarth, G. (1990). Type-1 diabetes: a chronic autoimmune disease of human, mouse, and rat. *Annu. Rev. Immunol.* **8**: 647–679.

Dalgleish, A., Beverley, P., Clapham, P. et al. (1984). The CD4(T4) antigen is an essential component of the receptor for AIDS retrovirus. *Nature* **312**: 763–767.

Dyson, P., Knight, A., Fairchild, S. et al. (1991). Genes encoding ligands for deletion of Vβ11 T cells cosegregate with mammary tumour virus genomes. *Nature* **349**: 530–531.

Frankel, W., Rudy, C., Coffin, J. et al. (1991). Linkage of Mls genes to endogenous mammary tumour viruses of inbred mice. *Nature* **349**: 526–527.

Hill, A.V.S., Elvin, J., Wills, A.C. et al. (1992). Molecular analysis of the association of HLA-B53 and resistance to severe malaria. *Nature* **360**: 434.

Jacob, C., Fronek, Z., Lewis, G. et al. (1990). Heritable major histocompatibility complex class II-associated differences in production of tumour necrosis factor α: relevance to systemic lupus erythematosus. *Proc. Natl Acad. Sci. USA* **87**: 1233–1237.

Jonsson, V., Wells, A. and Klein, G. (1982). Receptors for the complement C3d component and the Epstein–Barr virus are quantitatively co-expressed on a series of B cell lines and their derived somatic cell hybrids. *Cell Immunol.* **72**: 263–276.

Lund, T., O'Reilly, L., Hutchings, P. et al. (1990). Prevention of insulin-dependent diabetes mellitus in non-obese diabetic mice by transgenes encoding modified I-A β-chain or normal I-E α-chain. *Nature* **345**: 727–729.

Marrack, P., Kushnir, E. and Kappler, J. (1991). A maternally inherited superantigen encoded by a mammary tumour virus. *Nature* **349**: 524–525.

Miyazaki, T., Uno, M., Kikutani, H. et al. (1990). Direct evidence for the contribution of the unique I-A^{NOD} to the development of insulinitis in non-obese diabetic mice. *Nature* **345**: 722–724.

Nishimoto, H., Kikurani, H., Yamamura, K. et al. (1987). Prevention of autoimmune insulitis by expression of I-E molecules in NOD mice. *Nature* **328**: 432–434.

Olitsky, P. and Yager, R. (1949). Experimental disseminated encephalomyeltis in white mice. *J. Exp. Med.* **90**: 213–113.

Patarnen, J. and Koskimies, S. (1988). Low degree of DNA polymorphism in the HLA-linked lymphotoxin (Tumour Necrosis Factor β) gene. *Scand. J. Immunol.* **28**: 313–316.

Paterson, P. (1976). Experimental allergic encephalomyelitis and autoimmune disease. *Adv. Immunol.* **5**: 131–153.

Rivers, T. and Schwentker, F. (1935). Encephalomyelitis accompanied by myelin destruction experimentally produced in monkeys. *J. Exp. Med.* **61**: 689–702.

Sakai, K., Sinha, A., Mitchell, D. et al. (1988). Involvement of distinct T cell receptors in the autoimmune encephalitogenic response to nested epitopes of myelin basic protein. *Proc. Natl Acad. Sci. USA* **85**: 8608–8612.

Scott, C. and Steinman, L. (1990). The T lymphocyte in experimental allergic encephalomyelitis. *Annu. Rev. Immunol.* **8**: 579–621.

Signas, C., Katze, M., Persson, H. et al. (1982). An adenovirus glycoprotein binds heavy chains of class I transplantation antigens from man and mouse. *Nature* **299**: 175–178.

Slattery, R., Kjer-Nielsen, L., Allison, J. et al. (1990). Prevention of diabetes in non-obese diabetic I-A^k transgenic mice. *Nature* **345**: 724–727.

Steinman, L., Solomon, D., Zamvil, S. et al. (1983). Prevention of EAE with anti-I-A antibody: decreased accumulation of radiolabelled lymphocytes in the central nervous system. *J. Neuroimmunol.* **5**: 91–97.

Stuart, G. and Krikorian, K. (1928). The neuroparalytic accidents of anti-rabies treatment. *Ann. Trop. Med. Parasitol.* **22**: 327–377.

Todd, J., Fukui, Y. and Kitagawa, T. (1990). The A3 allele of the HLA-DQA1 locus is associated with susceptibility to type I diabetes in Japanese. *Proc. Natl Acad. Sci. USA* **87**: 1094–1098.

Turnley, A., Morahan, G., Okano, H. et al. (1991). Dysmyelination in transgenic mice resulting from expression of class I histocompatibility molecules in oligodendrocytes. *Nature* **353**: 566–567.

Vitiello, A., Ishioka, G, Gray, H.M. et al. (1995). Development of a lipopeptide-based therapeutic vaccine to treat chronic HBV infection. *J. Clin. Invest.* **95**: 341–349.

Waldor, M., Sriram, S., Hardy, R. et al. (1985). Reversal of experimental allergic encephalomyelitis with a monoclonal antibody to a T cell subset marker (L3T4). *Science* **227**: 415–417.

Walport, M.J. and Lachmann, P.J. (1990). Complement deficiencies and abnormalities of the complement system in systemic lupus erythematosus. *Curr. Opin. Rheumatol.* **2**: 661–663.

Williams, R. (1983). Rheumatic fever and the streptococcus. *Am. J. Med.* **75**: 727–730.

Woodlan, D., Happ, M., Gollob, K. et al. (1991). An endogenous retrovirus mediating deletion of αβ T cells? *Nature* **349**: 528–529.

Wraith, D., McDevitt, H., Steinman, L. et al. (1989). T cell recognition as the target for immune intervention in autoimmune disease. *Cell* **57**: 709–715.

Zamvil, S., Nelson, P., Trotter, J. et al. (1985a). T cell clones specific for myelin basic protein induce chronic relapsing EAE and demyelination. *Nature* **317**: 355–358.

Zamvil, S., Nelson, P., Mitchell, D. et al. (1985b). Encephalitogenic T cell clones specific for myelin basic protein: an unusual bias in antigen presentation. *J. Exp. Med.* **162**: 2107–2124.

CHAPTER 8

HLA Nomenclature: The Name of the Rose

Julia G. Bodmer, Walter F. Bodmer and Steven G.E. Marsh

INTRODUCTION

As William Shakespeare said, "What's in a name? That which we call a rose, by any other name would smell as sweet." That is true and there is no reason why a rose should not have another name, as long as everybody uses the same name. This is the basic *raison d'être* of a standardized nomenclature which should be universally known, universally accepted and universally used.

A common language is essential for understanding and working with highly complex phenomena, such as the HLA system. The definition of a nomenclature clarifies the concepts underlying the system to which it is being applied. In contrast, for example, to the development of the red cell blood groups, there were few simple monospecific serological reagents available to identify the white cell antigenic specificities that were to define the HLA system. Most of the initial serological reagents contained mixtures of many antibodies, and required extensive statistical analysis to draw out the definition of single specificities. Hence Race and Sanger's quaint reference to the "so splendidly complicated antigens of the white cells" (Race and Sanger, 1968). R.A. Fisher was probably the first to recognize, in the early 1930s, that serological assays may identify "the direct products of individual genes rather than have secondary reactions" (Bodmer, 1990). It was recognized almost immediately that there was a need to define specificities that could be identified reproducibly and unequivocally in different laboratories without prejudging the nature of the underlying genetics; and yet, right from the start, the geneticists primarily involved in developing the HLA nomenclature emphasized the importance of the principle that whatever nomenclature was introduced must in future be compatible with the underlying genetic system.

It is important to avoid backtracking in the establishment of nomenclature. This is often difficult to reconcile with forward compatibility as new discoveries are made and means that nomenclature should be assigned conservatively and always remain somewhat behind the most recent advances. In this way it is possible to wait to assign a definitive name until there is some reassurance that the entity being named will not disappear on further investigation. The only serological specificity to disappear was A20, which was, in fact, an unrecognized C locus specificity.

When reagents are hard to find and tests are relatively unreliable, as was the case with the initial HLA serology, it is essential that different laboratories compare results by exchanging reagents, typing a common panel of cells and comparing results using different types of assays. In this respect the International Histocompatibility Testing Workshops have been a spectacular success and essential to the orderly development of the HLA system and of its nomenclature. At first, the HLA typers were a small, dedicated band who could be brought together relatively easily. In spite of strong personalities, all realized the importance of a common basis for agreement of the definition of these new determinants that they were identifying. Once such a core group is established, it is very hard for any other group to compete. Thus, the core group expanded greatly, resulting in the current large-scale, international workshops. These involve hundreds of laboratories typing thousands of cells for hundreds of differences and looking at the influence of the HLA system in such diverse areas as transplantation, cancer and other diseases and population diversity.

HLA WORKSHOPS AND NOMENCLATURE REPORTS

The first HLA Workshop, organized by Bernard Amos in 1964, took place at the same time as Walter Bodmer was presenting the work done with Rose Payne, which established what came to be called HLA-A locus (Payne et al., 1964). The problem of nomenclature was already evident then and, in an appendix, an approach to naming antigens and genes was described using simple combinations of letters and numbers and avoiding superscripts and subscripts. This later came to be the basis for much of the subsequent nomenclature development. As we said then: "A system of nomenclature has to strike a compromise between the convenience and simplicity required for everyday use and the need for adequate definition of the concepts involved. The adoption of an unwieldy or inadequate notation may well retard advances in the understanding of a complex system." These are surely sentiments which apply today just as much, even though our knowledge of genetics has been revolutionized over these last 30 years.

A critical problem to be overcome was the overall naming of the system. Scientists have an extraordinary emotional attachment to names that they have chosen for phenomena, genes, products, etc. They no doubt feel that a name attached to something, which sticks, is a form of immortalization rather like that associated with a major donor attaching his or her name to a building or professorship in a university. We had staked a claim for LA, L standing for leucocyte and A, not as commonly assumed, for antigen, but for the first locus, anticipating the possibility that there would be LB, LC and so on. Dausset and his colleagues had assigned the name HU-1, meaning human system-1. The compromise that was reached at a meeting in America in 1968 and reported by Amos in a letter to *Science* (Amos, 1968), was HL-A, a sort of double recombinant between HU-1 and LA. It took 7 years and the recognition and naming of the A, B, C and D loci before it could be agreed that the hyphen should be dropped to simplify the nomenclature somewhat by leaving out at least one unnecessary symbol.

A verbatim account of the first substantive nomenclature committee meeting was provided by Roy Walford in Terasaki's *History of HLA: 10 Recollections* (Walford, 1990). This shows clearly the detailed discussions needed at that time to come to an agreement as to the criteria to be used for defining a serological determinant with the then available techniques. It also emphasizes the appropriately conservative strategy that was followed in naming essentially only those determinants on which everybody clearly agreed. A complete table was given of the equivalents between the newly established nomenclature and the designations used in the various laboratories that had taken part in the 1967 Turin Workshop on the basis of which the decisions were being made (WHO Nomenclature Committee, 1968). A simple consecutive numbering system was adopted and spaces were left at 4 and 6 for the later placing of van Rood's 4a and 4b which, to this day, remain the only epitopes that are separately defined. In retrospect, it is remarkable how robust these definitions were. The only confusion arose with "HL-A5", where use of cross-reacting sera had confused the definition of a broadly reactive antigen W5, which later became B35, and the narrower determinant which became B5 and which had earlier been called 4c. It is interesting that, at this time, when the red cell blood groupers still had a considerable influence, the emphasis was on the serological reagent as the standard. Every attempt should be made to obtain monospecific reagents and to define them as such by absorption, even though it was becoming clear that this was almost impossible, as was clearly established subsequently by the complex patterns of cross-reaction identified by monoclonal antibodies.

Until the DNA era, nomenclature committee meetings and reports followed each of the international workshops since, by virtue of common reagents, they were the basis on which new specificities and genes could be defined and the genetics of the system clarified by common agreement. Thus, the second Nomenclature Report in 1970 defined only four new specificities (WHO Nomenclature Committee, 1970), but introduced the concept of reference laboratories, presaging a gradual move from reagents to cells and, eventually, DNA as the standard. The third report, in 1972 (WHO Nomenclature Committee, 1972), introduced the concept of provisional definitions – the letter w before the number – and recognized the problem of broad and narrow specificities reflected in the split of A9 into A23 and A24. (Note that the *w* designation has now been dropped from every antigen except for HLA-Bw4 and -Bw6, HLA-C products – to differentiate them from complement alleles – and those antigens that are defined on the basis of cellular typing, i.e. HLA-D and -DP.) Originally, a decimal notation had been suggested for such splits, but this was soon seen to be unwieldy. It was also the 1972 report that first clearly recognized the need for genetics, e.g. family linkage studies, to ensure that a new determinant was at least closely linked to those that had been previously described. The Fourth Report in 1975 (WHO Nomenclature Committee, 1975), the first of which Walter Bodmer was rapporteur, clearly introduced, on a well-defined genetic basis, the system with which we are now familiar, the HLA region having just been assigned to chromosome 6. HLA was the name of the region or system and, following a hyphen, the letters A, B, C, D, etc. designated the genes within the system whose alleles were A1, A2, etc. Even at that time questions began to be raised as to which genes should be encompassed within the HLA nomenclature considerations, since some of the complement components had been mapped to the HLA region. Fortunately, although complement functions are immunological, the HLA Nomenclature Committee has never considered questions of complement nomenclature to be within its remit. Now for the first time specificities that were almost unique to non-Caucasoid populations were being identified, reflecting the inevitable Caucasoid bias of the initial serological reagents and specificity definitions. This was also the first time that strong emphasis was placed on urging workers in the field not to pre-empt the use of formal symbols, such as HLB or HLA-E (not defined at that time), and to allow the Nomenclature Committee to act as a central naming authority within a system that was becoming widely accepted.

The 7th Workshop in 1977, and its associated 5th Nomenclature Report (WHO Nomenclature Committee, 1978), had as the major advance, the naming of a DR gene for D-related, the HLA-D determinants having been established by cellular techniques, predominantly using the mixed lymphocyte culture assay. D itself was not used because there were still those then who argued that cellular determinants may not be recognizable by serology. It was only then, incidentally, that 4a and 4b became Bw4 and Bw6, respectively. A 6th Nomenclature Report, naming a few new specificities, was published following the 8th Workshop in 1980 (WHO Nomenclature Committee, 1980).

The next report in 1984 heralded the first major contribution of molecular biology (WHO Nomenclature Committee, 1985). The D-region genes had been cloned, their two-chain structure identified and a map of the region not too different from that which we now know had been established. The nomenclature DR, DP and DQ was introduced, and the suggestion was made that the genes for the separate chains should be called DRA, DRB, etc. By the time of the next report in 1987 (Bodmer et al., 1989), pseudogenes had been recognized and named and it was realized that many previously unitary antigens defined by serology would be "split" by DNA sequencing. Thus, in addition to the definition of several new genes, nomenclature to relate the DNA sequence to the predominant serological specificities was introduced so that, for example, subtypes or alleles related to B27 became B*2701 and B*2702. The emphasis was on the products that were expressed, but it was also recognized that there could be silent changes whose designation could be B*27011. At this stage, the asterisk between the gene name and the allele was introduced, since by that time many gene names ended with a number and the asterisk provided an appropriate spacer between the number defining a gene and that defining an allele. This is the only concession that the HLA community has made to the usage promoted by the larger human gene mapping community.

TABLE 8.1 Names for genes in the HLA region

Name	Previous equivalents	Molecular characteristics
HLA-A	–	Class I α-chain
HLA-B	–	Class I α-chain
HLA-C	–	Class I α-chain
HLA-E	E, '6.2'	expressed gene associated with class I 6.2-kB Hind III fragment
HLA-F	F, '5.4'	expressed gene associated with class I 5.4-kB Hind III fragment
HLA-G	G, '6.0'	expressed gene associated with class I 6.0-kB Hind III fragment
HLA-H	H, AR, '12.4'	Class I pseudogene associated with 5.4-kB Hind III fragment
HLA-J	cda12	Class I pseudogene associated with 5.9-kB Hind III fragment
HLA-K	HLA-70	Class I pseudogene associated with 7.0-kB Hind III fragment
HLA-L	HLA-92	Class I pseudogene associated with 9.2-kB Hind III fragment
HLA-DRA	DRα	DR α-chain
HLA-DRB1	DRβI, DR1B	DR β1-chain determining specificities DR1, DR2, DR3, DR4, DR5 etc.
HLA-DRB2	DRβII	pseudogene with DRB-like sequence
HLA-DRB3	DRβIII, DR3B	DR β3-chain determining DR52 and Dw24, Dw25, Dw26 specificities
HLA-DRB4	DRβIV, DR4B	DR β4-chain determining DR53
HLA-DRB5	DRβIII	DR β5-chain determining DR51
HLA-DRB6	DRBX, DRBσ	DRB pseudogene found on DR1, DR2 and DR10 haplotypes.
HLA-DRB7	DRBψ1	DRB pseudogene found on DR4, DR7 and DR9 haplotypes.
HLA-DRB8	DRBψ2	DRB pseudogene found on DR4, DR7 and DR9 haplotypes.
HLA-DRB9	M4.2 β exon	DRB pseudogene, isolated fragment
HLA-DQA1	DQα1, DQ1A	DQ α-chain
HLA-DQB1	DQβ1, DQ1B	DQ β-chain
HLA-DQA2	DXα, DQ2A	DQ α-chain-related sequence, not known to be expressed
HLA-DQB2	DXβ, DQ2B	DQ β-chain-related sequence, not known to be expressed
HLA-DQB3	DVβ, DQB3	DQ β-chain-related sequence, not known to be expressed
HLA-DOA	DZα, DOα, DNA	DO α-chain
HLA-DOB	DOβ	DO β-chain
HLA-DMA	RING6	DM α-chain
HLA-DMB	RING7	DM β-chain
HLA-DPA1	DPα1, DP1A	DP α-chain
HLA-DPB1	DPβ1, DP1B	DP β-chain
HLA-DPA2	DPα2, DP2A	DP α-chain-related pseudogene
HLA-DPB2	DPβ2, DP2B	DP β-chain-related pseudogene
TAP1	RING4, Y3, PSF1	ABC (ATP Binding Cassette) transporter
TAP2	RING11, Y1, PSF2	ABC (ATP Binding Cassette) transporter
LMP2	RING12	Proteasome-related sequence
LMP7	RING10	Proteasome-related sequence
MICA	MICA, PERB11.1	Class I chain-related gene
MICB	MICB, PERB11.2	Class I chain-related gene
MICC	MICC, PERB11.3	Class I chain-related pseudogene
MICD	MICD, PERB11.4	Class I chain-related pseudogene
MICE	MICE, PERB11.5	Class I chain-related pseudogene

The DOA gene has been recently renamed from its original designation of DNA; this is the only locus which has been renamed.

TABLE 8.2(a) HLA Class I alleles as named September 1999

A	B					C		E	F	G
A*0101	A*2416	B*07021	B*1536	B*3701	B*4703	Cw*0102	Cw*1602	E*0101	F*0101	G*01011
A*0102	A*2417	B*07022	B*1537	B*3702	B*4801	Cw*0103	Cw*16041	E*0102		G*01012
A*0103	A*2418	B*07023	B*1538	B*3801	B*4802	Cw*0104	Cw*1701	E*01031		G*01013
A*0104N	A*2419	B*0703	B*1539	B*38021	B*4803	Cw*02021	Cw*1702	E*01032		G*01014
A*0105N	A*2420	B*0704	B*1540	B*38022	B*4804	Cw*02022	Cw*1703	E*0104		G*01015
A*0106	A*2421	B*0705	B*1542	B*3803	B*4805	Cw*02023	Cw*1801			G*01016
A*02011	A*2422	B*0706	B*1543	B*39011	B*4806	Cw*02024	Cw*1802			G*01017
A*02012	A*2423	B*0707	B*1544	B*39013	B*4807	Cw*0203				G*01018
A*02013	A*2424	B*0708	B*1545	B*39021	B*4901	Cw*0302				G*0102
A*02014	A*2501	B*0709	B*1546	B*39022	B*5001	Cw*03031				G*0103
A*0202	A*2502	B*0710	B*1547	B*3903	B*5002	Cw*03032				G*01041
A*0203	A*2503	B*0711	B*1548	B*3904	B*5004	Cw*03041				G*01042
A*0204	A*2601	B*0712	B*1549	B*3905	B*51011	Cw*03042				G*01043
A*0205	A*2602	B*0713	B*1550	B*39061	B*51012	Cw*0305				G*0105N
A*0206	A*2603	B*0714	B*1551	B*39062	B*51021	Cw*0306				
A*0207	A*2604	B*0715	B*1552	B*3907	B*51022	Cw*0307				
A*0208	A*2605	B*0716	B*1553	B*3908	B*5103	Cw*0308				
A*0209	A*2606	B*0801	B*1801	B*3909	B*5104	Cw*0309				
A*0210	A*2607	B*0802	B*1802	B*3910	B*5105	Cw*0310				
A*0211	A*2608	B*0803	B*1803	B*3911	B*5106	Cw*0311				
A*0212	A*2609	B*0804	B*1804	B*3912	B*5107	Cw*04011				
A*0213	A*2610	B*0805	B*1805	B*3913	B*5108	Cw*04012				
A*0214	A*2611N	B*0806	B*1806	B*3914	B*5109	Cw*0403				
A*0215N	A*2612	B*0807	B*1807	B*3915	B*5110	Cw*0404				
A*0216	A*2613	B*0808N	B*1808	B*3916	B*5111N	Cw*0405				
A*02171	A*2901	B*0809	B*2701	B*3917	B*5112	Cw*0406				
A*02172	A*2902	B*0810	B*2702	B*40011	B*5113	Cw*0407				
A*0218	A*2903	B*1301	B*2703	B*40012	B*5114	Cw*0501				
A*0219	A*2904	B*1302	B*2704	B*4002	B*5115	Cw*0502				
A*0220	A*3001	B*1303	B*27052	B*4003	B*5116	Cw*0602				
A*0221	A*3002	B*1304	B*27053	B*4004	B*5117	Cw*0603				
A*0222	A*3003	B*1401	B*2706	B*4005	B*5118	Cw*0604				
A*0224	A*3004	B*1402	B*2707	B*4006	B*5119	Cw*0605				
A*0225	A*3006	B*1403	B*2708	B*4007	B*52011	Cw*07011				
A*0226	A*3007	B*1404	B*2709	B*4008	B*52012	Cw*07012				
A*0227	A*31012	B*1405	B*2710	B*4009	B*5301	Cw*0702				
A*0228	A*3102	B*14061	B*2711	B*4010	B*5302	Cw*0703				
A*0229	A*3103	B*14062	B*2712	B*4011	B*5303	Cw*0704				
A*0230	A*3104	B*15011	B*2713	B*4012	B*5304	Cw*0705				
A*0231	A*3201	B*1501102N	B*2714	B*4013	B*5401	Cw*0706				
A*0232N	A*3202	B*15012	B*2715	B*4014	B*5501	Cw*0707				
A*0233	A*3203	B*15013	B*2716	B*4015	B*5502	Cw*0708				
A*0234	A*3204	B*1502	B*3501	B*4016	B*5503	Cw*0709				
A*0235	A*3301	B*1503	B*3502	B*4018	B*5504	Cw*0710				
A*0236	A*3303	B*1504	B*3503	B*4019	B*5505	Cw*0711				
A*03011	A*3304	B*1505	B*3504	B*4020	B*5507	Cw*0712				
A*03012	A*3305	B*1506	B*3505	B*4021	B*5508	Cw*0713				
A*03013	A*3401	B*1507	B*3506	B*4022N	B*5601	Cw*0714				
A*0302	A*3402	B*1508	B*3507	B*4023	B*5602	Cw*0801				
A*0303N	A*3601	B*1509	B*3508	B*4024	B*5603	Cw*0802				
A*0304	A*4301	B*1510	B*35091	B*4025	B*5604	Cw*0803				
A*11011	A*6601	B*1511	B*35092	B*4101	B*5605	Cw*0804				
A*11012	A*6602	B*1512	B*3510	B*4102	B*5606	Cw*0805				
A*1102	A*6603	B*1513	B*3511	B*4103	B*5607	Cw*0806				
A*1103	A*68011	B*1514	B*3512	B*4201	B*5701	Cw*12021				
A*1104	A*68012	B*1515	B*3513	B*4202	B*5702	Cw*12022				

TABLE 8.2(a) *Continued.*

A	B					C	E	F	G
A*1105	A*6802	B*1516	B*3514	B*4402	B*5703	Cw*1203			
A*2301	A*68031	B*1517	B*3515	B*44031	B*5704	Cw*12041			
A*2302	A*68032	B*1518	B*3516	B*44032	B*5705	Cw*12042			
A*2303	A*6804	B*1519	B*3517	B*4404	B*5706	Cw*1205			
A*2402101	A*6805	B*1520	B*3518	B*4405	B*5801	Cw*1206			
A*2402102L	A*6806	B*1521	B*3519	B*4406	B*5802	Cw*1301			
A*24022	A*6807	B*1522	B*3520	B*4407	B*5901	Cw*14021			
A*24031	A*6808	B*1523	B*3521	B*4408	B*67011	Cw*14022			
A*24032	A*6809	B*1524	B*3522	B*4409	B*67012	Cw*1403			
A*2404	A*6810	B*1525	B*3523	B*4410	B*7301	Cw*1404			
A*2405	A*6811N	B*1526N	B*3524	B*4411	B*7801	Cw*15021			
A*2406	A*6812	B*1527	B*3525	B*4412	B*78021	Cw*15022			
A*2407	A*6813	B*1528	B*3526	B*4413	B*78022	Cw*1503			
A*2408	A*6814	B*1529	B*3527	B*4414	B*7803	Cw*1504			
A*2409N	A*6901	B*1530	B*3528	B*4415	B*7804	Cw*15051			
A*2410	A*7401	B*1531	B*3529	B*4501	B*8101	Cw*15052			
A*2411N	A*7402	B*1532	B*3530	B*4502	B*8201	Cw*1506			
A*2413	A*7403	B*1533	B*3531	B*4601		Cw*1507			
A*2414	A*8001	B*1534	B*3532	B*4701		Cw*1508			
A*2415		B*1535	B*3533	B*4702		Cw*1601			

TABLE 8.2(b) HLA alleles as named September 1999

DRA	DRB1			DRB2	DRB3	DRB4	DRB5	DRB6/7/8/9
DRA*0101	DRB1*0101	DRB1*0810	DRB1*1314	DRB2*0101	DRB3*01011	DRB4*01011	DRB5*01011	DRB6*0101
DRA*0102	DRB1*01021	DRB1*0811	DRB1*1315		DRB3*01012	DRB4*0102	DRB5*01012	DRB6*0201
	DRB1*01022	DRB1*0812	DRB1*1316		DRB3*01013	DRB4*0103101	DRB5*0102	DRB6*0202
	DRB1*0103	DRB1*0813	DRB1*1317		DRB3*01014	DRB4*0103102N	DRB5*0103	
	DRB1*0104	DRB1*0814	DRB1*1318		DRB3*0102	DRB4*01032	DRB5*0104	DRB7*01011
	DRB1*0105	DRB1*0815	DRB1*1319		DRB3*0103	DRB4*0104	DRB5*0105	DRB7*01012
	DRB1*0106	DRB1*0816	DRB1*1320		DRB3*0104	DRB4*0105	DRB5*0106	
	DRB1*03011	DRB1*0817	DRB1*1321		DRB3*0105	DRB4*0201N	DRB5*0107	DRB8*0101
	DRB1*03012	DRB1*0818	DRB1*1322		DRB3*0106	DRB4*0301N	DRB5*0108N	
	DRB1*03021	DRB1*0819	DRB1*1323		DRB3*0107		DRB5*0109	DRB9*0101
	DRB1*03022	DRB1*0820	DRB1*1324		DRB3*0201		DRB5*0110N	
	DRB1*0303	DRB1*0821	DRB1*1325		DRB3*02021		DRB5*0202	
	DRB1*0304	DRB1*09012	DRB1*1326		DRB3*02022		DRB5*0203	
	DRB1*0305	DRB1*1001	DRB1*1327		DRB3*0203		DRB5*0204	
	DRB1*0306	DRB1*11011	DRB1*1328		DRB3*0204			
	DRB1*0307	DRB1*11012	DRB1*1329		DRB3*0205			
	DRB1*0308	DRB1*11013	DRB1*1330		DRB3*0206			
	DRB1*0309	DRB1*1102	DRB1*1331		DRB3*0207			
	DRB1*0310	DRB1*1103	DRB1*1332		DRB3*0208			
	DRB1*0311	DRB1*11041	DRB1*1333		DRB3*0209			
	DRB1*0312	DRB1*11042	DRB1*1334		DRB3*0301			
	DRB1*0313	DRB1*1105	DRB1*1335		DRB3*0302			
	DRB1*0314	DRB1*1106	DRB1*1401		DRB3*0303			
	DRB1*0315	DRB1*1107	DRB1*1402					
	DRB1*04011	DRB1*11081	DRB1*1403					
	DRB1*04012	DRB1*11082	DRB1*1404					
	DRB1*0402	DRB1*1109	DRB1*1405					
	DRB1*04031	DRB1*1110	DRB1*1406					
	DRB1*04032	DRB1*1111	DRB1*1407					
	DRB1*0404	DRB1*1112	DRB1*1408					

TABLE 8.2(b) *Continued.*

DRA	DRB1			DRB2	DRB3	DRB4	DRB5	DRB6/7/8/9
	DRB1*04051	DRB1*1113	DRB1*1409					
	DRB1*04052	DRB1*1114	DRB1*1410					
	DRB1*0406	DRB1*1115	DRB1*1411					
	DRB1*0407	DRB1*1116	DRB1*1412					
	DRB1*0408	DRB1*1117	DRB1*1413					
	DRB1*0409	DRB1*1118	DRB1*1414					
	DRB1*0410	DRB1*1119	DRB1*1415					
	DRB1*0411	DRB1*1120	DRB1*1416					
	DRB1*0412	DRB1*1121	DRB1*1417					
	DRB1*0413	DRB1*1122	DRB1*1418					
	DRB1*0414	DRB1*1123	DRB1*1419					
	DRB1*0415	DRB1*1124	DRB1*1420					
	DRB1*0416	DRB1*1125	DRB1*1421					
	DRB1*0417	DRB1*1126	DRB1*1422					
	DRB1*0418	DRB1*1127	DRB1*1423					
	DRB1*0419	DRB1*1128	DRB1*1424					
	DRB1*0420	DRB1*1129	DRB1*1425					
	DRB1*0421	DRB1*1130	DRB1*1426					
	DRB1*0422	DRB1*1131	DRB1*1427					
	DRB1*0423	DRB1*1132	DRB1*1428					
	DRB1*0424	DRB1*1133	DRB1*1429					
	DRB1*0425	DRB1*1134	DRB1*1430					
	DRB1*0426	DRB1*1135	DRB1*1431					
	DRB1*0427	DRB1*1136	DRB1*1432					
	DRB1*0428	DRB1*1201	DRB1*1433					
	DRB1*0429	DRB1*12021	DRB1*15011					
	DRB1*0430	DRB1*12022	DRB1*15012					
	DRB1*0431	DRB1*12032	DRB1*15021					
	DRB1*0432	DRB1*1204	DRB1*15022					
	DRB1*0433	DRB1*1205	DRB1*15023					
	DRB1*07011	DRB1*1206	DRB1*1503					
	DRB1*07012	DRB1*1301	DRB1*1504					
	DRB1*0703	DRB1*1302	DRB1*1505					
	DRB1*0704	DRB1*13031	DRB1*1506					
	DRB1*0801	DRB1*13032	DRB1*1507					
	DRB1*08021	DRB1*1304	DRB1*1508					
	DRB1*08022	DRB1*1305	DRB1*16011					
	DRB1*08032	DRB1*1306	DRB1*16012					
	DRB1*08041	DRB1*13071	DRB1*16021					
	DRB1*08042	DRB1*13072	DRB1*16022					
	DRB1*08043	DRB1*1308	DRB1*1603					
	DRB1*0805	DRB1*1309	DRB1*1604					
	DRB1*0806	DRB1*1310	DRB1*1605					
	DRB1*0807	DRB1*1311	DRB1*1607					
	DRB1*0808	DRB1*1312	DRB1*1608					
	DRB1*0809	DRB1*1313						

TABLE 8.2(c) HLA alleles as named September 1999

DQA1	DQB1	DPA1	DPB1		DMA	DMB	DOA	DOB
DQA1*0101	DQB1*0201	DPA1*01031	DPB1*01011	DPB1*5401	DMA*0101	DMB*0101	DOA*01011	DOB*0101
DQA1*01021	DQB1*0202	DPA1*01032	DPB1*01012	DPB1*5501	DMA*0102	DMB*0102	DOA*0101201	DOB*0102
DQA1*01022	DQB1*0203	DPA1*0104	DPB1*02012	DPB1*5601	DMA*0103	DMB*0103	DOA*0101202	DOB*0103
DQA1*0103	DQB1*03011	DPA1*0105	DPB1*02013	DPB1*5701	DMA*0104	DMB*0104	DOA*0101203	
DQA1*0104	DQB1*03012	DPA1*0106	DPB1*0202	DPB1*5801		DMB*0105	DOA*01013	
DQA1*0105	DQB1*0302	DPA1*0107	DPB1*0301	DPB1*5901		DMB*0106	DOA*0101401	
DQA1*0106	DQB1*03032	DPA1*02011	DPB1*0401	DPB1*6001			DOA*0101402	
DQA1*0201	DQB1*03033	DPA1*02012	DPB1*0402	DPB1*6101N			DOA*01015	
DQA1*03011	DQB1*0304	DPA1*02013	DPB1*0501	DPB1*6201				
DQA1*0302	DQB1*0305	DPA1*02014	DPB1*0601	DPB1*6301				
DQA1*0303	DQB1*0306	DPA1*02015	DPB1*0801	DPB1*6401N				
DQA1*0401	DQB1*0307	DPA1*02021	DPB1*0901	DPB1*6501				
DQA1*05011	DQB1*0308	DPA1*02022	DPB1*1001	DPB1*6601				
DQA1*05012	DQB1*0309	DPA1*02023	DPB1*11011	DPB1*6701				
DQA1*0502	DQB1*0401	DPA1*0203	DPB1*11012	DPB1*6801				
DQA1*0503	DQB1*0402	DPA1*0301	DPB1*1301	DPB1*6901				
DQA1*0504	DQB1*05011	DPA1*0302	DPB1*1401	DPB1*7001				
DQA1*0505	DQB1*05012	DPA1*0401	DPB1*1501	DPB1*7101				
DQA1*06011	DQB1*0502		DPB1*1601	DPB1*7201				
DQA1*06012	DQB1*05031		DPB1*1701	DPB1*7301				
	DQB1*05032		DPB1*1801	DPB1*7401				
	DQB1*0504		DPB1*1901	DPB1*7501				
	DQB1*06011		DPB1*20011	DPB1*7601				
	DQB1*06012		DPB1*20012	DPB1*7701				
	DQB1*06013		DPB1*2101	DPB1*7801				
	DQB1*0602		DPB1*2201	DPB1*7901				
	DQB1*0603		DPB1*2301	DPB1*8001				
	DQB1*06041		DPB1*2401	DPB1*8101				
	DQB1*06042		DPB1*2501	DPB1*8201				
	DQB1*06051		DPB1*26011	DPB1*8301				
	DQB1*06052		DPB1*26012	DPB1*8401				
	DQB1*0606		DPB1*2701					
	DQB1*0607		DPB1*2801					
	DQB1*0608		DPB1*2901					
	DQB1*0609		DPB1*3001					
	DQB1*0610		DPB1*3101					
	DQB1*06111		DPB1*3201					
	DQB1*06112		DPB1*3301					
	DQB1*0612		DPB1*3401					
	DQB1*0613		DPB1*3501					
	DQB1*0614		DPB1*3601					
	DQB1*0615		DPB1*3701					
	DQB1*0616		DPB1*3801					
			DPB1*3901					
			DPB1*4001					
			DPB1*4101					
			DPB1*4401					
			DPB1*4501					
			DPB1*4601					
			DPB1*4701					
			DPB1*4801					
			DPB1*4901					
			DPB1*5001					
			DPB1*5101					
			DPB1*5201					
			DPB1*5301					

In virtually all other respects, the human gene mapping community has essentially adopted the approaches that came from the HLA Nomenclature Committee.

This has given rise to an alternative nomenclature for the polypeptides for which the HLA genes code. Thus, the heterodimeric protein product of the DRA and DRB1*0101 genes is called DR ($\alpha, \beta 1*0101$).

The 8th Nomenclature Report in 1989 (Bodmer et al., 1990), by which time Julia Bodmer had replaced Walter Bodmer as rapporteur, was the first to take place between workshops, recognizing the need for more rapid response, given the enormous rate of accumulation of sequence data. It had clearly become necessary to emphasize the need to deposit the sequences as they were discovered in an appropriate database and that, gradually, there would be a need for continuous updating. Nomenclature was turning into the management of a database and extensive communication with the HLA community and, thus, required significantly more resources than had been the case so far. These resources were provided by the Imperial Cancer Research Fund in the UK until September 1996. Since that date Steven Marsh has continued the work at the Anthony Nolan Research Institute, also in the UK.

The six most recent reports (Bodmer et al., 1991, 1992, 1994, 1995, 1997, 1999) have essentially all covered issues at the DNA level. Table 8.1 presents the list of genes, including expressed genes, pseudogenes, other genes not known to be expressed and fragments of genes, in the HLA region which had received official nomenclature by the time of the 1998 report. Table 8.2 shows the list of currently identified HLA alleles. There are currently recognized 151 HLA-A, 301 HLA-B, 83 HLA-C, 5 HLA-E, 1 HLA-F, 14 HLA-G, 2 HLA-DRA, 281 HLA-DRB, 20 HLA-DQA1, 43 HLA-DQB1, 18 HLA-DPA1, 87 HLA-DPB1, 4 HLA-DMA, 6 HLA-DMB, 8 HLA-DOA and 3 HLA-DOB alleles or 1027 in total. Since 1990, to cope with the increasing number of new HLA alleles that have been described, names have been assigned on a daily basis. The sequences are reported centrally to the HLA Sequence Database, now held at the Anthony Nolan Research Institute, and monthly nomenclature updates are published which give details on the newly defined alleles. Six updates have been published since the last full nomenclature report in 1998 (Marsh, 1999a–f). Figure 8.1 shows the numbers of alleles assigned each year since 1987. Currently, about 150 new HLA alleles are being described each year. Unfortunately, it is inevitable with this number of sequence submissions that occasionally sequences which contain errors are named and are later withdrawn. Table 8.3 lists all of the names that have been assigned to such sequences and have now been dropped, mostly because of sequencing errors in the early days of DNA sequencing of HLA alleles.

Serology has now been almost totally left behind, so that it even becomes inappropriate to recognize a serological specificity unless it has been defined by a DNA sequence. Table 8.4 shows the list of serological and cellular HLA specificities given in the 1996 Nomenclature Report. Questions have been raised concerning the definition of alleles due to silent base changes, the problem of partial sequences or sequences identified only by oligonucleotide annealing. The issue has also been raised as to whether a promoter region should be counted as part of an allele or considered somehow as a separate part of the gene. Unless the same promoter sequence is often associated with a variety of allele subsets, which because of linkage disequilibrium seems unlikely, such separate naming would be difficult to justify. Considerable emphasis is now placed on the requirement for making available DNA clones that lead to the definition of new sequences, and depositing, wherever possible, either DNA or cells in a central bank to allow others to access the material on the basis of which new sequences have been named. These are, in many respects, problems of quality control which are common to any DNA database. As the rate of acquisition of data increases, and many sequences are documented only in a database and not by publication, arguments continue over priority, although by this time it is not clear why it should matter whether one person or another first sequenced a new allele. Promoter and operator regions adjacent to the exons expressed in the protein should, it is now agreed, clearly be part of the definition of an allele. The time may come, however,

TABLE 8.3 Alleles which have been shown to be in error and renamed or abandoned

Deleted name	Sequence shown to be identical to
A*0223	A*0222
A*2401	–
A*2412	A*2408
A*3005	A*3004
A*31011	A*31012
A*3302	A*3303
B*0701	–
B*1305	B*1304
B*1541	B*1539
B*27051	B*27052
B*39012	B*39011
B*4017	B*4016
B*4203	Never assigned
B*4401	B*4402
B*5003	B*5002
B*5506	B*5504
B*5803	Never assigned
B*7901	Renamed B*1518
Cw*0101	Cw*0102
Cw*0201	Cw*02022
Cw*0301	Cw*0304
Cw*0402	Cw*04011
Cw*0601	Cw*0602
Cw*1101	–
Cw*1201	Cw*12022
Cw*1401	Cw*1402
Cw*1501	Cw*1502
Cw*1603	Cw*1403
Cw*16042	Cw*16041
Cw*1605	Cw*16041
DRB1*0702	DRB1*0701
DRB1*08031	DRB1*08032
DRB1*09011	DRB1*09012
DRB1*12031	DRB1*1201
DRB1*1606	DRB1*1605
DRB4*0101102N	DRB4*0103102N
DRB5*0201	DRB5*0202
DQA1*03012	DQA1*0302
DQA1*05013	Renamed DQA1*0505
DQB1*03031	DQB1*03032
DPA1*0101	DPA1*0103
DPA1*0102	DPA1*0103
DPB1*02011	DPB1*02012
DPB1*0701	Never assigned
DPB1*1201	Never assigned
DPB1*4201	DPB1*3101
DPB1*4301	DPB1*2801

when the end of a gene may not be clearly identified, although so far, fortunately, within the HLA genes, complex differential splicing has not been observed.

CURRENT NOMENCLATURE

Currently, over 150 new allele sequences are being described every year and this trend looks set to continue for the foreseeable future. However, most are rare, almost like describing new mutations, to which there is no limit. Long before all of the new mutations are described most of the population variance in major population groups will probably have been accounted for. When allele sequences were first named the designation was in accordance with the serological specificity associated with the expressed antigen. Hence, the first DR13 allele sequences were named DRB1*1301 and DRB1*1302, the "DRB1" indicating the locus, the asterisk as a separator between the locus name and the allele designation, the "13" indicating that the alleles belonged to the DR13 family, and the respective "01" and "02" indicating that these two alleles differed from each other by a differing amino acid sequence. However, with the continuing discovery of new allele sequences, it has become increasing difficult to maintain the link, given by the allele name, between the sequence and the serological profile of the antigen associated with it. In many cases, with researchers relying heavily on DNA technology, they are unable to provide a serological definition and, where this is available, they often cannot place the newly defined antigen within a known serological grouping. This particularly applies to the DRB1*03, *11, *13, *14 and *08 family of alleles, and the description of new alleles within the group illustrates that this is almost a continuum of allelic diversity. Although efforts are still made to indicate the serological grouping into which an allele will fall, this is not always possible. It is now, more than ever, most important that the name should be seen as no more than a unique designation.

In addition to the fifth digit which was introduced in 1990 to take account of alleles which differed only by silent, non-coding

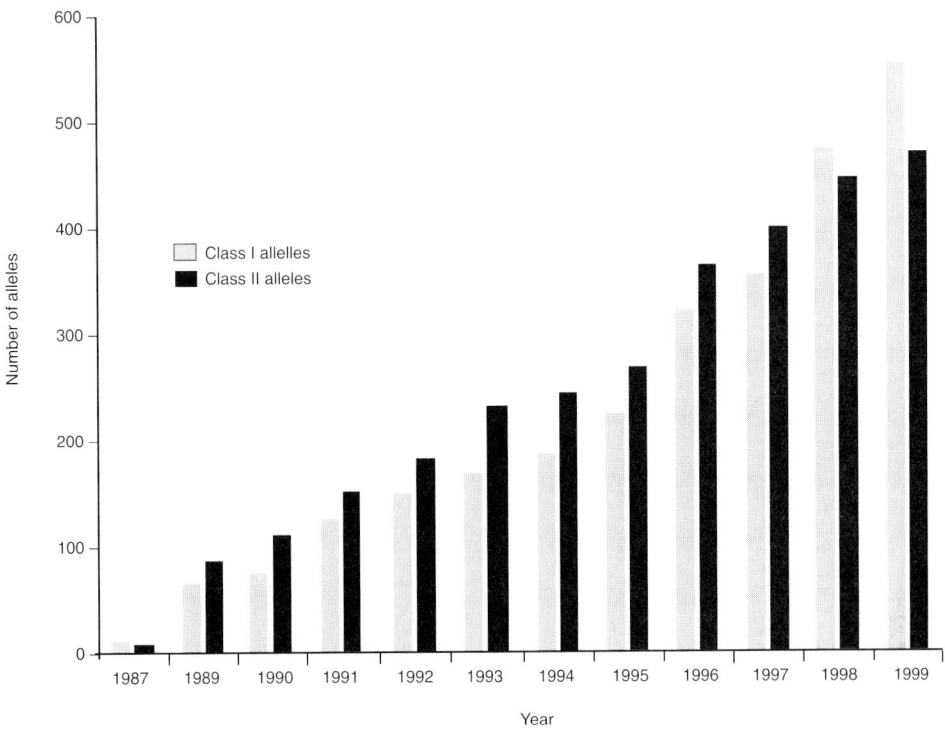

FIGURE 8.1 Numbers of officially recognized HLA alleles from 1987 to September 1999.

substitutions within exons of an allele, a further two digits (sixth and seventh) were added in the 1995 Nomenclature Report (Bodmer et al., 1995). The introduction of these digits has allowed for alleles to be named which have variation that lies outside the expressed regions of the sequence, such as polymorphism within the introns, 5′ or 3′ flanking sequences. Although most sequences are normally only assigned a four-digit allele designation, the extra three digits may be implied, because they have not yet been defined, so the DRB1*1301 allele sequence is actually DRB1*1301101. It is acceptable to use only DRB1*1301 rather than the full name DRB1*1301101, so long as this is not ambiguous in the context used. Similarly, where groups of alleles are referred to, terms such as the "DRB1*13 family" or "DRB1*13" are used.

Optional characters were also introduced for the first time in the 1995 report. Thus an "N" following a sequence name indicates that the allele is not expressed, namely it is a "null" allele. An "L" introduced in the 1996 report indicates that the allele sequence has a mutation which reduces the level of expression of an allele, to a level significantly below normal. These characters are optional as the allele name is always unique, regardless of the optional character. In addition, the character will always be given at the end of the allele name, which allows for the addition of extra digits in the future if it is necessary to expand the allele name further to code for additional polymorphisms. Table 8.5 indicates how a seven-digit allele name is comprised.

THE FUTURE FOR HLA NOMENCLATURE

The recent expansion of the Internet has allowed for greater communication and sharing of data on an international basis. Such access to

TABLE 8.4 Complete listing of recognized serological and cellular HLA specificities

A	B	C	D	DR	DQ	DP
A1	B5	Cw1	Dw1	DR1	DQ1	DPw1
A2	B7	Cw2	Dw2	DR103	DQ2	DPw2
A203	B703	Cw3	Dw3	DR2	DQ3	DPw3
A210	B8	Cw4	Dw4	DR3	DQ4	DPw4
A3	B12	Cw5	Dw5	DR4	DQ5(1)	DPw5
A9	B13	Cw6	Dw6	DR5	DQ6(1)	DPw6
A10	B14	Cw7	Dw7	DR6	DQ7(3)	
A11	B15	Cw8	Dw8	DR7	DQ8(3)	
A19	B16	Cw9(w3)	Dw9	DR8	DQ9(3)	
A23(9)	B17	Cw10(w3)	Dw10	DR9		
A24(9)	B18		Dw11(w7)	DR10		
A2403	B21		Dw12	DR11(5)		
A25(10)	B22		Dw13	DR12(5)		
A26(10)	B27		Dw14	DR13(6)		
A28	B2708		Dw15	DR14(6)		
A29(19)	B35		Dw16	DR1403		
A30(19)	B37		Dw17(w7)	DR1404		
A31(19)	B38(16)		Dw18(w6)	DR15(2)		
A32(19)	B39(16)		Dw19(w6)	DR16(2)		
A33(19)	B3901		Dw20	DR17(3)		
A34(10)	B3902		Dw21	DR18(3)		
A36	B40		Dw22			
A43	B4005		Dw23	DR51		
A66(10)	B41					
A68(28)	B42		Dw24	DR52		
A69(28)	B44(12)		Dw25			
A74(19)	B45(12)		Dw26	DR53		
A80	B46					
	B47					
	B48					
	B49(21)					
	B50(21)					
	B51(5)					
	B5102					
	B5103					
	B52(5)					
	B53					
	B54(22)					
	B55(22)					
	B56(22)					
	B57(17)					
	B58(17)					
	B59					
	B60(40)					
	B61(40)					
	B62(15)					
	B63(15)					
	B64(14)					
	B65(14)					
	B67					
	B70					
	B71(70)					
	B72(70)					
	B73					
	B75(15)					
	B76(15)					
	B77(15)					
	B78					
	B81					
	Bw4					
	Bw6					

TABLE 8.5 Structure of the nomenclature of HLA alleles

Nomenclature	Indicates
HLA	the HLA region and prefix for an HLA gene
HLA-DRB1	a particular HLA locus, i.e. DRB1
HLA-DRB1*13	a group of alleles which encode the DR13 specificity
HLA-DRB1*1301	a specific HLA allele
HLA-DRB1*1301N	a null allele
HLA-DRB1*13012	an allele which differs by a synonymous mutation
HLA-DRB1*1301102	an allele which contains a mutation outside the coding region
HLA-DRB1*1301102N	a null allele which contains a mutation outside the coding region
HLA-DRB1*1301103L	an allele which contains a mutation outside the coding region which leads to a lower level of expression

information has meant that it is now possible to publish nomenclature updates and the sequences to which they refer via the World Wide Web (WWW). This information is currently available from the URL, "www.anthonynolan.org.uk". An obvious expansion of this work is the construction of a large relational database, the IMGT/HLA database, which is currently being undertaken as part of a European initiative involving a collaboration among the Imperial Cancer Research Fund, the Anthony Nolan Research Institute and the European Bioinformatics Institute (Robinson et al., 2000). The database went live in December 1998 and can be accessed via the WWW at "www.ebi.ac.uk/imgt/hla". The database is being developed to address the problems currently experienced with the large sequence databases such as the EMBL Datalibrary and GenBank, which suffer both from a lack of detailed information pertaining to the HLA sequences they contain and from the many errors that are present in the sequence entries. This approach may be a model for the curation of other sequence families, such as the collagens or tyrosine kinase receptors.

Perhaps the most difficult questions in future will relate to the issue of which genes should be encompassed in HLA Nomenclature that are functionally related, but not necessarily homologous, such as the already named TAP1 and TAP2, and LMP2 and LMP7. Alleles of the class I-like genes, MICA were recently named by the committee (Bodmer et al., 1999), Table 8.6 shows the list of currently identified TAP and MICA alleles. It is true that DMA and DMB were named but although their sequence homology with other HLA D-region genes is low, their structure and function clearly imply functional relatedness and evolutionary homology. It is not clear that the same will be the case for the more distantly related class I genes being found telomeric to HLA; and, then, what should be done about CD1 and its relatives?

The need for strict adherence to nomenclature was well illustrated by the confusion over the name of the gene, now called HFE, which is thought to be mutated in haemochromatosis. Confusingly, this gene was called HLA-H in the original papers (Feder et al., 1997), by authors who were apparently unaware that this name had already been officially assigned to another gene in 1990. It is disconcerting that journals allow such lapses in the use of nomenclature. They are not, for example, so permissive in the organization of references! For an international nomenclature to remain useful, it is essential that it is adhered to, and this is the responsibility of both the involved scientific community, and the journals and their editors. The fact, however, that nomenclature reports are "hot", highly quoted

TABLE 8.6 TAP and MICA alleles as named September 1999

TAP1	TAP2	MICA
TAP1*0101	TAP2*0101	MICA*001
TAP1*0102N	TAP2*0102	MICA*002
TAP1*02011	TAP2*0103	MICA*004
TAP1*02012	TAP2*0201	MICA*005
TAP1*0301		MICA*006
TAP1*0401		MICA*007
		MICA*008
		MICA*009
		MICA*010
		MICA*011
		MICA*012
		MICA*013
		MICA*014
		MICA*015
		MICA*016

papers suggests that the majority of the scientific community welcomes the consistency and clarity of the HLA nomenclature. There is, unfortunately, no way that one can simplify the actual complexity of the system as it is found in nature.

HLA IS A MODEL FOR HUMAN GENOME PROJECT NOMENCLATURE

The HLA nomenclature has been a valuable prototype for the naming of genes by the human gene mapping community. To some extent this is due to the fortuitous overlap between some of those involved in the HLA field and in the Human Genome Project. The problems of gene nomenclature remain considerable and it is perhaps unfortunate that there has not been enough support for a consistent approach to dealing with this. At least the number of genes, whether it is 75 000 or 100 000, is finite and their naming is a much better defined problem than that of naming clones, contigs and various other aspects of DNA sequences. It seems clear that the large-scale databases such as GenBank and EMBL cannot, on their own, maintain and curate data with the detail and accuracy that can be achieved by a more specialist group such as the HLA community. There is now, as already mentioned, a collaboration with the European Bioinformatics Institute to explore the way in which specialist groups, such as those involved in HLA and immunoglobulin and T receptor nomenclature, could act as curators for their own particular families of sequences which are then deposited, after suitable annotation and checking. This is the only way in which to manage the detailed documentation that will be needed throughout the human genome, and in achieving this there is much to be learned from the continuing history of the development of HLA nomenclature.

ACKNOWLEDGEMENT

Part of this chapter was previously published in *Tissue Antigens* **49** (3 Pt 2), March 1997, © Munksgaard International Publishers.

REFERENCES

Amos, D.B. (1968). Nomenclature for factors of the HL-A system. *Science* **160**: 659–660.

Bodmer, J.G., Marsh, S.G.E., Parham, P. et al. (1990). Nomenclature for factors of the HLA system, 1989. *Tissue Antigens* **35**: 1–8.

Bodmer, J.G., Marsh, S.G.E., Albert, E. et al. (1991). Nomenclature for factors of the HLA system, 1990. *Tissue Antigens* **37**: 97–104.

Bodmer, J.G., Marsh, S.G.E., Albert, E. et al. (1992). Nomenclature for factors of the HLA system, 1991. In *HLA 1991* (eds T. Tsuji, M. Aizawa and T. Sasazuki), pp. 17–31, Oxford University Press, Oxford.

Bodmer, J.G., Marsh, S.G.E., Albert, E. et al. (1994). Nomenclature for factors of the HLA system, 1994. *Tissue Antigens* **44**: 1–18.

Bodmer, J.G., Marsh, S.G.E., Albert, E. et al. (1995). Nomenclature for factors of the HLA system, 1995. *Tissue Antigens* **46**: 1–18.

Bodmer, J.G., Marsh, S.G.E., Albert, E. et al. (1997). Nomenclature for factors of the HLA system, 1996. *Tissue Antigens* **49**: 297–321.

Bodmer, W.F. (1990). Genetics sequences. *Proc. R. Soc. Lond. B.* **241**: 85–92.

Bodmer, W.F., Albert, E., Bodmer, J.G. et al. (1989). Nomenclature for factors of the HLA system, 1987. In *Immunobiology of HLA* (ed. B. Dupont), pp. 72–79, Springer, New York.

Feder, J.N., Gnirke, A., Thomas, W. et al. (1997). A novel MHC class I-like gene is mutated in patients with hereditary haemochromatosis. *Nature Genet.* **13**: 399–408.

Marsh, S.G.E. (1999a). Nomenclature for factors of the HLA system, update January 1999. *Tissue Antigens* **54**: 106–107.

Marsh, S.G.E. (1999b). Nomenclature for factors of the HLA system, update February 1999. *Tissue Antigens* **54**: 108–109.

Marsh, S.G.E. (1999c). Nomenclature for factors of the HLA system, update March 1999. *Tissue Antigens* **54**: 110–111.

Marsh, S.G.E. (1999d). Nomenclature for factors of the HLA system, update April 1999. *Tissue Antigens* **54**: 209–211.

Marsh, S.G.E. (1999e). Nomenclature for factors of the HLA system, update May 1999. *Tissue Antigens* **54**: 212.

Marsh, S.G.E. (1999f). Nomenclature for factors of the HLA system, update June 1999. *Tissue Antigens* **54**: 322–324.

Payne, R., Tripp, M., Weigel, J. et al. (1964). A new leukocyte isoantigen system in man. *Cold Spring Harbor Symp. Quant. Biol.* **29**: 285–295.

Race, R.R. and Sanger, R. (1968). *Blood Groups in Man*, Blackwells, Oxford.

Robinson, J., Malik, A., Parham, P., Bodmer, J.G. and Marsh, S.G.E. (2000). IMGT/HLA Database – a sequence database for the Human Major Histocompatibilty Complex. *Tissue Antigens* **55**: (in press).

Walford, R.L. (1990). First meeting WHO Leucocyte Nomenclature Committee New York, September,

1968. In *History of HLA: Ten Recollections* (ed. P. I. Terasaki), pp. 121–149, UCLA Tissue Typing Laboratory, Los Angeles, LA.

WHO Nomenclature Committee (1968). *Bull. WHO* **39**: 483–486.

WHO Nomenclature Committee (1970). WHO terminology report. In *Histocompatibility Testing, 1970* (ed. P.I. Terasaki), p. 49, Munksgaard, Copenhagen.

WHO Nomenclature Committee (1972). *Bull. WHO* **47**: 659.

WHO Nomenclature Committee (1975). *Bull. WHO* **52**: 261.

WHO Nomenclature Committee (1978). *Bull. WHO* **56**: 461.

WHO Nomenclature Committee (1980). In *Histocompatibility Testing, 1980* (ed. P.I. Terasaki), pp. 18–20, UCLA Tissue Typing Laboratory, Los Angeles, CA.

WHO Nomenclature Committee (1985). Nomenclature for factors of the HLA system 1984. In *Histocompatibility Testing, 1984* (eds E.D. Albert, M.P. Baur and W.R. Mayr), pp. 4–8, Springer, Berlin.

CHAPTER 9

The Evolution of the Major Histocompatibility Complex: Insights from Phylogeny

Ronald E. Bontrop

INTRODUCTION

If the vertebrate immune system can be likened to a beautiful web woven to capture intruders, then the major histocompatibility complex (MHC) is clearly the spider that controls it. Comprising a cluster of linked genes, the MHC region controls which antigens are processed and presented to T cells for the purpose of initiating an antigen-specific immune response. The MHC developed relatively late in evolution. Most vertebrate species, with the possible exception of hagfish and lamprey, are thought to possess MHCs. During the 400 million years of vertebrate evolution, it seems that the MHCs of different species must have been subject to subtle modifications or even major changes, whereas other genetic regions have been conserved. In this chapter, the HLA system and its polymorphisms will be compared with the MHCs of other species and of non-human primates in particular.

CLASS I, II AND III

Located on the short arm of chromosome 6, the HLA system spans about 4 million base pairs of DNA and is one of the best characterized areas of the human genome. It is densely populated with genes and can be divided into three major sections, designated class I, class II and class III (Campbell and Trowsdale, 1997). The class I region encodes the highly polymorphic HLA-A, -B and -C transplantation antigens, which form dimers with β_2 microglobulin. The coding sequence for β_2 microglobulin is not on the same chromosome as the MHC. The MHC class I dimer is equipped with a peptide-binding site that can accommodate peptides of intracellular origin. These classical HLA class I transplantation antigens are normally expressed on the surface of all nucleated cells. MHC class I molecules complexed with a peptide, for instance, of viral origin, may be recognized by the T cell receptors of $CD8^+$ T cells. This recognition may eventually result in the lysis of the infected cell. In addition, some natural killer (NK) cells possess killer cell inhibitory receptors (KIR) which recognize self-MHC class I molecules (Leibson, 1995; Lanier, 1997). KIR are reactive with broad groups of HLA class I alleles but may even exhibit peptide specificity (Peruzzi et al., 1996; Rajagopalan and Long, 1997). NK cells can only lyse targets which fail to express self class I molecules, such as tumours and cells infected by viruses that interfere with MHC class I expression. Thus, classical HLA class I molecules may trigger cytotoxicity but also protect healthy cells from NK-mediated lysis.

The class I region also comprises the non-classical HLA-E, -F and -G genes, which exhibit low degrees of polymorphism and show restricted tissue distribution (Wei and Orr, 1990). Although HLA-E transcripts have been detected in many types of tissue, it is doubtful whether there is actual cell-surface expression of HLA-E molecules owing to insufficient peptide binding (Ulbrecht et al., 1992). However, the trophoblast is known to lack the expression of HLA-A and -B molecules but expresses various forms of HLA-G molecule. It has been speculated that HLA-G molecules may provide negative signals to NK cells.

The sequences from the MHC class I-related gene family (MIC), which seems to be conserved across mammalian evolution, also map within the HLA class I region (Bahram et al., 1994). MICA and MICB share a high degree of similarity but are distantly related (Bahram et al., 1996), whereas MICC, MICD and MICE are pseudogenes (Bahram and Spies, 1996). The precise function of these molecules and their ligands is not yet understood.

Not all class I-like molecules are part of the HLA region: for instance, the MHC class I-related CD1 gene family maps on chromosome 2 (Calabi and Milstein, 1986). These molecules have a specialized function and can present lipoglycan antigens to T cells (Prigozy et al., 1997).

The human class II region harbours the genes that encode the classical class II gene products such as HLA-DR, -DQ and -DP. These are transmembrane heterodimers composed of an α- and a β-chain encoded by A and B genes, respectively. Both genes map within the class II region and may, apart from DRA, display polymorphism. The class II transmembrane gene products are normally expressed on cells of the white blood group lineage. From a functional point of view the classical MHC class II molecules are involved in presenting peptides of extracellular origin, such as bacteria, to the T cell receptors on $CD4^+$ T cells. In the event of successful recognition, these latter cells respond by excreting cytokines which play a key role in tuning various complex types of immune reactions. In addition, HLA-DM molecules are involved in loading peptides on to class II molecules (Kelly et al., 1991; Sloan et al., 1995), whereas the TAP and LMP gene products are essential for producing the peptides that may complex with MHC class I molecules and assist their subsequent transport to the cell membrane.

The class III region lies interspersed between the class II and I regions. It contains a high number of genes, some of which are important for immune reactions, such as the complement factors C2, C4 and Bf and the cytokines TNFA and TNFB. Most of the genes that map within the class III region, however, have no obvious relationship with the immune system (Trowsdale, 1995).

WHY ARE THE MAJOR HISTOCOMPATIBILITY COMPLEX CLASS I AND II REGIONS GENETICALLY LINKED?

In most species studied, the classical MHC class I and II genes, as well as the LMP and TAP genes, are situated on the same chromosome. An exception to this rule may be seen in some species of fish where the MHC class I and II regions segregate independently from each other (Graser et al., 1998). The generic observation that the MHC class I and II regions are usually linked makes one wonder whether this confers some sort of advantage. Moreover, one should take into account that many MHC genes display polymorphism and that some alleles at different MHC loci occur more often on the same chromosome than expected by chance. This phenomenon is called linkage disequilibrium. Many pathogens are complex organisms that experience intracellular and extracellular life cycles. Polymorphism is thought to guarantee that at least some individuals have the capacity to mount protective MHC class I-mediated CTL responses, whereas others may possess the MHC class II alleles that co-ordinate the immune response and may generate other effect mechanisms such as the generation of antibody. Recombination may, just by chance, link certain MHC class I and II alleles. When such haplotypes control protective responses to particular parasites they may become subject to positive selection. Pathogenic pressures vary in time and appear to fluctuate among populations. The combination of these latter two phenomena may explain why linkage disequilibria differ among human ethnic populations.

THE ANTIQUITY OF HLA CLASS I AND II LOCI

The evolutionary equivalents of the HLA-A and -B loci have been well conserved during primate evolution, and representatives have been found to be present in great apes and Old World monkeys (Lawlor et al., 1988; Mayer et al., 1988; McAdam et al., 1994, 1995; Watkins 1995). Hence, these orthologues are considered to be at least 35 million years old. For example, in the orang utan and rhesus macaque, the A and/or B loci may have gone through several rounds of duplications (Chen et al., 1992; Watkins, 1995). Nucleotide sequence comparisons suggest that the HLA-B and -C loci also arose from a duplication event that took place somewhere during hominoid evolution. This is in line with the observation that the orthologues of HLA-C have thus far only been detected in gorillas and chimpanzees (Lawlor et al., 1988, 1991) and appear to be absent in orang utans, gibbons and macaques (Watkins, 1995). As both of these species have duplicated A and B loci, the function of HLA-C may have been taken over by one of the newly generated loci that arose from a duplication.

The ancestors of the non-classical HLA-E, -F and -G loci are at least 58 million years old, since orthologues have been detected in apes and Old and New World monkeys (Watkins, 1995). The E and F loci in particular have been remarkably well conserved during the evolution of primates (Otting and Bontrop, 1993; Boyson et al., 1995). The G locus appears to have been more subject to change. In the case of Old World monkeys, the G locus may be occupied by a pseudogene, since exon 3 bears stop codons (Boyson et al., 1996; Castro et al., 1996), whereas it appears to act as a classical transplantation antigen in New World monkeys (Watkins et al., 1990). Interestingly, in rhesus macaques the function of the inactivated equivalent of HLA-G seems to have been taken over by another locus designated Mamu-AG (Boyson et al., 1997).

The HLA class II loci are shared by rodents and primates and consequently are older than their HLA-class I counterparts. In humans the DQ subregion is occupied by two sets of genes arranged in the order HLA-DQA1, -DQB1, -DQA2 and -DQB2. Equivalents of the HLA-DQA1 and -DQB1 loci have been detected not only in primates, but in most mammals studied (Gyllensten and Erlich, 1989; Gyllensten et al., 1990; Kenter et al., 1992a; Otting et al., 1992). Equivalents of the HLA-DQA2 and -DQB2 genes are present in hominoids and New World monkeys but appear to have been deleted in Old World monkey species (Bontrop et al., 1995).

The primate MHC-DQ subregion is more stable than the DR subregion with respect to gene content. Owing to recombination-like processes, the arrangement and number of different types of MHC-DRB genes present per haplotype are not constant and may vary considerably between individuals, as well as between primate species (Böhme et al., 1985; Kasahara et al., 1992; Slierendregt et al., 1994). It should be noted, however, that the sharing of DR subregions between different primate species, or considerable segments thereof, has been reported (Brändle et al., 1992; Kasahara et al., 1992; Bontrop et al., 1995).

The equivalents of the HLA-DP genes appear to be present in at least some rodent species. The primate MHC-DP subregion is genetically very stable with regard to gene content, since a similar organization has been observed in humans, chimpanzees, orang utans and cotton-top tamarins (Grahovac et al., 1993). In each of these four species, the MHC-DP subregion consists of four loci arranged in the order MHC-DPB2, -DPA2, -DPB1 and -DPA1. The primate MHC-DPB2 and -DPA2 genes belong to a pseudogene locus that seems to have been inactivated long ago.

THE AGE OF LINEAGES AND ALLELES

A lineage represents a cluster of evolutionarily related alleles that evolved from a common ancestral sequence. Alleles are usually grouped or clustered by phylogenetic methods. In most primate species tested, both the MHC-DQA and -DQB genes exhibit polymorphism and a typical phylogenetic tree is shown for primate MHC-DQA1 and -DQA2 sequences. As can be seen, the different primate MHC-DQA1 alleles do not cluster according to the grouping of species but intermingle (Figure 9.1). For instance, the Patr-DQA1*0101 allele is more related to the HLA-DQA1*0101, -DQA1*0102 and -DQA1*0103 alleles than to any other known chimpanzee equivalent. This phenomenon indicates that the various types of MHC-DQA1 lineage predate

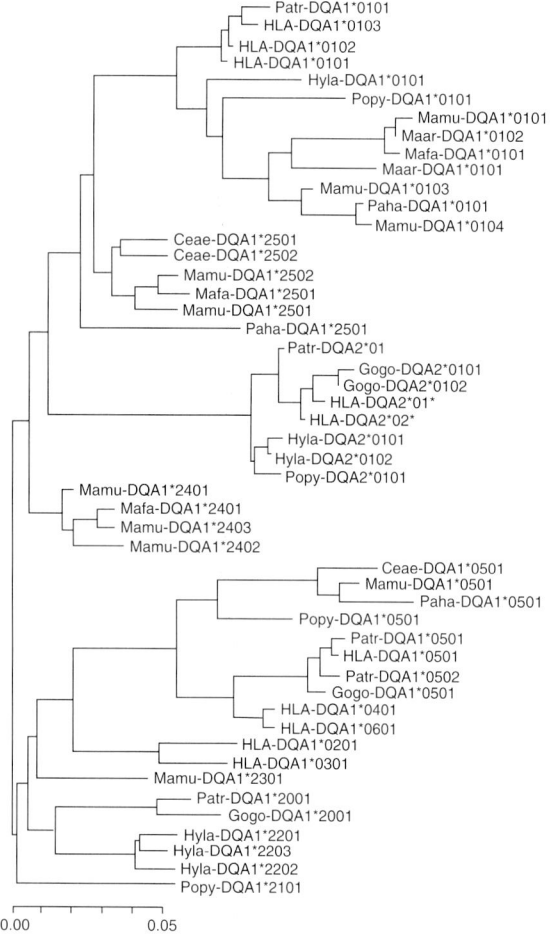

FIGURE 9.1 Phylogenetic tree of primate MHC-DQA1 and -DQA2 exon 2 sequences. The species prefixes are according to standardized MHC nomenclature.

speciation and were inherited in a trans-species mode of evolution (Klein et al., 1993). The MHC-DQA1*01, -DQA1*05 and -DQB1*06 lineages, which are common to humans, apes and monkeys, predate the speciation of higher primates and thus are at least 35 million years old (Gyllensten and Erlich 1989; Gyllensten et al., 1990; Kenter et al., 1992a; Otting et al., 1992).

MHC-DQB1*15 lineage members were discovered in chimpanzees and several Old World monkey species but appear to be absent in humans (Gyllensten et al., 1990; Otting et al., 1992). Theoretically, this lineage should thus also have been inherited in a trans-species manner, but its absence in the human population is a strong argument for the fact that humans or their progenitors appear to have lost this lineage. Most MHC-DQA1 and -DQB1 sequences are unique to a species and cluster at their own twigs in phylogenetic trees. Identical MHC-DQ alleles, as far as exon 2 sequences are concerned, have only been found in highly related macaque species which shared a common ancestor about 1 million years ago (Kenter et al., 1992a; Otting et al., 1992; Bontrop, 1994).

On the basis of nucleotide sequence comparisons, one might conclude that the MHC-DRB1*03, -DRB1*04 and -DRB1*10 lineages are entities that have existed for 55 million years or even longer (Bontrop et al., 1995). However, similar or identical motifs may have been generated *de novo* as a result of convergent evolution, as has been illustrated for various MHC-DRB1*03 lineage members detected in New World monkeys (Trtková et al., 1993).

New lineages can be generated by recombination. For instance, the HLA-DRB1*0901 gene is an apparent fusion product of genes belonging to the HLA-DRB1*07 and HLA-DRB5 lineages (Kenter et al., 1992b; Mayer et al., 1992), whereas the HLA-DRB1*08 specificity was generated by a recombination between MHC-DRB1*03 and MHC-DRB3 genes (Gorski, 1989; Kasahara et al., 1992).

Interestingly, pseudogenes such as HLA-DRB4, -DRB6, -DRB7, -DRB8 and -DRB9 are also shared among humans and various non-human primate species and have been conserved to a high degree (Klein et al., 1991; Corell et al., 1992; Kenter et al., 1992a; Mayer et al., 1993; Bontrop et al., 1995; Gongora et al., 1996).

The trans-species origin of lineages has been documented for the primate MHC-DR and -DQ loci, but no such relationship is observed for the MHC-DPB1 sequences. Phylogenetic analyses and sequence comparisons have demonstrated that the primate MHC-DPB1 locus evolves quickly owing to a frequent exchange of polymorphic segments between alleles and lineages (Slierendregt et al., 1995; Gyllensten et al., 1996). In contrast, comparison of HLA-DPA1 alleles and their non-human primate equivalents has shown that some MHC-DPA1 lineages are at least 35 million years old (Otting and Bontrop, 1994).

In the case of class I, all known chimpanzee MHC class I A-locus alleles were found to be related to the alleles of the HLA-A1, -A3, -A11 family (McAdam et al., 1995). Thus, this A-locus lineage predates at least the speciation of humans and chimpanzees. In sharp contrast, however, exon 2 sequences encoding the α_1 domains of HLA-B and Patr-B alleles may cluster in a lineage, whereas no such phenomenon is observed for the exon 3 to exon 8 gene segments. As found for MHC-DPB1, the primate MHC-B locus evolves quickly owing to a high frequency of interallelic shuffling of polymorphic stretches (Belich et al., 1992; Watkins et al., 1992; McAdam et al., 1994). Despite the trans-species inheritance of lineages, the MHC class I alleles are generally unique to a species. The sharing of full-length MHC class I sequences by different primate species has been documented in only a few cases (Cooper et al., 1998; Evans et al., 1998).

MECHANISMS THAT GENERATE POLYMORPHISM

Variation at the DNA level can, essentially, be created by point mutations and exchange of polymorphic segments by recombination. (This is discussed in greater detail in Chapter 10) The introduction of genetic variation, to some extent at least, is a random process. How different variants are selected during evolution is another question. Widely diverse theories have sought to explain the existence of MHC polymorphism. It is generally thought, however, that alleles which acquire disadvantageous mutations can be deleted from the gene pool as a result of negative selection. Theoretically, this would be the case for HLA alleles which control dominant and lethal susceptibility traits involving a particular infectious disease. Of course, such an allele would only be deleted from those populations coming into contact with the relevant pathogen. A complicating factor is that MHC antigens are expressed in a codominant fashion and are subject to balancing selection. Furthermore, dn/ds analyses have demonstrated that codons for the contact residues of the peptide-binding site of MHC class I and II gene products are more diverged than the codons that specify the molecule backbone residues (Hughes and Nei, 1988, 1989). Alleles with favourable mutations, which are able to cope with a particular infection, may become subject to frequency-dependent selection (Bodmer, 1972). This may be the case for HLA-B53, which was reported to protect some African human communities from severe malaria (Hill et al., 1991, 1992).

Another feature of the MHC is that it contains multiple highly related loci which encode proteins with similar but not identical functions. For instance, in the HLA system there are at least three classical class II loci (HLA-DR, -DQ and -DP) and three class I loci (HLA-A, -B and -C). These loci must have arisen from duplication events and may have acquired new specialized functions. That this may be true is exemplified by the observation that HLA-DR and -DQ molecules seem to use slightly different constraints for selecting the peptides that they bind (Raddrizzani et al., 1997). With regard to evolutionary stability, both the MHC class I and II regions encode different types of lineages. Some classical loci in humans encode lineages that are more or less conserved (HLA-A, -DR and -DQ), whereas others are subject to frequent change (HLA-B, -DR and -DP). The primate DR region contains multiple loci and has with regard to evolutionary stability an intermediate character. Some lineages appear to be extremely stable, whereas others appear to evolve more rapidly. On the one hand, it is possible to imagine that lineages which perform a successful function are well preserved, as demonstrated by some of the MHC-DRB1*03 lineage members, which seem to have been involved in mounting an immune response to bacterial antigens for millions of years (Geluk et al., 1993). On the other hand, parasites are known to change their antigenic appearance to escape immune recognition. To cope with these types of parasite, the MHC harbours loci that evolve much more quickly, such as MHC-B, -DPB1 and some -DRB lineages (Belich et al., 1992; Watkins et al., 1992; McAdam et al., 1995; Slierendregt et al., 1995).

ACKNOWLEDGEMENTS

The author wishes to acknowledge the expert assistance of Donna Devine, Mea van der Sman and Henk van Westbroek.

REFERENCES

Bahram, S. and Spies, T. (1996). The MIC gene family. *Res. Immunol.* **147**: 328–333.

Bahram, S., Bresnahan, M., Geraghty, D.E. and Spies, T. (1994). A second lineage of mammalian major histocompatibility complex class I genes. *Proc. Natl Acad. Sci. USA* **91**: 6259–6263.

Bahram, S., Mizuki, N., Inoko, H. and Spies, T. (1996). Nucleotide sequence of the human MHC class I *MICA* gene. *Immunogenetics* **45**: 80–81.

Belich, M.P., Madrigal, J.A., Hildebrand, W.H. et al. (1992). Unusual HLA-B alleles in two tribes of Brazilian Indians. *Nature* **357**: 326–330.

Bodmer, W.F. (1972). Evolutionary significance of the HL-A system. *Nature* **237**: 139–140.

Böhme, J., Andersson, M., Andersson, G. et al. (1985). HLA-DR beta genes vary in number between different DR specificities, whereas the number of DQ beta genes is constant. *J. Immunol.* **135**: 2149–2155.

Bontrop, R.E. (1994). Nonhuman primate *Mhc-DQA* and *-DQB* second exon nucleotide sequences: a compilation. *Immunogenetics* **39**: 81–92.

Bontrop, R.E., Otting, N., Slierendregt, B.L. and Lanchbury, J.S. (1995). Evolution of major histocompatibility complex polymorphisms and T cell receptor diversity in primates. *Immunol. Rev.* **143**: 33–62.

Boyson, J.E., McAdam, S.N., Gallimore, A. et al. (1995). The *MHC-E* locus in macaques is polymorphic and is conserved between macaques and humans. *Immunogenetics* **41**: 59–68.

Boyson, J.E., Iwanaga, K.K., Golos, T.G. and Watkins, D.I. (1996). Identification of the rhesus monkey HLA-G ortolog. Mamu-G is a pseudogene. *J. Immunol.* **157**: 5428–5437.

Boyson, J.E., Iwanaga, K.K., Golos, T.G. and Watkins, D.I. (1997). Identification of a novel MHC class I gene, Mamu-AG, expressed in the placenta of a primate with an inactivated G locus. *J. Immunol.* **159**: 3311–3321.

Brändle, U., Ono, H., Vincek, V. et al. (1992). Trans-species evolution of *Mhc-DRB* haplotype polymorphism in primates: organization of *DRB* genes in the chimpanzee. *Immunogenetics* **36**: 39–48.

Calabi, F. and Milstein, C. (1986). A novel family of human major histocompatibility related genes not mapping to chromosome 6. *Nature* **323**: 540–543.

Campbell, D. and Trowsdale, J. (1997). A map of the human major histocompatibility complex. *Immunol. Today* **43**: 18–19.

Castro, M.J., Morales, P., Fernandez-Soria, V. et al. (1996). Allelic diversity at the primate Mhc-G locus: exon 3 bears stop codons in all Cercophiecinae sequences. *Immunogenetics* **43**: 327–336.

Chen, Z.W., McAdam, S.N., Hughes, A.L. et al. (1992). Molecular cloning of orangutan and gibbon MHC class I cDNA: the HLA-A and -B loci diverged over 30 million years ago. *J. Immunol.* **148**: 2547–2551.

Cooper, S. Adams, E.J., Spencer Wells, R. et al. (1998). A major histocompatibility complex class I allele shared by two species of chimpanzee. *Immunogenetics* **47**: 212–217.

Corell, A., Morales, P., Varela, P. et al. (1992). Allelic diversity at the primate major histocompatibility complex *DRB6* locus. *Immunogenetics* **36**: 33–38.

Evans, D.T., Piekarczyk, M.S., Cadavid, L.F. et al. (1998). Two different primate species express an identical functional class I allele. *Immunogenetics* **47**: 206–211.

Geluk, A., Elferink, B.G., Slierendregt, B.L. et al. (1993). Evolutionary conservation of major histocompatibility complex-DR/peptide/T cell interactions in primates. *J. Exp. Med.* **177**: 979–987.

Gongora, R., Figueroa, F. and Klein, J. (1996). The HLA-DRB9 gene and the origin of HLA-DR haplotypes. *Hum. Immunol.* **51**: 23–31.

Gorski, J. (1989). The HLA-DRw8 lineage was generated by a deletion in the DRB region followed by first domain diversification. *J. Immunol.* **142**: 4041–4045.

Grahovac, B., Schönbach, C., Brändle, U. et al. (1993). Conservative evolution of the *Mhc-DP* region in anthropoid primates. *Hum. Immunol.* **37**: 75–84.

Graser, R., Vincek, V. and Klein, J. (1998). Analysis of Zebrafish *Mhc* using BAC clones. *Immunogenetics* **49**: 318–325.

Gyllensten, U.B. and Erlich, H.A. (1989). Ancient roots for polymorphism at the HLA-DQ alpha locus in primates. *Proc. Natl Acad. Sci. USA* **86**: 9986–9990.

Gyllensten, U.B., Lashkari, D. and Erlich, H.A. (1990). Allelic diversification at the class II DQB locus of the mammalian major histocompatibility complex. *Proc. Natl Acad. Sci. USA* **87**: 1835–1839.

Gyllensten, U., Bergstrom, T., Josefsson, A. et al. (1996). Rapid allelic diversification and intensified selection at antigen recognition sites of the Mhc class II DPB1 locus during hominoid evolution. *Tissue Antigens* **47**: 212–220.

Hill, A.V., Allsopp, C.E., Kwiatkowski, D. et al. (1991). Common west African HLA antigens are associated with protection from severe malaria. *Nature* **352**: 595–600.

Hill, A.V., Elvin, J., Willis, A.C. et al. (1992). Molecular analysis of the association of HLA-B53 and resistance to severe malaria. *Nature* **360**: 434–436.

Hughes, A.L. and Nei, M. (1988). Pattern of nucleotide substitution at major histocompatibility complex class I loci reveals overdominant selection. *Nature* **335**: 167–170.

Hughes, A.L. and Nei, M. (1989). Nucleotide substitution at major histocompatibility complex class II loci: evidence for overdominant selection. *Proc. Natl Acad. Sci. USA* **86**: 958–962.

Kasahara, M., Klein, D., Vincek, V. et al. (1992). Comparative anatomy of the primate major histocompatibility complex *DR* subregion: evidence for combinations of *DRB* genes conserved across species. *Genomics* **14**: 340–345.

Kelly, A.P., Monaco, J.J., Cho, S. and Trowsdale, J. (1991). A new human HLA class II related locus, DM. *Nature* **353**: 571–573.

Kenter, M., Otting, N., Anholts, J. et al. (1992a). Evolutionary relationships among the primate *Mhc-DQA1* and *DQA2* alleles. *Immunogenetics* **36**: 71–78.

Kenter, M., Otting, N., Anholts, J. et al. (1992b). *Mhc-DRB* diversity of the chimpanzee (*Pan troglodytes*). *Immunogenetics* **37**: 1–12.

Klein, D., Vincek, V., Kasahara, M. et al. (1991). Gorilla major histocompatibility complex-DRB pseudogene orthologous to HLA-DRBVIII. *Hum. Immunol.* **32**: 211–220.

Klein, J., O'hUigin, C., Figueroa, F. et al. (1993). Different modes of *Mhc* evolution in primates. *Mol. Biol. Evol.* **10**: 48–59.

Lanier, L.L. (1997). Natural killer cells: from no receptors to too many. *Immunity* **6**: 371–378.

Lawlor, D.A., Ward, F.E., Ennis, P.D. et al. (1988). HLA-A and -B polymorphisms predate the divergence of humans and chimpanzees. *Nature* **335**: 268–271.

Lawlor, D.A., Warren, E., Taylor, P. and Parham, P. (1991). Gorilla class I major histocompatibility complex alleles: comparison to human and chimpanzee class I. *J. Exp. Med.* **174**: 1491–1501.

Leibson, P.J. (1995). MHC-recognizing receptors: they're not just for T cells anymore. *Immunity* **3**: 5–8.

Mayer, W.E., Jonker, M., Klein, D. et al. (1988). Nucleotide sequence of chimpanzee MHC class I alleles: evidence for trans species mode of evolution. *EMBO J.* **7**: 2765–2774.

Mayer, W.E., O'hUigin, C., Zaleska, R.Z. and Klein, J. (1992). Trans-species origin of *Mhc-DRB* polymorphism in the chimpanzee. *Immunogenetics* **37**: 12–23.

Mayer, W.E., O'hUigin, C. and Klein, J. (1993). Resolution of the HLA-DRB6 puzzle: a case of grafting a de novo-generated exon on an existing gene. *Proc. Natl Acad. Sci. USA* **90**: 10 720–10 724.

McAdam, S.N., Boyson, J.E., Liu, X. et al. (1994). A uniquely high level of recombination at the HLA-B locus. *Proc. Natl Acad. Sci. USA* **91**: 5893–5897.

McAdam, S.N., Boyson, J.E., Liu, X. et al. (1995). Chimpanzee MHC class I A locus alleles are related to only one of the six families of human A locus alleles. *J. Immunol.* **154**: 6421–6429.

Otting, N. and Bontrop, R.E. (1993). Characterization of the rhesus macaque equivalent of HLA-F. *Immunogenetics* **38**: 141–145.

Otting, N. and Bontrop, R.E. (1994). Evolution of the major histocompatibility complex *DPA1* locus in primates. *Hum. Immunol.* **42**: 184–187.

Otting, N., Kenter, M., Weeren, P. van et al. (1992). *Mhc-DQB* repertoire variation in hominoid and Old World primate species. *J. Immunol.* **149**: 461–470.

Peruzzi, M., Parker, K.C., Long, E.O. and Malnati, M.S. (1996). Peptide requirements for the recognition of HLA-B*2705 by specific natural killer cells. *J. Immunol.* **157**: 3350–3356.

Prigozy, T.I., Sieling, P.A., Clemens, D. et al. (1997). The mannose receptor delivers lipoglycan antigens to endosomes for presentation to T cells by CD1b molecules. *Immunity* **6**: 187–197.

Raddrizzani, L., Sturniolo, T., Guenot, J et al. (1997). Different modes of peptide interaction enable HLA-DQ and HLA-DR molecules to bind diverse peptide repertoires. *J. Immunol.* **159**: 703–711.

Rajagopalan, S. and Long, E.O. (1997). The direct binding of a p58 killer cell inhibitory receptor to human histocompatibility leucocyte antigen HLA-Cw4 exhibits peptide selectivity. *J. Exp. Med.* **185**: 1523–1528.

Slierendregt, B.L., Otting, N., Van Besouw, N. et al. (1994). Expansion and contraction of rhesus macaque *DRB* regions by duplication and deletion. *J. Immunol.* **154**: 2298–2307.

Slierendregt, B.L., Otting, N., Kenter, M. and Bontrop, R.E. (1995). Allelic diversity at the *Mhc-DP* locus in rhesus macaques (*Macaca mulatta*). *Immunogenetics* **41**: 29–37.

Sloan, V.S., Cameron, P., Porter, G. et al. (1995). Mediation by HLA-DM of dissociation of peptides from HLA-DR. *Nature* **375**: 802–806.

Trowsdale, J. (1995). "Both man & bird & beast": comparative organization of MHC genes. *Immunogenetics* **41**: 1–17.

Trtková, K., Kupfermann, H., Grahovac, B. et al. (1993). *Mhc-DRB* genes of platyrrhine primates. *Immunogenetics* **38**: 210–222.

Ulbrecht, M., Kellerman, J., Johnson, J.P. and Weiss, E.H. (1992). Impaired intracellular transport and cell surface expression of non-polymorphic HLA-E: Evidence of insufficient peptide binding. *J. Exp. Med.* **176**: 1083–1090.

Watkins, D.I. (1995). The evolution of major histocompatibility class I genes in primates. *Crit. Rev. Immunol.* **15**: 1–29.

Watkins, D.I., Chen, Z.W., Hughes, A.L. et al. (1990). Evolution of the MHC class I genes of a New World primate from ancestral homologues of human classical genes. *Nature* **346**: 60–62.

Watkins, D.I., McAdam, S.N., Liu, X. et al. (1992). New recombinant HLA-B alleles in a tribe of South American Amerindians indicate rapid evolution of MHC class I loci. *Nature* **357**: 329–333.

Wei, X. and Orr, H.T. (1990). Differential expression of HLA-E, -F and -G transcripts in human tissue. *Hum. Immunol.* **29**: 131–142.

CHAPTER 10

The Generation and Selection of Major Histocompatibility Complex Polymorphisms

Jim Kaufman

INTRODUCTION

The most striking feature of the major histocompatibility complex (MHC) of most vertebrates, indeed the reason for its discovery in mice, is its amazingly high level of polymorphism (Klein, 1975). Now, after nearly a century of research, it is clear that the high polymorphism responsible for graft rejection in mice and humans is due to a few genes, the classical class I and class II genes, some of which have upwards of 200 alleles. High polymorphism is not a typical feature of vertebrate genes: most genes have few common alleles and often these have no perceptible differences in function (i.e. they are functionally neutral). In contrast, there are many data to indicate that alleles of the classical class I and class II genes of the MHC are selected, although the extraordinary level of polymorphism may be in part a consequence of population structure. In this short review, the generation of variation in classical MHC genes, the selection of polymorphism and the intensity of the selection will be considered.

GENERATION OF POLYMORPHISM IN CLASSICAL MAJOR HISTO-COMPATIBILITY COMPLEX GENES

There has been much controversy over the years as to the rate of mutation in classical MHC genes, with one explanation for the high level of polymorphism being a high mutation rate. However, the rate of change has been estimated for a number of genes and, even with the uncertainties and errors inherent in such tricky calculations, it seems clear that the range for such values is solidly in the middle of that found for vertebrate genes (Klein and Figueroa, 1986; Klein et al., 1993; Hughes and Hughes, 1995), and much lower than microsatellites and some genes whose alleles are generated from slippage of DNA polymerase. Thus, the polymorphism is not maintained by a high rate of mutation.

There appear to be three primary mechanisms for the generation of new classical MHC gene sequences: point mutation, recombination and microrecombination, with different relative rates in different gene loci (Bontrop et al., 1995; Parham et al., 1995). Single nucleotide differences between MHC sequences have been attributed to point mutation due to errors during DNA replication and repair (this is typical for all vertebrate genes) and such evidence has been uncontroversial. The evidence for the generation of new MHC alleles by recombination, particularly between exons, has also been straightforward. For instance, the human HLA-A69-coding allele has one exon that is identical to the HLA-A68-coding allele and the rest of the gene from the HLA-A2-coding allele, clearly resulting from a large-scale recombination event (Holmes and Parham, 1985).

In contrast, the interpretation of small stretches of sequence shared between alleles of a locus (or even between different gene loci) as evidence for microrecombination (or gene conversion) was not without controversy. A small shared stretch of sequence amid a lot of differences in genes can be explained in a number of ways, including divergence from a common sequence (in which the shared sequence does not diverge), convergent evolution (e.g. when two independent sequences mutate to a shared sequence owing to selection for that shared sequence) and microrecombination. There are two outcomes for a microrecombinational event, depending on which DNA strands are repaired. One outcome leads to two-way recombination (also called double-reciprocal recombination, in which the small sequences are simply exchanged between the two genes) and the other leads to one-way gene conversion (in which one donor gene remains unchanged but converts a stretch of sequence in the target gene). While two-way recombination leads to a scrambled sequence between two genes (and shared stretches of sequence between a number of genes), gene conversion can lead to a new target sequence (different from both the original target sequence and the donor sequence) or to homogenization of sequences (if a single donor gene consistently replaces large portions of the target sequence). Gene conversion was originally discovered in fungi, where all products of a meiosis (and thus both the donor and target gene) can be recovered and analysed (Holliday, 1964), and the evidence seems conclusive for a huge number of somatic gene conversion events during diversification of chicken antibody V genes in bursal B cells (Weill et al., 1987).

The interpretation of shared sequence stretches as the result of gene conversion was initially used to explain the bm mutants in mice (Nathenson et al., 1986) and stirred up a great deal of emotion (Klein, 1984; Klein and Figueroa, 1986; Klein et al., 1993; Hughes and Hughes, 1995; Klein and O'hUigin, 1995). In particular, those who have used (some would say misused) dendrograms of various kinds to estimate rates and times of divergence have resisted the idea with an uncommon vehemence, since it renders the conclusions of simplistic extrapolations from genetic distance absurd. The methodological problem is that simple two-dimensional trees work very well to show relationships between sequences that diverge by serial single changes from an ancestral sequence and can be used, with certain information and assumptions, to give rates and times of divergence. However, such trees are not good at showing the relationship of two sequences that have diverged far on to different branches of a tree and then become more related by exchanging or acquiring a stretch of sequence. Moreover, such an event confounds the rate and time interpretations, since a number of changes that might take place over a long period by point mutations happen instantaneously in a microrecombination event. The validity of many papers and certain oft-quoted conclusions depended on the resolution of this issue.

The sound and fury has mostly died down, following experiments at the DNA level that directly demonstrate reasonable rates of microrecombination between certain MHC genes and alleles in mouse and human sperm (Högstand and Boehme, 1994; Zangenberg et al., 1995), along with theoretical calculations to show their importance (Ohta, 1997; Satta, 1997). The mouse genes tested are some of the same class II gene alleles that were implicated in the original bm mutant series, which might be somehow special. In fact, the evidence suggests that some genes are more prone to be involved in such microrecombination than others, presumably based on some as yet undefined sequence features.

Comparison of the alleles at different MHC loci suggest that for each locus, variants are generated by a different mix of mechanisms (Bontrop et al., 1995; Parham et al., 1995). For human class I genes, the HLA-A gene seems to evolve primarily by point mutation with the occasional large-scale recombinational event (allowing fair use of trees), while HLA-B gene sequences show many segmental exchanges, and incidentally a higher rate of evolution (i.e. evidence for new alleles in isolated populations, short lifetimes for lineages) and a (consequent) larger number of alleles world-wide. For human class II alleles, DQA and -B genes appear to evolve primarily by point mutation, DPA and -B genes primarily by segmental exchange and DRB genes by a mixture. It is very striking that the DP genes appear as a patchwork of some six regions which mix and match from a limited number of sequence motifs. As for the class I genes, microrecombination apparently speeds up the rate of change, with the DQ genes appearing to evolve much more slowly than the DP genes. Not only the mutation rate but also the rise in gene frequency are necessary for a high level of polymorphism, which in classical MHC genes appears to be due to selection.

SELECTION OF POLYMORPHISM IN CLASSICAL MAJOR HISTO-COMPATIBILITY COMPLEX GENES

It is not enough for a new variant of a gene to appear; somehow it must rise in frequency enough to escape loss by drift. To illustrate how important this is, consider a fixed population of 100 males and 100 females, which replace themselves exactly every generation. Out of these 200 parents, one has a mutation. The progeny in that family each has a 50:50 chance of being either homozygous for wild-type or heterozygous for wild-type and the new mutation. That means, for the two progeny, there is only a 25% chance that both will be heterozygous (i.e. the gene frequency of the mutant will double), a 50% chance that one will be heterozygous and the other homozygous wild-type (i.e. the gene frequency of the mutant will stay low) and a 25% chance that both will be homozygous wild-type (i.e. the mutation will be lost from the gene pool by drift).

There are various ways by which a new mutation can rise in frequency: selection, genetic hitch-hiking, founder effect, high mutation rate and even drift. Selection, the "natural selection" of Darwin, comes about when a particular allele confers an advantage to an individual leading to more descendants: if the advantage is overwhelmingly better than that conferred by the old allele(s), the new allele will replace the old allele(s) completely (i.e. the new allele will "fix" in the population). In genetic hitch-hiking, if an allele of a particular gene rises in frequency because it is selected (i.e. individuals carrying that allele leave more descendants), then the closely linked allele of any gene that is not easily separated by recombination will be dragged up in frequency as well, including alleles that have neutral (i.e. functionally unimportant) or even deleterious changes. Founder effects are generally seen in populations isolated in some way; consider a population with many alleles, from which a pair of individuals moves to an isolated island and begins to breed: the few alleles (no matter how rare and/or deleterious in the original population) borne by those two individuals are now at a high frequency in the isolated population. Examples of high mutation rates were historically more difficult to find, but since the advent of DNA sequencing there are many examples; for instance, repeated slippage in different individuals by DNA polymerase replicating small nucleotide repeats are implicated in the high number of alleles in microsatellites and in some genes (including those used for "DNA typing"). Finally, the increase in gene frequency by the random process of drift, although unlikely, is possible; consider the situation for the population of 200 individuals described in the last paragraph, in which by chance both progeny in every generation always carry the mutant allele.

Classical MHC molecules show all of the signatures of selection. These include the large number of common alleles and a heterozygosity index in populations above that expected by neutrality alone. Perhaps the most compelling is based on the sequences themselves: the evidence that there is a greater number of differences leading to amino acid changes (replacement or non-synonymous changes) compared with those that do not lead to a change in amino acid (silent or synonymous changes) (Hughes and Nei, 1988, 1989; Hughes and Hughes, 1995). This is considered good evidence for selection at the level of protein rather than of DNA. Moreover, these changes are found only at certain sites: most are those directly involved in binding peptides, so the changes have a clear functional reason to be selected.

However, the basis of the selection pressure is still slightly contentious. Given that binding peptides derived from intracellular and extra-cellular proteins for presentation as antigens to T lymphocytes is the only function clearly shown at a molecular level for MHC molecules, the commonly accepted driving force for the large number of MHC alleles is an arms race between hosts and their pathogens. To see how this would work (Figure 10.1), consider a host population that is composed of individuals homozygous for one kind of MHC molecule with one peptide-binding specificity except for the rare individual heterozygous for a variant MHC molecule with a different peptide-binding specificity. Pathogens that bear peptides that can be bound by the common MHC molecule would not be able to replicate much in such a population, but a rare variant pathogen that did not have such peptides would not be recognized by the host immune system and so would increase in frequency (i.e. there is selection for a new pathogen variant). As this new pathogen variant became prominent, the host individuals homozygous for the common MHC molecule would be infected, and only the rare individual with the variant MHC molecule would resist the new variant pathogen well and engender offspring. As a result, the host population would change to have few homozygotes for the old common allele in favour of individuals homozygous or heterozygous for the new MHC allele, the latter being able to resist both the old and new pathogen variants. This would then select for a third pathogen variant that did not bear peptides for either the first or second MHC allele, and this pathogen variant would then select for yet another host MHC variant and so forth.

Such a scenario fits in well with the presence of virus isolates with changes in peptides bound by particular MHC alleles due evidently to direct selection. Examples include the paucity of certain Epstein–Barr virus variants in populations with a high proportion of HLA-A11 (Decamposlima et al., 1993) and the evolution of human immunodeficiency virus (HIV) sequences, apparently under the pressure of cytolytic T cells in certain infected individuals (McMichael and Phillips, 1997).

These selection pressures are the same as those identified for predator–prey relationships (in fact, pathogens are predators, in this and other senses): rare allele advantage (also known as frequency-dependent selection) and heterozygous advantage (also known as overdominant selection). In addition, if one protective MHC allele is enough for good survival, then even a totally unprotective MHC allele may only disappear slowly, since the lower the frequency of the unprotective allele, the fewer individuals homozygous for the unprotective allele exist to be eliminated (i.e. the unprotective MHC allele acts as a recessive lethal; Kaufman et al., 1990).

In addition, other biological functions may drive or at least contribute to the maintenance of polymorphism. The best developed of these ideas concerns sexual selection, which was first suggested on the basis of MHC-determined dif-

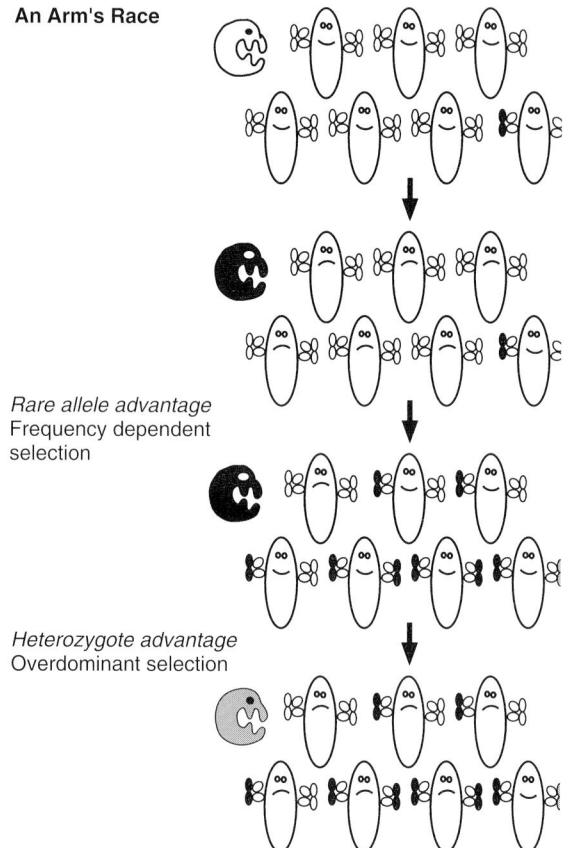

FIGURE 10.1 Coevolution of pathogens and their hosts. The population of host individuals starts with white MHC molecules that can bind and present peptides from white pathogens. This causes the selection of a variant pathogen (in this case, a black pathogen) which does not have the peptides that are bound by the white MHC molecules. This in turn selects for the rare host individual with a black MHC molecule that can bind and present peptides from black pathogens (rare allele advantage, frequency-dependent selection). As the number of host individuals bearing at least one black MHC molecule increases in the population, this leads to a mixture of individuals, some of whom are heterozygous and therefore resistant to both the white and black pathogens (heterozygote advantage, overdominant selection). This now selects for yet another pathogen variant (in this case, a grey pathogen) which in turn selects for a rare host with a grey MHC molecule, and so on.

ferences in mate choice between inbred lines of laboratory mice (Boyse et al., 1987), with investigations now extended to seminatural populations of mice in barns as well as natural populations of pheasants in the field (Potts and Wakeland, 1993; von Schantz et al., 1997). There is extensive data to show that mice and rats discriminate the MHC haplotype of other individuals by odours as well as evidence that the pheasants use visual correlates. The subject remains controversial on several levels (Penn and Potts, 1999), but even the strongest proponents suggest it only as a mechanism to boost a weak selection.

INTENSITY OF SELECTION ON CLASSICAL MAJOR HISTO-COMPATIBILITY COMPLEX GENES

While no one now doubts the importance of a molecular arms race between hosts and their pathogens in the selection for MHC polymorphism, there is little evidence for sufficient pressure to maintain the 200 common alleles found at some loci. For instance, population theory predicts that heterozygous advantage can maintain no more than roughly 10 alleles (since with 10 alleles there are virtually no homozygotes) and calculations (involving a number of assumptions) suggest a rather weak selection on individual MHC loci (Satta et al., 1994). Experimentally, monomorphic populations of mice on islands exist without evidence of frequent decimation by pathogens (Figueroa et al., 1986). Finally and most importantly, it was pointed out long ago (Klein and Figueroa, 1986; Klein, 1987) that contrary to expectations, there is very little evidence in well-studied mammals, such as humans, mice and rats, that MHC haplotypes confer differential resistance to infectious pathogens. Indeed, all of the strong associations of the human MHC are with autoimmune diseases (or biochemical defects), with the strongest associations with infectious diseases being much weaker (Tiwari and Terasaki, 1985; Hill et al., 1991; Hill, 1998).

This lack of evidence for strong selection has led to several suggestions to explain the extraordinary level of polymorphism at MHC loci, all of which may be operating simultaneously. One class of explanations involves population structure. For instance, if the species consists of small, isolated breeding populations, each one would originate and maintain a small number of alleles under the pressure of pathogens, but the alleles would be different in each population (even if the pathogens were the same, since there are many peptide-binding specificities which could recognize each pathogen). The discovery of completely new alleles among small isolated populations of Amerindians in South America has been interpreted in this way (Belich et al., 1992; Watkins et al., 1992). The total MHC polymorphism in the species would appear high, either because of sampling (as in different mouse demes) or because of historical coalescence (as in human conurbation). More panmictic species, such as rats, might be expected to have lower levels of total species polymorphism, as some have suggested is the case.

Another class of explanations is that the appearance of new alleles is most dramatic in times of new devastating diseases, such as zoonoses. For instance, it has been speculated that the black plague is responsible for the high level of HLA-A2 among northern Europeans, and that the prevalence of HLA-A1, B8 in the face of a large number of deleterious autoimmune associations must also have been due to a strong selection at some point. Changes in MHC gene frequencies among Dutch immigrants to Surinam in South America were also imputed to strong selection by pathogens in a new environment (de Vries et al., 1979). The recent evidence that certain HLA types have strong effects on the progression of HIV infection to lethal acquired immunodeficiency syndrome (AIDS) (Carrington et al., 1999) is now thought to be making a real impact on the genetic make-up of several human populations.

However, good examples of strong associations of the MHC with infectious diseases are found in chickens rather than mammals. Indeed, for certain pathogens, the haplotype of MHC (i.e. the B locus) borne by the individual chicken can mean life or death. The important feature appears to be the fact that many common chicken lines express only a single dominantly expressed classical class I (and class II) molecule, compared with the multigene family of classical MHC molecules expressed by well-studied mammals (Kaufman et al., 1995, 1999; Kaufman and Salomonsen, 1997).

Two pathogens have been particularly well examined. Rous sarcoma virus (RSV) is an acute transforming retrovirus (the first retrovirus to be characterized) with four genes (gag, pol, env and v-src) flanked by long terminal repeats (LTR). The chicken MHC is one of the major loci determining resistance (the others being the cellular receptors for the virus), based on studies with a number of MHC haplotypes and RSV strains. In the currently available system involving two MHC-congenic chicken lines (bred to be identical everywhere but the MHC), both lines become infected, replicate virus and have tumours transformed by the v-src gene, but in one line the tumours regress and all of the animals live, while in the other line the tumours progress and all of the animals die. This striking difference in response is due to the peptide-binding specificity of the single dominantly expressed class I molecule: a peptide from the v-src oncogene is presented to cytolytic T lymphocytes in the regressor line, whereas no peptide from the RSV strain binds to the dominantly expressed class I molecule in the progressor line (Kaufman et al., 1995). The peptide-binding specificity of the dominantly expressed MHC molecules seems likely to be responsible for the MHC-determined resistance to certain small pathogens as well as the response to killed vaccines.

The strongest association of the MHC with any phenomenon is the remarkable resistance of the B21 haplotype and the remarkable susceptibility of the B19 haplotype to the tumours resulting from transformation by Marek's disease virus (MDV). There is a rank hierarchy of susceptibility for all tested haplotypes, with B19 the most susceptible, B21 the least susceptible and the others ranged in between. Given that MDV expresses over 80 proteins, it seems unlikely that even the most fastidious MHC molecule would fail to find a peptide to bind and present to T cells. Instead, the level of cell-surface expression of class I molecules correlates well with the level of susceptibility: the resistant B21 haplotype has low cell-surface expression, the susceptible B19 haplotype has high expression and the other haplotypes are ranged in between. The mechanism of protection is unknown, but is likely to be due to natural killer (NK) cells. Such differences in cell-surface expression levels between haplotypes have not been reported for mammals, but there is some evidence for differences between alleles and loci. This leads to the suggestion that this phenomenon is another manifestation of having single, dominantly expressed MHC molecules in the chicken, so that the differences in expression are dramatic compared with mammals, where the multigene family of MHC molecules averages out the levels (Kaufman and Salomonsen, 1997).

Thus, it would appear that the typical mammalian MHC, which encodes a multigene family of both classical class I and class II molecules, confers more-or-less good protection to all pathogens (which, therefore, must use strategies other than lack of peptide binding to escape the attentions of the immune system). In contrast, the chicken (and perhaps most if not all non-mammalian vertebrates) have single MHC molecules whose

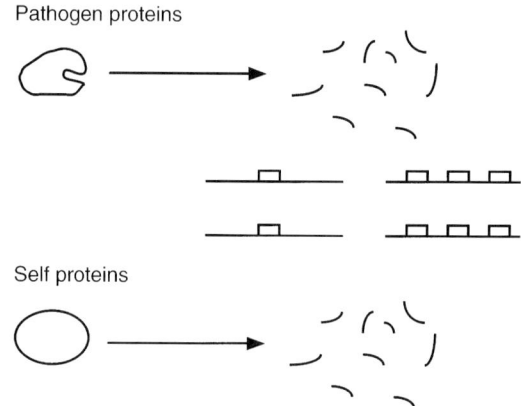

FIGURE 10.2 Immunity and autoimmunity are conversely affected by the numbers of classical MHC molecules. Both pathogen proteins and self-proteins are digested into peptides by normal cellular proteolysis and some of the peptides are bound by individual MHC molecules encoded by classical MHC genes. In vertebrates with single dominantly expressed MHC genes (e.g. one class I gene), there will be two MHC molecules found on the cell surface (since animals are diploid) and only those few peptides which fit the peptide-binding site of one or the other of the encoded molecules will be presented to T lymphocytes of the immune system. In vertebrates with a multigene family of well-expressed MHC genes (e.g. three class I genes), the peptides that can fit any of six MHC molecules will be presented. An animal with one class I molecule will have less chance of being responsive to a pathogen and also less chance of being responsive to a self-protein than an animal with multiple MHC molecules.

properties dominate the immune response and as a result suffer from pathogens that would not bother a typical mammal. Why the chicken pursues this seemingly suicidal strategy is unclear: various kinds of genome and chromosomal evolution have been invoked (Kaufman et al., 1995, 1999; Kaufman and Salomonsen, 1997). Another interesting possibility is that the propensity for autoimmunity rises with increasing number of genes (Figure 10.2): among other reasons, because the chance of cross-reaction between a pathogen peptide and a self-peptide to which the immune system is simply ignorant rather than fully tolerant increases. In any case, simple theory predicts that such strong selection pressures in chickens should lead to higher polymorphism than that found in mammals, a result not yet obvious from the existing data. Clearly, there is still much to learn about the generation and maintenance of polymorphism in MHC genes.

ACKNOWLEDGEMENTS

The author would like to thank Drs Tim Powell and Pete Kaiser for reading the manuscript.

REFERENCES

Belich, M., Madrigal, J.A., Hildebrand, W. et al. (1992). Unusual HLA-B alleles in two tribes of Brazilian Indians *Nature* **357**: 326–329.

Bontrop, R.E., Otting, N., Slierendregt, B.L. and Lanchbury, J.S. (1995). Evolution of major histocompatibility complex polymorphisms and T-cell receptor diversity in primates. *Immunol. Rev.* **143**: 33–62.

Boyse, E.A., Beauchamp, G.K. and Yamazaki, K. (1987). The genetics of body scent. *Trends Genet.* **3**: 97–102.

Carrington, M., Nelson, G.W., Martin, M.P. et al. (1999). HLA and HIV-1: heterozygote advantage and B*35-Cw*04 disadvantage. *Science* **283**: 1748–1752.

Decamposlima, P.O., Gavioli, R., Zhang, Q.J. et al. (1993). HLA-A11 epitope loss isolates of Epstein–Barr-virus from a highly A11+ population. *Science* **260**: 98–100.

Figueroa, F., Tichy, H., Berry, R.J. and Klein, J. (1986). Mhc polymorphism in island populations of mice. *Contemp. Top. Microbiol. Immunol.* **127**: 100–105.

Hill, A.V.S., Allsopp, C., Kwiatowski, D. et al. (1991). Common West African HLA antigens are associated with protection from severe malaria. *Nature* **352**: 595–600.

Hill, A.V.S. (1998). The immunogenetics of human infectious disease. *Annu. Rev. Immunol.* **16**: 593–617.

Högstand, K. and Boehme, J. (1994). A determination of gene conversion in unmanipulated mouse sperm. *Proc. Natl Acad. Sci. USA* **91**: 9921–9925.

Holliday, R. (1964). A mechanism for gene conversion in fungi. *Genet. Res.* **5**: 282–304.

Holmes, N. and Parham, P. (1985). Exon shuffling in vivo can generate novel HLA class I molecules. *EMBO J.* **4**: 2849–2854.

Hughes, A.L. and Nei, M. (1988). Pattern of nucleotide substitution at major histocompatibility complex class I loci reveals overdominant selection, *Nature* **335**: 167–170.

Hughes, A.L. and Nei, M. (1989). Nucleotide substitution at major histocompatibility complex class II loci: evidence for overdominant selection. *Proc. Natl Acad. Sci. USA* **86**: 958–962.

Hughes, A.L. and Hughes, M.K. (1995). Natural selection on the peptide-binding regions of major histocompatibility complex molecules. *Immunogenetics* **42**: 233–243.

Kaufman, J. and Salomonsen, J. (1997). The "Minimal Essential MHC" revisited: both peptide-binding and cell surface expression level of MHC molecules are polymorphisms selected by pathogens in chickens. *Hereditas* **127**: 67–73.

Kaufman, J., Skjoedt, K. and Salomonsen, J. (1990). The MHC molecules of nonmammalian vertebrates. *Immunol. Rev.* **113**: 83–117.

Kaufman, J., Völk, H. and Wallny H.J. (1995). A "minimal essential MHC" and an "unrecognized MHC": two extremes in selection for polymorphism. *Immunol. Rev.* **143**: 63–88.

Kaufman, J., Jacob, J., Shaw, I. et al. (1999). Gene organisation determines evolution of function in the chicken MHC. *Immunol. Rev.* **167**: 101–118.

Klein, J. (1975). *The Biology of the Mouse Histocompatibility-2 Complex: Principles of Immunogenetics Applied to a Single System*, Springer, Berlin.

Klein, J. (1984). Gene conversion in MHC genes. *Transplantation* **38**: 327–329.

Klein, J. (1987). Origin of major histocompatibility complex polymorphism: the trans-species hypothesis. *Hum. Immunol.* **19**: 155–162.

Klein, J. and Figueroa, F. (1986). Evolution of the major histocompatibility complex. *CRC Crit. Rev. Immunol.* **6**: 295–386.

Klein, J. and O'hUigin, C. (1995). Class II B Mhc motifs in an evolutionary perspective. *Immunol. Rev.* **143**: 89–112.

Klein, J., Satta, Y., O'hUigen, C. and Takahata, N. (1993). The molecular descent of the major histocompatibility complex. *Annu. Rev. Immunol.* **11**: 269–295.

McMichael, A.J. and Phillips, R.E. (1997). Escape of human immunodeficiency virus from immune control. *Annu. Rev. Immunol.* **15**: 271–296.

Nathenson, S.G., Geliebter, J., Pfaffenbach, G.M. and Zeff, R.A. (1986). Murine major histocompatibility complex class I mutants: molecular analysis and structure–function implications. *Annu. Rev. Immunol.* **4**: 471–502.

Ohta, T. (1997). Role of gene conversion in generating polymorphisms at major histocompatibility complex loci. *Hereditas* **127**: 97–103.

Parham, P., Adams, E.J. and Arnett, K.L. (1995). The origins of HLA-A, -B and -C polymorphisms. *Immunol. Rev.* **143**: 141–180.

Penn, D.J. and Potts, W.K. (1999). The evolution of mating preferences and major histocompatibility complex genes. *Amer. Nat.* **153**: 145–164.

Potts, W. and Wakeland, E. (1993). Evolution of MHC genetic diversity: a tale of incest, pestilence and sexual preference. *Trends Genet.* **9**: 408–412.

Satta, Y. (1997). Effects of intra-locus recombination on HLA polymorphism. *Hereditas* **127**: 105–112.

Satta, Y., O'hUigin, C., Takahata, N. and Klein, J. (1994). Intensity of natural selection at the major histocompatibility complex loci. *Proc. Natl Acad. Sci. USA* **91**: 7184–7188.

Schantz, T. von, Wittzell, H., Goransson, G. and Grahn, M. (1997). Mate choice, male condition-dependent ornamentation and MHC in the pheasant. *Hereditas* **127**: 133–140.

Tiwari, J. and Terasaki, P. (1985). *HLA and Disease Associations*, Springer, New York.

Vries, R.R. de, Meera-Khan, P., Bernini, L.F. et al. (1979). Genetic control of survival in epidemics. *J. Immunogenet.* **6**: 271–287.

Watkins, D., McAdam, S., Liu, X. et al. (1992). New recombinant HLA-B alleles in a tribe of South American Amerindians indicate rapid evolution of MHC class I loci. *Nature* **357**: 329–333.

Weill, J.C. and Reynaud, C.A. (1987). The chicken B cell compartment. *Science* **238**: 1094–1097.

Zangenberg, G., Huang, M., Arnheim, N. and Erlich, H.A. (1995). New HLA-DPB1 alleles generated by gene conversion detected by analysis of sperm. *Nature Genet.* **10**: 407–414.

SECTION II

HLA and Disease Associations

CHAPTER 11

HLA and Rheumatoid Arthritis

Gerald T. Nepom

INTRODUCTION

Genetic susceptibility is a major contributory factor in the pathogenesis of rheumatoid arthritis (RA). Multiple genes are likely to be involved; however, the principal known genetic locus contributing to susceptibility is HLA-DRB1 in the HLA class II region of chromosome 6, which encodes the β-chain of HLA-DR molecules.

HLA-DRB1 is a highly polymorphic locus, with over 150 allelic variants in the human population. Several of these DRB1 alleles are associated with susceptibility to RA (Nepom et al., 1989; Ollier and Thomson, 1992). Each of these susceptibility alleles has in common an important structural element known as the shared epitope (Gregersen et al., 1987; Winchester et al., 1992; Nepom and Nepom, 1993). This consists of a series of amino acids encoded by codons 67–74 of the DRB1 gene. When this sequence is LLEQKRAA or LLEQRRAA, the corresponding allele is highly associated with RA. This presence of the shared epitope sequence accounts for the association of five main alleles with RA in various ethnic groups: DRB1*0401, DRB1*0404 and DRB1*0101 in Caucasoid populations, DRB1*0405 in Asians and DRB1*1402 in native Americans (Willkens et al., 1991; Ollier and Thomson, 1992). A multitude of genetic typing studies in the 1980s and 1990s has documented the association of each of these alleles with RA in various populations. The presence of the shared epitope sequence in each case is the common link which distinguishes these alleles from closely related genes that are not associated with RA. For example, single codon substitutions within the DRB1*04 family of genes which modify the shared epitope sequence abrogate the associated disease susceptibility. DRB1*0402 and DRB1*0403, members of the HLA-DR4 family of alleles, contain polymorphisms within the shared epitope and are not associated with RA. Thus, the presence of the shared epitope is the primary unit of genetic susceptibility in patients with RA. Table 11.1 presents a summary of these associations (Ohta et al., 1982; Nepom et al., 1989; Wordsworth et al., 1989; Gao et al., 1990; Ronningen et al., 1990; Willkens et al., 1991; Nelson et al., 1992).

The HLA-DRB1 allelic markers used for HLA typing are specific for the susceptibility genes themselves and, as such, are more tightly associated with disease than any other anonymous or linked markers in the HLA region. Other linked genes in the HLA class II region, such as the DQ locus, are not independently associated with RA and do not contribute to disease susceptibility (Nepom et al., 1989; Fugger and Svejgaard, 1997). Interestingly, several reports analysing genetic polymorphisms around the tumor necrosis factor (TNF) locus telomeric to the class II region on chromosome 6 suggest the possibility of an independent susceptibility locus in that region, possibly modifying risk on haplotypes which also carry a DRB1 susceptibility allele (Mulcahy et al., 1996). Thus, it is possible that the major HLA susceptibility contribution of the DRB1 gene may be subject to epistatic effects from a remote locus, such as TNF.

GENE DOSAGE EFFECTS ON DISEASE SUSCEPTIBILITY

There is some heterogeneity in the relative contribution of the various RA susceptibility genes to disease susceptibility. In populations in which DRB1*0401 and DRB1*0404 are prevalent, these are usually more closely associated with disease than the DRB1*0101 gene. Conversely, in populations in which DRB1*0401 and DRB1*0404 are rare, other susceptibility genes, such as DRB1*0405 and DRB1*1402, account for the HLA-associated disease risk. DRB1*0101 is often associated with less severe forms of RA and, as a consequence, is more prevalent in patients studied in community practice than in patients studied in tertiary care centres. Most patients with RA inherit a single copy of one of these susceptibility alleles; thus, a single gene dose in a heterozygote individual (i.e. "dominant" mode of inheritance) is sufficient for increased genetic risk. However, in certain studies of patients in tertiary care centres reflecting severe long-standing disease, there is an increased association correlated with the inheritance of two of the susceptibility alleles (Wordsworth et al., 1992; Weyand and Goronzy, 1994). Thus, in some cases, there is a larger relative risk for susceptibility gene homozygotes, implying a gene dosage effect.

One particular combination of RA susceptibility alleles appears to be associated with a particularly high degree of risk. Heterozygous individuals expressing both DRB1*0401 and DRB1*0404 have a relative risk 5–10-fold higher than that of either of the single alleles alone (Nepom et al., 1984; Ollier and Thomson, 1992; Wordsworth et al., 1992). This observation has triggered intense study of the potential mechanism for this synergistic disease risk, which is still unknown. The extremely high association of DRB1*0401/DRB1*0404 heterozygotes with seropositive juvenile RA has led to the suggestion that this particular combination of susceptibility alleles also predisposes for earlier disease onset.

RELATIONSHIP TO DISEASE SEVERITY

Some of the conflicting issues regarding disease associations in RA are resolved by considering issues of disease severity. RA is a heterogeneous clinical syndrome, with some patients progressing through rapidly progressive, erosive synovitis to irreversible joint damage and others maintaining a relatively benign clinical course. Most studies which stratify patients based on severity of disease have correlated the DR4 susceptibility alleles (DRB1*0401, DRB1*0404, DRB1*0405) with severe erosive forms of disease, and most such studies have correlated the association of DRB1*0101 with a less severe disease course (Young et al., 1984; Olsen et al., 1988; van Zeben et al., 1991; Weyand et al., 1992; McMahon et al., 1993; Vehe et al., 1993). Stratification of these susceptibility alleles in this way marks the DR4 susceptibility genes as indicators of risk for severe forms of disease (Nepom et al., 1996).

Consideration of the DR4 susceptibility alleles as markers for risk for severe RA also helps to clarify the issue of gene dosage, discussed above. Thus, DR4/DR4 allelic combinations have a higher relative risk in patients with severe RA than DR4/DRx combinations or DR4/DR1. In reports from one center, individuals with two DR4 susceptibility alleles had the highest risk of extra-articular manifestations of RA, including vasculitis and Felty's syndrome (Weyand and Goronzy, 1990; Weyand et al., 1995).

UTILITY FOR DISEASE PROGNOSIS

The relative risk statistics outlined in Table 11.1 summarize the genetic association of each of the RA susceptibility genes, where relative risk represents an odds ratio expressing the risk to an individual with the susceptibility gene compared with an individual without this gene. From the perspective of an individual patient, however, an absolute risk statistic is more relevant than the relative risk. In other words, what is the risk of developing RA for an individual with one of the susceptibility genes? Table 11.2 summarizes the absolute risk ratios for susceptibility genes in RA, based on an estimated prevalence of RA in the Caucasoid population of 1%.

As illustrated in Table 11.2, the absolute risk to an individual in the population carrying one of the RA susceptibility genes is quite low, and is at worst one in seven for individuals heterozygous

TABLE 11.1 HLA-DRB1 susceptibility genes in rheumatoid arthritis

HLA class II gene	Relative risk	Shared epitope sequence (Residues 70–74)	Ethnic distribution
DRB1*0401	6–11	Q K R A A	Caucasian
DRB1*0404	5–14	Q R R A A	Caucasian
DRB1*0101	1–2	Q R R A A	Caucasian
DRB1*0405	6–10	Q R R A A	Oriental
DRB1*1402	1–2	Q R R A A	Native American

TABLE 11.2 Absolute risk attributable to the HLA-DRB1 susceptibility genes in rheumatoid arthritis

HLA class II gene	Approximate risk ratio
DRB1*0401	1 in 35
DRB1*0404	1 in 20
DRB1*0101	1 in 80
DRB1*0401/DRB1*0404	1 in 7

for both DRB1*0401 and DRB1*0404. This is an obvious consequence of the fact that the RA susceptibility genes are much more prevalent than is RA itself (Nepom and Nepom, 1992). Thus, HLA typing for RA susceptibility has no role for disease prediction in the general population.

Nevertheless, there is an important clinical utility for susceptibility genotyping. This clinical utility is based on the understanding, summarized above, that the DR4 susceptibility alleles are associated with susceptibility for severe, erosive forms of RA. Therefore, in a patient with early pre-erosive RA, the presence of the DR4 susceptibility genes can serve as a prognostic marker for the likelihood of progression to early erosive disease. Prognostic studies evaluating the clinical utility of early patient stratification are limited, but it has been estimated that the combination of HLA genotyping for DR4 susceptibility alleles, along with rheumatoid factor, will identify those patients likely to have a progressive erosive disease course (Emery, 1997). Figure 11.1 summarizes this estimate of erosive disease in patients with early RA. In studies performed to date, one copy of a DR4 susceptibility allele is sufficient to predict progression to erosive disease; it is not yet known whether there is a gene dosage effect which increases risk in this setting (Wagner et al., 1997). At any rate, from the point of view of clinical relevance, a single copy of a DR4 susceptibility gene is sufficient to stratify patients: In the early assessment of patients with RA, the presence

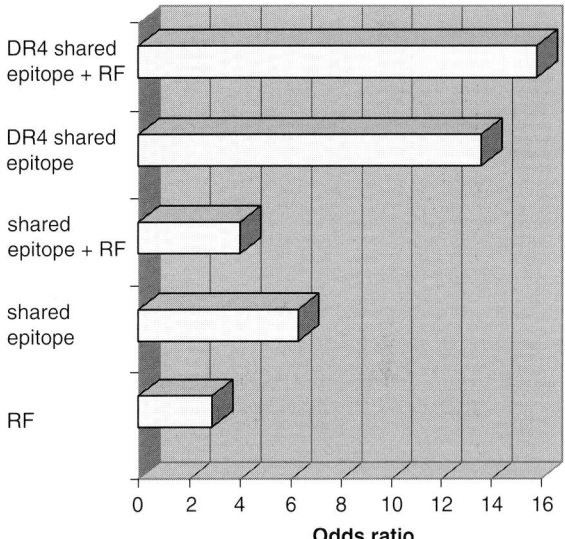

FIGURE 11.1 Prognosis for erosive disease in rheumatoid arthritis (RA): the impact of HLA-DR4 susceptibility. Patients with early polyarthritis can be assessed for the likelihood of progression to severe, erosive RA by a combination of clinical, laboratory and genetic parameters. Patients with the DR4 susceptibility alleles (containing the shared epitope) have a significantly increased risk of developing erosions (Wagner et al., 1997). Seropositivity for rheumatoid factor (RF) is an independent risk factor which, in this study, did not reach statistical significance. In this study, the probability of having erosive disease was increased to 0.91 in the presence of the DR4 susceptibility alleles, while the likelihood decreased to 0.35 when the shared epitope was absent. (Figure adapted from Wagner et al., 1997.)

of the DR4 susceptibility alleles may indicate a two in three risk for erosive disease, while the absence of HLA susceptibility alleles indicates an equally high probability of a fairly mild disease course. Thus, the use of HLA typing for prognosis in patients with pre-erosive early RA has an important potential clinical role. This has been referred to as clinical staging, in which patients with early RA are assessed for clinical parameters, rheumatoid factor and the presence of the shared epitope. This combination of predictive factors is very promising for predicting the likelihood of erosive disease outcome.

It should be emphasized that the high positive predictive value for DR4 susceptibility alleles pertains to the likelihood of the erosive disease in patients who do not yet have erosions and, as yet, only to the Caucasian population. As summarized previously, most patients who have erosive disease carry the RA susceptibility allele, so the positive predictive value from HLA typing patients who already have erosions is marginal.

What are the implications of predicting severe erosive disease in patients with early RA? If the early assessment of RA predicts for severe erosive disease, there is an opportunity for early therapeutic intervention (O'Dell et al., 1998). In the context of a decision regarding therapeutic options for patients with early RA, a number of studies has shown the benefit of aggressive therapy using combination drugs. It seems likely that the main role for HLA typing in RA will be to assist in making this therapeutic decision as early after the initial diagnosis as possible.

REFERENCES

Emery, P. (1997). Prognosis in inflammatory arthritis: the value of HLA genotyping and the oncological analogy. *J. Rheum.* **24**: 1436–1442.

Fugger, L. and Svejgaard, A. (1997). The HLA–DQ7 and –DQ8 associations in DR4-positive rheumatoid arthritis patients. *Tissue Antigens* **50**: 494–500.

Gao, X., Olsen, N.J., Pincus, T. and Stastny, P. (1990). HLA-DR alleles with naturally occurring amino acid substitutions and risk for development of rheumatoid arthritis. *Arthritis Rheum.* **33**: 939–946.

Gregersen, P.K., Silver, J. and Winchester, R.J. (1987). The shared epitope hypothesis: an approach to understanding the molecular genetics of susceptibility to rheumatoid arthritis. *Arthritis Rheum.* **30**: 1205–1213.

McMahon, M.J., Hillarby, M.C., Clarkson, R.W.E. et al. (1993). Major histocompatibility complex variants and articular disease severity in rheumatoid arthritis. *Br. J. Rheum.* **32**: 899–902.

Mulcahy, B., Waldron-Lynch, F., McDermott, M.F. et al. (1996). Genetic variability in the tumor necrosis factor–lymphotoxin region influences susceptibility to rheumatoid arthritis. *Am. J. Hum. Genet.* **59**: 676–683.

Nelson, J.L., Boyer, G., Templin, D. et al. (1992). HLA antigens in Tlingit Indians with rheumatoid arthritis. *Tissue Antigens* **40**: 57–63.

Nepom, G.T. and Nepom, B.S. (1992). Prediction of susceptibility to rheumatoid arthritis by human leukocyte antigen genotyping. In *Rheumatic Disease Clinics of North America* (ed. Nepom, G.T.), pp. 785–792, W.B. Saunders, Philadelphia, PA.

Nepom, B.S. and Nepom, G.T. (1993). Immunogenetics and the rheumatic diseases. In *Textbook of Rheumatology* (eds Kelley, W.N., Harris, E.D., Jr, Ruddy, S. and Sledge, C.B.), pp. 89–107, W.B. Saunders, Philadelphia, PA.

Nepom, B.S., Nepom, G.T., Mickelson, E. et al. (1984). Specific HLA-DR4-associated histocompatibility molecules characterize patients with seropositive juvenile rheumatoid arthritis. *J. Clin. Invest.* **74**: 287–291.

Nepom, G.T., Byers, P., Seyfried, C. et al. (1989). HLA genes associated with rheumatoid arthritis. *Arthritis Rheum.* **32**: 15–21.

Nepom, G.T., Gersuk, V. and Nepom, B.S. (1996). Prognostic implications of HLA genotyping in the early assessment of patients with rheumatoid arthritis. *J. Rheumatol.* **23**: 5–9.

O'Dell, J.R., Nepom, B.S., Haire, C. et al. (1998). DRBI typing in rheumatoid arthritis: predicting response to specific therapies. *Ann. Rheum. Dis.* **57**: 209–213.

Ohta, N., Nishimura, Y.K., Tanimoto, K. et al. (1982). Association between HLA and Japanese patients with rheumatoid arthritis. *Hum. Immunol.* **5**: 123–132.

Ollier, W. and Thomson, W. (1992). Population genetics of rheumatoid arthritis. In *Rheumatic Disease Clinics of North America* (ed. Nepom, G.T.), pp. 741–759, W.B. Saunders, Philadelphia, PA.

Olsen, N.J., Callahan, L.F., Brooks, R.H. et al. (1988). Associations of HLA-DR4 with rheumatoid factor and radiographic severity in rheumatoid arthritis. *Am. J. Med.* **84**: 257–264.

Ronningen, K.A., Spurkland, A., Egeland, T. et al. (1990). Rheumatoid arthritis may be primarily associated with HLA-DR4 molecules sharing a particular sequence at residues 67–74. *Tissue Antigens* **36**: 235–240.

Vehe, R.K., Nepom, G.T., Wilske, K.R. et al. (1993). Erosive rheumatoid factor positive and negative rheumatoid arthritis are immunogenetically similar. *J. Rheumatol.* **21**: 194–196.

Wagner, U., Kaltenhauser, S., Sauer, H. et al. (1997). HLA markers and prediction of clinical course and outcome in rheumatoid arthritis. *Arthritis Rheum.* **40**: 341–351.

Weyand, C.M. and Goronzy, J.J. (1990). Disease-associated human histocompatibility leukocyte antigen determinants in patients with seropositive rheumatoid arthritis. *J. Clin. Invest.* **85**: 1051–1057.

Weyand, C.M. and Goronzy, J.J. (1994). Disease mechanisms in rheumatoid arthritis: gene dosage effect of HLA-DR haplotypes. *J. Lab. Clin. Med.* **124**: 335–338.

Weyand, C.M., Hicok, K.C., Conn, D.L. and Goronzy, J.J. (1992). The influence of HLA-DRB1 genes on disease severity in rheumatoid arthritis. *Ann. Intern. Med.* **117**: 801–806.

Weyand, C.M., McCarthy, T.G. and Goronzy, J.J. (1995). Correlations between disease phenotype and genetic heterogeneity in rheumatoid arthritis. *J. Clin. Invest.* **95**: 2120–2126.

Willkens, R.F., Nepom, G.T., Marks, C.R. et al. (1991). The association of HLA-Dw16 with rheumatoid arthritis in Yakima Indians: further evidence for the "shared epitope" hypothesis. *Arthritis Rheum.* **34**: 43–47.

Winchester, R., Dwyer, E. and Rose, S. (1992). The genetic basis of rheumatoid arthritis: the shared epitope hypothesis. In *Rheumatic Disease Clinics of North America* (ed. Nepom, G.T.), pp. 761–783, W.B. Saunders, Philadelphia, PA.

Wordsworth, B., Lanchbury, J.S.S., Sakkas, L.I. et al. (1989). HLA-DR4 subtype frequencies in rhzeumatoid arthritis indicate that DRB1 is the major susceptibility locus within the HLA class II region. *Proc. Natl Acad. Sci. USA* **86**: 10 049–10 053.

Wordsworth, P., Pile, K.D., Buckley, J.D. et al. (1992). HLA heterozygosity contributes to susceptibility to rheumatoid arthritis. *Am. J. Hum. Genet.* **51**: 585–591.

Young, A., Jaraquemada, D., Awad, J. et al. (1984). Association of HLA-DR4/Dw4 and DR2/Dw2 with radiologic changes in a prospective study of patients with rheumatoid arthritis. *Arthritis Rheum.* **27**: 20–25.

Zeben, D. van, Hazes, J.M.W., Zwinderman, A.H. et al. (1991). Association of HLA-DR4 with a more progressive disease course in patients with rheumatoid arthritis. *Arthritis Rheum.* **34**: 822–830.

CHAPTER 12

HLA and the Spondylarthropathies

Paul Bowness

INTRODUCTION

The association of the human leucocyte antigen (HLA) class I allele HLA-B27 with the rheumatic disease ankylosing spondylitis (AS) has been recognized since the early 1970s (Brewerton et al., 1973; Schlosstein et al., 1973). AS is a relatively common inflammatory rheumatic disease, affecting approximately 0.1–0.5% of the UK population. Although the joints of the spine and axial skeleton are most commonly involved, AS is a multisystem disease and recognized features include asymmetrical peripheral arthritis, uveitis, aortic valve involvement and upper lobe pulmonary fibrosis. The association of AS with possession of HLA-B27 is amongst the strongest described for any HLA locus. A recent study found that 94% of AS patients are HLA-B27 positive, compared with 9.4% of controls, giving an odds ratio of 171, with 95% confidence interval 135–218 (Brown et al., 1996). HLA-B27 is also strongly associated with a number of other diseases, shown in Table 12.1.

These conditions have clinical features in common, including arthritis of the spine and large joints, and involvement of the skin, eye, genital mucosa and heart. Reactive arthritis (ReA) is less common than AS and less strongly associated with HLA-B27 (Brewerton et al., 1974; Laitinen et al., 1977). ReA is, however, an important model of autoimmune disease since it follows certain genital or gastrointestinal infections. Some, but not all patients, develop the classical triad of arthritis, urethritis and conjunctivitis described by Reiter. Only some infections predispose to ReA, with genitally acquired *Chlamydia* or gastrointestinal infection with *Salmonella*, *Yersinia*, *Shigella* or *Campylobacter* being most commonly implicated, all of which are Gram-negative bacteria that can live intracellularly. Spinal and peripheral joint involvement occurring together with psoriasis and inflammatory bowel disease are also HLA-B27 associated, as is isolated uveitis. A related but poorly understood clinical entity, "undifferentiated spondylarthritis", is also recognized. It is probable that all of these overlapping clinical

TABLE 12.1 HLA-B27-associated diseases

Disease	% HLA-B27+	Reference
Ankylosing spondylitis	96	Brewerton et al. (1973)
Reactive arthritis (includes "Reiter's disease")	60–70	Brewerton et al. (1974); Laitinen et al. (1977)
(Anterior) uveitis	40–69	Careless et al. (1997)
Aortic regurgitation together with cardiac conduction abnormality	67–88	Bergfeldt (1997)
Sacroileitis associated with inflammatory bowel disease	33–75	Calin and Taurog (1998)
Sacroileitis associated with psoriasis	40–50	Calin and Taurog (1998)
Undifferentiated oligoarthropathy		Calin and Taurog (1998)

conditions – the spondylarthropathies – have a related aetiology.

Whilst the pathogenetic role of HLA-B27 in the spondylarthropathies is unknown, numerous theories have been proposed. All such theories (see below) must explain not only the relationship with the specific triggering organisms found most clearly in ReA, but also the striking tissue distribution of these diseases.

This chapter will first describe the possible mechanisms by which HLA-B27 might be involved in disease pathogenesis, before presenting the data available from HLA-B27 transgenic animals, biochemical analysis of HLA-B27 function and molecular epidemiology studies. A concluding section will synthesize these lines of evidence and identify key lines of future research.

THEORIES EXPLAINING THE ASSOCIATION OF HLA-B27 WITH THE SPONDYLARTHROPATHIES

A number of different mechanisms has been put forward to explain the association of HLA-B27 with the spondylarthropathies. Some of these theories, summarized in Table 12.2, are also applicable to other HLA-associated autoimmune diseases.

It is also important to recognize not only that disease can occur in the absence of HLA-B27, but also that other HLA alleles may predispose to these conditions, either independently or together with HLA-B27 (e.g. HLA-B60 together with HLA-B27; Brown et al., 1996). Furthermore, family and twin studies have demonstrated that

TABLE 12.2 Hypotheses of HLA-B27 disease association

- Peptide presentation by HLA-B27 ("arthritogenic peptide")
- Thymic selection of T cell repertoire
- "Altered self" hypothesis
- Unusual cell biology of HLA-B27
- Presentation of HLA-B27-derived peptides by HLA class II molecules
- Interaction with superantigens
- Receptor hypothesis
- Linkage to a disease-associated gene
- Other mechanisms

other, as yet unidentified, non-HLA inherited genes make an even greater contribution to disease susceptibility than HLA-B27 (Calin et al., 1983).

HLA-B27 presents "arthritogenic" peptides to CD8 T cells

The finding that the natural role of HLA molecules is to bind and present peptide to T cells led to the suggestion that the spondylarthropathies result from an ability of HLA-B27 to bind a unique set of peptides (Benjamin and Parham, 1990). This "arthritogenic peptide" hypothesis proposes that disease results from an HLA-B27-restricted cytotoxic T cell (CTL) response to a peptide or peptides found only in joint tissues. This peptide, which would also be present in other affected tissues, would be bound and presented by all disease-associated HLA-B27 subtypes (see below), but not by other class I molecules. Pathogenic T cells might be primed in the joint or at other sites such as the genital or gut mucosa. A modification of the original hypothesis could entail a breakdown of self-tolerance by initial HLA-B27-restricted presentation of a peptide or peptides derived from one of the triggering pathogens.

Arthritogenic peptide models of disease causation remain attractive because of their simplicity, and are supported by the epidemiological and functional studies of HLA-B27 subtypes. Whilst the evidence from transgenic models is inconclusive (see below), recent data support models in which the binding of peptide to HLA-B27 is directly involved in disease (Zhou et al., 1998).

Thymic selection of T cell repertoire

Thymic expression of HLA-B27 will undoubtedly shape the peripheral T cell repertoire and could lead to selection of a population of arthritogenic T cells through either positive or negative selection. This hypothesis, which is related to the "arthritogenic peptide" theory [since thymic selection is in part dependent on the repertoire of major histocompatibility complex (MHC)-bound peptides], is supported by adoptive cell transfer studies in HLA-B27 transgenic rats which suggest that peripheral expression of HLA-B27 is not required for disease expression (see below).

Altered self and the HLA-B27 cysteine-67

The observation that HLA-B27 carries an unpaired cysteine residue at position 67 has led to the suggestion that this residue might be modified or form a disulfide bond with thiol-containing reagents. This might result in new immunological reactivity or "altered self" (Archer et al., 1990; Benjamin and Parham, 1990). Either an antibody or a cell-mediated (including natural killer) response could be elicited to this neoantigen. Alternatively, modification of Cys^{67} would almost certainly have a significant effect on peptide binding (although this has not been tested experimentally), and so could stimulate autoimmunity through the presentation of a different peptide repertoire. HLA-B27-restricted homocysteine-specific T cell responses have been demonstrated (Gao et al., 1996). It has recently been shown that HLA-B27 can form homodimers *in vitro* which are dependent on Cys^{67} (see below; Allen et al., 1999). This is of particular interest since rats transgenic for a variant of HLA-B27 mutated at position 67 apparently develop little arthritis (Taurog et al., unpublished data, discussed in Zhou et al., 1998).

It should be pointed out that a cysteine residue at position 67 is not unique to HLA-B27. It is also present in, for example, HLA-B14. However, no other alleles also have the same combination of amino acid residues at surrounding residues, in particular at positions 70 and 71, which are known to affect the redox potential of Cys^{67}.

HLA-B27 may have unique biological properties

Another possibility is that some feature of the biochemistry or cell biology of HLA-B27 predisposes to disease development. A number of lines of evidence suggests that HLA-B27 may not behave like other class I molecules. Benjamin and colleagues observed that "empty" cell-surface HLA-B27 molecules could present peptides to T cells and suggested that this might lead to the presentation of extracellular peptides not normally accessible to the class I processing pathway (Benjamin et al., 1991). There is also evidence that HLA-B27 is unusual in its interactions with some other molecules involved in antigen processing.

For example cell-surface expression can occur independently of tapasin (Peh et al., 1998). We have recently shown that HLA-B27 can form homodimers *in vitro* which are dependent on Cys^{67} but do not contain β_2-microglobulin. These B27 homodimers are stable and are capable of binding peptide (Allen et al., 1999). This finding, taken together with the data from transgenic rodents (see below), raises the possibility that HLA-B27 may be capable of peptide presentation in the absence of β_2-microglobulin, perhaps to $CD4^+$ T cells. There is, however, no direct evidence supporting this hypothesis.

HLA-B27-derived peptides may be presented by HLA class I or II molecules

The most abundant naturally processed peptides eluted from HLA class II molecules are those derived from (other) HLA molecules (Chicz et al., 1993). It is thus likely that HLA-B27-derived peptides are widely presented on the cell surface by class II molecules (Davenport, 1995). It has been suggested that HLA-B27 itself may be able to present B27-derived peptides (Scofield et al., 1995). Whilst tolerance to such epitopes should normally be present, it is possible that certain HLA class II-restricted T lymphocytes, stimulated by bacterial infection, might have cross-reactive specificity for an HLA-B27-derived peptide presented by host cells. Although a strong HLA class II association with disease might, therefore, be expected, the promiscuity of peptide binding of HLA class II molecules might obscure this (Davenport, 1995). However, it is not clear how such a mechanism could explain the tissue specificity of disease. In one study, proliferative T cell responses to a synthetic peptide encompassing HLA-B27 residues 60–72 were detected in 17 out of 55 patients with AS (Marker et al., 1997). Recent data from transgenic mice (see below) would, however, appear to rule out such a model (Khare et al., 1998).

HLA-B27 could interact with microbial superantigens

Superantigens are proteins produced by bacteria and viruses, including certain endogenous retroviruses, that are able to stimulate large numbers of T cells by cross-linking their T cell receptors with the MHC class II molecules of presenting cells (Marrack and Kappler, 1990). Although superantigens specific for MHC class I antigens have not been described, it is possible that an HLA-B27-specific superantigen could cause disease through immune stimulation in this manner.

Receptor hypothesis

This theory suggests that microorganisms could recognize specific MHC molecules and use them as a vehicle on which to enter the cell. Kapasi and Inman (1994) presented evidence that possession of HLA-B27 and serologically related class I molecules may alter the uptake of *Salmonella*, one of the bacterial species which can induce ReA.

Linkage to a disease-associated gene

It is possible that HLA-B27 is merely a marker for a linked disease-associated gene. For example, the association of HLA-B8 with the storage disease haemochromatosis (Yaounq et al., 1994) has been shown to be due to the presence of the linked HFE gene on chromosome 6. The strongest evidence that it is HLA-B27 itself which is directly involved in disease pathogenesis comes from studies of rodents transgenic for HLA-B27, which develop disease similar to the human spondylarthropathies (see below).

Other mechanisms

In addition to their role in antigen presentation and thymic selection, HLA molecules have other functions which might lead to disease, including interaction with CD8 and with natural killer cells (Lanier, 1998).

The original molecular mimicry hypothesis suggested that disease resulted from an antibody response cross-reactive between a unique portion of HLA-B27 and certain bacterial epitopes. Molecular mimicry between the HLA-B*2705 amino acid positions 72–77 (QTDRED) and both the *Klebsiella* nitrogenase reductase enzyme and a *Shigella flexneri* plasmid gene product have been described, and cross-reactive antibodies can be found in patients with AS (Ebringer et al., 1976; Tsuchiya et al., 1990). However, this exact amino acid sequence is not found in either HLA-B*2702 or HLA-B*2704, both disease-associated subtypes. Furthermore, whilst antibodies reacting with candidate regions undoubtedly exist, many specificities are found when appropriate controls are included (Lahesma et al., 1992).

ANIMAL MODELS Of HLA-B27-ASSOCIATED DISEASE

Strong evidence that HLA-B27 is directly involved in disease pathogenesis is provided by rats and mice carrying HLA-B27 as a transgene, in that they can develop illnesses similar to the spondylarthropathies. An increased frequency of the naturally occurring inflammatory joint disease ankylosing enthesopathy (ankent) has recently been reported in mice transgenic for HLA-B*2702 (Weinreich et al., 1995). Rats carrying a high copy number of HLA-B*2705 transgenes develop an illness characterized by peripheral and axial arthritis, gut inflammation and both genital and skin lesions (Hammer et al., 1990), whereas rats kept in germ-free conditions do not develop the inflammatory intestinal or peripheral joint disease (Taurog, 1994). Adoptive transfer experiments have been inconclusive, providing evidence of disease transfer by foetal liver cells (Breban et al., 1993), and by both $CD4^+$ and $CD8^+$ T cells (Breban et al., 1996). The significance of the latter experiments is uncertain since recipients developed predominantly gut inflammation and little peripheral arthritis. Recently, Taurog's group showed that the peptide repertoire bound by HLA-B27 is indeed critical for disease causation, since introduction of a strongly binding B27 epitope as a further transgene significantly reduces the incidence of arthritis in male rats (Zhou et al., 1998).

Mice rendered transgenic for HLA-B27, which is expressed on the cell surface together with murine β2m (Mβ2m), do not normally develop disease. However, spontaneous inflammatory arthritis develops in mice transgenic for HLA-B27 but lacking murine β2m ($B27^+Mβ2m^-$) when they are transferred from germ-free conditions (conventionalization). In the absence of β2m, these animals express very low levels of class I molecules. However, although normally conformed HLA-B27 is not expressed in these mice, HLA-B27 heavy chains, not associated with β2m, can be detected on the cell surface of concanavalin A-treated peripheral blood lymphocytes using the monoclonal antibody HC10 (Stam et al., 1986). Disease still occurs if human β2m is introduced as a further transgene (B27$^+$Mβ2m$^-$Hβ2m$^+$). These animals express both normally conformed HLA-B27 and the form recognized by HC10, which is expressed on peripheral blood lymphocytes and thymic epithelium. Significantly, disease can be prevented or delayed by treatment with HC10, but not with ME1, an antibody to normally conformed B27.

Recently, B27$^+$Mβ2m$^-$ transgenic mice have been further bred with mice lacking endogenous MHC class II molecules. These mice also develop disease, ruling out the possibility that HLA-B27-derived peptides are being presented by murine class II molecules to induce disease (Khare et al., 1998).

FUNCTIONAL STUDIES OF HLA-B27

Much is now known of the molecular structure, peptide-binding specificity and cell biology of HLA-B27. Solution of the crystal structure of HLA-B27, crystallized with a variety of self-peptides (Madden et al., 1991), showed that short peptides are bound in an extended conformation by a peptide-binding groove lying between the α_1 and α_2 helices. A common arginine was found at the second position of all bound peptides. The long side chain of this arginine was accommodated in the B or 45 pocket, comprising a unique combination of residues in HLA-B27: 45E, 67C, 34V, 26G and 24T. Amino acid analysis of self-peptides eluted from HLA-B27 has confirmed the presence of an anchor arginine residue at the second position (Jardetzky et al., 1991; Rotzschke et al., 1994). There also appear to be preferences for particular amino acids at other positions of bound peptides, which differ between HLA-B27 subtypes. Thus, HLA-B*2705 appears to bind peptides with C-terminal amino acids that are either aromatic, hydrophobic or positively charged, whereas HLA-B*2702 can probably only accommodate aromatic or hydrophobic residues at this position (Rotzschke et al., 1994). Finally, measurement of the ability of different peptides to bind to HLA-B27 has confirmed the importance of the P2 arginine (e.g. Bowness et al., 1994), and also shown that different subtypes probably bind different but overlapping subsets of peptides (Colbert et al., 1994; Tanigaki et al., 1994; D'Amato et al., 1995; Garcia et al., 1997). Given these differences in peptide binding, any arthritogenic peptide or peptides would be expected to bind with different affinity to different B27 subtypes, resulting in differences in the strength of disease association.

A subset of HLA-B27 molecules is capable of binding unusually long peptides (Urban et al., 1994). The significance of this finding is unknown but raises the possibility that HLA-B27 may be capable of adopting a class II-like function.

If the disease association of HLA-B27 is indeed a consequence of its physiological role in peptide selection, HLA-B27-restricted CTL, specific for self- or bacterial epitopes, should be demonstrable in the arthritic joints of patients with spondylarthropathies. Such responses have proved remarkably difficult to detect, in contrast to CD4$^+$ HLA class II-restricted cells (Viner et al., 1991; Hassell et al., 1992). However, *Yersinia* and *Salmonella*-specific clones have been grown from two patients with ReA (Hermann et al., 1993). Recently, several patients with ReA have been found to make CTL responses to a *Yersinia* heat shock protein 60 peptide (Ugrinovic et al., 1997), and a *Chlamydia* heat shock protein 70 peptide-specific T cell line from another ReA patient has been identified (P. Bowness, R.L. Allen and A.J. McMichael, unpublished observations). An autoreactive HLA-B27-restricted type II collagen-specific T cell line has also been isolated from the blood of a patient with ReA (Gao et al., 1994). Although nothing is known of the function of such T cells, a pathogenic role would be suggested if they could be shown to accumulate selectively in affected joints and disappear during disease remission. More direct evidence could be obtained either by direct immunological intervention in patients, e.g. with monoclonal antibodies directed against the T cell receptor, or by the use of animal models.

EPIDEMIOLOGY OF HLA-B27-ASSOCIATED DISEASE

At least 12 molecular subtypes of HLA-B27 have now been identified. Since these subtypes

TABLE 12.3 HLA-B27 subtypes associated with spondylarthropathy

	Reference
Disease-associated subtypes	
HLA-B*2702, HLA-B*2704, HLA-B*2705	Breur-Vriesendorp (1987); Brown et al. (1996)
Not associated	
HLA-B*2706, HLA-B*2709	Lopez-Larrea et al. (1995); D'Amato et al. (1995)
Uncertain association	
HLA-B*2703	Hill et al. (1991)

differ in their fine specificity for peptide, arthritogenic peptide models of disease would predict a differential association with disease. A key question is thus whether all subtypes are indeed disease associated. Recent studies using molecular typing methods confirm the disease association of HLA-B*2702, 04 and 05 (see Table 12.3).

Small, controlled studies suggest that the rare subtypes HLA-B*2706 and HLA-B*2709 are not associated with disease (D'Amato et al., 1995; Lopez-Larrea et al., 1995). It is also possible that HLA-B*2703, found in individuals of West African descent, may not be associated with spondylarthropathy (Hill et al., 1991), although at least one African–American patient with HLA-B*2703 has been identified. Larger confirmatory epidemiological studies are required and may also determine whether the common Caucasian subtypes, HLA-B*2705 and *02, are differentially associated with disease. Combining this preliminary epidemiological evidence with knowledge of peptide-binding differences, one might envisage a candidate arthritogenic peptide containing an arginine at P2 and a hydrophobic or an aromatic C-terminal residue, but lacking a positively charged N-terminal amino acid.

CONCLUSIONS

Although the role of HLA-B27 in the pathogenesis of the spondylarthropathies remains unclear, transgenic studies confirm that it has a key role. A great deal is now known about the molecular function of HLA-B27, with most available evidence consistent with a major role in peptide selection. In future, careful epidemiological studies should be able to delineate unequivocally subtype associations. Together with functional studies, this may narrow the search for arthritogenic peptides. It is likely that other features of the biology of HLA-B27 as an antigen-presenting molecule also contribute to disease pathogenesis. Studies exploring the possible role of free B27 heavy chains, the ability to bind long or extracellular peptides and the role of the free cysteine at position 67 are important in this respect.

REFERENCES

Allen, R.L., O'Callaghan, C.A., McMichael, A.J. and Bowness, P. (1999). HLA-B27 can form a novel $\beta(2)$-microglobulin-free heavy chain homodimer structure. *J. Immunol.* **162**: 5045–5048.

Archer, J.R., Whelan, M.A., Badakere, S.S. et al. (1990). Effect of a free sulphydryl group on expression of HLA-B27 specificity. *Scand. J. Rheumatol.* **87** Suppl.: 44–50.

Benjamin, R. and Parham, P. (1990). Guilt by association: HLA B27 and ankylosing spondylitis. *Immunol. Today* **11**: 137–142.

Benjamin, R.J., Madrigal, J.A. and Parham, P. (1991). Peptide binding to empty HLA-B27 molecules of viable human cells. *Nature* **351**: 74–77.

Bergfeldt, L. (1997). HLA-B27-associated cardiac disease. *Ann. Intern. Med.* **127**: 621–629

Bowness, P., Allen, R.L. and McMichael, A.J. (1994). Identification of T cell receptor recognition residues for a viral peptide presented by HLA B27. *Eur. J. Immunol.* **24**: 2357–2363.

Breban, M., Hammer, R.E., Ricardson, J.A. and Taurog, J.D. (1993). Transfer of the inflammatory disease of HLA-B27 transgenic rats by bone marrow engraftment. *J. Exp. Med.* **178**: 1607–1616.

Breban, M., Fernandez, S.J. Richardson, J.A. et al. (1996). T cells, but not thymic exposure to HLA-B27, are required for the inflammatory disease of HLA-B27 transgenic rats. *J. Immunol.* **156**: 794–803.

Breur-Vriesendorp, S., Dekker-Says, A. and Ivanyi, P. (1987). Distribution of HLA-B27 subtypes in patients with ankylosing spondylitis: the disease is associated with a common determinant of the various B27 molecules. *Ann. Rheum. Dis.* **46**: 353–356.

Brewerton, D.A., Caffrey, M., Hart, F.D. et al. (1973). Ankylosing spondylitis and HL-A27. *Lancet* **i**: 904–907.

Brewerton, D.A., Caffrey, M., Hart, F.D. et al. (1974). Reiters disease and HL-A27. *Lancet* **ii**: 996–998.

Brown, M.A., Pile, K.D., Kennedy, L.G. et al. (1996). HLA class I associations of ankylosing spondylitis in the white population in the United Kingdom. *Ann. Rheum. Dis.* **55**: 268–270.

Calin, A. and Taurog, J.D. eds (1998). *The Spondyl Arthritides*, Oxford University Press, Oxford.

Calin, A., Marder, A. and Becks, E. (1983). Genetic differences beweem B27 positive patients with ankylosing spondylitis and B27 positive healthy controls. *Arthritis Rheum.* **26**: 1460–1464.

Careless, D.J., Chiu, B., Rabinovitch, T. et al. (1997). Immunogenetic and microbial factors in acute anterior uveitis. *J. Rheumatol.* **24**: 102–108.

Chicz, R.M., Urban, R.G., Gorga, J.C. et al. (1993). Specificity and promiscuity among naturally processed peptides bound to HLA-DR alleles. *J. Exp. Med.* **178**: 27–47.

Colbert, R.A., Rowland-Jones, S.L., McMichael, A.J. and Frelinger, J.A. (1994). Differences in peptide presentation between B27 subtypes: the importance of the P1 side chain in maintaining high affinity peptide binding to B*2703. *Immunity* **1**: 121–130.

D'Amato, M., Fiorillo, M.T., Carcassi, C. et al. (1995). Relevance of residue 116 of HLA-B27 in determining susceptibility to ankylosing spondylitis. *Eur. J. Immunol.* **25**: 3199–3201.

Davenport, M. (1995). The promiscuous B27 hypothesis [Letter]. *Lancet* **346**: 500–501.

Ebringer, A., Cowling, P., Ngwa-Sah, N. et al. (1976). Cross reactivity between *Klebsiella aerogenes* species and B27 lymphocyte antigens. In *HLA and Disease* (eds J. Dausset and A. Svejgaard), p. 27, INSERM, Paris.

Gao, X.M., Wordsworth, P. and McMichael, A. (1994). Collagen-specific cytotoxic T lymphocyte responses in patients with ankylosing spondylitis and reactive arthritis. *Eur. J. Immunol.* **24**: 1665–1670.

Gao, X.M., Wordsworth, P., McMichael, A.J. et al. (1996). Homocysteine modification of HLA antigens and its immunological consequences. *Eur. J. Immunol.* **26**: 1443–1450.

Garcia, F., Marina, A. and Lopez, d.C.J. (1997). Lack of carboxyl-terminal tyrosine distinguishes the B*2706-bound peptide repertoire from those of B*2704 and other HLA-B27 subtypes associated with ankylosing spondylitis. *Tissue Antigens* **49**: 215–221.

Hammer, R.E., Maika, S.D., Richardson, J.A. et al. (1990). Spontaneous inflammatory disease in transgenic rats expressing HLA-B27 and human β2m: an animal model of HLA-B27-associated human disorders. *Cell* **63**: 1099–1112.

Hassell, A.B., Pilling, D., Reynolds, D. et al. (1992). MHC restriction of synovial fluid lymphocyte responses to the triggering organism in reactive arthritis: absence of a class I-restricted response. *Clin. Exp. Immunol.* **88**: 442–447.

Hermann, E., Yu, D.T., Meyer, z.B.K. and Fleischer, B. (1993). HLA-B27-restricted CD8 T cells derived from synovial fluids of patients with reactive arthritis and ankylosing spondylitis. *Lancet* **342**: 646–650.

Hill, A.V.S., Allsopp, C.E.M., Kwiatowski, D. et al. (1991). HLA class I typing by PCR: HLA-B27 and an African subtype. *Lancet* **337**: 640–642.

Jardetzky, T.S., Lane, W.S., Robinson, R.A. et al. (1991). Identification of self peptides bound to purified HLA-B27. *Nature* **353**: 326–329.

Kapasi, K. and Inman, R.D. (1994). ME1 epitope of HLA-B27 confers class I-mediated modulation of Gram-negative bacterial invasion. *J. Immunol.* **153**: 833–840.

Khare, S.D., Bull, M.J., Hanson, J. et al. (1998). Spontaneous inflammatory disease in HLA-B27 transgenic mice is independent of MHC class II molecules: a direct role for B27 heavy chains and not B27-derived peptides. *J. Immunol.* **160**: 101–106.

Lahesma, R., Skurnik, M., Granfors, K. et al. (1992). Molecular mimicry in the pathogenesis of spondyloarthropathies. A critical appraisal of croos-reactivity between microbial antigens and HLA B27. *Br. J. Rheumatol.* **31**: 221–229.

Laitinen, O., Leirisalo, M. and Skylv, G. (1977). Relation between HLA-B27 and clinical features in patients with *Yersinia* arthritis. *Arthritis Rheum.* **20**: 1121–1124.

Lanier, L.L. (1998). Follow the leader: NK cell receptors for classical and nonclassical MHC class I. *Cell* **92**: 705–707.

Lopez-Larrea, C., Sujirachato, K., Mehr, N. et al. (1995). HLA-B27 subtypes in Asian patients with ankylosing spondylitis. Evidence for new associations. *Tissue Antigens* **45**: 169–176.

Madden, D.R., Gorga, J.C., Strominger, J.L. and Wiley, D.C. (1991). The structure of HLA B27 reveals nonamer self-peptides bound in an extended conformation. *Nature* **353**: 321–325.

Marker, H.E., Meyer, z.B.K. and Wildner, G. (1997). HLA-B27-derived peptides as autoantigens for T lymphocytes in ankylosing spondylitis. *Arthritis Rheum.* **40**: 2047–2054.

Marrack, P. and Kappler, J. (1990). The staphylococcal enterotoxins and their relatives. *Science* **248**: 705–711.

Peh, C.A., Burrows, S.R., Barnden, M. et al. (1998). HLA-B27-restricted antigen presentation in the absence of tapasin reveals polymorphism in mechanisms of HLA class I peptide loading. *Immunity* **8**: 531–542.

Rotzschke, O., Falk, F., Stevanovic, S. et al. (1994). Dominant aromatic/aliphatic C-terminal anchor in HLA-B*2702 and B*2705 peptide motifs. *Immunogenetics* **39**: 74–77.

Schlosstein, L., Terasaki, P.I., Bluestone, J. et al. (1973). High association of an HL-A antigen, w27, with ankylosing spondylitis. *N. Engl. J. Med.* **288**: 704–706.

Scofield, R., Kuien, B., Gross, T. et al. (1995). HLA-B27 binding of peptide from its own sequence and similar peptides from bacteria: implications for spondylarthropathies. *Lancet* **345**: 1542–1544.

Stam, N.J., Spits, H. and Ploegh, H.L. (1986). Monoclonal antibodies raised against denatured HLA-B locus heavy chains permit biochemical characterization of certain HLA-C locus products. *J. Immunol.* **137**: 2299–2306.

Tanigaki, N., Fruci, D., Vigneti, E. et al. (1994). The peptide binding specificity of HLA-B27 subtypes. *Immunogenetics* **40**: 192–198.

Taurog, J.D., Richardson, J.A., Croft, J.T. et al. (1994). The germfree state prevents development of gut and joint inflammatory disease in HLA-B27 transgenic rats. *J. Exp. Med.* **180**: 2359–2364.

Tsuchiya, N., Husby, G., Williams, R.J. et al. (1990). Autoantibodies to the HLA-B27 sequence cross-react with the hypothetical peptide from the arthritis-associated Shigella plasmid. *J. Clin. Invest.* **86**: 1193–1203.

Ugrinovic, S., Mertz, A., Wu, P. et al. (1997). A single nonamer from the *Yersinia* 60-kDa heat shock protein is the target of HLA-B27-restricted CTL response in *Yersinia*-induced reactive arthritis. *J. Immunol.* **159**: 5715–5723.

Urban, R.G., Chicz, R.M., Lane, W.S. et al. (1994). A subset of HLA-B27 molecules contains peptides much longer than nonamers. *Proc. Natl Acad. Sci. USA* **91**: 1534–1538.

Viner, N.J., Bailey, L.C., Life, P.F. et al. (1991). Isolation of *Yersinia* specific T cell clones from the synovial membrane and synovial fluid of a patient with reactive arthritis. *Arthritis Rheum.* **34**: 1151–1157.

Weinreich, S., Euldrink, F., Capkova, J. et al. (1995). HLA-B27 as a relative risk factor in ankylosing enthesopathy in transgenic mice. *Hum. Immunol.* **42**: 103–115.

Yaounq, J., Perichon, M., Chorney, M. et al. (1994). Anonymous marker loci within 400 kb of HLA-A generate haplotypes in linkage disequilibrium with the hemochromatosis gene (HFE). *Am. J. Hum. Genet.* **54**: 252–263.

Zhou, M., Sayad, A., Simmons, W.A. et al. (1998). The specificity of peptides bound to human leukocyte antigen (HLA)-B27 influences the prevalance of arthritis in HLA-B27 transgenic rats. *J. Exp. Med.* **188**: 877–886.

CHAPTER 13

HLA and Renal Disease

Richard G. Phelps and Andrew J. Rees

INTRODUCTION

There has been considerable interest in the role of the immune system in renal disease, both for its role in initiating injury and because of the possibility that it perpetuates injury originally caused by other events. This has resulted in the search for associations between human leucocyte antigen (HLA) alleles and a wide range of renal diseases, but it is only in glomerulonephritis that there is a convincing rationale for such studies.

Glomerulonephritis is the generic name given to a group of immunologically mediated renal diseases, often associated with the deposition of immunoglobulins in glomeruli. The diseases differ in pathogenesis, morphology and prognosis, and are usually categorized into two broad groups: proliferative and non-proliferative (Table 13.1). The pathogenesis of most of these disorders is unknown and it is probable that there is heterogeneity even within individual clinicopathological entities. Nevertheless, analysis of the frequencies of HLA class I and II alleles has identified several striking associations with certain types of glomerulonephritis, including three which are amongst the strongest reported for diseases affecting any organ. The association with Goodpasture's disease is of particular interest because it is caused by autoimmunity to a well-characterized antigen. This has permitted analysis of the influence of the Goodpasture's disease-associated class II molecules on the processing of the Goodpasture autoantigen and its presentation to T cells, thus presenting a unique opportunity to study the mechanism by which inheritance of HLA alleles influences susceptibility to autoimmune disease.

This chapter will review the evidence that inheritance of HLA alleles influences the susceptibility to glomerulonephritis. The most recent data on the role of HLA molecules in Goodpasture's disease will be discussed first because the evidence is clearest, and because they provide a paradigm for how the HLA molecules might determine susceptibility. The chapter will then review in detail the data available for three other

TABLE 13.1 Histopathological classification of glomerulonephritides

	Histopathological description	Associated disease discussed in this chapter
Non-proliferative	Minimal change	Steroid-sensitive nephrotic syndrome
	Focal (and segmental) glomerulosclerosis (FSGS)	Steroid-resistant nephrotic syndrome
	Membranous	Idiopathic membranous nephropathy
Proliferative	Diffuse proliferative	Post-streptococcal glomerulonephritis
	Mesangial proliferative	IgA nephropathy
	Focal proliferative	
	Focal necrotizing/crescentic	Goodpasture's disease; ANCA-associated vasculitis
	Mesangiocapillary	

types of nephritis which have clearly been associated with HLA alleles. Finally, the other, more tentative HLA association with various types of nephritis will be mentioned. The role of the HLA complex in lupus nephritis will not be discussed because it is described together with other aspects of that disease in Chapter 21.

GOODPASTURE'S DISEASE

Goodpasture's disease is a cause of rapidly progressive glomerulonephritis often associated with lung haemorrhage. The disease is defined serologically by the presence of circulating antibodies specific for glomerular basement membrane (GBM), and immunopathologically by a linear deposition of immunoglobulin (usually IgG) along the GBM. Anti-GBM antibodies have a remarkably uniform specificity in Goodpasture's disease. The autoantibodies from all patients bind to the 230 amino acid carboxyl-terminal NC1 domain of the α3-chain of type IV collagen [α3(IV)NC1] (Saus, 1988; Turner, 1992). Furthermore, antibodies from most patients cross-inhibit each other, and their binding can be inhibited by a murine monoclonal antibody raised against α3(IV)NC1, suggesting that they recognize a single or very limited range of target epitopes (Levy et al., 1996). Anti-GBM antibodies do not occur in health and there is compelling evidence that they are pathogenic: for example, their titre is closely related to disease activity; the recurrence of Goodpasture's disease in renal transplants is closely related to the level of anti-GBM antibodies in the circulation and deposited in the graft; and, most importantly, antibodies eluted from the kidneys of Goodpasture patients rapidly cause nephritis when injected into monkeys (Lerner et al., 1967).

Goodpasture's disease is ideal for investigating the link between HLA alleles and autoimmune disease for several interrelated reasons: (i) it is the only nephritis in which the target of autoimmune attack has been defined; (ii) the disease is defined precisely by its immunopathogenesis, namely autoimmunity to the GBM; (iii) there is a much greater uniformity of the autoantibody response than is found in patients with other autoimmune diseases; (iv) the target antigen is a small relatively easily manipulated protein for which cDNA has been cloned and recombinant expression established; and, most importantly for this chapter, (v) inheritance of different HLA class II alleles appears to exert contrasting influences on susceptibility to the disease, as both strong positive and negative associations with disease are reported. It should also be mentioned that models of antibody-mediated nephritis in rats (Druet et al., 1972) and mice (Kalluri et al., 1997) are linked to the major histocompatibility complex (MHC).

Occurrence of HLA alleles in patients with Goodpasture's disease

The first indication of an association between histocompatibility locus antigens and susceptibility to Goodpasture's disease was reported in 1978 when Rees et al. (1978) described a high frequency of the HLA-DR2 specificity in small series of patients. A subsequent study extended the series to 38 patients, 34 of whom (88%) had inherited DR2 compared with 19% of 153 controls. There was also a weaker association with HLA-B7, but no stronger than expected from the known linkage disequilibrium between DR2 and B7 (Rees et al., 1984): confirmation of these associations came from small studies from Australia (Perl et al., 1981) and the USA (Garovoy, 1982).

At least 11 DRB1 alleles encode the β-chains of DR molecules carrying the DR2 specificity, namely DRB1*1501–1505 and DRB1*1601–1606. DRB1*1501 is overwhelmingly the most common in north European Caucasoids, accounting for 90% of these alleles. All of the alleles conferring the DR2 specificity occur on haplotypes with the functional DRB5 gene in very close linkage disequilibrium; so, for practical purposes, DRB1*1501 is always inherited with DRB5*0101, DRB1*1502 with DRB5*0102, and DRB1*1601 with DRB5*0201. DRB1*1501 is also in strong linkage disequilibrium with DQA1*0102 and DQB1*0602, which encode the α- and β-chains, respectively, of the DQ molecule with specificity DQ6, which is a subtype of DQ1. DRB1*1502 and the DR16 alleles are in linkage disequilibrium with different DQ alleles; the more common haplotypes are shown in Tables 13.2 and 13.3. Consequently, the association of the DR2 specificity with Goodpasture's disease could be explained by primary associations with any of these DRB1, DRB5, DQA or DQB alleles,

TABLE 13.2 Selected serologically defined HLA-DR,DQ haplotype frequencies

Haplotype	Haplotype frequency				
	Caucasoid			Non-Caucasoid	
	British	French	German	Japanese	Chinese Han
DR15,DQ1	10.1	11.2	8.8	17	17.2
DR16,DQ1		2.1	2.1	0.8	2.4
DR3,DQ2	12.4	10.4	9.4		2.3
DR4,DQ1			0.8		3.0
DR4,DQ3	6.4	5.8	6.9	5.8	
DR4,DQ4			1.0	12.1	2.0
DR4,DQ7	4.5	3.7	4.5	1.9	3.5
DR7,DQ2	12.4	10.4	9	0.4	10.1
DR7,DQ3	2.3	2.9	2		2.6
DR7,DQ7					2.3
DR8,DQ1				8.3	
DR8,DQ3				1.9	
DR8,DQ4	3	4	3.1	2.4	
DR9,DQ3		1.1	1	10.3	3.8

Data from the 11th International Histocompatibility Workshop.

or even with alleles at other more distant but still linked loci. These issues have now been partially resolved by analysis of the frequencies of alleles at DR and DQ loci in Goodpasture patients using sequence-specific oligonucleotides.

Association with DR alleles

Studies from the USA (Huey et al., 1993), Australia (Dunckley et al., 1991), France (Mercier et al., 1992) and the UK (Burns et al., 1995; Fisher

TABLE 13.3 Selected genetically defined HLA-DRB1,DQB1 haplotype frequencies

Haplotype			Haplotype frequency			
DRB1	DQB1		Caucasoid			Non-Caucasoid
			USA	French	German	Japanese
0101	0501		5	9.3	6.3	5.1
1501	0602	DRB5*0101	5.2	7.2	7.4	4.9
1502	0601	DRB5*0102	0.7[f]	0.3[f]	1.1[f]	9.6
160x[a]	0502	DRB5*020x[a]	1.1[f]	2.7		0.7
0301	0201		5.3	11	7.7	0.7
040x[b]	0302		2.5	7.2	6.3	5.7
0401	0301[c]		2.7	2.1	2.3	1.6
0405	0401[d]					11.9
0701	0201		10.4	11.2	8.5	0.7
0803	0601[e]					8.8
0801	0402			2.7	2.9	
0901	Various					9.7
1101	0301		3.6	8.6	9.5	2.3

Data from the 11th International Histocompatibility Workshop.
[a]$x = 1$ in Caucasoids and 2 in Japanese.
[b]$x = 1-4$ in Caucasoids and 3 or 6 in Japanese, associated with DQ8 and DQ3, respectively.
[c]DQ7 and DQ3 serologically.
[d]DQ4 serologically.
[e]DQ6 and DQ1 serologically.
[f]Frequency of the DRB1 allele is shown since haplotype data are not available and linkage disequilibrium is known to be very strong.

et al., 1997) have shown that almost all patients with Goodpasture's disease inherit a haplotype defined by the DRB1*1501, DRB5*0101 and DQB1*0602 alleles (Table 13.4). HLA-DRB1*1502 also occurred more frequently than expected, suggesting that it too confers susceptibility to the disease, but because DRB1*1502 is rare in Caucasoid populations this result was not statistically significant. It is notable that DRB1*16 alleles were not responsible for the DR2 specificity in any of the 153 patients reported in these studies, even though it would have been expected to occur in 3.5 patients, based on the frequencies in the control samples if randomly associated (Table 13.5).

The strength of the association between DR15 and Goodpasture's disease means that special statistical techniques are needed to analyse the other associations that have been suggested by various studies which, although weaker, are potentially important, e.g. the positive association with HLA-DR4 alleles and the negative associations with DR1 and DR7 (Table 13.5). In order to do this, Fisher et al. (1997) applied the technique of relative predispositional analysis (Payami et al., 1989) to a cohort of 82 British patients and confirmed that both the positive association with DR4 and the negative association with DR7 were highly significant (DR4, $p = 0.019$; DR7, $p = 0.02$): the number of patients studied was too small to judge whether the negative association with DR1 was significant. Similarly, the number of DR4 alleles was insufficient to identify significant associations with any single DRB1*04 allele.

To determine whether any other class II alleles have significant associations with Goodpasture's disease relative predispositional analysis was also applied to genotype data for 139 patients, assembled from the three largest reports on HLA in Goodpasture's disease (Dunckley et al., 1991; Huey et al., 1993; Fisher et al., 1997). This analysis confirmed the associations with DR15, DR4 and DR7 (Table 13.6) and suggested an additional weak but significant positive association with DR3. An association with DR3 had previously been suggested by one of the individual studies (Huey et al., 1993). Analysis of DR4 subtypes found that the most frequent DR4 allele in patients was DRB1*0401 (46% of DR4 alleles), but the only allele that occurred significantly more frequently in patients was DRB1*0404 (Table 13.7). The proposed negative association with DR1 was still not statistically significant when the combined data were analysed in this way. However, interpretation of the meta-analysis must be guarded as the control population data were necessarily even less satisfactory than usual. Nevertheless, taken together the data support ranking class II alleles as shown in Table 13.6. This ranking is confirmed and extended in a recent formal meta-analysis of genotype data (Phelps, 1999).

TABLE 13.4 Frequency of DRB1*1501/2 in Caucasoids with Goodpasture's disease

Reference	Country	Patients % (n)	Controls % (n)	OR	EF	p_c
Fisher et al. (1997)	UK	79 (82)	28 (177)	8.3	0.69	***
Huey et al. (1993)	USA	91 (23)	19.6[a]	41	0.89	***
Mercier et al. (1992)	France	92 (12)	21.2[a]	42	0.9	***
Dunckley et al. (1991)	Australia	63.9 (36)	26.8 (202)	4.8	0.51	***

[a]Frequency calculated from published allele frequency in USA population. ***$p_c < 0.0001$.
p_c: probability corrected for number of comparisons made.

TABLE 13.5 Summary of reported HLA-DRB1 associations with Goodpasture's disease

			DRB1 allele frequencies (phenotype)				
Reference	Country	Patients (n)	*15	*04	*16	*01	*07
Fisher et al. (1997)	UK	82	0.79[a]	0.43	0	0.1[c]	0.05[b]
Huey et al. (1993)	USA	23	0.91[a]	0.35	0	0.13	0.04
Mercier et al. (1992)	France	12	0.92[a]	0.58	0	0	0
Dunckley et al. (1991)	Australia	36	0.64[a]	0.58	0	0.06	0.03[d]

[a]$p < 0.0001$; [b]$p = 0.0002$; [c]$p = 0.0005$; [d]$p = 0.05$.

TABLE 13.6 Ranking of DRB1 alleles by association with Goodpasture's disease

HLA-DRB1 allele	Occurence of alleles		Relative predisposition		Strength of association
	Patients % (n)	Controls (%)	Denominator	p	
DRB1*1501 or 1502	43 (120)	16.3	278	7.6×10^{-34}	+++++
DRB1*04	44 (70)	26.6	158[a]	4.7×10^{-7}	+++
DRB1*0301	42 (37)	21.0	88[b]	1.2×10^{-6}	++
DRB1*1201	5.7 (5)	2.3	88	0.03	+
DRB1*08, 09, 11,13					Neutral
DRB1*16	0 (0)	1.2	88	0.3	−
DRB1*0101	13.7 (12)	21.0	88	0.09	−
DRB1*0701	8 (7)	22.0	88	0.0015	−

Genotype data for 139 patients (278 alleles) analysed by relative predisposition as described by Payami et al. (1989).
[a]Analysed after removing 120 DR15 alleles; [b]analysed after removing 120 DR15 and 70 DR4 alleles.

Association with DQ alleles

The DQB1*0602 allele is strongly associated with Goodpasture's disease and confers an odds ratio (OR) of 8.6, which slightly exceeds that conferred by inheritance of HLA-DRB1*1501 (OR 7.4), with which it is in linkage disequilibrium (Fisher et al., 1997). However, close inspection of the frequencies of other DRB1 and DQ alleles in Goodpasture's disease strongly suggests that the primary association is with DR alleles. Thus, DQB1*02 is commonly inherited with either DR7 or DR3, the former of which is negatively associated with the disease, whereas the latter is either neutral or positively associated. When the frequencies of DQ alleles remaining after DQB1*0602 alleles had

TABLE 13.7 Frequency of DR4 alleles in patients with Goodpasture's disease

DRB1*04	% (n[a])	Expected %
0401	9 (18)	10.7
0402	0 (0)	0.6
0403	2.5 (5)	2.04
0404	5.5 (11)	2.96
0405	0.5 (1)	0.56
0406	0.5 (1)	0.00
0407	1.5 (3)	3.04
0408	0 (0)	0.48
0409	0 (0)	0.00
0410	0 (0)	0.00
0411	0 (0)	0.00

Frequencies of DRB1*04 subtypes identified for 100 patients in two studies (Huey et al., 1993; Fisher et al., 1997) are compared with the frequencies of similar Caucasoid populations reported in the 11th HLA Workshop. Only the increased frequency of 0404 was statistically significant ($p = 0.03$), and then only before correction for the number of comparisons made.
[a]Out of 200 alleles.

been removed were analysed, only the negative association with DQB1*0501 remained significant, and this is attributable to that allele's linkage disequilibrium with DRB1*01 alleles. Therefore, the hierarchy of disease association observed for DRB1 alleles was not present for DQ alleles.

In summary, analysis of the frequency of HLA class II alleles in Goodpasture patients shows, first, that inheritance of certain alleles predisposes whereas other protect from the disease and, secondly, that alleles at the DRB1 locus rather than the DQ loci are responsible.

Relationship between the structure and function of class II molecules and their association with Goodpasture's disease

If the strong associations between HLA class II alleles and Goodpasture's disease are caused by differences in the properties of the encoded class II molecules then it would be expected that disease-associated and disease-protective class II molecules should have contrasting properties. Polymorphism of class II alleles largely influences residues flanking the class II peptide-binding groove, thereby determining peptide-binding preferences. It is therefore rational to compare the peptide-binding preferences of class II molecules with divergent influences on disease susceptibility. This can be done, to a degree, simply by comparing their gene sequences and predicted structure, as the relationship between structure and function is now fairly well understood, but direct analysis of function requires experimental determinations of the affinity of class II molecules for representative peptides.

Structure of disease-associated and disease-protective class II molecules

The first indication that class II alleles associated with Goodpasture's disease had significant structural similarity was the identification of a common motif (RFLDRYF) in parts of the sequences of the DR β-chains encoded by DRB1*1501 and DR4 alleles which were predicted to shape the class II peptide-binding groove (Burns et al., 1995). By contrast, the corresponding sequences of DR1 and DR7 were very different, suggesting that DR molecules encoded by disease-associated alleles might have similar peptide-binding specificities which were different from disease-protective class II molecules. The motif was subsequently shown to be present in 75/82 (91%) British patients but only in 62% of controls ($p < 0.0001$) (Fisher et al., 1997). Solutions of crystal structures of DR–peptide complexes have confirmed that this motif impinges on the peptide-binding groove (Stern et al., 1994), and binding studies have shown that at least one of the amino acids specified by the motif influences peptide binding to class II molecules (Krieger et al., 1991).

The accumulation of structural and functional data has led to a good understanding of the structural basis for class II–peptide interactions (reviewed in Madden, 1995), permitting rational comparison of the sequences of disease-associated and disease-protective alleles. Peptide-binding specificity is largely determined by the chemical character of "pockets" in the class II peptide-binding groove which accommodate side chains 1, 4, 6, 7 and 9 of the bound peptide, and which are referred to as pockets 1, 4, 6, 7 and 9, respectively, after the system proposed by Stern et al. (1994). A comparison of the amino acid residues flanking these pockets encoded by HLA-DRB alleles positively and negatively associated with Goodpasture's disease is shown in Table 13.8. Peptide side-chain/class II interactions at pockets 1 and 4 are particularly important for stable binding. Four DR β-chain residues flank pocket 1, of which two (β85 and β86) are polymorphic, but the amino acids encoded by disease-associated and -protective DRB1 alleles at either position do not correlate with disease susceptibility. Notably, β86 is the

TABLE 13.8 Residues flanking pockets in HLA-DR molecules encoded by DRB1 alleles associated with Goodpasture's disease

HLA class II allele	Residues (subdivided by pockets) in peptide-binding groove																
	1				4					6		7					9
	85	86	89	90	13	70	71	74	78	11	13	28	47	61	67	71	9	57
DRB1*1501	V	V	F	T	R	Q	A	A	Y	P	R	D	F	W	I	A	W	D
DRB1*0404	V	V	F	T	H	Q	R	A	Y	V	H	D	Y	W	L	R	E	D
DRB1*040x[a]	V	V,G	F	T	H	Q	R,K,E	A,E	Y	V	H	D	Y	W	I,L	R,K,E	E	D,S
DRB1*0301/2	V	V,G	F	T	S	Q	K	R	Y	S	S	D,E	F,Y	W	L	K	E	D
DRB1*1201	A	V	F	T	G	D	R	A	Y	S	G	E	F	W	I	R	E	V
DRB1*08	V	V,G	F	T	G	D	R	L,A	Y	S	G	D	Y	W	I,F	R	E	S,D,V,A
DRB1*09	V	G	F	T	F	R	R	E	V	D	F	H	Y	W	F	R	K	V
DRB1*13	V	V,G	F	T	S,G	D	E,K,RA	A,L	Y	S	S,G	D,E	F,Y	W	I,F,L	E,K,RA	E	D,S
DRB1*11	V,A	V,G	F	T	S,G,H	D,R	R,E,K	A,E	Y	S,V	S,H,G	D	F,Y	W	F,I,L	R,E,K	E	D
DRB1*0101	V	G	F	T	F	Q	R	A	Y	L	F	E	Y	W	L	R	W	D
DRB1*16	V	G	F	T	R	D	R	A,L	Y	P	R	D	Y	W	F	R	W	D
DRB1*0701	V	G	F	T	Y	D	R	Q	V	G	Y	E	Y	W	I	R	W	V
DRB4*0101	V	V	F	T	C	R	R	E	Y	A	C	I	Y	W	L	R	E	D
DRB5*0101	V	G	F	T	Y	D	R	A	Y	D	Y	H	Y	W	F	R	Q	D

Pocket numbering (bold numbers) and flanking residues (non-bold numbers) are taken from Stern et al. (1994). DRB1 alleles are shown ranked by influence on susceptibility to Goodpasture's disease as in Table 13.6. Alleles other than DR15, DR4, DR1 and DR7 are shaded as they have neutral or less firmly established weak influences on disease susceptibility. At the bottom of the table are shown two other alleles which may be relevant: DRB5*0101 is expressed on DR15 and DR16-bearing haplotypes and DRB4*0101 on DR7-bearing haplotypes.
[a]DRB1*04 alleles 0401–0411 except for DRB1*0404.

only position at which DRB1*1502 differs from DRB1*1501 and, as discussed earlier, it is probable that both alleles confer susceptibility to Goodpasture's disease. In contrast, there are striking differences between DRB1 alleles at residues flanking pocket 4. Disease-associated alleles encode positively charged residues at β13 (R or H) and a positively charged glutamine residue at β70, whereas disease-protective alleles encode hydrophobic residues at β13 (Y in DR7, F in DR1) and a negatively charged aspartic acid at β70. In addition, disease-associated alleles encode small residues (A) at β74 (and β71 in DRB1*1501), whereas DR7 has a bulky glutamine residue at β74. Residues flanking pocket 4 in DR16 alleles more closely resemble DR7 than DR15, consistent with DR16 having a negative influence on disease susceptibility. Of the secondary anchors, the combination of residues flanking pocket 6 in DRB1*1501 is also found in other DR15 and DR16 alleles, but in no other DRB1 allele, and at pockets 7 and 9 disease-associated and disease-protective DRB1 alleles differ only by conservative amino acid substitutions.

Thus, disease-associated alleles have structural similarities at a pocket important for peptide binding which are very different in disease-protective alleles. This correlation between structure and disease association is not seen if DRB5*0101 is considered in place of DRB1*1501 because the structure of DRB5*0101 is more similar to DR7 and DR1 than to DR4 alleles at primary anchor positions. Thus, consideration of structure strongly suggests that inheritance of DRB1*1501 and not the closely linked allele DRB5*0101 confers susceptibility to Goodpasture's disease. The observation that the most striking difference between disease-associated and protective class II molecules is at pocket 4 is reminiscent of the findings in rheumatoid arthritis and pemphigus vulgaris. In both of these cases the major difference relates to the charge of pocket 4, whereas here it relates to charge and size (Hammer et al., 1995; Wucherpfennig et al., 1995).

Functional differences between disease-associated and disease-protective class II molecules deduced from their structure

The differences in the predicted structures of disease-associated and disease-protective class II molecules suggest they should have very different affinities for some peptides, particularly peptides for which stable binding requires large side-chain residues to be accommodated in pocket 4. Such differences are interesting because they can be used to search the sequence of autoantigens for peptides with divergent predicted binding to disease-associated and disease-protective class II molecules which may be relevant in immunopathogenesis (Hammer et al., 1995). To identify peptides in α3(IV)NC1 likely to bind much better to the class II molecules comprising the gene products of DRA and DRB1*1501 (referred to throughout this chapter as DR15, although it represents only one form of DR15) than to DR7, the motif LIMVF,X,X,WYF was deduced (i.e. a four-residue motif in which the first residue is one of the residues L, I, M, V or F and the fourth one of W, Y or F; X represents any amino acid; derivation is discussed in the legend to Figure 13.1). The motif occurs seven times in the sequence of α3(IV)NC1, shown in Figure 13.1.

Functional differences between disease-associated and disease-protective class II molecules suggested by measurements of affinity for Goodpasture antigen peptides

Measurements of the affinity of α3(IV)NC1 peptides to DR15 confirm that the seven motif-containing peptides bind with intermediate to high affinity. However, peptides representative of all but two (peptides P3b, containing the motif VPLY, and P11, containing LEPY) of the seven bound equally or better to DR1 and DR7 molecules (Phelps et al., 1996a), possibly because the peptides examined were long enough to contain several alternative core binding sequences. Thus, analysis of structure and binding data show striking differences in the peptide-binding characteristics of disease-associated and protective class II molecules, and identify several peptides with divergent affinities which might be relevant in immunopathogenesis.

However, deducing autoantigen-contained peptides that might be relevant in immunopathogenesis from differences in the peptide-binding preferences of disease-associated class II molecules makes two major assumptions, first, that affinity for class II molecules relates to presentation to T cells. The validity of this assumption probably depends on the relative importance of class II peptide-binding preferences and antigen-

```
SPATWTTRGFVFTRHSQTTAIPSCPEGTVPLYSGFSFLFVQGNQRAHGQD

LGTLGSCLQRFTTMPFLFCNVNDVCNFASRNDYSYWLSTPALMPMNMAPI

TGRALEPYISRCTVCEGPAIAIAVHSQTTDIPPCPHGWISLWKGFSFIMF

TSAGSEGTGQALASPGSCLEEFRASPFLECHGRGTCNYYSNSYSFWLASL

NPERMFRKPIPSTVKAGELEKIISRCQVCMKKRH
```

FIGURE 13.1 Occurrence of a putative disease-associated motif within the sequence of α3(IV)NC1. The amino acid sequence of α3(IV)NC1 is shown in one-letter amino acid code. Occurrences of the motif LIMVF,X,X,WYF are underlined. The motif specifies an intermediate-sized amino acid at position 1 which should bind well in pocket 1 of any class II molecules. The size of hydrophobic side chains that can be accommodated in pocket 1 is determined by a glycine/valine polymorphism at DRβ86. Since this dimorphism does not correlate with disease susceptibility, it follows that disease-associated peptides are likely to bind with intermediate-sized hydrophobic side chains in pocket 1 (such as VILMF) and not larger residues such as WY which would bind much better to class II molecules with the small glycine residue at DRβ86. At pocket 4, peptides binding with bulky side chains would be expected to show the most divergent binding. Pocket 4 in DRB1*1501 should efficiently accommodate bulky hydrophobic residues, such as WY(F), which are too large to lodge stably in the smaller pocket 4 of DR7.

processing factors in determining how antigens are presented to T cells by antigen-presenting cells (APC), and probably depends on the antigen in question. In Goodpasture's disease this question has been directly examined by biochemically characterizing the major naturally processed α3(IV)NC1-derived peptides bound to DR15 molecules on the surface of human APC (Table 13.9) (Phelps et al., 1996b, 1998). They consist of three nested sets centred on core sequences with intermediate or better affinity for DR15 molecules, but importantly they are not the α3(IV)NC1 peptides with highest affinity for DR15 molecules, underscoring the importance of processing factors in determining how a self-protein is presented to T cells. Interestingly, one of the sets of naturally processed peptides is centred on the VPLY sequence identified by the motif above.

The second assumption that must be made is the relationship between how autoantigens are presented to T cells and susceptibility to autoimmunity. This relationship is not known and is very different in the competing hypotheses discussed in the next section.

Mechanisms by which HLA molecules might influence susceptibility to Goodpasture's disease

There are many hypotheses which seek to explain HLA class II associations with autoimmune diseases in terms of differences in class II–peptide interactions (reviewed in Theofilopoulos, 1995a, b). In Goodpasture's disease, knowledge of both strong positive and negative class II associations as well as the sequence of the autoantigen enables some of these hypotheses to be examined critically.

A widely advocated hypothesis (for example, see Hammer et al., 1995; Wucherpfennig and

TABLE 13.9 Sequences of naturally processed α3(IV)NC1 peptides eluted from DR15 molecules

Sequence
TGQALASPGSCLEEFRASPF
GQALASPGSCLEEFRASPFLECH
ALASPGSCLEEFRASPF
ALASPGSCLEEFRASPFLE
LEEFRASPFLECHGRGTCN
RFTTMPFLFCNVNDVCNF
PFLFCNVNDVCNFASR
LFCNVNDVCNFASRND
SCPEGTVPLYSGFSFLFVQ
PSCPEGTVPLYSGFSFLFVQG
GTVPLYSGFSFLFVQGNQRAHG

These 11 α3(IV)NC1 peptides were identified amongst peptides eluted from DR15 molecules (DR2α and DR2β) purified from α3(IV)NC1-pulsed EBV transformed DR15-homozygous APC. They are arranged into nested sets as is typical of class II bound peptides, reflecting the open ends of the class II peptide-binding groove and exoprotease trimming of bound polypeptides.

Strominger, 1995) is that class II molecules encoded by disease-associated alleles have high affinity for disease-associated antigen-derived peptides, whilst other class II alleles bind these peptides less well or not at all. When applied to Goodpasture's disease the hypothesis identifies seven peptides predicted to bind better to DR15 than to DR7 (Figure 13.2) as putative disease-associated peptides. Two of these have been confirmed to bind better to DR15 than to DR7, and failure to show this relationship for the other five may relate more to the length of the peptides examined than to the validity of the motif. However, this hypothesis can be attacked at several levels: (i) it ignores the higher tolerance established to better presented self-peptides; (ii) it requires unique peptide-presenting capabilities of one or a few disease-associated class II molecules where class II molecules are evidently good at binding a wide range of peptides; and (iii) it cannot account for the dominant protection conferred by class II alleles in Goodpasture's and other autoimmune diseases (Nepom, 1995).

An alternative hypothesis is that autoimmunity is directed at self-peptides with low class II affinity that are not normally presented to T cells and therefore not the target of self-tolerance; this appears to be the case in some experimental models (Fairchild et al., 1993; Fairchild and Wraith, 1996; Muraro et al., 1997). However, it is hard to reconcile this hypothesis with the striking hierarchical association between class II alleles in Goodpasture's and some other autoimmune diseases. In particular, it is hard to envisage how particular class II molecules could have unique disease-conferring or disease-protective effects, such as those seen in Goodpasture's disease, by making low-affinity peptide interactions which are generally also low specificity and likely to be just as poor for many other class II types.

It is possible to reconcile the specificity of the influence of class II alleles with self-tolerance by proposing that autoimmunity targets peptides which may have high class II affinity but are not usually presented to T cells, say, because of processing constraints (cryptic epitopes; see Sercarz et al., 1993). Applied to Goodpasture's disease, this model would predict that T cells in DR15 individuals should be tolerant to the major naturally presented α3(IV)NC1 peptides shown in Table 13.9, and that autoreactive T cells target other α3(IV)NC1 peptides. Intriguingly, peptides representing almost two-thirds of the sequence of α3(IV)NC1 bind to DR15 as well as the major naturally presented peptides but are not detectable bound to DR15 on α3(IV)NC1-pulsed APC, suggesting that α3(IV)NC1 contains at least 13 putative cryptic epitopes (Phelps et al., 1998). Abnormal processing of α3(IV)NC1, for example, by the extracellular digestion by neutrophil-derived collagenases during glomerular inflammation or at times of greatly increased collagen turnover, could result in presentation of these cryptic peptides, which might be expected to be powerful immunogens because of their high affinity for DRB1*1501. In this respect, it is noteworthy that the α3(IV)NC1 peptide with highest affinity for DR15 (P15) is rapidly cleaved by the endosomal proteases cathepsins E and D, possibly explaining its failure to become bound to DR15 (authors' unpublished observations).

Further work needs to be done, but there is the exciting prospect that, having partially defined the way in which α3(IV)NC1 is processed and presented to T cells, elucidation of the mechanism by which class II molecules influence susceptibility to Goodpasture's disease may only require characterization of the peptide specificities and HLA restriction of disease-causing T cells from patients.

IDIOPATHIC MEMBRANOUS NEPHROPATHY

Membranous nephropathy is the most common cause of nephrotic syndrome in adults. It is characterized by diffuse thickening of the GBM with deposition of immunoglobulins and complement along the basement membrane. The glomeruli have little or no leucocyte infiltration, and both morphology and immunohistology are identical to Heymann nephritis, an experimental renal disease of rats caused by autoimmunity to antigens expressed on the surface of glomerular epithelial cells (Kerjaschki and Farquhar, 1982; Salant et al., 1989). Membranous nephropathy can occur in isolation (idiopathic membranous nephropathy or IMN), as part of a systemic disease such as lupus erythematosus, or as a hypersensitivity reaction to drugs including gold, mercurials or penicillamine. It can also complicate infections, including

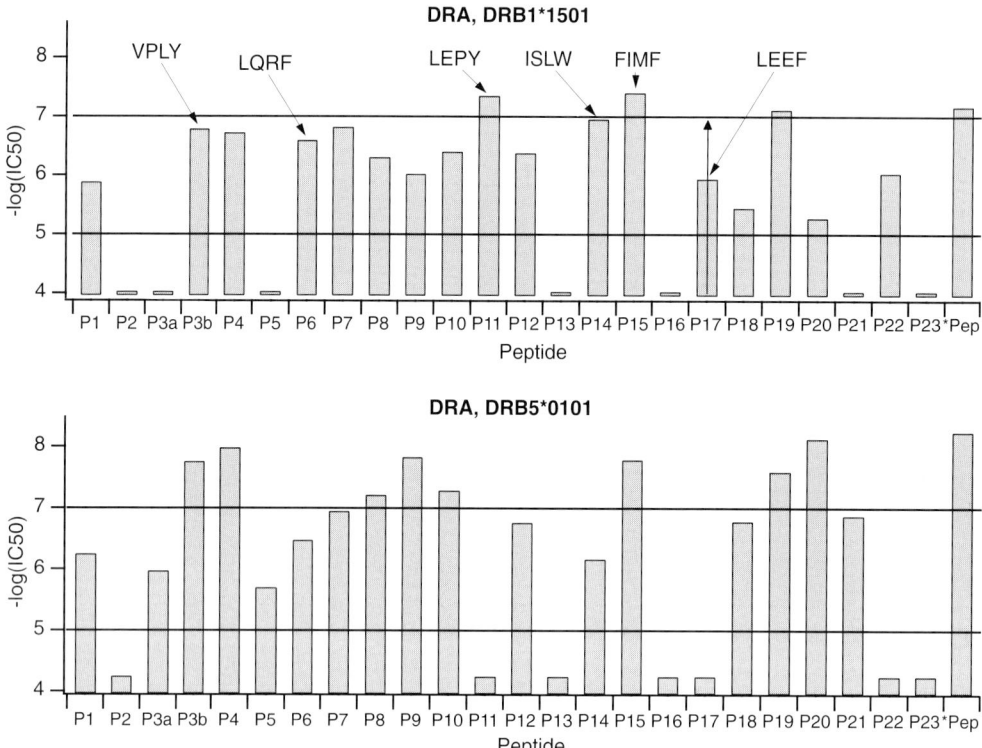

FIGURE 13.2 Affinity of α3(IV)NC1 peptides for DR15 and DR51 molecules. Binding (K_i, mean of at least three estimations) to DR15 (upper figure) and DR51 (lower) is shown for reference peptides (*Pep) and 24 overlapping peptides spanning the sequence of α3(IV)NC1. Horizontal lines demarcate high-affinity binding peptides (K_i < 100 nM, upper region), intermediate binders (K_i 100 nM–10 µM, middle region) and poor/non-binders (K_i > 10 µM, lower region). Reference peptides were *MBPP for DR15 and *HAP for DR51. Peptides containing the motif LIVMF,X,X,WYF are indicated. All have good affinity for DR2b. Peptide P17, containing LEEF, had the lowest affinity of this group but a shorter peptide (13 amino acids), also containing LEEF, had substantially higher affinity. Thus, the LEEF sequence has high affinity for DR51; presumably, the extra residues in P17 interfere with binding, perhaps by making unfavourable interactions outside the groove.

hepatitis B, or develop in association with diseases such as carcinomas and lymphomas.

There is little doubt that membranous nephropathy is immunologically mediated. The earliest pathological feature is subepithelial deposition of immunoglobulin (mainly IgG) and complement components. Despite this, the pathogenetic mechanism remains uncertain. However, it is now clear that idiopathic membranous nephropathy has a strong HLA association with HLA DR3–bearing haplotypes, as have gold- and penicillamine-induced disease (Hall, 1989).

Occurrence of HLA alleles in patients with idiopathic membranous nephropathy

Despite the possible heterogeneity amongst patients diagnosed as having IMN, there are very strong associations between IMN and the inheritance of certain HLA class II alleles (Klouda et al., 1979; Garovoy, 1980; Muller et al., 1981; Le Petit et al., 1982a; Rashid et al., 1983; Short et al., 1983; Berthoux et al., 1984; Papiha et al., 1987; Roccatello et al., 1987; Zucchelli et al., 1987; Vaughan et al., 1989). However, the relationship is much more complex than that so far identified for patients with Goodpasture's disease. In particular, IMN has different HLA associations in different racial groups.

Association with DR alleles

Numerous studies of HLA alleles or antigen frequencies in European Caucasoids have demonstrated a strong association between IMN and the inheritance of the HLA-DRB1*03 allele or expression of the DR3 antigen (Table 13.10). The strongest associations have been found in northern European Caucasoids, including populations from the UK (78–80% vs 27% of controls, OR 10; Cameron et al., 1990; Vaughan et al., 1995), Germany (76% vs 23% of controls, OR 10.7; Muller et al., 1981), France (62%, OR 7.6; Vaughan et al., 1992), Italy (55–61%, OR 3.7–7.6; Roccatello et al., 1987; Zucchelli et al., 1987) and Spain (67%, OR 7.5; Garovoy, 1980). Weaker but still significant associations have been found in Greek (34.5%, OR 2.97; Vaughan et al., 1995) and North American Caucasoids (33%, OR 1.9; Garovoy, 1980).

In striking contrast, Japanese patients with IMN have a high frequency of DRB1*1501 (65–80% vs ~32% in controls, OR 6.5–9.3; see Table 13.11) and a very low frequency of DRB1*03. The difference in the HLA association found in Japanese and Caucasoid populations is not simply a consequence of different allele frequencies in the two populations, because, although DRB1*03 is rare in the Japanese population, DRB1*1501 is common in Caucasoid populations. Moreover, DR2 alleles may be protective against IMN in Caucasoids (OR 0.21, $p = 0.00013$; Vaughan et al., 1995).

Association with DQ and DP alleles

One explanation for different associations with DR alleles in different races is that they are all due to a common susceptibility gene in linkage disequilibrium with different class II alleles in different populations. Vaughan et al. (1989) suggested that the HLA-DR3 association in UK Caucasoids

TABLE 13.10 Idiopathic membranous nephropathy: HLA-DR3 antigen frequency in Caucasoids

Reference	Country	Patients % (n)	Controls % (n)	OR	EF	p_c
Klouda et al. (1979)	UK	75 (32)	20 (60)	12	0.69	
Garovoy (1980)	UK	73 (51)	21 (3184)	8.7	0.66	**
Garovoy (1980)	Spain	67 (39)	21 (3184)	7.5	0.58	**
Garovoy (1980)	USA	33 (33)	21 (3184)	1.9	0.15	NS
Muller et al. (1988)	Germany	76 (21)	23 (122)	10.7	0.69	***
Le Petit et al. (1982a)	France	65 (26)	20 (74)	7.42	0.56	
Rashid et al. (1983)	UK	53 (35)	27 (325)	3.05	0.36	
Papiha et al. (1987)	UK	52 (55)	23 (70)	3.64	0.38	
Zucchelli et al. (1987)	Italy	55 (55)	25 (65)	3.67	0.40	
Roccatello et al. (1987)	Italy	61 (18)	17 (526)	7.6	0.53	**
Vaughan et al. (1989)	UK	65 (31)	37 (55)	2.9	0.44	*
Cameron et al. (1990)	UK	78 (78)	–	–		
Vaughan et al. (1992)	France	62 (37)	–	7.7		***
Vaughan et al. (1995)	UK	80 (52)	27 (100)	10.6	0.73	***
Vaughan et al. (1995)	Greece	33 (29)	15 (92)	3.0	0.21	NS

***$p_c < 0.0001$, **$p_c < 0.001$, *$p_c < 0.02$; NS: not significant.

$$EF : \text{aetiological fraction} = \frac{(\text{Relative risk} - 1) \times \text{Number of affected patients}}{\text{Relative risk} \times \text{Total number of patients}}$$

TABLE 13.11 Idiopathic membranous nephropathy: HLA-DR2 antigen frequency in Japanese

Reference	Country	Patients % (n)	Controls % (n)	OR	EF	p_c
Hiki et al. (1984)	Japan	80 (50)	36 (884)	7.12	0.69	***
Tomura et al. (1984)	Japan	74 (50)	30 (158)	6.5	0.63	***
Naito et al. (1987)	Japan	65 (52)	32 (106)	7.1	0.49	
Ogahara et al. (1992a)	Japan	80 (30)	30 (50)	9.3	0.71	***

***$p_c < 0.0001$.
EF: see legend to Table 13.10.

might reflect an underlying association with a DQA1 allele, and went on to study DR, DQ and DP alleles in 51 British and 29 Greek patients with IMN (Vaughan et al., 1995). The results showed a strong association with DQA1*0501 in British Caucasoid patients (85% vs 45%, OR 7.4) and Greek patients (96% vs 66%, OR 9.7), but because of the large number (80) of comparisons made, only the observation in British Caucasoids was statistically significant after correction ($p_c = 0.0056$, where p_c is the probability that an observation occurred by chance alone, corrected for the number of comparisons made).

Analysis of the alleles at the more centromerically located DQB1 locus in UK patients showed that the two DQB1 alleles commonly inherited with DQA1*0501 had contrasting frequencies; the frequency of DQB1*0201 was significantly raised (90% vs 44%, OR 10, $p_c = 0.0032$), whilst that of DQB1*0302 was reduced (OR 0.29, $p_c = 0.8$). A study of 37 French Caucasoid patients gave similar, but less striking, results (Vaughan et al., 1992). The frequency of DQB1*0201 was also increased in Greek patients (OR 3.6) but, as for DQA alleles, this was not significant after correction. Furthermore, there were no associations with alleles at the more distantly centromeric DPB1 locus (Vaughan et al., 1995). Analysis of alleles at DRB3 (telomeric to DRB1 on chromosome 6) again found contrasting associations, this time much stronger in Greek patients than in British Caucasoids (UK DRB3*0101, OR 4.2; Greek DRB3*0201/2, OR 11).

Very different DQ associations have been found in Japanese patients, most of whom (about 97%) carry DQA1*01 alleles and not DQA1*0501 (Vaughan et al., 1992). Ogahara et al. (1992b) analysed DQ frequencies in 30 Japanese patients and found that 80% (24/30) carried the extended haplotype DRB1*1501–DRB5*0101–DQA1*0102–DQB1*0602, but DQA1*0102 was also carried as part of alternative haplotypes by two patients and was the most frequent single DR or DQ allele (87% vs 26%, OR 18.6, $p = 1.1 \times 10^{-10}$).

Severity of disease and HLA alleles in idiopathic membranous nephropathy

The data on membranous nephropathy are contradictory. Zucchelli et al. (1987) suggested that patients with the DR3–B8 haplotype have a worse prognosis, whereas Short et al. (1983) reported that patients with HLA-DR3–BfF1–B18 did worse than those with other DR3-bearing haplotypes, including those bearing B8. Cameron et al. (1990) could not detect any effect of the HLA complex on prognosis, which is probably the correct answer.

Mechanisms by which HLA molecules might influence susceptibility to diopathic membranous nephropathy

The data on membranous nephropathy suggest that there is a primary association with DRB1*0301 in British and French Caucasoids, and secondary associations with alleles in linkage disequilibrium (most strongly DQA1*0501 and DQB1*0201, but also DRB3*0101 and B8). In Greek Caucasoids there is a trend towards a weak association with DRB1*0301 which is significant before correction for multiple comparisons. In Japanese patients there is a strong, but completely different, association with the haplotype DRB1*1501–DRB5*0101–DQA1*0102–DQB1*0602. Thus, analysis of the frequency of HLA class II alleles in patients with IMN identifies highly significant class II associations, but none that is common to the different racial groups (Table 13.12). There are at least three possible explanations for these findings:

- First, it is possible that the DR molecules themselves do not confer susceptibility, but are in linkage disequilibrium with the single true susceptibility gene and that the linkage is different between Japanese and Caucasoids. Three candidate loci have been studied. The TAP loci lie in the class II region between the DQA and DPB loci. There is limited polymorphism of these genes and differences in allele frequencies have been reported in membranous nephropathy but are attributable to linkage disequilibrium to DR (Chevrier et al., 1994). The C4A and tumour necrosis factor-β (TNFβ) loci are both located in the HLA class III region telomeric to DRA on chromosome 6. C4A null alleles and a 5.5 kb restriction fragment length polymorphism (RFLP) occur more frequently in Caucasoids with IMN than in control individuals (Sacks et al., 1992; Medcraft et al., 1993). However, neither allele occurs more frequently than expected through inheritance of the

TABLE 13.12 Strongest associations with idiopathic membranous nephropathy at *DR* and *DQ* loci (most frequent HLA class II alleles at *DRB* and *DQx1* loci in patients with IMN)

Population	Allele (% patients carrying allele, relative risk for IMN)			
	DQB1	DQA1	DRB1	DRB3/5
British	*0201* (94, **21.6**)	*0501* (85, 6.7)	*0301* (80, 9.8)	*0101* (66, 5.5)
French	*0201* (62, 2.5)	*0501* (84, 4.9)	*0301* (62, **7.6**)	*0101* (54, 3.6)
Greek	*0503* (25, 7.4)			
	0201 (54, 3.7)	*0501* (96, 14.2)	*0301* (35, 3)	*02xx* (97, **16.1**)
Japanese	*0602* (71, **16.8**)	*01xx* (97, 15.2)	*1501* (71, 14.2)	*0101*[a]

The strongest associations between IMN and HLA class II alleles at DQA1/DQB1 and DRB1/DRB3 loci are shown (data from Vaughan et al., 1992). Alleles associated with the greatest relative risk in each population are shown in bold. There is no tendency for more strongly associated alleles to occur at any of the loci shown, nor is there any tendency for the most strongly associated class II allele to occur at one end or other of the class II region of chromosome 6, which might suggest the presence of a common susceptibility factor on linkage disequilibrium with different class II alleles in different populations. The loci are shown in their order on chromosome 6, from centromeric to telomeric.
[a]Data not available but, as patients inheriting DRB1*1501 (71%) almost certainly also inherited DRB5*0101, the relative risk is probably similar.

DR3,B8,A3 extended haplotype, and Japanese DR2-bearing haplotypes do not have C4 null alleles at the C4A locus (Sacks et al., 1992). Another putative locus has been proposed to lie between DPA1 and DQB2, on the basis of differences between IMN-associated haplotypes from IMN patients and control individuals, identified by molecular mapping (Sacks et al., 1993). So far, confirmation and identification of the genes responsible have not been reported. Slightly against the presence of a common linked susceptibility locus is the position along chromosome 6 of the alleles most strongly associated with IMN in different populations, which does not point consistently to any one location that may harbour an unknown common susceptibility gene (see Table 13.12).

- Secondly, and equally difficult to disprove, is the possibility that the pathogenesis of IMN is different in different populations. For example, it has been reported that the prognosis of IMN in Japanese patients is different to that in Caucasoids (Kida et al., 1986). Clearly, this issue can only be tackled sensibly once the pathogenesis of IMN is understood.
- Thirdly, it is possible that the class II alleles most strongly associated with IMN in different populations share characteristics which predispose to IMN. It may be, for example, that they have similar peptide-binding preferences. The peptide-binding specificities of DR2 and DR3 molecules are fairly well characterized and quite different, but similarities between DQ molecules are more difficult to examine. This is because information on peptide-binding data to DQ molecules is generally sparse and extrapolations from structure (the DQ sequence superimposed on the known structure of DR1) are complicated by polymorphism at both DQA and DQB loci and the occurrence of DQ molecules composed of the gene products of DQA and DQB in *cis* and *trans*.

MINIMAL CHANGE NEPHROTIC SYNDROME

Minimal change nephrotic syndrome (MCNS) is the most common form of nephrotic syndrome in childhood, but can occur at any age. The syndrome is defined by the nephrotic syndrome and renal biopsy appearances that are essentially normal by light microscopy. Children with nephrotic syndrome and no identifiable associated disease (i.e. idiopathic nephrotic syndrome) who respond to treatment with corticosteroids almost always have MCNS, so renal biopsy is often not undertaken. For this reason it is not possible to undertake large studies of HLA allele frequencies in patients with renal biopsy-proven MCNS. Instead, children with idiopathic nephrotic syndrome are classified according to their response to treatment with corticosteroids into those who are steroid sensitive (SSNS), most of whom will have MCNS, and those who are steroid resistant (SRNS), who have a range of different renal pathologies.

Occurrence of HLA alleles in patients with minimal change nephrotic syndrome

Most studies of HLA antigens in MCNS have examined Caucasoid children with SSNS. Early reports identified associations with the class I antigens HLA-B12 (Thomson et al., 1976; Trompeter et al., 1980) and HLA-B8 (Lenhard et al., 1980; Noss et al., 1981; Ruder et al., 1982) in different Caucasoid populations. Later studies showed that these associations were due to strong associations with HLA-DR7 alleles (see Table 13.13). Studies in German and American Caucasoid populations have additionally identified a weaker association with DR3, which is in linkage disequilibrium with B8 (see Table 13.14). Many studies have also found significantly fewer DRB1*1501 alleles in patients, suggesting that this allele protects against SSNS (Konrad et al., 1994; Bouissou et al., 1995). Thus, the pattern of positive and negative class II associations in Caucasoids with SSNS is the exact opposite of that found in Caucasoids with Goodpasture's disease, and similarly strong [aetiological fraction (EF; defined in legend to Table 13.13) for positive association ~ 0.6 for both]. Some of the more recent studies have also examined the frequencies of DQ and DP alleles (Clark et al., 1990; Lagueruela et al., 1990; Bouissou et al., 1995; Haeffner et al., 1997). The findings for DQ reflect the strong linkage disequilibrium found between DR and DQ alleles (see Tables 13.2 and 13.3). Thus, there is a strong positive association with DQ2 (DQB1*0201 and DQA1*0201) (Table 13.15) and strong negative association with DQ6 (DQB1*0602 and DQA1*0102). Clark et al. (1990) suggested an additional weak association with DP–Cp63, but a large study by Haeffner (Haeffner et al., 1997) neither confirmed this nor found other DP associations. There are no consistent class II associations with steroid-resistant nephrotic syndrome.

The pattern of association in Japanese children with SSNS is very different. First, the only significant HLA-DR association is with DR8 (Komori et

TABLE 13.13 Occurrence of DR7 in children with steroid-sensitive nephrotic syndrome

		Patients	Controls			
Reference	Country	% (n)	% (n)	OR	EF	p_c
Alfiler et al. (1980)	Australia	71 (42)	30 (121)	5.9	0.59	**
Mouzon-Cambon et al. (1981)	France	74 (38)	30 (91)	6.3	0.63	**
Ruder et al. (1982)	Germany	59 (54)	18 (100)	6.3	0.50	***
Nunez-Roldan et al. (1982)	Spain	72 (50)	38 (179)	4.2	0.55	
McEnery and Welch (1989)	USA	44 (25)	24[b]	2.6	0.26	NS
Lagueruela et al. (1990)	USA[a]	17 (64)	5 (80)	4.1	0.17	
Ruder et al. (1990)	Germany	60 (91)	18 (100)	6.8	0.51	***
Mir et al. (1994)	Turkey	40 (30)	16 (630)	3.5	0.29	**
Zhou et al. (1994)	China	40 (40)	11 (285)	5.3	0.32	**
Bouissou et al. (1995)	Germany	55 (98)	29 (269)	3.2	0.37	***
Bouissou et al. (1995)	France	70 (142)	32 (300)	5.1	0.56	***
Clark et al. (1990)	UK	70 (40)	20 (60)	9.3	0.63	***
Jin et al. (1991)	Japan	5.6 (36)	0.9 (115)	6.5	0.05	NS
Komori et al. (1983)	Japan	3.6 (28)	2.6 (114)	1.4	0.01	NS

[a]Frequency of B44, DR7, DRw53, FC31 extended haplotype.
[b]Using allele frequencies published elsewhere.

$$\text{EF: aetiological fraction} = \frac{(\text{Relative risk} - 1) \times \text{Number of affected patients}}{\text{Relative risk} \times \text{Total number of patients}}$$

***p_c < 0.0001, **p_c < 0.001, *p_c < 0.02; NS: not significant.

TABLE 13.14 Other *DR* associations with steroid-sensitive nephrotic syndrome in children

Reference	Country	Patients % (n)	Controls % (n)	OR	EF	p_c
DR3						
Ruder et al. (1990)	Germany	37 (91)	14 (100)	3.6	0.3	*
Bouissou et al. (1995)	Germany	37 (98)	18 (269)	2.6	0.2	**
Bouissou et al. (1995)	France	26 (144)	22 (356)	1.3	0.1	NS
Lagueruela et al. (1990)	USA[a]	16 (64)	3.8 (80)		0.1	0.014
HLA-DR8						
Komori et al. (1983)	Japan[c]	46 (28)	8 (114)	10.1	0.4	*
Abe et al. (1995)	Japan	22 (27)	21 (116)	1.1	0.01	NS
Kobayashi et al. (1985)	Japan[c]	35 (40)	12.6 (40)	3.7	0.3	*
McEnery and Welch (1989)	USA	20 (25)	5[b]	4.8	0.2	NS

[a]Frequency of A1–B8–DR3–DRW52–SC01 extended haplotype.
[b]Using allele frequencies published elsewhere.
[c]Includes adults with MCNS.
*$p_c < 0.02$, **$p_c < 0.001$; NS: not significant.
EF: see legend to Table 13.13.

al., 1983; Jin et al., 1991), and in some studies no significant DR association has been found (Abe et al., 1995) (Table 13.14). DR7 is more common in Japanese with SSNS (3.6–5.6% vs 0.9–2.6%), but its rarity in the Japanese population means that this does not attain statistical significance (Table 13.13). Secondly, although DRB1*1501 is about half as common in Japanese populations as it is in Caucasoid populations (most Japanese carrying the DR2 specificity have DRB1*1502), it does not confer the protection from SSNS observed in Caucasoid children. Thirdly, there is a positive association with DQ3 (Kobayashi et al., 1985; Ogahara et al., 1992b; Abe et al., 1995), which in one study was due to a high frequency of DQB1*0302 (54% vs 16% of controls, OR 6.2, EF = 0.45, $p = 6.8 \times 10^{-6}$; Abe et al., 1995) (Table 13.15). Interestingly, several studies of Japanese adults with MCNS have also shown associations with DR8 and DQ3 (Komori et al., 1983; Kobayashi et al., 1985, 1995; Naito et al., 1987), in marked contrast to MCNS in Caucasoid adults, where a striking dearth of data suggests that only weak or no associations occur (Laurent et al., 1983, 1985).

TABLE 13.15 Positive associations between *DQ* alleles and steroid-sensitive nephrotic syndrome in children

Reference	Country	Patients % (n)	Controls % (n)	OR	EF	p_c
DQ2						
Clark et al. (1990)	UK	83 (40)	40 (60)	7.1	0.72	****
Bouissou et al. (1995)	France	81 (111)	48 (257)	4.7	0.63	****
Bouissou et al. (1995)	Germany	62 (29)	37 (97)	2.8[a]	0.40	0.03
Lagueruela et al. (1990)	USA	72 (32)	35 (†)	3.1	0.57	*
DQ3						
Kobayashi et al. (1985)	Japan	95 (40)	63 (40)	11.1	0.86	**
Abe et al. (1995)	Japan	82 (27)	44 (116)	5.6	0.68	**
Zhou et al. (1994)	China	85 (40)	77 (285)	1.6	0.35	NS

[a]But see text.
*$p_c < 0.02$, **$p_c < 0.001$; ****$p_c < 0.0001$; NS: not significant.
EF: see legend to Table 13.13.
†Allele frequency in USA population.

Chinese Han are the only other population in which HLA class II associations in childhood SSNS have been reported. Zhou et al. reported a strong association with DR7 (40% vs 11.2% of controls, OR 5.27, $p = 0.00025$) but, in contrast to Caucasoids, all of their patients carried DR7 on a haplotype including DQ3 and not DQ2 (Zhou et al., 1994). This difference is striking because DR7 is usually inherited with DQ2 in Chinese Han, as it is in Caucasoids (Table 13.2).

Pattern of disease and HLA alleles in minimal change nephrotic syndrome

Some children with SSNS enter long remissions after a single course of steroid treatment, others relapse during repeated attempts to withdraw steroids (steroid dependence) and many children relapse at some time after steroids have been withdrawn, in some cases on numerous occasions (frequent relapsers). The association with DR7 is strongest in Caucasoid children who are steroid dependent or frequent relapsers, less strong in children who relapse infrequently and, possibly, absent in the few children who never relapse (Bouissou et al., 1995). Inheritance of both DR7 and DR3 is particularly associated with frequent relapses (Ruder et al., 1990; Bouissou et al., 1995). Intriguingly, although SSNS in Chinese children is also associated with DR7, frequent relapses was associated with DR9, whilst DR7 was associated with no or few relapses (Zhou et al., 1994). The association of class II alleles with different patterns of disease may be of practical importance. Konrad has recently reported that the likelihood of further relapse in children with frequently relapsing SSNS treated with cytotoxic agents is strikingly increased in children who carry DR7 (Konrad et al., 1997), but whether knowledge of class II type will in future direct the type of treatment or just help to assess prognosis is not yet clear.

Primary HLA association with minimal change nephrotic syndrome

Analysis of class II frequencies in patients with SSNS from several populations suggests associations with HLA-DR and DQ but not DP. As with Goodpasture's disease, it is difficult to distinguish primary associations with DR and DQ because of the low frequency of recombination between the DRB1 and DQB1 loci.

In most Caucasoid populations the associations with DRB1*07 and DQB1*0201 (DQw2) are indistinguishably strong because of linkage disequilibrium. However, several factors support a primary association with DR7: (i) the disease is associated with DR7 but not DQw2 in Chinese Han patients; (ii) in some populations, DR3 is not more common in patients despite strong linkage disequilibrium with DQ2; (iii) DR7–DQw9 haplotypes (also identified as RFLP 7.2) are also elevated in some series (Clark et al., 1990); and (iv) in the German and US Caucasoid populations in which the patients have a higher frequency of DR3–DQ2 is similarly or less strongly associated than DR7, despite being the most frequent DQ molecules encoded on both DR3 and DR7 haplotypes. In contrast, the strongest association in Japanese patients is with DQB1*0302 (DQw3) and the inconsistent association with DR8 is attributable to linkage disequilibrium.

Several observations suggest that certain class II alleles inherited in combination have a greater influence on susceptibility to SSNS, or at least its severity, than do the alleles inherited in isolation: (i) Clark et al. (1990) examined the DR-DQ haplotypes of 40 British Caucasoids with SSNS and concluded that both DQw2 and DR7 independently contributed to susceptibility; (ii) inheritance of DR7 with DR3 in Caucasoids or DR9 in Chinese patients is associated with more severe disease; and (iii) in Chinese and Japanese patients, the presumed susceptibility class II alleles occur on haplotypes unusual for the respective populations. Thus, Chinese patients inherit DR7 with DQ3, not the more usual DR7–DQ2 haplotype, and Japanese patients have a high frequency of DQB1*0302, but not of the DR alleles (DRB1*0406, 0403, 0802 and 1201) normally in linkage disequilibrium with it.

Role of HLA alleles in pathogenesis of steroid-sensitive nephrotic syndrome

The striking associations between SSNS and HLA class II alleles suggest that the class II molecules, or proteins encoded by genes in close linkage disequilibrium, play a role in pathogenesis. However, the association with class II alleles is more complex than that described for Goodpasture's disease and the mechanism may also be more complex. Two observations must be explained: first, the association is different in different races; and secondly, disease susceptibility and severity appear to be influenced by the combination of class II alleles inherited and not simply the presence of a particular allele.

The data clearly indicate that the HLA associations with MCNS are different in different populations. There are at least four possible explanations. First, it is possible that the difference reflects a difference in haplotype frequencies in the different populations. For example, the data are consistent with DR7 conferring susceptibility to SSNS in all the populations studied (see relative risks in Table 13.13), with the lack of statistical significance in Japanese populations being attributed to the low frequency of DR7 in Japanese populations. Secondly, different class II molecules may confer susceptibility by the same pathogenetic mechanism in different populations. For example, it is conceivable that the allele most closely associated with SSNS in Japanese patients, DR8, is next best to DR7 at conferring susceptibility to the disease and substitutes because of the low frequency of DR7. In support of this, an increased frequency of DR8 has been reported occasionally in Caucasoid patients (McEnery and Welch, 1989). However, the relative peptide-binding characteristics of DR7 and DR8, although not known, are expected to be very different as they have very different amino acids flanking pockets 4 and 6. It is also unlikely that DR7 in Caucasoids and DQ3 in Japanese influence disease by the same mechanism, as there is no increase in DQ3 in Caucasoid patients (Bouissou et al., 1995). Thirdly, it may be that the DR and DQ alleles associated with MCNS in different populations are in linkage disequilibrium with a common susceptibility gene. Any such gene would be predicted to lie telomeric to DPA1, as DP alleles are not associated with the disease. However, the linkage between DR and DQ alleles is strong even across populations, and a study of microsatellite markers between DR and DQ did not yield a stronger association with disease than observed with DR or DQ alone (Mignot et al., 1995). Finally, it is possible that the pathogenesis of SSNS differs between populations.

Mechanisms by which HLA molecules might influence minimal change nephrotic syndrome

It is striking that the association with class II alleles is strongest in children who relapse frequently and very weak in children who never relapse (Bouissou et al., 1995). One explanation is that class II alleles are less influential in the initial disease process, which may have a non-immunological cause, but very important in determining whether subsequent immunologically mediated "relapses" occur. In this respect, it is intriguing that heterozygosity at DR influences the likelihood of frequent relapses in German (DR3 and DR7; Bouissou et al., 1995) and Chinese populations (DR9 and DR7; Zhou et al., 1994). This suggests that the tendency to relapse is related not to the gene dose of DR7 but to the simultaneous presence of DR3 in Caucasoids or DR9 in Chinese (DR9 is rare in Caucasoids), or a linked gene such as a DQ allele. This is reminiscent of reports in other autoimmune diseases where it is postulated that peptides derived from one HLA antigen are presented by another (Baum et al., 1996), but other mechanisms could account for these observations.

Finally, it may be relevant that the class II associations in Caucasoids with Goodpasture's disease and SSNS are similarly strong but almost exactly opposite. If class II alleles influence susceptibility to autoimmune diseases because they encode class II molecules with different peptide-binding specificities, and positive and negative disease associations reflect divergent peptide-binding specificities, then such reciprocal associations would be predicted.

MESANGIAL IMMUNOGLOBULIN A NEPHROPATHY

Mesangial IgA nephropathy (IgAN) is probably the most common chronic glomerulonephritis in the world and about 25% of affected patients develop end-stage renal failure. An autoimmune model has been postulated for the disease recently because IgG autoantibodies against specific determinants on mesangial cells have been found in sera from some patients (O'Donoghue et al., 1991), but patients with IgAN probably have a generalized disease of cellular and humoral immunity (Levy and Lesavre, 1992).

Occurrence of HLA alleles in patients with immunoglobulin A nephropathy

The familial incidence of IgAN is well described in both Caucasoid and Japanese populations and there are reports of disease in HLA-identical siblings and in twins. A number of studies, although not all, has linked familial IgAN to particular alleles. Levy and Lesarve (1992) analysed data from all reported families, including 10 affected sibs they had studied. There was a significant deviation from the expected Mendelian pattern of inheritance for HLA haplotypes but, as the authors emphasize, the use of reported families could have biased the results.

The possibility that sporadic IgAN was associated with HLA alleles in Caucasoids was first reported by Berthoux et al. (1978), who described an association with HLA-Bw35 in French Caucasoids. However, only two of 16 subsequent studies of class I frequencies have supported this observation (Brettle et al., 1978; Noel et al., 1978; Nagy et al., 1979; Savi et al., 1979; Bignon et al., 1980; Faucet et al., 1980; Macdonald et al., 1980; Arnaiz-Villena et al., 1981; Moutonen et al., 1981; Le Petit et al., 1982b; Rambausek et al., 1982; Julian et al., 1983; Feehally et al., 1984; Hanly et al., 1984; Waldo et al., 1986; Berthoux et al., 1988). Attention turned to class II alleles and in 1980 Faucet et al. reported that the frequency of DR4 was significantly increased in a group of 45 French patients with IgA disease. Again, with one exception (in French Caucasoids; Berger, 1984), subsequent studies in Caucasoids (see Table 13.16) have either failed to confirm this (Brettle et al., 1978; Bignon et al., 1980; Mouzon-Cambon et al., 1980; Le Petit et al., 1982b; Rambausek et al., 1982; Julian et al., 1983; Feehally et al., 1984; Hanly et al., 1984; Waldo et al., 1986; Berthoux et al., 1988; Moore et al., 1990; Freedman et al., 1994; Luger et al., 1994) or found weak associations without statistical significance (again in French Caucasoids; Raguenes et al., 1995). Studies of DQ and DP alleles have similarly shown no consistent association with IgA disease (Moore et al., 1990, 1992; Li et al., 1991; Moore, 1993; Rambausek et al., 1993; Luger et al., 1994; Fennessy et al., 1996). Moore et al. (1990) reported a significant association with the RFLP T2 (which identifies with DQB1*0302, 0303 and 0402 alleles) in British Caucasoids, but found no association in Italian or Finnish patients (Moore, 1993). Li et al. (1991) could not confirm this observation in British Caucasoids but described a significant association with DQw7 (DQB1*0301). Raguenes' study of 58 French Caucasoid patients found no significant association when the data were analysed for all the patients but, dividing the patients according to outcome, found a strong association between DQB1*0301 and progression to end-stage renal failure and between DQB1*0302 and stable renal function (Raguenes et al., 1995). Finally, a recent large study of British, Finnish and Italian Caucasoids found neither of the previously reported DQ associations but two negative associations: DQB1*0201 was significantly reduced in British Caucasoids and DQB1*0602 in Finnish patients (Fennessy et al., 1996).

The situation is quite different in Japanese, in which nine separate studies have shown an increased prevalence of HLA-DR4, with an incidence of about 60% in patients compared with 32–44% in healthy controls (Komori et al., 1979, 1983; Hiki et al., 1982, 1990, 1991; Kasahara et al., 1982; Kashiwabara et al., 1982; Kohara et al., 1985; Abe et al., 1993) (Table 13.17). Later studies of DQ frequencies showed a stronger association with DQ4, suggesting that the association with DR4 is due to its linkage disequilibrium with DQw4. A few other non-Caucasoid populations have been studied, generally without clear associations (Seedat et al., 1988; Huang et al., 1989; Chen, 1992). Li et al. (1994) found no DR or DQ allele that was significantly more frequent in a group of 79 Hong Kong Chinese patients, but homozygous DQ7 was slightly more frequent than in controls (16.4% vs 5.7%) and the DQA2 U allele (DXa U) was associated with an adverse prognosis.

In summary, no clear conclusions can be drawn from the data on European Caucasoids, but HLA-DQ4–DR4 appears to confer susceptibility for the disease in Japanese populations.

ANTI-NEUTROPHIL CYTOPLASMIC ANTIBODY (ANCA)-ASSOCIATED VASCULITIS

Small vessel vasculitis associated with anti-neutrophil cytoplasmic antibodies (ANCA) frequently presents with focal necrotizing glomerulonephritis

TABLE 13.16 Mesangial immunoglobulin A nephropathy: HLA-DR4 antigen frequency in Caucasoids

Reference	Country	Patients % (n)	Controls % (n)	OR	EF
Brettle et al. (1978)	UK	24 (17)	33 (208)	0.62	–0.13
Mouzon-Cambon et al. (1980)	France	0 (11)	12 (90)		
Bignon et al. (1980)	France	14 (35)	21 (56)	0.61	–0.09
Faucet et al. (1980)	France	49 (45)	19 (113)	3.96*	0.37
Rambausek et al. (1982)	FRG	17 (36)	26 (248)	0.56	–0.12
Le Petit et al. (1982b)	France	22 (49)	34 (74)	0.57	–0.18
Berger (1984)	France	43 (30)	13 (106)	5.02*	0.34
Feehally et al. (1984)	UK	37 (46)	32 (385)	1.01	0.07
Hanly et al. (1984)	Ireland	30 (46)	33 (212)	0.88	–0.04
Waldo et al. (1986)	USA	40 (27)	28		
Berthoux et al. (1988)	France	25 (1985)	32 (124)	0.69	–0.10
Luger et al. (1994)	USA	30 (56)	30 (102)	1	0.00
Moore et al. (1990)	UK	44 (73)	37 (149)	1.3	0.11
Raguenes et al. (1995)	France	29.3 (58)	19.7 (150)	1.7	0.12

*Significant result.
EF: see legend to Table 13.13.

with extensive crescent formation and scanty immune deposits. Clinically, many of the patients are diagnosed as having microscopic polyangitis or Wegener's granulomatosus, but some present with predominantly glomerular disease. Over 90% of untreated patients with active disease have detectable ANCA, usually directed against one of the neutrophil cytoplasmic antigens, myeloperoxidase (MPO) and proteinase-3 (Pr-3). The recognition that autoimmunity to neutrophil cytoplasmic antigens might be involved in the pathogenesis has led to a re-examination of the possible involvement of the HLA complex in these disorders.

These diseases have been reported in siblings, but identical twins discordant for the disease have also been observed (Muller et al., 1984). Early studies were performed before ANCA could be assayed and so are subject to some uncertainty. Nevertheless, Muller et al. (1984) reported that idiopathic rapidly progressive glomerulonephritis (of which ANCA-associated vasculitis is the major cause) was associated with HLA-DR2, MT3 and the complement allotype BfF, especially

TABLE 13.17 Mesangial immunoglobulin A nephropathy: HLA-DR4 and -DQ4 antigen frequency in Japanese

Reference	Patients % (n)	Controls % (n)	OR	EF
DR4				
Kashiwabara et al. (1982)	66 (42)	39 (158)	3.1*	0.44
Kasahara et al. (1982)	60 (104)	36 (147)	2.6*	0.38
Hiki et al. (1982)	66 (80)	41 (884)	2.78*	0.42
Komori et al. (1983)	55 (51)	44 (114)	1.6	0.20
Kohara et al. (1985)	58 (41)	32 (63)	3.03	0.38
Naito (1987)	58 (70)	34 (100)	2.7*	0.36
Hiki et al. (1990)	60 (130)	42 (472)	2.1*	0.31
Hiki et al. (1991)	50 (50)	42 (472)	1.4[NS]	0.14
Abe et al. (1993)	66 (32)	42 (124)	2.6*	0.41
DQ4				
Hiki et al. (1991)	36 (50)	10 (472)	4.97*	0.29
Abe et al. (1993)[a]	88 (56)	65 (121)	3.9*	0.66

*Significant result; [NS]not significant.
[a]Frequency of an RFLP identifying DQw4, 8 and 9.
EF: see legend to Table 13.13.

when inherited together as an extended haplotype. None of their patients was said to have had evidence of systemic disease, which is surprising in view of the high incidence of systemic symptoms in most series of RPGN. The 45 patients studied by Elkon et al. (1983) all presented with vasculitis and most had crescentic nephritis. The frequency of DR2 was significantly increased in patients diagnosed as having Wegener's granulomatosis (65% for the 17 patients vs 21% for 113 controls, OR 7), but not in the group as a whole. The hope that more precise definition of the patients by the specificity of their autoantibodies would lead to clearer HLA associations has proved overoptimistic. The frequency of HLA-DQB1*0301 was significantly increased in a group of 59 patients (Spencer et al., 1992), especially when found on a DR4-bearing haplotype. This association was independent of the clinical diagnosis and of the ANCA specificity. Zhang et al. (1995) did not confirm these findings in another study of British Caucasoids and did not identify any significant associations. A third study, this time of Dutch patients, failed to find any positive associations but reported a significant negative association with HLA-DR6 (Hagen et al., 1995). Each of these studies was relatively small and the results probably say more about ascribing statistical significance than about the nature of small-vessel vasculitis.

POST-STREPTOCOCCAL GLOMERULONEPHRITIS

The familial incidence of post-streptococcal glomerulonephritis is well recognized in epidemic outbreaks and is likely to reflect the presence of genetically determined susceptibility. However, there is no real evidence about the genes involved. Layrisse et al. (1983) were unable to find evidence of linkage to the HLA complex in 18 affected Venezuelan families, but reported a weak association with HLA-DR4 in 42 unrelated patients. Given the problems of small sample sizes, this suggestion will need to be confirmed before being accepted. Two small studies from Japan of class II associations are unconvincing (Sasajuki et al., 1979; Naito et al., 1987). Taken together, these results provide no evidence of association.

MESANGIOCAPILLARY GLOMERULONEPHRITIS

Although there are no definite associations between HLA antigens and susceptibility to mesangiocapillary nephritis, Welch et al. (1986) suggested that, despite treatment with steroids, patients with the extended haplotype HLA-B–DR3 are significantly more likely than patients with other haplotypes to develop renal failure.

CONCLUSION

The approach to the analysis of HLA associations with disease has been transformed since the late 1980s, mainly because of a much deeper understanding of the structure and function of class I and class II molecules. In some respects this has clarified interpretation of reported associations, for example through the identification of individual HLA alleles and by defining the topology of their peptide-binding grooves. However, other discoveries, such as the demonstration that the class II region contains genes which encode molecules that facilitate peptide loading of both class I and class II molecules, have added to the complexity of interpretation. These developments have emphasized the need for extremely rigorous mathematical treatment of the results obtained if spurious associations are to be avoided.

Two major difficulties hamper progress in understanding the role of HLA molecules in the development of nephritis. First, the rarity of these diseases increases the difficulty of assembling large studies. More important still is lack of understanding of the pathogenesis of most types of glomerulonephritis and, in particular, ignorance of the relevant renal autoantigens. The contrast between what is possible in Goodpasture's disease and other conditions with equally strong HLA associations highlights this problem.

Nevertheless, progress that has been made should not be underestimated, even though it is clear from clinical and experimental studies (Vyze and Kotzin, 1998) that we must now expect HLA associations to be complex and probably to involve epistatic interactions between alleles at different loci. Unravelling this presents a major challenge for the next decade.

REFERENCES

Abe, J., Kohsaka, T., Tanaka, M. et al. (1993). Genetic study on HLA class II and class III region in the disease associated with IgA nephropathy. *Nephron* **65**: 17–22.

Abe, K.K., Michinaga, I., Hiratsuka, T. et al. (1995). Association of DQB1*0302 alloantigens in Japanese pediatric patients with steroid-sensitive nephrotic syndrome. *Nephron* **70**: 28–34.

Alfiler, C.A., Roy, L.P., Doran, T. et al. (1980). HLA-DRw7 and steroid responsive-nephrotic syndrome of childhood. *Clin. Nephrol.* **14**: 71–74.

Arnaiz-Villena, A., Gonzalo, A., Mampaso, F. et al. (1981). HLA and IgA nephropathy in Spanish population. *Tissue Antigens* **17**: 549–550.

Baum, H., Davies, H., Peakman, M. et al. (1996). Molecular mimicry in the MHC: hidden clues to autoimmunity? *Immunol. Today* **17**: 64–70.

Berger, J. (1984). IgA mesangial nephropathy. 1968–1983. *Contrib. Nephrol.* **40**: 4–6.

Berthoux, F.C., Gagne, A., Sabatier, J.C. et al. (1978). HLA-Bw35 and mesangial IgA glomerulonephritis. *N. Engl. J. Med.* **298**: 1034–1035.

Berthoux, F.C., Genin, C., Le Petit, J.C. et al. (1984). Immunogenetics of mesangial IgA nephritis. *Contrib. Nephrol.* **40**: 118–123.

Berthoux, F.C., Alamartine, E., Pommier, G. et al. (1988). HLA and IgA nephritis revisited 10 years later: HLA B35 antigen as a prognostic factor. *N. Engl. J. Med.* **319**: 1609–1610.

Bignon, J.D., Houssin, A., Soulillou, J. et al. (1980). HLA antigens and Berger's disease. *Tissue Antigens* **16**: 108–111.

Bouissou, F., Meissner, I., Konrad, M. et al. (1995). Clinical implications from studies of HLA antigens in idiopathic nephrotic syndrome in children. *Clin. Nephrol.* **44**: 279–283.

Brettle, R., Peters, D.K., Batchelor, J.R. et al. (1978). Mesangial IgA glomerulonephritis and HLA antigens. *N. Engl. J. Med.* **299**: 200–201.

Burns, A.P., Fisher, M., Li, P. et al. (1995). Molecular analysis of HLA class II genes in Goodpasture's disease. *Q. J. Med.* **88**: 93–100.

Cameron, J.S., Healy, M.J. and Adu, D. (1990). The Medical Research Council trial of short-term high-dose alternate day prednisolone in idiopathic membranous nephropathy with nephrotic syndrome in adults. The MRC Glomerulonephritis Working Party. *Q. J. Med.* **74**: 133–156.

Chen, X. (1992). HLA-DR gene frequencies in IgA nephropathy patients obtained by oligonucleotide genotyping. *Chung Hua I Hsueh Tsa Chih (Taipei)* **72**: 206–209, 254.

Chevrier, D., Giral, M., Braud, V. et al. (1994). Membranous nephropathy and a TAP1 gene polymorphism [Letter]. *N. Engl. J. Med.* **331**: 133–134.

Clark, A., Vaughan, R., Stephens, H. et al. (1990). Genes encoding the beta-chains of HLA-DR7 and HLA-DQw2 define major susceptibility determinants for idiopathic nephrotic syndrome. *Clin. Sci. (Colch)* **78**: 391–397.

Druet, P., Bariety, J., Bellon, B. et al. (1972). Nephrotoxic serum nephritis in the rat: ultrastructural localization of nephrotoxic rabbit antibodies using peroxidase-labeled conjugates. *Lab. Invest.* **27**: 157–164.

Dunckley, H., Chapman, J.R., Burke, J. et al. (1991). HLA-DR and -DQ genotyping in anti-GBM disease. *Disease Markers* **9**: 249–256.

Elkon, K.B., Sutherland, D.C., Rees, A.J. et al. (1983). HLA antigen frequencies in systemic vasculites: increase in HLA-DR2 in Wegener's granulomatosis. *Arthritis. Rheum.* **26**: 98–101.

Fairchild, P.J. and Wraith D.C. (1996). Lowering the tone: mechanisms of immunodominance among epitopes with low affinity for MHC. *Immunol. Today* **17**: 80–85.

Fairchild, P., Wildgoose, R., Atherton, E. et al. (1993). An autoantigenic T cell epitope forms unstable complexes with class II MHC: a novel route for escape from tolerance induction. *Int. Immunol.* **5**: 1151–1158.

Faucet, R., Le Pogamp, P., Genetet, B. et al. (1980). HLA-DR4 antigen and IgA nephropathy. *Tissue Antigens* **16**: 405–410.

Feehally, J., Dyer, P.A., Davidson, J.A. et al. (1984). Immunogentics of IgA nephropathy: experience in a UK centre. *Dis. Markers* **2**: 493–500.

Fennessy, M., Hitman, G., Moore, R. et al. (1996). HLA-DQ gene polymorphism in primary IgA nephropathy in three European populations. *Kidney Int.* **49**: 477–480.

Fisher, M., Pusey, C.D., Vauhan, R.W. et al. (1997). Susceptibility to Goodpasture's disease is strongly associated with HLA-DRB1 genes. *Kidney Int.* **51**: 222–229.

Freedman, B., Spray, B. and Heise, E. (1994). HLA associations in IgA nephropathy and focal and segmental glomerulosclerosis. *Am. J. Kidney Dis.* **23**: 352–357.

Garovoy, M.R. (1980). Idiopathic membranous glomerulonephritis (IMGN): an HLA associated disease. In *Histocompatibility Workshop 1980* (ed. P. Terasaki), pp. 673–680, Los Angeles University Press, Los Angeles, CA.

Garovoy, M.R. (1982). Immunogenetic associations in nephrotic states. *Contemp.y Issues Nephrol.* **9**: 259–282.

Haeffner, A., Abbal, M., Mytilineos, J. et al. (1997). Oligotyping for HLA-DQA, -DQB, and -DPB in idiopathic nephrotic syndrome. *Pediatr. Nephrol.* **11**: 291–295.

Hagen, E.C., Stegeman, C.A., D'Amaro, J. et al. (1995). Decreased frequency of HLA-DR13DR6 in Wegener's granulomatosis. *Kidney Int.* **48**: 801–805.

Hall, C. (1989). The natural course of gold and penicillamine nephropathy: a longterm study of 54 patients. *Adv. Exp. Med. Biol.* **252**: 247–256.

Hammer, J., Gallazzi, F., Bono, E., et al. (1995). Peptide binding specificity of HLA-DR4 molecules: correlation with rheumatoid arthritis association. *J. Exp. Med.* **181**: 1847–1855.

Hanly, P., Garrett, P., Spencer, S. et al. (1984). HLA -A, -B and -DR antigens in IgA nephropathy. *Tissue Antigens* **23**: 270–273.

Hiki, Y., Kobayashi, Y., Tateno, S. et al. (1982). Strong association of HLA-DR4 with benign IgA nephropathy. *Nephron* **32**: 222–226.

Hiki, Y., Kobayashi, Y., Itoh, I. et al. (1984). Strong association of HLA-DR2 and MT1 with idiopathic membranous nephropathy in Japan. *Kidney Int.* **25**: 953–957.

Hiki, Y., Kobayashi, Y., Ookubo, M. et al. (1990). The role of HLA-DR4 in the long-term prognosis of IgA nephropathy. *Nephron* **54**: 264–265.

Hiki, Y., Kobayashi, Y., Ookubo, M. et al. (1991). Association of HLA-DQw4 with IgA nephropathy in the Japanese population. *Nephron* **58**: 109–111.

Huang, C., Hu, S., Lin, J. et al. (1989). HLA and Chinese IgA nephropathy in Taiwan. *Tissue Antigens* **33**: 45–47.

Huey, B., McCormack, K., Capper, J. et al. (1993). Associations of HLA-DR and HLA-DQ types with anti-GBM nephritis by sequence specific oligonucleotide probe hybridisation. *Kidney Int*. **44**: 307–312.

Jin, D., Kohsaka, T., Tanaka, M. et al. (1991). Human leukocyte antigens in childhood idiopathic nephrotic syndrome. *Acta Paediatr. Jpn*. **33**: 709–713.

Julian, B.A., Wyatt, R.J., McMorrow, R.G. et al. (1983). Serum complement proteins in IgA nephropathy. *Clin. Nephrol*. **20**: 251–258.

Kalluri, R., Danoff, T.M., Okada, H. et al. (1997). Susceptibility to anti-glomerular basement membrane disease and Goodpasture syndrome is linked to MHC class II genes and the emergence of T cell-mediated immunity in mice. *J. Clin. Invest*. **100**: 2263–2275.

Kasahara, M., Hamada, K., Kayama, T. et al. (1982). Role of HLA in IgA nephropathy. *Clin. Immunol. Immunopathol*. **25**: 189–195.

Kashiwabara, H., Shishido, H., Tomura, S. et al. (1982). Strong association between IgA nephropathy and HLA-DR4 antigen. *Kidney Int*. **22**: 377–382.

Kerjaschki, D. and Farquhar M.G. (1982). The pathogenic antigen of Heymann nephritis is a membrane glycoprotein of the renal proximal tubule brush border. *Proc. Natl Acad. Sci. USA* **79**: 5557–5581.

Kida, H., Asamoto, T., Yokoyama, H. et al. (1986). Long-term prognosis of membranous nephropathy. *Clin. Nephrol*. **25**: 64–69.

Klouda, P.T., Manos, J., Acheson, E.J. et al. (1979). Strong association between idiopathic membranous nephropathy and HLA-DRW3. *Lancet* **ii**: 770–771.

Kobayashi, T., Ogawa, A., Takahashi, K. et al. (1995). HLA-DQB1 allele associates with idiopathic nephrotic syndrome in Japanese children. *Acta Paediatr. Jpn*. **37**: 293–296.

Kobayashi, Y., Chen, X.M., Hiki, Y. et al. (1985). Association of HLA-DRw8 and DQw3 with minimal change nephrotic syndrome in Japanese adults. *Kidney Int*. **28**: 193–197.

Kohara, M., Naito, S., Arakawa, K. et al. (1985). The strong association of HLA-DR4 with spherical mesangial dense deposits in IgA nephropathy. *J. Clin. Lab. Immunol*. **18**: 157–160.

Komori, K., Nose, Y., Inouye, H. et al. (1979). Study of HLA system in IgA nephropathy. *Tissue Antigens* **14**: 32–36.

Komori, K., Nose, Y., Inouye, H. et al. (1983). Immunogenetical study in patients with chronic glomerulonephritis. *Tokai. J. Exp. Clin. Med*. **8**: 135–148.

Konrad, M., Mytilineos, J., Bouissou, F. et al. (1994). HLA class II associations with idiopathic nephrotic syndrome in children. *Tissue Antigens* **43**: 275–280.

Konrad, M., Mytilineos, J., Ruder, H. et al. (1997). HLA-DR7 predicts the response to alkylating agents in steroid-sensitive nephrotic syndrome. *Pediatr. Nephrol*. **11**: 16–19.

Krieger, J., Karr, R., Grey, H. et al. (1991). Single amino acid changes in DR and antigen define residues critical for peptide-MHC binding and T-cell recognition. *J. Immunol*. **146**: 2331–2340.

Lagueruela, C.C., Buettner, T.L., Cole, B.R. et al. (1990). HLA extended haplotypes in steroid-sensitive nephrotic syndrome of childhood. *Kidney Int*. **38**: 145–150.

Laurent, J., Ansquer, J.C., de Mouzon-Cambon, A. et al. (1983). Adult onset lipoid nephrosis is not DR7 associated. *Tissue Antigens* **22**: 229–230.

Laurent, J., Belghiti, D., Ansquer, J. et al. (1985). Idiopathic nephrotic syndrome and the HLA allele. Prevalence of age-related DR7. *Rev. Med. Interne*. **6**: 116–120.

Layrisse, Z., Rodriguez Iturbe, B., Garcia-Ramirez R. et al. (1983). Family studies of the HLA system in acute post-streptococcal glomerulonephritis. *Hum. Immunol*. **7**: 177–185.

Le Petit, J.C., Laurent, B., Berthoux, F.C. et al. (1982a). HLA-DR3 and idiopathic membranous nephritis (IMN) association. *Tissue Antigens* **20**: 227–228.

Le Petit, J.C., Cazes, M.H., Berthoux, F.C. et al. (1982b). Genetic investigation in mesangial IgA nephropathy. *Tissue Antigens* **19**: 108–114.

Lenhard, V., Dippel, J., Muller Wiefel, D.E. et al. (1980). HLA antigens in children with idiopathic nephrotic syndrome. *Proc. EDTA* **17**: 673–677.

Lerner, R., Glassock, R. and Dixon, F. (1967). The role anti-glomerular basement membrane antibody in the pathogenesis of human glomerulonephritis. *J. Exp. Med*. **126**: 989–1004.

Levy, J.B., Turner, A.N., George, A.J. et al. (1996). Epitope analysis of the Goodpasture antigen using a resonant mirror biosensor. *Clin. Exp. Immunol*. **106**: 79–85.

Levy, M. and Lesavre P. (1992). Genetic factors in IgA nephropathy (Berger's disease). *Adv. Nephrol. Necker Hosp*. **21**: 23–51.

Li, P.K., Burns, A.P., So, A.K. et al. (1991). The DQw7 allele at the HLA-DQB locus is associated with susceptibility to IgA nephropathy in caucsians. *Kidney Int*. **39**: 961–965.

Li, P.K., Poon, A.S. and Lai, K.N. (1994). Molecular genetics of MHC class II alleles in Chinese patients with IgA nephropathy. *Kidney Int*. **46**: 185–190.

Luger, A.M., Komathireddy, G., Walker, R.E. et al. (1994). Molecular and serologic analysis of HLA genes and immunoglobulin allotypes in IgA nephropathy. *Autoimmunity* **19**: 1–5.

Macdonald, I.M., Dumble, L.J. and Kincaid Smith, P.S. (1980). HLA Bw35, circulating immune complexes and IgA deposits in mesangial proliferative glomerulonephritis. *Aust. N.Z. J. Med*. **10**: 480–481.

McEnery, P.T. and Welch T.R. (1989). Major histocompatibility complex antigens in steroid-responsive nephrotic syndrome. *Pediatr. Nephrol*. **3**: 33–36.

Madden, D.R. (1995). The three-dimensional structure of peptide MHC complexes. *Annu. Rev. Immunol*. **13**: 587–622.

Medcraft, J., Hitman, G., Sachs, J. et al. (1993). Autoimmune renal disease and tumour necrosis factor beta gene polymorphism. *Clin. Nephrol*. **40**: 63–68.

Mercier, B., Bourbigot, O., Raguenes, O. et al. (1992). HLA class II typing of Goodpasture's syndrome affected patients. (abstract). *J. Am. Soc. Nephrol*. **3**: 658.

Mignot, E., Kimura, A., Abbal, M. et al. (1995). DQCAR microsatellite polymorphisms in three selected HLA class II-associated diseases. *Tissue Antigens* **46**: 299–304.

Mir, S., Kutukculer, N. and Kavakli, K. (1994). Major histocompatibility complex antigens and immune mechanisms in steroid-responsive nephrotic syndrome. *Acta Paediatr. Jpn.* **36**: 662–665.

Moore, R. (1993). MHC gene polymorphism in primary IgA nephropathy. *Kidney Int.* **39** Suppl.: S9–S12.

Moore, R.H., Hitman, G.A., Lucas, E. et al. (1990). HLA DQ region polymorphism associated with primary IgA nephropathy. *Kidney Int.* **37**: 991–995.

Moore, R., Hitman, G., Medcraft, J. et al. (1992). HLA-DP region gene polymorphism in primary IgA nephropathy: no association. *Nephrol. Dial. Transplant.* **7**: 200–204.

Moutonen, J., Pasternack, A., Helin, H. et al. (1981). Circulating immune complexes, the concentration of serum IgA and the distribution of HLA antigens in IgA nephropathy. *Nephron* **29**: 170–175.

Mouzon-Cambon, A., Ohayon, E., Bouissou, F. et al. (1980). HLA-DR typing in children with glomerular diseases. *Lancet* **ii**: 868.

Mouzon-Cambon, A., Bouissou, F., Dutau, G. et al. (1981). HLA DR-7 in children with idiopathic nephrotic syndrome. Correlation with atopy. *Tissue Antigens* **17**: 518–524.

Muller, G.A., Gebhardt, M., Kompf, J. et al. (1984). Association between rapidly progressive glomerulonephritis and the properdin factor BfF and different HLA-D region products. *Kidney Int.* **25**: 115–118.

Muller, G.A., Muller, C.A., Liebau, G. et al. (1981). Strong association of idiopathic membranous nephropathy (IMN) with HLA-DR3 and MT-2 without involvement of HLA-b18 and no association to BfF1. *Tissue Antigens* **17**: 332–337.

Muller, G.A., Muller, C.A., Lipovski, J. et al. (1988). Renal, major histocompatibility complex antigens and cellular components in rapidly progressive glomerulonephritis identified by monoclonal antibodies. *Nephron* **49**: 132–139.

Muraro, P.A., Vergelli, M., Kalbus, M. et al. (1997). Immunodominance of a low-affinity major histocompatibility complex-binding myelin basic protein epitope (residues 111–129) in HLA-DR4 (B1*0401) subjects is associated with a restricted T cell receptor repertoire. *J. Clin. Invest.* **100**: 339–349.

Nagy, J., Hamorri, A., Ambus, M. et al. (1979). More on IgA glomerulonephritis and HLA antigens. *N. Engl. J. Med.* **300**: 92.

Naito, S., Kohara, M. and Arakawa, K. (1987). Association of class II antigens of HLA with primary glomerulopathies. *Nephron* **45**: 111–114.

Nepom, G.T. (1995). Class II antigens and disease susceptibility. *Annu. Rev. Med.* **46**: 17–25.

Noel, L.H., Descamps, B., Jungers, P. et al. (1978). HLA antigens in three types of glomerulonephritis. *Clin. Immunol. Immunopathol.* **10**: 19–23.

Noss, G., Bachmann, H.J. et al. and Obling, H. Association of minimal change nephrotic syndrome (MCNS) with HLA-B8 and B13. *Clin. Nephrol.* **15**: 172–174.

Nunez-Roldan, A., Villechenous, E., Fernandez-Andrade, C. et al. (1982). Increased HLA-DR7 and decreased DR2 in steroid responsive nephrotic syndrome. *N. Engl. J. Med.* **306**: 366–367.

O'Donoghue, D.J., Darvill, A. and Ballardie, F.W. (1991). Mesangial cell autoantigens in immunoglobulin A nephropathy and Henoch–Schonlein purpura. *J. Clin. Invest.* **88**: 1522–1530.

Ogahara, S., Naito, S., Abe, K. et al. (1992a). Analysis of HLA class II genes in Japanese patients with idiopathic membranous glomerulonephritis. *Kidney Int.* **41**: 175–182.

Ogahara, S., Michinaga, I., Hiratsuka, T. et al. (1992b). DNA typing of HLA-class II genes in idiopathic nephropathy. *Nippon Rinsho* **50**: 3072–3078.

Papiha, S.S., Pareek, S.K., Rodger, R.S. et al. (1987). HLA-A,B,DR and Bf allotypes in patients with idiopathic membranous nephropathy(IMN). *Kidney Int.* **31**: 130–134.

Payami, H., Joe, S., Farid, N.R. et al. (1989). Relative predispositional effects (RPEs) of marker alleles with disease: HLA-DR alleles and Graves disease. *Am. J. Hum. Genet.* **45**: 451–546.

Perl, S.I., Pussell, B.A., Charlesworth, J.A. et al. (1981). Goodpasture's (anti-GBM) disease and HLA-DRw2 [Letter]. *N. Engl. J. Med.* **305**: 463–464.

Phelps, R.G., Jones, V.J., Turner, A.N. et al. (1996a). HLA-DR peptide binding capacity and susceptibility to Goodpasture's disease [Abstract]. *Immunology* **89** Suppl. 1: 79.

Phelps, R.G., Turner, N., Rees, A.J. et al. (1996b). Direct identification of naturally processed autoantigen derived peptides bound to HLA-DR15. *J. Biol. Chem.* **271**: 18 549–18 553.

Phelps, R.G., Jones, V., Coughlan, M. et al. (1998). Presentation of the Goodpasture autoantigen to CD4 T cells is influenced more by processing constraints than by HLA class II peptide binding preferences. *J. Biol. Chem.* **273**: 11440–11447.

Phelps, R.G. and Rees, A.J. (1999). The HLA complex in Goodpasture's disease: a model for analyzing susceptibility to autoimmunity. *Kidney Int.* **56**: 1638–1654.

Raguenes, O., Mercier, B., Cledes, J. et al. (1995). HLA class II typing and idiopathic IgA nephropathy (IgAN): DQB1*0301, a possible marker of unfavorable outcome. *Tissue Antigens* **45**: 246–249.

Rambausek, M., Seelig, H.P., Andressy, K. et al. (1982). Clinical and serological features of mesangial IgA glomerulonephritis. *Proc. EDTA* **19**: 663–668.

Rambausek, M.H., Waldherr, R. and Ritz, E. (1993). Immunogenetic findings in glomerulonephritis. *Kidney Int.* **39** Suppl.: S3–S8.

Rashid, H.U., Papiha, S.S., Agroyannis, B. et al. (1983). The associations of HLA and other genetic markers with glomerulonephritis. *Hum. Genet.* **63**: 38–44.

Rees, A.J., Peters, D.K., Compston, D.A. et al. (1978). Strong association between HLA-DRW2 and antibody-mediated Goodpasture's syndrome. *Lancet* **i**: 966–968.

Rees, A.J., Peters, D.K., Amos, N. et al. (1984). The influence of HLA-linked genes on the severity of anti-GBM antibody-mediated nephritis. *Kidney Int.* **26**: 445–450.

Roccatello, D., Coppo, R., Amoroso, A. et al. (1987). Failure to relate mononuclear phagocyte system function to HLA-A,B,C,DR,DQ antigens in membranous nephropathy. *Am. J. Kidney Dis.* **31**: 130–134.

Ruder, H., Scharer, K., Lenhard, V. et al. (1982). HLA phenotypes and idiopathic nephrotic syndrome in children. *Proc. Eur. Dial. Transplant. Assoc.* **19**: 602–606.

Ruder, H., Scharer, K., Opelz, G. et al. (1990). Human leucocyte antigens in idiopathic nephrotic syndrome in children. *Pediatr. Nephrol.* **4**: 478–481.

Sacks, S., Nomura, S., Warner, C. et al. (1992). Analysis of complement C4 loci in caucasoids and Japanese with idiopathic membranous nephropathy. *Kidney Int.* **42**: 882–887.

Sacks, S., Warner, C., Campbell, R. et al. (1993). Molecular mapping of the HLA class II region in HLA-DR3 associated idiopathic membranous nephropathy. *Kidney Int.* **39** Suppl.: S13–S19.

Salant, D.J., Quigg, R.J. and Cybulsky, M.V. (1989). Heymann nephritis: mechanisms of renal injury. *Kidney Int.* **35**: 976–984.

Sasajuki, T., Hayase, R., Iwanto, I. et al. (1979). HLA and acute post-streptococcal glomerulonephritis. *N. Engl. J. Med.* **301**: 1184–1185.

Saus, J., Wieslander, J., Langeveld, J. et al. (1988). Identification of the Goodpasture antigen as the α3(IV) chain of collagen IV. *J. Biol. Chem.* **263**: 13 374–13 380.

Savi, M., Neri, T.M., Silvestri, M.G. et al. (1979). HLA antigens and IgA mesangial glomerulonephritis. *Clin. Nephrol.* **12**: 45–46.

Seedat, Y., Nathoo, B., Parag, K. et al. (1988). IgA nephropathy in blacks and Indians of Natal. *Nephron* **50**: 137–141.

Sercarz, E.E., Lehman, P.V, Ametani, A. et al. (1993). Dominance and crypticity of T cell antigenic determinants. *Annu. Rev. Immunol.* **11**: 729–776.

Short, C.D., Dyer, P.A., Cairns, S.A. et al. (1983). A major histocompatibility system haplotype associated with poor prognosis in idiopathic membranous nephropathy. *Dis. Markers* **1**: 189–196.

Spencer, S.J., Burns, A., Gaskin, G. et al. (1992). HLA class II specificities in vasculitis with antibodies to neutrophil cytoplasmic antigens. *Kidney Int.* **41**: 1059–1063.

Stern, L.J., Brown, J.H., Jardetzky, T.S. et al. (1994). Crystal structure of the human class II MHC protein HLA-DR1 complexed with an influenza virus peptide. *Nature* **368**: 215–221.

Theofilopoulos, A. (1995a). The basis of autimmunity: Part I Mechanisms of aberrant self-recognition. *Immunol. Today* **16**: 90–97.

Theofilopoulos, A. (1995b). The basis of autimmunity: Part II Genetic predisposition. *Immunol. Today* **16**: 150–159.

Thomson, P.D., Barratt, T.M., Stokes, C.R. et al. (1976). HLA antigens and atopic features in steriod-responsive childhood nephrotic syndrome. *Lancet* **ii**: 765–768.

Tomura, S., Kashiwabara, H., Tuchida, H. et al. (1984). Strong association of idiopathic membranous nephropathy with HLA-DR2 and MT1 in Japanese. *Nephron* **36**: 242–245.

Trompeter, R.S., Barrett, T.M., Kay, R. et al. (1980). HLA atopy and cyclophosphamide in steroid responsive childhood nephrotic syndrome. *Kidney Int.* **17**: 113–117.

Turner, N., Mason, P.J., Brown, R. et al. (1992). Molecular cloning of the human Goodpasture antigen demonstrates it to be the alpha 3 chain of type IV collagen. *J. Clin. Invest.* **89**: 592–601.

Vaughan, R., Demaine, A. and Welsh, K. (1989). A DQA1 allele is strongly associated with idiopathic membranous nephropathy. *Tissue Antigens* **34**: 261–269.

Vaughan, R.W., Ogahara, S., Acquart, S. et al. (1992). Idiopathic membranous nephropathy. In *Proceedings of the Eleventh International Histocompatibility Workshop and Conference* (eds K. Tsuji, M. Aizawa and T. Sasazuki), Vol. 1, pp. 745–748, Oxford University Press, Oxford.

Vaughan, R., Tighe, M., Boki, K. et al. (1995). An analysis of HLA class II gene polymorphism in British and Greek idiopathic membranous nephropathy patients. *Eur. J. Immunogenet.* **22**: 179–186.

Vyze, T.J. and Kotzin, B.L. (1998). Genetic susceptibility to systemic lupus erythematosus. *Annu. Rev. Immunol.* **16**: 261–292.

Waldo, B.F., Beischel, L. and West, C.D. (1986). IgA synthesis by lymphocytes from patients with IgA nephropathy and their relatives. *Kidney Int.* **29**: 1229–1233.

Welch, T.R., Beischel, L., Balakrishnan, K. et al. (1986). Major-histocompatibility-complex extended haplotypes in membranoproliferative glomerulonephritis. *N. Engl. J. Med.* **314**: 1476–1481.

Wucherpfennig, K.W., Bei, Y., Bhol, K. et al. (1995). Structural basis for major histocompatibility complex (MHC)-linked susceptibility to autoimmunity: charged residues of a single MHC binding pocket confer selective presentation of self-peptides in pemphigus vulgaris. *Proc. Natl Acad. Sci. USA* **92**: 11 935–11 939.

Wucherpfennig, K.W. and Strominger, J.L. (1995). Selective binding of self-peptides to disease-associated major histocompatibility complex (MHC) molecules: a mechanism for MHC-linked susceptibilitty to human autoimmune diseases. *J. Exp. Med.* **181**: 1597–1601.

Zhang, L., Jayne, D.R., Zhao, M.H. et al. (1995). Distribution of MHC class II alleles in primary systemic vasculitis. *Kidney Int.* **47**: 294–298.

Zhou, G.P., Guo, Y.Q., Ji, Y.H. (1994). Major histocompatibility complex class II antigens in steroid-sensitive nephrotic syndrome in Chinese children. *Pediatr. Nephrol.* **8**: 140–141.

Zucchelli, P., Ponticelli, C., Cagnoli, L. et al. (1987). Genetic factors in the outcome of idiopathic membranous nephropathy [Letter]. *Nephrol. Dial. Transplant.* **1**: 265–266.

CHAPTER 14

HLA and Neurological Diseases

Jan Hillert and Anna Fogdell-Hahn

INTRODUCTION

The human leucocyte antigen (HLA) associations with disorders of the nervous and neuromuscular systems are quite extraordinary. They include the strongest HLA association known, in narcolepsy, as well as one of the best studied, in multiple sclerosis (MS). In myasthenia gravis (MG), it is uncertain whether class I or class II genes are primarily responsible for the association. In addition, the case of an HLA association in juvenile myoclonic epilepsy (JME) is unresolved and may, in any case, be due to a non-HLA allele in linkage disequilibrium. Thus, this group of conditions offers a good illustration of the basic problems and controversies regarding HLA associations and their interpretation.

DISEASES WITH AUTOIMMUNE CHARACTERISTICS

Multiple sclerosis

Background and summary

MS is a chronic demyelinating disease of the central nervous system (CNS) with inflammatory characteristics. It was first shown to be associated with HLA specificities as early as 1972, first with the class I alleles A3 and B7 (Jersild et al., 1972; Naito et al., 1972) and subsequently with the HLA class II specificity Dw2 (then named LD7a) (Jersild et al., 1973). The Dw2 or DR15, DQ6 association has since been confirmed in numerous studies, especially in populations of western European origin (Tiwari and Terasaki, 1985). Additional associations have repeatedly been described, but so far none of these has been sufficiently well corroborated for a consensus to have developed.

HLA-Dw2

With the ever-increasing recognition of alleles and improvements in HLA typing by genomic methods, the Dw2 association still holds true in most populations studied. With genomic nomenclature, the cellular specificity Dw2 corresponds to DRB1*1501, DRB5*0101, DQA1*0102, DQB1*0602 (Fogdell et al., 1995). To be adequate, HLA association studies in MS must use typing techniques with sufficient resolution to distinguish at least the DRB1*1501 from the other DR2 subtypes. Failure to do this, for example typing with the serological resolution of DR2, might give confusing results, especially in populations where DR2 haplotypes other than Dw2 are common. With cellular typing, however, it was possible to distinguish Dw2 from the other DR2-bearing specificities Dw12, Dw21 and Dw22, which all were typed as DR2 by serology. For these reasons, a general impression developed that the DR2 association was confined to Europeans. An interesting illustration of the difficulties involved is a study of Hungarian gypsies, who were shown to have a high frequency of DR2 and yet no association between DR2 and MS (Takács et al., 1990). However, genomic typing data provided in this study allow the conclusion to be drawn that most of the DR2-containing haplotypes expressed DRB1*1601 rather than DRB1*1501 (Olcrup and Hillert, 1991). A clear picture of the universal

TABLE 14.1 Phenotypic frequencies (%) of the Dw2 haplotype among multiple sclerosis patients and controls with various ethnic backgrounds, as determined by cellular or genomic methods

Country or ethnic group	Reference	% Dw2 in MS (n)	% Dw2 in controls (n)
USA	Gogolin et al. (1989)	52 (330)	18 (319)
USA: Black	Dupont et al. (1977)	35 (31)	0 (34)
Norway	Spurkland et al. (1991a)	72 (69)	33 (180)
Sweden	Hillert and Olerup (1993)	60 (148)	30 (158)
England	Francis et al. (1991)	63 (71)	32 (100)
N. Ireland	Cullen et al. (1991)	63 (54)	30 (150)
N. France	Marcadet et al. (1985)	70 (44)	? (44)
Sardinia	Marrosu et al. (1992)	5 (103)	7 (69)
Gypsies	Takács et al. (1990)	30 (10)	0 (21)
Israel	Brautbar et al. (1982)	14 (45)	6 (51)
Japan	Hao et al. (1992)	22 (60)	7 (44)
Hong Kong	Serjeantson et al. (1992)	50 (12)	12 (42)

increase in the DRB1*1501, DRB5*0101, DQA1*0102, DQB1*0602 haplotype among MS patients is seen in studies using cellular or genomic HLA class II typing techniques (Table 14.1). There are few populations where exceptions to this pattern occur (see below).

The normal Dw2 haplotype

DNA sequencing of the DR and DQ genes of the *Dw2* haplotype in MS patients shows identical sequences to those in Dw2-positive controls (Cowan et al., 1991). Similarly, analysis of flanking polymorphisms in Dw2-carrying MS patients and controls also shows close to identical results (Hillert and Olerup, 1993), indicating that it is the normal Dw2 haplotype, and not a mutated variant, that somehow increases the risk for MS.

Other DR associations

In Europeans, Dw2 is the primarily associated specificity/haplotype. Since around 60% of patients carry Dw2, this will have an influence on the frequencies of other alleles and haplotypes. If chromosomes with the Dw2 haplotype are subtracted in the analysis, however, some additional differences in allele frequencies may still be observed. One such possible additional class II association is with the DR17, DQ2 haplotype or, in sequence nomenclature, DRB1*0301, DRB3*0101, DQA1*0501, DQB1*0. If this haplotype were neutral in MS, a slight decrease in its frequency would be evident in most studies on account of the increased frequency of Dw2. However, several studies have observed a slight increase in DR3 among MS patients (Paty et al., 1977; de Moerloose et al., 1979; Cullen et al., 1991; Francis et al., 1991; Olerup and Hillert, 1991a; Spurkland et al., 1991).

Associations in MS with other specificities have been reported in different ethnic groups. DR4 has been suggested as the MS-associated haplotype in Arabs (Kurdi et al., 1977) and Sardinians (Marrosu et al., 1988). Subsequent studies of Sardinian MS patients have shown a stronger association with DRB1*0301 and a lesser association with DRB1*04, especially the subtype DRB1*0405 (Marrosu et al., 1997). DR6 has been reported to be associated with MS in Japan (Naito et al., 1978) and Mexico (Gorodezky et al., 1986). More recently, however, the DR6 association in the Japanese has not been confirmed, but instead an association with DR2 and DQB1*0602 has been observed (Hao et al., 1992).

Considering the peptide-binding properties of HLA molecules, that identical peptides can bind different alleles, it is not unlikely that alleles other than those of the associated Dw2 haplotype may contribute to disease susceptibility. One way to study this may be to look for sequence motifs shared between different associated alleles. These motifs, or single amino acid positions, would then be more strongly associated than the single alleles or haplotypes. Such sharing of sequences in MS has been suggested to occur for DQB1 alleles of the DR2, DR4 and DR6 haplotypes (Vartdal et al., 1989), and subsequently for DQA1 alleles (DQ α-chain, amino acid position 34) (Spurkland et al., 1991a) as well as for DR β-chain position 86 (Allen et al., 1994). However, none of these attempts to define a better marker for susceptibility than Dw2 has been confirmed (Francis et al., 1991; Olerup and Hillert, 1991; Sinha et al., 1991; Hillert et al., unpublished data).

Is HLA-DR or -DQ more important for the class II association in multiple sclerosis?

Owing to tight linkage disequilibrium, DR and DQ alleles are usually inherited together as conserved haplotypes. Therefore, it is difficult to determine whether either part of the haplotype accounts for the susceptibility alone. Alternatively, it may be the whole HLA-Dw2 haplotype as such that increases the risk for MS. One way to study this is to identify patients carrying only parts of this haplotype. A few patients positive for DQB1*0602 but not for DRB1*1501 have been found, which could indicate a primary importance of the DQ molecule (Serjeantson et al., 1992; Spurkland et al., 1997). However, patients carrying the DRB1*1501 allele in combination with DQB1*0603 (Spurkland et al., 1997) and DQB1*0601 (Serjeantson et al., 1992) have also been reported, although these DQB1 alleles are very similar to the DQB1*0602 allele. (Three nucleotides differ between *0602 and *0602, and 13 between *0601 and *0602.) Since the numbers of such patients are small, no definitive identification of which part of the associated haplotype more closely determines susceptibility can yet be made. If the association is due to the immunological function of the HLA molecules, it is also possible that the different heterodimers of the associated haplotype act in concert.

Association with DP

Genomic typing of reasonable resolution of HLA-DP alleles has not been available until recently. Three genomic techniques, sequencing-based typing (SBT) and polymerase chain reaction amplification with sequence-specific primers (PCR-SSP) or sequence-specific oligonucleotide probes (PCR-SSO), can identify most of the over 80 existing DPB1 and 10 DPA1 alleles (Aldener-Cannavá and Olerup, 1996). A previously reported association in MS with DPw4 using restriction fragment length polymorphism (RFLP) typing (Odum et al., 1988) was not confirmed in subsequent analysis of the same (Fugger et al., 1990) and other (Olerup et al., 1990; Cullen et al., 1991; Spurkland et al., 1991a) ethnic groups. With PCR-SSP typing, a recent study failed to find any associations with the DPA and DPB genes in MS independent of Dw2 (Fogdell et al., unpublished data).

HLA and clinical subtypes

The subdivision of MS into distinct clinical subgroups is controversial. However, two distinct clinical forms, primarily chronic progressive (PCP) and relapsing/remitting (RR) MS can be well characterized. Differences in HLA association in these forms have been suggested (Olerup et al., 1989; Hillert et al., 1992), but these findings have not been confirmed by other groups (Francis et al., 1991; Spurkland et al., 1991a). Very similar frequencies of HLA-Dw2 occur in both subgroups, which indicates similarity rather than distinction. More interestingly, in a recent Japanese study, patients were stratified into two clinical forms of MS, Western and Asian type, based on clinical features. The analysis showed that the DRB1*1501 and DRB5*0101 alleles were present in Western-type MS patients (41%), less commonly in healthy controls (14%) and not in Asian-type MS (0%) (Kira et al., 1996).

Reappreciation of HLA class I associations

The use of recently developed genomic HLA class I typing has revealed error rates as high as 10% in serological HLA typing, mostly due to overestimation of homozygosity or a failure to detect subtypes (Mytilineos et al., 1997). Using genomic class I typing by PCR-SSP, it was recently found that the originally described association with HLA-A3 is, surprisingly, independent of the Dw2 association, whereas the B7 association is not (Fogdell-Hahn et al., in press). No additional associations were found for the HLA-C gene. More interestingly, the A2 allele was shown to be decreased independently of Dw2 in the patient group, a tendency that may be noted in several previous studies. This suggests the existence of a protective influence of genes in the HLA class I region, a phenomenon previously described in experimental autoimmune encephalomyelitis (EAE) (Mustafa et al., 1994).

Linkage studies of HLA genes in multiple sclerosis

Several studies of genetic linkage between HLA genes and MS have been reported, by investigating either multiplex families in general or affected sib-pairs (Table 14.2). In general, non-parametric sib-pair analyses have provided less evidence for linkage than parametric analyses of multiplex families. This discrepancy may depend on study design or on the relative importance of these genes in the different populations investigated. Full genomic screening studies with microsatellite markers in MS-affected sib-pairs support linkage with the HLA region, although not fulfilling stringent criteria for genome-wide linkage (Ebers et al., 1996; Haines et al., 1996; Sawcer et al., 1996). At the same time, these studies have failed to identify any other major susceptibility genes in MS, the importance of which clearly exceeds that of the HLA genes.

Conclusion

Both association and linkage studies have repeatedly shown that HLA genes are important in MS. It is not clear whether this is due to the function of the HLA molecules themselves or to products of other genes in that region. Neither can we state that a single part of the associated HLA haplotype is more important than others. Genetic analysis of the HLA region suggests that HLA genes carried by MS patients are normal. Additional factors, such as interaction with the environment or the presence of other, non-HLA genes, are probably needed for disease development. Since the function of the HLA molecules is to present peptides, both self and foreign, this is likely to be an important aetiological factor. However, this remains to be shown.

Optic neuritis

Acute unilateral optic neuritis (ON) is a common disease manifestation in MS and a frequent presenting symptom. However, not all patients with monosymptomatic ON go on to develop signs or symptoms of a disseminated CNS demyelinating disease, e.g. MS. Thus, only a subgroup of ON patients is thought to have an MS-like disease. There has been a number of reports concerning the distribution of HLA alleles and haplotypes in ON. The consistent finding is that of a slight increase of Dw2, although at a lower level than seen in MS (Hely et al., 1986; Francis et al., 1987a; Sandberg-Wollheim et al., 1990). Most studies are limited in size, and the increase in Dw2 is only statistically significant in the largest study so far reported (Hillert et al., 1996). One interpretation of this finding, which is supported by findings in magnetic resonance tomography and cerebrospinal fluid analysis, is that the clinical syndrome of ON is heterogeneous.

Myasthenia gravis

Background and summary

MG is a neuromuscular disease characterized by muscular fatigue induced by autoantibodies directed against structures in the neuromuscular junction. The most prominent target for attack is the postsynaptic acetylcholine receptor (AChR), and particularly its α-subunit. Muscular weakness

TABLE 14.2 Linkage studies of HLA class II genes in multiple sclerosis

Country	Reference	Families (n)	Sib-pairs (n)	Results	Type of study
Canada	Ebers et al. (1982)	36	40	Negative	Sib-pair
Scotland	Francis et al. (1987)	34	31	Negative	Sib-pair
UK	Kellar-Wood et al. (1995)	115		Negative	Sib-pair
Australia	Stewart et al. (1981)	100		Positive	Families
Sweden	Hillert et al. (1994)	9	9	Positive	Families
Italy	Eoli et al. (1995)	28		Positive	Families
Finland	Tienari et al. (1993)	21		Positive	Families
Sweden	Fogdell et al. (1997)	49	19	Positive	Families
USA	Hauser et al. (1989)	33		Positive	Families
USA	Voskuhl et al. (1996)	9	29	Positive	Families and sib-pairs

may be generalized or limited to certain muscular groups, most typically ocular muscles, causing fluctuating ptosis and/or diplopia. Age of onset may vary, from adolescence to old age. There is a preponderance of females, especially in the younger age groups, where ocular symptoms are frequent. MG is usually associated with aberrations in thymus histology, most typically hyperplasia, sometimes involution and less commonly thymoma, which is usually seen in the oldest age group.

The existence of an HLA association in MG has been recognized since 1973 (Fritze et al., 1973; Säfwenberg et al., 1973). The haplotype A1, B8, DR3, DQ2 has drawn most attention, showing a strong association in populations of Western European origin. However, it remains unclear which part of this haplotype is most strongly associated, and it has even been suggested that non-HLA genes such as the tumour necrosis factor (TNF) genes may be of primary importance. In addition, several reports have indicated a heterogeneity, with the B8,DR3 association being most evident in young Caucasian females and other alleles or haplotypes being of importance in older patients and other ethnic groups.

B8,DR3

The early observations of an association in MG with the specificity then known as HL-A8 (Fritze et al., 1973; Säfwenberg et al., 1973), soon afterwards renamed HLA-B8, have been uniformly confirmed in later studies in Caucasian populations. In 1975, however, a similarly strong association was observed with what were later to be classified as HLA class II specificities (Kaakinen et al., 1975). Subsequently, it was realized that this depended on the linkage disequilibrium between B8 and DR3 (Christiansen et al., 1978; Naeim et al., 1978) which, in turn, is in even stronger linkage disequilibrium with the DQ genes. Thus, with the advent of genomic typing techniques increasing resolution, the haplotype has finally been identified as HLA-B*0801, DRB1*0301, DRB3*0101, DQA1*0501, DQB1*0201 (Spurkland et al., 1991b; Vieira et al., 1993; Hjelmström et al., 1995). In the telomeric direction, the haplotype includes HLA-Cw7 and HLA-A1, but with less strong associations. However, the association does not seem to extend centromerically to the TAP genes (Hjelmström et al., 1997) and clearly not to the DP genes (Ratanachaiyawong, 1994).

Much of the discussion has focused on the limitation of this association, but also on the identification of the exact part of this haplotype that is of primary importance. Since B8 and DR3, or more specifically the DRB1*0301, DQA1*0501, DQB1*0201 haplotype, or parts of it, may occur independently, there is an opportunity to investigate whether B8 and DR3 in isolation are also increased in frequency in MG. Since such a comparison would require a large number of individuals to enable a significant observation, such a study is yet to be performed. However, published data suggest that B8 in isolation is more common than DR3 in isolation (Degli-Esposti, 1992; Vieira et al., 1993), even after compensation for the background frequency of these alleles. Thus, it is possible that class I genes are of greater importance than class II genes in MG. Interestingly, Degli-Esposti et al. (1992) went one step further and identified a chromosomal region between the TNFα and HLA-B genes that was shared by 50 out of 51 patients positive for either part of the haplotype. Whether this locus represents the true site of the "MG gene" or not remains to be determined.

Whereas the B8,DR3 haplotype is clearly increased among MG patients of European origin, studies in other ethnic groups have shown that this may not be universal. In Chinese MG patients, B8 and/or DR3 are very infrequent in MG. Instead, other associations have been reported, particularly with HLA-B46 (Hawkins et al., 1984; Chiu, 1990). Likewise, in Japan, there is no clear increase in DR3 (Matsuki, 1990).

Disease heterogeneity

From the early studies, it was clear that the B8-DR3 association is most evident among patients with onset of disease in adolescence or early adulthood and particularly in females, who dominate this group of MG patients. In addition, MG in these patients is commonly confined to ocular muscles rather than generalized. With increasing age of onset, and with increasing numbers of men in the patient groups, the B8, DR3 association becomes weaker (Möller et al., 1976; Säfwenberg et al., 1978; Compston et al., 1980). In particular, patients with thymoma, a patient group often having onset of disease in old age, are rarely B8-DR3 carriers (Oosterhius et al., 1976; Compston et al., 1980; Carlsson et al., 1990). Further support for an immunogenetic heterogeneity in MG is provided by the observation that penicillamine-induced MG rarely is B8,DR3 associated, but instead is possibly associated with DR1 (Dawkins et al., 1983; Delamere et al., 1983). However, this latter association may merely reflect the well-established association with DR1 in rheumatoid arthritis.

Functional studies

In comparison with most autoimmune disorders, an evident candidate autoantigen has been available in MG since it became established that the antibody-mediated attack in MG was to a large extent directed against the AChR (for an overview, see Drachmann, 1994). More recently, the occurrence of a T cell response against AChR has been well established. Based on this knowledge, several investigators have focused on the immunological specificities of T cells from B8,DR3-positive patients as well as on the affinities of immunodominant peptides to B8 and DR3 molecules (Malcherek et al., 1993; Melms et al., 1993; reviewed by Hawke et al., 1996).

Conclusion

In summary, the HLA association in MG is of great interest, not least because of the presence of clear-cut immunogenetic heterogeneity. Together with the existence of a good candidate autoantigen, MG offers an exceptional opportunity to study the importance of HLA genes and molecules in an autoimmune disorder.

Peripheral nerve demyelination

Inflammatory demyelinating polyneuropathies

This group of diseases basically consists of the Guillain–Barré syndrome (GBS), also known as acute inflammatory demyelinating polyneuropathy (AIDP), and its chronic form chronic inflammatory demyelinating polyneuropathy (CIDP). Classical GBS is characterized by a subacutely ascending paresis usually beginning in the lower extremities. Sensory functions may or may not be involved. At its peak, total paralysis including respiratory muscles is common for a duration of up to several weeks, after which gradual recovery occurs. In CIDP, a more slowly evolving disease course is present and the course is often relapsing/remitting. Here, we have chosen also to include polyneuropathy occurring in presence of a monoclonal gammopathy, usually of immunoglobulin M (IgM) isotype.

GBS, with its dramatic clinical picture and clear autoimmune features, has inspired a number of HLA studies. In general, these studies have been of limited size and negative (Stewart et al., 1978; Winer et al., 1988; Hillert et al., 1991; Rees et al., 1995). It has been suggested that a reason for this could be the variety of triggering events in GBS, i.e. that a pathogenetically more distinct group of patients could potentially reveal associations. Kaslow et al. (1984) described an increase in HLA-A1 in patients with GBS following vaccination against swine influenza in the USA in 1976–1977. However, in a small number of patients with preceding *Campylobacter* infection, Winer et al. (1988) did not observe any differences from controls whereas, in a large study, Rees et al. (1995) describe an increase in DQB1*03 alleles in patients with a preceding *Campylobacter jejunii* infection. These results need further support.

Separate analyses of HLA antigens or genes in the CIDP syndrome have been reported. In two early studies of a limited numbers of patients, tentative increased frequencies of HLA-B8 and DR3/Dw3 were reported (Stewart et al., 1978; Adams et al., 1979). In 1990, a large study of 71 patients with CIDP indicated a possible association with the A3,B7,DR2 haplotype, but no association with B8-DR3 (Feeney et al., 1990). However, in a moderately sized study somewhat later, neither of these associations was confirmed (van Doorn et al., 1991). Thus, at present, there is still great hesitation as to the significance of HLA alleles in both the subacute (GBS/AIDP) and chronic (CIDP) forms of demyelinating polyneuropathy (Vaughan et al., 1990).

A somewhat different finding is described in a report by Vrethem et al. (1993) of patients with polyneuropathy associated with anti-myelin-associated glycoprotein (MAG) antibodies in the presence of a non-malignant paraproteinaemia. Several class II haplotypes were found to be increased in frequency among patients. A possible common denominator was identified as the presence of a tryptophan residue in amino acid position 9 of the DR β-chain, a position of theoretical importance for peptide-binding properties of the DR molecule. However, this hypothesis remains to be confirmed.

OTHER DISEASES

Narcolepsy

Background and summary

Narcolepsy is a condition characterized by irresistible daytime sleep attacks in combination with other symptoms of dysregulation of sleep, such as hypnagogic hallucinations, sleep paralysis and, most specifically, cataplexy (the abrupt loss of muscle tone upon sudden emotional stimulation such as fright or laughing). A striking feature in cataplectic narcolepsy is a near 100% HLA class II association, although the figures vary somewhat depending on the population studied. In Japan, 100% of narcoleptic patients carry the DR15, DQ6, Dw2 haplotype, whereas figures for Caucasians are 90–100% and for African–Americans 70% (see Table 14.3).

DR2/Dw2

The first associations reported for narcolepsy were with the HLA class I alleles A3 and B7 (Seignalet and Billiard, 1984; Poirier et al., 1986). However, these associations were not as strong as the 100% DR2/Dw2 association discovered shortly thereafter (Langdon et al., 1984; Juji et al., 1984; Mueller-Eckhardt et al., 1986). In addition, since Japanese narcoleptics carry the associated DR15, DQ6,Dw2 haplotype with different class I alleles (B15, B51 or B35) (Matsuki et al., 1985) than seen in Europeans (A3, B7), it is quite clear that the common denominator is in the class II region.

DR or DQ?

Of the class II genes of the DR2/Dw2 haplotype, DQ genes have been suggested to be more important than DR, since many of the DRB1*1501-negative patients carry the DQB1*0602 and DQA1*0102 alleles (Table 14.4). DQ is certainly a better marker for narcolepsy in African–Americans, where only 75% carried DRB1*1501 whereas 27/28 (96%) carried DQB1*0602 (Mignot et al., 1994). However, in Caucasians, 18/19 (95%) were DRB1*1501 positive and the Caucasian DR2-negative narcoleptic patient did not carry DQB1*0602 (Mignot et al., 1994). Therefore, even though the HLA association is strong, occasional patients carry neither DRB1*1501 nor DQB1*0602 (Mignot et al., 1997a). In addition, since these alleles are carried by up to 30% of the healthy population, the penetrance of narcolepsy remains low.

Interpretations

The aetiology of narcolepsy is unknown, but the strong association with HLA implies that narcolepsy might have an autoimmune component (Mignot et al., 1995). However, spontaneous narcolepsy occurring naturally in dogs is caused by a single autosomal recessive trait encoded in the immunoglobulin heavy-chain complex (Mignot, 1997b), and lacks a major histocompatibility complex (MHC) association (Dean et al., 1989). In

TABLE 14.3 Frequency of *HLA-Dw2* association in cataplectic narcolepsy in various ethnic groups

Country or ethnic group	Reference	Frequency and genes		n
UK	Langdon et al. (1984)	100%	DR2	37
Japan	Juji et al. (1984)	100%	DR2, not B52	40
Germany	Mueller-Eckhardt et al. (1986)	98.3%	DR2,DQ1	58
France	Billiard et al. (1986)	100%	DR2,DQ1	23
Japan	Honda et al. (1986)	100%	DR2,DQ1	135
Japan	Juji et al. (1988)	100%	Dw2	190
Czechoslovakia	Roth et al. (1988)	98%	DR2	124
Israel	Kwon et al. (1995)	100%	DRB1*1501,	8
African–Americans	Neely et al. (1987)	58%	DR2	12
African–Americans	Mignot et al. (1994)	75%	DRB1*1501	28

TABLE 14.4 HLA-DRB1 alleles present in narcolepsy patients who carry HLA-DQ6 in the absence of DRB1*1501

DRB1 allele	Reference
DRB1*1202 with DQA1*0102,DQB1*0602	Behar et al. (1995)
DR1,DQ1/DR5,DQ3	Douglass et al. (1989)
DRB1*08 deleted	Grumet and Mignot (1994)
36% of non-DR15 were DQA1*0102,DQB1*0602	Mignot et al. (1997)
DR8-related with deletion, DQ1	Roushdy et al. (1993)

addition, familial narcolepsy may occur in the absence of DR2/Dw2 (Ditta et al., 1992). Hypothetically, the products of non-HLA narcolepsy genes may interact with specific alleles of genes in the HLA region. Another possibility is that sleep-regulating genes are in linkage disequilibrium with the associated HLA class II haplotype. Attempts to identify such genes have been made. The alleles of the Dw2 haplotype have been analysed by direct sequencing (Lock et al., 1988; Uryu et al., 1989) and by RFLP (Olerup et al., 1990), although without showing differences from the Dw2 alleles carried by healthy individuals. Both groups also carry the same DRB5 allele (Fogdell et al., 1995). Studies with microsatellite markers indicates that the normal DQB1*0602 is present in all narcoleptic patients (Ellis et al., 1997). This supports the hypothesis that it is the HLA genes themselves that are the predisposing factor in narcolepsy.

Juvenile myoclonic epilepsy (JME)

Background and summary

Epilepsy is genetically, as well as aetiologically, heterogeneous. Whereas most cases of epilepsy are non-genetic and secondary to other conditions, there is also a number of clearly genetic epilepsy disorders of more or less distinct phenotype. Linkage for such conditions has been reported with a number of loci on chromosomes 1q, 6p, 8q, 16p, 20q, 21q and 22q (for review, see Delgado-Escueta et al., 1994).

The most common genetic form of epilepsy is JME, which comprises 5–10% of all cases of epilepsy. This is a form of generalized epilepsy where myoclonic jerks occur in isolation or preceding a generalized seizure. JME usually starts in adolescence, is characterized by a synchronous spike-and-wave electroencephalographic pattern and often responds well to sodium valproate.

Greenberg et al. (1988) reported significant linkage (lod score > 3) with genetic markers in the HLA gene complex in 24 JME families. This observation was later confirmed, although somewhat modified. At present, it seems clear that this locus, referred to as EJM 1, confers disease in some, but not all, families with JME. The exact location of the gene is still unclear, but it is likely to be centromeric to the HLA class II locus. Thus, JME seems to be an example of a disorder where linkage and association with the HLA gene complex are explained by flanking genes.

Linkage studies

In the original study by Greenberg et al. (1988), the highest lod scores with HLA markers were obtained for a recessive model. Weissbecker et al. (1991), in an analysis of 23 families of probands with JME, reported a significant lod score for a dominant model. However, in this study, a number of different disease phenotype definitions was used and results varied somewhat depending on definition. In some, recessive models resulted in indicative lod scores. Similarly, Durner et al. also reported significant linkage with a dominant model (Durner et al., 1991; Sander et al., 1995). More recently, further support for the importance of a genetic factor in or at the centromeric side of the HLA gene complex has been reported (Liu et al., 1995; Sander et al., 1997).

In contrast, Whitehouse et al. (1993), analysing families of JME patients, failed to find evidence for linkage with several markers on chromosome 6. A subsequent study from the same group, now confined to JME families using stringent definitions, again failed to provide evidence for linkage to chromosome 6 (Elmslie et al., 1996). Instead, they recently reported significant linkage with a locus on chromosome 15 (Elmslie et al., 1997).

Association studies

Findings in linkage studies naturally prompted the search for possible associations with HLA markers. Here, the picture is similar to that of the linkage data. Thus, whereas several studies have identified an association with HLA-DR6 (Durner et al., 1992) or its split DR13 (Obeid et al., 1994; Greenberg et al., 1996), other investigators have reported a similar distribution of alleles in cases and controls (Sander et al., 1997).

Conclusion

JME is likely to be a genetically heterogeneous condition where some genetic susceptibility is conferred by genes within or at the centromeric side of the HLA class II region. The observation of an association with HLA alleles in several ethnic groups may indicate that the true location of the responsible gene is not far away from the HLA gene region. In this respect, JME may display, on the centromeric side of the HLA complex, an HLA association due to linkage disequilibrium, as does haemochromatosis on the telomeric side.

REFERENCES

Adams, D., Festenstein, H., Gibson, J.D. et al. (1979). HLA antigens in chronic relapsing idiopathic inflammatory polyneuropathy. *J. Neurol. Neurosurg. Psychiat.* **42**: 184–186.

Aldener-Cannava, A. and Olerup, O. (1996). HLA-DPA1 typing by PCR amplification with sequence-specific primers (PCR-SSP) and distribution of DPA1 alleles in Caucasian, African and Oriental populations. *Tissue Antigens* **48**: 153–160.

Allen, M., Sandberg-Wollheim, M., Sjögren, K. et al. (1994). Association of susceptibility to multiple sclerosis in Sweden with HLA class II DRB1 and DQB1 alleles. *Hum. Immunol.* **39**: 41–48.

Behar, E., Lin, X., Grumet, F.C. and Mignot, E. (1995). A new DRB1*1202 allele (DRB1*12022) found in association with DQA1*0102 and DQB1*0602 in two black narcoleptic subjects. *Immunogenetics* **41**: 52.

Billiard, M., Seignalet, J., Besset, A. and Cadilhac, J. (1986). HLA-DR2 and narcolepsy. *Sleep* **9**: 149–152.

Brautbar, C., Amar, A., Cohen, N. et al. (1982). HLA-D typing in multiple sclerosis: Israelis tested with European homozygous typing cells. *Tissue Antigens* **19**: 189–197.

Carlsson, B., Wallin, J., Pirskanen, R. et al. (1990). Different HLA DR-DQ associations in subgroups of idiopathic myasthenia gravis. *Immunogenetics* **31**: 285–290.

Chiu, H.C., Hsieh, R.P. and Hsieh, K.H. (1990). Association of HLA antigens with myasthenia gravis in Chinese on Taiwan. *Chin. J. Microbiol. Immunol.* **23**: 12–18.

Christiansen, F.T., Houliston, J.B. and Dawkins, R.L. (1978). HLA, anti-DNA, and complement in myasthenia gravis. *Muscle Nerve* **1**: 467–470.

Compston, D.A.S., Vincent, A., Newsom-Davies, J. and Batchelor, J.R. (1980). Clinical, pathological, HLA antigen and immunological evidence for disease heterogeneity in myasthenia gravis. *Brain* **103**: 579–601.

Cowan, E.P., Pierce, M.L., McFarland, H.F. and McFarlin, D.E. (1991). HLA-DR and -DQ allelic sequences in multiple sclerosis patients are identical to those found in the general population. *Hum. Immunol.* **32**: 203–210.

Cullen, C.G., Middleton, D., Savage, D.A. and Hawkins, S. (1991). HLA-DR and DQ DNA genotyping in multiple sclerosis patients in Northern Ireland. *Hum. Immunol.* **30**: 1–6.

Dawkins, R.L., Christiansen, F.T., Kay, P.H. et al. (1983). Disease associations with complotypes, supratypes and haplotypes. *Immunol. Rev.* **70**: 1–22.

Dean, R.R., Kilduff, T.S., Dement, W.C. and Grumet, F.C. (1989). Narcolepsy without unique MHC class II antigen association: studies in the canine model. *Hum. Immunol.* **25**: 27–35.

Degli-Esposti, M.A., Andreas, A., Christiansen, F.T. et al. (1992). An approach to the localization of the susceptibility genes for generalized myasthenia gravis by mapping recombinant ancestral haplotypes. *Immunogenetics* **35**: 355–364.

Delamere, J.P., Jobson, S., Mackintosh, L.P. et al. (1983). Penicillamine-induced myasthenia in rheumatoid arthritis: its clinical and genetic features. *Ann. Rheum. Dis.* **42**: 500–504.

Delgado-Escueta, A.V., Serratosa, J.M., Liu, A. et al. (1994). Progress in mapping human epilepsy genes. *Epilepsia* **35** Suppl. 1: S29–S40.

Ditta, S.D., George, C.F. and Singh, S.M. (1992). HLA-D-region genomic DNA restriction fragments in DRw15 (DR2) familial narcolepsy. *Sleep* **15**: 48–57.

Doorn, P.A. van, Schreuder, G.M., Vermeulen, M. et al. (1991). HLA antigens in patients with chronic inflammatory demyelinating polyneuropathy. *J. Neuroimmunol.* **32**: 133–139.

Douglass, A.B., Harris, L. and Pazderka, F. (1989). Monozygotic twins concordant for the narcoleptic syndrome. *Neurology* **39**: 140–141.

Drachman, D.B. (1994). Myasthenia gravis. *N. Engl. J. Med.* **330**: 1797–1810.

Dupont, B., Lisak, R.P., Jersild, C. et al. (1977). HLA antigens in black American patients with multiple sclerosis. *Transplant. Proc.* **9**: 181–185.

Durner, M., Sander, T., Greenberg, D.A. et al. (1991). Localization of idiopathic generalized epilepsy on chromosome 6p in families of juvenile myoclonic epilepsy patientspatients. *Neurology* **41**: 1651–1655.

Durner, M., Janz, D., Zingsem, J. and Greenberg, D.A. (1992). Possible association of juvenile myoclonic epilepsy with HLA-DRw6. *Epilepsia* **33**: 814–816.

Ebers, G.C., Pathy, D., Stiller, C. et al. (1982). HLA typing in sibs with multiple sclerosis. *Lancet* **ii**: 1278.

Ebers, G.C., Kukay, K., Bulman, D.E. et al. (1996). A full genome search in multiple sclerosis. *Nature Genet.* **13**: 472–476.

Ellis, M.C., Hetisimer, A.H., Ruddy, D.A. et al. (1997). HLA class II haplotype and sequence analysis support a role for DQ in narcolepsy. *Immunogenetics* **46**: 410–417.

Elmslie, F.V., Rees, M., Williamson, M.P. et al. (1996). Linkage analysis of juvenile myoclonic epilepsy and microsatellite loci spanning 61 cM of human chromosome 6p in 19 nuclear pedigrees provides no evidence for a suscpetibility locus in this region. *Am. J. Hum. Genet.* **59**: 653–663.

Elmslie, F.V., Williamson, M.P., Rees, M. et al. (1997). Genetic mapping of a major susceptibility locus for juvenile myoclonic epilepsy on chromosome 15q. *Hum. Molec. Genet.* **6**: 1329–1334.

Eoli, M., Pandolfo, M., Amoroso, A. et al. (1995). Evidence of linkage between susceptibility to multiple sclerosis and HLA-class II loci in Italian multiplex families. *Eur. J. Hum. Genet.* **3**: 303–311.

Feeney, D.J., Pollard, J.D., McLeod, J.G. et al. (1990). HLA antigens in chronic inflammatory demyelinating polyneuropathy. *J. Neurol. Neurosurg. Psychiat.* **53**: 170–172.

Fogdell, A., Hillert, J., Sachs, C. and Olerup, O. (1995). The multiple sclerosis- and narcolepsy-associated HLA class II haplotype includes the DRB5*0101 allele. *Tissue Antigens* **46**: 333–336.

Fogdell, A., Olerup, O., Fredrikson, S. et al. (1997). Linkage analysis of HLA class II genes in Swedish multiplex families with multiple sclerosis. *Neurology* **48**: 758–762.

Fogdell-Hahn, A., Ligers, A., Gronning, M. et al. (1999). Multiple sclerosis: a modifying influence of HLA class I genes in an HLA class II associated disease. *Tissue Antigens* (in press).

Francis, D.A., Batchelor, J.R., McDonald, W.I. et al. (1987a). HLA genetic determinants in familial MS. A study from the Grampian region of Scotland. *Tissue Antigens* **29**: 7–12.

Francis, D.A., Compston, D.A.S., Batchelor, J.R. and McDonald, W.I. (1987b). A reassessment of the risk of multiple sclerosis developing in patients with optic neuritis after extended follow-up. *J. Neurol. Neurosurg. Psychiat.* **50**: 758–765.

Francis, D.A., Thompson, A.J., Brookes, P. et al. (1991). Multiple sclerosis and HLA: is the susceptibility gene really HLA-DR or -DQ? *Hum. Immunol.* **32**: 119–124.

Fritze, D., Herrmann, C., Jr, Smith, G.S. and Walford, R.L. (1973). HL-A types in myasthenia gravis. *Lancet* **ii**: 211.

Fugger, L., Ryder, L.P., Morling, N. et al. (1990). DNA typing for HLA-DPB1*02 and -DPB1*04 in multiple sclerosis and juvenile rheumatoid arthritis. *Immunogenetics* **32**: 150–156.

Gogolin, K.J., Kolaga, V.J., Baker, L. et al. (1989). Subtypes of HLA-DQ and -DR defined by DQB1 and DRB1 RFLPs: allele frequencies in the general population and in insulin-dependent diabetes (IDDM) and multiple sclerosis patients. *Ann. Hum. Genet.* **53**: 327–338.

Gorodezky, C., Najera, R., Rangel, B.E. et al. (1986). Immunogenetic profile of multiple sclerosis in Mexicans. *Hum. Immunol.* **16**: 364–374.

Greenberg, D.A., Delgado-Escueta, A.V., Widelitz, H. et al. (1988). Juvenile myoclonic epilepsy (JME) may be linked to the BF and HLA loci on human chromosome 6. *Am. J. Med. Genet.* **31**: 185–192.

Greenberg, D.A., Durner, M., Shinnar, S. et al. (1996). Association of HLA class II alleles in patients with juvenile myoclonic epilepsy compared with patients with other forms of adolescent-onset generalized epilepsy. *Neurology* **47**: 750–755.

Grumet, F.C. and Mignot, E. (1994). A non-DR2 narcoleptic patient with a unique DRB1-08del, DQB1–0602 haplotype having a new DRB1 allele with a condon deletion. *Hum. Immunol.* **39**: 302.

Haines, J.L., Ter-Minassian, M., Bazyk, A. et al. (1996). A complete genomic screen for multiple sclerosis underscores a role for the major histocompatability complex. The Multiple Sclerosis Genetics Group. *Nature Genet.* **13**: 469–471.

Hao, Q., Saida, T., Kawakami, H. et al. (1992). HLAs and genes in Japanese patients with multiple sclerosis: evidence for increased frequencies of HLA-Cw3, HLA-DR2, and HLA-DQB1*0602. *Hum. Immunol.* **35**: 116–124.

Hauser, S.L., Fleischnick, E., Weiner, H.L. et al. (1989). Extended major histocompatibility complex haplotypes in patients with multiple sclerosis. *Neurology* **39**: 275–277.

Hawke, S., Matsuo, H.. Nicolle, M. et al. (1996). Autoimmune T cells in myasthenia gravis: heterogeneity and potential for specific immunotargeting. *Immunol. Today* **17**: 307–311.

Hawkins, B.R., Chan-Lui, W.Y., Choi, E.K. and Ho, A.Y. (1984). Strong association of HLA BW46 with juvenile onset myasthenia gravis in Hong Kong Chinese. *J. Neurol. Neurosurg. Psychiat.* **47**: 555–557.

Hely, M.A., McManis, P.G., Doran, T.J. et al. (1986). Acute optic neuritis: a prospective study of risk factors for multiple sclerosis. *J. Neurol. Neurosurg. Psychiat.* **49**: 1125–1130.

Hillert, J. and Olerup, O. (1993). Multiple sclerosis is associated with genes within or close to the HLA-DR-DQ subregion on a normal DR15,DQ6,Dw2 haplotype. *Neurology* **43**: 163–168.

Hillert, J., Osterman, P.O. and Olerup, O. (1991). No association with HLA-DR, -DQ or -DP alleles in Guillain–Barré syndrome. *J. Neuroimmunol.* **31**: 67–72.

Hillert, J., Gronning, M., Nyland, H. et al. (1992). An immunogenetic heterogeneity in multiple sclerosis. *J. Neurol. Neurosurg. Psychiat.* **55**: 887–890.

Hillert, J., Kall, T., Vrethem, M. et al. (1994). The HLA-Dw2 haplotype segregates closely with multiple sclerosis in multiplex families. *J. Neuroimmunol.* **50**: 95–100.

Hillert, J., Käll, T., Olerup, O., Söderström, M. (1996). Distribution of HLA-Dw2 in optic neuritis and multiple sclerosis indicates heterogeneity. *Acta Neurol. Scand.* **94**: 161–166.

Hjelmström, P., Giscombe, R., Lefvert, A.K. et al. (1995). Different HLA-DQ are positively and negatively associated in Swedish patients with myasthenia gravis. *Autoimmunity* **22**: 59–65.

Hjelmström, P., Giscombe, R., Lefvert, A.K. et al. (1997). TAP polymorphisms in Swedish myasthenia gravis patients. *Tissue Antigens* **49**: 176–179.

Honda, Y., Juji, T., Matsuki, K. et al. (1986). HLA-DR2 and Dw2 in narcolepsy and in other disorders of excessive somnolence without cataplexy. *Sleep* **9**: 133–142.

Jersild, C., Svejgaard, A. and Fog, T. (1972). HL-A antigens and multiple sclerosis. *Lancet* **i**: 1240–1241.

Jersild, C., Fog, T., Hansen, G.S. et al. (1973). Histocompatibility determinants in multiple sclerosis, with special reference to clinical course. *Lancet* **ii**: 1221–1225.

Juji, T., Satake, M., Honda, Y. and Doi, Y. (1984). HLA antigens in Japanese patients with narcolepsy. All the patients were DR2 positive. *Tissue Antigens* **24**: 316–319.

Juji, T., Matsuki, K., Tokunaga, K. et al. (1988). Narcolepsy and HLA in the Japanese. *Ann. N.Y. Acad. Sci.* **540**: 106–114.

Kaakinen, A., Pirskanen, R. and Tiilikainen, A. (1975). LD antigens associated with HL-A8 and myasthenia gravis. *Tissue Antigens* **6**: 175–182.

Kaslow, R.A., Sullivan-Bolyai, J.Z. and Hafkin, B. (1984). HLA antigens in Guillain–Barré syndrome. *Neurology* **34**: 240–242.

Kellar-Wood, H.F., Wood, N.W., Holmans, P. et al. (1995). Multiple sclerosis and the HLA-D region: linkage and association studies. *J. Neuroimmunol.* **58**: 183–190.

Kalman, B., Takacs, K. and Gyodi, E. (1991). Sclerosis multiplex in gypsies. *Acta Neurol. Scand.* **84**: 181–185.

Kira, J., Kanai, T., Nishimura, Y. et al. (1996). Western versus Asian types of multiple sclerosis: immunogenetically and clinically distinct disorders. *Ann. Neurol.* **40**: 569–574.

Kurdi, A., Ayesh, I., Abdallat, A. and Maayta, U. (1977). Different B lymphocyte alloantigens associated with multiple sclerosis in Arabs and North Europeans. *Lancet* **i**: 1123–1125.

Kwon, O.J., Peled, N., Miller, K. et al. (1995). HLA class II analysis in Jewish Israeli narcoleptic patients. *Hum. Immunol.* **44**: 199–202.

Langdon, N., Welsh, K. I., Dam, M. van et al. (1984). Genetic markers in narcolepsy. *Lancet* **2**: 1178–1180.

Liu, A.W., Delgado-Escueta, A.V., Serratosa, J.M. et al. (1995). Juvenile myoclonic epilepsy locus in chromosome 6p21.2-p11: linkage to convulsions and electroencephalography trait. *Am. J. Hum. Genet.* **57**: 368–381.

Lock, C.B., So, A.K., Welsh, K.I. et al. (1988). MHC class II sequences of an HLA-DR2 narcoleptic. *Immunogenetics* **27**: 449–455.

Malcherek, G., Falk, K., Rotzschke, O. et al. (1993). Natural peptide ligand motifs of two HLA molecules associated with myasthenia gravis. *Int. Immunol.* **5**: 1229–1237.

Marcadet, A., Massart, C., Semana, G. et al. (1985). Association of class II HLA-DQ beta chain DNA restriction fragments with multiple sclerosis. *Immunogenetics* **22**: 93–96.

Marrosu, M.G., Muntoni, F., Murru, M.R. et al. (1988). Sardinian multiple sclerosis is associated with HLA-DR4: a serologic and molecular analysis. *Neurology* **38**: 1749–1753.

Marrosu, M.G., Muntoni, F., Murru, M.R. et al. (1992). HLA-DQB1 genotype in Sardinian multiple sclerosis: evidence for a key role of DQB1 *0201 and *0302 alleles. *Neurology* **42**: 883–886.

Marrosu, M.G., Murru, M.R., Costa, G. et al. (1997). Multiple sclerosis in Sardinia is associated and in linkage disequilibrium with HLA–DR3 and -DR4 alleles. *Am. J. Hum. Genet.* **61**: 454–457.

Matsuki, K., Juji, T., Tokunaga, K. et al. (1985). Human histocompatibility leukocyte antigen (HLA) haplotype frequencies estimated from the data on HLA class I, II, and III antigens in 111 Japanese narcoleptics. *J. Clin. Invest.* **76**: 2078–2083.

Matsuki, K., Juji, T., Tokunaga, K. et al. (1990). HLA antigens in Japanese patients with myasthenia gravis. *J. Clin. Invest.* **86**: 392–399.

Melms, A., Malcherek, G., Schoepfer, R. et al. (1993). Acetylcholine receptor-specific T cells are present in the normal immune repertoire. A study with recombinant polypeptides of the human acetylcholine receptor alpha-subunit. *Ann. N.Y. Acad. Sci.* **681**: 310–312.

Mignot, E. (1997). Genetics of narcolepsy and other sleep disorders. *Am. J. Hum. Genet.* **60**: 1289–1302.

Mignot, E., Lin, X., Arrigoni, J. et al. (1994). DQB1*0602 and DQA1*0102 (DQ1) are better markers than DR2 for narcolepsy in Caucasian and black Americans. *Sleep* **17**: S60–67.

Mignot, E., Tafti, M., Dement, W.C. and Grumet, F.C. (1995). Narcolepsy and immunity. *Advan. Neuroimmunol.* **5**: 23–37.

Mignot, E., Kimura, A., Latterman, A. et al. (1997a). HLA class II studies in non DRB1*1501 patients with narcolepsy-cataplexy. In *Genetic Diversity of HLA – Functional and Medical Implications* (ed. D. Charron), pp. 436–440, Medical and Scientific International, Paris.

Mignot, E., Kimura, A., Lattermann, A. et al. (1997b). Extensive HLA class II studies in 58 non-DRB1*15 (DR2) narcoleptic patients with cataplexy. *Tissue Antigens* **49**: 329–341.

Moerloose, P. de, Jeannet, M., Martins-da-Silva, B. et al. (1979). Increased frequency of HLA—DRw2 and DRw3 in multiple sclerosis. *Tissue Antigens* **13**: 357–360.

Möller, E., Hammarström, L., Smith, E. and Matell, G. (1976). HL-A8 and LD-8a in patients with myasthenia gravis. *Tissue Antigens* **7**: 39–44.

Mueller-Eckhardt, G., Meier-Ewert, K., Schendel, D.J. et al. (1986). HLA and narcolepsy in a German population. *Tissue Antigens* **28**: 163–169.

Mustafa, M., Vingsbo, C., Olsson, T. et al. (1994). Protective influences on experimental autoimmune encephalomyelitis by MHC class I and cIIIeles. *J. Immunol.* **153**: 3337–3344.

Mytilineos, J., Lempert, M., Middleton, D. et al. (1997). HLA class I DNA typing of 215 HLA-A, -B, -DR zero mismatched kidney transplants. *Tissue Antigens* **50**: 355–358.

Naeim, F., Keesey, J.C., Herrmann, C. et al. (1978). Association of HLA-B8, DRw3, and anti-acetylcholine receptor antibodies in myasthenia gravis. *Tissue Antigens* **12**: 381–386.

Naito, S., Namerow, N., Mickey, M.R. and Terasaki, P.I. (1972). Multiple sclerosis: association with HL-A3. *Tissue Antigens* **2**: 1–4.

Naito, S., Kuroiwa, Y., Itoyama, T. et al. (1978). HLA and Japanese MS. *Tissue Antigens* **12**, 19–24.

Neely, S., Rosenberg, R., Spire, J.P. et al. (1987). HLA antigens in narcolepsy. *Neurology* **37**: 1858–1860.

Obeid, T., el Rab, M.O., Daif, A.K. et al. (1994). Is HLA-DRW13 (W6) associated with juvenile myoclonic epilepsy in Arab patients? *Epilepsia* **35**: 319–321.

Odum, N., Hyldig-Nielsen, J.J., Morling, N. et al. (1988). HLA-DP antigens are involved in the susceptibility to multiple sclerosis. *Tissue Antigens* **31**: 235–237.

Olerup, O. and Hillert, J. (1991). HLA class II-associated genetic susceptibility in multiple sclerosis: a critical evaluation. *Tissue Antigens* **38**: 1–15.

Olerup, O., Hillert, J., Fredrikson, S. et al. (1989). Primarily chronic progressive and relapsing/remitting multiple sclerosis: two immunogenetically distinct disease entities. *Proc. Nat. Acad. Sci. USA* **86**: 7113–7117.

Olerup, O., Schaffer, M., Hillert, J. and Sachs, C. (1990). The narcolepsy-associated DRw15,DQw6,Dw2 haplotype has no unique HLA-DQA or -DQB restriction fragments and does not extend to the HLA-DP subregion. *Immunogenetics* **32**: 41–44.

Oosterhuis, H.J., Feltkamp, T.E., Rossum, A.L. van et al. (1976). HL-A antigens, autoantibody production, and associated diseases in thymoma patients, with and without myasthenia gravis. *Ann. N.Y. Acad. Sci.* **274**: 468–474.

Paty, D.W., Dossetor, J.B., Stiller, C.R. et al. (1977). HLA in multiple sclerosis. Relationship to measles antibody, mitogen responsiveness and clinical course. *J. Neurol. Sci.* **32**: 371–379.

Poirier, G., Montplaisir, J., Decary, F. et al. (1986). HLA antigens in narcolepsy and idiopathic central nervous system hypersomnolence. *Sleep* **9**: 153–158.

Ratanachaiyavong, S., Fleming, D., Janer, M. et al. (1994). HLA-DPB1 polymorphisms in patients with hyperthyroid Graves' disease and early onset myasthenia gravis. *Autoimmunity* **17**: 99–104.

Rees, J.H., Vaughan, R.W., Kondeatis, E. and Hughes, R.A. (1995). HLA-class II alleles in Guillain–Barré syndrome and Miller Fisher syndrome and their association with preceding *Campylobacter jejuni* infection. *J. Neuroimmunol.* **62**: 53–57.

Roth, B., Nevsimalova, S., Sonka, K. et al. (1988). A study of the occurrence of HLA DR2 in 124 narcoleptics: clinical aspects. *Schweizer Archiv. Neurol. Psychiat.* **139**: 41–51.

Roushdy, J., Santoso, S., Kalb, R. et al. (1993). A deletion in the second exon of an HLA-DRB1 allele found in a DR2-negative narcolepsy patient. *Hum. Immunol.* **37**: 1–6.

Säfwenberg, J., Lindblom, J.B. and Osterman, P.O. (1973). HL-A frequencies in patients with myasthenia gravis. *Tissue Antigens* **3**: 465–469.

Säfwenberg, J., Hammarstrom, L., Lindblom, J.B. et al. (1978). HLA-A, -B, -C and -D antigens in male patients with myasthenia gravis. *Tissue Antigens* **12**: 136–142.

Sandberg-Wollheim, M., Bynke, H., Cronquist, S. et al. (1990). A long-term prospective study of optic neuritis: evaluation of risk factors. *Ann. Neurol.* **27**: 386–393.

Sander, T., Hildmann, T., Janz, D. et al. (1995). The phenotypic spectrum related to the human epilepsy susceptibility gene "EJM1". *Ann. Neurol.* **38**: 210–217.

Sander, T., Bockenkamp, B., Hildmann, T. et al. (1997). Refined mapping of the epilepsy susceptibility locus EJM1 on chromosome 6. *Neurology* **49**: 842–847.

Sawcer, S., Jones, H.B., Feakes, R. et al. (1996). A genome screen in multiple sclerosis reveals susceptibility loci on chromosome 6p21 and 17q22. *Nature Genet.* **13**: 464–468.

Seignalet, J. and Billiard, M. (1984). Possible association between HLA-B7 and narcolepsy. *Tissue Antigens* **23**: 188–189.

Serjeantson, S.W., Gao, X., Hawkins, B.R. et al. (1992). Novel HLA-DR2-related haplotypes in Hong Kong Chinese implicate the DQB1*0602 allele in susceptibility to multiple sclerosis. *Eur. J. Immunogenet.* **19**: 11–19.

Sinha, A.A., Bell, R.B., Steinman, L. and McDevitt, H.O. (1991). Oligonucleotide dot-blot analysis of HLA-DQ beta alleles associated with multiple sclerosis. *J. Neuroimmunol.* **32**: 61–65.

Spurkland, A., Celius, E.G., Knutsen, I. et al. (1997). The HLA-DQA1*0102,DQB1*0602 heterodimer may confer susceptibility to multiple sclerosis in the absence of the HLA-DRA1*01,DRB1*1501 heterodimer. *Tissue Antigens* **50**: 15–22.

Spurkland, A., Ronningen, K.S., Vandvik, B. et al. (1991a). HLA-DQA1 and HLA-DQB1 genes may jointly determine susceptibility to develop multiple sclerosis. *Hum. Immunol.* **30**: 69–75.

Spurkland, A., Gilhus, N.E., Ronningen, K.S. et al. (1991b). Myasthenia gravis patients with thymus hyperplasia and myasthenia gravis patients with thymoma display different HLA associations. *Tissue Antigens* **37**: 90–93.

Stewart, G.J., Pollard, J.D., McLeod, J.G. and Wolnizer, C.M. (1978). HLA antigens in the Landry–Guillain–Barré syndrome and chronic relapsing polyneuritis. *Ann. Neurol.* **4**: 285–289.

Stewart, G.J., McLeod, J.G., Basten, A. and Bashir, H.V. (1981). HLA family studies and multiple sclerosis: a common gene, dominantly expressed. *Hum. Immunol.* **3**: 13–29.

Takács, K., Kálmán, B., Gyódi, E. et al. (1990). Association between the lack of HLA-DQw6 and the low incidence of multiple sclerosis in Hungarian gypsies. *Immunogenetics* **31**: 383–385.

Tienari, P.J., Wikstrom, J., Koskimies, S. et al. (1993). Reappraisal of HLA in multiple sclerosis: close linkage in multiplex families. *Eur. J. Hum. Genet.* **1**: 257–268.

Tiwari, J.L. and Terasaki, P.I. (1985). *HLA and Disease Associations*, Springer, New York.

Uryu, N., Maeda, M., Nagata, Y. et al. (1989). No difference in the nucleotide sequence of the DQ beta beta 1 domain between narcoleptic and healthy individuals with DR2,Dw2. *Hum. Immunol.* **24**: 175–181.

Vartdal, F., Sollid, L.M., Vandvik, B. et al. (1989). Patients with multiple sclerosis carry DQB1 genes which encode shared polymorphic amino acid sequences. *Hum. Immunol.* **25**: 103–110.

Vaughan,R.W., Adam, A.M., Gray, I.A. et al. (1990). Major histocompatibility complex class I and class II polymorphism in chronic idiopathic demyelinating polyradiculoneuropathy. *J. Neuroimmunol.* **27**: 149–153.

Vieira, M.L., Caillat-Zucman, S., Gajdos, P. et al. (1993). Identification by genomic typing of non-DR3 HLA class II genes associated with myasthenia gravis. *J. Neuroimmunol.* **47**: 115–122.

Voskuhl, R.R., Goldstein, A.M., Simonis, T. et al. (1996). DR2/DQw1 inheritance and haplotype sharing in affected siblings from multiple sclerosis families. *Ann. Neurol.* **39**: 804–807.

Vrethem, M., Ernerudh, J., Cruz, M. et al. (1993). Susceptibility to demyelinating polyneuropathy in plasma cell dyscrasia may be influenced by amino acid position 9 of the HLA-DR beta chain. *J. Neuroimmunol.* **43**: 139–144.

Weissbecker, K.A., Durner, M., Janz, D. et al. (1991). Confirmation of linkage between juvenile myoclonic epilepsy locus and the HLA region of chromosome 6. *Am. J. Med. Genet.* **38**: 32–36.

Winer, J.B., Hughes, R.A., Anderson, M.J. et al. (1988). A prospective study of acute idiopathic neuropathy. II. Antecedent events. *J. Neurol. Neurosurg. Psych.* **51**: 613–618.

Whitehouse, W.P., Rees, M., Curtis, D. et al. (1993). Linkage analysis of idiopathic generalised epilepsy (IGE) and marker loci on chomosome 6p in families of patients with juvenile myoclonic epilepsy: no evidence for an epilepsy locus in the HLA region. *Am. J. Hum. Genet.* **53**: 652–662.

CHAPTER 15

HLA and Type I Diabetes

Gerald T. Nepom

INTRODUCTION

Human leucocyte antigen (HLA) associations with autoimmune diabetes have been analysed extensively and there are numerous excellent published summaries (Wassmuth and Lernmark, 1989; Lernmark et al., 1991; Nepom and Erlich, 1991; Ronningen et al., 1991; Nepom, 1993; Reijonen and Nepom, 1996). The main principles from these studies include the following.

- HLA class II genes are responsible for the strong genetic association between autoimmune (type I) diabetes and the major histocompatibility complex.
- Specific alleles at the HLA-DQ locus account for these associations, and therefore HLA-DQ molecules are implicated as risk factors in the immunopathogenesis of diabetes.
- There are multiple HLA-DQ alleles which correlate with susceptibility, and the hierarchy of allelic associations varies, influenced by ethnicity and age of the population tested.
- In most ethnic groups tested, the predominant susceptibility allele is the HLA-DQB1*0302 gene.
- The degree of genetic risk conferred by the presence of the HLA-DQB1*0302 gene is modulated by other genes, including other major histocompatibility complex (MHC) genes which either increase or decrease the relative risk of disease.
- The strongest "protective" genetic influence encoded in the MHC is due to the presence of the HLA-DQB1*0602 allele; heterozygotes with both this protective allele and the DQB1*0302 susceptibility allele are at very low risk for diabetes.

Table 15.1 lists the major disease-associated HLA-DQB1 alleles and indicates the HLA class II haplotypes which contain these susceptibility genes. The risk associated with each of these haplotypes is given as an average odds ratio.

HLA-DQ ASSOCIATIONS IN TYPE I DIABETES

The association between DQB1*0302 and type I or insulin-dependent diabetes mellitus (IDDM) has been studied extensively, and illustrates

TABLE 15.1 HLA-DQB1 susceptibility genes in autoimmune diabetes

HLA class II susceptibility gene	DQ-DR haplotype	Relative risk
DQB1*0302	DQB1*0302–DQA1*0301–DRB1*04	~8[a]
DQB1*0201	DQB1*0201–DQA1*0501–DRB1*03	~4
DQB1*04	DQB1*04–DQA1*0301–DRB1*04	~4
DQB1*0303	DQB1*0303–DQA1*0301–DRB1*09	~4

[a]Risk varies depending on subtype of DRB1*04 allele (see Table 15.3).
Adapted from Reijonen and Nepom (1996).

several distinctive features with practical implications for disease prediction. First, there is a strong correlation between age of disease onset and the presence of HLA susceptibility genes, in that older-onset patients (after the age of approx. 15 years) have a much lower frequency of the "disease-associated" alleles (Tait et al., 1995). This may imply that the younger-onset patients have a major genetic contribution to pathogenesis which precipitates early disease onset, while patients without these susceptibility genes have a more indolent disease process which takes time to evolve into clinical pathology. A second important feature of the HLA-DQ associations is the striking synergy between compound heterozygotes, in which genetic susceptibility is markedly increased, as illustrated in Table 15.2.

In addition to this genetic interaction between different haplotypes in heterozygous individuals, a genetic interaction occurs between different class II molecules on the same haplotype, a phenomenon referred to as epistasis. The DQB1*0302–DQA1*0301–DRB1*04 haplotype is associated with the susceptibility to develop IDDM in most populations studied (e.g. Wassmuth and Lernmark, 1989; Baisch et al., 1990; Erlich et al., 1990; Todd et al., 1990; Ronningen et al., 1992), but only certain DR4-associated DRB1 alleles are prevalent among IDDM patients (Table 15.3). DRB1*0401 and *0402 are positively associated with IDDM, while DRB1*0403 and *0406 are not, and DRB1*0404 is intermediate with respect to disease association (Sheehy et al., 1989; Erlich et al., 1993; Ronningen et al., 1992; Caillat-Zucman et al., 1992; Mijovic et al., 1993; Awata and Kanazawa, 1994; Van der Auwera et al., 1995; Ilonen et al., 1996; Reijonen et al., 1997; Undlien et al., 1997). Thus, the range of disease susceptibility associated with DQB1*0302 varies from a relative risk of 2 up to a relative risk of 10 or more, depending on the associated DRB1*04 allele (Nepom and Erlich, 1991).

The role of the DQB1*0201–DQA1*0501–DRB1*03 haplotype in IDDM susceptibility is complex. There are several distinctive features of the disease associations with this haplotype, which suggest the possibility of multiple genetic contributions. First, as noted above, heterozygotes with this haplotype in combination with a DQB1*0302-positive haplotype have a synergistic, rather than an additive risk for IDDM. This has frequently been interpreted as being consistent with the formation of *trans*-encoded DQ heterodimers, in which the DQ molecules expressed in such heterozygotes combine an α-chain from one haplotype with a β-chain from the other haplotype. Such *trans*-dimers have indeed been found in DQB1*0201/DQB1*0302 heterozygous cells (Nepom et al., 1987) and have been shown to have unique functional peptide-binding properties (Kwok et al., 1995).

However, neither the DQA1*0501 nor the DQB1*0201 allele is invariably associated with IDDM. Both of these genes occur on other common haplotypes, such as those containing DRB1*11 and DRB1*07 alleles, respectively, which are not associated with diabetes and which do not synergize with DQB1*0302-positive haplo-

TABLE 15.2 Heterozygous HLA class II genetic synergy for type I diabetes risk

DQ-DR haplotypes	Relative risk
DQB1*0302–DQA1*0301–DRB1*04 and DQB1*0201–DQA1*0501–DRB1*03	~20
DQB1*0201–DQA1*0501–DRB1*03 and DQB1*0303–DQA1*0301–DRB1*09	~8
DQB1*0402–DQA1*0301–DRB1*08 and DQB1*0302–DQA1*0301–DRB1*04	~12

TABLE 15.3 Epistatic influence of *DRB1*04* alleles on type I diabetes susceptibility

DQ-DR haplotypes	Relative risk
DQB1*0302–DQA1*0301–DRB1*0401	~8
DQB1*0302–DQA1*0301–DRB1*0402	~10
DQB1*0302–DQA1*0301–DRB1*0403	~2
DQB1*0302–DQA1*0301–DRB1*0404	~4

types. In addition, the DQA1*0301–DQB1*0201 heterodimer occurs in heterozygotes with the haplotype DR4–DQA1*0301–DQB1*0301 in *trans* with DR3 or DR7 haplotypes, but is not associated with IDDM in these cases, making it unlikely that this particular *trans*-dimer is involved in disease susceptibility.

The association between DQB1*0201–DQA1*0501–DRB1*03 haplotypes and IDDM varies in different populations and in several studies this haplotype lacks an independent risk for IDDM, although it still synergizes with the DQB1*0302–DQA1*0301–DRB1*04 haplotype in heterozygotes (Ronningen et al., 1989). This contrasts with a study from Sardinia, where the major susceptibility haplotype is DQB1*0201–DQA1*0501–DRB1*03, with increased risk associated in homozygotes (Cucca et al., 1993). One possible reconciliation of these contrasting results is the recognition that the DQB1*0201–DQA1*0501–DRB1*03 haplotype is not the same in southern and northern Europe, and differences at the HLA class I loci (and probably at other MHC-linked genes as well) may be important factors in contributing to disease risk. The HLA-A30–B18–DR3 haplotype is frequent in Southern Europe and is associated with a higher disease risk than the HLA-A1–B8–DR3 haplotype prevalent in northern Europe (Deschamps et al., 1988; Vicario et al., 1992). This HLA-A30–B18–DR3 haplotype is also more prevalent in childhood-onset IDDM (Deschamps et al., 1988; Tait et al., 1995). Another HLA class I association has been reported in Japanese, with HLA-A24 (Nakanishi et al., 1993). One of the implications of these observations is that the DQB1*0201–DQA1*0501–DRB1*03 haplotypes are likely to contain more than one genetic factor associated with IDDM, only one of which is due to the DQB1 locus. Whether the additional susceptibility genes are encoded by class I alleles themselves, or are located in the large interval on chromosome 6 between the HLA class II and class I genes, remains to be clarified.

In addition to these various HLA susceptibility genes, a negative association between IDDM and HLA-DR2 has been recognized since the late 1980s (Maclaren et al., 1988; Thomson et al., 1988). Different DR2-positive haplotypes contain different linked DQA1 and DQB1 alleles, and it is primarily the DQB1*0602–DQA1*0102–DRB1*1501 haplotype which is negatively associated with IDDM. This negative correlation occurs in multiple ethnic groups and can be quite profound: the relative risk of diabetes with DQB1*0602 is approximately 0.05, even in heterozygotes who carry a disease-susceptible DQB1*0302–DQA1*0301–DRB1*04 haplotype (Baisch et al., 1990). HLA-DR6 haplotypes containing DQB1*0603–DQA1*0103–DRB1*1301 may also be negatively associated with diabetes (Gjertsen et al., 1990; Khalil et al., 1990; Vallet-Colom et al., 1990; Heimberg et al., 1992; de Vries et al., 1993).

MOLECULAR PROPERTIES OF HLA-DQ

The multiple genetic factors which participate in the genesis or modulation of diabetes susceptibility provide a challenge for understanding the genetic mechanisms of disease. Since the principal known function of HLA class II genes is to encode the class II heterodimer which presents peptide antigens to specific T lymphocytes, understanding HLA contributions to IDDM have more recently focused on the peptide interactions of specific DQ molecules.

HLA-DQ3.2, the most highly IDDM-associated HLA molecule, is encoded by the DQA1*0301 and DQB1*0302 genes and has a peptide binding motif which is influenced by both the α- and the β-chain (Kwok et al., 1995). There are several polymorphisms on the β-chain which distinguish the DQ3.2 molecule from other DQ molecules, and these genetic polymorphisms account for the characteristic properties of peptides which efficiently bind to DQ3.2: there are four predominant amino acid residues on the peptide which appear to be crucial for binding, at relative positions 1, 4, 6 and 9, similar to peptides which bind other class II molecules (Stern and Wiley, 1992; Stern et al., 1994; Kwok et al., 1995). Peptides which have permissive residues at these positions can bind to DQ3.2, thereby orientating amino acids at other peptide positions towards the T cell receptor. Of the four crucial peptide binding positions, residues 4 and 9 participate in binding interactions which are

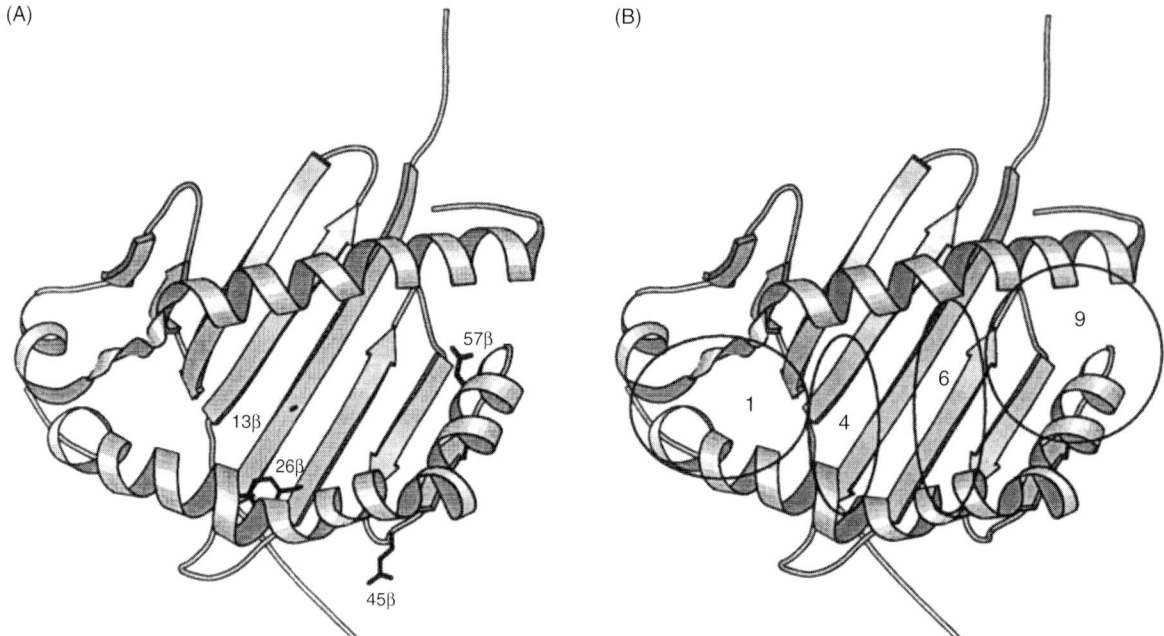

FIGURE 15.1 Structural model of the DQ class II molecule, highlighting the four positions (A) that distinguish the diabetes-associated DQ3.2 molecule from the closely related DQ3.1 molecule. These differences lead to changes in the interactions between peptides and the DQ binding "pockets" (B), which correspond to positions 1, 4, 6, and 9 of the bound antigenic peptide. Changes in pockets 4 and 9, due to the structural features of residues 13, 26 and 57, account for the unique binding properties of the DQ3.2 molecule. (Adapted from DeWeese et al., 1996.)

specific for the DQ3.2 molecule; interactions with residues 1 and 6, while important for binding, are not allele specific. Therefore, the disease-associated properties of the DQ3.2 molecule are largely influenced by two site-specific interactions, corresponding to the binding pockets for residues 4 and 9 of antigenic peptides. A structural model of the DQ class II molecule is shown in Figure 15.1.

The DQ3.2 binding pocket for residue 4 accommodates aliphatic and hydrophobic side-chains, while the binding pocket for residue 9 prefers to bind negatively charged amino acids. This preference for negatively charged residues in pocket 9 is of particular interest, since it corresponds to a conserved polymorphism at residue 57 of the DQ β-chain which has sometimes been used as a marker for the alleles (DQB1*0302 and DQB1*0201) most often associated with IDDM. DQ β residue 57 is located near one end of the peptide-binding groove of class II molecules, where it functions as a molecular gate-keeper for the peptide side chain at position 9. When DQ β residue 57 is a small non-charged residue, such as alanine, large peptide side chains are able to occupy pocket 9. There is a positively charged arginine residue at position 79 of the DQ α-chain adjacent to pocket 9, and if the peptide side chain carries a negative charge (such as Asp or Glu), there is a strong attractive interaction which leads to a stable, high-avidity binding interaction at this site. When DQ β residue 57 is an aspartic acid, as in DQB1*0301, DQB1*0602 and several other DQB1 alleles, there appears to be a native "salt-bridge" interaction between the DQ β Asp^{57} and the Arg^{79} of the DQ α-chain. By analogy with a similar structure in the DRB1 molecules resolved by crystallography (Stern et al., 1994), this salt bridge forms a structural constraint on the peptide residues which can be accommodated in pocket 9. Peptides with an Asp or Glu at residue 9 do not bind but small aliphatic residues are sufficient to stabilize the binding interaction (Kwok et al., 1996a, b; Nepom et al., 1996). The most likely interpretation for this binding pattern is that the positively charged arginine residue on the DQ α-chain determines the character of the peptide residue at pocket 9, either by interaction with the Asp^{57} residue of the β-chain or by direct interaction with the side chain contributed by the peptide. Although this polymorphic interaction with DQ β residue 57 is critical for peptide-binding specificity, it is not an explanation for the genetic complexity of HLA associations with IDDM. Several of the disease-associated alleles, such as HLA-DQB1*0303 and HLA-DQB1*04, are Asp57 positive, while others, such as DQB1*0302 and DQB1*0201, are Asp^{57} negative. Similarly, there is a wide spectrum of relative protection from IDDM among different alleles which are Asp^{57} positive, ranging from a strongly negative association to a neutral genetic influence.

HLA-DQ IN THE PREDICTION OF TYPE I DIABETES

In spite of the strong genetic associations between specific HLA-DQB1 alleles and IDDM, especially in childhood, a number of confounding issues complicates the use of HLA testing for disease prediction.

Autoimmune (type I) diabetes is a chronic autoimmune disease. Commonly, autoantibodies to islet autoantigens are present in individuals months or years before the clinical onset of diabetes. Thus, screening programmes have been developed to identify these "prediabetics", using a combination of antibody testing and HLA typing (Eisenbarth et al., 1993). When these screening programmes are used to evaluate individuals within families where one family member already has diabetes, the antibody testing is very accurate for predicting the likelihood of disease (Bingley et al., 1992). In this clinical setting, the likelihood is that such antibody-positive family members have inherited the same HLA susceptibility alleles as the affected proband, and therefore the additional predictive value from using HLA-DQ typing as a screening tool is limited. In other words, since such families are ascertained because there is a diabetic member, there is a high probability that multiple family members will carry the HLA-DQ susceptibility alleles, and discriminating between the multiple family members who are at genetic risk is facilitated by the antibody testing. Probably the most useful HLA analysis in this family situation is accomplished by screening for the protective HLA-DQB1*0602 allele, which can identify family members unlikely to progress to disease, even if they have autoantibodies to islet antigens.

In contrast to the situation where a family member is diabetic, screening for diabetes risk in the general population is more difficult. At the present time, cost and technology to screen the general population for autoantibodies are limiting, and there is a sufficiently high false-positive rate in such screening tests to make a universal screening programme problematic. It is possible, however, to use HLA testing as a method to reduce the size of the population subjected to autoantibody testing and to increase the specificity of the screening procedure. Although only 5% of the general population is heterozygous for the high-risk diabetes haplotypes DQB1*0302–DQA1*0301–DRB1*04 and DQB1*0201–DQA1*0501–DRB1*03, at least 30% of childhood-onset diabetics will carry this genotype. Therefore, screening populations for this heterozygous HLA type will identify a high-risk population that can be monitored for autoantibodies to islet antigens, with a high specificity for the testing procedure. However, since the DQB1*0302–DQA1*0301–DRB1*04 haplotype is associated with diabetes risk even in the absence of the DQB1*0201–DQA1*0301–DRB1*03 haplotype in 30–50% of patients (depending on their age), screening for high-risk heterozygotes is not a very sensitive testing approach. If the HLA screening procedure is expanded from heterozygote testing to including all DQB1*0302-positive individuals, there is increased sensitivity but decreased specificity, since DQB1*0302 is a common allele in the unaffected population. Current studies, along with technological improvements in screening methods, are underway to address these issues. It is likely that the ability to predict diabetes using a combination of HLA and autoantibody testing will become a reality in the near future, awaiting application when specific therapeutic intervention is proven to arrest the autoimmune process in prediabetic individuals.

REFERENCES

Awata, T. and Kanazawa, Y. (1994). Genetic markers for insulin-dependent diabetes mellitus in Japanese. *Diabetes Res. Clin. Pract.* **24** Suppl.: S83–S87.

Baisch, J.M., Weeks, T., Giles, T. et al. (1990). Analysis of HLA-DQ genotypes and susceptibility in insulin-dependent diabetes mellitus. *N. Engl. J. Med.* **322**: 1836–1841.

Bingley, P.J., Colman, P., Eisenbarth, G.S. et al. (1992). Standardization of IVGTT to predict IDDM. *Diabetes Care* **15**: 1313–1316.

Caillat-Zucman, S., Garchon, H.J., Timsit, J. et al. (1992). HLA genetic heterogeneity of insulin dependent diabetes mellitus. Indication for an age dependent susceptibility gradient. *J. Clin. Invest.* **90**: 2242–2250.

Cucca, F., Muntoni, F., Lampis, R. et al. (1993). Combinations of specific DRB1, DQA1, DQB1 haplotypes are associated with insulin-dependent diabetes mellitus in Sardinia. *Human Immunol.* **37**: 85–94.

Deschamps, I., Marcelli-Barge, A., Poirier, J.C. et al. (1988). Two distinct HLA-DR3 haplotypes are associated with age related heterogeneity in type I (insulin-dependent) diabetes. *Diabetologia* **1988**: 896–901.

Eisenbarth, G.S., Verge, C.F., Allen, H. and Rewers, M.J. (1993). The design of trials for prevention of IDDM. *Diabetes* **42**: 941–947.

Erlich, H.A., Bugawan, T.L., Scharf, S. et al. (1990). HLA-DQb sequence polymorphism and genetic susceptibility to IDDM. *Diabetes* **39**: 96–103.

Erlich, H.A., Zeidler, A., Chang, J. et al. (1993). HLA class II alleles and susceptibility and resistance to insulin dependent diabetes mellitus in Mexican–American families. *Nature Genet.* **3**: 358–364.

Gjertsen, H.A., Lundin, K.E.A., Ronningen, K.S. et al. (1990). T cells recognizing an HLA-DQ αβ heterodimer encoded in cis by the DR4DQw4 haplotype and in trans by DR4DQw8/DRw8DQw4 heterozygous cells. *Human Immunol.* **30**: 226–232.

Heimberg, H., Nagy, Z.P., Somers, G. et al. (1992). Complementation of HLA-DQA and -DQB genes confers susceptibility and protection to insulin-dependent diabetes mellitus. *Human Immunol.* **33**: 10–17.

Ilonen, J., Reijonen, H., Herva, E. et al. and the Childhood Diabetes in Finland Study Groups (1996). Rapid HLA-DQB1 genotyping for four alleles in the assessment of risk for IDDM in the Finnish population. *Diabetes Care* **19**: 795–800.

Khalil, I., D'Auriol, L., Gobet, M. et al. (1990). A combination of HLA-DQb Asp57-negative and HLA DQa Arg52 confers susceptibility to insulin-dependent diabetes mellitus. *J. Clin. Invest.* **85**: 1315–1319.

Kwok, W.W., Nepom, G.T. and Raymond, F.C. (1995). HLA-DQ polymorphisms are highly selective for peptide binding interactions. *J. Immunol.* **155**: 2468–2476.

Kwok, W.W., Domeier, M.E., Johnson, M.L. et al. (1996a). HLA-DQB1 codon 57 is critical for peptide binding and recognition. *J. Exp. Med.* **183**: 1253–1258.

Kwok, W.W., Domeier, M.E., Raymond, F.C. et al. (1996b). Allele-specific motifs characterize HLA-DQ interactions with a diabetes-associated peptide derived from glutamic acid decarboxylase. *J. Immunol.* **156**: 2171–2177.

Lernmark, A., Barmeier, H., Dube, S. et al. (1991). Autoimmunity of diabetes. *Endocrinol. Metab. Clin.* **20**: 589–617.

Maclaren, N., Riley, W., Skordis, N. et al. (1988). Inherited susceptibility to insulin-dependent diabetes is associated with HLA-DR1, while DR5 is protective. *Autoimmunity* **1**: 197–205.

Mijovic, C.H., Jenkins, D. and Penny, M.A. (1993). Susceptibility to type I (insulin-dependent) diabetes mellitus is determined by MHC class II molecules. What about DR4? (Review). *Diabetologia* **36**: 1210–1211.

Nakanishi, K., Kobayashi, T., Murase, T. et al. (1993). Association of HLA-A24 with complete beta-cell destruction in IDDM. *Diabetes* **42**: 1086–1093.

Nepom, G.T. (1993). Immunogenetics and IDDM. *Diabetes Rev.* **1**: 93–103.

Nepom, G.T. and Erlich, H. (1991). MHC class II molecules and autoimmunity. *Annu. Rev. Immunol.* **9**: 493–525.

Nepom, B.S., Schwarz, D., Palmer, J.P. and Nepom, G.T. (1987). Transcomplementation of HLA genes in IDDM. HLA-DQ alpha- and beta-chains produce hybrid molecules in DR3/4 heterozygotes. *Diabetes* **36**: 114–117.

Nepom, B.S., Nepom, G.T., Coleman, M. and Kwok, W.W. (1996). Critical contribution of β chain residue 57 in peptide binding ability of both HLA DR and DQ molecules. *Proc. Natl Acad. Sci. USA* **93**: 7202–7206.

Reijonen, H. and Nepom, G.T. (1996). The role of HLA susceptibility in predisposing to IDDM. In *The Molecular Pathogenesis of Diabetes Mellitus* (ed. R.D.G. Leslie), Karger, Basel.

Reijonen, H., Nejentsev, S., Tuokko, J. et al. and the Childhood Diabetes in Finland Study Groups (1997). HLA-DR4 subtype and -B alleles in DQB1*0302-positive haplotypes associated with IDDM. *Eur. J. Immunogen.* **24**: 357–363.

Ronningen, K.S., Markussen, G., Iwe, T. and Thorsby, E. (1989). An increased risk of insulin-dependent diabetes mellitus (IDDM) among HLA-DR4,DQw8/ DRw8,DQw4 heterozygotes. *Hum. Immunol.* **24**: 165–173.

Ronningen, K.S., Spurkland, A., Iwe, T. et al. (1991). Distribution of HLA-DRB1, -DQA1 and -DQB1 alleles and DQA1-DQB1 genotypes among Norwegian patients with insulin-dependent diabetes mellitus. *Tissue Antigens* **37**: 105–111.

Ronningen, K.S., Spurkland, A., Tait, B.D. et al. (1992). HLA class II associations in insulin-dependent diabetes mellitus among Blacks, Caucasoids, and Japanese. In *HLA 1991* (ed. K. Tsuji, M. Aizawa and T. Sasazuki), pp. 713–722, Oxford University Press, Oxford.

Sheehy, M.J., Scharf, S.J., Rowe, J.R. et al. (1989). A diabetes-susceptible HLA haplotype is best defined by a combination of HLA-DR and -DQ alleles. *J. Clin. Invest.* **83**: 830–835.

Stern, L.J. and Wiley, D.C. (1992). The human class II MHC protein HLA-DR1 assembles as empty ab heterodimers in the absence of antigenic peptide. *Cell* **68**: 465–477.

Stern, L.J., Brown, J.H., Jardetzky, T.S. et al. (1994). Crystal structure of the human class II MHC protein HLA-DR1 complexed with an influenza virus peptide. *Nature* **368**: 215–221.

Tait, B.D., Harrison, L.C., Drummond, B.P. et al. (1995). HLA antigens and age at diagnosis of insulin-dependent diabetes mellitus. *Hum. Immunol.* **42**: 116–122.

Thomson, G., Robinson, W.P., Kuhner, M.K. et al. (1988). Genetic heterogeneity, modes of inheritance, and risk estimates for a joint study of Caucasians with insulin-dependent diabetes mellitus. *Am. J. Hum. Genet.* **43**: 799–816.

Todd, J.A., Fukui, Y., Kitagawa, T. and Sasazuki, T. (1990). The A3 allele of the HLA-DQA1 locus is associated with susceptibility to type I diabetes in Japanese. *Proc. Natl Acad. Sci. USA* **87**: 1094–1098.

Undlien, D.E., Friede, T., Rammensee, H.G. et al. (1997). HLA encoded genetic predisposition in insulin-dependent diabetes mellitus (IDDM). DR4 subtypes may be associated with different degrees of protection. *Diabetes* **46**: 143–149.

Vallet-Colom, I., Levy-Marchal, C., Zarrouk, D. et al. (1990). HLA-DQB1 codon 57 and genetic susceptibility to type I (insulin-dependent) diabetes mellitus in French children. *Diabetologia* **33**: 174–176.

Van der Auwera, B., Van Waeyenberge, C., Schuit, F. et al. (1995). DRB1*0403 protects against IDDM in Caucasians with the high-risk heterozygous DQA1*0301–DQB*0302/DQA1*0501–DQB1*0201 genotype. *Diabetes* **44**: 527–530.

Vicario, J.L., Martinez-Laso, J., Corell, A. et al. (1992). Comparison between HLA-DRB and DQ DNA sequences and classic serological markers as type-1 (insulin-dependent) diabetes-mellitus predictive risk markers in the Spanish population. *Diabetologia* **35**: 475–481.

Vries, N. de, Ronningen, K.S., Tilanus, M.G. et al. (1993). HLA-DR1 and rheumatoid arthritis in Israeli Jews: sequencing reveals that DRB1*0102 is the predominant HLA-DR1 subtype. *Tissue Antigens* **41**: 26–30.

Wassmuth, R. and Lernmark, A. (1989). The genetics of susceptibility to diabetes (Review). *Clin. Immunol.* **53**: 358–399.

CHAPTER 16

HLA and Endocrine Disease

Anthony P. Weetman

INTRODUCTION

Autoimmune hypothyroidism is the prototype of organ-specific autoimmune diseases and similar T cell-mediated pathogenic mechanisms are likely to underlie the endocrine failure in Addison's disease and some cases of premature ovarian failure and hypophysitis. By contrast, the proximal cause of Graves' disease is stimulation of the thyroid-stimulating hormone receptor (TSH-R) by antibodies, although patients also have an underlying thyroiditis which can produce spontaneous hypothyroidism later in the course of the disease (Weetman and McGregor, 1994). These endocrinopathies, and type I diabetes mellitus, vitiligo and pernicious anaemia, occur together in individuals and families more frequently than expected, constituting the autoimmune polyglandular syndromes (APS). Human leucocyte antigen (HLA) associations have therefore been sought both with the isolated diseases and in the context of APS. Two other endocrine conditions discussed in this chapter have HLA associations but are not due to autoimmunity. Congenital adrenal hyperplasia is caused by adrenal enzyme defects and the HLA association has its basis in the location of the adrenal 21-hydroxylase gene in the class III region, while subacute thyroiditis is due to a viral infection and is associated with particular class I alleles.

GRAVES' DISEASE

Graves' disease is the most common form of hyperthyroidism, affecting approximately 1% of Caucasian women and 0.1% of men. Over 90% of patients have subclinical evidence of thyroid-associated ophthalmopathy (TAO), although clinical signs are less common, ranging from lid lag and retraction (50%) through proptosis (30%) to diplopia or visual field impairment (5%).

Population studies

Caucasians

Many population studies have examined HLA associations with Graves' disease. In Caucasians living in Europe, Australia and North America, HLA-A1 and B8 were the first associations to be reported in the 1970s and early 1980s, with a mean relative risk of 1.7 and 2.3, respectively (Tiwari and Terasaki, 1986). Thereafter, attention has focused on the class II region and, in terms of the highest relative risk, HLA-DR3 has proved to be more strongly associated than HLA-A1 or B8. Representative results are shown in Table 16.1.

The development of DNA-based genotyping has led to the exploration of other class II associations which could be still stronger. HLA-DQ2 is increased because of its linkage disequilibrium with DR3, but both of the DR3 subtypes (DR17 and DR18) have been associated with Graves' disease (Mangklabruks et al., 1991), which at first sight is surprising, given the fact that A1 and B8 are in linkage disequilibrium with DR17. However, as DR18 has no such class I linkage, these results are compatible with the idea that it is DR3 in general which is important in Caucasians. Results from other methods of testing extended DR3-associated haplotypes have

TABLE 16.1 HLA-DR3 associations with Graves' disease in Caucasians

Study	Country	Patients (n)	Relative risk
Farid et al. (1980)	Canada	83	3.1
McGregor et al. (1980)	England	65	3.2
Dahlberg et al. (1981)	Sweden	78	3.9
McKenna et al. (1982)	Ireland	86	2.6
Allannic et al. (1983)	France	94	4.2
Stenszky et al. (1985)	Hungary	256	4.8
Weetman et al. (1986)	Wales	65	2.9
Schleusener et al. (1989)	Germany	253	2.5

given similar information (Weetman and McCorkle, 1990) and a report of associations between a polymorphism in the heat shock protein 70 locus and Graves' disease may well reflect linkage disequilibrium with the C4A gene deletion, which is part of the HLA-B8–DR3 haplotype (Ratanachaiyavong et al., 1991).

More recently, HLA-DQA1*0501 has been found to confer an additional susceptibility to DR3 alone, giving a relative risk of 3.4 after exclusion of DR3-positive subjects (Yanagawa et al., 1993). This association was particularly strong in men with Graves' disease, but these cases are around 10-fold less frequent than in women. A larger series has confirmed the increased frequency of DQA1*0501 in DR3-negative Graves' disease but failed to show the male excess for this allele (Barlow et al., 1996).

Non-Caucasians

Fewer studies have been performed in non-Caucasians but widely different HLA associations have emerged. Asian Indians with Graves' disease have the closest pattern to European Caucasians, with a significant increase in HLA-B8 and DQ2, having relative risks of 4.1 and 5.5, respectively (Tandon et al., 1990). The strongest association in Japanese and Korean patients is with HLA-DR5, giving relative risks of 8.1 and 4.4, while DR8 is less strongly associated (Sasazuki et al., 1983; Cho et al., 1987). More recently, susceptibility and resistance to Graves' disease have been associated with DPB1*0501 and DQB1*0501, respectively, in Japanese (Dong et al., 1992; Tamai et al., 1994). Partial confirmation of these results comes from the demonstration that HLA-DPB1*0501 is increased in early-onset Graves' disease (below the age of 20 years) in Japan, but not in those with later onset (Onuma et al., 1994). The pathophysiological basis for this split by age of onset is unclear. Overall in this study, however, there was only a modest association with DR8 (relative risk 2.1) and the strongest relative risks were with B46 (4.8) and *C1* (2.8).

Similar discrepancies exist in Black populations, which could be related to either ethnic variations or the influence of environmental factors. No overall DR or DQ associations have been found in US Blacks with Graves' disease (Sridama et al., 1987; Yanagawa and DeGroot, 1996) whereas in South African Blacks of Zulu descent, HLA-DR1 and DR3 were significantly increased in patients with Graves' disease, giving relative risks of 3.5 and 2.4, respectively (Omar et al., 1990).

Family studies

The concordance of Graves' disease is 3–9% in dizygotic and 30–60% in monozygotic twins (Volpé et al., 1972); these early reports are almost certainly subject to ascertainment bias. In particular, a recent survey in Denmark has shown a concordance rate of only 22% in 18 monozygotic twin pairs (Brix et al., 1997). The risk in siblings is 3% and in HLA-identical siblings is 7% (Stenszky et al., 1985). Therefore, other genetic factors besides HLA, as well as environmental factors, must contribute to susceptibility. Polymorphisms in the CTLA-4 gene have recently been associated with Graves' disease, as well as diabetes mellitus and autoimmune hypothyroidism in population-based surveys (Yanagawa et al., 1995; Nisticò et al., 1996; Kotsa et al., 1997), but a limited search in 19 Graves' families revealed negative lod scores for markers in the region of CTLA-4 and other candidate genes (Tomer et al., 1997). A weakly positive score (<3) was found for markers on 14q31, studied because the TSH-R is encoded on this chromosome. However, based on population and family studies the TSH-R gene itself is unlikely to be a candidate (Watson et al., 1995; DeRoux et al., 1996).

Specific linkage to HLA was not found in a study of 15 Caucasian Graves' families and 12 other families with autoimmune hypothyroidism, with or without Graves' disease (Roman et al., 1992). However, this approach has been criticized, as linkage disequilibrium was assumed, yet there is known to be a clear association between Graves' disease and HLA-DR3 (Shields et al., 1994). By combined linkage and segregation analysis of Graves' disease with a thyroid antibody diathesis, evidence for linkage to HLA-DR was obtained when the effect of DR3 was considered (lod score 6.6).

Associations with clinical features

Attempts to find additional susceptibility factors for TAO have been dogged by differences in defining this complication; subclinical disease is a near universal feature of Graves' disease and there is disagreement over what constitutes severe TAO. Early studies in Germany and Hungary found an increase in HLA-B8,DR3 in patients with clinically obvious TAO plus Graves' disease, compared with those with Graves' disease alone, and an additional influence of HLA-DR7 (Schleusener et al., 1983; Frecker et al., 1986). However, others in Denmark, France and the UK found no difference in HLA-DR types between the two groups (Bech et al., 1977; Allannic et al., 1983; Kendall-Taylor et al., 1988; Weetman et al., 1988). HLA-DR4 and DR6 were increased in American Blacks with Graves' disease plus clinical evidence of TAO, compared with controls (Sridama et al., 1987), but as there was no difference from concurrently studied patients with Graves' disease alone, who did not differ from controls in HLA-DR4 or DR6, the significance of these results is debatable. HLA-DPB2.1/8, defined by restriction fragment length polymorphism (RFLP) analysis, was significantly less frequent in Caucasian patients with severe TAO than in Graves' disease without severe TAO or in controls, suggesting that this may have a protective effect for the development of severe eye signs (Weetman et al., 1990). This protective effect of DP2 has also been documented in Japanese patients in whom HLA-A11, B5, D12 and DQ4 were positively associated with TAO (Inoue et al., 1992).

The other major area of investigation has been to determine whether outcome after a course of antithyroid drugs has any HLA associations, because the ability to predict which patients will enter permanent remission after such treatment would be valuable. Despite several initial studies reporting a worse prognosis in individuals who were HLA-DR3 positive, subsequent observations have shown no effect of HLA-DR3 on outcome after antithyroid drugs (Allannic et al., 1983; Weetman et al., 1986; Young et al., 1988) and so far there are no factors, alone or combined, which are sufficiently discriminating to predict outcome after such medication.

One recent report from Japan has demonstrated a positive association between HLA-DRB1*08032 and the likelihood of developing agranulocytosis after antithyroid drugs (Tamai et al., 1996). The relative risk was 5.4 but the number of subjects was small ($n = 24$). Further confirmation is required, especially in other ethnic groups, and this complication is so rare that the clinical implications of the observation are limited, but it does suggest a T cell-mediated aetiology for such drug-induced agranulocytosis. Another rare complication of antithyroid drug therapy is the development of the insulin autoimmune syndrome, characterized by spontaneous hypoglycaemia, a high endogenous insulin level and the presence of autoantibodies to insulin. The condition is most frequent in Japan and over one-third of such patients have taken methimazole (although none has used propylthiouracil). The syndrome is generally associated with DRB1*0406–DQA1*0301–DQB1*0302 and this is preserved in Graves' patients who develop the syndrome while taking methimazole (Uchigata et al., 1993). Methimazole may predispose to the syndrome by cleaving insulin at its disulfide bonds, allowing presentation of immunogenic peptides in genetically predisposed individuals.

AUTOIMMUNE HYPOTHYROIDISM

Autoimmune hypothyroidism affects around 1% of women and 0.1% of men and is by far the most common cause of primary thyroid failure in iodine-sufficient areas. Thyroid damage is slowly progressive over months to years; individuals at risk have circulating antibodies to thyroglobulin or thyroid peroxidase. Both cell-mediated and humoral immunity contribute to thyroid failure (Weetman and McGregor, 1994). Histological and clinical subdivisions have been made, notably into goitrous (Hashimoto's thyroiditis) and non-goitrous (primary myxoedema) forms, but accurate classification is generally unavailable as biopsy is rarely indicated, and thyroid size not only requires ultrasound for precise measurement but also is not constant through the course of the disease.

Population studies

Caucasians

A summary of population surveys from several different countries is given in Table 16.2. Initial studies suggested a clear genetic distinction between Hashimoto's thyroiditis and primary myxoedema, HLA-DR5 being associated with the former and HLA-B8,DR3 with the latter (Moens and Farid, 1978; Weissel et al., 1980). However, this has not been confirmed in subsequent studies (Table 16.2), possibly because of changes in typing reagents and regional variations. A meta-analysis found evidence for an association between Hashimoto's thyroiditis and HLA-DR3 and DR4, with respective relative risks of 2.3 and 1.6 (Jenkins et al., 1992). This has been confirmed by the rigorous approach of genotyping thyroid tissue blocks from patients with histologically proven Hashimoto's thyroiditis; DQA1*0301 (in linkage disequilibrium with DR4) and DQB1*0201 (in linkage disequilibrium with DR3) were increased in frequency with similar relative risks (Shi et al., 1992).

Overall, these results indicate a relatively weak and heterogeneous association between HLA-DR3, DR4 and possibly DR5, and autoimmune hypothyroidism, irrespective of goitre. Other candidate susceptibility genes have been explored without any consensus (Mangklabruks et al., 1991). However, polymorphisms in the CTLA-4 gene are associated with autoimmune hypothyroidism (Kotsa et al., 1997), suggesting a shared genetic background with Graves' disease and compatible with the clinical observation that the two disorders frequently occur together in families.

Non-Caucasians

Studies in other racial groups have also given conflicting results. In Chinese patients with Hashimoto's thyroiditis there is an association with HLA-B46 and DR9 (Hawkins et al., 1987; Wang et al., 1988), whereas HLA-DR53 (and, equivocally, DR9) are associated in Japanese (Honda et al., 1989; Onuma et al., 1993). Japanese and Korean hypothyroid patients have been classified into those with and those without thyroid-blocking anti-

TABLE 16.2 HLA-DR associations with Hashimoto's thyroiditis and primary myxoedema in Caucasians

Study	Country	Patients (n)	HLA	Relative risk
Hashimoto's thyroiditis				
Weissel et al. (1980)	Austria	39	DR5	3.2
Farid et al. (1981)	Canada	40	DR5	3.1
Thompson and Farid (1985)	Canada	21	DR4	5.0
Stenszky et al. (1987)	Hungary	60	DR3	3.3
Vargas et al. (1988)	Canada	60	DR5	4.2
Badenhoop et al. (1990)	Canada/England	64	DR4/DR5	2.9/3.8
Tandon et al. (1991)	England	86	DR3	2.2
Mangklabruks et al. (1991)	USA	63	None	None
Primary myxoedema				
Farid et al. (1981)	Canada	50	DR3	5.7
Bogner et al. (1992)	Germany	36	DR5	4.6

bodies. These antibodies operate at the level of the TSH-R and are particularly associated with primary myxoedema in these racial groups, but not in Caucasians in whom blocking antibodies also occur with Hashimoto's thyroiditis (Weetman and McGregor, 1994). Those with blocking antibodies have HLA associations similar to local Graves' disease patients, whereas primary myxoedema without blocking antibodies has similar associations to local Hashimoto's thyroiditis (Cho et al., 1993; Inoue et al., 1993). Similar studies are awaited in non-Oriental groups to determine whether this reflects an underlying association between certain HLA alleles and a generalized autoimmune response against the TSH-R.

Family studies

Monozygotic twin pairs with autoimmune hypothyroidism have been identified but there are no reliable estimates of concordance (Hoffman, 1966). Studies in a limited number of families with Hashimoto's thyroiditis have found no evidence for linkage to HLA (Roman et al., 1992; Tomer et al., 1997). Autosomal dominant transmission of autoantibodies to thyroglobulin and thyroid peroxidase, with incomplete penetrance in men, has been reported in families with autoimmune thyroid disease but there was no linkage of this diathesis to HLA (Phillips et al., 1990). Unfortunately, a subsequent analysis could not distinguish between a single locus and a multifactorial model for the inheritance of thyroid antibodies, despite a large sample size (Phillips et al., 1993). This will make it difficult to ascertain the true role of HLA and other genes in susceptibility by genetic linkage strategies.

Associated disorders

There is a high prevalence of autoimmune hypothyroidism in Down's syndrome, and those with DQA1*0301 (in linkage with DR4 and DR7) are particularly at risk, implying an interaction between effects encoded on chromosome 21 and by a particular HLA locus in susceptibility (Nicholson et al., 1994). The likely importance of chromosome 21 is reflected by the high prevalence of autoimmune thyroid disease in familial Alzheimer's disease kindreds (Ewins et al., 1991).

POSTPARTUM THYROIDITIS

This is defined as the occurrence of transient dysfunction in the year after delivery and occurs in around 5% of women postpartum (Lazarus et al., 1996). These women usually have thyroid antibodies antepartum and this mild autoimmune response is believed to be enhanced in the year after delivery, leading to thyroid damage. In around one-quarter of such women permanent hypothyroidism supervenes after recovery.

Several studies have reported, in common with autoimmune hypothyroidism, associations with HLA-DR3, DR4 and DR5, and these are reviewed in detail elsewhere (Stagnaro-Green, 1993). Many of these involved small samples and all were based on serological typing, which has caused misinterpretations (Thompson and Farid, 1985). Recently, RFLP-based typing has been conducted on a reasonably sized sample ($n = 86$) of women with carefully characterized postpartum thyroiditis (Parkes et al., 1996). Only HLA-DR5 (both DR11 and DR12 subtypes) was significantly associated with postpartum thyroiditis (relative risk 2.4); there was also a weak association with DR3 (DR17) that did not survive correction, and a reduced frequency of DR15. Unfortunately, local patients with autoimmune hypothyroidism were not included and it is therefore difficult to know how much overlap there is between the two disorders.

Overall, it appears that all three types of thyroid autoimmune disease share a weak HLA association with HLA-DR3 in Caucasians, and autoimmune hypothyroidism and postpartum thyroiditis are associated with DR4 and DR5 in some populations. However, the major genetic susceptibility for all three disorders lies outside the HLA complex; CTLA-4 is one candidate gene, but its role will be, at best, a modest one.

SUBACUTE THYROIDITIS

Subacute (de Quervain's) thyroiditis is believed to be the result of a viral infection, although a wide variety of viruses has been implicated (Volpé, 1979). Clinically, the condition is characterized by a tender, enlarged thyroid, high erythrocyte sedimentation rate, reduced radioisotope uptake and a destructive thyroiditis. Circulating thyroid hormone levels are initially raised, followed by a phase of hypothyroidism and usually restoration of euthyroidism after several months. An association with HLA-B35 was first reported from Czechoslovakia (Nyulassy et al., 1975). This has been confirmed by many other groups in Caucasian, Chinese and Japanese patients and in a pair of monozygotic twins (Rubin and Guay, 1991).

The identification of new HLA-B specificities has led to a reappraisal of patients with subacute thyroiditis, in whom, again, HLA-B35 gave the strongest relative risk (18), but an association with B67 was also found (relative risk 11.2); no HLA-B35/B67 heterozygotes were identified (Ohsako et al., 1995). The B67-positive patients were more likely to have the classical picture of thyroid dysfunction and a summer or an autumn onset, whereas thyroiditis in the B35-positive subjects occurred throughout the year and the hypothyroid phase was often absent. Whether these associations reflect the ability of B35 and B67 to present clinically different thyrotropic viral epitopes, or are related to a more complex type of pathogenesis such as that proposed for HLA-B27 and *Klebsiella* infection in ankylosing spondylitis (Section II, Chapter 1), is unknown.

AUTOIMMUNE ADDISON'S DISEASE

In this rare disorder, affecting around 1 in 25 000 of the population, patients have autoantibodies against P450c21 and, to a lesser extent, other cytochrome enzymes in the steroid synthetic pathway, but the mechanisms resulting in adrenal cortical destruction remain unclear (Weetman, 1997). Addison's disease is a component of both types of APS defined in Table 16.3 and discussed further below. While APS type 1 is very rare, APS type 2 is much more common: around half of all patients with autoimmune Addison's disease have an associated endocrinopathy as part of this syndrome (Papadopoulos and Hallengren, 1990), which is clearly important in population surveys (there have been no family studies of Addison's disease, owing to its rarity) as HLA associations may reflect the influence of the associated disorders.

The first such association identified was with HLA-B8 in Caucasians (Thomsen et al., 1975) and this was more pronounced in those patients with Addison's disease as part of APS type 2 (see below). Subsequently, a stronger association with HLA-DR3 was reported in 34 North American patients (relative risk 12.1), but DR4 was also associated (relative risk 8.9), even in the absence of concurrent type I diabetes mellitus, for which this specificity is a risk factor (Maclaren and Riley, 1986). Three further studies on a total of 139 European patients found only an association with HLA-DR3 and more modest relative risks of 3.4–3.6 (Latinne et al., 1987; Boehm et al., 1991; Weetman et al., 1991). Associations between DR4 haplotypes and concurrent diabetes followed the expected pattern in Addison's patients (Boehm et al., 1991). It is unclear from these reports whether there is a stronger association between HLA-DR3 and the subset of Addison's disease with APS type 2, because the numbers are too small.

The enzyme P450c21 is encoded by the CYP21B gene, located in tandem with a pseudogene (CYP21A) in the HLA class III region, and there is linkage disequilibrium between this and the class II region. However, there is no obvious difference between patients and controls in their CYP21B alleles beyond that expected by linkage

TABLE 16.3 Main characteristics of the autoimmune polyglandular syndromes (APS)

	APS 1	APS 2
Inheritance	Autosomal recessive	Autosomal dominant with variable penetrance
Frequency	Rare	Common
Major components[a]	Addison's disease, hypoparathyroidism, chronic mucocutaneous candidiasis	Autoimmune thyroid disease, type I diabetes mellitus, Addison's disease
Other components	Ovarian failure, alopecia, vitiligo, diabetes, pernicious anaemia, ectodermal dysplasia	Ovarian failure, vitiligo, alopecia, pernicious anaemia, coeliac disease, myasthenia gravis
HLA association	May modify phenotype	HLA-A1, B8, DR3

[a]At least two major components are needed for diagnosis.

disequilibrium with the HLA-B8,DR3 haplotype (Peterson et al., 1995).

PREMATURE OVARIAN FAILURE

Defined as the onset of menopause before the age of 40 years, this affects around 1% of women, in up to one-third of whom there is a history of thyroid or other autoimmune disorders (Hock et al., 1997). Those cases with associated Addison's disease clearly have an autoimmune basis but constitute only a small proportion of women with premature ovarian failure. It has been widely assumed that the isolated idiopathic form also has an autoimmune aetiology, but the evidence for this remains equivocal. In a limited study of 19 Canadian patients an association with HLA-DR3 was found (Walfish et al., 1983). However, no HLA class I or II associations were found in 27 UK women (Jaroudi et al., 1994) and a much larger study of 102 US women also found no difference from local controls in HLA-DR frequencies determined serologically (Anasti et al., 1994). A weak association with HLA-DR4 was noted when patients were compared with a large reference control population, but this was not significant when the smaller sample of local controls we used. The present author found no HLA-DR associations in 40 patients using polymerase chain reaction (PCR)-based typing (unpublished observations). HLA associations could, none the less, exist in a small subset of patients with idiopathic premature ovarian failure whose disease has an autoimmune basis, but these have yet to be properly defined.

LYMPHOCYTIC HYPOPHYSITIS

This is a rare inflammatory condition, probably of autoimmune origin. Fewer than 100 cases have been reported and HLA associations remain unclear, with only isolated patients having had typing performed (Thoudou et al., 1995).

AUTOIMMUNE POLYGLANDULAR SYNDROME TYPES 1 AND 2

APS type 1 (Table 16.3) is a rare autosomal recessive disorder and the disease locus has been assigned to chromosome 21q22.3 (Aaltonen et al., 1994). However, HLA genes may modify the expression of the syndrome, as HLA-A28 is more frequent in patients with hypoparathyroidism and diabetes mellitus, A3 in ovarian failure and A2 in keratopathy (Ahonen et al., 1988). Further, larger studies are needed to confirm these observations.

There is nothing unique about the autoimmune responses against the various target organs in patients with APS type 2 compared with those in the individual sporadic forms, but the HLA-B8, DR3 haplotype is more frequent, depending on the components of the syndrome. This haplotype conferred a relative risk of 2.8 in pernicious anaemia plus one other endocrinopathy, 7.1 in vitiligo plus one other endocrinopathy, 16 in type I diabetes mellitus plus Graves' disease and 48.3 in type I diabetes plus Addison's disease (Eisenbarth and Jackson, 1981). In a more recent survey using PCR-based typing, the importance of DR3, DQB1*0201 in APS type 2 was confirmed and HLA-DR4, DQB1*0302 was associated only with type I diabetes mellitus (Huang et al., 1996).

Finally, a unique form of APS has been reported in Persian Jews, not associated with HLA-DR3, in which there is hypoparathyroidism and hypogonadism, as well as Addison's disease, diabetes and hypothyroidism, less commonly (Shapiro et al., 1987). All subjects were HLA-A26 or A27 positive.

CONGENITAL ADRENAL HYPERPLASIA

Congenital adrenal hyperplasia is a group of autosomal recessive diseases, the most common of which is 21-hydroxylase deficiency (95% of cases). A severe (classical) form of this disorder presents in infancy with ambiguous genitalia in females, postnatal virilization in both genders and, in 75% of cases, salt wasting. The classical form occurs in 1 in 12 000 births, whereas a mild (non-classical) form is present in between 1 in 100 and 1 in 1000 people in various populations. There is often late-onset virilization in women with the non-classical disorder, as well as insulin resistance, reduced fertility and lower than expected height in both genders (New and White, 1995).

As mentioned above, P450c21 is encoded by the CYP21B (or CYP21) gene, which is 30 kb from the CYP21A pseudogene (or CYP21P) and 3 kb downstream of each C4 gene (Section I, Chapter 2). There is a high degree of homology between CYP21 and CYP21P, which allows (i) unequal crossing-over in meiosis, resulting in CYP21 deletions or duplications, and (ii) gene conversion, in which base changes on CYP21P are transferred to CYP21 (Miller, 1994; New and White, 1995). Random gene deletions and point mutations are rare. Because there are many different CYP21 defects, most patients are compound heterozygotes rather than truly homozygous. Around 15% of classical 21-hydroxylase deficiency is due to gene deletions and the remainder, plus all non-classical forms, is caused by gene conversion. The severity of disease depends on the least affected allele in compound heterozygotes but, even in true homozygotes, phenotypic expression is variable, probably because other non-adrenal enzymes can mediate 21-hydroxylation (Miller, 1994). Salt wasting is generally associated with complete abolition of P450c21 activity and simple virilization at birth with 1–2% of enzymatic activity, while missense mutations that reduce enzymatic activity to 20–50% of normal are associated with the non-classical form of the disease. A comprehensive list of mutations is given elsewhere (New and White, 1995).

REFERENCES

Aaltonen, J., Björses, P., Sandkuijl, L. et al. (1994). An autosomal locus causing autoimmune disease: autoimmune polyglandular disease type I assigned to chromosome 21. *Nature Genet.* **8**: 83–87.

Ahonen, P., Koskimies, S., Lokki, M.L. et al. (1988). The expression of autoimmune polyglandular disease type I appears associated with several HLA-A antigens but not with HLA-DR. *J. Clin. Endocrinol. Metab.* **66**: 1152–1157.

Allannic, H., Fauchet, R., Lorcy, Y. et al. (1983). A prospective study of the relationship between relapse of hyperthyroid Graves' disease after antithyroid drugs and HLA haplotype. *J. Clin. Endocrinol. Metab.* **57**: 719–722.

Anasti, J.N., Adams, S., Kimzey, L.M. et al. (1994). Karyotypically normal spontaneous premature ovarian failure: evaluation of association with the class II major histocompatibility complex. *J. Clin. Endocrinol. Metab.* **78**: 722–723.

Badenhoop, K., Schwarz, G., Walfish, P.G. et al. (1990). Susceptibility to thyroid autoimmune disease: molecular analysis of HLA-D region genes identifies new markers for goitrous Hashimoto's thyroiditis. *J. Clin. Endocrinol. Metab.* **71**: 1131–1137.

Barlow, A.B.T., Wheatcroft, N. and Weetman, A.P. (1996). Association of HLA-DQA1*0501 with Graves' disease in English Caucasian men and women. *Clin. Endocrinol.* **44**: 73–77.

Bech, K., Lumholz, B., Nerup, J. et al. (1977). HLA antigens in Graves' disease. *Acta Endocrinol.* **86**: 510–515.

Boehm, B.O., Manfras, B., Seidl, S. et al. (1991). The HLA-DQβ non-Asp-57 allele: a predictor of future insulin-dependent diabetes mellitus in patients with autoimmune Addison's disease. *Tissue Antigens* **37**: 130–132.

Bogner, U., Badenhoop, K., Peters, H. et al. (1992). HLA-DR/DQ gene variation in nongoitrous autoimmune thyroiditis at the serological and molecular level. *Autoimmunity* **14**: 155–158.

Brix, T.H., Christensen, K., Holm, N.V. et al. (1998). A population-based study of Graves' disease in Danish twins. *Clin. Endocrinol.* **48**: 397–400.

Cho, B.Y., Chung, J.H., Shong, Y.K. et al. (1993). A strong association between thyrotropin receptor-blocking antibody-positive atrophic autoimmune thyroiditis and HLA-DR8 and HLA-DQB1*0302 in Koreans. *J. Clin. Endocrinol. Metab.* **77**: 611–615.

Cho, B.Y., Rhee, B.D., Lee, D.S. et al. (1987). HLA and Graves' disease in Koreans. *Tissue Antigens* **31**: 119–121.

Dahlberg, P.A., Holmlund, G., Karlsson, F.A. and Säfwenberg, J. (1981). HLA-A, -B, -C and -DR antigens in patients with Graves' disease and their correlation with signs and clinical course. *Acta Endocrinol.* **97**: 42–47.

De Roux, N., Shields, D.C., Misrahi, M. et al. (1996). Analysis of the thyrotropin receptor as a candidate gene in familial Graves' disease. *J. Clin. Endocrinol. Metab.* **81**: 3483–3486.

Dong, R.-P., Kimura, A., Okubo, R. et al. (1992). HLA-A and DPB1 loci confer susceptibility to Graves' disease. *Human Immunol.* **35**: 165–172.

Eisenbarth, G.S. and Jackson, R.A. (1981). Immunogenetics of polyglandular failure and related diseases. In *HLA in Endocrine and Metabolic Disorders* (ed. N.R. Farid), pp. 235–264, Academic Press, London.

Ewins, D.L., Rossor, M.N., Butler, J. et al. (1991). Association between autoimmune thyroid disease and familial Alzheimer's disease. *Clin. Endocrinol.* **35**: 93–95.

Farid, N.R., Stone, E. and Johnson, G. (1980). Graves' disease and HLA – clinical and epidemiological associations. *Clin. Endocrinol.* **13**: 535–544.

Farid, N.R., Sampson, L., Moens, H. and Barnard, J.M. (1981). The association of goitrous autoimmune thyroiditis with HLA-DR5. *Tissue Antigens* **17**: 265–268.

Frecker, M., Stenszky, V., Balázs, C. et al. (1986). Genetic factors in Graves' ophthalmopathy. *Clin. Endocrinol.* **25**: 479–485.

Hawkins, B.R., Lam, K.S.L., Ma, J.T.C. et al. (1987). Strong association between HLA-DRw9 and Hashimoto's thyroiditis in southern Chinese. *Acta Endocrinol.* **114**: 543–546.

Hock, A., Schoemaker, J. and Drexhage, H.A. (1997). Premature ovarian failure and ovarian autoimmunity. *Endocr. Rev.* **18**, 107–134.

Hoffman, E. (1966). Hashimoto's thyroiditis (struma lymphomatosa). Case report of a mother and three daughters, two of whom are monovular twins. *Arch. Surg.* **92**: 865–867.

Honda, K., Tamai, J., Morita, T. et al. (1989). Hashimoto's thyroiditis and HLA in Japanese. *J. Clin. Endocrinol. Metab.* **69**: 1268–1273.

Huang, W., Connor, E., Dela Rosa, T. et al. (1996). Although DR3-DQB1*0201 may be associated with multiple component diseases of the autoimmune polyglandular syndromes, the human leukocyte antigen DR4-DQB1*0302 haplotype is implicated only in β-cell autoimmunity. *J. Clin. Endocrinol. Metab.* **81**: 2559–2563.

Inoue, D., Sato, K., Enomoto, T. et al. (1992). Correlation of HLA types and clinical findings in Japanese patients with hyperthyroid Graves' disease: evidence indicating the existence of four subpopulations. *Clin. Endocrinol.* **36**: 75–82.

Inoue, D., Sato, K., Sugawa, H. et al. (1993). Apparent genetic difference between hypothyroid patients with blocking-type thyrotropin receptor antibody and those without, as shown by restriction fragment length polymorphism analyses of HLA-DP loci. *J. Clin. Endocrinol. Metab.* **77**: 606–610.

Jaroudi, K.A., Arora, M., Sheth, K.V. et al. (1994). Human leukocyte antigen typing and associated abnormalities in premature ovarian failure. *Hum. Reprod.* **9**: 2006–2009.

Jenkins, D., Penny, M.A., Fletcher, J.A. et al. (1992). HLA class II gene polymorphism contributes little to Hashimoto's thyroiditis. *Clin. Endocrinol.* **37**: 141–145.

Kendall-Taylor, P., Stephenson, A., Stratton, A. et al. (1988). Differentiation of autoimmune ophthalmopathy from Graves' hyperthyroidism by analysis of genetic markers. *Clin. Endocrinol.* **28**: 601–610.

Kotsa, K., Watson, P.F. and Weetman, A.P. (1997). A CTLA-4 gene polymorphism is associated with both Graves' disease and autoimmune hypothyroidism. *Clin. Endocrinol.* **46**: 551–554.

Latinne, D., Vandeput, Y., De Bruyere, M. et al. (1987). Addison's disease: immunological aspects. *Tissue Antigens* **30**: 23–24.

Lazarus, J.H., Hall, R., Othman, S. et al. (1996). The clinical spectrum of postpartum thyroid disease. *Q. J. Med.* **89**: 429–435.

McGregor, A.M., Rees Smith, B., Hall, R. et al. (1980). Prediction of relapse in hyperthyroid Graves' disease. *Lancet* **i**: 1101–1103.

McKenna, R., Kearns, M., Sugrue, D. et al. (1982). HLA and hyperthyroidism in Ireland. *Tissue Antigens* **19**: 97–99.

Maclaren, N.K. and Riley, W.J. (1986). Inherited susceptibility to autoimmune Addison's disease is linked to human leukocyte antigens DR3 and/or DR4, except when associated with type I autoimmune polyglandular syndrome. *J. Clin. Endocrinol. Metab.* **62**: 455–459.

Mangklabruks, A., Cox, N. and DeGroot, L.J. (1991). Genetic factors in autoimmune thyroid disease analyzed by restriction fragment length polymorphisms of candidate genes. *J. Clin. Endocrinol. Metab.* **73**: 236–244.

Miller, W.L. (1994). Genetics, diagnosis, and management of 21-hydroxylase deficiency. *J. Clin. Endocrinol. Metab.* **78**: 241–246.

Moens, H. and Farid, N.R. (1978). Hashimoto's thyroiditis is associated with HLA-DRw3. *N. Engl. J. Med.* **299**: 133–134.

New, M.I. and White, P.C. (1995). Genetic disorders of steroid hormone synthesis and metabolism. *Baillière's Clin. Endocrinol. Metab.* **9**: 525–554.

Nicholson, L.B., Wong, F.S., Ewins, D.L. et al. (1994). Susceptibility to autoimmune thyroiditis in Down's syndrome is associated with the major histocompatibility class II DQA 0301 allele. *Clin. Endocrinol.* **41**: 381–383.

Nisticò, L., Buzzetti, R., Pritchard, L.E. et al. (1996). The *CTLA-4* gene region of chromosome 2q33 is linked to, and associated with, type 1 diabetes. *Hum. Mol. Gen.* **5**: 1075–1080.

Nyulassy, S., Hnilica, P. and Stefanovic, J. (1975). The HL-A system and subacute thyroiditis: a preliminary report. *Tissue Antigens* **6**: 105–106.

Ohsako, N., Tamai, H., Sudo, T. et al. (1995). Clinical characteristics of subacute thyroiditis classified according to human leukocyte antigen typing. *J. Clin. Endocrinol. Metab.* **80**: 3653–3656.

Omar, M.A.K., Hammond, M.G., Desai, R.K. et al. (1990). HLA class I and II antigens in South African Blacks with Graves' disease. *Clin. Immunol. Immunopathol.* **54**: 98–102.

Onuma, H., Ota, M., Sugenoya, A. et al. (1993). Association of HLA-DR53 and lack of association of DPB1 alleles with Hashimoto's thyroiditis in Japanese. *Tissue Antigens* **42**: 150–152.

Onuma, H., Ota, M., Sugenoya, A. and Inoko, H. (1994). Association of HLA-DBP1*0501 with early-onset Graves' disease in Japanese. *Hum. Immunol.* **195**: 195–201.

Papadopoulos, K.L. and Hallengren, B. (1990). Polyglandular autoimmune syndrome type II in idiopathic Addison's disease. *Acta Endocrinol.* **124**: 472–478.

Parkes, A.B., Darke, C., Othman, S. et al. (1996). Major histocompatibility complex class II and complement polymorphisms in postpartum thyroiditis. *Eur. J. Endocrinol.* **134**: 449–453.

Peterson, P., Partanen, J., Aavik, E. et al. (1995). Steroid 21-hydroxylase gene polymorphism in Addison's disease patients. *Tissue Antigens* **46**: 63–67.

Phillips, D., McLachlan, S., Stephenson, A. et al. (1990). Autosomal dominant transmission of autoantibodies to thyroglobulin and thyroid peroxidase. *J. Clin. Endocrinol. Metab.* **70**: 742–746.

Phillips, D.I.W., Shields, D.C., Dougoujon, J.M. et al. (1993). Complex segregation analysis of thyroid autoantibodies: are they inherited as an autosomal dominant trait? *Hum. Hered.* **43**: 141–146.

Ratanachaiyavong, S., Demaine, A.G., Campbell, R.D. and McGregor, A.M. (1991). Heat shock protein 70 (HSP70) and complement C4 genotypes in patients with hyperthyroid Graves' disease. *Clin. Exp. Immunol.* **84**: 48–52.

Roman, S.H., Greenberg, D., Rubinstein, P. et al. (1992). Genetics of autoimmune thyroid disease: lack of evidence for linkage to HLA within families. *J. Clin. Endocrinol. Metab.* **74**: 496–503.

Rubin, R.A. and Guay, A.T. (1991). Susceptibility of subacute thyroiditis is genetically influenced: familial occurrence in identical twins. *Thyroid* **1**: 157–159.

Sasazuki, T., Nishimura, Y., Muto, M. and Ohta, N. (1983). HLA-linked genes controlling immune response and disease susceptibility. *Immunol. Rev.* **70**: 51–73.

Schleusener, H., Schernthaner, G., Mayr, W.R. et al. (1983). HLA-DR3 and HLA-DR5 associated thyrotoxicosis – two different types of toxic diffuse goiter. *J. Clin. Endocrinol. Metab.* **56**: 781–785.

Schleusener, H., Schwander, J., Fischer, C. et al. (1989). Prospective multicentre study on the prediction of relapse after anti-thyroid drug treatment in patients with Graves' disease. *Acta Endocrinol.* **120**: 689–701.

Shapiro, M.S., Zamir, R., Weiss, E. et al. (1987). The polyglandular deficiency syndrome: a new variant in Persian Jews. *J. Endocrinol. Invest.* **10**: 1–7.

Shi, T., Zou, M., Robb, D. and Farid, N.R. (1992). Typing for major histocompatibility complex class II antigens in thyroid tissue blocks: association of Hashimoto's throiditis with HLA-DQA1*0501 and DRB*0201 alleles. *J. Clin. Endocrinol. Metab.* **75**: 943–946.

Shields, D., Ratanachaiyavong, S., McGregor, A.M. et al. (1994). Combined segregation and linkage analysis of Graves' disease with a thyroid autoantibody diathesis. *Am. J. Hum. Genet.* **55**: 540–554.

Sridama, V., Hara, Y., Fauchet, R. and DeGroot, L.J. (1987). HLA immunogenetic heterogeneity in Black American patients with Graves' disease. *Arch. Int. Med.* **147**: 229–231.

Stagnaro-Green, A. (1993). Postpartum thyroiditis: prevalence, etiology, and clinical implications. *Thyroid Today* **16**: 1–11.

Stenszky, V., Kozma, L., Balázs, C. et al. (1985). The genetics of Graves' disease: HLA and disease susceptibility. *J. Clin. Endocrinol. Metab.* **61**: 735–740.

Stenszky, V., Balázs, C., Kraszits, E. et al. (1987). Association of goitrous autoimmune thyroiditis with HLA-DR3 in eastern Hungary. *J. Immunogenet.* **14**: 143–148.

Tamai, H., Kimura, A., Dong, R.-P. et al. (1994). Resistance to autoimmune thyroid disease is associated with HLA-DQ. *J. Clin. Endocrinol. Metab.* **78**: 94–97.

Tamai, H., Sudo, T., Kimura, A. et al. (1996). Association between the DRB1*08032 histocompatibility antigen and methimazole-induced agranulocytosis in Japanese patients with Graves' disease. *Ann. Int. Med.* **124**: 490–494.

Tandon, N., Mehra, N.K., Taneja, V. et al. (1990). HLA antigens in Asian Indian patients with Graves' disease. *Clin. Endocrinol.* **33**: 21–26.

Tandon, N., Zhang, L. and Weetman, A.P. (1991). HLA associations in Hashimoto's thyroiditis. *Clin. Endocrinol.* **34**: 383–386.

Thodou, E., Asa, S.L., Kontogeorgos, G. et al. (1995). Clinical case seminar: lymphocytic hypophysitis; clinicopathological findings. *J. Clin. Endocrinol. Metab.* **80**: 2302–2311.

Thompson, C. and Farid, N.R. (1985). Post-partum thyroiditis and goitrous (Hashimoto's) thyroiditis are associated with HLA-DR4. *Immunol. Letts.* **11**: 301–303.

Thomsen, M., Platz, P., Ortved Anderson, O. et al. (1975). MLC typing in juvenile diabetes mellitus and idiopathic Addison's disease. *Transplant. Rev.* **22**: 125–147.

Tiwari, J.L. and Terasaki, P.I. (1986). *HLA and Disease Associations*, pp. 214–220, Springer, New York.

Tomer, Y., Barbesino, G., Keddache, M. et al. (1997). Mapping of a major susceptibility locus for Graves' disease (GD-1) to chromosome 14q31. *J. Clin. Endocrinol. Metab.* **82**: 1645–1649.

Uchigata, Y., Kuwata, S., Tsushima, T. et al. (1993). Patients with Graves' disease who developed insulin autoimmune syndrome (Hirata disease) possess HLA-B62/Cw4, DR4 carrying DRB1*0406. *J. Clin. Endocrinol. Metab.* **77**: 249–254.

Varga, M.T., Briones–Urbina, R., Gladman, D. et al. (1988). Antithyroid microsomal antibodies and HLA-DR5 are associated with postpartum thyroid dysfunction: evidence supporting autoimmune pathogenesis. *J. Clin. Endocrinol. Metab.* **67**: 327–333.

Volpé, R., Edmonds, M., Lamki, L. et al. (1972). The pathogenesis of Graves' disease. A disorder of delayed hypersensitivity? *Mayo Clin. Proc.* **47**: 824–834.

Walfish, P.G., Gottesman, I.S., Shewchuk, A.B. et al. (1983). Association of premature ovarian failure with HLA antigens. *Tissue Antigens* **32**: 235–236.

Wang, F.W., Yu, Z.Q., Xy, J.J. et al. (1988). HLA and hypertrophic Hashimoto's thyroiditis in Shanghai Chinese. *Tissue Antigens* **32**: 235–236.

Watson, P.F., French, A., Pickerill, A.P. et al. (1995). Lack of association between a polymorphism in the coding region of the thyrotrophin receptor gene and Graves' disease. *J. Clin. Endocrinol. Metab.* **80**: 1032–1035.

Weetman, A.P. (1997). Autoantigens in Addison's disease and associated syndromes. *Clin. Exp. Immunol.* **6**: 227–229.

Weetman, A.P. and McCorkle, R. (1990). Evidence against extended DR3-related haplotypes in Graves' disease. *J. Immunogenet.* **17**: 403–407.

Weetman, A.P. and McGregor, A.M. (1994). Autoimmune thyroid disease: further developments in our understanding. *Endocr. Rev.* **15**: 788–830.

Weetman, A.P., Ratanachaiyavong, S., Middleton, G.W. et al. (1986). Prediction of outcome in Graves' disease after carbimazole treatment. *Q. J. Med.* **59**: 409–419.

Weetman, A.P., So, A.K., Warner, C.A. et al. (1988). Immunogenetics of Graves' ophthalmopathy. *Clin. Endocrinol.* **28**: 619–628.

Weetman, A.P., Zhang, L., Webb, S. and Shine, B. (1990). Analysis of HLA-DQB and HLA-DPB alleles in Graves' disease by oligonucleotide probing of enzymatically amplified DNA. *Clin. Endocrinol.* **33**: 65–71.

Weetman, A.P., Zhang, L., Tandon, N. and Edwards, O.M. (1991). HLA associations with autoimmune Addison's disease. *Tissue Antigens* **81**: 31–33.

Weissel, M., Hofer, R., Zasmeta, H. and Mayr, W.R. (1980). HLA-DR and Hashimoto's thyroiditis. *Tissue Antigens* **16**: 256–259.

Yanagawa, T. and DeGroot, L.J. (1996). HLA class II associations in African–American female patients with Graves' disease. *Thyroid* **6**: 37–39.

Yanagawa, T., Mangklabruks, A., Chang, Y.B. et al. (1993). Human histocompatibility leucocyte antigen-DQA*0501 allele associated with genetic susceptibility to Graves' disease in a Caucasian population. *J. Clin. Endocrinol. Metab.* **76**: 1569–1574.

Yanagawa, T., Hidaka, Y., Guimaraes, V. et al. (1995). CTLA-4 gene polymorphism associated with Graves' disease in Caucasian population. *J. Clin. Endocrinol. Metab.* **80**: 41–45.

Young, E.T., Steel, N.R., Taylor, J.J. et al. (1988). Prediction of remission after antithyroid drug treatment in Graves' disease. *Q. J. Med.* **66**: 175–189.

CHAPTER 17

HLA and Gastrointestinal Diseases

Ludvig M. Sollid, A. Spurkland and Erik Thorsby

HLA ASSOCIATIONS IN GASTROINTESTINAL DISEASES

Of the gastrointestinal disorders, coeliac disease (CD) demonstrates the strongest and most clear-cut human leucocyte antigen (HLA) association, as is treated in the next section. In this section, other gastrointestinal diseases where HLA associations have been reported are discussed. The diseases and their HLA associations are summarized in Table 17.1.

Inflammatory bowel disease demonstrates variable HLA associations

There have been numerous attempts to identify HLA antigens that are positively or negatively associated with inflammatory bowel disease (IBD). The results have been variable, both in ulcerative colitis (UC) and in Crohn's disease, and most reported HLA associations are quite weak (Table 17.1).

Several studies have reported a slight increase in DR2 (DRB1*15) among UC patients (Toyoda et al., 1993; Bouma et al., 1997; de la Concha et al., 1997; Roussomoustakaki et al., 1997), and there are also several reports of a reduced frequency of DR4 (Toyoda et al., 1993; Bouma et al., 1997; Roussomoustakaki et al., 1997) and DR6 (DRB1*13) (Toyoda et al., 1993; Bouma et al., 1997). However, in a large study of UC, a reduced frequency of DR4 could only be demonstrated among patients with extensive disease (Satsangi et al., 1996). In the latter study a reduced frequency of DR3 was also observed, especially in females with only distal colitis (Satsangi et al., 1996).

There are few reports of significant HLA associations in Crohn's disease. Most studies conclude that this disease is not associated with HLA (Wassmuth et al., 1993; Satsangi et al., 1996; Bouma et al., 1997). In two large, independent European studies, however, where genomic typing of DR alleles was performed, a significantly increased frequency of DR7 and a reduced frequency of DR3 among patients were reported (Danzé et al., 1996; Reinshagen et al., 1996). An increase in DR7 was also observed in patients with lesions in the small intestine (Heresbach et al., 1996b). One study reported a positive association with the DRB3*0301 allele (Forcione et al., 1996). Among the Japanese, a positive association with DR4 (Fujita et al., 1984; Kobayashi et al., 1990) and the closely linked DQB1*04 allele has been reported (Nakajima et al., 1995).

The complex and heterogeneous picture emerging from the attempts to identify HLA associations in IBD may reflect heterogeneity of these diseases. It may be that given HLA alleles influence disease susceptibility in only a subgroup of patients and that different forms of the disease are associated with different HLA alleles. Subgrouping of the patients with IBD may, however, be difficult, owing to the fluctuating course of the disease. Attempts to relate HLA genotype to various clinical subgroups have yielded conflicting results (Duerr and Neigut, 1995; Heresbach et al., 1996a; Satsangi et al., 1996; Bouma et al., 1997; Roussomoustakaki et al., 1997). One small subgroup of patients with IBD developed primary sclerosing cholangitis (PSC), which demonstrates a convincing HLA association (see next

TABLE 17.1 Gastrointestinal diseases demonstrating significant and consistent HLA associations

	HLA association	Odds ratio	p-Value	Country	Reference
Coeliac disease[a]	B8				
	DR3				
	DQ2				
	DQA1*0501,DQB1*02				
	DQA1*03,DQB1*0302				
Ulcerative colitis (UC)	DRB1*0103	5.0	$< 1 \times 10^{-5}$	UK	Roussomoustakaki et al. (1997)
		2.8	< 0.001	UK	Satsangi et al. (1996)
	DR2	1.6	< 0.05	UK	Roussomoustakaki et al. (1997)
		2.6	< 0.01	USA	Toyoda et al. (1993)
		2.1	< 0.01	Netherlands	Bouma et al. (1997)
		1.2	ns	USA	Dueir and Neigut (1995)
	DR15	1.9	0.02	Spain	de la Concha et al. (1997)
		1.6	ns	France	Heresbach et al. (1996a)
	DRB1*1502	4.6	< 0.0001	Japan	Futami et al. (1995)
	DR3	0.4	< 0.005	Spain	de la Concha et al. (1997)
		0.2	< 0.01	France	Heresbach et al. (1996a)
	DR4	0.5	< 0.01	UK	Roussomoustakaki et al. (1997)
		0.4	< 0.02	USA	Toyoda et al. (1993)
		0.7	< 0.05	UK	Satsangi et al. (1996)
	DR13	0.5	< 0.05	USA	Toyoda et al. (1993)
		0.5	< 0.05	Netherlands	Bouma et al. (1997)
Crohn's disease	No association	–		UK	Satsangi et al. (1996)
		–		Sweden	Wassmuth et al. (1993)
		–		Netherlands	Bouma et al. (1997)
	DQB1*0501	1.6	0.01	France	Danzé et al. (1996)
	DR1,DQB1*0501	2.5	0.02	USA	Toyoda et al. (1993)
	DR3	0.5	< 0.003	Germany	Reinshagen et al. (1996)
		0.5	$< 1 \times 10^{-5}$	France	Danzé et al. (1996)
	DR7	1.9	< 0.0001	Germany	Reinshagen et al. (1996)
		1.6	< 0.01	France	Danzé et al. (1996)
	DR4	4.8	< 0.001	Japan	Fujita et al. (1984)
		6.8	< 0.01	Japan	Kobayashi et al. (1990)
	DQB1*04	3.3	$< 1 \times 10^{-5}$	Japan	Nakajima et al. (1995)
Crohn's disease, small intestine	DR7	>1	< 0.05	France	Heresbach et al. (1996b)
	DRB3*0301	6.2	< 0.0005	USA	Forcione et al. (1996)
Primary sclerosing cholangitis (PSC)	B8	12	< 0.0005	Norway	Schrumpf et al. (1982)
		4.4	< 0.001	UK	Chapman et al. (1983)
	DR3	9.6	< 0.0005	Norway	Schrumpf et al. (1982)
		4.6	$< 1 \times 10^{-5}$	UK	Donaldson et al. (1991)
		4.3	$< 1 \times 10^{-5}$	Norway	Spurkland et al. (1999)
	DR13,DQB1*0603	4.8	< 0.001	UK	Farrant et al. (1992)
		3.6	$< 1 \times 10^{-5}$	Sweden	Olerup et al. (1995)
		2.2	< 0.001	Norway	Spurkland et al. (1999)
	DR4	0.4	< 0.01	UK	Mehal et al. (1994)
		0.3	< 0.01	UK	Farrant et al. (1992)
		0.3	< 0.01	Sweden	Olerup et al. (1995)
		0.3	< 0.001	Norway	Spurkland et al. (1999)

TABLE 17.1 Continued.

	HLA association	Odds ratio	p-Value	Country	Reference
Primary biliary cirrhosis (PBC)	DRB1*150l	0.3	0.01	USA	Begovich et al. (1994)
	DR8	8.3	$\ll 1 \times 10^{-5}$	USA	Gores et al. (1987)
		15	0.001	Germany	Manns et al. (1991)
		2	< 0.005	UK	Gregory et al. (1993)
	DRB1*08,DQB1*04	5.5	< 0.001	UK	Underhill et al. (1992)
		3.8	< 0.005	USA	Begovich et al. (1994)
	DPB1*0501	1.5	<0.001	Japan	Seki et al. (1993)
	DPB1*0301	6.8	< 0.015	Germany	Mella et al. (1995)
				UK	Underhill et al. (1995)
	C4AQ0	4.7	< 0.001	Germany	Manns et al. (1991)
	BAT2/Rsa 2.7-kb			Denmark	Fugger et al. (1991)
Autoimmune hepatitis (AIH)	DR3	5.5	$\ll 1 \times 10^{-5}$	UK	Donaldson et al. (1991)
		4.6	< 0.00005	USA	Strettell et al. (1997)
	DR4	5.6	< 0.0001	UK	Donaldson et al. (1991)
		148	$< 1.10^{-5}$	Japan	Seki et al. (1990)
		12.8	$< 1 \times 10^{-5}$	Japan	Ota et al. (1992)
		2.1	0.02	Argentina	Marcos et al. (1994)
	DRB1*0401	6.0	0.0001	USA	Strettell et al. (1997)
	DRB1*150l	0.3	0.02	USA	Strettell et al. (1997)
Child-onset AIH	DR4	0.3	0.02	UK	Donaldson (1996)
	DRB1*1301	16.2	< 0.0001	Argentina	Fainboim et al. (1994)

[a]For references and discussion see text (*Coeliac disease is primarily associated with DQ2 or DQ8*).
ns: not significant.

section). The frequency of liver pathology in IBD patients is, however, unknown and may be underestimated. In addition, not all patients with PSC have IBD.

Primary sclerosing cholangitis is positively associated with DRB1*03 and DRB1*1301

Primary sclerosing cholangitis (PSC) is a chronic liver disease often found in patients with IBD. The disease leads to chronic liver failure, and 10–20% of the patients die from cholangiocarcinoma. In several studies a positive association of PSC with DR3 has been established (Schrumpf et al., 1982; Donaldson et al., 1991; Noguchi et al., 1992; Leidenius et al., 1995; Olerup et al., 1995; Spurkland et al., 1999). However, only 50% of the PSC patients carry DR3. In the remaining patients an association with the DRB1*1301–DQB1*0603 haplotype has repeatedly been observed (Farrant et al., 1992; Olerup et al., 1995; Spurkland et al., 1999). A reduced frequency of DR4 in patients with PSC has also consistently been found (Donaldson et al., 1991; Farrant et al., 1992; Mehal et al., 1994; Olerup et al., 1995; Spurkland et al., 1998), indicating a protective effect of alleles on this haplotype in PSC. It is of interest to note, as mentioned above, that the frequency of DR4 has also been reported to be reduced in UC, which is found in 70–80% of patients with PSC (see Table 17.1 and previous section). Thus, it is possible that the negative association of DR4 with PSC is secondary to that found in UC. An influence of certain HLA alleles on the prognosis of the disease has also been reported (Mehal et al., 1994; Leidenius et al., 1995), but this remains controversial (Olerup et al., 1995).

Primary biliary cirrhosis is positively associated with DRB1*08

Primary biliary cirrhosis (PBC) is another liver disease characterized by inflammatory scarring of the liver. The disease has consistently been found to be positively associated with DR8 (Gores et al., 1987; Manns et al., 1991; Underhill et al., 1992; Gregory et al., 1993; Seki et al., 1993; Begovich et al., 1994). Only 20–30% of the patients, however, carry DR8. Seki et al. (1993) have suggested that DPB1*0501, which is in close linkage disequilibrium with DRB1*08 in the Japanese, is more closely associated with PBC than is DRB1*08. Studies of patients of Caucasoid origin have suggested an association to the DPB1*0301 allele (Mella et al., 1995; Underhill et al., 1995). An increased frequency of the HLA-associated C4 (complement factor 4) null allele (C4AQ0), independent of linkage disequilibrium with DR8, has also been found among PBC patients (Manns et al., 1991). In a study of HLA-B-associated transcripts (BAT), which are transcribed from genes of unknown function telomeric to the class III region, Fugger et al. (1991) found a polymorphism that was increased in patients with PBC. However, this finding was not significant when corrected for the number of comparisons. The HLA-linked genetic susceptibility to PBC may thus be determined by genes outside the HLA class II region.

Autoimmune hepatitis is associated with the A1,B8,DR3 haplotype

Autoimmune hepatitis (AIH) is a heterogeneous disorder characterized by chronic hepatitis in the absence of markers for hepatitis B or C viral infection. The disease may present in both children and adults. Owing to the heterogeneity of the disease, there has been much controversy as to possible HLA associations. As diagnostic tools have become available to allow an ever larger percentage of virally induced hepatitis to be excluded, the HLA associations have become more consistent. It appears that AIH is positively associated with the A1–B8–DR3 haplotype, especially in early-onset disease (Donaldson et al., 1991). A second HLA association has been found with DR4, which is increased in Caucasoid AIH patients not carrying DR3 (Donaldson et al., 1991), and which is also found in most Japanese AIH patients (Seki et al., 1990; Ota et al., 1992). Among children with AIH in Argentina, a strong positive association with DRB1*1301 was observed (Fainboim et al., 1994), whereas among adults a positive association with DR4 was observed (Marcos et al., 1994).

Conclusions

Except in CD, where a strong and consistent HLA association has been found in all populations studied (see next section), most gastrointestinal diseases generally display variable or weak HLA associations. Some of this variability may be explained by the heterogeneity of these diseases.

COELIAC DISEASE: A MODEL DISEASE FOR HLA-DISEASE ASSOCIATIONS

CD is a malabsorptive disorder of the small intestine which is precipitated by ingestion of wheat gluten and related proteins of other cereals (barley, rye and, possibly, oats). The typical CD patient has malabsorption, diarrhoea and weight loss, but a wide variety of disease manifestations exists, from asymptomatic patients detected by serological screening to severe steatorrhoea, fatigue and anaemia (Trier, 1991; Marsh, 1992). The mucosal lesions are characterized by villous atrophy, hyperplastic crypts and T cell infiltration in the epithelium and in the lamina propria (Scott et al., 1997). The disease is believed to be caused by an abnormal immune response to gluten, which most probably is initiated by the activation of T cells in the intestinal mucosa to gluten-derived peptides (Scott et al., 1997). The disease is treated with a gluten-free diet, on which the gut mucosa of most patients normalizes completely. The gut lesion reappears when gluten is reintroduced into the diet (gluten challenge).

Dermatitis herpetiformis (DH) is closely related to CD. This dermatological disorder is characterized by symmetrical skin blisters, urticarial patches and papules on the buttocks, knees, elbows, etc., in addition to a variable enteropathy. Both the rash and the enteropathy of DH are gluten dependent (Fry, 1992).

The development of CD (and DH) is strongly associated with specific genes in the HLA complex, and CD offers a unique model for studies of HLA-associated diseases in general for several reasons: (i) the primary HLA associations have been established; (ii) the disease-inducing agent (gluten) is known; (iii) gluten-specific T cells from the intestinal mucosa are accessible from biopsies for *in vitro* studies; and (iv) the initiation of the disease can be mimicked by challenging treated patients with gluten.

Coeliac disease is primarily associated with DQ2 or DQ8

The concordance rate for monozygotic twins to develop CD is approximately 70% (Polanco et al., 1981; Strober, 1992). For HLA identical sibs the figure is about 30% (Mearin et al., 1983; Hernández et al., 1991), while for HLA-disparate sibs the concordance rate is about 4% (Hernández et al., 1991). These figures, which are not very different from those found in other strongly HLA-associated diseases (e.g. type I diabetes), demonstrate that both genetic and environmental factors contribute to the development of CD, where genes in the HLA complex are responsible for a major part of the genetic predisposition. Attempts to map predisposing non-HLA genes by linkage analysis have so far not revealed unambiguous candidate genes or chromosomal regions (Zhong et al., 1996; Houlston et al., 1997).

CD was first found to be associated with the HLA class I molecule B8 (Falchuk et al., 1972; Stokes et al., 1972). Later, a stronger association with the HLA class II molecule DR3 was found (Keunig et al., 1976; Solheim et al., 1976), and this association is present in all populations studied (Tiwari and Terasaki, 1985; Strober, 1992). Both the B8–DR3 and the B18–DR3 haplotypes are associated with CD (Coniga et al., 1992). CD is also associated with DR7 (Tiwari and Terasaki, 1985; Strober, 1992; Sollid and Thorsby, 1993), but only or mainly when DR7 occurs together with DR3 or DR5 (Sollid and Thorsby, 1993). This is in contrast to the susceptibility associated with DR3, which is seen irrespective of the accompanying DR allele The distribution of patients with CD expressing DR3, DR5/DR7 or other alleles in various populations is illustrated in Figure 17.1. It appears that in all populations studied most CD patients either carry DR3 or are DR5/DR7 heterozygous, and that only a few (usually < 10%) are neither DR3 nor DR5/DR7. The latter mainly carry the DR4–DQ8 haplotype (see later).

In 1983 Tosi and co-workers provided evidence for an even stronger association of CD with DQ2, which is in strong linkage disequilibrium with

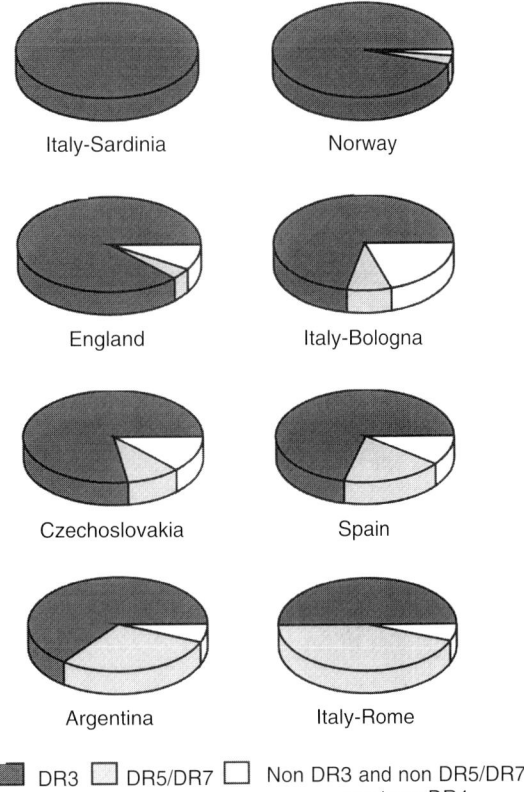

FIGURE 17.1 Schematic representation of the distribution of HLA-DR phenotypes in CD patients of various Caucasoid populations. (Figure adapted from Sollid and Thorsby, 1993, by copyright permission of W.B. Saunders Co.)

both DR3 and DR7 (Tosi et al., 1983). The DR3–DQ2 haplotype demonstrates the strongest association with CD.

Strong evidence suggests that CD patients who carry the DR3–DQ2 haplotype or are DR5–DQ7/DR7–DQ2 heterozygous share the genetic information conferring CD susceptibility (Sollid and Thorsby, 1993). Sequencing of DQ genes revealed that the DR3–DQ2 haplotype carries the DQA1*0501 and DQB1*02 alleles (Boss and Strominger, 1984; Schenning et al., 1984). Sequencing also demonstrated that the DR5–DQ7 haplotype carries the DQA1*0501 and DQB1*0301 alleles (Schiffenbauer et al., 1987) and that the DR7–DQ2 haplotype carries the DQA1*0201 and DQB1*02 alleles (Chang et al., 1983; Karr et al., 1986). Thus, the DQA1 allele of the DR3–DQ2 and DR5–DQ7 haplotypes and the DQB1 allele of the DR3–DQ2 and DR7–DQ2 haplotypes are identical (except for the codon of residue 135 in the membrane-proximal domain of DQ chain). As a result of this, a DR5–DQ7/DR7–DQ2 heterozygous individual would carry in the *trans* position the same DQA1 and DQB1 genes as those found in the *cis* position in DR3–DQ2 individuals, and both types of individual might therefore express a common DQ molecule (Figure 17.2). DNA typing using oligonucleotide probes demonstrated that 93 of 94 Norwegian CD patients had both the DQA1*0501 and DQB1*02 alleles either in the *cis* or the *trans* position. This strongly indicates that CD is primarily associated with a particular variant of DQ2, i.e. DQ(α1*0501–β1*02), encoded either in *cis* or in *trans* (Sollid et al., 1989). The DQ(α1*0501–β1*02) heterodimer is expressed when encoded both in *cis* and in *trans*, as demonstrated with alloreactive T cell clones (TCCs) (Lundin et al., 1990).

This concept has since been corroborated by many studies (see Sollid and Thorsby, 1993, for references). Approximately 90% of CD patients carry the DQ(α1*0501–β1*02) heterodimer, where the fraction of CD patients in different populations having this heterodimer encoded in *cis* or in *trans* is dependent on the haplotype frequencies of DR3–DQ2, DR5–DQ7 and DR7–DQ2, respectively in the population (Sollid and Thorsby, 1990). In a few patients the DQ(α1*0501–β1*02) heterodimer may be found to be encoded in *cis* by haplotypes other than DR3–DQ2 or in *trans* by individuals heterozygous for combinations other than DR5–DQ7/DR7–DQ2 (Sollid and Thorsby, 1993). There is no increase in the DQ(α1*0501–β1*0301) or DQ(α1*0201–β1*02) heterodimers in CD, demonstrating that susceptibility is dependent on the particular combination of the DQα- and DQβ-chains in the DQ(α1*0501–β1*02) heterodimer.

Approximately 2–10% of CD patients do not carry the DQ(α1*0501–β1*02) heterodimer. Most of them instead carry different subtypes of DR4, but share the DQ(α1*03–β1*0302) heterodimer (= DQ8), suggesting that susceptibility to develop CD in these patients is primarily associated with DQ8 (Spurkland et al., 1992; Tighe et al., 1993; Lundin et al., 1994a).

Associations with particular DP alleles have also been reported in different populations, but many of these associations may be explained by linkage disequilibrium between the DP allele(s) involved and DQA1*0501 and DQB1*02 (Sollid and Thorsby, 1993). Furthermore, no independent associations with alleles at the TAP1 and TAP2 loci have been found (Colonna et al., 1992; Powis et al., 1993; Meddeb-Garnaoui et al., 1995). Recently, an additional predisposing role of tumour necrosis factor (TNF) genes or genes in close linkage disequilibrium has been suggested. An association independent of DQ2 has been demonstrated with a microsatellite polymorphism near the TNF genes (McManus et al., 1996a) and a polymorphism of the gene promoter has been demonstrated to be a component of the DR3–DQ2 haplotype (McManus et al., 1996b). None of these observations has been replicated yet. It has also been suggested that a gene on the DR7–DQ2 haplotype confers an additive effect to that of the DQA1*0501 and DQB1*02 genes (Fernandez-Arquero et al., 1995), but this finding needs confirmation.

Roep et al. (1988) reported that particular DQ polymorphisms were more often found among CD patients than among DR- and DQ-matched healthy controls, suggesting that CD patients may carry unique DQ alleles. Mantovani et al. (1991) could not, however, confirm this finding, and sequencing of class II alleles has not demonstrated any disease-specific sequences (Kagnoff et al.,

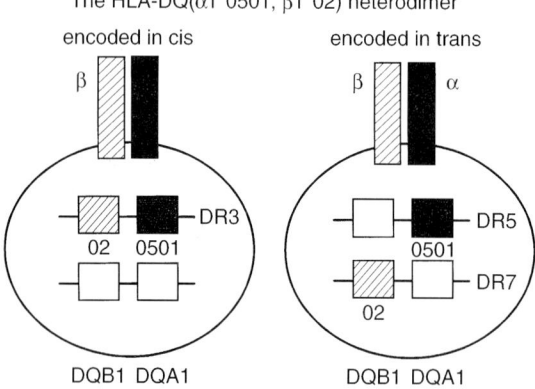

FIGURE 17.2 The DQ(α1*0501–β1*02) heterodimer may be encoded in the *cis* position by the *(DR3)*, DQA1*0501,DQB1*0201 haplotype and in the *trans* position in (DR5), DQA1*0501,B1*0301/(DR7), DQA1*0201,B1*0202 heterozygotes. (Figure adapted from Sollid and Thorsby, 1993, by copyright permission of W.B. Saunders Co.)

1989; Jacob and McDevitt, 1989). Thus, susceptibility to develop CD is associated with normal allelic variants of HLA molecules, a notion which seems to be true for most HLA-associated diseases (see Thorsby, 1997, and elsewhere in this volume).

Taken together, available data thus strongly suggest that susceptibility to develop CD is primarily associated with two conventional peptide-presenting DQ molecules: DQ($\alpha1^*0501$–$\beta1^*02$) (=DQ2) or, to a lesser extent, DQ($\alpha1^*03$–$\beta1^*0302$) (= DQ8). The elucidation of these primary HLA associations clearly illustrates the difficulties imposed by the strong linkage disequilibrium of HLA complex genes on unravelling which genes are primarily involved in HLA-associated diseases. Only a minority of CD patients (generally less than 5%) carries neither of these DQ molecules (Sollid and Thorsby, 1993). Interestingly, the HLA associations in DH are very similar, if not identical to those found in CD (Spurkland et al., 1997).

HLA restriction of gluten-specific T cells

Several mechanisms could explain a primary involvement of given peptide-presenting HLA molecules in disease susceptibility (see Chapter 7). The most straightforward explanation would be effects of the disease-associated HLA molecules on selection of the T cell receptor (TCR) repertoire during thymic development of T cells and/or preferential peptide presentation by the involved HLA molecules.

This issue has been addressed by studying gluten-specific T cells from the intestinal mucosa of CD patients. Stimulation of small-intestinal biopsy specimens with a peptic/tryptic digest of gluten induces rapid activation [i.e. expression of CD25, the interleukin-2 (IL-2) receptor] of the T cells in the lamina propria from CD patients, but not from non-CD control subjects (Halstensen et al., 1993). Such activated T cells can be isolated and propagated in the absence of further stimulation with gluten (Lundin et al., 1993). Gluten-reactive T cell lines (TCLs) have been established from several CD patients (Lundin et al., 1993; Molberg et al., 1997, 1998a) but cannot be isolated from intestinal biopsies of non-CD controls (Molberg et al., 1997).

Gluten-specific TCLs have been isolated from biopsy specimens of a number of patients carrying the DR3–DQ2 haplotype. Inhibition studies with anti-HLA antibodies indicated preferential DQ restriction for the majority of these TCLs (Lundin et al., 1993, 1994b; Molberg et al., 1997, 1998a). Predominant DQ restriction was also observed in a large number of TCCs (all CD4$^+$ TCR α/β^+) which were established from the gluten-reactive TCLs. Almost all of the TCCs were completely inhibited in their gluten-specific responses by anti-DQ, but not by anti-DR monoclonal antibodies (mAbs) (Lundin et al., 1993, 1994b; Molberg et al., 1997, 1998a). Panel studies with allogeneic B cell lines as antigen-presenting cells (APC) confirmed that DQ2 was the dominant restricting element (Lundin et al., 1993, 1994b; Molberg et al., 1997, 1998a). Be it encoded by genes in *cis* or in *trans* position, the DQ($\alpha1^*0501$–$\beta1^*02$) molecule could present gluten peptides to TCCs (Lundin et al., 1993, 1994b). Thus, in patients carrying the DR3–DQ2 haplotype the gluten-reactive T cells from the intestinal mucosa mainly recognize gluten-derived peptides when presented by the CD-associated DQ($\alpha1^*0501$–$\beta1^*02$) heterodimer, i.e. DQ2.

When gluten-specific T cells were isolated from small intestinal biopsies of heterozygous DR4–DQ7/DR4–DQ8 CD patients, the TCLs and the TCCs mainly responded to gluten when presented by APCs carrying DR4–DQ8, and the responses were strongly inhibited by anti-DQ but not by anti-DR mAbs (Lundin et al., 1994a, b). In fact, while 14 DQ8-restricted TCCs were found, only a single DR4-restricted clone was detected. Thus, in these patients gluten-specific CD4$^+$ T cells from the intestinal mucosa recognized gluten-derived peptides predominantly when presented by the other CD associated DQ heterodimer, DQ($\alpha1^*03$–$\beta1^*0302$) or DQ8.

DQ2 or DQ8 molecules are not preferential antigen-presenting molecules in the intestinal mucosa, irrespective of antigen. The responses of gut-derived TCLs and TCCs specific for astrovirus (a common gastroenteritis virus) were predominantly found to be DR restricted, in contrast to the predominant DQ2 restriction found for gluten-specific T cells (Molberg et al., 1998a, b).

Gluten-specific T cells may be found in the peripheral blood of both CD patients and healthy controls. Interestingly, gluten-reactive T cells derived from peripheral blood can be restricted either by DR, DP or DQ molecules, including DQ2 (Gjertsen et al., 1994; Jensen et al., 1995), but the predominant DQ2 or DQ8 restriction seen for gluten-specific T cells from the intestinal mucosa of CD patients is not observed with peripheral blood T cells.

The gut-derived, gluten-specific TCCs uniformly secrete interferon-γ (IFNγ) at high concentrations (Nilsen et al., 1995). In addition, many TCCs secrete one or several of the cytokines IL-4, IL-5, IL-6, IL-10, TNFα and transforming growth factor-β (TGFβ), suggesting that a major proportion of these gluten-specific T cells is of the Th0/Th1 type (Nilsen et al., 1995).

Taken together, the reported data demonstrate that (i) gluten-specific T cells from the intestinal mucosa of CD patients usually belong to the Th0/Th1 type of CD4$^+$ T cells and they mainly recognize gluten when presented by the disease-associated DQ2 or DQ8 molecules; (ii) the preferential DQ2 and DQ8 restriction is typical of gluten-specific and not of other antigen-specific T cell in the intestinal mucosa; and (iii) this restriction pattern is much less pronounced or not seen at all in gluten-specific T cells in the peripheral blood. The last observation is important to keep in mind when using T cells from peripheral blood to investigate the specificity of T cells in target organs (e.g. type I diabetes).

The data point to preferential binding and presentation of gluten-derived peptides by the CD-associated DQ2 or DQ8 molecules in the intestinal mucosa as the mechanism behind the HLA association, but do not exclude thymic effects of the same DQ molecules on the development of the TCR repertoire. However, we believe that the latter mechanism is less likely since there is no TCR usage by gluten-specific mucosal T cells (Lundin et al., 1993).

The peptide-binding motif of the coeliac disease-associated DQ2 heterodimer

Since preferential binding and presentation of gluten-derived peptides by the CD-associated DQ2 (or DQ8) molecules to T cells in the intestinal mucosa is a likely mechanism underlying the HLA association in CD, it is important to establish the peptide-binding motif of DQ(α1*0501–β1*02). This motif has been characterized by sequencing peptides eluted from affinity-purified molecules, and by binding experiments using truncated and substituted variants of high-affinity binding peptides (Johansen et al., 1996a; Vartdal et al., 1996; van de Wal et al., 1996, 1997). The anchor residues for peptides binding to DQ2 are found in the positions P1, P4, P6, P7 and P9 (Table 17.2). This is the same spacing as previously found for DR molecules, suggesting that DQ2 bound peptides adopt to a similar conformation as peptides bound to DR molecules. The peptide-binding motif of DQ2 is, however, quite different from other class II binding motifs described in the literature. Notably, the preference for negatively charged residues in the three anchor positions in the middle seems to be unique for DQ2.

The DQ2 peptide-binding motif is different from the motifs of the closely related DQ(α1*0501–β1*0301) and DQ(α1*0201–β1*02) molecules (Johansen et al., 1996b; van de Wal et

TABLE 17.2 Peptide binding motif of the DQ(α1*0501,β1*02) molecule[a]

Residues	Relative position								
	1	2	3	4	5	6	7	8	9
Preferred	F, Y, W, L, I, V			E, D,		E, D, P	E, D		L, I, F, Y, W, M, V
Not preferred	R, K, H, A, G, S, D			P, R, K, P, W		K, W	R, H, K, P		K, H, R

[a]The motif description is based on compilation of results reported by Johansen et al. (1996a), Vartdal et al. (1996), van de Wal et al. (1996, 1997) and Quarsten et al. (1998).

al., 1997), which do not confer susceptibility to CD (see above). In particular, the binding specificity of the DQ($\alpha1^*0501$–$\beta1^*0301$) molecule is different from that of DQ($\alpha1^*0501$–$\beta1^*02$). Binding experiments and computer modelling of the three-dimensional structures indicate that the binding sites of the DQ($\alpha1^*0501$–$\beta1^*02$) and the DQ($\alpha1^*0501$–$\beta1^*0301$) molecules differ at the P4, P7 and P9 pockets (Johansen et al., 1996b). The knowledge of the binding specificities of the three above-mentioned DQ molecules should allow a reverse immunogenetic approach to describe important features of peptides being able to bind only to DQ($\alpha1^*0501$–$\beta1^*02$) and thereby provoke a disease-causing immune response. This approach can be fully exploited, however, only when more detailed knowledge of the peptide-binding motifs of these three DQ molecules exists.

Antigen specificity of the gluten-reactive T cells

Gluten is a mixture of glutenin and gliadin proteins. Both types of protein serve as a source of nitrogen and carbon for the growing seedling during germination. The gliadin proteins are known to precipitate CD, whereas the role of glutenins is still elusive. All of the gut-derived TCLs investigated in the present authors' studies recognize gliadin proteins (Lundin et al., 1993, 1994a, b).

In a single wheat variety there are approximately 45 different gliadins (Wrigley and Shephard, 1973), which may be subclassified according to their N-terminal amino acid sequence as being either α-type, γ-type or ω-type gliadins (Shewry et al., 1992). The various gliadins are rich in glutamine and proline, but display slight differences in amino acid sequence and molecular mass within each type or family. Interestingly, when testing gut-derived TCCs against some purified α-type and γ-type gliadins, a heterogeneous response pattern was observed (Lundin et al., 1997), indicating that gliadin-specific T cells from the intestinal mucosa of CD patients recognize diverse gliadin epitopes. It remains to be established whether these gliadin epitopes share any common features.

Nature of gliadin epitopes which may be recognized by DQ2-restricted T cells

Negatively charged residues (aspartic acid and glutamic acid) are preferred as anchor residues in positions P4, P6 and P7 in peptides binding DQ2 with high affinity (see above). Proteins with a high content of negatively charged amino acids are therefore particularly likely sources of DQ2-binding epitopes. Paradoxically, gliadins have a very low content of negatively charged amino acids (usually <2% residue usage). Notably, deamidation (e.g. glutamine $\rightarrow$ glutamic acid) could transform gliadins into proteins which are exceptionally rich in glutamic acid, as gliadins have a very high content of glutamine (usually >35% residue usage). In this context, it is intriguing that an increase in antigenic potency of gliadin was observed following treatment known to cause deamidation (Lundin et al., 1997). Further support for a permissive role of deamidation for binding of gliadin peptides to DQ2 has been obtained in peptide-binding experiments. Replacement of some glutamine with glutamic acid residues, in particular gliadin peptides, led to a significant increase in the affinity for binding to DQ2 (Sollid et al., 1997a).

A crucial question is whether all DQ2-restricted T cells of the CD lesion recognize deamidated, modified gliadins. A reliable answer to this question must await a further characterization of the epitopes recognized by additional gut-derived TCCs.

Gliadin-specific T cells can provide help for autoantibody production

Anti-endomysial antibodies are a hallmark of CD and the demonstration of these antibodies is used for diagnostic purposes. Gliadin seems to drive this antibody secretion as anti-endomysial antibodies are produced in biopsies cultured with a peptic/tryptic digest of gliadin (Picarelli et al., 1996). Recently, Dieterich and colleagues identified tissue transglutaminase (TGase) as the antigen reactive with antiendomysial antibodies, and their data indicated that TGase forms complexes with gliadin (Dieterich et al., 1997).

As discussed elsewhere (Sollid et al., 1997b), complex formation between TGase and gliadin may create a hapten–carrier complex where gliadin acts as a carrier and TGase as the hapten

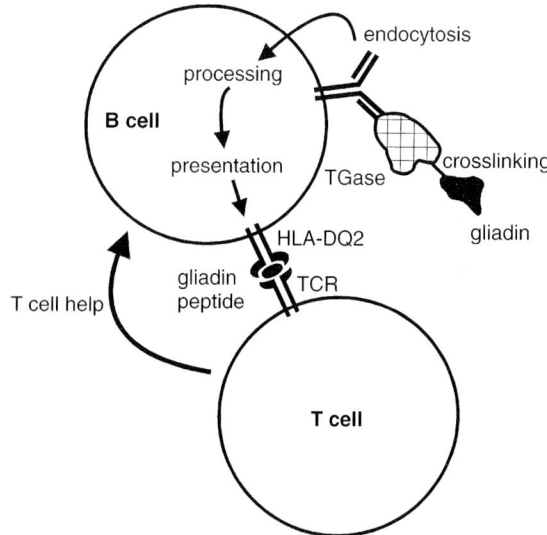

FIGURE 17.3 Gliadin-specific T cells can aid the antibody production of TGase-specific B cells by a carrier effect of gliadin in complex with TGase. (Figure adapted from Sollid et al., 1997a, by copyright permission of BMJ Publishing Group.)

(Figure 17.3). Gliadin-specific CD4$^+$ helper T cells exist in the small-intestinal mucosa of CD disease patients (Lundin et al., 1993) and may provide help for TGase-specific B cells and thus explain the production of anti-endomysial antibodies in CD.

The anti-endomysial antibodies, as suggested by Mäki and co-workers (Halttunen and Mäki, 1996), can potentially be involved in disease development by blocking the mesenchymal cell–epithelial cell cross-talk in the jejunal crypt villous axis and hence interfere with normal villous formation. Thus, there might be an autoimmune component in the immunopathogenesis of CD.

The molecular basis of the HLA association with coeliac disease

As discussed earlier in this chapter, both genetic and environmental factors contribute to the development of CD. Certain HLA molecules (DQ2 or DQ8) are almost necessary, but not sufficient for CD to develop. A similar picture is seen for other HLA-associated diseases. The nature of the environmental factors necessary for development of CD is not known (as is also the case for other HLA-associated diseases). In CD the environmental factors (together with other non-HLA genetic factors) may form a basis for a local immune response to gliadin-derived peptides to take place in the intestinal mucosa. For example, an inflammatory response in the intestinal mucosa (possibly secondary to bacterial or viral infection) may lead to maturation of local dendritic cells so that they may directly activate gliadin-reactive intestinal T cells (see Sollid et al., 1997c).

The molecular basis for the HLA association in CD is most probably preferential binding by DQ2 (or DQ8) of possibly modified (i.e. deamidated) gliadin peptides. Presentation of (modified) gliadin peptides by these HLA molecules expressed by mature dendritic cells to CD4$^+$ T cells in the intestinal mucosa might induce a gliadin-specific T cell immune response which can cause local immunopathology. The effector mechanism(s) responsible for the pathology are poorly understood. It could be a direct effect of cytokines (e.g. IFNγ) on the mucosal epithelial cells. Alternatively, the autoantibodies specific for TGase may be involved. If the latter is the case and the formation of these antibodies is a result of gliadin being a carrier protein (see above), this could explain why gliadin and not most other edible proteins cause this immunopathology in the gut.

CONCLUSION

Chronic inflammatory disorders such as rheumatoid arthritis, type I diabetes and multiple sclerosis are major health concerns in Western countries, for which better treatment regimes are needed. Given the increasing knowledge of the molecular basis of the HLA association in CD, this disease emerges as a model disorder not only for examining the basis for HLA associations in other diseases, but also for studying the interplay between genes and environmental factors leading to HLA-associated diseases. Further understanding of these factors will hopefully lead to better treatment strategies for these associated diseases.

Since the finalization of this manuscript, several important findings have been made about the immunopathogenesis of CD. Further evidence has been obtained for a preferential recognition of deamidated gliadins by gut-derived T cells (Sjöström et al., 1998). The sequence of three DQ2 restricted gliadin T cell epitopes (one from a γ-type gliadin and two from α-type gliadins; Sjöström et al., 1998; Arentz-Hansen et al., 2000) and two DQ8 restricted epitopes (one from an α-gliadin and one from a glutenin; van de Wal et al., *Proc. Natl. Acad. Sci. USA* 1998, **95**: 10050–10054 and van de Wal et al., 1998a; 1999) have been identified. With the exception of the DQ8 restricted glutenin epitope, all these epitopes contain deamidated glutamins that are important for binding to the DQ2 and/or DQ8 molecules and T cell recognition. The deamidation of gluten may take place in the acidic environment in the stomach, but increasing evidence indicates that the enzyme tissue transglutanminase, which in the gut is expressed just beneath the epithelium and is involved in the deamidation of the gluten-derived peptides (Molberg et al., 1998c; van de Wal et al., 1998b; Quarsten et al., 1999).

REFERENCES

Arentz-Hansen, H., Korner, R., Molberg, O. et al. (2000). The intestinal T cell response to α-gliadin in adult celiac disease is focused on a single deamidated glutamine targeted by tissue transglutaminase. *J. Exp. Med.* **191**: 603–612.

Begovich, A.B., Klitz, W., Moonsamy, P.V. et al. (1994). Genes within the HLA class II region confer both predisposition and resistance to primary biliary cirrhosis. *Tissue Antigens* **43**: 71–77.

Boss, J.M. and Strominger, J.L. (1984). Cloning and sequence analysis of the human major histocompatibility complex genes DC-3. *Proc. Natl Acad. Sci. USA* **81**: 5199–5203.

Bouma, G., Oudkerk Pool, M., Crusius, J.B.A. et al. (1997). Evidence for genetic heterogeneity in inflammatory bowel disease (IBD); HLA genes in the predisposition to suffer from ulcerative colitis (UC) and Crohn's disease (CD). *Clin. Exp. Immunol.* **109**: 175–179.

Chang, H.-C., Moriuchi, T. and Silver, J. (1983). The heavy chain of human B-cell alloantigen HLA-DS has a variable N-terminal region and a constant immunoglobulin-like region. *Nature* **305**: 813–815.

Chapman, R.W., Varghese, Z., Gaul, R. et al. (1983). Association of primary sclerosing cholangitis with HLA-B8. *Gut* **24**: 38–41.

Colonna, M., Bresnahan, M., Bahram, S. et al. (1992). Allelic variants of the human putative peptide transporter involved in antigen processing. *Proc. Natl Acad. Sci. USA* **89**: 3932–3936.

Concha, E.G. de la, Fernandez-Arquero, M., Santa-Cruz, S. et al. (1997). Positive and negative associations of distinct HLA-DR2 subtypes with ulcerative colitis (UC). *Clin. Exp. Immunol.* **108**: 392–395.

Coniga, M., Frau, F., Lampis, R. et al. (1992). A high frequency of the A30, B18, DR3, DRw52,DQw2 extended haplotype in Sardinian celiac disease patients; further evidence that the disease is conferred by DQA1*0501, B1*0201. *Tissue Antigens* **39**: 78–83.

Danzé, P.M., Colombel, J.F., Jacquot, S. et al. (1996). Association of HLA class II genes with susceptibility to Crohn's disease. *Gut* **39**: 69–72.

Dieterich W., Ehnis, T., Bauer, M. et al. (1997). Identification of tissue transglutaminase as the auto antigen of coeliac disease. *Nat. Med.* **3**: 797–801.

Donaldson, P.T. (1996). Immunogenetics in liver disease. *Baillières Clin. Gastroenterol.* **10**: 533–549.

Donaldson, P.T., Doherty, D.G., Hayllar, K.M. et al. (1991). Susceptibility to autoimmune chronic active hepatitis: human leukocyte antigens DR4 and A1-B8-DR3 are independent risk factors. *Hepatology* **13**: 701–706.

Duerr, R.H. and Neigut, D.A. (1995). Molecularly defined HLA-DR2 alleles in ulcerative colitis and an antineutrophil cytoplasmic antibody-positive subgroup. *Gastroenterology* **108**: 423–427.

Fainboim, L., Marcos, Y., Pando, M. et al. (1994). Chronic active autoimmune hepatitis in children. Strong association with a particular HLA-DR6 (DRB1*1301) haplotype. *Hum. Immunol.* **41**: 146–150.

Falchuck, Z.M., Rogentine, G.N. and Strober, W. (1972). Predominance of histocompatibility antigen HL-A8 in patients with gluten-sensitive enteropathy. *J. Clin. Invest.* **51**: 1602–1605.

Farrant, J.M., Doherty, D.G., Donaldson, P.T. et al. (1992). Amino acid substitutions at position 38 of the DR beta polypeptide confer susceptibility to and protection from primary sclerosing cholangitis. *Hepatology* **16**: 390–395.

Fernandez-Arquero, M., Figueredo, M.A., Maluenda, C. et al. (1995). HLA-linked genes acting as additive susceptibility factors in celiac disease. *Hum. Immunol.* **42**: 295–300.

Forcione, D.G., Sands, B., Isselbacher, K.J. et al. (1996). An increased risk of Crohn's disease in individuals who inherit the HLA class II DRB3*0301 allele. *Proc. Natl Acad. Sci. USA* **93**: 5094–5098.

Fry, L. (1992). Dermatitis herpetiformis. In *Coeliac Disease* (ed. M.N. Marsh), pp. 81–104, Blackwell, Oxford.

Fugger, L., Morling, N., Ryder, L.P. et al. (1991). Restriction fragment length polymorphism of two HLA-B-associated transcripts genes in five autoimmune diseases. *Hum. Immunol.* **30**: 27–31.

Fujita, K., Naito, S., Okabe, N. et al. (1984). Immunological studies in Crohn's disease. I. Association with HLA systems in the Japanese. *J. Clin. Lab. Immunol.* **14**: 99–102.

Futami, S., Aoyama, N., Honsako, Y. et al. (1995). HLA-DRB1*1502 allele, subtype of DR15, is associated with susceptibility to ulcerative colitis and its progression. *Dig. Dis. Sci.* **40**: 814–818.

Gjertsen, H.A., Sollid, L.M., Ek, J. et al. (1994). Both HLA-DR, -DQ, and -DP restricted gluten-specific T cell clones can be isolated from the peripheral blood of celiac disease patients. *Scand. J. Immunol.* **39**: 567–574.

Gores, G.J., Moore, S.B., Fisher, L.D. et al. (1987). Primary biliary cirrhosis: associations with class II major histocompatibility complex antigens. *Hepatology* **7**: 889–892.

Gregory, W.L., Mehal, W., Dunn, A.N. et al. (1993). Primary biliary cirrhosis: contribution of HLA class II allele DR8. *Q. J. Med.* **86**: 393–399.

Halstensen, T.S., Scott, H., Fausa, O. et al. (1993). Glutens stimulation of coeliac mucosa *in vitro* induces activation (CD25) of lamina propria CD4$^+$ T cells and macrophages but no crypt-cell hyperplasia. *Scand. J. Immunol.* **38**: 581–590.

Halttunen, T. and Mäki, M. (1999). Serum immunoglobin A from patients with celiac disease inhibits human T84 intestinal crypt epthilial cell differentiation. *Gastroenterology* **116**: 566–572.

Heresbach, D., Alizadeh, M., Reumaux, D. et al. (1996a). Are HLA-DR or TAP genes genetic markers of severity in ulcerative colitis? *J. Autoimmun.* **9**: 777–784.

Heresbach, D., Alizadeh, M., Bretagne, J.F. et al. (1996b). Investigation of the association of major histocompatibility complex genes, including HLA class I, class II and TAP genes, with clinical forms of Crohn's disease. *Eur. J. Immunogenet.* **23**: 141–151.

Hernández, J.L., Michalski, J.P., McCombs, C.C. et al. (1991). Evidence for a dominant gene mechanism underlying coeliac disease in the West of Ireland. *Genetic Epidemiol.* **8**: 13–27.

Houlston, R.S., Tomlinson, I.P.M., Ford, D. et al. (1997). Linkage analysis of candidate regions for coeliac disease. *Hum. Mol. Genet.* **6**: 1335–1339.

Jacob, C.O. and McDevitt, H.O. (1989). Absence of polymorphism between DR and DQ sequences isolated from celiac disease patients and normals. In *Immunobiology of HLA* (ed. B. Dupont), pp. 448–449, Springer, New York.

Jensen, K., Sollid, L.M., Scott, H. et al. (1995). Gliadin-specific T cell responses in peripheral blood of healthy individuals involve T cells restricted by the coeliac disease associated DQ2 heterodimer. *Scand. J. Immunol.* **42**: 166–170.

Johansen, B.H., Vartdal, F., Eriksen, J.A. et al. (1996a). Identification of a putative motif for binding of peptides to HLA-DQ2. *Int. Immunol.* **8**: 177–182.

Johansen, B.H., Jensen, T., Thorpe, C.J. et al. (1996b). Both α- and β-chain polymorphisms determine the specificity of the coeliac disease associated HLA-DQ2 molecule with β-chain residues being most influential. *Immunogenetics* **45**: 142–159.

Kagnoff, M.F., Harwood, J.I., Bugawan, T.L. et al. (1989). Structural analysis of the HLA-DR, -DQ, and -DP alleles on the celiac disease associated HLA-DR3 (DRw17) haplotype. *Proc. Natl Acad. Sci. USA* **86**: 6274–6278.

Karr, R.W., Gregersen, P.K., Obata, F. et al. (1986). Analysis of DR and DQ chain cDNA clones from a DR7 haplotype. *J. Immunol.* **137**: 2886–2890.

Keuning, J.J., Peña, A.S., van Leuwen, A. et al. (1976). HLA-Dw3 associated with coeliac disease. *Lancet* **i**: 506–508.

Kobayashi, K., Atoh, M., Yagita, A. et al. (1990). Crohn's disease in the Japanese is associated with the HLA-DRw53. *Exp Clin Immunogenet.* **7**: 101–108.

Leidenius, M.H., Koskimies, S.A., Kellokumpu, I.H. et al. (1995). HLA antigens in ulcerative colitis and primary sclerosing cholangitis. *APMIS* **103**: 519–524.

Lundin, K.E.A., Sollid, L.M., Qvigstad, E. et al. (1990). T lymphocyte recognition of a celiac disease-associated *cis*- or *trans*-encoded HLA-DQ α/β-heterodimer. *J. Immunol.* **145**: 136–139.

Lundin, K.E.A., Scott, H., Hansen, T. et al. (1993). Gliadin specific HLA-DQ(α1*0501,β1*0201) restricted T cells isolated from the small intestinal mucosa of celiac disease patients. *J. Exp. Med.* **178**: 187–196.

Lundin, K.E.A., Scott, H., Fausa, O. et al. (1994a). T cells from the small intestinal mucosa of a DR4, DQ7/DR4, DQ8 celiac disease patient preferentially recognize gliadin presented by DQ8. *Hum. Immunol.* **41**: 285–291.

Lundin, K.E.A., Gjertsen, H.A., Scott, H. et al. (1994b). Function of DQ2 and DQ8 as HLA susceptibility molecules in celiac disease. *Hum. Immunol.* **41**: 24–27.

Lundin, K.E.A., Sollid, L.M., Norén, O. et al. (1997). Heterogeneous reactivity patterns of HLA-DQ-restricted small intestinal T cell clones from patients with celiac disease. *Gastroenterology* **112**: 752–759.

Manns, M.P., Bremm, A., Schneider, P.M. et al. (1991). HLA DRw8 and complement C4 deficiency as risk factors in primary biliary cirrhosis. *Gastroenterology* **101**: 1367–1373.

Mantovani, V., Corazza, C.R., Angelini, G. et al. (1991). Molecular analysis of HLA-DQ A alleles in coeliac disease; lack of a unique disease-associated sequence. *Tissue Antigens* **83**: 74–78.

Marcos, Y., Fainboim, H.A., Capucchio, M. et al. (1994). Two-locus involvement in the association of human leukocyte antigen with the extrahepatic manifestations of autoimmune chronic active hepatitis. *Hepatology* **19**: 1371–1374.

Marsh, M.N. (1992). Gluten, major histocompatibility complex, and the small intestine. A molecular and immunobiological approach to the spectrum of gluten sensitivity ("celiac sprue"). *Gastroenterology* **102**: 330–354.

McManus, R., Moloney, M., Borton, M. et al. (1996a). Association of celiac disease with microsatellite polymorphisms close to the tumour necrosis factor genes. *Hum. Immunol.* **45**: 24–31.

McManus, R., Wilson, A.G., Mansfield, J. et al. (1996b). TNF2, a polymorphism of the tumour necrosis-α gene promotor, is a component of the celiac disease major histocompatibility complex haplotype. *Eur. J. Immunol.* **26**: 2113–2118.

Mearin, M.L., Biemond, I., Peña, A.S. et al. (1983). HLA-DR phenotypes in Spanish coeliac children; their contribution to the understanding of the genetics of the disease. *Gut* **24**: 532–537.

Meddeb-Garnaoui, A., Zeliszewski, D., Mougenot, J.F. et al. (1995). Reevalution of the relative risk for susceptibility to celiac disease of HLA-DRB1, -DQA1, -DQB1, -DPB1 and -TAP2 alleles in a French population. *Hum. Immunol.* **43**: 190–199.

Mehal, W.Z., Lo, Y.-M.D, Wordsworth, B.P. et al. (1994). HLA DR4 is a marker for rapid disease progression in primary sclerosing cholangitis. *Gastroenterology* **106**: 160–167.

Mella, J.G., Roschmann, E., Maier, K.-P. et al. (1995). Association of primary biliary cirrhosis with the allele HLA-DPB 1*0301 in a German population. *Hepatology* **21**: 398–402.

Molberg, Ø., Kett, K., Scott, H. et al. (1997). Gliadin specific, HLA DQ2-restricted T cells are commonly found in small intestinal biopsies from coeliac disease patients, but not from controls. *Scand. J. Immunol.* **46**: 103–109.

Molberg, Ø., Lundin, K.E.A., Nilsen, E.M. et al. (1998a). HLA restriction patterns of gliadin- and astrovirus-specific CD4$^+$ T cells isolated in parallel from the small intestine of celiac disease patients. *Tissue Antigens* **52**: 407–415.

Molberg, Ø., Nilsen, E.M., Sollid, L.M. et al. (1998b). CD4$^+$ T cells with specific reactivity against astrovirus isolated from normal human small intestine. *Gastroenterology* **114**: 115–122.

Molberg, Ø., McAdam, S.N., Korner, R. et al. (1998). Tissue transglutaminase selectively modifies gliadin peptides that are recognized by gut derived T cells in celiac disease. *Nat. Med.* **4**: 413–717.

Nakajima, A., Matsuhashi, N., Kodama, T. et al. (1995). HLA-linked susceptibility and resistance genes in Crohn's disease. *Gastroenterology* **109**: 1462–1467.

Nilsen, E.M., Lundin, K.E.A., Krajci, P. et al. (1995). Gluten-specific, HLA-DQ restricted T cells from coeliac mucosa produce cytokines with Th1 or Th0 profile dominated by interferon γ. *Gut* **37**: 766–776.

Noguchi, K., Kobayashi, M., Yagihashi, A. et al. (1992). HLA antigens in primary sclerosing cholangitis. *Transplant Proc.* **24**: 2775–2776.

Olerup, O., Olsson, R., Hultcrantz, R. et al. (1995). HLA-DR and HLA-DQ are not markers for rapid disease progression in primary sclerosing cholangitis. *Gastroenterology* **108**: 870–878.

Ota, M., Seki, T., Kiyosawa, K. et al. (1992). A possible association between basic amino acids of position 13 of DRB1 chains and autoimmune hepatitis. *Immunogenetics* **36**: 49–55.

Picarelli, A., Maiuri, L., Frate A. et al. (1996). Production of antiendomysial antibodies after in vitro gliadin challenge of small intestine biopsy samples from patients with coeliac disease. *Lancet* **348**: 1065–1067.

Polanco, I., Biemond, I., van Leeuwen, A. (1981). Gluten sensitive enteropathy in Spain: genetic and environmental factors. In *The Genetics of Coeliac Disease* (ed. R.B. McConnell), pp. 211–231, MTB, Lancaster.

Powis, S.H., Rosenberg, W.M.C., Hall, M. et al. (1993). TAP1 and TAP2 polymorphisms in coeliac disease. *Immunogenetics* **38**: 345–350.

Quarsten, H., Paulsen, G. Johansen, B. et al. (1998) The P9 pocket of HLA-DQ2 (non-Aspβ57) has no particular preference for negatively charged anchor residues found in other type 1 diabetes predisposing non-Aspβ57 MHC class II molecules. *Int. Immunol.* **10**: 1229–1236.

Reinshagen, M., Loeliger, C., Kuehnl, P. et al. (1996). HLA class II gene frequencies in Crohn's disease: a population based analysis in Germany. *Gut* **38**: 538–542.

Roep, B.O., Bontrop, R.E., Peña, A.S. et al. (1988). An HLA-DQ alpha allele identified at DNA and protein level is strongly associated with celiac disease. *Hum. Immunol.* **23**: 271–279.

Roussomoustakaki, M., Satsangi, J., Welsh, K. et al. (1997). Genetic markers may predict disease behavior in patients with ulcerative colitis. *Gastroenterology* **112**: 1845–1853.

Satsangi, J., Welsh, K.I., Bunce, M. et al. (1996). Contribution of genes of the major histocompatibility complex to susceptibility and disease phenotype in inflammatory bowel disease. *Lancet* **347**: 1212–1217.

Schenning, L., Larhammar, D., Bill, P. et al. (1984). Both α and β chains of HLA-DC class II histocompatibility antigens display extensive polymorphism in their amino-terminal domains. *EMBO J.* **3**: 447–452.

Schiffenbauer, J., Didier, D.K. and Klearman, M. (1987). Complete sequence of the HLA DQα and DQβ cDNA from a DR5/DQw3 cell line. *J. Immunol.* **139**: 228–233.

Schrumpf, E., Fausa, O., Førre, O. et al. (1982). HLA antigens and immunoregulatory T cells in ulcerative colitis associated with hepatobiliary disease. *Scand. J. Gastroenterol.* **17**: 187–191.

Scott, H., Nilsen, E., Sollid, L.M. et al. (1997). Immunopathology of gluten-sensitive enteropathy. *Springer Semin. Immunopathol.* **18**: 535–553.

Seki, T., Kiyosawa, K., Inoko, H. et al. (1990). Association of autoimmune hepatitis with HLA-Bw54 and DR4 in Japanese patients. *Hepatology* **12**: 1300–1304.

Seki, T., Kiyosawa, K., Ota, M. et al. (1993). Association of primary biliary cirrhosis with human leukocyte antigen DPB1*0501 in Japanese patients. *Hepatology* **18**: 73–78.

Shewry, P.R., Tatham, A.S. and Kasarda, D.D. (1992). Cereal proteins and coeliac disease. In *Coeliac Disease* (ed. M.N. Marsh), pp. 305–348, Blackwell, Oxford.

Sjöström, H., Lundin, K.E., Molberg, Ø. et al. (1998). Identification of a gliadin T-cell epitope in coeliac disease: general importance of gliadin deamidation for intestinal T-cell recognition. *Scand. J. Immunol.* **48**: 111–115.

Solheim, B.G., Ek, J., Thune, P.O. et al. (1976). HLA antigens in dermatitis herpetiformis and coeliac disease. *Tissue Antigens* **7**: 57–59.

Sollid, L.M. and Thorsby, E. (1990). The primary association of celiac disease to a given HLA-DQ α/β heterodimer explains the divergent HLA-DR associations observed in various Caucasian populations. *Tissue Antigens* **36**: 136–137.

Sollid, L.M. and Thorsby, E. (1993). HLA susceptibility genes in celiac disease: genetic mapping and role in pathogenesis. *Gastroenterology* **105**: 910–922.

Sollid, L.M., Markussen, G., Ek, J. et al. (1989). Evidence for a primary association of celiac disease to a particular HLA-DQ α/β heterodimer. *J. Exp. Med.* **169**: 345–350.

Sollid, L.M., Lundin, K.E.A., Sjöström, H. et al. (1997a). HLA-DQ molecules, peptides and T cells in coeliac disease. In *Coeliac Disease. Proceedings of the 7th International Symposium on Coeliac Disease* (eds M. Mäki, P. Collin and J.K. Visakorpi), pp. 265–277, Coeliac Disease Study Group, Tampere.

Sollid, L.M, Molberg, Ø., McAdam, S. et al. (1997b). Autoantibodies in coeliac disease: tissue transglutaminase – guilt by association. *Gut* **41**: 851–852.

Sollid, L.M., Johansen, B.H., Lundin, K.E.A. et al. (1997c). The molecular basis of the HLA association in celiac disease: a model for other HLA-associated diseases. In *HLA and Disease – The Molecular Basis. Alfred Benzon Symposium 40* (eds A. Svejgaard, S. Buus and L. Fugger), pp. 362–372, Munksgaard, Copenhagen.

Spurkland, A., Sollid, L.M. Polanco, I. et al. (1992). HLA-DR and –DQ genotypes of celiac disease patients serologically typed to be non-DR3 or non-DR5/7. *Hum. Immunol.* **35**: 188–192.

Spurkland, A., Ingvarsson, G., Falk, E.S. et al. (1997). Dermatitis herpetiformis and celiac disease are both primarily associated with the HLA-DQ (α1*0501,β1*02) or the HLA-DQ (α1*03,β1*0302) heterodimers. *Tissue Antigens* **49**: 29–34.

Spurkland, A., Saarinen, S., Boberg, K.M. et al. (1999). Primary sclerosing cholangitis displays different HLA associations in Northern and Southern Europeans. *Tissue Antigens* **53**: 459–469.

Stokes, P.L., Asquith, P., Holmes, G.K.T. et al. (1972). Histocompatibility antigens associated with adult coeliac disease. *Lancet* **ii**: 162–164.

Strettell, M.D.J., Donaldson, P.T., Thomson, L.J. et al. (1997). Allelic basis for HLA-encoded susceptibility to type 1 autoimmune hepatitis. *Gastroenterology* **112**: 2028–2035.

Strober, W. (1992). Gluten sensitive entropathy. In *The Genetic Basis of Common Diseases* (eds R.A. King, J.I. Rotter and A.G. Motulsky), pp. 279–304, Oxford University Press, Oxford.

Thorsby, E. (1997). Invited anniversary review: HLA associated diseases. *Hum. Immunol.* **53**: 1–11.

Tighe, M.R., Hall, M.A., Ashkenazi, A. et al. (1993). Celiac disease among Ashkenazi Jews from Israel. A study of the HLA class II alleles and their associations with disease susceptibility. *Hum. Immunol.* **38**: 270–276.

Tiwari, J.L. and Terasaki, P.I. (1985). *HLA and Disease Associations*, Springer, New York.

Tosi, R., Vismara, D., Tanigaki, N. et al. (1983). Evidence that celiac disease is primarily associated with a DC locus allelic specificity. *Clin. Immunol. Immunopathol.* **28**: 395–404.

Toyoda, H., Wang, S.-J., Yang, H.-Y. et al. (1993). Distinct associations of HLA class II genes with inflammatory bowel disease. *Gastroenterology* **104**: 741–748.

Trier, J.S. (1991). Celiac sprue. *N. Engl. J. Med.* **325**: 1709–1719.

Underhill, J., Donaldson, P., Bray, G. et al. (1992). Susceptibility to primary biliary cirrhosis is associated with the HLA-DR8-DQB1*0402 haplotype. *Hepatology* **16**: 1404–1408.

Underhill, J.A., Donaldson, P.T., Doherty, D.G. et al. (1995). HLA DPB polymorphism in primary sclerosing cholangitis and primary biliary cirrhosis. *Hepatology* **21**: 959–962.

Vartdal, F., Johansen, B.H., Friede, T. et al. (1996). The peptide binding motif of the disease associated HLA-DQ (α1*0501,β1*0201) molecule. *Eur. J. Immunol.* **26**: 2764–2772.

Wal, Y. van de, Kooy, Y.M.C., Drijfhout, J.W. et al. (1996). Peptide binding characteristics of the coeliac disease-associated DQ(α*0501,β1*0201) molecule. *Immunogenetics* **44**: 246–253.

Wal, Y. van de, Kooy, Y.M.C., Drijfhout, J.W. et al. (1997). Unique peptide binding characteristics of the disease-associated DQ(α*0501,β1*0201) vs the non-disease-associated DQ(α1*0201,β1*0202) molecule. *Immunogenetics* **46**: 484–492.

Wal, Y. van de, Kooy, Y.M.C., van Veelen, P.A. et al. (1998a). Small intestinal T cells of celiac disease patients recognize a natural pepsin fragment of gliadin. *Proc. Natl. Acad. Sci. USA* **95**: 10050–10054.

Wal, Y. van de, Kooy, Y.M.C., van Veelen, P.A. et al. (1998b). Selective deamidation by tissue transglutaminase strongly enhances gliadin-specific T cell reactivity. *J. Immunol* **161**: 1585–1588.

Wal, Y. van de, Kooy, Y.M.C., van Veelen, P.A. et al. (1999). Glutenin is involved in the gluten-driven mucosal T cell response. *Eur. J. Immunol.* **29**: 3133–3139.

Wassmuth, R., Eastman, S., Kockum, I. et al. (1993). HLA DR and DQ RFLP analysis in Crohn's disease. *Eur. J. Immunogenet.* **20**: 429–433.

Wrigley, C.W. and Shepard, K.W. (1973). Electrofocusing of grain proteins from wheat genotypes. *Ann. N.Y. Acad. Sci.* **209**: 154–162.

Zhong, F., McCombs, C.C., Olson J.M. et al. (1996). An autosomal screen for genes that predispose to celiac disease in the western counties of Ireland. *Nature Genet.* **14**: 329–333.

CHAPTER 18

HLA and Respiratory Disease

Cesare Saltini, Luca Richeldi and Massimo Amicosante

THE LUNG IMMUNE SYSTEM

As the site of gas exchange, the epithelial surface of the respiratory tract, with an area of 50–100 m^2, is constantly burdened with antigenic and non-antigenic particles present in the 7000–10 000 l of air that are inhaled daily. The lungs have two general mechanisms by which they dispose of these particles: bronchial clearance systems to remove them and inflammatory and immunocompetent cells to inactivate and destroy them. The evidence from experimental animal and human studies suggests that the lung contains a compartmentalized mucosal immune system. In this regard, the density of inflammatory and immune cells recovered from the lower respiratory tract is several-fold greater than that of the blood. Lung T lymphocyte density is twice that of the blood, whereas B lymphocytes are virtually absent on the alveolar surface but appear in the bronchial mucosal lymphoid aggregates. Since the lung lymphocyte population is almost entirely comprised of memory T cells, it is likely that the lymphocytes in the alveolar and bronchial associated lymphoid tissue have the ability to mount strong immune responses (Saltini et al., 1991). The lung, like the gut and the skin, is a site where foreign antigens may initiate abnormal, damaging immune reactions; it is therefore an ideal organ model for the study of the immunogenetic basis of hypersensitivity and inflammatory reactions to environmental agents.

HYPERSENSITIVITY DISORDERS OF THE LOWER RESPIRATORY TRACT

Berylliosis

Berylliosis is probably the best understood of the environmental chronic inflammatory disorders. It is characterized by the accumulation of CD4$^+$ T cells and macrophages in the lower respiratory tract, non-caseating granuloma formation and, eventually, fibrosis in response to beryllium (Be) inhalation. Pathologically, it is indistinguishable from aluminium- and titanium-induced granulomas and from sarcoidosis.

Exposure to Be can cause acute and chronic disorders of the skin and the lung. Exposure to high levels of Be salts can cause acute dermatitis, conjunctivitis, rhinitis and a pneumonitis similar to the metal fume fever syndrome induced by aluminium and other metals. Exposure to lower levels of insoluble Be dust, if protracted, may cause berylliosis, which affects 2–5% of exposed workers (Kreiss et al., 1994). Since the immunotoxicity of Be was recognized in the late 1930s and hypersensitivity suspected as the important mechanism in disease incidence, the levels of Be in the air in the industrial environment have been reduced by over 100-fold and strict industrial hygiene measures implemented. While acute disease has been eliminated, the chronic condition has not, suggesting that prevention may need to be targeted to the more sensitive segment of the exposed population.

In individuals affected by berylliosis, Be-sensitized lung T cells are dominated by CD4+ CD45R0+ memory T cells (Saltini et al., 1990a), which are polyclonal (Saltini et al., 1989), although they may express a limited set of T cell antigen receptor variable region genes (Rossman et al., 1992) and recognize Be as a specific major histocompatibility complex (MHC)-restricted antigen/hapten (Saltini et al., 1989). Consistent with the disease immunopathology, T cells in the lower respiratory tract of patients with berylliosis are TH_1 T cells, i.e. they produce interleukin-2 (IL-2) and interferon-γ (IFN-γ), the macrophage-activating cytokine driving the granulomatous reaction (Tinkle et al., 1997). The finding that Be is recognized as a specific HLA class II-restricted hapten/antigen prompted the search for immune response genes associated with Be hypersensitivity. Richeldi et al. (1993) found in a retrospective study of 32 cases that the human leucocyte antigen (HLA) class II HLA-DP allele HLA-DPB1*0201 was associated with disease risk and the allele HLA-DPB1*0401 with disease protection (Table 18.1). Sequence analysis showed that the susceptibility was associated with a polymorphic sequence coding for a lysine to glutamic acid change in the fourth variable domain D (residues 67–69) of the β-chain of the HLA-DP molecule (HLA-DPβGlu^{69}) (Figure 18.1). This association was confirmed in a cohort study of 137 Be ceramics workers, where an 80% frequency of HLA-DPβGlu^{69} was found among individuals with berylliosis compared with 35% in exposed controls (Richeldi et al., 1997a). Similar results were recently obtained in a larger cohort of 650 Be-exposed workers (Richeldi and Saltini, manuscript in preparation). Stubbs et al. (1996) confirmed the association of berylliosis with HLA-DPβGlu^{69} and also found that HLA-DR allelic variants were associated with disease.

What is the function of HLA-DP in the pathogenesis of berylliosis? In the context of current knowledge of the interactions of metals with the human immune system, several suggestions may be offered. First, the DPBI gene may be a disease marker resulting from linkage to some as yet unidentified nearby gene(s) influencing

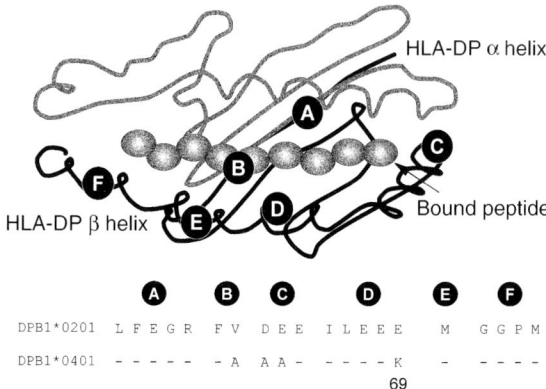

FIGURE 18.1 Structure of the HLA-DP molecule. Shown below the diagram are the amino acid sequence changes in the five variable domains of the HLA-DP β-chain gene.

TABLE 18.1 HLA association with hypersensitivity lung disorders[a]

Disease	HLA allele associated with		Other associated features
	Susceptibility	Protection	
Beryllosis	DPβ1 Glu^{69}	DPβ1 Lys^{69}	
Gold-induced HP	DR1		
Bird breeder's disease	B8, B40, DR3, DW6		
	DR3		Acute disease
	DR7		Mexicans
Farmer's lung	B7, DR2		
Summer-type HP	HLA-DQ3		Japanese
Sarcoidosis	A1, B8, DR3		Acute
	DR4, DR5		Chronic
	DPβ1 Glu^{69}		

[a]Data sources are quoted in the text.
HP: hypersensitivity pneumonitis.

the individual's immune responsiveness. The available evidence seems to support a specific immune response gene effect, although an effect of linkage or a combination of both effects cannot be ruled out. Secondly, certain DPB alleles may be directing the immune response toward the preferential expression of a TH_1 response, i.e. towards chronic inflammation and autoimmunity. Thirdly, certain DBP alleles may have higher affinity for Be, resulting in the ability to select Be as a specific antigen/hapten for T cell presentation and thus cause an allergic response.

HLA-DPβGlu^{69} might simply be a DNA marker linked to other nearby immune response genes with direct influence on disease susceptibility. In this regard, studies on the association of the tumour necrosis factor-α (TNFα) gene with chronic inflammatory and autoimmune disorders such as rheumatoid arthritis (RA) and diabetes may suggest that in berylliosis the association of HLA-DP with disease is also the consequence of linkage disequilibrium with the TNFα or other cytokine genes. However, the HLA-DP gene is not linked, or it is only weakly linked, to the other HLA locus genes, including the TNFoC gene. It cannot be excluded that other chromosome 6 genes, linked to the HLA-DP gene and strongly associated with susceptibility to berylliosis, will be identified.

Some observations support the hypothesis that HLA-DP functions as a TH_1 immune response gene. Individuals with Be hypersensitivity identified through screening programmes using an immunological blood test to identify that those with Be sensitization have less severe disease than patients identified because of clinical signs and symptoms (Pappas et al., 1993). Furthermore, some hypersensitive individuals may not progress towards lung disease at all. HLA-DPβGlu^{69} has a higher prevalence among subjects with clinical features of disease (Richeldi et al., 1993) than in those whose sensitization was picked up by laboratory screening (Richeldi et al., 1997a). Taken together with the observation that HLA-DPβGlu^{69} was also significantly more prevalent among individuals with disease than in those with lone hypersensitivity (Richeldi, Kreiss, Amicosante, Wiedemann and Saltini, manuscript in preparation), these data support the notion that HLA-DPβGlu^{69} confers upon exposed individuals a greater susceptibility to granulomatous inflammation. However, while HLA-DPβGlu^{69} is also associated with the TH_1 reactions of sarcoidosis (Lympany et al., 1996) and the generation of alloreactive type 1 T cells in transplant rejection (Potolocchio, 1997), it is also associated with TH_2 type responses such as the T cell response to albumin in egg allergy (Shinibara et al., 1995), arguing against a general association between DPβGlu^{69} and a tendency to generate a TH_1 type response.

The suggestion that the HLA-DP gene may function as the primary immune response gene is supported by the observation of Lombardi et al. (1997) who, consistent with earlier T cell studies (Saltini et al., 1989, 1990b), found that Be-specific T cell clones from individuals with berylliosis were activated by Be presented by HLA-DP molecules expressing glutamic acid in position 69 of the β chain. Recent data, indicating that T cell production of IFNγ in response to Be is also restricted by HLA-DP, lend further support to this hypothesis (Franchi, Wiedemann and Saltini, manuscript in preparation) (Figure 18.2). Current concepts of metal recognition by T cells are that metals can haptenize peptides, thereby generating neoantigens which can be bound by HLA molecules and induce allergy or can bind directly to the HLA molecule and generate an altered self-HLA, thereby inducing hypersensitivity. In nickel-allergic individuals, nickel binds to histidine residues on peptides, making them antigenic. In gold hypersensitivity, gold binds directly to the HLA molecule (Sinigaglia, 1994). The nature of the Be antigen has not yet been clarified. It may bind directly to the HLA-DP molecule, thereby generating an "altered self" and an immune reaction. In this context, in gold and Be-induced lung hypersensitivity, specific HLA alleles have a direct role in generating restricted neoantigens. Thus, HLA genes may function as primary susceptibility genes or specific immune response genes for these hypersensitivity diseases. Under these circumstances, HLA genes may be useful as epidemiological probes in identifying populations at higher risk and preventing disease associated with occupation, manufactured products and the environment.

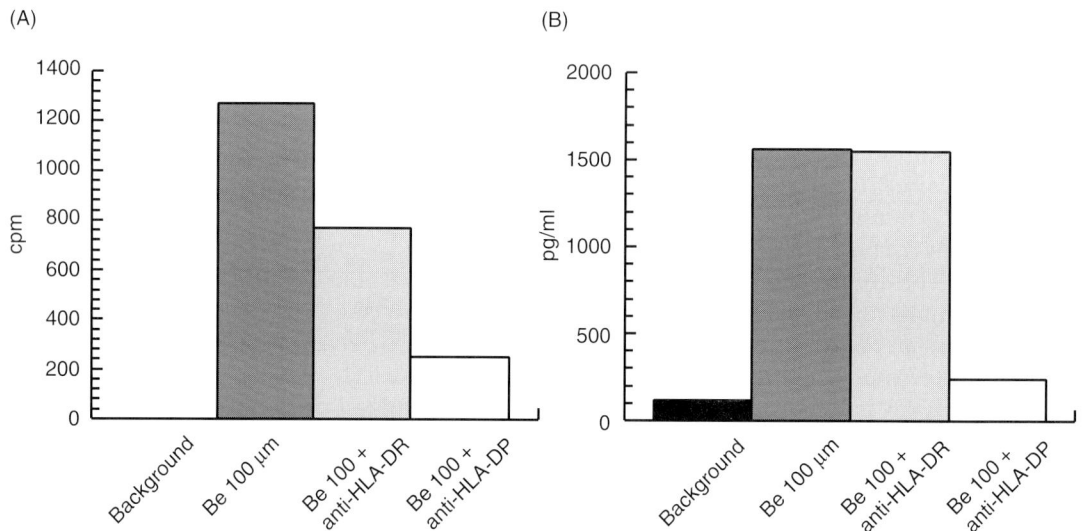

FIGURE 18.2 Role of HLA-DP in beryllium presentation and T cell activation. The diagram summarizes the results of the effect of the monoclonal antibodies L243 (anti-DR) and B7/21 (anti-DP) on T cell proliferation and production of granulocyte macrophage–colony-stimulating factor (GM-CSF), upon the stimulation with beryllium of lung T cells from patients with berylliosis. Shown on the ordinate are (A) [^{3}H]thymidine incorporation in counts per minute and (B) GM-CSF production in pg/ml of stimulated supernatant. The experimental conditions are shown on the abscissae.

Gold hypersensitivity pneumonitis

The administration of colloidal gold to patients with RA may cause skin and lung hypersensitivity. Studies in Sinigaglia's laboratory have shown that the response to gold of individuals with this allergy is maintained by T cells recognizing gold in an antigen-specific fashion. Unlike nickel, which can haptenize any peptide carrying a histidine residue and can therefore react with a large repertoire of antigenic moieties, gold is thought to bind directly to the HLA molecule. Sinigaglia's studies indicate that this very property may make gold both immunosuppressive and immunogenic at the same time. By binding to the HLA molecule, it may block the responses to the autoantigen(s) of RA. However, for the same reason it may transform certain HLA molecules into neoantigens (Sinigaglia, 1994).

Hakala et al. (1986) showed that gold-induced hypersensitivity pneumonitis (HP) is associated with HLA-DR1 in Finnish RA patients and the same association was found by Partanen et al. (1987). Strikingly, in the RA patients with gold hypersensitivity described by Romagnoli et al. (1992), the T cell response to gold was restricted by HLA-DR1, leading to the hypothesis that gold may be able to haptenize or "neoantigenize" the HLA-DR1 molecule.

Organic dust-induced hypersensitivity pneumonitis

Exposure to a variety of organic dusts causes chronic hypersensitivity lung reactions. Extrinsic allergic alveolitis, or HP, is caused by repeated exposure to organic dusts, most of which are products from animals, vegetables or microorganisms. A few small organic compounds can also cause HP. The best-known examples of HP are farmer's lung, bird breeder's disease and, in Japan, the summer-type HP. Farmer's lung is caused by the inhalation of *Actimomyces thermophilus rectivirgula* grown in mouldy hay, bird breeder's disease by exposure to bird droppings and summer-type HP by exposure to the microorganism *Trichosporon cutaneum*. The typical HP patient presents with flu-like symptoms developing 4–6 h after exposure. Respiratory symptoms may worsen insidiously, with repeated exposure leading to severe respiratory dysfunction, anorexia and weight loss. HP patients have positive skin tests and serum precipitating antibodies against the offending agent. HP is an immunologically mediated disorder: exposure to the dusts triggers macrophage activation and the release of granulocyte chemotactic factors leading to the accumulation of granulocytes in the lower respiratory tract during the first few hours after exposure. A T cell alveolitis follows, which is dominated by $CD4^+$ T cells in the acute stage and by $CD8^+$ cytotoxic T cells in the chronic phase. Tissue lesions are characterized by (i) interstitial infiltrates of mononuclear cells without eosinophilia, (ii) scattered poorly formed granulomas without necrosis, and (iii) bronchiolitis. That HP lesions are the consequence of an exaggerated reaction to specific antigen(s) or superantigen(s) has been suggested by the observation of expansion of both $CD8^+$ and $CD4^+$ T cells bearing selected T cell antigen receptor (TCR) α and β chain variable region genes (Wahlstorm et al., 1997).

The disease affects fewer than 10% of exposed individuals and it has been suggested that individual susceptibility factors may play a role in its pathogenesis. A number of studies has addressed the association of HP with HLA genes. Allen et al. (1977) described the inhibition of antigen-specific T cell responses by anti-HLA antiserum, postulating an immune response gene for HP. They evaluated the association between HP and HLA genes and found an increased frequency of HLA-B40 in bird breeder's disease. Sennekamp et al. (1978) found that the HLA-B8 gene was strongly associated with bird breeder's disease. Non-significant associations with hypersensitivity to avian antigens were reported by Berrill and van Rood (1977) for HLA-Dw6 and by Muers et al. (1982) for HLA-DR3. This observation was later confirmed by Rittner et al. (1983), who found a strong association between bird breeder's disease and HLA-DR3 in a study of 116 German pigeon breeders. Selman et al. (1987) described an association with HLA-DR7 and the haplotype HLA-B35–DR4, the postulated Hispanic equivalent of the Caucasian HLA-B8–DR3, reported in Mexican bird breeders.

In farmer's lung, Khomenko et al. (1985) observed an increased frequency of HLA-B7 and disease and Wahlstorm et al. (1997) of HLA-DR2. By contrast, no association was found in a study of 37 unselected, end-stage HP patients by Nowack et al. (1987). In summer-type HP, a Japanese study by Ando et al. (1989) showed an association with HLA-DQ3.

GRANULOMATOUS LUNG DISORDERS OF UNKNOWN ORIGIN

Sarcoidosis

Sarcoidosis is a chronic multisystem granulomatous disorder of unknown origin, possibly caused by a number of offending agents including mycobacteria, affecting primarily the lung. The immunopathology of pulmonary sarcoidosis is characterized by (i) a macrophage–lymphocyte alveolitis; (ii) the formation of non-caseating granulomas in the peribronchial and perivascular interstitium of the lung; and (iii) fibrosis. The alveolitis of sarcoidosis is dominated by activated macrophages releasing T cell activating lymphokines such as TNFα (Bost et al., 1994) and by the accumulation of activated type-1 $CD4^+$ T cells expressing IL-2 receptors and HLA-DR surface molecules and releasing IL-2 and IFNγ at exaggerated levels (Saltini et al., 1986). These T cells express biased repertoires of T cell antigen receptor rearranged genes (see below), suggesting an exaggerated response to a single antigen.

The presentation and clinical course of sarcoidosis vary. Stage I is characterized by prominent mediastinal lymph node reaction and rapid resolution, stage II by diffuse lung infiltration with granulomatous lesions, and stage III by diffuse fibrosis and severe lung dysfunction. Markers of exuberant local T cell activation, such as the increase in the number and proportions of the $CD4^+$ T cells, are associated with stage I disease and with rapid resolution.

Several lines of evidence suggest that susceptibility to sarcoidosis is inherited. First, there is a definite prevalence of sarcoidosis in certain ethnic groups such as the Irish, the Swedish and the North American Black population. Secondly, familial clusters of sarcoidosis are described in the literature. Finally, there is an excess of concordance for sarcoidosis in monozygotic twins. However, susceptibility to sarcoidosis is likely to involve multiple genes, rather than a single gene, since no clear inheritance pattern has been identified in the affected families. In the context of the accumulating evidence that local T cell activation, strongly resembling that of an antigen-driven process, is among the dominant features of sarcoidosis, a number of studies has attempted to link sarcoidosis susceptibility with HLA genes.

Hedfors and Lindstrom (1983) showed that the HLA-B8–DR3 haplotype was over-represented among acute-onset, stage I resolving sarcoidosis patients with arthritis. The association between acute, rapidly resolving sarcoidosis and HLA-B8–DR3 was confirmed by Gardner et al. (1984) and by Smith et al. (1981) in Caucasians but not in West Indians (Gardner et al., 1984). Studies on German and Japanese patient populations (Abe et al., 1987; Nowack et al., 1987) have instead shown an association with HLA-DR5. Abe et al. (1987) found that HLA-DR5, not HLA-DR3, was associated with non-resolving sarcoidosis. Similar conclusions were reached by Ikeda et al. (1992), also in a Japanese study. In this regard, they hypothesized that sarcoidosis is associated with the HLA-DR antigen group, including DR3, DR5, DR6 and DR8. Consistent with the notion that HLA-DR3 is very common among Caucasians but very rare among the Japanese, sarcoidosis is associated with the group of HLA antigens including DR3, DR5 and DR6 in Caucasians, and HLA-DR5, DR6 and DR8 in the Japanese (Ishihara et al., 1994).

Looking at the HLA-DP gene, Lympany et al. (1996) found an association between the berylliosis-associated HLA-DPβ Glu^{69} and sarcoidosis in a small group of patients. This was not confirmed by Richeldi et al. (1996) in patients of different ethnicity.

To address the question of whether HLA genes play a role in disease course, a collaborative study was carried out in Luisetti's laboratory on 233 European patients from Italy and Czechoslovakia categorized by disease onset, stage, extrapulmonary spread and disease course. As predicted, sarcoidosis was associated with HLA-A1, -B8 and -D3. In addition, it was associated with HLA-DR4 in women, -DR5 in men, -B22 in Italians and -B12 in Czechs. HLA-A1, -B8, -B27 and -DR3 were associated with stage I and -DR3 with spontaneous resolution. In contrast, HLA-B12 and -DR4 were associated with stage III disease (Martinetti et al., 1995).

What is the role of HLA genes in the pathogenesis of sarcoidosis? As in berylliosis, where certain HLA-DP alleles seem to function to restrict the presentation of Be, HLA genes may function in sarcoidosis as immune response genes restricting the presentation of specific antigen(s) or superantigen(s). Grunewald et al. (1992, 1994) observed in Scandinavian sarcoid patients carrying the HLA-DR3(DR17)–DQ2 haplotype that CD4$^+$ T cells expressing the Vα2.3 TCR gene accumulated in the lung with a greater frequency than was expected. Similarly, Usui et al. (1996) found that in Japanese patients carrying HLA-DR12, T cells accumulating in the lung expressed the Vβ6 gene. Grunewald et al. (1995) found few T cells expressing Vα2.3 in Japanese patients who did not express HLA-DR3. In contrast, they found an exaggerated expression of Vβ19 and Vβ22 TCR on CD8$^+$ cells in patients carrying the class I genes HLA-B51 or -B61. Although these findings were not confirmed in a subsequent Japanese study (Hoshino et al., 1997), they suggest the presence of specific antigen(s) or superantigen(s) with high affinity for the disease-associated HLA gene(s), and hence a direct role for both the HLA and the TCR genes in disease susceptibility.

Alternatively, since HLA-B8 and HLA-DR3, both of which have been associated with sarcoidosis in many studies, are in positive linkage disequilibrium with the TNF gene, it may be hypothesized that TNF is the primary sarcoidosis gene. However, Ishihara et al. (1995) showed that the TNFβ gene NcoI polymorphism is more weakly associated with disease than HLA-DR alleles, leaving open the question of whether the disease course, rather than the disease itself, is associated with the TNF gene. In this regard, Richeldi et al. (1997b) extended Marinetti's study and looked at the TNFα promoter region –308 biallelic A/G gene polymorphism. They found that stage I sarcoidosis was associated with the TNF2 allele (associated with high TNFα production) and not with TNF1 (associated with low TNFα production). Since TNF2 is in positive linkage disequilibrium with HLA-DR3 (Deng et al., 1996) and in negative disequilibrium with -DR5, it may be hypothesized that in stage I sarcoidosis, which is characterized by exuberant CD4$^+$ T cell activation, the expression of a "high TNF" genotype may play a role in the genesis of immunopathological manifestations. The opposite may happen with HLA-DR5 and stage III sarcoidosis.

ALLERGIES OF THE UPPER RESPIRATORY TRACT

Asthma

Asthma is a disease of the airways characterized by chronic inflammation with infiltration of lymphocytes, eosinophils and mast cells leading to epithelial desquamation and thickening of the bronchial mucosa. Clinically, the disease is characterized by airway narrowing and wheezing in response to a variety of stimuli. Asthma is caused by, or associated with (i) atopy, a condition involving exaggerated amounts of immunoglobulin E (IgE) antibodies in response to aeroallergens, and/or (ii) bronchial hyper-responsiveness (BHR), an exaggerated narrowing of the airways in response to non-antigenic stimuli. Family and population studies have strongly indicated that asthma is a complex genetic disorder.

Four phenotypic markers have been used in the search for the asthma gene(s): (i) total serum IgE levels; (ii) immediate reactivity in skin tests to aeroallergens and specific IgE levels; (iii) BHR to physical and pharmacological stimuli; and (iv) a history of wheezing. The genetic loci tentatively associated with asthma are: a locus on chromosome 11q13 in close proximity to the IgE receptor and a locus on chromosome 5q31 located near a cytokine gene cluster comprising the IL-4, IL-5 and IL-13 genes and close to the β_2-adrenergic receptor gene. These associations, however, have not been confirmed in all studies, possibly owing to varying diagnostic criteria and differences in the ethnic groups tested (Sandford et al., 1996). The association of HLA genes with asthma has been explored by many investigators, with conflicting results. Consistent with the antigen-presenting function of HLA molecules, HLA genes have not been associated with high levels of IgE, BHR, wheezing or the diagnosis of asthma per se, contrary to earlier reports (Rachelefsky et al., 1976). Among others, a British and a Greek study found a weak association of atopic asthma with the haplotype HLA-A1–B8–DR3 (Turner et al., 1977; Apostolakis et al., 1996) and a French family study, with the DR53 group of DRBI alleles (HLA-DR4–DR7–DR9) (Aron et al., 1995). However, no associations were found in American, British and Chinese studies (Amelung et al., 1992; Li et al., 1995; Holloway et al., 1996). Work from several laboratories has indicated that, in contrast to atopy or asthma per se, HLA genes may be associated with responsiveness to specific protein or peptide allergens (Marsh and Huang, 1991) (Table 18.2).

In allergy to ragweed, sensitization to the specific allergen Amb aV is associated with HLA-DR2 and to Amb aVI with HLA-DR5 (Marsh and Huang, 1991; Blumenthal et al., 1992). Among subjects with allergy to rye-grass pollen, sensitization to the allergens Lol pI and Lol pII is associated with HLA-DR3 and to Lol pIII with both HLA-DR3 and -DR5. Antigen-specific T cells are restricted by the respective allelic variants of the

TABLE 18.2 HLA association with obstructive lung disorders[a]

Disease	HLA allele associated with susceptibility	Other associated features
Ragweed asthma	DR2	Amb aV fragment
	DR5	Amb aVI fragment
Rye-grass asthma	DR3	Lol pI, Lol pII fragment
	DR3, DR5	Lol pIII fragment
House dust mite asthma	DQ7	Der p2
	DQA0301	Dermatophagoides farinae
Anhydride asthma	DR3	Trimellitic anhydride
Isocyanate asthma	HLA-DQB0503, 0201/0301	Aspartate-57

[a] Data sources are quoted in the text.

HLA-DR molecule (Ansari et al., 1989). Unlike grass-pollen allergy, allergy to house dust mites (HDM) seems to be associated with the HLA-DQ gene. Skin-prick test reactivity to *Dermatophagoides farinae* has been found to be associated with *HLA-DQA1*0301* but not with the DR53 haplotype identified in the French study quoted above (Holloway et al., 1996). O'Brien et al. (1995) found an association between *in vitro* T cell reactivity to *Dermatophagoides pteronissimus* Der p2 antigen-derived peptides and the HLA-DQ7 antigen. In the context of the observation that Der p2 is recognized by T cells within the HLA-DR molecule, but not HLA-DQ or HLA-DP (Van Neerven et al., 1993), it has been hypothesized that HLA-DQ functions as a T cell suppressor antigen in HDM-induced asthma (O'Brien et al., 1995; Holloway et al., 1996). HLA-DQ has also been implicated in aspirin-induced asthma in a small American study (Mullarkey et al., 1986), but the data were not confirmed in a study of British and German aspirin-sensitive asthmatics (Lympany et al., 1993).

Occupational asthma

Occupational asthma has been defined by the Industrial Injuries Advisory Council of Great Britain as "variable air flow limitation caused by sensitization to a specific agent encountered at work and excluding other occupational causes of variable air flow limitations not due to sensitization". The definition of occupational asthma must include: (i) reversible air flow impairment; (ii) work relatedness; (iii) documented sensitization if allergic; or (iv) non-specific hyper-responsiveness (Brooks, 1992). Occupational asthma affects hundreds of thousands of workers throughout the industrialized world. More than 200 agents causing asthma in the work environment have been identified, including a number of reactive small chemicals such as metals, anhydrides and isocyanates which may react as haptens with endogenous proteins, generating neoantigens.

It is thought that, in occupational asthma, as in contact dermatitis, a subpopulation of exposed individuals may be more sensitive to environmental agents since they have "leaky" airways as a result of atopy or injured bronchial epithelium (Brooks, 1992). Since the identification of causative agents is critical to the diagnosis, occupational asthma is an excellent model to test individual susceptibility to allergy. It has been shown recently that HLA genes are a major component of susceptibility to chemicals inducing occupational asthma, suggesting that they may function as specific immune response genes in asthma as well as in delayed-type hypersensitivity disorders of the lung.

Isocyanate-induced asthma has been considered by many authorities to be an irritant-induced type of asthma. However, it has been shown that T cells from diisocyanate-induced asthmatics react *in vitro* against the offending agent, thus indicating a specific immune reaction. Consistent with this notion, diisocyanate-induced asthma has been associated with HLA-DQB1*0503, heterozygosity for DQB1*0201 and *0301 (Bignon et al., 1994). Based on this finding Balboni et al. (1996) suggested that disease susceptibility may be associated with the expression of a specific amino acid residue (aspartate 57) on the HLA-DQ molecule β chain. A direct role for HLA-DQ in antigen presentation of these chemicals was thus suggested. Consistent with a direct role of the TCR and HLA in sensitization to isocyanates, Bernstein et al. (1997) showed that T cells from diisocyanate-induced asthmatics expressed a biased TCR repertoire consistent with the expansion of Vβ1 and Vβ5 genes in isocyanate-stimulated T cells. However, they did not find any association with HLA.

In anhydride-induced asthma, Young et al. (1995) showed a significant excess of HLA-DR3 in 30 asthma cases with a positive radioallergosorbent test compared with 30 exposed controls without specific IgE. The excess of HLA-DR3 was particularly associated with the presence of IgE against trimellitic anhydride, indicating that HLA class II proteins may be an important determinant in the generation of a specific IgE response to inhaled haptens.

FIBROTIC LUNG DISORDERS

Several dusts are capable of inducing fibrotic lung disorders. The most notable examples are asbestos, coal, silica and hard metals. Current concepts of the pathogenesis of dust-induced fibrotic lung disorders are that dusts may directly stimulate the activation of macrophages, inducing the production of inflammatory cytokines and resulting in chronic inflammation, activation and proliferation of fibroblasts, and progressive fibrosis and lung dysfunction. Although immunological mechanisms have been hypothesized in these diseases, they have not yet been convincingly substantiated.

Hard-metal lung disease

Exposure to a mixture of tungsten carbide and cobalt, or to cobalt alone, may be the cause of allergic dermatitis, asthma, pulmonary fibrosis and, possibly, of HP. Cobalt allergic dermatitis is thought to affect fewer than 5% of exposed individuals, suggesting that individual susceptibility might play a role in the prevalence of allergy. There is uncertainty as to the epidemiology of pulmonary fibrosis. Although the disease has been considered a rare event, two reports have claimed a prevalence between 20 and 30% (Churg and Green, 1995). Cobalt-induced fibrosis is characterized by the accumulation of large numbers of macrophages and giant cells in the alveolar spaces with diffuse interstitial fibrosis. Lymphoid infiltrates are not prominent and the role of T cells is uncertain. In this regard, the observation that exposed individuals develop either fibrosis or asthma might negate a role of an immune reaction to cobalt in pulmonary fibrosis. A study evaluating HLA-A, -B, -C and -DR genes in 853 hard-metal workers, 39 of whom had skin sensitization to cobalt, found no association with any of the genes tested (Fisher et al., 1984). Another study evaluated DQA, DQB, DRB, DPA and DPB polymorphisms in a group of patients including 38-cobalt sensitive, 26 chromium-sensitive and 70 nickel-sensitive individuals (Emtestam et al., 1993) and found no associations. With regard to cobalt-induced lung fibrosis in hard-metal workers, Potolicchio et al. (1997) showed that the disease is very strongly associated with the HLA-DPB1 gene, as all patients in the study expressed allelic variants of HLA-DP coding for the lysine-to-glutamic acid substitution at position 69 of the β chain (Table 18.3), a finding even more impressive given the disparate ethnic origin of the study subjects. Such a finding calls into question the current concept of disease pathogenesis, since such a strong association would imply a direct role of HLA-DP in the interaction with the offending agent. Alternatively, as a recent phenotypic study based on TNF localization in the affected tissue has claimed that hard-metal lung disease may be associated with exaggerated TNFα production (Rolfe et al., 1992), it is important to consider that HLA-DP is in tight linkage disequilibrium with a nearby gene regulating the fibrogenic response.

Asbestosis

Asbestos is a fibrous material widely used in industry because of its insulating properties

TABLE 18.3 HLA association with fibrotic lung disorders[a]

	HLA allele associated with		
Disease	Susceptibility	Protection	Other associated features
HMLD (cobalt)	DPβ1Glu^{69}		
Asbestosis	B27, DR5, DR53, DQ2		
	B12 (1)		Severe fibrosis (1)
CWP, silicosis	A19, B21, B29		CWP
	A1, B8, B21, B45		Complicated
	B18, DR8		PMF
	B44		Late onset
	B54, DR4, DR53	DR2, DR52	RA+, Japanese
IPF	B15-DR6, DR2		Diffuse pattern

[a]Data sources are quoted in the text.
HMLD: hard-metal lung disease; CWP: coal worker's pneumoconiosis; IPF: idiopathic pulmonary fibrosis; PMF: progressive massive fibrosis; RA: rheumatoid arthritis.

against heat and cold, incombustibility, great tensile strength and flexibility. Exposure to asbestos is the cause of pulmonary fibrosis in 4–15% of individuals with consistent exposure, with pleural fibrosis, pleural mesothelioma and lung cancer occurring in 3–50% of exposed workers.

Although there is no demonstrable lymphocyte reaction against asbestos, asbestos-exposed workers have been reported to have altered immune reactions characterized by an increased prevalence of rheumatoid factor and antinuclear antibodies, and elevated levels of circulating immunoglobulins and IgA immune complexes (Zone and Rom, 1985).

A pilot study on asbestos-exposed workers with or without asbestosis found an association between the risk of asbestosis and HLA-B27 (Merchant et al., 1975). The finding was felt to be important since the same antigen is strongly associated with ankylosing spondilitis, an autoimmune disorder frequently complicated by pulmonary fibrosis (Ferdoutsis et al., 1995). The association between HLA-B27 and asbestosis was not confirmed in a study of 37 patients and 37 matched exposed controls (Evans et al., 1977). Evans et al. found no association even when they combined their data with those of Matej et al. (1977). Evans et al. found instead that subjects with the HLA-B12 allele had more severe fibrosis. With the background that the B12 antigen has been associated with cryptogenic fibrosing alveolitis, a disease similar to pulmonary asbestosis in the absence of asbestos exposure, they suggested that a gene in positive linkage disequilibrium with HLA-B12 could be responsible for the weak association observed. In contrast, Huuskonen et al. (1979) found a protective effect of the HLA-A18 and HLA-B27 alleles in a small population of asbestos-exposed workers with and without asbestosis. Darke et al. (1979), in a study of 78 diseased and 92 unaffected asbestos workers, found an increase in the frequency of the B27 antigen, although this was not statistically significant. Begin et al. (1987) found no association with class I or class II genes in a study of 72 French Canadian asbestos workers, 40 of whom had disease. In a small study, Shih et al. (1993) evaluated HLA-DR and -DQ genes in asbestos-induced pulmonary fibrosis and found an association with HLA-DR53 and HLA-DQ2. In contrast, Al Jarad et al. (1992), in a study of 99 asbestos workers, did not find any association with class I HLA-A and HLA-B or with class II HLA-DR and HLA-DQ genes, with the exception of a weak negative association of HLA-DR5 when their data were combined with those of Begin et al. (1987).

Based on a meta-analysis of the literature, Al Jarad et al. (1992) attributed the weak associations found in the previous studies to the lack of stringent analytical criteria. They were thought more likely to be type I errors due to the high number of comparisons rather than type II errors, or missed associations due to the small size of samples. Thus, even considering that diagnostic standards for asbestosis, i.e. positive chest X-rays and bibasilar rales at auscultation, are insensitive and lead to an underestimation of the prevalence of disease (as shown using the more sensitive high-resolution computed tomography), a role for HLA genes in susceptibility to asbestos-induced pulmonary and pleural fibrosis is far from established.

Coal worker's pneumoconiosis and silicosis

Coal mining has been associated with a number of fibrotic lung disorders of varying severity, including (i) chronic bronchitis and emphysema, (ii) uncomplicated coal worker's pneumoconiosis, (iii) massive pulmonary fibrosis and rheumatoid pneumoconiosis (or Caplan's syndrome), and (iv) silicosis. Uncomplicated pneumoconiosis, affecting 20% of miners, is the most prevalent, while silicosis (12%) and progressive pulmonary fibrosis (5%) are less common. It has been proposed that uncomplicated pneumoconiosis develops when the lungs' capacity to eliminate coal dusts is overwhelmed. Progressive pulmonary fibrosis, which may develop after cessation of exposure, is often associated with immunological abnormalities and severe lung damage and disability. Exposure to silica in a variety of occupations, including sandblasting and grinding, mining, quarrying, tunnelling and foundry work, is the cause of "classic" silicosis, a fibrotic lung condition similar to coal worker's pneumoconiosis, but characterized by a typical silica-induced lesion called the silicotic nodule. Silica toxicity is, at least in part, due to the ability of this material to trigger oxygen radical reactions, which can damage cellular structures.

Individual susceptibility to coal worker's pneumoconiosis has been suggested by (i) the lack of a uniform relationship between exposure dose and disease incidence and (ii) the association of pneumoconiosis with immunological abnormalities, such as an increased prevalence of antinuclear antibodies and rheumatoid factor, as well as increased levels of immunoglobulins and immune complexes. Heise et al. (1979) found no strong association between progressive massive fibrosis, coal worker's pneumoconiosis and HLA-A and HLA-B genes in a study of 358 northeastern American miners. In contrast, Wagner and Darke (1979), in a study of over 400 Welsh coal workers with or without pneumoconiosis, found an association between HLA-B21 and pneumoconiosis and between HLA-B45 and complicated pneumoconiosis. In a follow-up study, the same authors confirmed the presence of HLA-B45 in complicated pneumoconiosis with rheumatoid factor. They also found a weak association between HLA-A1 and -B8 with complicated silicosis without rheumatoid factor (Darke et al., 1983). Koskinen et al. (1983) found an increased frequency of the HLA-A19 allele among Finnish silicosis patients compared with silica-exposed unaffected subjects. They were able to demonstrate an increased frequency of the haplotype HLA-A19–B18 among those developing progressive fibrosis, suggesting a role for this haplotype in the pathogenesis of lung fibrosis. Kreiss et al. (1989) found a statistically significant increase in the frequency of the HLA-A29 and -B44 alleles among 59 western American coal miners with silicosis compared with controls. They also found that B44-positive subjects were older at the time of diagnosis and were less symptomatic than other subjects, while A29-positive subjects were more likely to have pneumoconiosis-associated immune abnormalities such as high levels of IgG and immune complexes.

While Kreiss et al. (1989) found no association with HLA-DR and HLA-DQ, Rihs et al. (1994) found an increased frequency of HLA-DR8 in German miners developing silicosis with a rapid disease course, while the frequency of HLA-DR1 was increased and that of -DR52 decreased in those not developing silicosis. Consistent with the notion that silicosis patients have more severe disease in the presence of RA or of rheumatoid factor (Sluis-Cremer et al., 1986), Honda et al. (1988) reported similarly increased frequencies of the HLA antigens B54, DR4 and DR53 in silicosis and in RA patients. The same authors were later able to extend and confirm their observation in the same population of 46 Japanese silicosis patients, showing that the frequencies of HLA-B54, -DR4 (Dw15), -DR53 and -DQ4, were elevated. No associations were found with the alleles of the HLA-A or HLA-DP genes (Honda et al., 1993). Overall, the data suggest that HLA genes are variably associated in different geographical groups with disease presentation rather than disease per se.

Do HLA genes play a role in susceptibility to coal worker's pneumoconiosis and silicosis? In a critique of the published literature, Sluis-Cremes and Maier (1984) pointed to the inconsistency of the associations found in different populations and the weakness of statistical results obtained to conclude, consistent with the notion that the basic mechanism of silica toxicity is independent of T cell recognition and reaction, that it was unlikely that one or more HLA genes were involved in disease susceptibility. However, these weak associations may suggest linkage disequilibrium with other genes in the HLA region. As Honda et al. (1988) suggested, this gene is likely to be in positive linkage disequilibrium with HLA-B54–DR4–DR53 and in negative linkage disequilibrium with HLA-B52–DR2. Zhai et al. (1997) presented preliminary evidence on 13 pneumoconiotic miners suggesting that the TNF2 allele of the TNFα gene is associated with susceptibility to pneumoconiosis, thus making the TNFα gene the leading silicosis susceptibility candidate gene.

FIBROTIC LUNG DISORDERS OF UNKNOWN ORIGIN

Pulmonary fibrosis

Pulmonary fibrosis, also called cryptogenic fibrosing alveolitis (CFA) or idiopathic pulmonary fibrosis (IPF), is a chronic disorder of the lung characterized by the accumulation of inflammatory cells in the lower respiratory tract: macrophages accumulate in the alveolar spaces and the interstitium of the lung, accompanied by lymphocytes and polymorphonuclear leucocytes. Exaggerated amounts of oxygen toxic compounds are released in the alveolar spaces, causing tissue damage. Exuberant fibroblast proliferation and collagen production, smooth-muscle cell hyperplasia and squamous metaplasia of the bronchial epithelium ensue, resulting in distortion of the lung architecture and derangement of gas exchange. Several lines of evidence implicate a genetic component in susceptibility to pulmonary fibrosis. Pulmonary fibrosis is a frequent complication of a number of autoimmune disorders, including ankylosing spondilitis, a disorder closely associated with HLA-B27 (Hillerdal et al., 1983). A familial form of idiopathic pulmonary fibrosis has been described and pulmonary fibrosis is often observed in the course of the Hermanski–Pudlack syndrome, a genetic disorder dominated by albinism and blood-cell maturation abnormalities.

The association between pulmonary fibrosis and HLA genes has been examined in several studies. Turton et al. (1978) found a significant association between CFA and the HLA-B8 antigen in patients presenting with the disease at a young age. Varpela et al. (1979) found a higher prevalence of HLA-B15 and HLA-Dw6 in patients with cryptogenic fibrosing alveolitis. Neither finding was confirmed by Fulmer et al. (1978) in 33 American patients with IPF. In contrast, Libby et al. (1983) identified a strong association between HLA-DR2 and IPF in 20 patients. They also found a weaker association with HLA-B7 and -A3, which was attributed to linkage disequilibrium with HLA-DR2 in Caucasian patients. The association was particularly strong in those with the diffuse form of IPF.

CONCLUSION

In summary, HLA associations have been clearly demonstrated for hypersensitivity and allergic disorders caused by specific antigen or haptens such as berylliosis, gold hypersensitivity, certain forms of HP, occupational asthma and respiratory allergy to specific antigens. A role for HLA genes has not been convincingly shown in fibrotic lung disorders. However, the current evidence leads one to hypothesize on the presence of non-HLA susceptibility gene(s) located in chromosome 6. TNFα is a promising candidate. As the lung is continuously exposed to airborne agents, and since distinct HLA alleles are associated with sensitization to specific respiratory antigens, it is possible that HLA markers may be used to assess more accurately the risk associated with environmental and occupational respiratory hazards and to improve preventive measures.

ACKNOWLEDGEMENTS

Cesare Saltini was supported in part by grants from the Department of Energy of the United States of America, contract no. DE-FG02-93ER61714: "Structural and functional basis of genetic susceptibility to chronic beryllium disease".

REFERENCES

Abe, S., Yamaguchi, E., Makimura, S. et al. (1987). Association of HLA-DR with sarcoidosis. Correlation with clinical course. *Chest* **92**: 488–490.

Al Jarad, N., Uthayakumar, S., Buckland, E.J. et al. (1992). The histocompatibility antigen in asbestos related disease. *Br. J. Ind. Med.* **49**: 826–831.

Allen, D.H., Basten, A., Woolcock, A.J. and Guinan, J. (1977). HLA and bird breeder's hypersensitivity pneumonitis. *Monogr. Allergy* **11**: 45–54.

Amelung, P.J., Panhuysen, C.I.M., Postma, D.S. et al. (1992). Atopy and bronchial hyperresponsiveness: exclusion of linkage to markers on chromosome 11q and 6p. *Clin. Exp. Allergy* **22**: 1077–1084.

Ando, M., Hirayama, K., Soda, K. et al. (1989). HLA-DQw3 in Japanese summer-type hypersensitivity pneumonitis induced by *Trichosporon cutaneum*. *Am. Rev. Respir. Dis.* **140**: 948–950.

Ansari, A.A., Friedhoff, L.R. and Marsh, D.G. (1989). Molecular genetics of human immune responsiveness to Lolium perenne (rye) allergen Lol pIII. *Int. Arch. Allergy Appl. Immunol.* **88**: 164–169.

Apostolakis, J., Toumbis, M., Kostantinopoulos, K et al. (1996). HLA and asthma in Greeks. *Respir. Med.* **90**: 201–204.

Aron, Y., Swierczewski, E. and Lockhart, A. (1995). HLA class II haplotypes in atopic asthmatic and non-atopic control subjects. *Clin. Exp. Allergy* **25**: 65–67.

Balboni, A., Baricordi, O.R., Fabbri, L.M. et al. (1996). Association between toluene diisocyanate-induced asthma and DQB1 markers: a possible role for aspartic acid at position 57. *Eur. Respir. J.* **9**: 207–210.

Begin, R., Menard, H., Decarie, F. and St Sauveur, A. (1987). Immunogenetic factors as determinants of asbestosis. *Lung* **165**: 159–163.

Berrill, W.T. and van Rood, J.J. (1977). HLA-DW6 and avian hypersensitivity [Letter]. *Lancet* **ii**: 248–249.

Bignon, J.S, Aron, Y., Ju, L.Y. et al. (1994). HLA class II alleles in isocyanate-induced asthma. *Am. J. Respir. Crit. Care Med.* **149**: 71–75.

Bernstein, J.A., Munson, J., Lummus, Z.L. et al. (1997). T-cell receptor Vb gene segment expression in diisocyanate-induced occupational asthma. *J. Allergy Clin. Immunol.* **99**: 254–250.

Blumenthal, M., Marcus-Bagley, D., Awdeh, Z. et al. (1992). HLA-DR2, [HLA-B7, SC31, DR2], and [HLA-B8, SC01, DR3] haplotypes distinguish subjects with asthma from those with rhinitis only in ragweed pollen allergy. *J. Immunol.* **148**: 411–416.

Bost, T.W., Riches, D.W.H., Schumacher, B. et al. (1994). Alveolar macrophages from patients with beryllium disease and sarcoidosis express increased levels of mRNA for tumor necrosis factor-a and interleukin-6 but not interleukin-1b. *Am. J. Respir. Cell Mol. Biol.* **10**: 506–513.

Brooks, S.M. (1992). Occupational and environmental asthma. In *Environmental and Occupational Medicine* (ed. W.N. Rom), pp. 393–446, Little Brown, Boston, MA.

Darke, C., Wagner, M.M. and McMillan, G.H. (1979). HLA-A and B antigen frequencies in an asbestos exposed population with normal and abnormal chest radiographs. *Tissue Antigens* **13**: 228–232.

Darke, C., Wagner, M.M., Nuki, G. and Dyer, P.A. (1983). HLA-A, B and DR antigens and properdin factor B allotypes in Caplan's syndrome. *Br. J. Dis. Chest* **77**: 235–242.

Deng, G.Y., Maclaren, N.K., Huang, H.S. et al. (1996). No primary association between the 308 polymorophism in the tumor necrosis factor a promoter region and insulin dependent diabetes mellitus. *Hum. Immunol.* **45**: 137–142.

Churg, A.M. and Green, F.H.Y. (1995). In *Occupational Lung Disease* (eds W.M. Thurlbeck and A.M. Churg), pp. 851–930, Thieme, New York.

Emtestam, L., Zetterquist, H. and Olerup, O. (1993). HLA-DR, -DQ and -DP alleles in nickel, chromium, and/or cobalt-sensitive individuals: genomic analysis based on restriction fragment length polymorphisms. *J. Invest. Dermatol.* **100**: 271–274.

Evans, C.C., Lewinsohn, H.C. and Evans, J.M. (1977). Frequency of HLA antigen in asbestos workers with and without pulmonary fibrosis. *BMJ* **i**: 603–605.

Ferdoutsis, M., Bouros, D., Meletis, G. et al. (1995). Diffuse interstitial lung disease as an early manifestation of ankylosing spondylitis. *Respiration* **62**: 286–289.

Fisher, T., Rystedt, I. and Safwenberg, E.I. (1984). HLA-A, -B, -C and -DR antigens in individuals with sensitivity to cobalt. *Acta Derm. Venereol.* **64**: 121–124.

Fulmer, J.D., Sposovska, M.S., Gal, E.R. von et al. (1978). Distribution of HLA antigens in idiopathic pulmonary fibrosis. *Am. Rev. Respir. Dis.* **118**: 141–147.

Gardner, J., Kennedy, H.G., Hamblin, A. and Jones, E. (1984). HLA association in sarcoidosis: a study of two ethnic groups. *Thorax* **39**: 19–22.

Gossart, S., Cambon, C., Orfila, C. et al. (1996). Reactive oxygen intermediates as regulators of TNF-alpha production in rat lung inflammation induced by silica. *J. Immunol.* **156**: 1540–1548.

Grunewald, J., Janson, C.H., Eklund, A. et al. (1992). Restricted V alpha 2.3 gene usage by CD4$^+$ T lymphocytes in bronchoalveolar lavage fluid from sarcoidosis patients correlates with HLA-DR3. *Eur. J. Immunol.* **22**: 129–135.

Grunewald, J., Olerup, O., Persson, U. et al. (1994). T-cell receptor variable region gene usage by CD4$^+$ and CD8$^+$ T cells in bronchoalveolar lavage fluid and peripheral blood of sarcoidosis patients. *Proc. Natl Acad. Sci. USA* **91**: 4965–4969.

Grunewald, J., Shigematsu, M., Nagai, S. et al. (1995). T-cell receptor V gene expression in HLA-typed Japanese patients with pulmonary sarcoidosis. *Am. J. Respir. Crit. Care Med.* **151**: 151–156.

Hakala, M., Assendelft, A.H., Ilonen, J. et al. (1986). Association of different HLA antigens with various toxic effects of gold salts in rheumatoid arthritis. *Ann. Rheum. Dis.* **45**: 177–182.

Hedfors, E. and Lindstrom, F. (1983). HLA-B8 /DR3 in sarcoidosis. *Tissue Antigens* **22**: 200–203.

Heise, E.R., Mentnech, M.S., Olenchock, S.A. et al. (1979). HLA-A1 and coalworker pneumoconiosis. *Am. Rev. Respir. Dis.* **119**: 903–908.

Hillerdal, G. (1983). Ankylosing spondylitis lung disease – an underdiagnosed entity? *Eur. J. Respir. Dis.* **64**: 437–441.

Holloway, J.W., Doull. I., Begishvili, B. et al. (1996). Lack of evidence of a significant association between HLA-DR, DQ and DP genotypes and atopy in families with HDM allergy. *Clin. Exp. Allergy* **26**: 1142–1149.

Honda, K., Hirayama, K., Kikuchi, I. et al. (1988). HLA and silicosis in Japan. *N. Engl. J. Med.* **318**: 1610.

Honda, K., Kimura, A., Dong, R.P. et al. (1993). Immunogenetic analysis of silicosis in Japan. *Am. J. Respir. Cell Mol. Biol.* **8**: 106–111.

Hoshino, T., Suzuki, R., Tsuruta, Y. et al. (1997). Non-restricted T cell receptor (TCR)-V alpha and -V beta gene usage in patients with pulmonary sarcoidosis. *Clin. Exp. Immunol.* **108**: 529–538.

Huang, S.K. and Marsh, D.G. (1991). Human T-cell responses to ragweed allergens: A mb V homologues. *Immunology* **73**: 363–365.

Huuskonen, M.S., Tiilikainen, A. and Alanko, K. (1979). HLA-B18 antigens and protection from pulmonary fibrosis in asbestos workers. *Br. J. Dis. Chest* **73**: 253–259.

Ikeda, T., Hayashi, S., Kamikawaji, N. et al. (1992). Adverse effect of chronic tonsillitis on clinical course of sarcoidosis in relation to HLA distribution. *Chest* **101**: 758–762.

Ishihara, M., Ohno, S., Ishida, T. et al. (1994). Molecular genetic analysis of HLA class II genes in sarcoidosis. *Tissue Antigens* **43**: 238–241.

Ishihara, M., Ohno, S., Ishida, T. et al. (1995). Genetic polymorphism of the TNFB and HSP70 genes located in the human major histocompatibility complex in sarcoidosis. *Tissue Antigens* **46**: 59–62.

Khomenko, A.G., Pospelov, L.E., Malenko, A.F. et al. (1985). HLA antigens in lung diseases. *Ter. Arkh.* **57**: 77–80.

Koskinen, H., Tiilikainen, A. and Nordman, H. (1983). Increased prevalence of HLA-Aw19 and of the phenogroup Aw19,B18 in advanced silicosis. *Chest* **83**: 848–852.

Kreiss, K., Danilovs, J.A. and Newman, L.S. (1989). Histocompatibility antigens in a population based silicosis series. *Br. J. Ind. Med.* **46**: 364–369.

Kreiss, K., Miller, F., Newman, L.S. et al. (1994). Chronic beryllium disease – from the work place to cellular immunology, molecular immunogenetics, and back. *Cell Immunol. Immunopathol.* **71**: 123–129.

Li, P.K.T., Lai, C.K.W., Poon, A.S.Y. et al. (1995). Lack of association between HLA-DQ and -DR genotypes and asthma in southern Chinese patients. *Clin. Exp. Allergy* **25**: 323–331.

Libby, D.M., Gibofsky, A., Fotino, M. et al. (1983). Immunogenetic and clinical findings in idiopathic pulmonary fibrosis. Association with the B-cell alloantigen HLA-DR2. *Am. Rev. Respir. Dis.* **127**: 618–622.

Lombardi, G., Uren, J., Jones-Williams, W., Saltini, C. and Lechler, R. (1997). Molecular basis of HLA-DP associated susceptibility to beryllium disease. In *Genetic Diversity of HLA. Functional and Medical Implications* (ed. D. Charron), pp. 709–711, EDK, Paris.

Lympany, P.A., Welsh, K.I., Christie, P.E. et al. (1993). An analysis with sequence specific oligonucleotide probes of the association between aspirin-induced asthma and antigens of the HLA system. *J. Allergy Clin. Immunol.* **92**: 114–123.

Lympany, P.A., Petrek, M., Southcott, A.M. et al. (1996). HLA-DPB polymorphisms: Glu 69 association with sarcoidosis. *Eur. J. Immunogenet.* **23**: 353–359.

Martinetti, M., Tinelli, C., Kolek, V. et al. (1995). "The sarcoidosis map": a joint survey of clinical and immunogenetic findings in two European countries. *Am. J. Respir. Crit. Care Med.* **152**: 557–564.

Matej, H., Lange, A. and Smolik, R. (1977). HLA antigens in asbestosis. *Arch. Immunol. Ther. Exp. Warsz.* **25**: 489–491.

Merchant, J.A., Klouda, P.T., Soutar, C.A. et al. (1975). The HL-A system in asbestos workers. *BMJ* **i**: 189–191.

Muers, M.F., Faux, J.A., Ting, A. and Morris, P.J. (1982). HLA-A, B, C and HLA-DR antigens in extrinsic allergic alveolitis (budgerigar fancier's lung disease). *Clin. Allergy* **12**: 47–53.

Mullarkey, M.F., Thomas, P.S., Hansen, J.A. et al. (1986). Association of aspirin-sensitive asthna with HLA-DQw2. *Am. Rev. Respir. Dis.* **133**: 261–263.

Nowack, D. and Goebel, K.M. (1987). Genetic aspects of sarcoidosis. Class II histocompatibility antigens and a family study. *Arch. Intern. Med.* **147**: 481–483.

O'Brien, R.M., Thomas, W.R., Nicholson, I. et al. (1995). An immunogenetic analysis of the major house dust mite allergen Der p2; identification of high- and low-responder HLA-DQ alleles and localization of T-cell epitopes. *Immunology* **86**: 176–182.

Pappas, G.P. and Newman, L.S. Early pulmonary physiologic abnormalities in beryllium disease. *Am. Rev. Respir. Dis.* **148**: 661–666.

Partanen, J., Assendelft, A.H.W., Koskimies, S. et al. (1987). Patients with rheumatoid arthritis and gold-induced pneumonitis express two high-risk major histocompatibility complex patterns. *Chest* **92**: 277–281.

Potolicchio, I., Mosconi, G., Forni, A. et al. (1997). Susceptibility to hard metal lung disease is strongly associated with the presence of glutamate 69 in HLA-DPβ chain. *Eur. J. Immunol.* **27**: 2741–2743.

Rachelefsky, G., Park, M.S., Siegel, S. et al. (1976). Strong association between B-lymphocyte group-2 specificity and asthma. *Lancet* **ii**: 1042–1044.

Richeldi, L., Sorrentino, R. and Saltini, C. (1993). HLA-DPB1 glutamate 69: a genetic marker of beryllium disease. *Science* **262**: 242–244.

Richeldi, L., Losi, M., Bacchelli, B. et al. (1996). Role of two HLA class II and class III genetic polymorphisms in a population of northern Italian patients with pulmonary sarcoidosis. *Eur. Respir. J.* **9**: 171s.

Richeldi, L., Kreiss, K., Mroz, M.M. et al. (1997a). Interaction of genetic and exposure factors in the prevalence of chronic beruyllium disease among beryllium ceramic workers. *Am. J. Ind. Med.* **32**: 337–340.

Richeldi, L., Losi, M., Pelori, F. et al. (1997b). Role of the tumor necrosis factor α-308 polymorphism in pulmonary sarcoidosis. *Eur. Respir. J.* **10**: 98s.

Rihs, H.P., Lipps, P., May-Taube, K. et al. (1994). Immunogenetic studies on HLA-DR in German coal miners with and without coal worker's pneumoconiosis. *Lung* **172**: 347–354.

Rittner, C., Sennekamp, J., Mollenhauer, E. et al. (1983). Pigeon breeder's lung: association with HLA-DR 3. *Tissue Antigens* **21**: 374–379.

Rolfe, M.W., Paine, R., Davenport, R.B. and Strieter, R.M. (1992). Hard metal pneumoconiosis and the association with tumor necrosis factor-alpha. *Am. Rev. Respir. Dis.* **146**: 1600–1602.

Romagnoli, P., Spinas, G.A. and Sinigaglia, F. (1992). Gold-specific T-cells in rheumatoid arthritis patients treated with gold. *J. Clin. Invest.* **89**: 254–258.

Rossman, M.D., Yang, H.C., Murrray, R.K. et al. (1992). Chronic beryllium disease: an immunne response by restricted families of T-cells. *Am. Rev. Respir. Dis.* **145**: A415.

Saltini, C., Spurzem, J.R., Lee, J.L. et al. (1986). Spontaneous release of interleukin-2 by lung T-lymphocytes in active pulmonary sarcoidosis is primarily from the Leu3$^+$DR$^+$ T-cell subset. *J. Clin. Invest.* **77**: 1962–1970.

Saltini, C., Winestock, K., Kirby, M. et al. (1989). Maintenance of alveolitis in patients with chronic beryllium disease by beryllium-specific helper T-cells. *N. Engl. J. Med.* **320**: 1103–1109.

Saltini, C., Pinkston, P. and Crystal, R.G. (1990a). The specific immunity of chronic beryllium disease. In *Neutrophils, Lymphocytes and Lung* (eds M. Baggiolini, E. Pozzi, and G. Semenzato), pp. 249–260, Masson Italia, Milano.

Saltini, C., Kirby, M., Trapnell, B. et al. (1990b). Biased accumulation of T-lymphocytes with "memory"-type CD45 leukocyte common antigen gene expression on the epithelial surface of the human lung. *J. Exp. Med.* **171**: 1123–1140.

Saltini, C., Kirby, M., Bisetti, A. and Crystal, R.G. (1991).The lung epithelial immune system. *J. Immunol. Res.* **3**: 43–48.

Sandford, A, Weir, T. and Paré, P. (1996). The genetics of asthma. *Am. J. Respir. Crit. Care Med.* **153**: 1749–1765.

Selman, M., Teran, L., Mendoza, A. et al. (1987). Increase of HLA-DR7 in pigeon breeder's lung in a Mexican population. *Clin. Immunol. Immunopathol.* **44**: 63–70.

Sennekamp, J., Rittner, C., Vogel, F. and Tauberecht, I. (1978). Distribution of the HL-A antigens in patients with pigeon breeder's lung. *Schweiz. Med. Wochenschr.* **108**: 315–317.

Shih, J.F., Hunninghake, G.W., Goeken, N.E. et al. (1993). The relationship between HLA-A, B, DQ, and DR antigens and asbestos-induced lung disease. *Chest* **104**: 26–31.

Shinbara, M., Kondo, N., Agata, H. et al. (1995). T cell proliferation restricted by HLA class II molecules in patients with hen's egg allergy. *Exp. Clin. Immunogenet.* **12**: 103–110.

Sinigaglia, F. (1994). The molecular basis of metal recognition by T-cells. *J. Invest. Dermatol.* **102**: 398–401.

Sluis-Cremer, G. and Maier, G. (1984). HLA antigens of the A and B locus in relation to the development of silicosis. *Br. J. Ind. Med.* **41**: 417–418.

Sluis-Cremer, G., Hessel, P.A., Hnzido, E. and Churchill, A. (1986). Relationship between silicosis and rheumatoid arthritis. *Thorax* **41**: 596–601.

Smith, M.J., Turton, C.W., Mitchell, D.N. et al. (1981). Association of HLA B8 with spontaneous resolution in sarcoidosis. *Thorax* **36**: 296–298.

Stubbs, J., Argyris, E., Lee, C.W. et al. (1996). Genetic markers in beryllium hypersensitivity. *Chest* **109** (Suppl. 3): 45S.

Tinkle, S.S., Kittle, L.A., Schumacher, B.A. and Newman, L.S. (1997). Beryllium induces IL-2 and IFN-γ in berylliosis. *J. Immunol.* **158**: 518–526.

Turner, M.W., Brostoff, J., Wells, R.S. et al. (1977). HLA in eczema and hay fever. *Clin. Exp. Immunol.* **27**: 43–47.

Turton, C.W.G., Morris, L.M., Lawler, S.D. and Turner Warwick, M. (1978). HLA in criptogenic fibrosing alveolitis. *Lancet* **i**: 507–508.

Usui, Y., Kohsaka, H., Eishi, Y. et al. (1996). Shared amino acid motifs in T-cell receptor beta junctional regions of bronchoalveolar T cells in patients with pulmonary sarcoidosis. *Am. J. Respir. Crit. Care Med.* **154**: 50–56.

Van Neerven, R.J., Van-t'Hof, W., Ringrose, J.H. et al. (1993). T-cell epitopes of house dust mite major allergen Der pII. *J. Immunol.* **151**: 2326–3235.

Varpela, E., Tiilikainen, A., Varpela, M. and Tukiainen, P. (1979). High prevalences of HLA-B15 and HLA-Dw6 in patients with cryptogenic fibrosing alveolitis. *Tissue Antigens* **14**: 68–71.

Wagner, M.M. and Darke, C. (1979). HLA-A and B antigen frequencies in Welsh coalworkers with pneumoconiosis and Caplan's syndrome. *Tissue Antigens* **14**: 165–168.

Wahlstorm, J., Berlin, M., Lundgren, R. et al. (1997). Lung and blood T-cell receptor repertoire in extrinsic allergic alveolitis. *Eur. Respir. J.* **10**: 772–779.

Young, R.P., Barker, R.D., Pile, K.D. et al. (1995). The association of HLA-DR3 with specific IgE to inhaled acid anhydrides. *Am. J. Respir. Crit. Care Med.* **151**: 219–221.

Zhai, R., Franssen, H., Schins, R.P.F. and Borm, P.J.A. (1997). TNF-α polymorphisms in ex-coal workers with and without coal worker's pneumoconiosis. *Eur. Respir. J.* **10**: 234s.

Zone, I.J. and Rom, W.N. (1985). Circulating immune complexes in asbestos workers. *Environ. Res.* **37**: 383–389.

CHAPTER 19

HLA and Eye Disease

Jose S. Pulido and J. Wayne Streilein

IMMUNE PRIVILEGE AND THE EYE

Definition and description of immune privilege

The usual fate of organs or tissues transplanted from one individual to another is that the graft is destroyed. Normally, T cells of the recipient's immune system recognize antigens on the foreign transplant and mount an attack that leads to graft rejection. However, there are certain tissues and sites in the body where a different fate is observed. On the one hand, grafts placed in these special sites are often not rejected (or only very slowly); on the other hand, grafts prepared from these special tissues often succeed even when placed at conventional sites. Immune privilege was the term first advanced by Sir Peter Medawar in 1948 to identify this curious phenomenon. In truth, the phenomenon of immune privilege was discovered experimentally towards the end of the nineteenth century, when it was observed that foreign tumor cells placed in the anterior chamber of the eye of rabbits grew inexorably, whereas similar cells placed subcutaneously did not (Barker and Billingham, 1977). Until the 1940s, the reasons for the difference in tumor graft survival at these two sites were unknown. In describing the fundamental principles of transplantation immunobiology, Medawar and his colleagues demonstrated that rejection of skin grafts placed orthotopically (on the body wall) was mediated by lymphocytes of the recipient that recognized histocompatibility antigens on cells of the graft. Medawar was struck by the importance of lymphatic drainage pathways in graft-induced immunity. He believed that antigens from the graft escaped via lymphatics and were first detected when they reached the lymph node containing antigen-reactive lymphocytes. Medawar was aware that the eye was deficient in lymphatic drainage and he knew that the eye (like the brain) existed behind a blood–tissue barrier. He postulated, therefore, that immune privilege in the eye was predicated on the inability of graft-derived antigens to escape or immune effectors to gain access to intraocular compartments. In other words, foreign grafts placed in the anterior chamber were tolerated out of "immunologic ignorance".

This view was accepted until the mid-1970s, when three separate lines of evidence called into question the "immune ignorance" hypothesis. First, certain privileged sites were found to display readily detectable lymphatic drainage pathways: the testis and the brain are good examples (Head, 1983). In fact, even the eye possesses a potential route for lymphatic drainage, known as the uveo-scleral pathway. Secondly, antigenic material placed in privileged sites has now been detected in the blood and in distant organs (Niederkorn and Streilein, 1983; Wilbanks and Streilein, 1989). Thirdly, the systemic immune apparatus does detect and respond to antigens placed in privileged sites such as the eye. However, the nature of that response is deviant, giving rise to the phenomenon of anterior chamber-associated immune deviation (ACAID) (Streilein et al., 1980; Streilein, 1987; Niederkorn, 1990; Wilbanks and Streilein, 1990). In ACAID, the systemic immune response to ocular antigens is devoid of T cells that mediate delayed hypersensitivity or B cells that secrete complement-fixing antibodies. These three lines of evidence indicate that immune privilege results from active, rather than passive, mechanisms in which both systemic and local immune responses are modified and regulated (Streilein, 1995).

Mechanisms responsible for immune privilege

Within the eye, immune privilege has been defined for the cornea, the anterior chamber, the lens, the vitreous cavity and the subretinal space (Streilein, 1995). Outside the eye, immune privilege has been recognized for the brain, the testis and ovary, the liver, cartilage and certain tumors (Streilein, 1995). Organs and tissues that possess immune privilege display an array of special features that are thought to be important in the phenomenon. Some of the features are anatomical arrangements or constitutively expressed molecules that erect physical and/or chemical barriers to immune cells and molecules. These features can be considered to be "passive". Other features of privileged tissues and sites are more dynamic and actively contribute to creation of the privileged state by regulating one or another aspect of the immune response. These features can be referred to as "active". Table 19.1 summarizes the active and passive features now considered to be important in ocular immune privilege.

TABLE 19.1 Mechanisms responsible for ocular immune privilege

Passive	Active
Blood–tissue barrier	Cell-surface expression
Deficient lymphatics	of Fas ligand, CD59,
Tissue fluids drain	MCP, DAF
intravascularly	Soluble factors in
Reduced expression	immunosuppressive
of MHC class I	microenvironment:
and II molecules	TGFβ, α-MSH, VIP,
Deficit of professional	CGRP, MIF, IL-1
antigen-presenting	receptor antagonist,
cells	free cortisol

TGF: transforming growth factor; MSH: melanocyte stimulating hormone; VIP: vasoactive intestinal peptide; CGRP: calcitonin gene-related peptide; MIF: migration inhibition factor; IL: interleukin.

HLA expression and ocular immune privilege

It is axiomatic that T cells recognize antigens in the context of major histocompatibility complex (MHC) class I or class II molecules. Moreover, graft rejection is mediated primarily, although not exclusively, by T lymphocytes. Therefore, it is of no surprise that expression (or lack thereof) of class I and II molecules on cells of the eye can contribute to the existence of immune privilege. In the cornea, class I molecules are highly expressed on cells of the surface epithelium and on keratocytes (stromal fibroblasts), but class I expression on the corneal endothelium is significantly reduced (Wang et al., 1987; Abi-Hanna et al., 1988; Liversedge et al., 1988). Similarly, class I molecules are constitutively expressed at low levels by other ocular parenchymal cells: the epithelium of the iris and ciliary body, retinal pigment epithelium and cells of the neuronal retina. In addition, none of these cells expresses class II MHC molecules under normal conditions. When treated with interferon-γ (IFNγ), ocular parenchymal cells can upregulate both class I and class II expression, but the response is sluggish and requires heroic doses of the cytokine. Thus, the MHC antigenic "load" of ocular parenchymal cells is low.

As in other organs and tissues, bone marrow-derived cells of the dendritic cell and macrophage lineages infiltrate ocular tissues under normal conditions. Cells of this type have been described in the conjunctiva, in association with the epithelia and the stroma of the iris and ciliary body, in the space between retinal pigment epithelium and the choriocapillaries and, as microglia, within the neuronal retina (McMenamin, 1994; Thanos et al., 1996). While intraocular dendritic cells are routinely class I positive, a significant fraction fails to express class II MHC molecules constitutively. Upon treatment with granulocyte macrophage–colony-stimulating factor (GM-CSF) *in vitro*, or during intraocular inflammation *in vivo*, intraocular dendritic cells become intensely class II positive.

Unlike virtually any other tissue in the body, the cornea is unique in possessing no (or extremely few) bone marrow-derived dendritic cells (including Langerhans cells) (Streilein et al.,

1979). In all other tissues, dendritic cells and macrophages expressing high levels of MHC class I and II molecules function as mobile antigen-presenting cells (APC). In the special context of solid tissue allografts, cells of this type (referred to as passenger leucocytes) confer high levels of immunogenicity upon the graft (Steinmuller, 1967; Lafferty et al., 1983). In the cornea, the absence of passenger leucocytes has a profound effect on the tissue's immunogenicity as an allograft (see section on MHC expression and experimental models of corneal transplantation, p. 283).

The logic of immune privilege

All organs and tissues are susceptible to exogenous and endogenous pathogens for which immunity affords protection, but the range of pathogens that threatens one tissue (e.g. the skin) may be quite different from the range of pathogens that affects another (e.g. the gastrointestinal tract). As the immune system has coevolved with pathogens, it has been necessary to create diverse immune effector mechanisms to meet the challenge of diverse pathogens. The isotype of antibodies that is particularly useful on mucosal surfaces (IgA) is of little value in protecting against cutaneous pathogens. These considerations have led to the realization that the immune system is, in a sense, "balkanized" into regions; the term regional immunology refers to a branch of study devoted to understanding regional specializations in immune responses (Streilein, 1997). Effector T cells and antibodies procure pathogen elimination by direct binding and elimination, and by enlisting non-specific host-defense mechanisms. In the latter case, typically mediated by delayed-type hypersensitivity (DTH) or complement, there is considerable tissue injury and damage associated with the immune response. If this non-specific damage occurs amidst the cells of a vital organ, disease may result.

In this context, immune privilege represents the most extreme form of regional immunity. The eye is highly vulnerable to the ravages of immunopathogenic injury evoked by, for example, delayed hypersensitivity and complement activation, and immune privilege appears to be designed to limit or prevent the intraocular expression of immunity that is attended by intense inflammation. The delicate visual apparatus is exceedingly sensitive to minor adjustments, since an accurate image falls on the retina only when the apparatus is perfectly aligned. Inflammation disrupts the integrity of the visual axis and blindness is the inevitable result. In an evolutionary sense, immune privilege appears to have been created as a means by which the eye can receive certain forms of immune protection (cytotoxic T cells, non-complement fixing antibodies) that do not disrupt vision, while it suppresses the other immune effectors that potentially threaten sight. The classic description of immune privilege, in which a foreign graft survives beyond expectations, reflects the important role that DTH and complement fixing antibodies play in graft rejection. The central role that $CD4^+$ T cells play in fomenting delayed hypersensitivity and in promoting complement-fixing antibodies correlates with restricted expression of class II MHC molecules on ocular cells and indicates that regulation of MHC gene expression in the eye is likely to be important in the pathogenesis of ocular inflammatory diseases and transplantation. In a very real sense, the existence of ocular immune privilege protects the eye from inappropriate immune responses that may damage vision. When privilege is lost, the eye becomes vulnerable to immunogenic inflammation and vision suffers.

ANIMAL MODELS OF EYE DISEASES AND THE MHC

Experimental models of uveitis and the MHC

Experimental autoimmune uveitis

Extraocular introduction of several retinal antigens can cause uveitis in animal models. This includes recoverin, S-antigen, rhodopsin and interphotoreceptor-binding protein (IRBP). Recoverin and S-antigen, also called arrestin, are involved in phototransduction and can cause uveitis. Rhodopsin, which has a molecular weight of 40 000, is the photoreceptor molecule in rod cells. IRBP is a major protein in the interphotoreceptor space, is a glycoprotein of 1262 amino acids and has a molecular weight of 148 000. Phosducin has also been used to cause uveitis.

The clinical and histological findings following sensitization depend on the animal species and the dose of the uveitogenic challenge. Overall, however, this model is used to obtain a granulomatous uveitis similar to that seen in Vogt–Koyanagi–Harada's syndrome or sympathetic ophthalmia and has been termed experimental autoimmune uveitis (EAU). For instance, injection of IRBP with complete Freund's adjuvant systemically caused a granulomatous infiltration of the retina and choroid with lymphocytic infiltration of the vitreous and anterior chamber in mice (Nakamura, 1994). This reaction is most severe by day 12–14 following immunization (Konda et al., 1994).

EAU is caused by a DTH reaction to the retinal antigens and can be adaptively transferred to other animals by T helper lymphocytes. There is a role for the MHC because anti-Ia antibodies ameliorate the uveitis (Wetzig et al., 1988a; Rao et al., 1989). In addition, studies with different animal strains have shown that there is a strong role of genetics in the development of EAU. Two strains of congenic mice can have a markedly distinct rate of development of EAU (Nakamura, 1994). Caspi et al. (1992) showed that in a mouse model using IRBP, EAU developed more often in mice with the $H-2^k$ haplotype than other haplotypes. The I-a subregion seemed to be the most important MHC region in determining susceptibility. The final expression of EAU in the most susceptible MHC haplotype depended on the non-MHC background. Therefore, both non-MHC and MHC antigens are important in the development of EAU. A study using either the entire S-antigen or 18 amino acids which correspond to amino acids 303–320 of S-antigen and which are very immunogenic in a rat model showed that the non-MHC antigens were important in development of EAU in the absence of concurrent injection of *Bordetella pertussis* (Hirose et al., 1991). At this time, therefore, it is difficult to be sure of the relative importance of MHC and non-MHC antigens in EAU but it appears that both are important.

Other models of uveitis

Uveitis can be induced by the injection of lipopolysaccharide S (endotoxin) into the eye or by systemic injection. The uveitis is an anterior uveitis manifested ophthalmologically by iris hyperaemia, miosis, and flare and cells in the anterior chamber. Histopathologically, there is leucocytic infiltration of the anterior chamber, iris, ciliary body and vitreous, which is mainly neutrophilic. There can be some infiltration of the retina and choroid as well (Yang et al., 1997). The onset of disease is within 4 h of endotoxin injection and the neutrophilic infiltration is maximal at 18–24 h. T cells appear to be involved in this disease as well, but their exact role is as yet unknown (McMenamin and Crewe, 1995). At this time it appears that the lipopolysaccharide response gene, the presence of mast cells and the non-histocompatibility background are the important factors in developing endotoxin-induced uveitis (Li et al., 1995).

EAU can be induced by systemic injection of detergent-insoluble melanin-associated antigens isolated from bovine retinal pigment epithelium (RPE) or bovine uvea. The uveitis was first produced in the presence of complete Freund's adjuvant (CFA) but recently it has been reproduced in the absence of CFA (Broekhuyse et al., 1991; Bora et al., 1997). The onset of disease occurs in 12 weeks and there is marked infiltration of the iris and ciliary body with lymphocytes and neutrophils. There may also be mild infiltration of the choroid. The disease can be adaptively transferred by sensitized $CD4^+$ T cells.

Lens-induced uveitis can be induced by subcutaneous injection of xenogeneic lens protein with CFA (Augustin, 1994). This model causes a severe anterior uveitis with marked neutrophilic and lymphocytic infiltration of the anterior chamber, iris and ciliary body. It is an antigen–antibody mediated reaction but granulomata are also noted, consistent with a cell-mediated reaction. The role of T cells is not well elucidated.

Models of ocular infection and the MHC

Herpetic corneal disease and herpetic retinal disease are severe blinding diseases. There are differences in development of ocular disease among different strains of herpes simplex virus (HSV) but there also are differences in susceptibilty among different animal strains. At least in rats, genetic susceptibility to corneal herpetic disease appears to be dependent on non-MHC loci (Nicholls et al., 1994). Susceptibility to central nervous system (CNS) disease is also related to non-MHC loci. The role of MHC in herpetic retinitis is not known but it appears that CD4$^+$ T cells contribute to contralateral retinitis (Azumi and Atherton, 1994; Berra et al., 1994). This raises the possibility that the MHC is involved in disease. In addition, β_2 microglobulin-deficient mice, which cannot develop a CD8 response, still had a cytotoxic HSV-specific CD4$^+$ response (Niemialtowski et al., 1994).

MHC expression and experimental models of corneal transplantation

Transplantation of allogeneic cornea represents the most successful solid organ transplant conducted in humans (Council on Scientific Affairs, 1988). Similarly, orthotopic corneal transplants carried out in experimental animals often succeed, even in the absence of local or systemic immunosuppressive therapy (Maumanee, 1951; Katami et al., 1989; Sonoda and Streilein, 1992). These two statements dramatically reveal the immune privileged nature of the cornea as a graft. Moreover, they also make an important qualification of immune privilege: it is neither necessarily absolute nor permanent. Not only does a significant number of corneal transplants fail in experimental animals but, all too frequently, orthotopic corneal transplants fail in humans, especially when the recipient eye has suffered from an inflammatory or a scarring disease. Eyes with inflamed and/or scarred corneas in which there is a high probability of corneal allograft failure are termed high risk (Wilson and Kaufman, 1990; Mader and Stulting, 1991). The failure rate of grafts placed in these eyes exceeds 50%, even when very high doses of local and systemic immune suppression are used. Thus, ocular immunologists study orthotopic corneal transplants (i) in order to understand how immune privilege works, and (ii) to discover the factors that limit privilege and confer high-risk status on a corneal graft bed.

Several features of the cornea contribute to its immune privileged nature. First, the cornea lacks blood vessels and lymphatic channels. Secondly, the cornea (except at its limbic attachments) is devoid of bone marrow-derived cells that could function in antigen presentation. Thirdly, cells of the cornea secrete immunosuppressive factors: interleukin-1 (IL-1) receptor antagonist, transforming growth factor-β (TGFβ) and other factors yet to be described (Streilein et al., 1996). Fourthly, corneal endothelial cells (as well as other corneal cells) express Fas ligand constitutively (Griffith et al., 1995; Yamagami et al., 1997) and they express complement inhibitors in their surface membranes (Bora et al., 1993). Directly or indirectly, each of these features has been demonstrated experimentally to contribute to corneal allograft survival in normal eyes.

However, none of these features explains the most surprising aspect of orthotopic corneal transplantation in mice: minor histocompatibility antigens, rather than MHC-encoded antigens, offer the more formidable barrier to successful engraftment (Sonoda and Streilein, 1992). That is, corneal allografts that confront their recipients with multiple minor H incompatibilities are rejected more often and more promptly than allografts which display class I and/or class II MHC disparities in the absence of minor H incompatibility. This outcome is equally true whether the corneal allografts are placed in normal recipient eyes or in eyes rendered high-risk experimentally. The reasons why the traditional immunogenetic rules of transplantation immunology are inverted in the case of orthotopic corneal transplantation have now been elucidated, and they involve (i) the reduced MHC expression on corneal cells, and (ii) the absence of bone marrow-derived dendritic cells in the normal cornea.

While T cells recognize antigenic peptides in the context of MHC class I and II molecules, the T cells that recognize MHC alloantigens have two possible avenues of recognition available, termed the direct and indirect pathways of allorecognition (Lechler and Batchelor, 1982 a, b; Lee et al., 1994). First, one set of alloreactive T cells can recognize allogeneic class I or class II molecules directly, without the benefit of antigen processing. This is the direct pathway of antigen recognition; a remarkably high frequency of such T cells exists in the blood of normal individuals. The ability of passenger leucocytes to enhance the immunogenicity of solid tissue allografts is directly related to the ability of these mobile APC to migrate from the graft into the draining lymphatics of the recipient. When these cells reach the lymph node, they are recognized directly by T cells of this type. Secondly, another set of alloreactive T cells can recognize class I and II alloantigens when peptides cleaved from these molecules are processed and presented on recipient MHC molecules. In this case, because the alloantigens are being recognized on recipient, rather than donor, APC, the recognition is said to be indirect. While the indirect pathway was initially described for presentation of peptides derived from MHC alloantigens, this pathway is precisely the same as the pathway that enables peptides from minor H alloantigens to be presented to recipient T cells. Indeed, minor H antigens are presented exclusively by the indirect pathway.

Since the cornea lacks passenger leucocytes, recipient T cells are largely prevented from detecting donor alloantigens via the direct pathway. As a consequence, the recipient immune response is directed predominately at alloantigenic peptides that are prepared from the donor graft by infiltrating, recipient APC. Since constitutive expression of class I and II molecules on corneal cells is quite reduced compared with cells from other solid tissues, minor H alloantigens make a quantitatively greater contribution to the alloantigenic peptide pool that is processed from donor corneal cells by recipient APC. This appears to be the reason why minor H alloantigens offer a greater hurdle to corneal allograft acceptance than MHC-encoded antigens. The muted importance of MHC-encoded antigens in rejection of experimental corneal allografts may help to explain why it has been so difficult to prove that human leucocyte antigen (HLA) matching improves corneal transplants in humans (see *Corneal transplantation and HLA expression*, p. 284).

THE HLA LOCUS AND SPECIFIC OCULAR CONDITIONS

There are many diseases that affect the eye and have an association with specific products of the HLA loci. Some of these are associated with infectious causes and others appear to have an autoimmune origin. In addition to diseases that have an HLA association and affect only the eye, the eye can be secondarily involved in HLA-related diseases that affect other organ systems. The most common HLA-associated disease that affects the eye is diabetes mellitus (DM). Type I DM is an autoimmune disease that destroys β-cells of the pancreatic islets. The ensuing hyperglycemia induces microangiopathic changes, neovascularization and vitreous changes that ultimately cause diabetic retinopathy (Pulido, 1996). In this section, however, only primary ocular conditions will be discussed.

Disorders of the optic nerve and orbit

Optic neuritis

Optic neuritis is the most common inflammatory condition of the optic nerve. It can be caused by infections, including syphilis, Lyme's disease, *Bartonella henselae* (cat scratch disease), giant cell arteritis, other collagen vascular diseases and pharmacological agents. Most cases of optic neuritis are not associated with an infectious cause. These cases are often idiopathic or occur in patients with a history of multiple sclerosis (MS). Idiopathic cases are usually associated with decreased color perception and visual acuity. The affected patients have scotomata and may have pain with ocular movement. An afferent pupillary defect is usually present. The optic nerve is swollen and splinter hemorrhages may be present.

The Optic Neuritis Treatment Trial (ONTT) evaluated the treatment and subsequent development of MS in patients who had suffered an episode of optic neuritis (Beck et al., 1993a, b). Of the cases enrolled, 85% were Caucasian women with a median age of 31.8 years. Initially 49% had magnetic resonance imaging (MRI) evidence consistent with asymptomatic demyelinating disease. A Kaplan–Meier curve showed that over 3.5 years, about 25% of patients with optic neuritis would subsequently develop MS. The strongest predictor of developing MS in patients with optic neuritis was patients with two or more abnormal periventricular signals on MRI at the initial examination: 36% of patients who met that criterion developed clinically definite MS within 2 years. Patients who were treated with intravenous methylprednisolone for 3 days followed by a further 11 days of prednisone had a statistically lower rate of subsequent development of MS. The ONTT therefore showed a very strong association between idiopathic optic neuritis and MS.

HLA associations with MS have been examined in multiple studies. The results have been variable. Original studies showed an association with HLA-A3 and -B7. Subsequent studies showed that the associations with these two class I HLA loci resulted from linkage disequilibrium with HLA-DR2. This locus has been shown to be associated with MS among Caucasians but originally not in Asians. The DR/DQ haplotype associated with the HLA-DR2 locus has been shown by DNA-based techniques to be DRB1*1501–DQA1*0102–DQB1*0602 (Hogancamp et al., 1997). Possible mechanisms underlying the association include mimicry epitope spreading, the encoding of an antigen-recognition molecule which binds myelin-associated proteins, the failure to delete autoreactive T cells, or that the HLA alleles are merely markers of a closely linked gene within the region and that they are not actually associated with the disease themselves (Oksenberg, 1996). The molecular mimicry and epitope spreading concept is bolstered by a murine model of demyelinating disease seen after inoculation of mice with Theiler's murine encephalomyelitis virus (Miller et al., 1997).

Epplen (1997) showed that, in Germans, positivity for the appropriate HLA-DRB1 allele had a relative risk of 3.64. A recent result by Barcellos (1997) again demonstrated an association with HLA-DRB1*1501–DQA1*0102–DQB*0602 in 30.4% of cases with MS compared with 17% in controls ($p < 0.001$, odds ratio 2.1) in patients and controls from California.

Kira et al. (1996) segregated Japanese patients into two forms of MS, a Western type of MS which had disseminated CNS findings and an Asian type which had selective involvement of the optic nerve and spinal cord and a relapsing-remitting course. Examining HLA loci following this stratification, they showed that the HLA-DRB1*1501 allele was seen in 41% of Western-type MS, in only 14% of controls (relative risk = 3) and in no patients with Asian-type MS. Barcellos (1997) also showed that in a northern Chinese cohort there was a higher number of patients with the same DRB1 haplotype with MS than in the controls, but this was not statistically significant. These studies show that the DRB1*1501 locus is important but provides only a small contribution to the overall risk of developing MS.

Evaluation of the myelin basic protein (MBP) locus on chromosome 18 to date has not been associated with disease (Barcellos, 1997). MBP can be used in animal models to cause the development of experimental allergic encephalomyelitis (EAE). EAE is mediated by effector T cells that respond to myelin antigens. Another locus in the HLA class I region near the HLA-F locus encodes myelin oligodendrocyte glycoprotein (MOG). Injection of MOG has also been shown to cause EAE (Genain et al., 1996). Evaluation of an association with the MOG locus showed a barely statistically significant association ($p = 0.043$). This may be because of linkage disequilibrium between the MOG locus and DRB1*1501 (Barcellos, 1997). Further studies evaluating this locus are needed.

About 10% of patients with MS develop intermediate uveitis (see below). The patients may have inflammatory cells in the vitreous, cyclitis, retinal phlebitis and large aggregates of cells near the pars plana inferiorly (Bramford et al., 1987; Kerrison et al., 1994). The clinical appearance is similar to the inflammation seen in patients with pars planitis; however, a definite snowbank (fibrovascular tissue overlying the inferior pars plana of at least one eye) is rarely seen in patients that were initially diagnosed with MS. In some cases of EAE, iridocyclitis has been noted in animal studies (Bullington and Waksman, 1958).

Dysthyroid ophthalmopathy

Dysthyroid ophthalmopathy is characterized by ocular and orbital findings, usually in association with present or previous thyroid disease. Proptosis secondary to thyroid ophthalmopathy is the most common cause of both unilateral and bilateral proptosis. Graves' disease or Basedow's disease is the eponym for dysthyroid ophthalmopathy in association with hyperthroidism and diffuse goitre. The clinical features of Graves' disease include eyelid retraction, lagophthalmos, exophthalmos (proptosis), strabismus, and optic neuropathy. Other findings include dermopathy (pretibial myxedema) and acropachy.

Bartley (1996) characterized the clinical findings in 120 patients with dysthyroid ophthalmopathy. Women were affected in 86% of the cases. There were two peaks of incidence: one between the ages of 40 and 49 and another between 60 and 69 years. Eyelid retraction was noted in 91%, proptosis was seen in 62% and 6% had optic neuropathy. Extraocular muscle restriction was noted in 50%.

In this study, 90% of patients with dysthyroid ophthalmopathy had autoimmune hyperthroidism (Graves' disease), 1% had primary hypothyroidism, 3% had Hashimoto's thyroiditis and 6% were euthyroid. Fewer than 1% of patients had evidence of any other autoimmune disease. Interestingly, patients with other autoimmune diseases, e.g. myasthenia gravis, have a 5% incidence of having thyroid disease (Sahay et al., 1965).

Autoimmune hyperthyroidism is an autoimmune disease in which there are antibodies against the thyroid-stimulating hormone (TSH) receptor on thyroid follicular cells. This causes the cells to be stimulated to secrete thyroid hormone. Circulating T cells also react to antigens expressed on thyroid follicular cells (Heufelder, 1995).

Hashimoto's thyroiditis is an autoimmune disease in which T cells that react to thyroid follicular cells are present. These T cells react with Fas, which is expressed on the thyrocytes of patients with Hashimoto's thyroiditis (Giordano et al., 1997). FasL is constitutively expressed on thyrocytes and with expression of Fas on the thyrocytes of patients with Hashimoto's mutual destruction of thyrocytes can occur, obviating the need for cytotoxic T cells. Autoantibodies to thyroglobulin are also involved in the disease.

How or why the orbit and specifically the extraocular muscles are involved in dysthyroid ophthalmopathy is unknown. Histopathology shows infiltration of extraocular muscles by leucocytes and deposition of mucopolysaccharides, edema and fatty infiltration. There is still dispute as to whether the inflammation is antibody associated or cell mediated. Arnold et al. (1994) evaluated whether there was proliferation of circulating T cells from Graves' patients and from control normals to preparations of human and porcine extraocular muscle antigens. Only two out of 10 patients responded to extraocular muscle proteins in the 25–50 kDa range. They concluded that extraocular muscle antigens are unlikely candidate autoantigens to T cells. Other studies have shown that these patients have antibodies against antigens in the extraocular muscles. Some studies have shown antibodies against a 64 kDa antigen isolated from extraocular muscles which cross-react with 64 kDa proteins. Other cells including fibroblasts or adipocytes may be the source of autoantigens (Heufelder, 1995).

Determination of HLA alleles in Graves' disease would allow for determination of HLA-associated alleles with dysthyroid ophthalmopathy, since 90% of patients with dysthyroid ophthalmopathy have a history of Graves' disease. Studies showing a relationship with HLA loci have had variable results and those which have have demonstrated a low one. Baldini et al. (1995) showed that HLA-B8 was seen in 24% of Graves' patients compared with 12% of controls (odds ratio 1.98) and HLA-DR3 was seen in 31% of patients compared with 18% of controls (odds ratio 1.75). Others have not shown an HLA association (Tomer et al., 1997).

A murine model for autoimmune thyroiditis can be induced by injecting thyroglobulin, causing findings similar to those seen in Hashimoto's thyroiditis (Kong et al., 1996). Immunization of HLA-DRB1*0301 (DR3) transgenic mice with thyroglobulin-induced autoimmune thyroiditis, while injecting HLA-DRB1*1502 (DR2) transgenic mice with thyroglobulin did not cause autoimmune thyroiditis. This shows that at least for a possible animal model of Hashimoto's thyroiditis, the MHC is important in disease. The HLA associations of thyroid disease are discussed in Chapter 16.

Conjunctival and corneal disorders

Trachoma

It is estimated by the World Health Organization that 500 million people have trachoma and that five million are blind because of complications from this disease. Trachoma is caused by *Chlamy-*

dia trachomatis, a Gram-negative, intracellular bacterium. The elementary bodies are the extracellular infectious forms, while the reticular bodies are the intracellular form which generate the inclusion bodies seen by microscopy.

Trachoma is a chronic follicular keratoconjunctivitis. The chronic inflammation causes conjunctival scarring which can cause trichiasis and entropion. In addition, corneal vascularization occurs secondary to the chronic inflammation as well as from the secondary lid changes.

In a single community, there may be a marked difference between findings among families of similar socioeconomic backgrounds. This has raised the issue of whether there is a genetic predisposition for the more severe forms of the disease. In addition, there is some evidence that leprosy and tuberculosis, two other diseases caused by intracellular organisms, may also have an HLA predisposition. In a study evaluating patients with trachoma in Gambia there appeared to be an association with HLA-A28, specifically the HLA-A*6802 subtype (Conway et al., 1996). However, a similar study among the Omanis did not show an association with HLA-A28, but rather with HLA-DR2 (DR16), with a relative risk of 3.82 (White et al., 1997). Fewer patients appeared to have the HLA-DR53 locus (which is in linkage disequilibrium with DR4 and DR7) compared with controls. This was consistent with a possible protective effect. Considering that *Chlamydia* is an intracellular organism, it may be reasonable to expect that there is a relation with class II MHC loci, but further studies are needed to determine whether these relations hold among different ethnic groups.

Mooren's ulcer

Mooren's ulcer is a painful, relentless, chronic ulcerative keratitis that begins peripherally and progresses circumferentially and centripetally. Men are more commonly affected than women. It has been considered that there may be a more severe and painful bilateral form which occurs in younger, black males and another more benign and unilateral form which occurs in older, white patients. This epidemiological difference is under dispute.

Recent studies have shown a relationship with a chronic hepatitis C infection in some cases. Treatment with IFNα causes remission of ulceration in cases where hepatitis is documented (Moazami et al., 1995). Several studies have shown a relationship between HLA-DR3, -DR4 or -DR7 and hepatitis C-associated infections (Czaja, 1997; Jurado et al., 1997; Mann, 1997).

Molecular mimicry may be an important aspect of this disease. Gottsch et al. (1995) showed that patients with Mooren's ulcers have high antibody levels to a corneal antigen that is a protein of a molecular weight of 30 000. It is a tetramer consisting of monomer of 7000 monomers. Subsequent purification and sequencing of the corneal antigen revealed strong homology to calcium-binding proteins of the S-100 family (Liu and Gottsch, 1996). There may be some homology of a sequence of this protein to sequences of a helminthic or hepatitis C protein.

Cogan's syndrome

Cogan's syndrome is a rare disease that manifests as interstitial keratitis. Audiovestiblular dysfunction is common and manifests as hearing loss, tinnitus and vertigo. Approximately one-half have a history of a recent upper respiratory infection. Systemic manifestations may include neuropathy in 50%; other organ systems may be involved as well. An elevated erythrocyte sedimentation rate occurs in 50% of cases. Caseating granulomas can be seen by histopathology.

There have been reports of association of HLA-A9, -B17, -Bw35 and -Cw4 with Cogan's syndrome (Del Carpio et al., 1976; Kaiser-Kupfer et al., 1978), but the number of cases has been too small to confirm these findings.

Other conjunctival and corneal disorders associated with HLA loci

Rheumatoid arthritis (RA) can be associated with dry eyes, peripheral corneal ulcerations, scleritis, iritis and retinal vasculitis. There is an association of RA with HLA-DR4 and -DR1 (Auger et al., 1997). These share a similar amino acid sequence in positions 70–74 of the β_1 chain (QKRAA is the shared sequence). This association is also seen with Vogt–Koyanagi–Harada disease (see later section). The HLA-DRB*0401 allele may be found more often in patients with extra-articular manifestations (Perdriger et al., 1997). The authors did not, however, specifically evaluate ocular manifestations.

Cicatricial pemphigoid is usually recognized in patients in their seventh and eighth decade of life. It is a systemic disease which produces scarring of the affected surfaces. If the conjunctiva is involved, it can become foreshortened. There is an association between HLA-B12 or -DR4 and cicatricial pemphigoid.

Thygeson's superficial punctate keratopathy is a chronic epithelial keratitis which occurs in patients with a mean age of 29 years. Symptoms include photophobia and sensations of tearing or of a foreign body. The deposits appear as grouped punctate white-to-gray dot-like opacities. There may be an association with HLA-DR3 with a relative risk of 5.3. Further studies are needed to substantiate this (Darrell, 1981).

Corneal transplantation and HLA expression

The overall results of corneal transplantation in humans are among the best of any solid tissue transplants performed (Volker-Dieben et al., 1982; Council on Scientific Affairs, 1988; Price et al., 1991; Eye Bank, 1992; Stark et al., 1992). One-year survival rates of 89% are superior to recent experience with kidney, liver, heart and other organ transplants. As mentioned previously, allogeneic corneas grafted into high-risk eyes suffer a much lower success rate (35–65%). When grafts of this type are excluded, the success rate for corneas transplanted into low-risk human eyes exceeds 95%. Considering the fact that systemic immunosuppression is almost never used in this situation, the success of corneal transplants in uncomplicated eyes is simply unparalleled. With an acceptance rate that exceeds 95%, it is no surprise that HLA matching has been found to have no effect on graft outcome. However, one might have expected that corneas grafted into high-risk eyes would have benefited from HLA matching. To that end, several studies have reported that HLA matching improved graft outcome in the high-risk setting (Ehlers and Ahrons, 1971; Allansmith et al., 1974; Batchlelor et al., 1976; Stark et al., 1978). In general, these studies were conducted in Europe, among populations considerably less genetically heterogeneous than American and Australian populations. When similar, highly controlled studies have been performed in North America, HLA matching has not been found to improve graft survival in high-risk eyes (Stark et al., 1978). Recently, a study conducted in Europe came to the same conclusion (Vail et al., 1994). This study showed an increased risk of corneal allograft failure when donor and recipient displayed *no* mismatches with respect to HLA-DR antigens. Curiously, one American study which, upon meta-analysis, suggested that HLA-DR mismatching is associated with increased risks of graft failure (Gore et al., 1995), also reported that ABO mismatching of donor and recipient prejudiced corneal allograft survival.

It may be possible to make sense of this confusing set of findings by referring back to the recent studies conducted in mice. In the murine system, deficient MHC class I and II expression in the donor cornea, combined with lack of indigenous donor APC, results in recipient T cell responses directed predominantly against minor H antigens presented via the indirect allorecognition pathway. For mice, whether donor and recipient are matched at the rodent MHC (H-2) is largely irrelevant to graft outcome. Minor transplantation antigens are more important and MHC matching necessarily excludes these disparities. In the human system, disparity at ABO appears to be an important antigenic marker for corneal graft failure. In a sense, the alloenzymes (glycosyltransferases) that create the A and B isohemagglutinin antigens from the O antigen have the potential to be minor histocompatibility antigens. It has been proposed that when donor cornea expresses A or B, and the recipient lacks these antigens, peptides derived from the I_A or I_B glycosyltransferase are processed by recipient APC that infiltrate the graft, and these peptides (bound to recipient MHC molecules) are presented to the T cells that will eventually cause graft rejection (indirect pathway). If donor and recipient share one or more HLA class II alleles, then T cells sensitized in this fashion can find a ready target among cells of the cornea graft.

At present, HLA matching seems an unnecessary burden to be placed on orthotopic corneal grafting in human beings. There is good experimental evidence in mice to suggest that MHC-encoded antigens are not so important in the corneal graft setting. Reports of deleterious consequences of HLA class II matching between donor and recipient should bring considerable caution to this discussion. If and when matching for minor H antigens becomes more sophisticated, matching for these determinants may improve the survival of allografts placed in high-risk eyes. At the very least, matching for ABO compatibility is more likely to be cost-effective.

Disorders affecting the anterior segment and vitreous cavity

Acute anterior uveitis

Iritis or anterior uveitis is an inflammation of the anterior chamber of the eye. About 50% of cases of uveitis seen in a uveitis clinic are cases of anterior uveitis (Pivetti-Pezzi et al., 1996). Anterior uveitis can be divided into acute or chronic and granulomatous or non-granulomatous cases. About 55% of cases of acute anterior uveitis (AAU) are associated with an HLA-B27-positive serotype. About 70% of patients with recurrent episodes of acute anterior uveitis are HLA-B27 positive. HLA-B27-associated AAU is characterized by an abrupt onset and a fulminant course with a marked fibrinous response and a marked white cell response which can be so dense that a hypopyon (a layering of white cells in the anterior chamber) can occur.

About 1% of people who are HLA-B27 positive develop AAU (Linssen et al., 1991). The relative risk of developing AAU in an HLA-B27 patient is about 8 (Derhaag, 1988a, b). This compares with a relative risk of 70 for ankylosing spondylitis and 26 for Reiter's syndrome. Knowing that a patient with AAU is HLA-B27 positive is important since 84% of these patients have other associated HLA-B27 diseases: specifically, Reiter's syndrome, ankylosing spondylitis or psoriatic arthritis. Conversely, about 20–30% of patients with ankylosing spondylitis or Reiter's syndrome develop AAU (Rosenbaum, 1992).

There are probably genetic factors other than HLA-B27 in the development of AAU. AAU is more common in males than females and the prevalence of AAU in first degree relatives of HLA-B27-positive patients is 13%.

How the HLA-B27 genotype predisposes to AAU is unknown. Speculation includes molecular mimicry with certain Gram-negative bacteria including *Yersinia* and *Klebsiella*, or *Chlamydia*.

The development of transgenic HLA-B27-positive rodents that also express human β_2 microglobulin is proving helpful in elucidating the pathogenesis of the spondyloarthropathy. Rats with high gene copies spontaneously develop joint and intestinal inflammation. If they are kept in a germ-free environment, they do not develop inflammation. Attempts at causing AAU in these rats have not been very successful to date.

In one study, the transgenic mice were treated with subcutaneous intraretinal binding protein (IRBP) in complete Freund's adjuvant (Wilbanks et al., 1997). (IRBP injection is known to cause EAU in susceptible animals.) In the control non-transgenic mice, 50% developed EAU while only 20% of B27-positive animals developed uveitis.

In a study by Baggia et al. (1997), control rats and transgenic rats were inoculated with live *Salmonella enteritidis* and *Yersinia enterocolitica* intravenously or by gavage. At a dose of 10^5 colony-forming units, *Salmonella* given by gavage caused AAU in 17% of B27-positive transgenics compared with 50% of control animals. If the bacteria were administered intravenously, all non-transgenic animals developed AAU, compared with 80% of B27-positive animals. This study confirmed that systemic Gram-negative infections can cause AAU but did not establish the relationship between Gram-negative infections and HLA-B27-associated AAU. The mechanism for development of AAU in association with HLA-B27 is still unknown. Whether it is a direct consequence of the antigen-presenting function of the HLA-B27 or whether it is induced by HLA class II presentation of a B27-derived peptide remains unclear (Parham, 1996).

Uveitis associated with juvenile rheumatoid arthritis

Juvenile rheumatoid arthritis (JRA; also called juvenile chronic arthritis) occurs in three clinical forms, a systemic, a polyarticular and a pauciarticular form. JRA is defined as arthritis of at least 3 months' duration in a child younger than 16 years after other potential causes of arthritis have been excluded. The exact incidence of uveitis in patients with JRA is unknown (Kanski, 1990).

The systemic form is associated with fever, hepatosplenomegaly, lymphadenopathy and symmetrical polyarticular arthritis. It is rarely associated with antinuclear antibodies (ANA) or uveitis. The polyarticular form affects five or more joints, has only mild constitutional systemic findings and can involve ANA in 25% of cases. Over 75% of patients with the polyarticular form are girls. This form is associated with about a 5% incidence of uveitis.

The pauciarticular form occurs predominantly in girls. It is seen in at least 50% of children with JRA. ANA occurs in at least 75% of the patients and is seen very often, possibly in up to 100% of patients with uveitis (Malleson et al., 1992). The uveitis is a chronic insidious form of iridocyclitis and occurs in about 20% of cases. It rarely causes symptoms or conjunctival redness. Band keratopathy, cataracts, glaucoma and cystoid macular edema are common complications in untreated cases. These cases then progress to phthisis. Early diagnosis and aggressive treatment involving steroids and in some cases methotrexate are critical to avoid blindness.

There is a strong relationship between certain HLA alleles and uveitis in cases of JRA (Kanski, 1990; Fiannini et al., 1991). Children with the highest risk have HLA-DR5 and -DP2 but are HLA-DR1 negative. The risk is approximately 75% after 5 years of having arthritis. Those who are HLA-DR5 negative, -DP2 negative and -DR1 positive are at the lowest risk for development of uveitis, at around 6–7%.

Inflammatory bowel disease

Inflammatory bowel disease is also associated with anterior uveitis. Ocular symptoms of inflammatory bowel disease usually manifest as subacute or chronic iridocyclitis or scleritis. It is more common with ulcerative colitis than with Crohn's disease. There have been conflicting studies showing possible associations between inflammatory bowel disease and HLA loci.

Studies from Japan have shown a possible association with the HLA-DR2 allele and specifically the DRB*1502 subtype (Asakura et al., 1982; Toyoda, 1993). Others have not found an association with HLA-DR2 in other populations (Satsangi et al., 1996). A recent study has shown an association between DRB1*0103 and extraintestinal manifestations (Roussomoustakaki et al., 1997). Patients with ulcerative colitis and uveitis were positive for this allele in 35.7% of cases, while it was present in 3.2% of controls ($p < 0.0001$).

An association with antineutrophil cytoplasmic antibodies (ANCA) has been shown to exist in patients with ulcerative colitis. The association is stronger with perinuclear localization of ANCA (pANCA) positivity. This association has been shown to be present in 50–75% of patients with ulcerative colitis (Roussomoustakaki et al., 1997).

Fuchs' heterochromic iridocyclitis

Most cases of Fuchs' heterochromic iridocyclitis are diagnosed in patients in their third or fourth decade of life (Schwab, 1990). The disease is more often recognized in lighter colored irides. The affected patients have a low-grade chronic inflammation. The patients may present with reduced vision secondary to posterior subcapsular cataracts. Occasionally, they present because of the heterochromia or from a spontaneous hyphema.

Heterochromia is a hallmark of this disease. There may also be fine radial or neovascular vessels. Glaucoma occurs in 20–60% of patients. Small stellate keratic precipitates may be present and these are distributed uniformly throughout the entire cornea. A small number of anterior chamber and vitreous cells is present.

There is some evidence of genetic clustering of this disease. There has been a report of Fuchs' in identical twins (Makley). An epidemiological study comparing Belgian patients with Fuchs' heterochromic iridocyclitis to a pool of Belgian blood donors showed that there was a decreased frequency of HLA-DRw53 in the patients with Fuchs' iridocyclitis ($p = 0.03$) (De Bruyere et al., 1986). This has not been subsequently confirmed.

Intermediate uveitis

Intermediate uveitis is manifested by infiltration of the vitreous by inflammatory cells. The patients complain of decreased vision, usually due to cystoid macular edema, or of floaters. Rarely, they can develop vitreous hemorrhages or retinal detachments. Intermediate uveitis is a clinical descriptive diagnosis and can be caused by many diseases. Among the diseases to be considered are sarcoidosis, syphilis, tuberculosis,

Lyme's disease, Whipple's disease, *Toxocara canis* and non-Hodgkin's lymphoma.

One particular form of intermediate uveitis has been described which usually presents in patients in their second or third decade of life. It is usually bilateral and the patients often have a dense accumulation of white cells in the vitreous, retinal vasculitis, which is usually a phlebitis, and fibrovascular tissue overlying the inferior pars plana of at least one eye (snowbank). This has been termed pars planitis or idiopathic intermediate uveitis. Separating this form of intermediate uveitis from secondary causes, especially sarcoid, has been difficult because there is no definitive test for either diagnosis (Brockhurst, 1960). Indeed, some of the cases of pars planitis that were described by Brockhurst had fundoscopic findings of peripheral punched-out lesions that would now be ascribed to presumed sarcoidosis or multifocal choroiditis and panuveitis instead of to intermediate uveitis.

Malinowski et al. (1993a) evaluated the long-term outcomes in patients with pars planitis. Overall, the visual acuity at presentation was about the same after 7.5 years. Cataracts requiring surgery and retinal detachments occurred in 15% and 8% of patients, respectively. Of particular interest was that optic neuritis or MS developed in 20% of patients after 5 years and in 30% after 10 years. This showed that there was a strong relation between the diagnosis of pars planitis and MS. There was a family history of MS in 11% of patients. As noted in the Optic neuritis section (p. 284), a form of intermediate uveitis has been seen in 10% of patients who had an initial diagnosis of MS (Bramford et al., 1987; Kerrison et al., 1994).

Studies in the 1980s had shown a possible genetic predisposition to intermediate uveitis/pars planitis in certain cases where several family members had developed the disease (Augsburger et al., 1981; Culbertson et al., 1983; Wetzig et al., 1988b). Attempts at finding a relationship with MHC class I loci were unsuccessful (Ohno et al., 1977). Some of the older studies may not have separated the different etiologies or there may be differences in genetic predisposition among different ethnic groups (Arocker-Mettinger et al., 1992). The study by Arocker-Mettinger included patients with early-onset disease, with a mean age of 17.5 years, and another group with late-onset disease, with a mean age of 47.6 years and a range of 31–74 years. There was an equal number of patients in both groups. This may have confounded the results, since pars planitis is not a disease that begins in the sixth decade.

Malinowski et al. (1993b) developed very strict inclusion criteria prior to accepting patients with a diagnosis of pars planitis for HLA typing. Patients had to be under the age of 35 years at the time of initial diagnosis and other possible causes of intermediate uveitis had to be ruled out. Only patients with snowbanks in at least one eye were accepted. HLA-B8 and -B51 were found more often in affected patients than in controls (relative risks of 2.44 and 2.12, respectively). The HLA-DR antigens were examined and there was a very strong association with the HLA-DR2 phenotype. This phenotype was seen in 67% of patients and 28% of controls (relative risk, 5.32; $p < 0.0001$). Because the patients accepted into the study were cases of only classic pars planitis and the HLA determination was done by microlymphocytotoxicity, a subsequent study was performed by Tang et al. (1997) to evaluate the HLA association of all cases of idiopathic intermediate uveitis once all other causes including sarcoidosis had been reasonable excluded. This study evaluated the HLA association using both serology and polymerase chain reaction (PCR) techniques. Of 18 patients with intermediate uveitis, 13 were HLA-DR15 positive, with a relative risk of 6.36 ($p < 0.001$). The allele involved was DRB1*1501. Three patients (23%) had a family history of MS.

Considering the studies showing a similar HLA phenotype as in patients with MS, as well as the long-term outcome studies showing that a high percentage of patients with pars planitis will ultimately develop MS, there is a clear association between these two diseases. Whether it is that they share the same genetic predisposition to acquire two separate diseases or that MS and pars planitis are two manifestations of the same disease is, as yet, unknown. Studies by Bora et al. (1996) showed a possible circulating antigen seen only in patients with active pars planitis and which has homology to nucleopore proteins. If subsequent studies verify this finding, this will become a helpful serological test in further separating this disease from other causes of intermediate uveitis and will allow better understanding of the genetic risks for this disease.

Retinal and choroidal disorders

Birdshot chorioretinopathy and retinal vasculitis

Ryan and Maumanee (1980) first described a series of patients who had multiple depigmented spots under the retina which radiated from the optic disc. There was vitritis, retinal vasculitis and cystoid macular edema. They termed this disease birdshot chorioretinopathy (BSCR). Other terms have been used, including vitiliginous chorioretinitis (Gass, 1981). The radiating spots are quite characteristic and the vitritis is seen in about 83% of cases, cystoid macular edema in 62%, retinal arteriolar narrowing in 61%, retinal vasculitis in 40% and disc swelling in 38% (Priem and Oosterhuis, 1988). It usually affects middle-aged patients with an average age of 50 years. Women may have a slightly higher incidence. About one-half of the patients have an abnormal electroretinogram.

Shortly after the initial description of the disease, Nussenblatt found a strong association between BSCR and the HLA-A29 antigen (Nussenblatt et al., 1982). This antigen was seen in 80% of patients, with a relative risk of 50. Subsequent studies have shown an even higher incidence of the antigen and a higher relative risk. These subsequent studies have been from Europe and it may be that there is a stronger association with this antigen in Europe or that better understanding of the disease has allowed better segregation of the affected patients from other diseases prior to HLA typing. This disease has one of the highest associations with the HLA of any known diseases. Baarsma et al. (1990) showed that 96% of their cases of BSCR were positive compared with 20% of controls, with a relative risk of 224. Bloch-Michel and Frau (1991) found the HLA-A29 phenotype in 95% of their patients compared with 11% of controls, with a relative risk of 156. Since the serologically defined HLA-A29 phenotype can be divided into A29.1 and 29.2 subtypes, LeHoang et al. (1992) evaluated which one was associated with BSCR. The HLA-A29 antigen was found in 93% of affected patients and 8% of controls, with a relative risk of 157. The HLA-A29.2 was found in 100% of the affected patients who were HLA-A29 positive. There may be some cases that are associated with the HLA-A29.1 subtype, but this has not been completely verified.

There is only a single amino acid difference between the two subtypes: at position 102 outside the peptide-binding groove (Boisgerault et al., 1996). Boisgerault et al. evaluated the peptide-binding site of HLA-A29.2. The peptide motif of HLA-29.2 is nine or 10 amino acids long. Two amino acid nonamers located near the C-terminal end of the retinal S-antigen bound the peptide-binding site of HLA-29.2 with high affinity. The retinal S-antigen is highly uveitogenic in animal models of uveitis (see *Experimental autoimmune uveitis*, p. 281). This study is one of the first to evaluate the T cell epitopes that could be important in ocular disease. It is reasonable that this should have first been conducted with BSCR since it has such a high association with the HLA compared with other ocular diseases.

Retinal vasculitis may be seen in multiple diseases. Systemic causes include Behçet's disease, sarcoidosis and MS. Syphilis, cytomegalovirus and toxoplasmic retinochoroiditis are a few of the infectious causes of retinal vasculitis. BSCR, pars planitis and multifocal choroiditis with panuveitis are primary ocular diseases associated with retinal vasculitis. Idiopathic retinal vasculitis is diagnosed once the other systemic and ocular diseases associated with retinal vasculitis have been excluded. It usually presents in patients between the ages of 15 and 40 years. There may be several different diseases that have been categorized under the term of idiopathic retinal vasculitis. The HLA association with this entity was HLA-A29 in 61% of cases in one series, with a relative risk of 12.65; HLA-B44 was also strongly associated (Bloch-Michel and Frau, 1991). There has been an association between Behçet's disease and HLA-B51, with a relative risk of 8 in the Japanese population and 6.3 in an Irish population (Kilmartin et al., 1997). It may be that some cases which have been categorized as retinal vasculitis may actually be a forme fruste of Behçet's disease and therefore may be positive for this locus. Finally, some cases of retinal vasculitis may be positive for the HLA-B8 locus, which has been reported in some cases to be associated with sarcoidosis.

Vogt–Koyanagi–Harada disease and sympathetic ophthalmia

Vogt–Koyanagi–Harada (VKH) disease is an acute bilateral diffuse panuveitis. Ocular findings include iritis, vitritis, cystoid macular edema and exudative retinal detachments. There is a granulomatous infiltration of the ciliary body and the choroid. The choroidal infiltrate was shown to be principally T cells (Kahn et al., 1993). Macrophages were also present in the granulomas. Similar clinical and pathological findings can be seen in sympathetic ophthalmia and in EAU. Systemic findings include sensory hearing loss, alopecia, poliosis, vitiligo, headaches and other diffuse, non-specific CNS symptoms.

VKH is more common in Asians. In the western hemisphere, the disease is more commonly seen in patients with American Indian ancestry (Davis et al., 1990). The fact that there are ethnic differences in prevalence and that, even among the Japanese, there were cases of familial clustering added to the possibility that there are genetic associations with the disease (Ishikawa et al., 1994). Among the Japanese the HLA-DR53, -DR4 and -DQ4 loci were found in 98%, 93% and 82% of patients, with relative risks of 27, 17 and 10, respectively (Islam, 1994). At the genomic level all patients had HLA-DQA1*0301 with a relative risk of 56. HLA-DRB4*0101 had a relative risk of 30. HLA-DRB4*0101 is serologically defined as HLA-DR53. The twins with VKH reported by Ishikawa also showed the presence of HLA-DR4 and -DR53. Of importance is that HLA-DR53 and -DQA1 are in strong positive linkage disequilibrium with -DR4. In a Han Chinese population the HLA-DR4 and -Dw53 loci were also strongly associated with relative risks of 16 and 34, respectively (Zhao et al., 1991).

In a group of Hispanics of Mexican and Central American ancestry with American Indian roots who had VKH, the HLA-DR4 locus was present in 56% of cases compared with 29% of controls, with a relative risk of 2 (Weisz, 1995). The other loci found to be associated in the Japanese population were not associated. The DR1 phenotype was present in 36% of patients compared with 9% of controls for a relative risk of 4.1. The authors postulated that since HLA-DR1 and HLA-DR4 share very similar amino acid sequences near their peptide-binding site, they both could present similar antigens. Of concern is that the DR1 locus was not found to be increased in the VKH Japanese population. In a study of Italians with VKH, the DR4 antigen was the only one strongly associated with disease, with 50% of VKH patients expressing this antigen compared with 10% of controls ($p = 0.063$) (Pivetti-Pezzi et al., 1996). At this point there appears to be an association between the DR4 antigen and VKH that crosses over racial groups. There may be other loci that may be important depending on the racial background of patients.

Sympathetic ophthalmia (SO) is a rare uveitic syndrome seen in patients who have had a prior history of penetrating trauma or surgery. Ocular findings are similar to those in VKH but the systemic findings are less frequently encountered in SO. HLA typing in Japan of cases with SO again showed a strong correlation with HLA-DR4 and DRw53, with relative risks of 19 and 15, respectively (Shindo et al., 1997). The association with DRw53 may be because of the strong linkage of DRw53 with DR4 and again it is the DR4 association that carries the risk. A study from the USA of eight cases of SO did not include enough patients to show statistical significance, but five out of eight cases were DR4 positive (Davis et al., 1990).

Acute posterior multifocal placoid pigment epitheliopathy

Acute posterior multifocal placoid pigment epitheliopathy (APMPPE) occurs in young, previously healthy patients. There is usually a history of a prodromal upper respiratory tract illness. Patients report visual loss and a classic fundoscopic finding of circumscribed yellow–white postequatorial lesions is noted. These lesions characteristically block fluorescence early during a fluorescein angiogram and in the later stages of the angiogram stain intensely with fluorescein. The optic nerve may be swollen and mild vitritis may be present.

Rarely, the patients may have CNS manifestations as well. The disease is usually self-limiting with improvement of vision and mild scarring of the lesions. A study of 30 patients with APMPPE showed that the HLA-B7 antigen was present in 40% of cases compared with 17% of controls and the DR2 serotype was present in 57% of cases compared with 28% of controls, with relative risks of 3.4 and 3.3, respectively (Wolf et al., 1990). It should be noted that the HLA-B7 and -DR2 serotypes are in linkage disequilibrium, probably accounting for both being associated with APMPPE. A similar disease, serpiginous choroiditis, has been shown to be associated with HLA-B7. This study did not include class II antigens and therefore the B7 association may result from the associated DR2 linkage.

ACKNOWLEDGEMENTS

Jose Pulido and J. Wayne Streilein were supported in part by an unrestricted grant from Research to Prevent Blindness, New York, a core grant from NEI and a Wm Pendill grant.

REFERENCES

Abi-Hanna, D., Wakefield, D. and Watkins, S. (1988). HLA antigens in ocular tissues. *Transplantation* 45: 610–613.

Allansmith, M.R., Fine, M. and Payne, R. (1974). Histocompatibility typing and corneal transplantation. *Trans. Am. Acad. Ophthalmol. Otolaryngol.* 78: 445–460.

Arnold, K., Tandon, N., McIntosh, R.S. et al. (1994). T cell responses to orbital antigens in thyroid-associated ophthalmopathy. *Cin. Exp. Immunol.* 96: 329–334.

Arocker-Mettinger, E., Mayr, W.R., Huber-Spitzy, V. et al. (1992). Do HLA antigens play a role in intermediate uveitis? *Vox Sang.* 63: 282–284.

Asakura, H., Tsuchiya, M., Aiso, S. et al. (1982). Assoication of the human leukocyte DR2 antigen with Japanese ulcerative colitis. *Gastroenterology* 82: 413–418.

Auger, I., Toussirot, E. and Roudier, J. (1997). Molecular mechanisms involved in the association of HLA-DR4 and rheumatoid arthritis. *Immunol. Rev.* 16: 121–126.

Augustin, A.J., Boker, T., Blumenroder, S.H. et al. (1994). Free radical scavenging and antioxidant activity of allopurinol and oxypurinol in experimental lens-induced uveitis. *Invest. Ophthalmol. Vis. Sci.* 35: 3897–3904.

Augsburger, J.J., Annesley, W.H., Sergott, R.C. et al. (1981). Familial pars planitis. *Ann. Ophthalmol.* 13: 533–537.

Azumi, A. and Atherton, S.S. (1994). Sparing of the ipsilateral retina after anterior chamber inoculation of HSV-1: requirement for either $CD4^+$ or $CD8^+$ T cells. *Invest. Ophthalmol. Vis. Sci.* 35: 3251–3259.

Baarsma, G.S., Priem, H.A. and Kijlstra, A. (1990). Association of birdshot retinochoroidopathy and HLA-A29 antigen. *Curr. Eye Res.* 9 (Suppl.): 63–68.

Baggia, S., Lyons, J.L., Angell, E. et al. (1997). A novel model of bacterially-induced acute anterior uveitis in rats and the lack of effect from HLA-B27 expression. *J. Invest. Med.* 45: 295–301.

Baldini, M., Papplettera, M., Lecchi, L. et al. (1995). Human lymphocyte antigens in Graves' disease: correlation with persistent course of disease. *Am. J. Med. Sci.* 309: 43–48.

Barcellos, L.F., Thomson, G., Carrington, M. et al. (1997). Chromosome 19 single-locus and multilocus haplotype associations with multiple sclerosis. Evidence of a new susceptibility locus in Caucasian and Chinese patients. *JAMA* 278: 1256–1261.

Barker, C.F. and Billingham, R.E. (1977). Immunologically privileged sites. *Adv. Immunol.* 25: 1–54.

Bartley, G.B., Fatourechi, V., Kadrmas, E.F. et al. (1996). Clinical features of Graves' ophthalmopathy in an incidence cohort. *Am. J. Ophthalmol.* 121: 284–290.

Batchelor, J.R., Casey, T.A., Gibbs, D.C. et al. (1976). HLA matching and corneal grafting. *Lancet* i: 551–554.

Beck, R.W., Cleary, P.A. and the Optic Neuritis Study Group (1993b). Optic neuritis treatment trial: one-year follow-up results. *Arch. Ophthalmol.* 11: 773–775.

Beck, R.W., Cleary, P.A., Trobe, J.D. et al. (1993a). The effect of corticosteroids for acute optic neuritis on the subsequent development of multiple sclerosis. *N. Engl. J. Med.* 329: 1764–1769.

Berra, A., Rodriguez, A., Heiligenhaus, A. et al. (1994). The role of macrophages in the pathogenesis of HSV-1 induced chorioretinitis in BALB/c mice. *Invest. Ophthalmol. Vis. Sci.* 35: 2990–2998.

Bloch-Michel, E. and Frau, E. (1991). Birdshot retinochoroidopathy and $HLA-A29^+$ and HLA-A29- idiopathic retinal vasculitis: comparative study of 56 cases. *Can. J. Ophthalmol.* 26: 361–366.

Boisgerault, F., Khalil, I., Tieng, V. et al. (1996). Definition of the HLA-A29 peptide ligand motif allows predictioin of potential T-cell epitopes from the retinal soluble antigen, a candidate autoantigen in birdshot retinopathy. *Proc. Natl Acad. Sci. USA* 93: 3466–3470.

Bora, N.S., Gobleman, Cl., Atkinson, J.P. et al. (1993). Differential expression of the complement regulatory proteins in the human eye. *Invest. Ophthalmol. Vis. Sci.* 34: 3579–3584.

Bora, N.S., Bora, P.S. and Kaplan, H.J. (1996). Identification, quantitation, and purification of a 36Kda circulating protein associated with pars planitis. *Invest. Ophthalmol. Vis. Sci.* 37: 1870–1876.

Bora, N.S., Woon, M.D., Tandhasetti, M.T. et al. (1997). Induction of experimental autoimmune anterior uveitis by a self-antigen. Melanin complex without adjuvant. *Invest. Ophthalmol. Vis. Sci.* 38: 2171–2175.

Bramford, C.R., Ganley, J.P., Sibley, W.A. and Laguna, J.F. (1987). Uveitis, perivenous sheathing and multiple sclerosis. *Neurology* 28: 119–124.

Brockhurst, R.J., Schepens, C.L. and Okamura, I.D. (1960). Uveitis. II. Peripheral uvetis: Clinical description, complications and differential diagnosis. *Am. J. Ophthalmol.* 49: 1257–1266.

Broekhuyse, R.M., Kuhlman, E.D., Winkens, H.G. and Van Vugt, A.H.M. (1991). Experimental autoimmune anterior uveitis (EAAU), a new form of experimental uveitis. I. Induction by a detergent-insoluble intrinsic protein fraction of the retina pigment epithelium. *Exp. Eye Res.* 52: 465–474.

Bullington, R.J. and Waksman, B.H. (1958). Uveitis in rabbits with experimental allergic encephalomyelitis. *Arch. Ophthalmol.* **59**: 435–455.

Caspi, R.R., Grubbs, B.G., Chan, C.C. et al. (1992). Genetic control of susceptibility to experimental autoimmune uveoretinitis in the mouse model. *J. Immunol.* **148**: 2384–2389.

Conway, D.J., Holland, M.J., Campbell, A.E. et al. (1996). HLA classs I and II polymorphisms and trachomatous scarring in a *Chlamydia trachomatis*-endemic population. *J. Infect. Dis.* **174**: 643–646.

Council on Scientific Affairs (1988). Report on the Organ Transplant Panel: Corneal Transplantation. *JAMA* **259**: 719–722.

Culbertson, W.W., Giles, C.L., West, C. and Stafford, T. (1983). Familial pars planitis. *Retina* **3**: 179–181.

Czaja, A.J. (1997). Extrahepatic immunologic features of chronic viral hepatitis. *Dig. Dis.* **15**: 125–144.

Darrell, R.W. (1981). Thygeson's superficial punctate keratitis: natural history and association with HLA-DR3. *Trans. Am. Ophthalmol. Soc.* **74**: 486–516.

Davis, J.L., Mittal, K.K., Freidlin, V. et al. (1990). HLA associations and ancestry in Vogt–Koyanagi–Harada disease and sympathetic ophthalmia. *Ophthalmol.* **97**: 1137–1142.

De Bruyere, M., Dernouchamps, J.P. and Sokal, G. (1986). HLA antigens in Fuchs' heterochromic iridocyclitis. *Am. J. Ophthalmol.* **102**: 392–393.

Del Carpio, J., Espinoza, L.R. and Osterland, C.K. (1976). Cogan's syndrome and HLA BW1. *N. Engl. J. Med.* **295**: 1262–1263.

Derhaag, P.J.F.M., Linssen, A., Broekema, N. et al. (1988a). A familial study of the inheritance of HLA-B27-positive acute anterior uveitis. *Am. J. Ophthalmol.* **105**: 603–606.

Derhaag, P.R.F.M., Waal, L.P. de, Linssen, A. and Feltkamp, T.E.W. (1988b). Acute anterior uveitis and HLA-B27 subtypes. *Invest. Ophthalmol. Vis. Sci.* **29**: 1137–1140.

Ehlers, N. and Ahrons, S. (1971). Corneal transplantation and histocompatibility. *Acta Ophthalmol.* **49**: 513–527.

Epplen, C., Jackel, S., Santos, E.J. et al. (1997). Genetic predisposition to multiple sclerosis as revealed by immunoprinting. *Ann. Neurol.* **41**: 341–352.

Eye Bank Association of America (1992). *Activity Report, 1992,* Eye Bank Association of America, Washington, DC.

Feltkamp, T.E.W. (1990). HLA-B27 and acute anterior uveitis. *Curr. Eye Res.* **9** (Suppl.): 213–214.

Fiannini, E.H., Malagon, C.N., Van Kerhove, C. et al. (1991). Longitudinal analysis of HLA-associated risks for iridocyclitis in juvenile rheumatoid arthritis. *J. Rheumatol.* **18**: 1394–1397.

Gass, J.D.M. (1981). Vitiliginous chorioretinitis. *Arch. Ophthalmol.* **99**: 1778–1787.

Genain, C.P., Abel, K., Belmar, N. et al. (1996). Late complications of immune deviation therapy in a nonhuman primate. *Science* **274**: 2054–2057.

Giordano, C., Stassi, G., De Maria, R. et al. (1997). Potential involvement of Fas and its ligand in the pathogenesis of Hashimoto's thyroiditis. *Science* **275**: 960–963.

Gore, S.M., Vail, A., Bradley, B.A. et al. (1995). HLA-DR matching in corneal transplantation. Systemic review of published evidence. *Transplantation* **60**: 1033–1039.

Gottsch, J.D., Liu, S.H., Minkovitz, J.B. et al. (1995). Autoimmunity to a cornea-associated stromal antigen in patients with Mooren's ulcer. *Invest. Ophthalmol. Vis. Sci.* **36**: 1541–1547.

Griffith, T.S., Brunner, T., Fletcher, S.M. et al. (1995). Fas ligand-induced apoptosis as a mechanism of immune privilege. *Science* **270**: 1189–1192.

Hammer, R.E., Maika, S.D., Richardson, J.A. et al. (1990). Spontaneous inflammatory disease in trangenic rats expressing HLA-B27 and human beta-2 microglobulin: an animal model of HLA-B27-associated human disorders. *Cell* **63**: 1099–1112.

Head, J.R., Neaves, W.B. and Billingham, R.E. (1983). Reconsideration of the lymphatic drainage of the rat testis. *Transplantation* **35**: 91–98.

Heufelder, A.E. (1995). Involvement of the orbital fibroblast and TSH receptor in the pathogenesis of Graves' ophthalmopathy. *Thyroid* **5**: 331–340.

Hirose, S., Ogasawara, K., Natori, T. et al. (1991). Regulation of experimental autoimmune uveitis in rats – separation of MHC and non-MHC gene effects. *Clin. Exp. Immunol.* **86**: 419–425.

Hogancamp, W.E., Rodriguez, M. and Weinshenker, B.G. (1997). Identification of multiple sclerosis-associated genes. *Mayo. Clin. Proc.* **72**: 965–976.

Ishikawa, A., Shiono, T. and Uchida, S. (1994). Vogt–Koyanagi–Harada disease in identical twins. *Retina* **14**: 435–437.

Islam, S.M.M., Numaga, J., Fujino, Y. et al. (1997). HLA Class II genes in Vogt-Koyanagi-Harada disease. *Invest. Opthalmol. Vis. Sci.* **35**: 3890–3896.

Jurado, A., Cardaba, B., Jara, P. et al. (1997). Autoimmune hepatitis type 2 and hepatitis C virus infection: study of HLA antigens. *J. Hepatol.* **26**: 983–991.

Kahn, M., Pepose, J.S., Green, W.R. et al. (1993). Immunocytologic findings in a case of Vogt–Koyanagi–Harada syndrome. *Ophthalmology* **100**: 1191–1198.

Kaiser-Kupfer, M.I., Mittal, K.K., Del Valle, L.A. and Haynes, B.F. (1978). The HLA antigens in Cogan's syndrome. *Am. J. Ophthalmol.* **86**: 314–316.

Kanski, J.J. (1990). Juvenile arthritis and uveitis. *Surv. Ophthalmol.* **34**: 253–267.

Katami, M., Madden, P.W., White, D.J.G. et al. (1989). The extent of immunological privilege of orthotopic corneal grafts in the inbred rat. *Transplantation* **48**: 371–376.

Kerrison, J.B., Flynn, T. and Green, W.R. (1994). Retinal pathologic changes in multiple sclerosis. *Retina* **14**: 445–451.

Kilmartin, D.J., Finch, A. and Acheson, R.W. (1997). Primary association of HLA-B51 with Behçet's disease in Ireland. *Br. J. Ophthalmol.* **81**: 649–653.

Kira, J., Kanai, T., Nishimura, Y. et al. (1996). Western versus Asian types of muliple sclerosis: immunogenetically and clinically distinct disorders. *Ann. Neurol.* **40**: 569–574.

Konda, B.R., Pararajasegaram, G., Wu, G.S. et al. (1994). Role of retinal pigment epithelium in the development of experimental autoimmune uveitis. *Invest. Ophthalmol. Vis. Sci.* **35**: 40–47.

Kong, Y.C., Lomo, L.C., Motte, R.W. et al. (1996). HLA-DRB1 polymorphism determines susceptibility to auoimmune thyroiditis in transgenic mice: definitive association with HLA-DRB1*0301 (DR3) gene. *J. Exp. Med.* **184**: 1167–1172.

Lafferty, K.J., Prowse, S.J., Simeonovic, C.J. and Warren, H.S. (1983). Immunobiology of tissue transplantation: a return to the passenger leukocyte concept. *Annu. Rev. Immunol.* **1**: 142–173.

Lechler, R.I. and Batchelor, J.R. (1982a). Restoration of immunogenicity to passenger cell-depleted kidney allografts by the addition of donor strain dendritic cells. *J. Exp. Med.* **155**: 31–41.

Lechler, R.I. and Batchelor, J.R. (1982b). Immunogenicity of retransplanted rat kidney allografts: effects of including chimerism in the first recipient and quantitative studies on immunosuppression of the second recipient. *J. Exp. Med.* **156**: 1835–1841.

Lee, R.S., Grusby, M.J., Glimcher, L.H. et al. (1994). indirect recognition by helper cells can induce donor-specific cytotoxic T lymphocytes *in vivo*. *J. Exp. Med.* **179**: 865–872.

LeHoang, P., Ozdemir, N., Benhamou, et al. (1992). HLA-A29.2 subtype associated with birdshot retinochoroidopathy. *Am. J. Ophthalmol.* **113**: 33–35.

Li, Q., Peng, B., Whitcup, S.M. et al. (1995). Endotoxin induced uveitis in the mouse: susceptibility and genetic control. *Exp. Eye Res.* **61**: 629–632.

Linssen, A., Rothova, A. and Valkenburg, H.A. (1991). The lifetime cumulative incidence of acute anterior uveitis in a normal population and its relation to ankylosing spondylitis and histocompatibility antigen HLA-B27. *Invest. Ophthalmol. Vis. Sci.* **32**: 2568–2578.

Liu, S.H. and Gottsch, J.D. (1996). Amino acid sequence of an immunogenic corneal stromal protein. *Invest. Ophthalmol. Vis. Sci.* **37**: 944–948.

Liversedge, J.M., Sewell, H.F. and Forrester, J.V. (1988). Human retinal pigment epithelial cells differentially express MHC class II (HLA DP, DR and DQ) antigens in response to *in vitro* stimulation with lymphokine or purified IFN-γ. *Clin. Exp. Immunol.* **73**: 489–494.

Mader, T.H. and Stulting, R.D. (1991). The high-risk penetrating keratoplasty. *Ophthalmol. Clin. North Am.* **4**: 411–426.

Makley, T.A., Jr. (1956). Heterochronic cyclitis in identical twins. *Am. J. Opthalmol.* **41**: 768–772.

Malinowski, S.M., Pulido, J.S. and Folk, J.C. (1993a). Long-term visual outcome and complications associated with pars planitis. *Ophthalmology* **100**: 818–825.

Malinowski, S.M., Pulido, J.S., Goeken, N.E. et al. (1993b). The association of HLA-B8, B51, DR2, and multiple sclerosis in pars planitis. *Ophthalmology* **100**: 1199–1205.

Malleson, P.N., Fung, M.Y., Petty, R.E. et al. (1992). Autoantibodies in chronic arthritis of childhood: Relations with each other and with histocompatibility antigens. *Ann. Rheum. Dis.* **51**: 1301–1306.

Manns, M.P. (1997). Hepatotropic viruses and autoimmunity 1997. *J. Viral Hepatitis* **4** (Suppl.): 7–10.

Maumanee, A.E. (1951). The influence of donor–recipient sensitization on corneal grafts. *Am. J. Ophthalmol.* **34**: 142–152.

McMenamin, P.G. (1994). Immunocompetent cells in the anterior segment. *Prog. Retinal Eye Res.* **13**: 555–591.

McMenamin, P.G. and Crewe, J. (1995). Endotoxin-induced uveitis. Kinetics and phenotype of the inflammatory cell infiltrate and the response of the resident tissue macrophages and dendritic cells in the iris and ciliary body. *Invest. Ophthalmol. Vis. Sci.* **36**: 1949–1959.

Medawas, P.B. (1948). Immunity to homologous grafted skin. III. The fate of the skin homografts transplanted to the brain, to subcutaneous tissue, and to the anterior chamber of the eye. *Bi. J. Exp. Pathol.* **29**: 58–69.

Miller, S.D., Vanderlugt, C.L., Begolka, W.S. et al. (1997). Persistent infection with Theiler's virus leads to CNS autoimmunity via epitope spreading. *Nature Med.* **10**: 1133–1136.

Moazami, G., Auran, J.D., Florakis, G.J. et al. (1995). Interferon treatment of Mooren's ulcers associated with hepatitis C. *Am. J. Ophthalmol.* **119**: 365–366.

Nakamura, S., Yamakawa, T., Sugita, M. et al. (1994). The role of tumor necrosis factor-alpha in the induction of experimental autoimmune uveoretinitis in mice. *Invest. Opthalmol. Vis. Sci.* **35**: 3884–3889.

Nicholls, S.M., Benylles, A., Shimeld, C. et al. (1994). Ocular infection with herpes simplex virus in several strains of rat. *Invest. Ophthalmol. Vis. Sci.* **35**: 3260–3267.

Niederkorn, J.Y. (1990). Immune privilege and immune regulation in the eye. *Adv. Immunol.* **48**: 199–208.

Niederkorn, J.Y. and Streilein, J.W. (1983). Intracamerally-induced concomitant immunity: mice harboring progressively growing intraocular tumors are immune to spontaneous metastases and secondary tumor challenge. *J. Immunol.* **131**: 2587–2594.

Niemialtowski, M.G., Godfrey, V.L. and Rouse, B.T. (1994). Quantitative studies on CD4[+] and CD8[+] cytotoxic T lymphocyte responses against herpes simplex virus type 1 in normal and beta 2-m deficient mice. *Immunobiology* **190**: 183–194.

Nussenblatt, R.B., Mittal, K.K., Ryan, S. et al. (1982). Birdshot retinochoroidopathy associated with HLA-A29 antigen and immune responsiveness to retinal S-antigen. *Am. J. Ophthalmol.* **94**: 147–158.

Ohno, S., Kimura, S.J., O'Connor, G.R. and Char, D.H. (1977). HLA antigens and uveitis. *Br. J. Ophthalmol.* **61**: 62–64.

Oksenberg, J.R. (1996). Immunogenetics and heterogeneity in multiple sclerosis. *Ann. Neurol.* **40**: 557–558.

Parham, P. (1996). Presentation of HLA class I-derived peptides: potential involvement in allorecognition and HLA-B27-associated arthritis. *Immunol. Rev.* **154**: 137–149.

Perdiger, A., Chales, G., Semana, G. et al. (1997). Role of HLA-DR-DR and DR-DQ associations in the expression of extrarticular manifestations and rheumatoid factor in rheumatoid arthritis. *J. Rheumatol.* **24**: 1272–1276.

Pivetti-Pezzi, P., Accorinti, M., La Cava, M. et al. (1996). Endogenous uveitis: an analysis of 1,417 cases. *Ophthalmologica* **210**: 234–238.

Price, F.W., Whitson, W.E. and Marks, R.G. (1991). Progression of visual acuity after penetrating keratoplasty. *Ophthalmology* **98**: 1177–1185.

Priem, H.A. and Oosterhuis, J.A. (1988). Birdshot chorioretiniopathy: clinical characteristics and evolution. *Br. J. Ophthalmol.* **72**: 646–659.

Pulido, J.S. (1996). Experimental nonenzymatic glycosylation of vitreous collagens occurs by two pathways. *Trans. Am. Ophthalmol. Soc.* **94**: 1029–1072.

Rao, N.A., Atalla, L., Linker-Israeli, M. et al. (1989). Suppression of experimental uveitis in rats by anti-I-A antibodies. *Invest. Ophthalmol. Vis. Sci.* **30**: 2348–2355.

Rosenbaum, J.T. (1992). Acute anterior uveitis and spondyloarthropathies. *Rheum. Dis. Clin. N. Am.* **18**: 143–150.

Roussomoustakaki, M., Satsangi, J., Welsh, K. et al. (1997). Genetic markers may predict disease behaviour in patients with ulcerative colitis. *Gastroenterology* **112**: 1845–1853.

Ryan, S.J. and Maumenee, A.E. (1980). Birdshot retinochoroidopathy. *Am. J. Ophthalmol.* **89**: 31–45.

Sahay, B.M., Blendis, L.M. and Grene. R, (1965). Relation between myasthenia gravis and thyroid disease. *BMJ* **i**: 762–765.

Satsangi, J., Welsch, K.I., Bunce, M. et al. (1996). Contribution of genes of the major histocompatibility complex to susceptibility and disese phenotype in inflammatory bowel diease. *Lancet* **347**: 1212–1217.

Schwab, I.R. (1990). Fuchs' heterochromic iridocyclitis. *Int. Ophthalmol. Clin.* **30**: 252–256.

Shindo, Y., Ohno, S., Usui, M. et al. (1997). Immunogenetic study of sympathetic ophthalmia. *Tissue Antigens* **49**: 111–115.

Sonoda, Y. and Streilein, J.W. (1992). Orthotopic corneal transplantation in mice. Evidence that the immunogenetic rules of rejection do not apply. *Transplantation* **54**: 694–703.

Stark, W., Stulting, D., Maguire, M. and Streilein, J.W. (1992). The Collaborative Corneal Transplantation Studies (CCTS): effectiveness of histocompatibility matching of donors and recipients in high risk corneal transplantation. *Arch. Ophthalmol.* **110**: 1392–1403.

Stark, W.J., Taylor, H.R., Bias, W.B. and Maumenee, A.E. (1978). Histocompatibility (HLA) antigens and keratoplasty. *Am. J. Ophthalmol.* **86**: 595–604.

Steinmuller, D. (1967). Immunization with skin isografts taken from tolerant mice. *Science* **158**: 127–129.

Streilein, J.W. (1987). Immune regulation and the eye: a dangerous compromise. *FASEB J.* **1**: 199–208.

Streilein, J.W. (1995). Unraveling immune privilege. *Science* **270**: 1158–1159.

Streilein, J.W. (1997). Regional immunology. In *Encyclopedia of Human Biology*, 2nd edn (ed. R. Dulbecco), Vol. 4, pp. 767–776, Academic Press, San Diego, CA.

Streilein, J.W., Toews, G.B. and Bergstresser, P.R. (1979). Corneal allografts fail to express Ia antigens. *Nature* **282**: 325–327.

Streilein, J.W., Niederkorn, J.Y. and Shadduck, J.A. (1980). Systemic immune unresponsiveness induced in adult mice by anterior chamber presentation of minor histocompatibility antigens. *J. Exp. Med.* **152**: 1121–1125.

Streilein, J.W., Bradley, D. and Sano, Y. (1996). Immunosuppressive properties of tissues of the ocular anterior segment. *Ocular Immunol. Inflamm.* **4**: 57–68.

Tang, W.M., Pulido, J.S., Eckels, D.D. et al. (1997). The association of HLA-DR15 and intermediate uveitis. *Am. J. Ophthalmol.* **123**: 70–75.

Thanos, S., Moore, S. and Hong, Y. (1996). Retinal microglia. *Prog. Retinal Eye Res.* **15**: 331–361.

Tomer, Y., Barbesino, G., Keddache, M. et al. (1997). Mapping of a major susceptibility locus for Graves' disease (GD-1) to chromosome 14q31. *J. Clin. Endocrinol. Metab.* **82**: 1645–1648.

Toyoda, H., Wang, S.J., Yang, H.Y. et al. (1993). Distinct associations of HLA class II genes with inflammatory bowel disease. *Gastroenterology* **104**: 741–748.

Vail, A., Gore, S.M., Bradley, B.A. et al. (1994). Influence of donor and histocompatibility factors on corneal graft outcome. *Transplantation* **58**: 1210–1216.

Volker-Dieben, H.J.M., Kok-van, Alphen, C.C., Lansbergen, Q. and Persijn, G.G. (1982). Different influences on corneal graft survival in 539 transplants. *Acta Ophthalmol.* **60**: 190–202.

Wang, H.M., Kaplan, H.J., Chan, W.C. and Johnson, M. (1987). The distribution and ontogeny of MHC antigens in murine ocular tissues. *Invest. Ophthalmol. Vis. Sci.* **28**: 1383–1389.

Weisz, J.M., Holland, G.N., Roer, L.N. et al. (1995). Association between Vogt-Koyanagi-Harada syndrome and HLA-DR1 and -DR4 in Hispanic patients living in southern California. *Opthalmology* **102**: 1012–1015.

Wetzig, R., Hooks, J.J., Percopo, C.M. et al. (1988a). Anti-Ia antibody diminishes ocular inflammmation in experimental autoimmune uveitis. *Curr. Eye Res.* **7**: 809–818.

Wetzig, R.P., Chan, C.-C., Nussenblatt, R.B. et al. (1988b). Clinical and immunopathological studies of pars planitis in a family. *Br. J. Ophthalmol.* **72**: 5–10.

White, A.G., Bogh, J., Leheny, W. et al. (1997). HLA antigens in Omanis with blinding trachoma: markers for disease susceptibility and resistance. *Br. J. Ophthalmol.* **81**: 431–434.

Wilbanks, G.A. and Streilein, J.W. (1989). The differing patterns of antigen release and local retention following anterior chamber and intravenous inoculation of soluble antigen. Evidence that the eye acts as an antigen depot. *Regional Immunol.* **2**: 390–398.

Wilbanks, G.A. and Streilein, J.W. (1990). Distinctive humoral responses following anterior chamber and intravenous administration of soluble antigen. Evidence for active suppression of IgG2a-secreting B cells. *Immunology* **71**: 566–572.

Wilbanks, G.A., Rootman, D.S., Jay, V. et al. (1997). Experimental autoimmune uveitis in HLA-B27 transgenic mice. *Hum. Immunol.* **53**: 188–194.

Wilson, S.E. and Kaufman, H.E. (1990). Graft failure after penetrating keratoplasty. *Surv. Ophthalmol.* **34**: 325–356.

Wolf, M.D., Folk, J.C., Panknen, C.A. and Goeken, N.E. (1990). HLA-B7 and HLA-DR2 antigens and acute posterior multifocal placoid pigment epitheliopathy. *Arch. Ophthalmol.* **108**: 698–700.

Yamagami, S., Kawashima, H., Tsuru, T. et al. (1997). Role of Fas–Fas ligand interactions in the immunorejection of allogeneic mouse corneal transplants. *Transplantation* **64**: 1107–1111.

Yang, P., Vos, A.F. de and Kijlstra, A. (1997). Macrophages and MHC class II positive cells in the choroid during endotoxin induced uveitis. *Br. J. Ophthalmol.* **81**: 396–401.

Zhao, M., Jiang, Y. and Abrahams, W. (1991). Association of HLA antigens with Vog–Koyanagi–Harada syndrome in a Han Chinese population. *Arch. Ophthalmol.* **109**: 368–370.

CHAPTER 20

HLA and Infectious Diseases

Paul A. Glynne and Nicholas M. Price

INTRODUCTION TO HLA AND INFECTIOUS DISEASES

Overview

The fundamental role of the human immune system is to protect against infectious pathogens. Impairment of the host immune response to infections may lead to overt clinical disease. Other factors, such as virulence properties of the pathogen and socioenvironmental factors, are also important determinants of disease susceptibility.

In many cases, virulence and social factors alone are not sufficient to explain the observed differences in infectious disease patterns between different populations. In addition, within a population the number of individuals exposed to a particular pathogen far exceeds the number that develops clinical disease. It is clear that the host immune response is a critical factor in disease outcome and the level of this response is determined predominantly by genetic factors.

The 1990s have seen an enormous expansion of research into the role of both the major histocompatibility complex (MHC) genes and non-MHC genes in the immunopathogenesis of human infectious diseases. Evidence from human leucocyte antigen (HLA) disease-association studies has shown that the MHC and its gene products have a major impact on infectious disease outcome through their central role in the regulation of the immune response. Evolutionary pressure coming from an enormous variety of antigens of infectious origin has resulted in a balance in the frequency of the highly polymorphic alleles of the MHC. Populations of certain geographical areas appear to have selected a particular set of MHC molecules or HLA haplotypes, which confer protection from certain pathogens by presenting antigens in an optimal manner.

The aim of this chapter is to discuss critically those infectious diseases in which the MHC has been shown to influence significantly both the susceptibility to and the outcome of infection. In addition to the specific diseases detailed in the sections below, a number of isolated clinical reports has suggested possible MHC associations with other pathogens. Where not discussed in the text, these associations have been incorporated into Table 20.1.

Approach to the study of HLA and infectious diseases

Recent trends in the study of infectious disease immunogenetics have seen an expansion from traditional HLA disease-association studies to a more global approach to identifying a wide range of immune response (Ir) genes. A variety of approaches have been used to study the association of HLA with infectious diseases (Hill, 1997, 1998). The aims of such studies are to determine which HLA polymorphisms affect differential susceptibility to infectious diseases in humans and whether these variations account for the heterogeneity in immune responses observed following infection. Large population studies also provide

Table 20.1 Examples of reported HLA allelic associations with infectious diseases

Infectious disease	HLA allele	Population	Study
Parasitic infections			
Malaria	B53↓ DRB1*1302↓	Gambia	Hill et al. (1991)
Mucocutaneous leishmaniasis	A28↑ B22↑ DQ28↑	Venezuela	Lara et al. (1991)
	DR2↓ DQ3↑	Brazil	Petzl-Erler et al. (1991)
	A11↑ B5↑ B7↑	Egypt	Morsy et al. (1990)
Visceral leishmaniasis	A26↑	Iran	Faghiri et al. (1995)
Filariasis	DQB1*0201↑	Liberia	Meyer et al. (1994)
	DPB1*0402↑		
	DQA1*0501↓		
	DQB1*0301↓		
	DQA1*0501↑	Nigeria	Murdoch et al. (1997)
	DQB1*0301↑		
Viral infections			
Hepatitis B persistence	DRB1*1302↓	Gambia	Thursz et al. (1995, 1997)
	DRB1*1302↓	Germany	Hohler et al. (1997)
	DR6↓	Holland	Van Huttum et al. (1987)
	DR2↓ DR7↑	Qatar	Almari and Batchelor (1994)
Hepatitis C persistence	DR5↓	Italy	Peano et al. (1994); Zavaglia et al. (1996)
	DR5↓	England	Minton et al. (1998)
	DQB1*0301↓	England	Tibbs et al. (1996)
	DQB1*0301↓	France	Alric et al. (1997); Cramp et al. (1998); Minton et al. (1998)
	B54↑ DQB1*0601↓	Japan	Kuzushita et al. (1998)
Hantavirus	B8↑ B27↓	Finland	Mustonen et al. (1996, 1998)
Bacterial infections			
Chronic Lyme disease	DRB1*0401↑	USA	Steere et al. (1990)

Human immunodeficiency virus and mycobacterial HLA associations are listed in Tables 20.2 and 20.3, respectively.
↑: susceptibility allele; ↓: protective allele.

insights into the role of selection by various pathogens in giving rise to genetic polymorphisms that determine susceptibility to infection.

An estimate of the magnitude of the host genetic component can be made from family studies by analysis of pedigrees. The most important epidemiological parameter is the relative risk, λ_S, which describes the increased risk of disease in the sibling of an affected individual compared with the general population risk. However, this value may overestimate the host genetic contribution to infectious disease susceptibility because other family members are subject to the same environmental factors. Twin studies reduce these confounding environmental factors.

In case-controlled population studies, the distribution of HLA haplotypes is compared between unrelated patients with disease and healthy controls, selected according to age, ethnic group and other demographic factors. These studies require prior knowledge of candidate genes of interest and their associated polymorphisms. This approach has been widely used to determine HLA associations with many infectious diseases. Case–control studies have the disadvantage that they cannot exclude bias resulting from population heterogeneity. This problem can be overcome by using as the control group those alleles of their parents which the cases under study did not inherit, in family-based association studies. Family studies exclude these confounding variables as well as demonstrating associations not observable at the population level.

In families with many individuals affected by the disease, genetic linkage studies can analyse non-random segregation of HLA haplotypes in diseased children compared with disease-free siblings. These techniques are able to identify HLA association with susceptibility to disease. Segregation patterns are analysed by the lod score statistic (the log odds ratio between the observed and expected distribution of the genetic markers).

For infectious diseases, where patterns of allele segregation and inheritance are complex, simpler pedigrees are often preferred consisting of two affected siblings and their parents. The affected sib-pair method compares the observed and expected distribution of parental alleles at marker loci. At any locus, a pair of siblings may share zero, one or two alleles (in the ratio 25:50:25%) by random segregation. If the locus is genetically linked to disease, affected siblings will share a higher number of alleles at that locus than expected. Using microsatellite genetic markers it is possible to examine the whole genome for evidence of linkage, although this carries with it a high risk of false-positive associations (Todd, 1992). This necessitates having large numbers of families to generate the required statistical power to define accurately minor immunogenetic factors.

The genome-wide approach offers exciting possibilities through the identification of novel immune response genes that determine susceptibility to infectious diseases. Ultimately, this may lead to the discovery of novel therapeutic interventions and vaccine strategies.

PARASITIC INFECTIONS

Malaria

HLA class I and II associations with severe falciparum malaria

The largest HLA association studies of an infectious disease have been of malaria caused by *Plasmodium falciparum*. There are significant differences in the frequencies of HLA genes between malaria-exposed and -unexposed populations, suggesting that the human MHC may protect populations in endemic disease areas who are exposed to malaria parasites.

A case–control study of over 2000 individuals in the Gambia showed that the HLA class I allele, HLA-B*5301, and the HLA class II haplotype, HLA-DRB1*1302–DQB1*0501, were independently associated with a reduced risk of severe malaria (defined as cerebral malaria and severe anaemia) in childhood (Hill et al., 1991). The protective effect of HLA-B53 was equally strong for heterozygotes and homozygotes. The HLA-B*5301 association was most marked in very young children (Gupta and Hill, 1995) and more marked for severe malaria than other manifestations of malaria infection. This protective effect might be mediated by HLA class I-restricted $CD8^+$ cytotoxic T lymphocytes (CTL) during the liver stage of the parasite's life-cycle. Evidence for this comes from animal models of malaria infection in which a protective role for MHC class I-restricted CTL has been demonstrated (Weiss et al., 1988, 1990; Romero et al., 1988). In humans, there is indirect evidence for an MHC class I-restricted response: in the malaria-immune population, HLA-B53-restricted CTL recognized a conserved nonapeptide from a liver stage-specific antigen, but no HLA-B53-restricted epitopes were identified in antigens from other stages (Hill et al., 1992). These data provide a molecular basis for this HLA disease association.

In a similar, recently reviewed case–control study of severe malaria in Kenya, a protective association with HLA-DRB1*0101 but not HLA-B*5301 and HLA DRB1*1302 was found (Hill, 1998). The large sample sizes in both the Gambian and Kenyan studies demonstrated a statistically significant difference in HLA association between the two populations. It is not clear whether such heterogeneity was due to differences in malarial parasite CTL epitopes between the two regions or to other factors such as transmission intensity (Gupta and Hill, 1995). Studies of CTL responses to variant epitopes in other pathogens, e.g. human immunodeficiency virus (HIV) and hepatitis B virus (HBV), have shown that some naturally occurring variants of these epitopes may specifically downregulate the CTL response through altered peptide ligand antagonism (Bertoletti et al., 1994; Klenerman et al., 1994). The simultaneous presence of the variant antagonist epitope delivers an altered signal to the responding CTL that may result in its impaired responsiveness to the target agonist epitope. HLA-B35-restricted CTL in the Gambia often recognize a polymorphic epitope in a variable region of the malarial circumsporozoite (CS) protein, of which two allelic variants are CTL epitopes that bind HLA-B35 (cp27 and cp29) (Hill et al., 1992). Recently, it has been demonstrated that the distribution of these variants was significantly different between individuals with and without this class I type (Gilbert et al., 1998). Furthermore, cp26 and cp29 can mutually interfere with the *in vitro* induction of primary CTL responses by the other peptide epitope; i.e. the CTL response to malaria is itself subject to altered peptide ligand antagonism. From an evolutionary perspective, cohabitation of parasite strains that can down-regulate cellular immune responses may indeed enhance each strain's ability to survive. From the above evidence, it seems likely that the striking heterogeneity observed in the immune response to *P. falciparum* infection may be determined in part by the interaction of HLA phenotype, polymorphisms in the malaria parasite and immunological antagonism of malaria CTL epitopes.

Tumour necrosis factor-α and cerebral malaria

Clinical studies have demonstrated an association between tumour necrosis factor-α (TNFα) levels and the severity of malaria infection (Grau et al., 1989; Kern et al., 1989), and experimental work has suggested a pathogenic role for TNFα in the aetiology of cerebral malaria (Grau et al., 1987). These observations raised the possibility that cerebral malaria may be related to a genetic predisposition of the host to produce TNFα, the gene for which resides within the MHC class III region. Several studies have shown that individual differences in TNFα production can be linked to HLA type (Molvig et al., 1988; Pociot et al., 1993). Studies of polymorphisms in and near to the TNF gene have shown striking associations with differential susceptibility to malaria infection (McGuire et al., 1994). A polymorphism directly affecting the regulation of TNFα is located at −308 nucleotides relative to the transcriptional start site of the gene (Wilson et al., 1992, 1993). There are two allelic forms, TNF1 (the common allele) and TNF2. They are identical apart from a single base transition (G to A) at position −308. The −308 allele promoter variant (TNF2 allele) is associated with higher constitutive and inducible levels of transcription than the common (TNF1) allele and has been demonstrated to be a stronger transcriptional activator than the common allele in reporter gene assays (Wilson et al., 1997). In addition, there is a very strong association between the uncommon TNF allele and HLA-A1, B8 and DR3 alleles.

McGuire et al. (1994) found an increased prevalence of homozygotes for the TNF2 allele in a large case–control study of children with cerebral malaria in the Gambia, but there was no association with severe malarial anaemia. In this study population, the TNF2 allele was associated with HLA-B70, B50, Cw2, DRβ1*1302 and DRβ1*1101 but none of these HLA types was associated with susceptibility to cerebral malaria, i.e. the TNF association is independent of HLA class I and II variation. It was also shown that there was no relationship between TNF2 and HLA-B53, which is protective against severe malaria, and indeed these two alleles probably affect disease outcome by very different mechanisms.

HLA and mild falciparum malaria

Although previous HLA studies have identified class I, II and III genes that are associated with altered susceptibility to severe malaria (Hill et al., 1991; McGuire et al., 1994), they did not show clear associations with uncomplicated clinical malaria. Other studies have also failed to detect an association between mild malaria and individual alleles (Bennett et al., 1993). A more recent genetic linkage study reported a small, affected sib-pair study of Gambian families where two dizygotic twins had mild clinical malaria during a defined study period (Jepson et al., 1997). Using microsatellite markers within the MHC, linkage was demonstrated using only 22 sibling pairs. The data showed significant non-random sharing of alleles, suggesting that the MHC as a whole had a greater effect on malaria susceptibility than specific resistance alleles, at particular loci such as HLA-B, HLA-DRB1 and TNF.

Other immunogenetic factors and malaria infection

In addition to the compelling evidence supporting the role of the MHC in determining susceptibility to malaria infection, many other important immunogenetic factors are clearly involved. An epidemiological study in west Africa compared malaria infection rates, morbidity and antibody levels in three distinct ethnic groups living in the same endemic region (Modiano et al., 1996). The Fulani ethnic group showed lower parasite prevalence and fewer episodes of clinical infection. The interethnic differences could not be accounted for by differences in either haemoglobin S or HLA-B53 frequencies and it is suggested that the differences have another immunogenetic basis. Indeed, a diverse number of other malaria resistance genes has been characterized through case–control studies; these include the genes coding for glucose-6-phosphate dehydrogenase, intracellular adhesion molecule-1, α-thalassaemia, and complement receptor-1 (Hill, 1992; Williams et al., 1996; Fernandez-Reyes et al., 1997; Rowe et al., 1997).

HLA and malaria vaccine strategies

Malaria is a world-wide cause of morbidity and mortality, and an effective vaccine is urgently required in regions where drug resistance is increasing.

HLA-based vaccine strategies require knowledge of the following variables: (i) the common HLA molecules expressed in the population to be immunized; (ii) the HLA-restricted T cell epitopes present in the immunogen; and (iii) the epitope variants present in the population to be vaccinated. The discovery of HLA-B53-restricted CTL responses against epitopes expressed by the liver-stage or pre-erythrocytic parasite has made these epitopes potential targets for vaccination programmes. In animal models, CTL against liver-stage epitopes can confer resistance to infection by eliminating liver-stage parasites (Nardin and Nussenzweig, 1993). Mice and humans immunized with irradiated sporozoites can be protected from malaria (Nussenzweig and Nussenzweig, 1989).

Human malaria vaccine designs require a range of different HLA-restricted CS epitopes that reflect the diversity of HLA allele expression between different populations. Preliminary studies have demonstrated protection against experimental malaria challenges from vaccinations designed to induce antibodies directed against the CS protein present on the sporozoite surface (Stoute et al., 1997). There are now exciting new strategies based on the development of novel CTL-inducing recombinant vaccine constructs containing strings of *P. falciparum*-derived CTL epitopes restricted through common HLA alleles (Plebanski et al., 1998).

Leishmaniasis

Leishmaniasis is caused by a protozoan parasite of the genus *Leishmania*. In humans, several species cause infection which can be broadly subdivided into either cutaneous (*L. major, L. amazonensis*) or visceral (*L. donovani*) disease. Extensive information is available on the genetic regulation of cutaneous leishmaniasis in murine models but much less is known about human disease. MHC class II-restricted $CD4^+$ T cells play a major role in host immune response against the parasite, which resides as an intracellular amastigote in the phagolysosome of macrophages (Chakkalath et al., 1995). There is also evidence in murine models that MHC class I-restricted $CD8^+$ T cells prevent reinfection (da Conceição-Silva et al., 1994).

The mechanism by which parasites evade $CD4^+$ T cell recognition is poorly understood. *Leishmania* parasites reduce exogenous antigen presentation by macrophages *in vitro* (Prina et al., 1993) and downregulation of MHC class II molecules has been suggested as an explanation for this (Reiner et al., 1987). Kima et al. (1996) have proposed that leishmania antigens have reduced access to the MHC class II processing pathway. Interestingly, de Souza Leao et al. (1995) demonstrated that MHC class II molecules were endocytosed and degraded by amastigotes within the phagolysosome of mouse bone marrow-derived macrophages. However, the immunological significance of this mechanism is unclear, since other work has demonstrated that MHC class II expression by macrophages is unchanged 48 h postinfection (Prina et al., 1993; Lang et al., 1994).

There are few patient data on HLA associations with leishmaniasis. In a multicase family study in Venezuela, localized cutaneous leishmaniasis (*L. braziliensis, L. mexicana*) was associated with HLA-A28, -B22 and -DQ8 (Lara et al., 1991). In Brazil, mucocutaneous leishmaniasis (*L. braziliensis*) has been associated with a decrease in HLA-DR2 and an increase in HLA-DQ3 (Petzl-Erler et al., 1991). In Venezuelan patients with mucocutaneous leishmaniasis (*L. braziliensis*), a higher frequency of the –308 TNFα promoter polymorphism was observed (Cabrera et al., 1995). High TNFα levels correlate with severe mucocutaneous disease and the functional significance of the −308 allele promoter variant has been discussed above. In an Iranian population study, visceral leishmaniasis (*L. donovani*) has been associated with HLA-A26 (Faghiri et al., 1995).

Filariasis

Little information is available on the role of HLA molecules in onchocerciasis and earlier associations with Brugian (or lymphatic) filariasis have not been confirmed (Yazdanbakhsh et al., 1997).

Onchocerciasis, or "river blindness", is caused by the nematode *Onchocerca volvulus* and results in a spectrum of ocular and skin pathology in humans. In a large population study in west Africa, HLA class II haplotypes were identified in patients with onchocerciasis compared with asymptomatic controls (Meyer et al., 1994). The HLA haplotype, DQA1*0101–DQB*0501, and the HLA allele, DQB1*0201, were found more frequently in generalized onchocerciasis (characterized by skin atrophy, ocular disease, blindness, high cutaneous microfilaria counts and low proliferative T cell responses to *O. volvulus*). HLA-DPB1*0402 was associated with localized disease (characterized by focal dermal pathology, low microfilaria counts and high proliferative T cell responses). The HLA haplotype DQA1*0501–DQB1*0301 was more frequent among asymptomatic subjects who were considered to be "putatively immune". Notably, the HLA-DQB*0501 allele is also associated with protection from severe malaria anaemia (Hill et al., 1991). The authors speculated that greater protection against severe malaria might be at the expense of an increased susceptibility to river blindness.

More recently, HLA-DQA1*0501 and -DQB1*0301 were also found to be associated with skin depigmentation, a chronic feature of onchocerciasis, compared with individuals with normal skin but high microfilarial loads (Murdoch et al., 1997). A decreased frequency of HLA-DQA1*0101 was also reported in subjects with depigmentation. It was proposed that immunological tolerance to *O. volvulus* degenerates over time in individuals with the HLA alleles DQA1*0501 and DQB1*0301, resulting in clinical disease.

HLA AND VIRAL INFECTIONS

Introduction to HLA and viral infections

Host genetic factors are important determinants of susceptibility to viral infections in humans. Zinkernagel and Doherty (1974) discovered that MHC class I-restricted CTL responses to viral antigens play a major role in the destruction of virally infected cells. However, there have been surprisingly few documented HLA class I alleles associated with protection against human viral infections. The most convincing associations with viral infections have been with HLA class II molecules. For example, certain HLA class II antigens increase disease susceptibility in infection with human lymphotropic virus-I (HTLV-I) (Jeffery et al., 1999) and human papillomavirus (HPV), and certain HLA class II antigens are associated with clearance of hepatitis B and C viruses. Following the introduction of more sensitive molecular typing techniques for identification of low-frequency HLA class I variants, class I disease associations have become less difficult to establish. Indeed, certain HLA class I alleles have recently been shown to be protective against the progression of disease in patients infected with HTLV-I (HLA-A*02; Jeffery et al., 1999) and HIV-1. HLA class I heterozygosity has been demonstrated to be protective against rapid progression to acquired immunodeficiency syndrome (AIDS) after HIV-1 infection (Carrington et al., 1999). Heterozygote advantage, also documented for HLA class II-restricted T cell responses against hepatitis B infection (Thursz et al., 1997), probably results from an increase in the number of MHC-restricted epitopes that can be presented to T cells and subsequently a stronger immune response. Ultimately, identification of protective HLA class I associations offers exciting prospects for the development of effective vaccines that elicit strong antiviral CTL responses.

Human immunodeficiency virus

HLA associations with HIV-1 infection have been studied extensively, but few data are available for HIV-2. The clearest evidence for HLA involvement in determining the speed of HIV-1 disease progression is from a study of 95 HIV-1 seropositive sibling pairs with haemophilia. Sibling pairs sharing one or two, but not zero, MHC haplotypes were found to be significantly concordant in CD4$^+$ T cell decline and AIDS status within 5 years of seroconversion (Kroner et al., 1995). There is good evidence that HLA class I haplotypes are associated with the rate of disease progression but HLA class II associations are inconsistent. Recently, Carrington et al. (1999) reported a selective advantage and delayed AIDS onset among patients with maximum HLA heterozygosity of class I loci (A, B and C). Non-HLA genes, located within the MHC complex and in linkage disequilibrium with HLA genes, may also have important effects.

HLA class I and human immunodeficiency virus

In several studies, the extended haplotype HLA-A1–B8–DR3 has been associated with accelerated HIV-1 progression using various measures of outcome including CD4$^+$ T cell decline, HIV-1-positive versus AIDS, clinical stage and time to AIDS (Just, 1995; Rowland-Jones et al., 1997; Roger, 1998) (see Table 20.2). Interestingly, HLA-B8 is known to bind selectively highly variable viral epitopes, such as Gag, that fail to stimulate protective CD8$^+$ T cell responses (McAdam et al., 1995). In addition, it is notable that no epitope has yet been described that is presented by HLA-A1, despite the relative frequency of this HLA type (McMichael and Walker, 1994). Furthermore, when the alleles HLA-A1–B8 are split from DR3, the association with rapid disease progression segregates with HLA-A1–B8 and not HLA-DR3 (Mann et al., 1994).

HLA-B35 has also been found to be associated with rapid HIV-1 progression in several studies using similar outcome measures (see Table 20.2). Most recently, HLA-B35 was reported to be associated with the rapid development of AIDS in Caucasians (Carrington et al., 1999). Tomiyama et al. (1997) reported that natural mutations of HLA-B35-restricted epitopes could result in failure of peptide binding and/or CTL receptor recognition. In contrast, a protective role for HLA-B35 (the most common Gambian HLA class I molecule) has been suggested in Gambian female prostitutes, apparently resistant to HIV-1 and -2 infection (Rowland-Jones et al., 1995). In this study, cross-reactivity was demonstrated between HIV-1 and -2 epitopes presented by HLA-B35-restricted CTL responses from HIV-infected donors. Initial exposure to HIV-2 (which is less pathogenic than HIV-1) might therefore also confer immunity to HIV-1 in these high-risk individuals. This may be an example of where conclusions drawn from one study population are not applicable to another because of differences in both the prevalent HLA haplotypes and virus type, i.e. HIV-1 or -2.

Slower disease progression has been consistently associated with HLA-B27 and -B57, which are rare alleles in most populations. Although the reason for this is unclear, these molecules may selectively bind more highly conserved immunodominant HIV peptide sequences more efficiently than other HLA types (McMichael and Walker, 1994).

HLA class II and human immunodeficiency virus

Some associations have been described for HLA class II alleles but reports are conflicting. Certain HLA-DR5 and -DR6 alleles have primarily been associated with accelerated disease but correlated with slow HIV progression in one study (Itescu et al., 1994). One proposed mechanism is molecular mimicry, since considerable homology between specific amino acids of the HIV gp120 region and HLA class I and II molecules has been reported in several studies. In particular, Itescu et al. (1994) found remarkable similarity between HLA-DR5 and -DR6 and the C-terminal region of the HIV-1 envelope V3 loop. It has previously been suggested that AIDS may essentially be an autoimmune disease. If this is true, the authors speculated that this form of HLA mimicry might protect the host by preventing the destruction of HIV-infected CD4$^+$ T cells by deletion of self-reactive CTL.

HLA-DR1 and -DR3 haplotypes are more commonly associated with rapidly progressive disease. However, these genes may represent surrogate markers for closely linked disease genes. For example, HLA-DR3 is in linkage disequilibrium with TNFα- and lymphotoxin (LT or TNFβ)-coding genes (Pociot et al., 1993), as well as HLA-A1–B8 (Mann et al., 1994).

TABLE 20.2 MHC associations with human immunodeficiency virus (HIV) progression

MHC allele	Cases (controls)*	Speed of progression	Risk group	Population	Reference
HLA-B35	51	Rapid	Mixed	Italian	Scorza-Smeraldi et al. (1986)
HLA-A1–B8–DR3	9(7)[a]	Rapid	Haemophiliac	Scottish	Steel et al. (1988)
HLA-DR2, B35, Cw4	51	Rapid	Mixed	Swiss	Jeannet et al. (1989)
HLA-A1–Cw7–B8–DR3, A24	49(59)[a]	Rapid	Homosexual	European	Kaslow et al. (1990)
HLA-DR3–DQ2	76	Rapid	Haemophiliac	Italian	Fabio et al. (1990)
HLA-A1–B8–DR3, B21, B35	180	Rapid	Mixed	European	Kaplan et al. (1990)
HLA-DR1, DR4		Slow			
HLA-B35	46	Rapid	Homosexual	Cauc. American	Itescu et al. (1992)
HLA-DR3, DR5	65	Rapid	Haemophiliac	Cauc. Australian	Donald et al. (1992)
HLA-DR4		Slow			
HLA-B35	144	Rapid	Haemophiliac	French	Sahmoud et al. (1993)
HLA-B35	106	Rapid	Homosexual	Dutch	Klein et al. (1994)
HLA-B*35, Cw*04	498	Rapid	Mixed	Cauc. American	Carrington et al. (1999)
HLA-A1–B8–DR3	313	Rapid	i.v. drug users	Scottish	McNeil et al. (1996)
HLA-B27		Slow			
HLA-DQB1*0301	81	Rapid	Perinatal trans.	Spanish	Just et al. (1996)
HLA-B37, B49	241	Rapid	Homosexual	Cauc. American	Kaslow et al. (1996)
HLA-A28 and TAP2.3					
HLA-A29 and TAP2.1					
HLA-B8 and TAP2.1					
HLA-B35 and C4					
HLA-DRB1*0401–DQA1*0300–DQB*0301 and TAP1.2					
HLA-DRB1*1200–DQA1*0501–DQB*0301 and TAP1.2					
HLA-DRB1*1300–DQA1*0102–DQB*0604 and TAP1.2					
HLA-DRB1*1400–DQA1*0101–DQB*0503 and TAP1.2					
HLA-B27, B51, B57		Slow			
HLA-(A25 or A26 or A32 or B18) and TAP2.3					
HLA-A1–B8–C4AQ*0–DR3	26	Rapid	Homosexual	Cauc. Australian	Cameron et al. (1990)
HLA-A11–B35–C4BQ*0–DR1					
HLA-C4AQ*0, C4BQ*0	54	Rapid	Homosexual	Luxembourg	Hentges et al. (1992)
TNFβ polymorphism (TNF-c2 microsatellite)	44	Slow	Homosexual/ haemophiliac	England	Khoo et al. (1997)

[a] Case–control study. i.v.: intravenous; trans.: transmission; Cauc.: Caucasian.

HLA class III and human immunodeficiency virus

Deletion or non-functional mutations of the complement gene C4A is associated with a more rapid clinical course to AIDS (Hentges et al., 1992). Although reduced complement levels may result, the mechanism is unclear since plasma C4 concentration does not correlate with disease status in HIV-infected individuals.

Other MHC genes and human immunodeficiency virus

Non-HLA genes, such as the TAP (transporter associated with antigen processing)-1 and -2 genes, in linkage disequilibrium with HLA genes may also affect HIV progression. TAP1/TAP2 proteins transport peptides from the cytosol to the endoplasmic reticulum, where they bind to HLA class I molecules and export class I molecule/peptide complexes to the cell surface. Mutations or polymorphisms in TAP genes, which are located in the MHC class II region, may disrupt antigen processing and have been implicated in HIV infection. Using a novel study design, Kaslow et al. (1996) identified specific combinations of TAP gene variants with several class I and II haplotypes which were associated with the speed of progression to AIDS (Table 20.2). Although this mechanism may be important, HIV has evolved TAP-independent pathways for processing certain endogenously expressed viral proteins. HIV-1 envelope protein, a molecule that naturally translocates to the endoplasmic reticulum, is one of only a small number of viral proteins that are processed in this manner (Hammond et al., 1993, 1995).

The class III region contains the TNFα and LT genes, which are interposed between the HLA class I and II regions. Both of these are potent inducers of HIV replication and may accelerate disease progression and $CD4^+$ T cell depletion. Microsatellite polymorphisms (TNF-c2) within the first intron of the LT gene were recently identified in patients with slow rates of $CD4^+$ T cell depletion, however, it was not reported whether this resulted in reduced LT production (Khoo et al., 1997). In contrast, no association was found between progression to AIDS and several TNFα promoter region polymorphisms in another study (Brinkman et al., 1997).

HLA and human immunodeficiency virus vaccine strategies

A major objective of current HIV vaccination strategies is the induction of HIV-1-specific HLA class I-restricted CTL responses. The most formidable obstacle to this is viral variability, particularly the hypervariability of the envelope protein. HIV-1 can be divided into two distinct groups, O and M, based on phylogenetic analysis of the envelope gene. The major group, M, can be further subdivided into at least nine different clades (designated A–I); B is the most prevalent clade in the West.

A potential way of overcoming the hypervariability of HIV is to develop vaccines using immunodominant viral epitopes that are well conserved within and between clades. There is encouraging evidence suggesting that this approach may be possible. For example, Africans infected with HIV clade A, C and G subtypes demonstrate cross-recognition of clade B virus Gag, Pol and Nef proteins (Betts et al., 1997; Cao et al., 1997). However, another major factor which determines the immunological breadth of the CTL response is the HLA class I haplotype of the host. In a recent study, low CTL reactivity following vaccination with gp120 was attributed to the possession of HLA-A11 by the host (Ferrari et al., 1997). Multiple peptide vaccines may therefore have the best chance of effectively immunizing diverse populations, although HLA-specific vaccines may need to be offered to certain groups.

Another exciting vaccine strategy currently being explored stems from the discovery that certain antigenic peptides have the ability to bind to multiple HLA class I molecules ("degenerate" binding). Furthermore, recognition of cells presenting the same peptide in the context of a different class I molecule has also been reported ("promiscuous" recognition) (Oldstone et al., 1992). Another approach to overcome the problem resulting from HLA restriction might therefore be to use such peptides as vaccines. Recently, Threlkeld et al. (1997) reported that a variety of HIV-1 peptide epitopes, in the context of both HLA-A3 and -A11, could elicit such degenerate and promiscuous CTL recognition.

Hepatitis B virus

HLA and hepatitis B infection

Infection with HBV results in a wide spectrum of disease, ranging from asymptomatic carriage to fulminant hepatic failure. The majority of infected individuals will develop acute hepatitis and clear the virus, but approximately 10–15% of people become persistent carriers. Chronic HBV carriage, which predisposes patients to chronic liver disease and hepatocellular carcinoma, is responsible for significant world-wide morbidity. The fact that outcome of HBV infection appears unrelated to the virulence properties of HBV strains suggests that the host immune response is likely to play an important role in determining disease manifestation.

There is much evidence to suggest that both cellular and humoral responses to HBV-encoded antigens are required for HBV clearance. While the humoral antibody response to viral envelope antigens contributes to the clearance of circulating virus particles, HBV-specific CTL play a major role in controlling infection through lysis of infected cells and local cytokine release (Chisari and Ferrari, 1995; Guidotti et al., 1996). Indeed, the generation of adequate HLA class I- and class II-restricted T cell responses to the virus is a critical determinant of HBV infection outcome (Chisari and Ferrari, 1995). Patients with transient infection develop strong and multispecific HLA class I- and class II-restricted T-cell responses (Ferrari et al., 1991; Penna et al., 1991; Nayersina et al., 1993; Rehermann and Chisari, 1995), in contrast to weak responses from patients who become persistent HBV carriers (Ferrari et al., 1990; Jung et al., 1995).

HLA class I and hepatitis B infection

Patients with transient hepatitis develop a vigorous, polyclonal, HLA class I-restricted CTL response against multiple epitopes in the HBV envelope, nucleocapsid and polymerase proteins (Penna et al., 1991; Nayersina et al., 1993; Chisari and Ferrari, 1995). Failure to clear the virus is associated with weak HBV-specific CTL responses. The majority of CTL epitopes is HLA-A2 restricted; there is a single nucleocapsid epitope (Penna et al., 1991), 11 epitopes in the HBV envelope protein (Nayersina et al., 1993) and five in the polymerase protein (Rehermann and Chisari, 1995). In addition, an HLA-A31- and HLA-A68-restricted CTL response to a single HBV nucleocapsid epitope occurs during acute HBV infection (Missale et al., 1993).

Among the different CTL epitopes so far identified, sequence 18–27 of the HBV core molecule (HBc18–27) is widely recognized by CTL of HLA-A2-positive patients with transient infection (Vitiello et al., 1995). HBc18–27 has been extensively characterized and can be recognized by specific CTL when presented in the context of a broad range of HLA-A2 subtypes. HBc18–27 can also stimulate HLA class II-restricted T cell responses (Bertoletti et al., 1993, 1997). It represents the main component of peptide-based therapeutic vaccination strategies designed to stimulate the HBV-specific CTL immune response in patients with chronic HBV infection.

HLA class II and hepatitis B infection

A vigorous HLA class II-restricted, $CD4^+$ helper T cell response occurs in almost all patients with acute self-limited hepatitis. In contrast to the multispecific class I-restricted CTL responses, there is marked dominance of the class II response to HBV by the viral nucleocapsid, with correspondingly fewer HBV-specific $CD4^+$ helper T cell epitopes (Chisari and Ferrari, 1995).

In west Africa, 90% of the population have been infected with HBV before adulthood. In a large two-stage (children and adults) case–control study in the Gambia, the frequency of MHC class I antigens and class II haplotypes was studied in patients with either transient or persistent infection (Thursz et al., 1995a). HLA-DRB1*1302 was associated with clearance of HBV infection; individuals with this HLA type were less likely to become persistent carriers, who are at high risk of chronic liver disease. A smaller case–control study of Caucasian patients also demonstrated that the HLA-DRB1*1302 allele was protective against chronic hepatitis associated with HBV persistence (Hohler et al., 1997). As discussed previously, HLA-DRB1*1302 is also associated with a reduced risk of severe malaria in Gambian children (Hill et al., 1991). Malaria and HBV are both major causes of premature death in west Africa. Indeed, HBV surface antigen carriage was significantly increased amongst cases of severe malaria in a case–control study of over 1200 Gambian children (Thursz et al., 1995b). It would therefore seem likely that there is a selection advantage for the protective HLA-DRB1*1302 allele in the Gambia associated with both of these infections.

A large study of Gambian subjects with HBV infection not only confirmed the lower frequency of HLA-DRB1*1302 in persistent carriers, but also demonstrated that heterozygosity at the MHC class II loci afforded protection from persistent infection (Thursz et al., 1997). Heterozygote advantage was seen for all class II haplotypes, making this effect independent of the protective association with HLA-DRB1*1302. The equivalent differences were not observed at the MHC class I loci, for which there is a far greater number of HBV-specific T-cell epitopes. The authors suggested that heterozygosity may increase the number of MHC class II-restricted epitopes that can be presented to T cells, so increasing the patient's ability to clear the virus.

HLA-DR6, the serological supertype of DRB1*1302 and DRB1*1302, has been identified as being protective against persistent HBV infection. van Hattum and Schreuder (1987) found significantly more HLA-DR6-positive north European patients with transient HBV infection than with persistent infection. In addition, HLA-DR6 was associated with a more favourable response of chronic HBV infection to interferon (IFN) therapy (Scully et al., 1990). In two consecutive and independent studies of Qatari patients with persistent HBV, HLA-DR2 was associated with significant protection from both acute and persistent infection. HLA-DR7 was associated with increased susceptibility to persistent infection. It should be noted that the sample size in the second-stage study was small (Almarri and Batchelor, 1994).

HLA class III and hepatitis B infection

Few studies have examined the role of other immunogenetic factors in the pathogenesis of chronic HBV infection. A recent report suggests that there might be an association between a TNFα promoter polymorphism at position –238 and the development of chronic HBV infection. This was unexplained by linkage disequilibrium to HLA class I or II alleles (Hohler et al., 1998c) and patient numbers were small.

HLA and hepatitis B vaccine strategies

HBV vaccinations have been highly effective in preventing disease, probably through $CD4^+$ helper T cell-dependent production of anti-hepatitis B surface antigen antibodies (Egea et al., 1991; Deulofeut et al., 1993; Lemon and Thomas, 1997). Non-responsiveness to vaccination (5% of individuals) is significantly influenced by the MHC and is subject to considerable interethnic variation: 20% of oriental patients are non-responders; this has been associated with the common haplotype, HLA-B54–DR4–DR53 (Watanabe et al., 1988), while Caucasian non-responders show an increased frequency of HLA-DR3 and DR7 (Craven et al., 1986) and of the extended MHC haplotypes HLA-B8–SCO1–DR3 and B44–FC31–DR7 (Alper et al., 1989). There is a pattern of inheritance of non-responsiveness that suggests a recessive trait with an increased frequency of HLA-DR3 and -DR7 homozygotes in the non-responders (Kruskall et al., 1992). In a recent two-stage study (adults and infants), non-responsiveness was associated with HLA-DRB1*3 and -DRB1*7; DRB1*13 was increased among vaccine responders (Hohler et al., 1998a). In a large series of haemodialysis patients, a population known to have high rates of non-responsiveness, HLA-DR3 and -DR14 were associated with a poor response to vaccination, whereas -DR1 and -DR15 were associated with a good response (Caillat-Zucman et al., 1998).

Hepatitis C virus

Over 70% of patients infected with the hepatitis C virus (HCV) develop chronic infection with persistent viraemia and detectable HCV RNA, potentially leading to cirrhosis and hepatocellular carcinoma (Alter et al., 1992). Susceptibility to chronic HCV infection is determined largely by the route of infection, inoculum size and viral genotype (Dusheiko et al., 1994; Tong et al., 1995). However, a small percentage of patients clears the virus (with loss of detectable HCV RNA) independently of these factors, the mechanism for which is unclear. There is increasing evidence that MHC class I- and particularly class II-restricted T cell responses play a central role in determining outcome of HCV infection and the response of HCV infection to treatment.

HLA class I and hepatitis C infection

Recent reports have identified strikingly broad and heterogeneous HLA class I-restricted (A2, A3, A11, A23, B7, B8, B53) CTL responses specific for HCV proteins among liver infiltrating lymphocytes of patients with chronic HCV infection (Koziel et al., 1995; Wong et al., 1998). However, there was no correlation between the presence of a detectable CTL response and viral load or eradication. Therefore, the role of HLA class I-restricted CTL responses in the pathogenesis of chronic HCV infection remains to be clarified.

HLA class II and hepatitis C infection

There are associations between class II alleles and protection from, or susceptibility to, chronic HCV infection. A lower frequency of HLA-DR5 was found in Italian HCV-infected individuals with chronic liver disease than those with asymptomatic infection or in healthy controls (Peano et al., 1994; Zavaglia et al., 1996), although neither study defined patient groups according to HCV RNA status. This association was maintained for HLA-DR5-positive haplotypes in strong linkage disequilibrium with HLA-DQB1*0301, but was lost for those carrying HLA-A9 or HLA-B35, which usually cosegregates with DR5 in Italian populations. In a more recent study of Italian patients with chronic HCV infection as defined by the presence of detectable HCV RNA, the *trans*-dimer DQA1*0201–DQB1*0201 predisposed to chronic HCV infection, whereas the DRB1*1104–DQA1*0501–DQB1*0301 haplotype conferred protection (Zavaglia et al., 1998). The involvement of the DQA1 allele in viral clearance is doubtful because the tight linkage disequilibrium within the DR/DQ region dictates that the DQA1*0501 allele cosegregates with DRB1 haplotypes.

Minton et al. (1998) found that HLA-DRB1*1101, which is the main DR5 allele in Europeans, was associated with clearance of HCV infection in England. A protective association with HLA-DQB1*0301 in this study was consistent with two smaller studies of English HCV-infected individuals (Tibbs et al., 1996; Cramp et al., 1998) and another series of French patients (Alric et al., 1997). The strong association with the DQB1*0301 allele in three different populations suggests that this is an important HLA class II allele in promoting viral clearance. As discussed for *DQA1*, it would seem likely that any association between the *DRB1* allele and viral clearance has arisen as a result of tight linkage with DQB1*0301.

In contrast to the above reports, a study of Japanese patients showed that extended HLA haplotypes including the HLA-B54 allele were associated with progressive HCV-induced liver injury, whereas extended haplotypes including class II DRB1*1302–DQB1*0604 were associated with enhanced viral clearance (Kuzushita et al., 1998).

Preliminary data suggest that MHC class II alleles may play a role in determining the response of patients with chronic HCV to IFNα therapy (Almarri et al., 1998).

HLA class III and hepatitis C infection

The precise mechanisms underlying viral clearance are not understood. The protective MHC class II alleles may function through their role in optimal presentation of HCV-specific antigens, but it is also possible that these associations might arise through linkage disequilibrium with other genes.

TNFα may play an important pathogenic role in HCV infection. Elevated plasma levels of TNFα have been found in patients with acute and chronic HCV (Tilg et al., 1992; Torre et al., 1994) and are associated with non-responsiveness to IFNα therapy (Gonzalez-Amaro et al., 1994). Liver-infiltrating CTL have been shown to produce TNFα (Koziel et al., 1995). Hohler et al. (1998b) demonstrated that the TNFα promoter polymorphism at position –238 was associated with an increased susceptibility to chronic HCV infection, which could not be explained by linkage disequilibrium with HLA-B or –DR genes.

Epstein–Barr virus

HLA, Epstein–Barr virus and lymphoblastoid cell lines

Epstein–Barr virus (EBV), a human gamma herpesvirus, is carried by approximately 90% of the population as an asymptomatic infection. In developing countries primary infection occurs in the first few years of life and clinically manifest infectious mononucleosis (IM) is rare. In western populations, primary infection is delayed until adolescence in approximately 20% of individuals, half of whom develop IM. The host immune response is critical in determining outcome of viral infection and controlling virus carrier status. Disturbances of this host–virus balance with loss of host immune surveillance by EBV-specific CTL can lead to a diverse spectrum of clinical disease, including EBV-associated malignancies. The relatively high frequency of EBV-driven lymphoproliferative tumours in T cell-immuno-compromised patients (Craig et al., 1993) suggests that cell-mediated immune responses play a major role in the control of EBV infection. Indeed, donor CTL infusions have resulted in dramatic resolution of EBV-induced lymphomas in recipients of bone-marrow transplants (Papadopoulos et al., 1994; Rooney et al., 1995).

EBV causes a markedly immunogenic immortalizing infection of B lymphocytes *in vitro* to produce lymphoblastoid cell lines (LCL). LCL have two distinct cycles: a latent cycle characterized by virus-driven expansion of infected B cell clones and a lytic cycle characterized by viral replication. These stages are equivalent to the EBV carrier state and primary infection, *in vivo*, respectively (Sixbey et al., 1984; Tierney et al., 1994).

Latently infected LCL express six EBV-coded nuclear antigens (EBNAs 1, 2, 3A, 3B, 3C, -LP) and two latent membrane proteins (LMP1 and 2). LMP1 is thought to play a major role in the growth-transforming ability of the virus (Rickinson and Moss, 1997). These proteins elicit primary and memory $CD8^+$ T cell responses that display strong HLA restriction in the EBV system and maintain a stable host–virus balance. The LCL lytic phase is characterized by viral replication and production of virus-coded immediate-early (BZLF1) and early (BHRF1) proteins.

HLA and Epstein–Barr virus infection

Many of the EBV-specific CTL responses have been mapped to defined epitopes and are strongly influenced by HLA class I alleles (Rickinson and Moss, 1997). The influence of HLA class II-restricted $CD4^+$ T cell responses is smaller and less clearly defined (Misko et al., 1984). Studies of Caucasoid donors, using type I virus isolates (the prevalent virus strain in this population), have shown a hierarchy of immunodominant EBV-specific CTL responses. The dominant responses are to EBNA3A, 3B or 3C proteins, with subdominant responses recognizing other latent proteins including LMP2 greater than LMP1, EBNA2 and EBNA-LP (Gavioli et al., 1992; Khanna et al., 1992; Murray et al., 1992). The differences in immunogenicity of particular epitopes in this population probably reflect the very strong influence of the common Caucasoid HLA-A3, A11, B7, B8, B44 alleles which optimally present EBNA3A, 3B and 3C epitopes (Khanna et al., 1992; Gavioli et al., 1993; Burrows et al., 1994; Hill et al., 1995).

The EBV system has evolved other pathways of antigen processing separate from the typical TAP-dependent HLA class I processing route. *In vitro* $CD8^+$ CTL responses to EBNA1 are extremely rare, which appears to be the result of a protective internal glycine–alanine repeat that appears to render the protein inaccessible to HLA class I processing (Levitskaya et al., 1995), possibly by restricting proteasome-mediated proteolytic cleavage of antigen (Rickinson and Moss, 1997a). However, both HLA class I and II-restricted T lymphocyte responses have been detected from human isolates. HLA class II-restricted EBNA1-specific CTL responses occur against LCL supplied with exogenous antigen (Khanna et al., 1995, 1997a). Blake et al. (1997) isolated human $CD8^+$ CTL clones recognizing EBNA1-specific peptides in the context of the allele products at HLA-B*3501 and HLA-A*203. Results from this study showed that EBNA1, supplied as exogenous antigen, was presented on HLA class I molecules by a TAP-independent pathway. The authors suggested that this was a potential mechanism for *in vivo* priming of EBNA1-specific CTL.

The second TAP-independent pathway for antigen presentation involves the LMP2 antigen. TAP-negative LCL are unable to present the immunodominant epitopes, but are still able to present CTL epitopes within endogenously expressed LMP2, the mechanism for which is not understood (Khanna et al., 1996; Lee et al., 1996). This is analogous to the processing of HIV-1 envelope protein (see above), although the transmembranous LMP2 protein does not have a cytoplasmic tail that extends into the endoplasmic reticulum. More recently, a $CD8^+$ T-cell clone isolated from two siblings with a peptide TAP deficiency was found to recognize LMP2 presented in the context of HLA-B molecules, which were the predominant cell-surface expressed HLA class I molecules (de la Salle et al., 1997).

Epitope choice is highly allele specific and individual alleles present only two or three epitopes, one of which is immunodominant (Rickinson and Moss, 1997). For example, HLA-A11 induces a response to EBNA3B 416–424 (immunodominant) and to EBNA3B 399–408 (subdominant) in HLA-A11-positive people (Gavioli et al., 1993); HLA-B8 induces a response to EBNA3A 325–333 (immunodominant) and to EBNA3A 158–166 (subdominant) in HLA-B8-positive individuals (Burrows et al., 1994; Rickinson and Moss, 1997). Immunodominance of one epitope over another restricted through the same alleles may occur as a result of differential peptide/MHC complex expression (Levitsky et al., 1996). HLA-restricted epitopes are highly conserved for particular virus strains within certain populations. HLA-A11 epitopes are highly conserved for type I EBV strains from Caucasoid and African populations, whereas the epitope sequences are different in Chinese and Papua New Guinea isolates, two populations with a very high prevalence of the HLA-A11 allele (de Campos-Lima et al., 1993, 1994; Burrows et al., 1996). The sequence changes involved selective mutation of epitope anchor residues affecting peptide/A11 binding, resulting in loss of CTL recognition. These findings suggested that such mutations occurred as a result of selection pressure in those particular communities. In HLA-B8-positive individuals, the HLA-B8 restricted CTL memory response to the immunodominant EBNA epitope has shown to be composed of clones with highly conserved T cell receptors (TCR) (Argaet et al., 1994). These findings suggested that following long-term exposure to EBV, there is selection for memory CTL that optimally bind peptide/HLA complexes.

HLA and infectious mononucleosis

Primary infection with EBV can cause IM and is associated with detectable HLA-restricted EBV-specific $CD8^+$ T cell responses against both latent (EBNA3A, 3B, 3C) and lytic cell (BZLF1) antigens (Steven et al., 1996, 1997). The best-characterized CTL response against both latent and lytic cell antigens has been in the context of HLA-B8 (Steven et al., 1996). HLA-B8-positive IM patients with a primary CTL response to EBNA3A show clonal expansion of T lymphocytes with a highly conserved TCR region that is also found on the memory CTL of viral carriers (Silins et al., 1996). This suggests that these particular clones might be preferentially selected into the memory CTL response, thus optimizing long-term CTL surveillance and the maintenance of the viral carrier state.

HLA and Epstein–Barr virus-associated malignancies

It is clear that in all previously EBV-infected individuals, the virus persists for life as a latent infection, kept under check by a population of EBV-specific CTL. Loss of immune surveillance can lead to uncontrolled virus-driven cell proliferation and the development of EBV-driven neoplasia.

There are two well-defined immunopathogenetic pathways underlying EBV-driven malignancies. First, impairment of the EBV-specific CTL response, as occurs in patients on immunosuppressive therapy, may result in lymphoproliferative disease. For example, the incidence of post-transplant lymphomas correlates with the intensity of immunosuppression. Secondly, EBV-associated malignancies can develop despite ongoing CTL-effected immune surveillance. This usually occurs in the context of phenotype changes of infected cells whereby the normal MHC-restricted immune responses are evaded, e.g. Burkitt's lymphoma (BL) and nasopharyngeal carcinoma (NPC).

HLA and Burkitt's lymphoma

BL is an aggressive, EBV-driven childhood tumour. In striking contrast to LCL, BL cell lines display an immunologically silent phenotype such that this particular EBV-driven malignancy has escaped HLA class I-restricted immune recognition (Rooney et al., 1985, 1986). The reasons for this are twofold: (i) the viral genome is down-regulated and EBNA1 is the only viral protein detectable in BL cells (Rowe et al., 1987); and (ii) BL lines demonstrate very low levels of expression of HLA class I and TAP1/TAP 2 gene products such that there is a marked antigen-processing defect (Khanna et al., 1994; Rowe et al., 1995). IFNγ and LMP1 are able to restore endogenous antigen processing in BL cell lines by upregulating TAP and HLA class I proteins, so restoring accessibility to CTL surveillance (Rowe et al., 1995).

While attention has largely been focused on the role of HLA class I-restricted CD8$^+$ CTL responses and recognition of EBV-associated malignancies, there is growing evidence in support of a role for CD4$^+$ T cells in determining outcome of EBV infection (White et al., 1996). Khanna et al. (1997b) demonstrated that BL cells expressed high levels of HLA-DR and HLA-DQ molecules, and were recognized by HLA class II-restricted EBV-specific CTL. The identification of HLA class II antigen processing in BL cells has clear implications for future immunotherapeutic strategies against EBV-driven malignancies.

HLA, nasopharyngeal carcinoma and Hodgkin's disease

EBV is associated with all cases of NPC and 40% of cases of Hodgkin's disease (HD) (Rickinson and Moss, 1997). In contrast to BL, these tumours grow in patients with intact antigen-processing pathways (TAP-positive, HLA class I-positive phenotype) and normal systemic CTL surveillance (Agathanggelou et al., 1995; Frisan et al., 1995; Oudejans et al., 1996). EBV-specific CTL can be reactivated from the blood of patients, but not from tumour-infiltrating lymphocytes, suggesting local tumour-mediated CTL suppression (Frisan et al., 1995). Similar to BL, NPC and HD express LMP1 but the remaining immunodominant antigens are downregulated. However, unlike BL, EBV within these tumours expresses LMP1 and also LMP2. The CTL responses to LMP1 are poorly defined, but there is an increasing number of CTL epitopes recognized in LMP2 restricted for a number of HLA class I alleles, including HLA-A2.01, A2.06, A11, A24 and B60 (Lee et al., 1993, 1997). These alleles are common in Caucasoid and south-east Asian populations. It is likely that weak CTL responses against LMP2 epitopes and local tumour-derived cytokine suppression of CTL favour tumorigenesis. The fact that the LMP2 epitopes are highly conserved between populations has implications for the design of CTL-based antitumour therapy.

HLA and Epstein–Barr virus vaccine strategies

A key strategy for an EBV vaccine that will afford protection against IM and development of post-transplant lymphoproliferative disease in EBV seronegative allograft recipients is based on induction of EBV-specific CTL. Trials are underway vaccinating HLA-B8-positive, EBV-negative individuals with a construct consisting of the HLA-B8-restricted epitope sequence FLRGRAYL (from EBNA3A) and tetanus toxoid emulsified in the water-in-oil adjuvant Montanide ISA 720 (Moss et al., 1996, 1998). This phase I study should accurately determine the safety and immunogenicity of EBV CTL epitope vaccines. In the long term, the vaccines will require multiple CTL epitopes to accommodate the enormous diversity of HLA alleles expressed across different populations. Indeed, this may well be achieved since it has been found that vaccine constructs, containing multiple synthetic epitope peptides, are able to elicit effective CTL responses against all epitopes in a murine model (Thomson et al., 1995, 1996). More recently, Thomson et al. (1998a) described a DNA plasmid encoding a polyepitope or "polytope" protein, containing mixtures of minimal murine CTL epitopes, that elicited HLA-restricted CTL responses to each of the epitopes.

Although current immunological strategies are aimed at eliciting HLA class I-restricted CD8$^+$ CTL responses, it is likely that vaccines that include both CD4$^+$- and CD8$^+$-T-cell epitopes will be more effective in controlling viral infections. Lysosome-targeted (i.e. targeted to HLA class II processing compartments) recombinant polytope constructs have been shown to elicit EBV-specific CD4$^+$ CTL responses *in vitro*, findings that have important implications for future vaccine design (Thomson et al., 1998b).

As already discussed in the previous sections, other EBV-driven malignancies (BL, NPC and HD) arise as a result of various escape mechanisms which evade the existing EBV-specific CTL responses. Therapeutic vaccines designed to target EBV and non-EBV-encoded tumour antigens are therefore likely to be the most successful treatment strategies.

Human papillomavirus

HPV play a major aetiological role in the development of cervical intraepithelial neoplasia (CIN) (Koutsky et al., 1992), a disease that may progress to invasive carcinoma. HPV infection is relatively common in healthy young women, suggesting that factors other than virulence properties of the virus influence the progression of cervical disease. Of the 70 HPV types which have been identified to date, HPV16 and HPV18 carry the highest risk with regard to the development of cervical carcinoma.

Han et al. (1992) showed that an increased risk of oncogenic transformation with the related cottontail rabbit papillomavirus was associated with a rabbit MHC (RLA) class II DQα restriction fragment length polymorphism (RFLP). The same study also demonstrated a protective association with an RLA-DRα RFLP.

A study of molecular DQB1 typing of Caucasoid cervical carcinoma patients demonstrated an increase in DQB1*0301 allele frequency and a decrease in the allele frequency of DQB1*0302 (Wank and Thomssen, 1991; Wank et al., 1992). An increased allele frequency of DQB1*0301 was also observed in a study of British women with CIN (Odunsi et al., 1996). Apple et al. (1994) conducted a case–control analysis using HPV and HLA class II DRB1 and DQB1 polymerase chain reaction (PCR)/sequence specific oligonucleotide probe typing of cervical cancer tissue from Hispanic patients. The Hispanic population was derived from European Caucasoids and native Americans. This study showed a significant association between the HLA class II haplotype DRB1*1501–DQB1*0602 (haplotypes of Caucasian origin) and cervical carcinoma; a negative association was observed for DR13 haplotypes. All of the HLA associations were specific for HPV16. A similar case–control study of Hispanic patients demonstrated identical HLA class II associations with advanced CIN and carcinoma *in situ* (Apple et al., 1995). In contrast to the earlier Caucasian study (Wank et al., 1992), an increase in DQB1*0301 was not observed in the Hispanic population, nor was there a protective effect of DQB1*0302. This probably reflects interethnic differences in the DRB1 alleles encoding the DR4 serotype. In this Hispanic population, many DQB1*0301 alleles are coupled to DQA1*0501 on DRB1*1402 and *1602 haplotypes of native American origin, whereas in Caucasians, most of the DQB1*0301 alleles are on DRB1*11 haplotypes. A recent Norwegian case–control study also observed an increased allele frequency of DQB1*0602 associated with increased risk of CIN in HPV16-infected women (Helland et al., 1998).

More recently, HLA-A*0201-restricted CTL specific for the E6 and E7 proteins of HPV16 have been found in the peripheral blood of patients with CIN and cervical carcinoma (Nakagawa et al., 1997; Nimako et al., 1997). These CTL may play a role in restricting disease progression. There are now exciting prospects for the development of novel immunotherapeutic strategies based on vaccination using constructs with E6/7 epitopes to elicit HLA class I-restricted HPV-specific CTL-mediated protective immunity to HPV16-induced malignancies (Borysiewicz et al., 1996; De Bruijn et al., 1998).

Hantavirus

The Puumala Hantavirus is an enveloped RNA virus, carried by rodents, which causes a wide spectrum of human disease. Epidemics in the USA are of Hantaviruses that cause predominantly respiratory disease, whereas in northern Europe, renal disease, known as nephropathia epidemica (NE), is the major clinical manifestation. There is significant variability in the clinical severity of NE; although most cases of infection are either subclinical or mild, approximately 10% of patients develop shock and approximately 5% develop dialysis-dependent acute renal failure (McNicholl, 1998). Recent evidence supports an important role for the MHC in determining the outcome of NE. In a north European series of adult patients with NE, those with the most severe disease course had a very high frequency of HLA-B8 DRB1*0301 alleles (Mustonen et al., 1996). The HLA-B8 haplotype was associated with the worst clinical disease and with higher levels of viral RNA in blood and urine (Plyusnin et al., 1997). Conversely, as for HIV infection, HLA-B27 was associated with a benign clinical course of NE (Mustonen et al., 1998).

Dengue fever virus

The arthropod-borne dengue viruses, of which four distinct serotypes exist, are a major cause of morbidity world-wide. The most severe clinical manifestation, dengue haemorrhagic fever

(DHF), occurs following secondary infection only in a minority of pre-exposed patients. The factors determining susceptibility to DHF following secondary infection have not been fully elucidated, but there is some evidence to suggest that host genetic factors may play a role.

All studies to date have investigated HLA-restricted, virus-specific responses of dengue virus serotype cross-reactive T-cell clones isolated from small numbers of individual donors following secondary infection (either naturally dengue immune or subject to sequential immunization with vaccines). HLA-DR-15- and DP2-restricted $CD4^+$ T-cell responses, and HLA-B35-restricted $CD8^+$ CTL responses to the immunodominant non-structural dengue virus protein-3 (NS3) have been reported (Livingston et al., 1995; Zeng et al., 1996; Kurane et al., 1998; Okamoto et al., 1998).

BACTERIAL INFECTIONS

HLA and mycobacterial infections

Mycobacterium tuberculosis and *Mycobacterium leprae* are the most important mycobacterial infections in humans and are the causative organisms of tuberculosis (TB) and leprosy, respectively. The spectrum of clinical manifestations produced by these infections is determined primarily by the host immune response to these organisms. HLA-restricted responses may cause pathology as well as protective immunity and have been studied extensively in leprosy and to a lesser extent in TB (Meyer et al., 1998). Several HLA class II associations with TB and leprosy have been found, but class I associations are inconsistent.

HLA haplotypes, however, are not the only immunogenetic factors influencing the host immune response to mycobacteria. Jepson et al. (1997) compared the immune responses to *M. tuberculosis* purified protein derivative and malarial antigens between adult Gambian dizygotic twins. HLA typing enabled the genetic factors involved in the immune responses to be divided between HLA and non-HLA genes. Although there were HLA associations for most humoral and cellular immune responses, non-HLA genes contributed the greater proportion of the overall genetic effect. The precise nature of these non-HLA genes is unclear.

Leprosy

The clinical spectrum of leprosy spans tuberculoid (TT) and lepromatous forms (LL). TT is characterized by focal skin and nerve lesions, a vigorous cell-mediated host immune response and an absence of bacilli. At the other extreme, LL is a systemic infection, host cell-mediated immunity is poorly developed and bacilli are found in abundance. Intermediate forms are classified as borderline tuberculoid, mid-borderline and borderline lepromatous.

HLA class I and leprosy

Various class I associations have been suggested but none has emerged as clear or consistent.

HLA class II and leprosy

Class II associations, such as that with HLA-DR2, have been consistently observed in leprosy patients from ethnically diverse populations (see Table 20.3). However, attempts to correlate consistently specific clinical forms of disease (i.e. TT or LL) with HLA type have not been convincing. For example, in a study of families with multiple cases of leprosy, preferential segregation of HLA-DR2 was observed in children with TT leprosy (van Eden et al., 1980). The HLA-DR2 association with TT was also observed in population studies in India (de Vries et al., 1980) and Thailand (Shauf et al., 1985). In contrast, there was no difference in HLA-DR2 allele frequency between TT and LL in a Korean study (Kim et al., 1987). More recently, HLA-DR2 (specifically DRB1*1501) has been found to be more strongly associated with LL in Indian populations (Rani et al., 1992, 1993).

An association between HLA-DR3 and TT has also been reported in a multicase family study in Venezuela (van Eden et al., 1985) and in a population study in Surinam (van Eden et al., 1981). HLA-DQ1 has also been associated with LL (Rani et al., 1992) and TT in Thailand (Shauf et al., 1985).

The functional significance of these HLA associations in leprosy is poorly understood. The *M. leprae* heat shock protein (HSP)-65-derived peptide 3-13 is presented exclusively in the context of the HLA-DR3 (DR17) molecule (Geluk et al.,

TABLE 20.3 MHC class II genes and mycobacterial infections

Genetic factors	Clinical form	Population	Reference
Leprosy			
↑ HLA-DR2	TT	India	van Eden et al. (1980)
↑ HLA-DR2	TT	India	de Vries et al. (1980)
↑ HLA-DR2, DQ1	TT	Thailand	Shauf et al. (1985)
↑ HLA-DR5	LL		
↑ HLA-DR3	TT	Venezuela	van Eden et al. (1985)
↑ HLA-DR3	TT	Surinam	van Eden et al. (1981)
↑ HLA-DR1, DR2, DR9, DQ1	TT and LL	Korea	Kim et al. (1987)
↓ HLA-DR4, DR3, DQ3	TT and LL		
↑ HLA-DR2, DQ8, DQ1	LL	India	Rani et al. (1992)
↑ HLA-DRB1*1501, DRB*1502, DQB1*0601	TT and LL	India	Rani et al. (1993)
↑ HLA-DQA1*0103	LL		
↓ HLA-DQB1*0503	TT		
−308 TNFα polymorphism	TT	India	Roy et al. (1997)
HLA-DR1 and TAP2B	TT	India	Rajalingam et al. (1997)
Pulmonary tuberculosis			
↑ HLA-DR2		India	Singh et al. (1983a)
↑ HLA-DR2		India	Bramajothi et al. (1991)
↑ HLA-DR2, DQ1		Indonesia	Bothamley et al. (1989)
↑ HLA-DR2		Former USSR	Khomenko et al. (1990)
↓ HLA-DR3			
↑ HLA-DRB1*1501, DRB1*1502		India	Rajalingam et al. (1996)
↓ HLA-DR3			
↑ HLA-DQB1*0503		Cambodia	Goldfeld et al. (1998)
TAP2-A/F		India	Rajalingam et al. (1997)

1992). Moreover, this peptide/HLA complex has been shown to select specific TCR Vβ rearrangements in a patient with TT (Struyk et al., 1995). Variation of specific residues in HLA-DR17 does not decrease peptide binding but may alter the conformation of the peptide/HLA complex (Geluk et al., 1994). More recently, molecular modelling has suggested that arginine at positions 30, 70 and 71 of the HLA-DRB1 product in patients with TT also modifies peptide binding. This may stimulate particular T cell clones which result in the detrimental immune response of TT leprosy (Zerva et al., 1996).

In contrast, a notable feature of LL is that most patients fail to respond to challenges from mycobacterial antigens. This anergic response may partly be explained by CD8$^+$ T cells which appear to act as suppressor cells in LL patients (Modlin et al., 1986), although the existence of such T suppressor cells is controversial. There is evidence that this mechanism may also be HLA restricted. Using a panel of *M. leprae*-specific CD4$^+$ and CD8$^+$ T cell clones restricted for different HLA class II haplotypes, suppression of CD4$^+$ T cell responses was found to be restricted to HLA-DQ but not HLA-DR2 (Salgame et al., 1991). Furthermore, anti-HLA-DQ antibodies could inhibit this suppression.

There is also evidence that HLA polymorphisms may modulate the cytokine profile of *M. leprae* reactive CD4$^+$ T cells and hence influence the T helper (Th) response. Mitra et al. (1997) found that HSP65 and HSP18 induced a Th1 cytokine profile [IFNγ and interleukin-2 (IL-2)] in the context of all HLA-DR restricted alleles except for HLA-DR1 and -DR7, which induced a Th2 (IL-4) type response. This is important since in mycobacterial infections strong Th1 responses are thought to be desirable and Th2 responses ineffective and associated with the LL or disseminated TB.

Other MHC genes and leprosy

It remains unclear whether the clinical manifestations of infection directly involve class II genes or a combined effect involving a nearby non-HLA genetic locus. For example, polymorphisms within the TNF gene in TT leprosy patients, which is located within 1 Mb of the MHC class II region, may lead to variability in TNFα secretion during infection (Roy et al., 1997). This is significant since, in mycobacterial infections, TNFα promotes host defence mechanisms and granuloma formation, but high concentrations of TNFα are associated with immunopathology.

HLA and tuberculosis

Several studies have shown that pulmonary tuberculosis (PTB) is associated with certain HLA class II alleles. In north India, using multicase family studies, HLA-DR2 was found to be more common in children with PTB than their healthy siblings (Singh et al., 1983a). The HLA-DR2 association has also been reported in populations from south India, Indonesia and the former Soviet Union (Bothamley et al., 1989; Khomenko et al., 1990; Bramajothi et al., 1991). Rajalingam et al. (1996) reported the association between HLA-DR2 (DRB1*1501, DRB1*1502) and failure to respond to antituberculous chemotherapy in north Indian patients. In contrast, in Mexican–American, north and south Indian population studies, HLA-DR2 was not significantly increased in PTB patients compared with control subjects (Cox et al., 1988; Singh et al., 1983b; Sanjeevi et al., 1992).

A negative association between HLA-DR3 and PTB was also reported in PTB patients in Mexico and the former Soviet Union (Cox et al., 1988; Khomenko et al., 1990). Very recently, HLA-DQ*0503 was found to be significantly associated with susceptibility to TB in Cambodian patients (Goldfeld et al., 1998). This allele may affect the electrical charge of the putative peptide-binding pocket (Stern et al., 1994) and decrease the efficiency of antigen binding. In the same study, no difference was found in the distribution of two TNFα variant alleles compared with controls.

In addition, polymorphisms within TAP genes which are located within the MHC class II region have been found in patients with PTB and TT leprosy (Rajalingam et al., 1997). It has been suggested that mutant TAP allele products may modify antigen processing and presentation by HLA molecules, but the precise mechanism causing disease is unclear. These and other observations require confirmation in larger studies.

HLA and mycobacterial vaccine strategies

HLA restriction is an important consideration in HLA-based antimycobacterial vaccine design. HSP derived from *M. leprae* and *M. tuberculosis* are potential candidate peptide subunit vaccines, since they contain multiple T cell determinants (Oftung et al., 1994) rather than a few immunodominant epitopes, and induce vigorous CD4$^+$ T cell proliferative responses restricted by many HLA-DR alleles (Mustafa et al., 1993). There is also evidence that peptides that stimulate class I restricted CD8$^+$ T cell responses might be effective vaccines (Lalvani et al., 1998). In the future, "peptide cocktails" might be used to induce a wide variety of HLA restricted protective T cell responses in diverse populations.

Staphylococcal and streptococcal infections and superantigens

Staphylococcal and streptococcal superantigens (Sags), which include staphylococcal enterotoxins, toxic shock syndrome toxin-1 and streptococcal pyrogenic exotoxins, are potent stimulators of T cell proliferation and cytokine production. This effect is thought to be responsible for the clinical phenotype of diseases such as staphylococcal and streptococcal toxic shock syndromes (Johnson et al., 1992; Kotzin et al., 1993; Schlievert, 1993). In addition, susceptibility to lethal Gram-negative endotoxic shock is enhanced by a number of bacterial Sags (Stiles et al., 1993). In contrast to conventional antigens, Sags do not require processing because they can bind directly to MHC class II molecules at residues outside the peptide-binding groove and interact with the Vβ region of the TCR (Mollick et al., 1989; Dellabona et al., 1990; Marrack and Kappler, 1990; Herman et al., 1991; Irwin et al., 1993).

T cell recognition of Sag is largely determined by the Vβ elements of the TCR (Herman et al., 1991). Although Sags depend on MHC class II molecules for functional expression, recognition of Sag/MHC complexes is not typically MHC restricted, i.e. one Sag can be recognized in the context of multiple MHC class II alleles (Herman et al., 1991; Janeway, 1991; Woodland and

Blackman, 1993). However, MHC class II molecules vary in their ability to present Sag, probably as a result of differences in the affinity of the Sag for the class II molecule. *In vitro* studies have demonstrated that MHC class II isotypic diversity and allelic polymorphism influence the avidity of binding of Sags to MHC/TCR complex (Herman et al., 1990; Scholl et al., 1990). The streptococcal pyrogenic exotoxin SpeA1 has a greater affinity for HLA-DQ than it does for HLA-DR or HLA-DP; in contrast, the staphylococcal enterotoxin SEB has a greater affinity for HLA-DR (Papageorgiou et al., 1999). Indeed, HLA-DQ transgenic mice have an increased susceptibility to infection with SpeA-producing streptococci compared with HLA-DR transgenic animals and controls (Sriskandan, S. and Altmann, D., personal communication). Data from human subjects with invasive streptococcal disease showed that the HLA-DRB1*02, DQA1*03 alleles and the HLA-DRB1*02–DQA1*0102 haplotype were associated with protection against severe disease (Guedez et al., 1997). These preliminary results suggested that HLA class II haplotypes may play a role in determining the outcome of streptococcal infection.

Lyme disease

The vector-borne spirochaete *Borrelia burgdorferi* causes a multisystem disease that may affect the skin, nervous system, heart or joints. Chronic Lyme arthritis (LA) is a late manifestation of the disease in approximately 10% of patients, despite antibiotic therapy, and joint inflammation may persist for months or even years after initial therapy. In patients with chronic LA, *B. burgdorferi* DNA is often undetectable in synovial fluid after antibiotic treatment, suggesting that there is ongoing inflammation despite eradication of the pathogen.

Studies of MHC class II alleles have shown that HLA-DRB1*0401 alleles are associated with chronic LA and refractoriness to antibiotic therapy (Steere et al., 1990). This allele is associated with an increased risk of rheumatoid arthritis (Gregerson et al., 1987). Patients with LA often develop an immunoglobulin G response to the C-terminal fragment of the outer surface protein A (OspA) of *B. burgdorferi* near the beginning of prolonged episodes of arthritis. The duration of arthritis is longer in patients with HLA-DR4 and OspA immunoreactivity than in those who lack responses to these surface proteins (Kalish et al., 1993). More recently, HLA class I-restricted, *B. burgdorferi*-specific CTL responses have been demonstrated in peripheral blood from patients with LA (Busch et al., 1996). These data suggest a possible role for HLA class I-restricted CTL responses in disease control and have implications for the design of vaccines against Lyme borreliosis.

ACKNOWLEDGEMENT

The authors would like to thank Professor Adrian V.S. Hill, Wellcome Trust Principal Research Fellow, Wellcome Trust Centre for Human Genetics, University of Oxford for his critical review of the manuscript and valuable suggestions.

REFERENCES

Agathanggelou, A., Niedobitek, G., Chen, R. et al. (1995). Expression of immune regulatory molecules in Epstein–Barr virus-associated nasopharyngeal carcinomas with prominent lymphoid stroma: evidence for a functional interaction between epithelial tumour cells and infiltrating lymphoid cells. *Am. J. Pathol.* **147**: 1152–1160.

Almarri, A. and Batchelor, J.R. (1994). HLA and hepatitis B infection. *Lancet* **344**: 1194–1195.

Almarri, A., Dwick, N., Kabi, S. et al. (1998). Interferon-α therapy in HCV hepatitis: HLA phenotype and cirrhosis are independent predictors of clinical outcome. *Hum. Immunol.* **59**: 239–242.

Alper, C.A., Kruksall, M.S., Marcus-Bagley, D. et al. (1989). Genetic prediction of nonresponse to hepatitis B vaccine. *N. Engl. J. Med.* **321**: 708–712.

Alric, L., Fort, M., Izopet, J. et al. (1997). Genes of the major histocompatibility complex class II influence the outcome of hepatitis C virus infection. *Gastroenterology* **113**: 1675–1681.

Alter, M.J., Margolis, H.S., Krawczynski, K. et al. for the Sentinel Counties Chronic Non-A, Non-B Hepatitis Study Team (1992). The natural history of community-acquired hepatits C in the United States. *N. Engl. J. Med.* **327**: 1899–1905.

Apple, R.J., Erlich, H.A., Klitz, W. et al. (1994). HLA DR-DQ associations with cervical carcinoma show papillomavirus-type specificity. *Nat. Genet.* **6**: 157–162.

Apple R.J., Becker, T.M., Wheeler, C.M. and Erlich, H.A. (1995). Comparison of human leukocyte antigen DR-DQ disease associations found with cervical dysplasia and invasive cervical carcinoma. *J. Natl Cancer. Inst.* **87**: 427–436.

Argaet, V.P., Schmidt, C.W., Burrows, S.R. et al. (1994). Dominant selection of an invariant T-cell antigen receptor in response to persistent infection by Epstein–Barr virus. *J. Exp. Med.* **180**: 2335–2340.

Bennett, S., Allen, S.J., Olerup, O. et al. (1993). Human leucocyte antigen (HLA) and malaria morbidity in a Gambian community. *Trans. R. Soc. Trop. Med. Hyg.* **87**: 286–287.

Bertoletti, A., Chisari, F.V., Penna, A. et al. (1993). Definition of a minimal optimal cytotoxic T-cell epitope within the Hepatitis B virus nucleocapsid protein. *J. Virol.* **67**: 2376–2380.

Bertoletti, A., Sette, A., Chisari, F.V. et al. (1994). Natural variants of cytotoxic epitopes are T-cell receptor antagonists for antiviral cytotoxic T cells. *Nature* **369**: 407–410.

Bertoletti, A., Southwood, S., Chesnut, R. et al. (1997). Molecular features of the hepatitis B virus nucleocapsid T-cell epitope 18–27: interaction with HLA and T-cell receptor. *Hepatology* **26**: 1027–1034.

Betts, M.R., Krowka, J., Santamaria, C. et al. (1997). Cross-clade human immunodeficiency virus (HIV)-specific cytotoxic T-lymphocyte responses in HIV-Infected Zambians. *J. Virol.* **71**: 8908–8911.

Blake, N., Lee, S., Redchenko, I. et al. (1997). Human CD8+ T cell responses to EBV EBNA1: HLA class I presentation of the (Gly-Ala)-containing protein requires exogenous processing. *Immunity* **7**: 791–802.

Borysiewicz, L.K., Fiander, A., Nimako, M. et al. (1996). A recombinant vaccinia virus encoding human papillomavirus types 16 and 18, E6 and E7 proteins as immunotherapy for cervical cancer. *Lancet* **347**: 1523–1527.

Bothamley, G.H., Beck, J.S., Schreuder, G.M. Th. et al. (1989). Association of tuberculosis and *M. tuberculosis*-specific antibody. *J. Infect. Dis.* **159**: 549–555.

Brahmajothi, V., Pitchappan, R.M., Kakkahaiah, V.N. et al. (1991). Association of pulmonary tuberculosis and HLA in South India. *Tubercule* **72**: 123–132.

Brinkman, B.M.N., Keet, I.P.M., Miedema, F. et al. (1997). Polymorphisms within the human tumor necrosis factor-α promoter region in human immunodeficiency virus type 1-seropositive persons. *J. Infect. Dis.* **175**: 188–190.

Burrows, S.R., Gardener, J., Khanna, R. et al. (1994). Five new cytotoxic T-cell epitopes identified within Epstein–Barr virus nuclear antigen 3. *J. Gen. Virol.* **75**: 2489–2493.

Burrows, J.M., Burrows, S.R., Poulsen, L.M. et al. (1996). Unusually high frequency of Epstein–Barr virus genetic variants in Papua New Guinea that can escape cytotoxic T cell recognition: implications for virus evolution. *J. Virol.* **70**: 2490–2496.

Busch, D.H., Jassoy, C., Brinckmann, U. et al. (1996). Detection of *Borrelia burgdorferi*-specific CD8$^+$ cytotoxic T cells in patients with Lyme arthritis. *J. Immunol.* **157**: 3534–3541.

Cabrera, M., Shaw, M.A., Sharples, C. et al. (1995). Polymorphism in tumour necrosis factor genes associated with mucocutaneous leishmaniasis. *J. Exp. Med.* **182**: 259–1264.

Caillat-Zucman, S., Gimenez, J.J., Wambergue, F. et al. (1998). Distinct HLA class II alleles determine antibody response to vaccination with hepatitis B surface antigen. *Kidney Int.* **53**: 1626–1630.

Callan, M.F.C., Steven, N., Krausa, P. et al. (1996). Large clonal expansions of CD8$^+$ T cells in acute infectious mononucleosis. *Nature Med.* **2**: 906–911.

Cameron, P.U., Mallal, S.A., French, M.A.H. and Dawkins, R.L. (1990). Major histocompatibility complex genes influence the outcome of HIV infection: ancestral haplotypes with C4 null alleles explain diverse HLA associations. *Hum. Immunol.* **29**: 282–295.

Campos-Lima, P.O. de, Gavioli, R., Zhang, Q.-J. et al. (1993). HLA-A11 epitope loss isolates of Epstein–Barr virus from a highly A11+ population. *Science* **260**: 98–100.

Campos-Lima, P.O. de, Levitsky, V., Brooks, J. et al. (1994). T cell responses and virus evolution: loss of HLA-A11-restricted CTL epitopes in Epstein–Barr virus isolates from high A11-positive populations by selective mutation of anchor residues. *J. Exp. Med.* **179**: 1297–1305.

Cao, H., Kanki, P., Sankale, J.-L. et al. (1997). Cytotoxic T-lymphocyte cross-reactivity among different human immunodeficiency virus type 1 clades: implications for vaccine development. *J. Virol.* **71**: 8615–8623.

Carrington, M., Nelson, G.W., Martin, M.P. et al. (1999). HLA and HIV-1: heterozygote advantage and *B*35-Cw*04* disadvantage. *Science* **283**: 1748–1752.

Chakkkalath, H.R., Theodos, C.M., Markowitz, J.S. et al. (1995). Class II major histocompatibility complex-deficient mice initially control an infection with *Leishmania major* but succumb to disease. *J. Infect. Dis.* **171**: 1302–1308.

Chisari, F.V. and Ferrari, C. (1995). Hepatitis B virus immunopathogenesis. *Annu. Rev. Immunol.* **13**: 29–60.

Conceição-Silva, F. da, Perlaza, B.L., Louis, J.A. and Romero, P. (1994). *Leishmania major* infection in mice primes for specific major histocompatibility complex class I-restricted CD8$^+$ cytotoxic T cell responses. *Eur. J. Immunol.* **24**: 2813–2817.

Cox, R.A., Downs, M., Neimes, R.E. et al. (1998). Immunogenetic analysis of human tuberculosis. *J. Infect. Dis.* **158**: 1302–1308.

Craig, F.E., Gulley, M.L. and Banks, P.M. (1993). Post-transplantation lymphoproliferative disorders. *Am. J. Clin. Pathol.* **99**: 265–276.

Cramp, M.E., Carucci, P., Underhill, J. et al. (1998). Association between HLA class II genotype and spontaneous clearance of hepatitis C viraemia. *J. Hepatol.* **29**: 207–213.

Craven, D.E., Awdeh, Z.L., Kunches, L.M. et al. (1986). Nonresponsiveness to hepatitis B vaccine in healthcare workers. Results of revaccination. *Ann. Intern. Med.* **105**: 356–360.

De Bruijn, M.L., Schuurhuis, D.H., Vieboom, M.P. et al. (1998). Immunization with human papillomavirus type 16 (HPV16) oncoprotein-loaded dendritic cells as well as protein in adjuvant induces MHC class I-restricted protection to HPV16-induced tumor cells. *Cancer Res.* **58**: 724–731.

Dellabona, P., Peccoud, J., Kappler, J. et al. (1990). Superantigens interact with MHC class II molecules outside the antigen groove. *Cell* **62**: 1115–1121.

Deulofeut, H., Nagy, M., Bing, D.H. et al. (1993). Cellular recognition and HLA restriction of a midsequence HBsAg in hepatitis B vaccinated individuals. *Mol. Immunol.* **30**: 941–948.

Donald, J.A., Rudman, K., Cooper, D.W. et al. (1992). Progression of HIV related disease is associated with HLA-DQ and -DR alleles defined by restriction fragment length polymorphisms. *Tissue Antigens* **39**: 241–248.

Dusheiko, G., Schmilovitz-Weiss, H., Brown, D. et al. (1994). Hepatitis C virus genotypes: an investigation of type-specific differences in geographic origin and disease. *Hepatology* **19**: 13–18.

Eden, W. van, Vries, R.R.P. de, Mehra, N.K. et al. (1980). HLA segregation of tuberculoid leprosy: confirmation of the DR2 marker. *J. Infect. Dis.* **141**: 693–701.

Eden, W. van, Vries, R.R.P. de, D'Amaro, J. et al. (1981). HLA-DR-associated genetic control of the type of leprosy in a population from Surinam. *Hum. Immunol.* **4**: 343–350.

Eden, W. van, Gonzalez, N.M., Vries, R.R.P. de, et al. (1985). HLA-linked control of predisposition to lepromatous leprosy. *J. Infect. Dis.* **151**: 9–15.

Egea, E., Iglesias, A., Slazar, M. et al. (1991). The cellular basis for lack of antibody response to hepatitis B vaccine in humans. *J. Exp. Med.* **173**: 531–538.

Fabio, G., Scorza-Smeraldi, R., Gringeri, A et al. (1990). Susceptibility to HIV infection and AIDS in Italian haemophiliacs is HLA-associated. *Br. J. Haematol.* **75**: 531–536.

Faghiri, Z., Tabei, S.Z. and Taneri, F. (1995). Study of the association of HLA class I antigens with Kala-Azar. *Hum. Hered.* **45**: 258–261.

Fernandes-Reyes, D., Craig, A.G., Kyes, S.A. et al. (1997). A high frequency African coding polymorphism in the N-terminal domain of ICAM-1 predisposing to cerebral malaria in Kenya. *Hum. Mol. Genet.* **112**: 1357–1360.

Ferrari, C., Penna, A., Bertoletti, A. et al. (1990). Cellular immune response to hepatitis B virus-encoded antigenas in acute and chronic hepatitis B virus infection. *J. Immunol.* **145**: 3442–3449.

Ferrari, C., Bertoletti, A., Penna, A. et al. (1991). Identification of immunodominant T cell epitopes of the hepatitis B virus nucleocapsid antigen. *J. Clin. Invest.* **88**: 214–222.

Ferrari, G., Humphrey, W., McElrath, M.J. et al. (1997). Clade B-based HIV-1 vaccines elicit cross-clade cytotoxic T lymphocyte reactivities in uninfected volunteers. *Proc. Natl Acad. Sci. USA* **94**: 1396–1401.

Frisan, T., Sjoberg, J., Dolcetti, R. et al. (1995). Local suppression of Epstein–Barr virus (EBV)-specific cytotoxicity in biopsies of EBV-positive Hodgkin's disease. *Blood* **86**: 1493–1501.

Gavioli, R., Campos-Lima, P.O. de, Kurilla, M.G. et al. (1992). Recognition of the Epstein–Barr virus encoded nuclear antigens EBNA4 and EBNA6 by HLA-A11-restricted cytotoxic T lymphocytes: implications for down-regulation of HLA-A11 in Burkitt lymphoma. *Proc. Natl Acad. Sci. USA* **89**: 5862–5866.

Gavioli, R., Kurilla, M.G., Campos-Lima, P.O. de al. (1993). Multiple HLA-A11-restricted cytotoxic T-lymphocyte epitopes of different immunogenicities in the Epstein–Barr virus-encoded nuclear antigen 4. *J. Virol.* **67**: 1572–1578.

Geluk, A., Meijgaarden, K.E. van, Janson, A.A.M. et al. (1992). Functional analysis of DR17(DR3)-restricted mycobacterial T cell epitopes reveals DR17-binding motiing motif and enables the design of allele-specific competitor peptides. *J. Immunol.* **149**: 2864–2871.

Geluk, A., Fu, X-T., Meijgaarden, K.E. van et al. (1994). T cell receptor and peptide-contacting residues in the HLA-DR17|(3)β1 chain. *Eur. J. Immunol.* **24**: 3241–3244.

Gilbert, S.C., Plebanski, M., Gupta, S. et al. (1998). Association of malaria parasite population structure, HLA, and immunological antagonism. *Science* **279**: 1173–1177.

Goldfeld, A.E., Delgado, J.C., Thim, S. et al. (1998). Association of an HLA-DQ allele with clinical tuberculosis. *JAMA* **279**: 226–228.

Gonzalez-Amarro, R., Garcia-Monzon, C., Garcia-Buey, L. et al. (1994). Induction of tumour necrosis factor α production by human hepatocytes in chronic viral hepatitis. *J. Exp. Med.* **179**: 841–848.

Goulder, P.J.R., Sewell, A.K., Lalloo, D.G. et al. (1997). Patterns of immunodominance in HIV-1-specific cytotoxic T lymphocyte responses in two human histocompatibility leukocyte antigens (HLA)-identical siblings with HLA-A*0201 are influenced by epitope mapping. *J. Exp. Med.* **185**: 1423–1433.

Grau, G.E., Fajardo, L.F., Piguet, P.F. et al. (1987). Tumour necrosis factor (cachectin) as an essential mediator in murine cerebral malaria. *Science* **237**: 1210–1212.

Grau, G.E., Taylor, T.E., Molyneux, M.E. et al. (1989). Tumour necrosis factor and disease severity in children with falciparum malaria. *N. Engl. J. Med.* **320**: 1586–1591.

Gregerson, P.K., Silber, J. and Winchester, R.J. (1987). The shared epitope hypothesis: an approach to understanding the molecular genetics of rheumatoid arthritis. *Arthritis Rheum.* **30**: 1205–1213.

Guedez, Y., Norby-Teglund, A., Low, D.E. et al. (1997). HLA class II alleles associated with outcome of invasive Group A streptococcal infections (Abstract). *Conference Proceedings from the 37th Interscience Conference on Antimicrobial Agents and Chemotherapy*, September 28–October 1, Toronto, Canada.

Guidotti, L.G., Ishikava, T., Hobbs, M.V. et al. (1996). Intracellular inactivation of the hepatitis B virus by cytotoxic T lymphocytes. *Immunity* **4**: 25–36.

Gupta, S. and Hill, A.V.S. (1995). Dynamic interactions in malaria: host heterogeneity meets parasite polymorphism. *Proc. R. Soc. London B. Biol. Sci.* **261**: 271–277.

Hammond, S.A., Bollinger, R.C., Tobery, T.W. and Siliciano, R.F. (1993). Transporter-independent processing of HIV-1 envelope protein for recognition by $CD8^+$ T-cells. *Nature* **364**: 158–161.

Hammond, S.A., Johnson, R.P., Kalams, S.A. et al. (1995). An epitope-selective transporter associated with antigen presenatation (TAP)-1/2-independent pathway and a more general TAP-1/2-dependent antigen-processing pathway allow recognition of the HIV-1 envelope glycoprotein by $CD8^+$ CTL. *J. Immunol.* **154**: 6140–6156.

Han, R., Breitburd, F., Marche, P.N. and Orth, G. (1992). Linkage of regression and malignant conversion of rabbit viral papillomas to MHC class II genes. *Nature* **356**: 66–68.

Hattum, J. van and Schreuder, G.M. (1987). HLA antigens in patients with various courses after hepatitis B virus infection. *Hepatology* **7**: 11–14.

Helland, A., Olsen, A.O., Gjoen, K. et al. (1998). An increased risk of cervical intra-epithelial neoplasia grade II-III among human papillomavirus positive patients with the *HLA-DQA1*0102-DQB1*0602* haplotype: a population-based case–control study of Norwegian women. *Int. J. Cancer* **76**: 19–24.

Hentges, F., Hoffmann, A., Oliveira de Araujo, F. and Hemmer, R. (1992). Prolonged clinically asymptomatic evolution after HIV-1 infection is marked by the absence of complement C4 null alleles at the MHC. *Clin. Exp. Immunol.* **88**: 237–242.

Herman, A., Croteau, G., Sekaly, R.P. et al. (1990). HLA-DR alleles differ in their ability to present staphylococcal enterotoxins to T cells. *J. Exp. Med.* **172**: 709–717.

Herman, A., Jappler, J., Marrack, P. and Pullen, A. (1991). Superantigens: mechanism of T-cell stimulation and role in immune responses. *Annu. Rev. Immunol.* **9**: 745–772.

Hill, A.V.S. (1992). Malaria resistance genes: a natural selection. *Trans. R. Soc. Trop. Med. Hyg.* **86**: 225–226.

Hill, A.B., Worth, A., Elliot, T. et al. (1995). Characterisation of two Epstein–Barr virus epitopes restricted by HLA-B7. *Eur. J. Immunol.* **25**: 18–24.

Hill, A.V.S. (1997). Genetic susceptibility to multifactorial diseases. *Trans. R. Soc. Trop. Med. Hyg.* **91**: 369–371.

Hill, A.V.S. (1998). The immunogenetics of human infectious diseases. *Annu. Rev. Immunol.* **16**: 593–617.

Hill, A.V.S., Allsopp, C.E., Kwiatkowski, D. et al. (1991). Common west African HLA antigens are associated with protection from severe malaria. *Nature* **352**: 595–600.

Hill, A.V.S., Elvin, J., Wills, A.C. et al. (1992). Molecular analysis of the association of HLA-B53 and resistance to severe malaria. *Nature* **360**: 434–439.

Hohler, T., Gerken, G., Notghi, A. et al. (1997). *HLA-DRB1*1301* and **1302* protect against chronic hepatitis B. *J. Hepatol.* **26**: 503–507.

Hohler T., Meyer, C.U., Notghi, A. et al. (1998a). The influence of major histocompatibility complex class II genes and T-cell Vβ repertoire on response to immunization with HBsAg. *Hum. Immunol.* **59**: 212–218.

Hohler, T., Kruger, A., Gerken, G. et al. (1998b). Tumour necrosis factor alpha promoter polymorphism at position -238 is associated with chronic active hepatitis C infection. *J. Med. Virol.* **54**: 173–177.

Hohler, T., Kruger, A., Gerken, G. et al. (1998c). A tumour necrosis factor-alpha (TNF-alpha) promoter polymorphism is associated with chronic hepatitis B infection. *Clin. Exp. Immunol.* **111**: 579–582.

Irwin, M.J., Hudson, K.R.., Ames, K.T. et al. (1993). T cell receptor beta-chain binding to enterotoxin superantigens. *Immunol. Rev.* **131**: 61–78.

Itescu, S., Mathur, Wagh, U. et al. (1992). HLA-B35 is associated with accelerated progression to AIDS. *J. Aids* **5**: 37–45.

Itescu, S., Rose, S., Dwyer, E. and Winchester, R. (1994). Certain HLA-DR5 and -DR6 major histocompatibility complex class II alleles are associated with a CD8 lymphocytic host response to human immunodeficiency virus type 1 characterized by low lymphocyte viral strain heterogeneity and slow disease progression. *Proc. Natl Acad. Sci. USA* **91**: 11 472–11 476.

Janeway, C.A., Jr (1991). Selective elements for the Vβ region of the T cell receptor: Mls and the bacterial toxic mitogens. *Adv. Immunol.* **50**: 1–53.

Jeannet, M., Sztajzel, R., Carpentier, N. et al. (1989). HLA antigens are risk factors for the development of AIDS. *J. Aids* **2**: 28–32.

Jeffery, K.J.M., Usuku, K., Hall, S.E. et al. (1999). HLA alleles determine human T-lymphotropic virus-I (HTLV-I) proviral load and the risk of HTLV-I-associated myelopathy. *Proc. Natl Acad. Sci. USA* **96**: 3848–3853.

Jepson, A., Sisay-Joof, F., Banya, W. et al. (1997). Genetic linkage of mild malaria to the MHC in Gambian children. *BMJ* **315**: 96–97.

Johnson, H., Russel, R. and Pontzer, C. (1992). Superantigens in human disease. *Sci. Am.* **266**: (92–95), 98–101.

Jung, M.C., Diepolder, H.M., Spengler, U. et al. (1995). Activation of a heterogeneous hepatitis B (HB) core and an antigen-specific $CD4^+$ T-cell population during seroconversion to anti-HBe and anti-HBs in hepatitis B virus infection. *J. Virol.* **69**: 3358–3368.

Just, J.J. (1995). Genetic predisposition to HIV-1 infection and acquired immune deficiency virus syndrome. *Hum. Immunol.* **44**: 156–169.

Just, J.J., Casabona, J., Bertran, J. et al. (1996). MHC class II alleles associated with clinical and immunological manifestations of HIV infection among children in Catalonia, Spain. *Tissue Antigens* **47**: 313–318.

Kalish, R.A., Leong, J.M. and Steere, A.C. (1993). Association of treatment-resistant chronic lyme arthritis with HLA-DR4 and antibody reactivity with OspA and OspB of *Borrelia burgdorferi*. *Infect. Immun.* **61**: 2774–2779.

Kaplan, C., Muller, J.Y., Doinel, C. et al. (1990). HLA-associated susceptibility to acquired immune deficiency syndrome in HIV-1-seropositive subjects. *Hum. Hered.* **40**: 290–298.

Kaslow, R.A., Duquesnoy, R., VanRaden, M. et al. (1990). A1, Cw7, B8, DR3 HLA antigen combination associated with rapid decline of T-helper lymphocytes in HIV-1 infection. *Lancet* **335**: 927–930.

Kaslow, R.A., Carrington, M., Apple, R. et al. (1996). Influence of combinations of human major histocompatibility complex genes on the course of HIV-1 infection. *Nature Med.* **2**: 405–411.

Kern, P., Hemmer, C.J., Van Damme, J. et al. (1989). Elevated tumour necrosis factor alpha and interleukin-6 serum levels as markers for complicated *Plasmodium falciparum* malaria. *Am. J. Med.* **87**: 139–143.

Khanna, R., Burrows, S.R., Kurilla, M.G. et al. (1992). Localisation of Epstein–Barr virus cytotoxic T cell epitopes using recombinant vaccinia: implications for vaccine development. *J. Exp. Med.* **176**: 169–178.

Khanna, R., Burrows, S.R., Argaet, V. and Moss, D.J. (1994). Endoplasmic reticulum signal sequence facilitated transport of peptide epitopes restores immunogenicity of an antigen processing defective tumour cell line. *Int. Immunol.* **6**: 639–645.

Khanna, R., Burrows, S.R., Steigerwald-Mullen, P.M. et al. (1995). Isolation of cytotoxic T lymphocytes from healthy seropositive individuals specific for peptide epitopes from Epstein–Barr virus nuclear antigen 1: implications for viral persistence and tumour surveillance. *Virology* **214**: 633–637.

Khanna, R., Burrows, S.R., Moss, D.J. and Silins, S.L. (1996). Peptide transporter (TAP-1/2)-independent endogenous processing of Epstein–Barr virus (EBV) latent membrane protein 2A: implications for CTL control of EBV-associated malignancies. *J. Virol.* **70**: 5357–5362.

Khanna, R., Burrows, S.R., Steigerwald-Mullen, P.M. et al. (1997a). Targeting Epstein–Barr virus nuclear antigen 1 (EBNA1) through HLA class II pathways restores immune recognition by EBNA1-specific cytotoxic T lymphocytes: evidence for HLA-DM independent processing. *Int. Immunol.* **9**: 1537–1543.

Khanna, R.B., S.R., Thomson, S.R., Moss, D.J. et al. (1997b). Class I processing-defective Burkitt's lymphoma cells are recognized efficiently by CD4$^+$ EBV-specific CTLs. *J. Immunol.* **158**: 3619–3625.

Khomenko, A.G., Litvinov, V.I., Chukanova, V.P. and Pospelov, L.E. (1990). Tuberculosis in patients with various HLA phenotypes. *Tubercle* **71**: 187–192.

Khoo, S.H., Pepper, L., Snowden, N. et al. (1997). Tumour necrosis factor c2 microsatellite allele is associated with the rate of HIV disease progression. *AIDS* **11**: 423–428.

Kim, S.J., Choi, I.H., Dahlberg, S. et al. (1987). HLA and leprosy in Koreans. *Tissue Antigens* **29**: 146–153.

Kim, J., Sette, A., Rodda, S. et al. (1997). Determinants of T cell reactivity to the *Mycobacterium leprae* GroES homologue. *J. Immunol.* **159**: 335–343.

Kima, P.E., Soong, L., Chicharro, C. et al. (1996). Leishmania-infected macrophages sequester endogenously synthesized parasite antigens from presentation to CD4$^+$ T cells. *Eur. J. Immunol.* **26**: 3163–3169.

Klein, M.R., Keet, I.P.M., D'Amaro, J. et al. (1994). Associations between HLA frequencies and pathogenic features of human immunodeficiency virus type 1 infection in seroconverters from the Amsterdam cohort of homosexual men. *J. Infect. Dis.* **169**: 1244–1249.

Klenerman, P., Rowland-Jones, S., McAdam, S. et al. (1994). Cytotoxic T-cell activity antagonized by naturally occurring HIV-1 Gag variants. *Nature* **369**: 403–407.

Kotzin, B., Leung, D., Kappler, J. and Marrack, P. (1993). Superantigens and their potential role in human disease. *Adv. Immunol.* **54**: 99–166.

Koutsky, L.A., Holmes, K.K., Critchlow, C.W. et al. (1992). A cohort study of the risk of cervical intraepithelial neoplasia grade 2 or 3 in relation to papillomavirus infection. *N. Engl. J. Med.* **327**: 1272–1278.

Koziel, M.J., Dudley, D., Afdhal, N. et al. (1995). HLA class I-restricted cytotoxic T lymphocytes specific for hepatitis C virus. Identification of multiple epitopes and characterisation of patterns of cytokine release. *J. Clin. Invest.* **96**: 2311–2321.

Kroner, B.L., Goedert, J.J., Blattner, W.A. et al. (1995). Concordance of human leukocyte antigen haplotype-sharing, CD4 decline and AIDS in hemophiliac siblings. *AIDS* **9**: 275–280.

Kruksall, M.S., Alper, C.A., Awdeh, Z.L. et al. (1992). The immune response to hepatitis B vaccine in humans: inheritance patterns in families. *J. Exp. Med.* **175**: 495–502.

Kurane, I., Zeng, L., Brinton, M.A. and Ennis, F.A. (1998). Definition of an epitope on NS3 recognized by human CD4$^+$ cytotoxic T lymphocyte clones cross-reactive for dengue virus types 2, 3 and 4. *Virology*. **240**: 169–174.

Kuzushita, N., Hayashi, N., Moribe, T. et al. (1998). Influence of HLA haplotypes on the clinical courses of individuals infected with hepatitis C virus. *Hepatology* **27**: 240–244.

Lalvani, A., Brooks, E.S.R., Wilkinson, R.J. et al. (1998). Human cytolytic and interferon-secreting CD8$^+$ T lymphocytes specific for *Mycobacterium tuberculosis*. *Proc. Natl Acad. Sci. USA* **95**: 270–275.

Lang, T., Chastellier, C. de, Frehel, C. et al. (1994). Distribution of MHC class I and of MHC class II molecules in macrophages infected with *Leishmania amazonensis*. *J. Cell Sci.* **107**: 69–82.

Lara, M.L., Layrisse, Z., Scorza, J.V. et al. (1991). Immunogenetics of human American cutaneous leishmaniasis: study of HLA haplotypes in 24 families from Venezuela. *Hum. Immunol.* **30**: 129–135.

Lee, S.P., Thomas, W.A., Murray, R.J. et al. (1993). HLA-A2.1-restricted cytotoxic T cells recognizing a range of Epstein–Barr virus isolates through a defined epitope in latent membrane protein LMP2. *J. Virol.* **67**: 7428–7435.

Lee, S.P., Thomas, W.A., Blake, N.W. and Rickinson, A.B. (1996). Transporter (TAP)-independent processing of a multiple membrane-spanning protein, the Epstein–Barr virus latent membrane protein 2. *Eur. J. Immunol.* **26**: 1875–1883.

Lee, S.P., Tierney, R.J., Thomas, W.A. et al. (1997). Conserved cytotoxic T lymphocyte (CTL) epitopes within Epstein–Barr virus (EBV) latent membrane protein 2, a potential target for CTL-based tumour therapy. *J. Immunol.* **158**: 3325–3334.

Lemon, S.M. and Thomas, D.L. (1997). Vaccines to prevent viral hepatitis. *N. Engl. J. Med.* **336**: 196–204.

Levitskaya, J., Coram, M., Levitsky, V. et al. (1995). Inhibition of antigen processing by the internal repeat region of the Epstein–Barr virus nuclear antigen 1. *Nature* **375**: 685–688.

Levitsky, V., Zhang, Q.-J., Levitskaya, J. and Masucci, M.G. (1996). The life span of major histocompatibility complex-peptide complexes influences the efficiency of presentation and immunogenicity of two class I-restricted cytotoxic T lymphocyte epitopes in the Epstein–Barr virus nuclear antigen 4. *J. Exp. Med.* **183**: 915–926.

Livingston, P.G., Kurane, I., Dai, L.C. et al. (1995). Dengue virus-specific, HLA-B35-restricted, human CD8$^+$ cytotoxic T lymphocyte (CTL) clones. Recognition of NS3 amino acids 500 to 508 by CTL clones of two different serotype specificities. *J. Immunol.* **154**: 1287–1295.

McAdam, S., Klenerman, P., Tussey, L. et al. (1995). Immunogenic HIV variant peptides that bind to HLA-B8 can fail to stimulate cytotoxic T lymphocyte responses. *J. Immunol.* **155**: 2729–2736.

McGuire, W., Hill, A.V., Allsopp, C.E. et al. (1994). Variation in the TNF alpha promoter region associated with susceptibility to cerebral malaria. *Nature* **371**: 508–510.

McMichael, A.J. and Walker, B.D. (1994). Cytotoxic T lymphocyte epitopes: implications for HIV vaccines. *AIDS* **8** (Suppl. 1): S155–S173.

McNeil, A.J., Yap, P.L., Gore, S.M. et al. (1996). Association of HLA types A1-B8-DR3 and B27 with rapid and slow progression of HIV disease. *Q. J. Med.* **89**: 177–185.

McNicholl, J. (1998). Host genes and infectious diseases. *Emerg. Infect. Dis.* **4**: 423–426.

Mann, D.L., Carrington, M.N. and Kroner, B.L. (1994). The human major histocompatibility complex in HIV-1 pathogenesis. *AIDS* **8** (Suppl. 1): S53–S60.

Marrack, P. and Kappler, J. (1990). The staphylococcal enterotoxins and their relatives. *Science* **248**: 705–711.

Meyer, C.G., Gallin, M., Erttmann, K.D. et al. (1994). HLA-D alleles associate with generalized disease, localized disease, and putative immunity in *Onchocerca volvulus* infection. *Proc. Natl Acad. Sci. USA* **91**: 7515–7519.

Meyer, C.G., May, J. and Stark, K. (1998). Human leukocyte antigens in tuberculosis and leprosy. *Trends Microbiol.* **6**: 148–153.

Minton, E.J., Smillie, D., Neal, K. et al. and the Trent Hepatitis C virus study group. (1998). Association between MHC class II alleles and clearance of circulating hepatitis C virus. *J. Infect. Dis.* **178**: 39–44.

Misko, I.S., Pope, J.H., Hutter, P. et al. (1984). HLA-DR-antigen-associated restriction of EBV-specific cytotoxic T-cell colonies. *Int. J. Cancer* **33**: 239–243.

Missale, G., Redeker, A., Person, J. et al. (1993). HLA-A31- and HLA-Aw68-restricted cytotoxic T cell responses to a single hepatitis B virus nucleocapsid epitope during acute viral hepatitis. *J. Exp. Med.* **177**: 751–762.

Mitra, D.K., Rajalingam, R., Taneja, V. et al. (1997). HLA-DR polymorphism modulates the cytokine profile of *Mycobacterium leprae* HSP-reactive $CD4^+$ T cells. *Clin. Immunol. Immunopathol.* **82**: 60–67.

Modiano, D., Petrarca, V., Sirima, B.S. et al. (1996). Different response to *Plasmodium falciparum* malaria in west African sympatric ethnic groups. *Proc. Natl Acad. Sci. USA* **93**: 13 206–13 211.

Modlin, R.L., Kato, H., Mehra, V. et al. (1986). Genetically restricted suppressor T-cell clones derived from lepromatous leprosy lesions. *Nature* **322**: 459–461.

Mollick, J., Cook, R. and Rich, R. (1989). Class II MHC molecules are specific receptors for staphylococcal enterotoxin A. *Science* **244**: 817–820.

Molvig, J., Baek, L., Christensen, P. et al. (1988). Endotoxin-stimulated human monocyte secretion of interleukin 1, tumour necrosis factor alpha, and prostaglandin E2 shows stable inter-individual differences. *Scand. J. Immunol.* **27**: 705–716.

Morsy, T.A., Romia, S.A., al-Ganayni, G.A. et al. (1990). Histocompatibility antigens (HLA) in Egyptians with two parasitic skin diseases (scabies and leishmaniasis). *J. Egypt. Soc. Parasitol.* **20**: 565–572.

Moss, D.J., Schmidt, C., Elliot, S. et al. (1996). Strategies involved in developing an effective vaccine for EBV-associated diseases. *Adv. Cancer Res.* **69**: 213–244.

Moss, D.J., Suhrbier, A. and Elliot, S.L. (1998). Candidate vaccines for Epstein–Barr virus. *BMJ* **317**: 423–424.

Murdoch, M.E., Payton, A., Abiose, A. et al. (1997). HLA-DQ alleles associate with cutaneous features of onchocerciasis. *Hum. Immunol.* **55**: 46–52.

Murray, R.J., Kurilla, M.G., Brooks, J.M. et al. (1992). Identification of target antigens for the human cytotoxic T cell response to Epstein–Barr virus (EBV): implications for the immune control of EBV-positive malignancies. *J. Exp. Med.* **176**: 157–168.

Mustafa, A.S., Lundin, K.E.A. and Oftung, F. (1993). Human T cells recognize mycobacterial heat shock proteins in the context of multiple HLA-DR molecules: studies with healthy subjects vaccinated with *Mycobacterium bovis* BCG and *Mycobacterium leprae*. *Infect. Immun.* **61**: 5294–5301.

Mustonen, J., Partanen, J., Kanerva, M. et al. (1996). Genetic susceptibility to severe course of nephropathia epidemica caused by Puumala Hantavirus. *Kidney Int.* **49**: 217–221.

Mustonen, J., Partanen, J., Kanerva, M. et al. (1998). Association of HLA-B27 with benign clinical course of nephropathia epidemica caused by Puumala Hantavirus. *Scand. J. Immunol.* **47**: 277–279.

Nakagawa, M., Stites, D.P., Farhat, S. et al. (1997). Cytotoxic T lymphocyte responses to E6 and E7 proteins of human papillomavirus type 16: relationship to cervical intraepithelial neoplasia. *J. Infect. Dis.* **175**: 927–931.

Nardin, E.H. and Nussenzweig, R.S. (1993). T cell responses to pre-erythrocytic stages of malaria: role in protection and vaccine development against pre-erythrocytic stages. *Annu. Rev. Immunol.* **11**: 686–727.

Nayersina, R., Fowler, P., Guilhot, S. et al. (1993). HLA A2 restricted cytotoxic T lymphocyte responses to multiple hepatitis B surface antigen epitopes during hepatitis B virus infection. *J. Immunol.* **150**: 4659–4671.

Nimako, M., Fiander, A.N., Wilkinson, G.W. et al. (1997). Human papillomavirus-specific cytotoxic lymphocytes in patients with cervical intraepithelial neoplasia grade III. *Cancer Res.* **57**: 4855–4861.

Nussenzweig, V. and Nussenzweig, R.S. (1989). Rationale for development of an engineered sporozoite malaria vaccine. *Adv. Immunol.* **45**: 283–334.

Odunsi, K., Terry, G., Ho, L. et al. (1996). Susceptibility to human papillomavirus-associated cervical intra-epithelial neoplasia is determined by specific HLA DR-DQ alleles. *Int. J. Cancer* **67**: 595–602.

Oftung, F., Geluk, A., Lundin, K.E.A. et al. (1994). Mapping of multiple HLA class II-restricted T-cell epitopes of the mycobacterial 70-kilodalton heat shock protein. *Infect. Immun.* **62**: 5411–5418.

Okamoto, Y., Kurane, I., Leporati, A.M. and Ennis, F.A. (1998). Definition of the region which contains multiple epitopes recognized by dengue virus serotype-cross reactive and flavivirus cross-reactive, HLA-DPw2-restricted $CD4^+$ T cell clones. *J. Gen. Virol.* **79**: 697–704.

Oldstone, M.B.A., Tishon, A., Geckeler, R. et al. (1992). A common antiviral cytotoxic T-lymphocyte epitope for diverse major histocompatibility complex haplotypes: implications for vaccination. *Proc. Natl Acad. Sci. USA* **89**: 2752–2755.

Oudejans, J.J., Jiwa, N.M., Kummer, J.A. et al. (1996). Analysis of MHC class I expression on Reed–Sternberg cells in relation to the cytotoxic T-cell response in Epstein–Barr virus positive and negative Hodgkin's disease. *Blood* **87**: 3844–3851.

Papadopoulos, E.B., Ladanyi, M., Emanuel, D. et al. (1994). Infusions of donor leukocytes to treat Epstein–Barr virus-associated lymphoproliferative disorders after allogeneic bone marrow transplantation. *N. Engl. J. Med.* **330**: 1185–1191.

Papageorgiou, A.C., Collins, C.M., Gutman, D.M. et al. (1999). Structural basis for the recognition of superantigen streptococcal pyrogenic exotoxin A (SpeA1) by MHC class II molecules and T-cell receptors. *EMBO J.* **18**: 9–21.

Peano, G., Menardi, G., Ponzetto, A. and Fenoglio, L.M. (1994). HLA DR5 antigen – a genetic factor influencing the outcome of hepatitis C virus infection? *Arch. Intern. Med.* **154**: 2733–2736.

Penna, A., Chisari, F.V., Bertoletti, A. et al. (1991). Cytotoxic T lymphocytes recognize an HLA-A2-restricted epitope within the Hepatitis B virus nucleocapsid antigen. *J. Exp. Med.* **174**: 1565–1570.

Petzl-Erler, M.L., Belich, M.P. and Queiroz-Telles, F. (1991). Association of mucosal Leishmaniasis with HLA. *Hum. Immunol.* **32**: 254–260.

Plebanski, M., Gilbert, S.C., Schneider, J. et al. (1998). Protection from *Plasmodium berghei* infection by priming and boosting T cells to a single class I-restricted epitope with recombinant carriers suitable for human use. *Eur. J. Immunol.* **28**: 4345–4355.

Plyusnin, A., Horling, J., Kanerva, M. et al. (1997). Puumala Hantavirus genome in patients with nephropathia epidemica: correlation of PCR positivity with HLA haplotype and link to viral sequences in local rodents. *J. Clin. Microbiol.* **35**: 1090–1096.

Pociot, F., Briant, L., Jongeneel, C.V. et al. (1993). Association of tumour necrosis factor (TNF) and class II major histocompatibility complex alleles with the secretion of TNF-α and TNF-β by human mononuclear cells: a possible link to insulin-dependent diabetes mellitus. *Eur. J. Immunol.* **23**: 224–231.

Prina, E., Jouanne, C., Souza Leao, S. de et al. (1993). Antigen presentation capacity of murine macrophages infected with *Leishmania amazonensis* amastigotes. *J. Immunol.* **151**: 2050–2061.

Rajalingam, R., Mehra, N.K., Jain, R.C. et al. (1996). Polymerase chain reaction-based sequence-specific oligonucleotide hybridisation analysis of HLA class II antigens in pulmonary tuberculosis: relevance to chemotherapy and disease severity. *J. Infect. Dis.* **173**: 669–676.

Rajalingam, R., Singal, D.P. and Mehra, N.K. (1997). Transporter associated with antigen-processing (TAP) genes and susceptibility to tuberculoid leprosy and pulmonary tuberculosis. *Tissue Antigens* **49**: 168–172.

Rani, R., Zaheer, A. and Mukherjee, R. (1992). Do human leukocyte antigens have a role to play in differential manifestation of multibacillary leprosy: A study on multibacillary leprosy patients from North India. *Tissue Antigens* **40**: 124–127.

Rani, R., Fernandez-Vina, M.A., Zaheer, S.A. et al. (1993). Study of HLA class II alleles by PCR oligotyping in leprosy patients from North India. *Tissue Antigens* **42**: 133–137.

Rehermann, B. and Chisari, F.V. (1995). The cytotoxic T lymphocyte response to multiple hepatitis B virus polymerase epitopes during and after acute viral hepatitis. *J. Exp. Med.* **181**: 3442–3449.

Reiner, N.E., Ng, W. and McMaster, W.R. (1987). Parasite-accessory cell interactions in murine leishmaniasis. *J. Immunol.* **138**: 1926–1932.

Rickinson, A.B. and Moss, D.J. (1997). Human cytotoxic T lymphocyte responses to Epstein–Barr Virus infection. *Annu. Rev. Immunol.* **15**: 405–431.

Roger, M. (1998). Influence of host genes on HIV-1 disease progression. *FASEB J.* **12**: 625–632.

Romero, P., Maryanski, J.L., Corradin, G. et al. (1988). Cloned cytotoxic T cells recognise an epitope in the circumsporozoite protein and protect against malaria. *Nature* **341**: 323–326.

Rooney, C.M., Rowe, M., Wallace, L.E. and Rickinson, A.B. (1985). Epstein–Barr virus-positive Burkitt's lymphoma cells not recognised by virus-specific T cell surveillance. *Nature* **317**: 629–631.

Rooney, C.M., Edwards, C.F., Lenoir, G.M. et al. (1986). Differential activation of cytotoxic responses by Burkitt's lymphoma (BL)-cell lines: relationship to the BL-cell surface phenotype. *Cell. Immunol.* **102**: 99–112.

Rooney, C.M., Smith, C.A., Ng, C.Y.C. et al. (1995). Use of gene-modified virus-specific T lymphocytes to control Epstein–Barr virus-related lymphoproliferation. *Lancet* **345**: 9–13.

Rowe, J.A., Moulds, J.M., Newbold, C.I. and Miller, L.M. (1997). *P. falciparum* rosetting mediated by a parasite variant erythrocyte membrane protein and complement receptor 1. *Nature* **388**: 292–295.

Rowe, M., Khanna, R., Jacob, C.A. et al. (1995). Restoration of endogenous processing in Burkitt's lymphoma cells by Epstein–Barr virus latent membrane protein-1: coordinate upregulation of peptide transporters and HLA class I antigen expression. *Eur. J. Immunol.* **25**: 1374–1384.

Rowe, M., Rowe, D.T., Gregory, C.D. et al. (1987). Differences in B cell growth phenotype reflect novel patterns of Epstein–Barr virus latent gene expression in Burkitt's lymphoma cells. *EMBO J.* **6**: 2743–2751.

Rowland-Jones, S.L., Powis, S.H., Sutton, J. et al. (1993). An antigen processing polymorphism revealed by HLA-B8-restricted cytotoxic T lymphocytes which does not correlate with TAP gene polymorphism. *Eur. J. Immunol.* **23**: 1999–2004.

Rowland-Jones, S., Sutton, J., Ariyoshi, K. et al. (1995). HIV-specific cytotoxic T-cells in HIV-exposed but uninfected Gambian women. *Nature Med.* **1**: 59–64.

Rowland-Jones, S., Tan, R. and McMichael, A. (1997). Role of cellular immunity in protection against HIV infection. *Adv. Immunol.* **65**: 277–346.

Roy, S., McGuire, W., Mascie-Taylor, C.G.N. et al. (1997). Tumor necrosis factor promoter polymorphism and susceptibility to lepromatous leprosy. *J. Infect. Dis.* **176**: 530–532.

Sahmoud, T., Laurian, Y., Gazengel, C. et al. (1993). Progression to AIDS in French haemophiliacs: association with HLA-B35. *AIDS* **7**: 497–500.

Salgame, P., Convit, J. and Bloom, B.R. (1991). Immunological suppression by human $CD8^+$ T cells is receptor dependent and HLA-DQ restricted. *Proc. Natl Acad. Sci. USA* **88**: 2598–2602.

Salle, H. de la, Houssaint, E., Peyrat, M.A. et al. (1997). Human peptide transporter deficiency: importance of HLA-B in the presentation of TAP-independent EBV antigens. *J. Immunol.* **158**: 4555–4563.

Sanjeevi, C.B., Narayanan, P.R., Prabakar, R. et al. (1992). No association or linkage with HLA-DR or -DQ genes in South Indians with pulmonary tuberculosis. *Tubercle Lung Dis.* **73**: 280–284.

Schauf, V., Ryan, S., Scollard, D. et al. (1985). Leprosy associated with HLA-DR2 and DQw1 in the population of northern Thailand. *Tissue Antigens* **26**: 243–247.

Schlievert, P. (1993). Role of superantigens in human disease. *J. Infect. Dis.* **167**: 997–1002.

Scholl, P.R., Diez, A., Karr, R. et al. (1990). Effect of isotypes and allelic polymorphism on the binding of staphylococcal exotoxins to MHC class II molecules. *J. Immunol.* **144**: 226–230.

Scorza-Smeraldi, R., Fabio, G., Lazzarin, A. et al. (1986). HLA-associated susceptibility to acquired immunodeficiency syndrome in Italian patients with human-immunodeficiency-virus infection. *Lancet* **ii**: 1187–1189.

Scully, L., Brown, D., Shein, R. and Thomas, H.C. (1990). Immunological studies before and during interferon therapy in chronic HBV infection: identification of factors predicting response. *Hepatology* **12**: 1111–1117.

Silins, S.L., Cross, S.M., Elliot, S.L. et al. (1996). Development of Epstein–Barr virus-specific memory TCR clonotypes in acute infectious mononucleosis. *J. Exp. Med.* **184**: 1815–1824.

Singh, S.P.N., Mehra, N.K., Dingley, H.B. et al. (1983a). Human leukocyte antigen (HLA)-linked control of susceptibility to pulmonary tuberculosis and association with HLA-DR types. *J. Infect. Dis.* **148**: 676–681.

Singh, S.P.N., Mehra, N.K., Dingley, H.B. et al. (1983b). HLA-A, -B, -C and -DR antigen profile in pulmonary tuberculosis in North India. *Tissue Antigens* **21**: 380–384.

Sixbey, J.W., Nedrud, J.G., Raab-Traub, N. et al. (1984). Epstein–Barr virus replication in oropharyngeal epithelial cells. *N. Engl. J. Med.* **310**: 1225–1230.

Souza Leao, S. de, Lang, T., Prina, E. et al. (1995). Intracellular *Leishmania amazonensis* amastigotes internalize and degrade MHC class II molecules of their host cells. *J. Cell. Sci.* **108**: 3219–3231.

Steel, C.M., Beatson, D., Cuthbert, R.J.D. et al. (1988). HLA haplotype A1 B8 DR3 as a risk factor for HIV-related disease. *Lancet* **i**: 1185–1188.

Steere, A.C., Dwyer, E. and Winchester, R. (1990). Association of chronic Lyme arthritis with HLA-DR4 and HLA-DR2 alleles. *N. Engl. J. Med.* **323**: 219–223.

Stern, L.J., Brown, J.H., Jardetzky, T.S. et al. (1994). Crystal structure of the human class II MHC protein HLA-DR1 complexed with an influenza virus peptide. *Nature* **368**: 215–221.

Steven, N.M., Leese, A.M., Annels, N.E. et al. (1996). Epitope focussing in the primary cytotoxic T cell response to Epstein–Barr virus and its relationship to T cell memory. *J. Exp. Med.* **184**: 1801–1813.

Steven, N.M., Annels, N.E., Kumar, A. et al. (1997). Immediate early and early lytic cycle proteins are frequent targets of the Epstein–Barr virus-induced cytotoxic T cell response. *J. Exp. Med.* **185**: 1605–1617.

Stiles, B.G., Bavari, S., Krakauer, T. and Ulrich, R.G. (1993). Toxicity of staphylococcal enterotoxins potentiated by lipopolysaccharide: major histocompatibility complex class II molecule dependency and cytokine release. *Infect. Immun.* **61**: 5333–5338.

Stoute, J.A., Slaoui, M., Heppner, G. et al. (1997). A preliminary evaluation of a recombinant circumsporozoite protein vaccine against *Plasmodium falciparum* malaria. *N. Engl. J. Med.* **336**: 86–91.

Struyk, L., Hawes, G.E., Haanen, J.B.A.G. et al. (1995). Clonal dominance and selection for similar complementarity determining region 3 motifs among T lymphcytes responding to the HLA-DR3-associated *Mycobacterium leprae* heat shock protein 65-kd peptide 3-13. *Hum. Immunol.* **44**: 220–227.

Thomson, S.A., Khanna, R., Gardener, J. et al. (1995). Minimal epitopes expressed in a recombinant polyepitope protein are processed and presented to $CD8^+$ cytotoxic T cells: implications for vaccine design. *Proc. Natl Acad. Sci.* **92**: 5845–5849.

Thomson, S., Elliot, S., Sherritt, M. et al. (1996). Recombinant polyepitope vaccines for the delivery of multiple $CD8^+$ cytotoxic T cell epitopes. *J. Immunol.* **157**: 822–826.

Thomson, S.A., Sherrit, M.A., Medveczky, J. et al. (1998a). Delivery of multiple CD8 cytotoxic T cell epitopes by DNA vaccination. *J. Immunol.* **160**: 1717–1723.

Thomson, S.A., Burrows, S.R., Misko, I.S. et al. (1998b). Targeting a polyepitope protein incorporating multiple class II-restricted viral epitopes to the secretory/endocytic pathway facilitates immunorecognition by $CD4^{(+)}$ cytotoxic T lymphocytes: a novel approach to vaccine design. *J. Virol.* **72**: 2246–2252.

Threlkeld, S.C., Wentworth, P.A., Kalms, S.A. et al. (1997). Degenerate and promiscuous recognition by CTL of peptides presented by the MHC class 1 A3-like superfamily. *J. Immunol.* **159**: 1648–1657.

Thursz, M., Kwiatkowski, D., Allsopp, C.E.M. et al. (1995a). Association between an MHC class II allele with clearance of hepatitis B virus infection in the Gambia. *N. Engl. J. Med.* **332**: 1065–1069.

Thursz, M.R., Kwiatkowski, D, Estee Torok, M. et al. (1995b). Association of hepatitis B surface antigen carriage with severe malaria in Gambian children. *Nature Med.* **1**: 374–375.

Thursz, M.R., Thomas, H.C., Greenwood, B.M. and Hill, A.V.S. (1997). Heterozygote advantage for HLA class II type in hepatitis B virus infection [Letter]. *Nat. Genet.* **17**: 11–12.

Tibbs, C., Donaldson, P., Underhill, J. et al. (1996). Evidence that the *HLA-DQA*03* allele confers protection from chronic HCV infection in Northern European Caucasoids. *Hepatology* **24**: 1342–1345.

Tierney, R.J., Steven, N., Young, L.S. and Rickinson, A.B. (1994). Epstein–Barr virus latency in blood mononuclear cells: analysis of viral gene transcription during primary infection and in the carrier state. *J. Virol.* **68**: 7374–7385.

Tilg, H., Wilmer, A., Vogel, W. et al. (1992). Serum levels of cytokines in chronic liver diseases. *Gastrenterology* **103**: 264–273.

Todd, J.A. (1992). Le carte des microsatellites est arrivee! *Hum. Mol. Genet.* **5**: 1515–1519.

Tomiyama, H., Miwa, K., Shiga, H. et al. (1997). Evidence of presentation of multiple HIV-1 cytotoxic T lymphocyte epitopes by *HLA-B*3501* molecules that are associated with the accelerated progression of AIDS. *J. Immunol.* **158**: 5026–5034.

Tong, M.L., Farra, N.S. el, Reikes, A.R. and Co, R.L. (1995). Clinical outcomes after transfusion-associated hepatitis C. *N. Engl. J. Med.* **332**: 1463–1466.

Torre, D., Zeroli, C., Giola, M. et al. (1994). Serum levels of interleukin-1α, interleukin-1β, interleukin-6 and tumour necrosis factor in patients with acute viral hepatitis. *Clin. Infect. Dis.* **18**: 194–198.

Vitiello, A., Ishioka, G., Grey, H.M. et al. (1995). Development of a lipopeptide-based therapeutic vaccine to treat chronic HBV infection. *J. Clin. Invest.* **95**: 341–349.

Vries, R.R.P. de, Mehra, N.K., Vaidya, M.C et al. (1980). HLA-linked control of susceptibility to tuberculoid leprosy and association with HLA-DR types. *Tissue Antigens* **16**: 294–304.

Wank, R. and Thomssen, C. (1991). High risk of squamous cell carcinoma of the cervix for women with HLA-DQw3. *Nature* **352**: 723–725.

Wank, R., Schendel, D.J. and Thomssen, C. (1992). HLA antigens and cervical carcinoma. *Nature* **356**: 22–23.

Watanabe, H., Matsushita, S., Nobuhiro, K et al. (1988). Immunosuppressor gene on HLA-Bw54, DR4, DRw53 haplotype controls nonresponsiveness in humans to hepatitis B surface antigen via $CD8^+$ suppressor T cells. *Hum. Immunol.* **22**: 9–14.

Weiss, W.R., Sedegah, M., Beaudoin, R.L. et al. (1988). $CD8^+$ T cells are required for protection in mice immunized with irradiated sporozoites. *Proc. Natl Acad. Sci. USA* **88**: 573–576.

Weiss, W.R., Mellouk, S., Houghten, R.A. et al. (1990). Cytotoxic T cells recognize a peptide from the circumsporozoite protein on malaria-infected hepatocytes. *J. Exp. Med.* **171**: 763–773.

White, C.A., Cross, S.M., Kurilla, M.G. et al. (1996). Recruitment during infectious mononucleosis of $CD3^+CD4^+CD8^+$ virus-specific cytotoxic T cells which recognize Epstein–Barr virus lytic antigen BHRF1. *Virology* **219**: 489–492.

Williams, T.N., Maitland, K., Bennett, S. et al. (1996). High incidence of malaria in alpha thalassaemic children. *Nature* **383**: 522–525.

Wilson, A.G., Giovine, F.S. di, Blakemore, A.I. and Duff, G.W. (1992). Single base polymorphism in the human tumour necrosis factor alpha (TNF alpha) gene detectable by NcoI restriction of PCR product. *Hum. Mol. Genet.* **1**: 353.

Wilson, A.G., Vries, N. de, Pociot, F. et al. (1993). An allelic polymorphism within the human tumour necrosis alpha promoter region is strongly associated with HLA A, B8 and DR3 alleles. *J. Exp. Med.* **177**: 557–560.

Wilson, A.G., Symons, J.A., McDowell, T.L. et al. (1997). Effects of a polymorphism in the human tumour necrosis factor a promoter on transcriptional activation. *Proc. Natl Acad. Sci. USA* **94**: 3195–3199.

Wong, D.K.H., Dudley, D.D., Afdhal, N.H. et al. (1998). Liver-derived CTL in hepatitis C virus infection: breadth and specificity of responses in a cohort of persons with chronic infection. *J. Immunol.* **160**: 1479–1488.

Woodland, D.L. and Blackman, M.A. (1993). How do T-cell receptors, MHC molecules and superantigens get together? *Immunol. Today* **14**: 208–212.

Yazdanbakhsh, M., Abadi, K., Roo, M. de, et al. (1997). HLA and elephantiasis revisited. *Eur. J. Immunogen.* **24**: 439–442.

Zavaglia, C., Bortolon, C., Ferrioli, G. et al. (1996). HLA typing in chronic type B, D and C hepatitis. *J. Hepatol.* **24**: 658–665.

Zavaglia, C., Martinetti, M., Silini, E. et al. (1998). Association between HLA class II alleles and protection from or susceptibility to chronic hepatitis C. *J. Hepatol.* **28**: 1–7.

Zeng, L., Kurane, I., Okamoto, Y. et al. (1996). Identification of amino acids involved in recognition by dengue virus NS3-specific, HLA-DR15-restricted cytotoxic $CD4^+$ T-cell clones. *J. Virol.* **70**: 3108–3117.

Zerva, L., Cizman, B., Mehra, N.K. et al. (1996). Arginine at positions 13 or 70-71 in pocket 4 of HLA-DRB1 alleles is associated with susceptibility to tuberculoid leprosy. *J. Exp. Med.* **183**: 829–836.

Zinkernagel, R.M. and Doherty, P.C. (1974). Restriction of *in vitro* T cell-mediated cytotoxicity in lymphocytic choriomeningitis within a syngeneic or semiallogeneic system. *Nature* **248**: 701–702.

CHAPTER 21

HLA and Systemic Vasculitides, Systemic Lupus Erythematosus and Sjögren's Syndrome

Matthew C. Pickering, Mohini Perraudeau and Mark J. Walport

INTRODUCTION

In this chapter the human leucocyte antigen (HLA) associations with the systemic vasculitides, systemic lupus erythematosus (SLE) and primary Sjögren's syndrome are discussed. The systemic vasculitides are rare conditions characterized histologically by inflammation and necrosis of blood vessel walls. There is considerable overlap in the clinical expression of the different vasculitic syndromes and in the majority of conditions the underlying cause is unknown. Classification criteria proposed by the American College of Rheumatology in 1990 (Hunder et al., 1990) and the Chapel Hill consensus conference in 1994 (Jeanette et al., 1994) have enabled selection of relatively homogeneous groups of patients, which in turn has allowed valid comparisons to be made between studies. Behçet's syndrome, which is not part of either classification but is a condition in which vasculitis plays a central role, has also been included.

For each of the diseases considered in this chapter there is evidence that disease-susceptibility and/or severity genes are located within the major histocompatibility complex (MHC). In almost every case the precise identity of the relevant disease-susceptibility gene within the MHC is not known because of the phenomenon of linkage disequilibrium. As is discussed elsewhere in this volume (see Chapter 6), the study of disease associations in ethnically diverse populations is one important approach to identifying a disease-susceptibility gene. In the case of SLE, discussed in detail below, the strong association of the disease with homozygous classical pathway complement deficiencies, such as C1q, encoded outside the MHC, points to the candidature of C4 and C2 null alleles as the relevant disease-susceptibility genes within the MHC. Other approaches to identifying the relevant disease-susceptibility genes include animals bearing the candidate human disease-susceptibility gene (for example, see Chapter 12).

SYSTEMIC VASCULITIDES

Large vessel vasculitis

Polymyalgia rheumatica and giant cell arteritis

Giant cell arteritis (GCA) and polymyalgia rheumatica (PMR) are overlapping diseases, characterized by a prominent acute-phase response. GCA is a medium and large vessel vasculitis affecting mainly cranial vessels and vessels arising from the aortic arch, which occurs almost exclusively in patients over the age of 55 years. Clinical features include malaise, lethargy, temporal headache, jaw claudication, scalp and temporal artery tenderness. PMR is characterized by limb-girdle stiffness and pain associated with a dramatic improvement following low-dose corticosteroid therapy. PMR and GCA often occur in the same individual. A role for genetic influences in these two conditions is suggested by reports of familial cases (Liang et al., 1974; Granato et al., 1980) and by the finding that the majority of cases occur in Caucasoid individuals from the UK, Scandinavia and North America. The condition is rare in Black populations (Love et al., 1986).

The majority of early studies showed weak or no significant HLA class I associations in patients with either GCA or PMR (Table 21.1). However, several later studies demonstrated a significant association with HLA-DR4. In addition to an increase in the frequency of HLA-DR4 [67.4% vs 30.3%, relative risk (RR) 4.8], one study of 44 patients with PMR also detected an increase in the number of HLA-DR4 homozygous patients compared with controls ($p = 0.00003$, RR 8.7) (Sakkas et al., 1990). One French study detected an increase in HLA-DR7 but not HLA-DR4 (Gouet et al., 1985) and an association with both HLA-DR4 and HLA-DR3 was seen in another study (Lowenstein et al., 1983). The HLA-DR4 association appears to be strongest for PMR (Richardson et al., 1987). One study found that the prevalence of the HLA-DR4 allele in Spanish patients with GCA was similar to controls (Cid et al., 1988). However, this has not been a consistent finding, with other studies demonstrating associations between HLA-DR4 and the presence of biopsy-proven GCA (Ninet et al., 1987; Weyand et al., 1992). The association with HLA-DR4 implies that this allele or a closely linked gene is related to disease susceptibility. Ethnic studies are consistent with a role for HLA-DR4 as the relevant disease-susceptibility gene. PMR and GCA are uncommon in Black populations (Love et al., 1986) and in Northern Italians (Salvarani et al., 1991), where the prevalence of HLA-DR4 is low.

HLA-DRβ, DQα and DQβ hybridizations of class II restriction fragments from 27 Caucasoid HLA-DR4-positive biopsy-proven GCA patients did not reveal any differences compared with HLA-DR4-positive controls (Bignon et al., 1988). Subsequent analysis of HLA-DRB1 alleles in patients with biopsy-proven GCA has demonstrated an increased frequency of HLA-DRB1*0401 and DRB1*0404/8 (Weyand et al., 1992). Furthermore, in patients without these alleles, HLA-DRB1*03, DRB1*08 and DRB1*13 were increased compared with controls. These findings suggested the presence of a disease-associated sequence motif common to these DRB1 alleles. Sequence analysis confirmed homology within the second hypervariable region of these alleles. This sequence motif (amino acids 28–31 of the DRβ1-chain, DRYF) maps to the antigen-binding site of the class II molecule. A later study confirmed that the distribution of HLA-DRB1 alleles described above was also seen in patients with PMR and that both conditions share the sequence motif DRYF (Weyand et al., 1994). In contrast, a recent study of Swiss patients with PMR (with or without GCA) found only a non-significant increase in HLA-DR4 and no increase in the DRYF sequence motif compared with controls (Guerne et al., 1997).

No differences in HLA-DQ alleles have been demonstrated serologically (Cid et al., 1988; Salvarani et al., 1991) or using DNA typing (Bignon et al., 1988; Sakkas et al., 1990).

In conclusion, both PMR and GCA are associated with HLA-DR4 and recent molecular analysis suggests the presence of a sequence motif common to the disease-associated HLA-DR4 subtypes. However, further confirmation in large, clearly defined patient populations is needed.

Takayasu arteritis

Takayasu arteritis is an inflammatory, obliterative vasculitis of large and medium-sized arteries with a predilection for the aorta and its major branches (Takayasu, 1908). In addition to the American College of Rheumatology (Arend et al., 1990) and the Chapel Hill nomenclature, an alternative classification exists based on the first classification criteria proposed in 1988 (Ishikawa, 1988). Ishikawa's criteria were modified in 1995 to improve diagnostic sensitivity (Sharma et al., 1995). The disease primarily affects young women from Asia and South America and only rarely affects Caucasoids.

TABLE 21.1 Class I and II allele associations in different populations of patients with polymyalgia rheumatica (PMR) and giant cell arteritis (GCA)

Reference	Ethnic group	Study number Patients	Controls	Class I Allele	Class I RR or p-value*	Class II Allele	Class II RR or p-value*
Rosenthal et al. (1975)[a]	Caucasoid	PMR 15	1142	A8	$p < 0.05$	Not tested	—
				A10	$p < 0.001$		
Terasaki et al. (1976)[a]	Caucasoid	56[c] (PMR 47, GCA 9)	3895	None	—	Not tested	—
Hazleman et al. (1977)[a]	Caucasoid	PMR 27	216	A10	$p < 0.05$	Not tested	—
				B8	$p < 0.05$		
				B8	$p < 0.05$		
Hunder et al. (1977)[a]	Caucasoid	GCA 30	700	None	—	Not tested	—
		PMR 12		None	—		
		GCA 43					
Siegnalet et al. (1977)[a]	Caucasoid	GCA 61	300	B14	$p = 0.05$	Not tested	—
Mattingly et al. (1978)[a]	Caucasoid	PMR 44	100	None	—	Not tested	—
Malmvall et al. (1980)[a]	Caucasoid	GCA 63	101	None	—	Not tested	—
Bridgeford et al. (1980)[a]	Caucasoid	PMR 36	100	None	—	Not tested	—
		GCA 14		None			
		Both 8		None			
Calamia et al. (1981)[a]	Caucasoid	69[c] (PMR 18, GCA 24, both 27)	113	Not tested	—	DR4	$p = 0.03$
Barrier et al. (1981)[a]	Caucasoid	GCA 50	284	None	—	DR4	2.76
						DR3	2.3
Lowenstein et al. (1983)[a]	Caucasoid	78[c] (PMR 61, GCA 7, both 10)	495	Not tested	—	DR4	4.5
Armstrong et al. (1983)[a]	Caucasoid	55[c] (PMR 35, GCA 20)	153	A1	$p = 0.019$	DR4	$p = 0.0082$
				Cw3	$p = 0.012$		
				Cw6	$p = 0.003$		
Bignon et al. (1984)[a]	Caucasoid	GCA 40	146	Not tested	—	DR4	3.20
Armstrong et al. (1984)[a]	Caucasoid	PMR 34	157	A1	$p < 0.05$	DR4	$p < 0.02$
Gouet et al. (1985)[a]	Caucasoid	66[c] (PMR 30, GCA 20, both 16)	116	Not tested	—	DR7	$p = 0.025$
Hansen et al. (1985)[a]	Caucasoid	GCA 35	219	A31	3.09	DR4	2.70
				B40	3.00		
				Cw3	5.65		
Richardson et al. (1987)[a]	Caucasoid	PMR 16	243	None	—	DR4	NS
		GCA 21		None		None	—
		Both 22		None		DR4	$p = 0.035$
Ninet et al. (1987)[a]	Caucasoid	PMR 37	124	None	—	DR4	NS
		GCA 37				DR4	$p = 0.05$
Cid et al. (1988)[a]	Caucasoid	GCAS 31 both 34	200	None	—	DR4	NS
						DR4	5.71
Sakkas et al. (1990)[b]	Caucasoid	PMR 43	132	Not tested	—	DR4	4.8
Salvarani et al. (1991)[a]	Caucasoid	99[c] (PMR 56, GCA 23, both 20)	242	None	—	DR4	NS
Weyand et al. (1992)[b]	Caucasoid	GCA 42	63	Not tested	—	DR4 (DRB1*0401–0404/8 in 60%)	$p = 0.003$; $p = 0.03$
Weyand et al. (1994)[b]	Caucasoid	PMR 46	72	Not tested	—	DRB1*04 (67.4%)	$p = 0.0001$
						DRB1*01	$p = 0.03$
		GCA 52				DRB1*04 (61.5%)	$p = 0.0001$
						DRB1*01	$p = 0.02$
Guerne et al. (1997)[b]	Caucasoid	PMR 100	200	Not tested	—	DR4	NS

[a]Serological typing; [b]DNA typing; [c]alleles apply to total patient group. *p-Value if no relative risk (RR) given; NS: not significant.

Early studies of HLA associations in Japanese patients showed significant increases in the frequency of HLA-B5 (Table 21.2). However, a later study of 82 Japanese patients demonstrated that the association was significant only for the B52 subtype of HLA-B5 (43.9% vs 12.9%, RR 5.5) (Isohisa et al., 1978). More recently, DNA analysis of class I alleles confirmed an increase in HLA-B52 and an increase in HLA-B39.2, a serological subtype of HLA-B39 (probably the product of HLA-B*39021), was also seen (Yoshida et al., 1993). This subtype was not present in the control population and conferred a RR of 20.9. Sequencing of HLA-B52 and HLA-39.2 alleles from patients showed amino acid homology (glutamine and serine at positions 63 and 67, respectively), leading to the hypothesis that this shared epitope is relevant to disease susceptibility (Yoshida et al., 1993; Kimura et al., 1996). Ethnic differences do exist. Serological typing of Indian patients showed an association with HLA-B5 but no association with its subtypes, B51 and B52 (Mehra et al., 1996). In Mexican patients an association with HLA-DR6 and HLA-B62 was evident (Girona et al., 1996), whilst no association was seen in a small study of North American patients (Khraishi et al., 1992).

DNA typing of class II alleles has shown a significant excess of the HLA-DRB1*1502–DRB5*0102 (DR2,Dw12) and HLA-DQA1*0103–DQB1*0601 (DQ6) alleles in Japanese patients (Dong et al., 1992). The HLA-DPB1*0901 allele was also increased significantly. From studies of family members, the disease-associated haplotype was demonstrated to be: HLA-B52–DRB1*1502–DRB5*0102–DQA1*0103–DQB1*0601–DPA1*02–DPB1*0901 (Dong et al., 1992). However, the strongest RR for disease-susceptibility alleles within the MHC in Japanese patients are with HLA-B52 (RR = 3.52) and HLA-39.2 (RR = 20.9). The presence of the DR and DP alleles on the disease-susceptibility haplotype seem most likely to be a consequence of linkage disequilibrium with HLA-B52 (Kimura et al., 1996).

The presence of the HLA-B52 allele is associated with more severe disease in Japanese patients. Compared with HLA-B52-negative patients, those with HLA-B52 have more severe left ventricular involvement (Kasuya et al., 1992), more severe inflammatory manifestations requiring steroid therapy at higher doses and for longer periods (Moriwaki and Numano, 1992) and an increased frequency of mural thrombi (Hata and Numano, 1995).

In summary, Takayasu arteritis in Japanese patients is strongly associated with HLA-B52 and HLA-39.2 alleles which share homology within the peptide-binding region of the HLA-B allele. The presence of this shared epitope leads to the hypothesis that these alleles predispose an individual to the efficient presentation of a disease-inciting antigen.

Medium vessel vasculitis

Polyarteritis nodosa

Polyarteritis nodosa (PAN) was classified by the Chapel Hill consensus conference, as a necrotizing vasculitis involving medium-sized arteries without glomerulonephritis or small-vessel vasculitis. This distinguishes it from the antineutrophil cytoplasmic antibody (ANCA)-positive vasculitis, microscopic polyarteritis (discussed below). In this latter condition, necrotizing vasculitis affects both small and medium-sized vessels and necrotizing glomerulonephritis is very common. Prior to the Chapel Hill meeting, many cases which would now be classified as microscopic polyarteritis were included under the definition of PAN.

Few studies have reported HLA associations in PAN. In a serological study of 15 patients no significant differences in the frequencies of class I and class II alleles were found (Elkon et al., 1983). Furthermore, no HLA associations were demonstrated in a study of 18 unrelated patients (Reveille et al., 1989b). HLA-DQ7 was detected in two siblings, one with classical PAN and another with microscopic polyarteritis (Mason et al., 1994). However, another study, which included two family members who developed PAN following hepatitis B infection, did not find a common HLA type (Reveille et al., 1989b). Other familial reports of PAN are limited by the lack of histopathological evidence of disease but show no convincing HLA association (Harris and Jones, 1970; Leff et al., 1971).

TABLE 21.2 Class I and II allele associations in different populations of patients with Takayasu arteritis

Reference	Ethnic group	Study number		Class I		Class II	
		Patients	Controls	Allele	RR or p-value*	Allele	RR or p-value*
Isohisa et al. (1978)[a]	Japanese	65	128	B5	$p < 0.0001$	Not tested	—
Naito et al. (1978)[a]	Japanese	82	128	B52	5.5	Not tested	—
Naito et al. (1978)[a]	Japanese	38	160	B5	4.1	Not tested	—
Numano et al. (1979)[a]	Japanese	65	128	A10	$p < 0.001$	Not tested	—
				B5	$p < 0.001$		
Volkman et al. (1982)[a]	Caucasoid 8 Korean 2 Mixed 1	11	—	None	—	DR4	$p < 0.015$
						MB3	$p < 0.006$
Moriuchi et al. (1982)[a]	Japanese	47	76	A9	2.4	DR2	6.0
				B52	7.8	DQ1	12.6
Park and Park (1992)[a]	Korean	59	100	B52	$p < 0.02$	DR7	$p < 0.04$
				Cw6	$p < 0.05$	DQ2	$p < 0.01$
Dong et al. (1992)[a,b]	Japanese	64	317	B52	3.59	DRB1*1502	2.80
				B39	2.66	DRB5*0102	2.80
						DQA1*0103	2.27
						DQB1*0601	2.16
						DPB1*0901	3.02
Kraishi et al. (1992)[a]	North American	21	243	None	—	None	—
Yoshida et al. (1993)[b]	Japanese	64	156	B52	3.52	Not tested	—
				B39.2	20.9		
Kimura et al. (1996)[b]	Japanese	138	492	B52	3.7	DRB1*1502	2.7
				B39.2	10.9	DPB1*0901	3.0
Mehra et al. (1996)[a]	Indian	80	289	B5	4.3	DR8	4.7
Girona et al. (1996)[a]	Mexican	24	200	B62	3.13	DR6	5.08

[a]Serological typing, [b]DNA typing. *p-Value if no relative risk (RR) given; NS: not significant.

Kawasaki disease

Kawasaki disease or mucocutaneous lymph node syndrome was first described in Japanese children in 1974 (Kawasaki et al., 1974). It is a multisystem inflammatory illness affecting children and is characterized by acute onset of fever (lasting for more than 5 days), conjunctival congestion, dry lips, erythematous tongue and oral mucosa, cervical lymphadenopathy, polymorphous skin eruption, and redness of the palms and soles with subsequent desquamation. Coronary artery aneurysms are a well-recognized serious complication occurring in up to 20% of affected children (Kato et al., 1975).

The disease has a high incidence in Japan (Yanagawa et al., 1988) and among Japanese Americans (Morens et al., 1980), implying the presence of predisposing genetic factors within the Japanese population. Nation-wide epidemiological surveys of Kawasaki disease in Japan since 1970 have shown three epidemic years (1979, 1982 and 1986) and age-specific incidence curves show a unimodal peak between 9 and 11 months of age (Yanagawa et al., 1988). Furthermore, the proportion of sibling cases among the total number of reported cases was raised. These findings suggest an infectious aetiology triggering the disease in genetically predisposed individuals.

Early studies in Japanese patients suggested that the disease was linked with a Japanese-specific class I allele, Bw22J (now designated B54), which is also associated with Takayasu arteritis and thromboangiitis obliterans (Buerger's disease) (Ohtawa, 1976). In a study of 32 Japanese children, the Bw22J allele was non-significantly increased compared with controls (20.7% vs 13.1%) (Matsuda et al., 1977). However, a subsequent large study of 205 Japanese patients did detect a significant association with Bw22J (25.4% vs 11.8%, $p < 0.0005$) (Kato et al., 1978). In addition, the HLA-Bw15 allele was significantly associated with the presence of coronary artery aneurysms compared with patients without this complication (27.3% vs 11.2, $p < 0.02$). Conversely, a study of predominantly Caucasoid American patients found an association with HLA-B51 which conferred a RR of 14.8 (Krensky et al., 1981). HLA-B51 was also significantly raised among a cohort of Israeli patients with a RR of 19.6 (Keren et al., 1982). In Caucasoid patients, different class I associations during epidemic and endemic cases of the disease have been demonstrated (Krensky et al., 1983). During endemic periods an association with HLA-B51 was demonstrated (70% vs 3%, RR 80), whilst epidemic cases were associated with a non-significant increase in HLA-B44. A further study of epidemic cases in Caucasoid patients found that the strongest association was with HLA-A2–B44–Cw5 (Kaslow et al., 1985). These studies could be explained on the basis that different disease-inciting antigens are responsible for epidemic and endemic cases.

This condition is not associated with particular class II alleles. DNA analysis did not reveal any significant class II associations in a study of 25 Caucasoid and two Black patients (Fildes et al., 1992) and in a study of 44 Caucasoid, 13 Asian and five African–American patients (Barron et al., 1992).

In summary, Kawasaki disease is linked to HLA-B54 in Japanese individuals and HLA-B51 in Caucasoid patients. The size and number of studies are limited by the rarity of this disease and the significance of these associations remains unclear.

Small vessel vasculitis

Wegener's granulomatosis and microscopic polyangiitas

Wegener's granulomatosis is characterized by granulomatous inflammation of the respiratory tract, glomerulonephritis and often vasculitis involving other organs. Wegener's granulomatosis and microscopic polyarteris (previously called microscopic polyangiitis) are both characterized by vasculitis involving small and medium-sized arteries, capillaries and veins. The common clinical, histological and serological manifestations suggest that these conditions are related and belong to the same spectrum of disease. Both conditions are associated with ANCA. The pathogenesis is not known. The HLA associations of Wegener's granulomatosis are shown in Table 21.3. There are no consistent findings between different studies, which suggests that there are no powerful disease-susceptibility genes for this condition located within the MHC.

TABLE 21.3 Class I and II allele associations in different populations of patients with Wegener's granulomatosis

Reference	Ethnic group	Study number		Class I		Class II	
		Patients	Controls	Allele	RR or p-value*	Allele	RR or p-value*
Strimlan et al. (1978)[a]	Caucasoid	31	–[d]	None	–	Not tested	–
Katz et al. (1979)[a]	Caucasoid	31	418	B8	p < 0.01	Not tested	–
Elkon et al. (1983)[a]	Caucasoid	17	113	None	–	DR2	p < 0.008
Murty et al. (1991)[a,b]	Caucasoid	41	200	None	–	None	–
Papiha et al. (1992)[a]	Caucasoid	27	105	None	–	DR1	2.3
						DQ1	5.7
Spencer et al. (1992)[b]	Caucasoid	34	1103	Not tested	–	DQ7	2.9
Cotch et al. (1995)[c]	Not specified	83	4039	B50	4.9	DR9	6.7
Hagen et al. (1995)[a]	Caucasoid	118	2443	None	–	DR6	0.4
						DR13	0.37
	Caucasoid	106	2443	None	–	DR6	0.47
						DR13	0.29
Zhang et al. (1995)[b]	Caucasoid	56	90	Not tested	–	None	–

[a]Serological typing; [b]DNA typing; [c]typing method not specified; [d]control group number not given. *p-Value if no relative risk (RR) given.

In an early study, an association with HLA-B8 was demonstrated (Katz et al., 1979). HLA-B8-positive patients could not be distinguished from HLA-B8-negative individuals on the basis of organ involvement or severity of disease. Although the increase in this antigen was significant, the association between this antigen and Wegener's granulomatosis was not strong. Furthermore, other studies have not confirmed this association (see Table 21.3). One study reported an association with HLA-B50 but this allele was detected in only 8.4% of the patients studied (Cotch et al., 1995).

In a study of 27 patients treated in the north of England, a highly significant increase in HLA-DR1 (RR 2.3) was demonstrated (Papiha et al., 1992). The percentage combined frequency of HLA-DR1 and HLA-DQ1 alleles was significantly higher than in controls. However, the patient group was similar to that used in an earlier study where no significant increase in HLA-DR alleles was seen (Murty et al., 1991). In this study no association was demonstrated for HLA-B8 or HLA-DR2 (DR15) alleles, although the largest difference between the patient and control groups was found for HLA-DR15 (43.9% patients compared with 30% of controls). As 98.6% of HLA-DR2-positive Caucasoid individuals in the UK possess the HLA-DR15 subtype, this difference may reflect the fact that the sample size was not large enough to pick up skewing of such a common allele. A significant increase in the frequency of HLA-DR2 has been reported (Elkon et al., 1983). Both HLA-B7–DR2 and HLA-DR2 alone were present in significantly higher frequency than in controls; 47% of the patients with Wegener's granulomatosis had HLA-B8 compared with 22% of the controls but, unlike the earlier study by Katz, the difference was not statistically significant. A significant decrease in the frequency of HLA-DR13 and HLA-DR6 was demonstrated in one study (Hagen et al., 1995). The frequency of these antigens was also low in the microscopic polyangiitis group but this did not reach statistical significance. In a study of 34 patients with Wegener's granulomatosis and 25 patients with microscopic polyarteritis, there was a significant increase in the frequency of DQ7 at the DQB locus when all 59 patients were analysed together (Spencer et al., 1992); 47% of those with Wegener's granulomatosis inherited DQ7, as did 62% of those with microscopic polyangiitis. Similarly, no difference was found when patients were classified according to the pattern of ANCA staining: DQ7 was found in 51% patients with cANCA (cytoplasmic) and 50% of those with pANCA (perinuclear). Forty-nine patients were also analysed to determine the effects of alleles on the persistence of ANCA following completion of induction immunosuppressive therapy. It was found that HLA-DR4 was associated with transient and HLA-DR2 with persistent ANCA synthesis. The frequency of DQ7 was equal in both groups. In contrast, the DR3–DQw2a haplotype, as defined by restriction fragment length polymorphism (RFLP) analysis, was less common in the patients than the controls. HLA-DQw2a is a serotype which is invariably inherited with HLA-DR3 in Caucasoid individuals and so it was impossible to separate the effects of the two alleles. The HLA-DQ7 specificity was inherited on a variety of haplotypes in these patients which included the HLA-DR4 and HLA-DR5 alleles. This study mapped the putative susceptibility gene for Wegener's granulomatosis to the DQB1 locus, as the DQA alleles which encode the DQα of the DQ7 molecules are different on the HLA-DR4- and HLA-DR5-bearing haplotypes. However, DNA typing of 94 patients with Wegener's granulomatosis, microscopic polyangiitis or renal limited vasculitis showed no significant differences in the distribution of DRB1, DQB1 and DPB1 alleles (Zhang et al., 1995). In addition, no differences were evident when they divided the patients according to the specificities of their ANCA, antiproteinase-3 (PR3) and antimyeloperoxidase (MPO).

The variation in findings may be due to the small sample size of the studies and possibly differences in ascertaining the HLA type. Ultimately, the fact that there appears to be no consistent association suggests that any such association is weak if it exists at all.

Mixed essential cryoglobulinaemia

Mixed essential cryoglobulinaemia is commonly associated with chronic hepatitis C infection (Agnello et al., 1992). Many patients are asymptomatic and those that develop vasculitis usually have evidence of type II cryoglobulins and hypocomplementaemia (Wong et al., 1996) The most common clinical manifestations are cutaneous, with the development of palpable purpura on the lower extremities, which histologically represents a leucocytoclastic vasculitis (Brouet et al., 1974). Other important manifestations include polyarthralgia, renal disease and peripheral neuropathy. Inhibition of viral replication with α-interferon therapy results in clinical improvement and falls in both cryocrit and hepatitis C RNA levels (Misiani et al., 1994). Clearly, the vasculitis is closely related to the viraemia.

Few studies have analysed HLA associations in patients with mixed essential cryoglobulinaemia. In an early study of 36 Italian patients with this condition, only non-significant increases in HLA-DR3 and HLA-DR8 were seen (Migliorini et al., 1981). Following the discovery that this disease is strongly associated with chronic hepatitis C infection, it became clear that only a minority of patients with chronic hepatitis C infection develops overt clinical signs of cryoglobulinaemia. It has been hypothesized that this could be due to genetic differences between these patient groups. DNA analysis of Sardinian, thalassaemic patients found a significant increase in HLA-DR2 (DRB1*1601, DQB1*0502) among the 30 patients who were hepatitis C negative, suggesting that this association was protective (Congia et al., 1996). Conversely, a serological study of 25 patients with chronic hepatitis C infection and cryoglobulinaemia demonstrated a significant increase in HLA-B8 (40% vs 10.1%, RR 5.9) and HLA-DR3 (40% vs 15.1%, RR 3.7) compared with hepatitis C patients without cryoglobulinaemia (Lenzi et al., 1998). However, another study involving 16 patients detected an increase in HLA-A9, -B51 and -B35 alleles (Ossi et al., 1995).

Further studies are needed, but the presence of protective and disease-associated HLA alleles in this condition is plausible, particularly as specific hepatitis C genotypes correlate with the presence of mixed cryoglobulinaemia in hepatitis C infection.

Hypersensitivity vasculitis

Cutaneous leucocytoclastic vasculitis refers to leucocytoclastic vasculitis confined exclusively to the skin and, following the Chapel Hill consensus conference, replaces the term "hypersensitivity vasculitis". Previously, hypersensitivity vasculitis was either primary (confined to the skin) or occurred in the setting of systemic vasculitis or connective tissue disease. In a study of 31 patients with primary and secondary leucocytoclastic vasculitis, no HLA association was found in those with primary leucocytoclastic vasculitis (Glass, 1976a). However, HLA-A11 and HLA-B35 were found in 5 of 19 patients with secondary leucocytoclastic vasculitis (RR 10.92). The significance of this finding is uncertain.

Henoch–Schönlein purpura

The Chapel Hill consensus conference defines Henoch–Schönlein purpura as a small vessel vasculitis with predominant IgA vascular deposition. It is the most common small-vessel vasculitis syndrome in children. The classic triad of arthritis, abdominal pain and purpura occurs in the majority of cases.

Studies to examine a possible influence of the MHC have focused on complement deficiency as a contributing cause. An increased prevalence of C4 null alleles (McLean et al., 1984; Abe et al., 1993) and of homozygous C2 deficiency have been reported in Henoch–Schönlein purpura (Sussman et al., 1973; Gelfand et al., 1975; D'Cruz et al., 1992). In a study of 24 Japanese patients with Henoch–Schönlein purpura, a C4 gene deletion was found in 16.7% of the patients compared with only 2.8% of controls ($p = 0.002$, RR 7.0). However, the frequency of the null allele for both C4A and C4B among patients was not different to controls. Conversely, in a study of 19 unrelated Caucasoid patients with Henoch–Schönlein purpura, the presence of homozygous C4A or C4B null alleles was significantly raised in the patient group compared with controls (26.3% vs 3.9%, $p = 0.003$) (McLean et al., 1984). The relative risk for the development of Henoch–Schönlein purpura was greatest in C4AQ*0 homozygotes (RR 11.9) compared with that for C4BQ*0 homozygotes (RR 6.2). No significant increase in either class I or class II antigens was found in this study. Finally, a 17-year-old patient with Henoch–Schönlein purpura who was homozygous for the HLA haplotype, A30–B18–DR7–C4AQ0–C4BQ0, has been reported (Lhotta et al., 1990).

Churg–Strauss syndrome

The Chapel Hill consensus conference classifies Churg–Strauss vasculitis as a necrotizing vasculitis affecting small to medium-sized vessels with asthma, peripheral blood eosinophilia and eosinophilic-rich inflammation. The condition is rare and only one study has analysed the frequency of class I and II alleles (Elkon et al., 1983). No significant associations were found in the 14 patients studied.

Behçet's syndrome

Behçet's syndrome was classified by the international study group criteria as recurrent oral ulceration plus two of the following: recurrent genital ulceration, eye lesions, skin lesions or a positive pathergy test (Anonymous, 1990). It is one of the vasculitides that can affect the venous as well as the arterial side of the vascular tree (Lie, 1992). Vessels of all sizes may be affected, with superficial thrombophlebitis being the most frequent type of vascular involvement. Epidemiological studies have demonstrated that the disease is prevalent amongst populations living alongside the old Silk Road (stretching from Far East Asia, China, the Middle East and along to the Eastern Mediterranean). Behçet's syndrome is also seen less commonly outside the silk route in America and Europe.

Behçet's syndrome is strongly associated with the class I antigen HLA-B51 (Ohno et al., 1978). In Japanese patients HLA-B51 is seen in 57%, significantly higher than the frequency of this allele in the Japanese population (16%) (Mizuki et al., 1997a). This HLA association has been noted in many different ethnic groups. However, in a community study in rural Turkey, patients did not show any association with HLA-B51, suggesting that the association may be linked with severe disease (Yurdakul et al., 1988). The association with HLA-B51 is especially marked in patients with eye involvement (Mizuki et al., 1992a). Of the five HLA-B51 alleles, HLA-B*5101 is most commonly found in this condition. The frequency of HLA-B*5101 and HLA-B*5102 was 91.8% and 1.9%, respectively, in HLA-B51-positive Japanese patients (Mizuki et al., 1993). However, the frequency of these alleles was not different compared with healthy HLA-B51-positive Japanese controls.

Peripheral blood neutrophils in patients with Behçet's syndrome demonstrate increased chemotactic activity compared with healthy controls (Matsumura and Mizushima, 1975). This increased neutrophil activity has been linked to HLA-B51 (Takeno et al., 1995). Both HLA-B51-positive patients and HLA-B51-positive healthy individuals demonstrated significantly higher neutrophil superoxide production compared with their respective HLA-B51-negative controls. No difference was seen in superoxide production between HLA-B51-negative patients and HLA-B51 negative healthy controls. The same study demonstrated that neutrophils from HLA-B51 transgenic mice produce excessive amounts of superoxide after stimulation with N-formyl-methionyl-leucyl-phenylalanine. However, these transgenic mice do not have any other phenotypic changes comparable with those seen in Behçet's syndrome.

A lower frequency of a tumour necrosis factor-β (TNFβ) RFLP in patients with ocular disease has been demonstrated (Mizuki et al., 1992b). This was consistent with the genetic susceptibility being linked to a locus close to, but centromeric to, HLA-B51. Polymorphism in the Tau-a microsatellite region located between HLA-B and TNF genes on chromosome 6 has been demonstrated in Japanese patients with Behçet's syndrome (Mizuki et al., 1995). This polymorphism was significantly more frequent in HLA-B51-positive and HLA-B51-negative patients than in healthy HLA-B51-positive and HLA-B51 negative controls, respectively. Thus, it appears that the genetic susceptibility in Behçet's syndrome may not be the HLA-B51 antigen but possibly a non-HLA gene located centromeric to HLA-B51 and in linkage disequilibrium with the Tau-a microsatellite. Many novel and unidentified genes exist between the HLA-B locus and the genes encoding for TNF. One of these candidate genes is MICA (MHC class I chain-related gene A).

MICA is an MHC class I chain-related gene located 40 kb centromeric to the HLA-B gene on the short arm of chromosome 6. The function of MICA is unknown but it is likely to play a role in antigen presentation, possibly to γδ T cells. Using a triplet repeat microsatellite polymorphism in the transmembrane region of the MICA gene, 57 out of 77 (74%) of Japanese patients were found to have the microsatellite allele (consisting of six repetitions of GCT/AGC) at a significantly higher frequency than controls (Mizuki et al., 1997b). This allele (A6) was present in all patients who were HLA-B51 positive and in 13 out of 33 (39%) who were HLA-B51 negative. The genetic susceptibility to Behçet's syndrome may therefore be causally linked to the MICA locus rather than to HLA-B51.

SYSTEMIC LUPUS ERYTHEMATOSUS

Three groups of genes within the MHC are candidates for disease-susceptibility genes for SLE. These are the HLA class II genes, complement genes within the class III region and other genes within the class III region, e.g. TNF. This section reviews the current knowledge of haplotype associations and candidate genes within the MHC. The relevance of HLA alleles to autoantibody production and disease subsets is also discussed.

Haplotype associations

The most common extended haplotype in Caucasoid, western European lupus patients is HLA-A1–B8–Cw–DR3, C4AQ*0–C4*B1–C2*C–B*fS (Fielder et al., 1983; Schur et al., 1990; So et al., 1990; Hartung et al., 1992a; Truedsson et al., 1995). In a study of 10 families with two or more patients with lupus (27 patients among 66 individuals) only the HLA B8–DR17(DR3)–SCO1(C4AQ*0–C4*B1–C2*C–B*fS) haplotype was more common in patients than their healthy relatives (Truedsson et al., 1995). In a large multicentre European study, MHC haplotypes were determined in 155 families by familial segregation analysis. Two disease-susceptibility haplotypes were found: B8–DR3–C4AQ*0–C4B1 and B7–DR2–C4A3–C4B1 (Hartung et al., 1992a). This was the first study to demonstrate an association between the HLA-B7–DR2 haplotype and lupus. However, although the frequency of HLA-DR2 was increased independently of the HLA-B7-DR2 haplotype, the frequency of the HLA-B7 allele was not. It is therefore likely that the disease-susceptibility locus on this haplotype is nearer to the DR allele than the B7 allele.

However, in some Caucasoid populations the association of SLE with these haplotypes has not been found. The extended haplotype, HLA-B8–DR3–C4QA*0, was found in lupus patients of English/Irish ancestry, but not in a sample of other Europeans (Schur et al., 1990). Similarly, these haplotypes were not detected in French–Canadians, amongst whom the disease was associated with HLA-DQ6 (Goldstein and Sengar, 1993). Amongst non-Caucasoid populations an association with HLA-DR2 has been described in Japanese (Hashimoto et al., 1985, 1994), Southern Chinese (Hawkins et al., 1987; Doherty et al., 1992), Malayans (Kong et al., 1994) and in some studies of African–Americans (Kachru et al., 1984; Barron et al., 1993; Reveille et al., 1998).

The HLA-DR3 association with lupus is predominantly associated with class II alleles, haplotype HLA-DRB1*0301–DQA1*0501–DQB1*0201 (Skarsvag, 1995a; Yao et al., 1993a) and the HLA-DR2 association with the class II haplotype, HLA-DRB1*1501–DQA1*0102–DQB1*0602. This latter association has been shown in Caucasoids (Bettinotti et al., 1993a; Yao et al., 1993a), African–Americans (Reveille et al., 1991b; Barron et al., 1993), Chinese (Doherty et al., 1992) and Japanese (Hashimoto et al., 1994) lupus patients. This implies the presence of a disease-susceptibility gene within this particular haplotype. However, studies of DNA polymorphisms have not been able to demonstrate a disease-specific HLA-DR2 subtype. Amongst Caucasoid patients participating in the European multicentre trial, the haplotype HLA-DRB1*1501–DQA1*0102–DQB1*0602, was identified in 81% of the 62 HLA-DR2-positive patients, but this haplotype was also present in 81.8% of HLA-DR2-positive controls (Bettinotti et al., 1993a).

Candidate genes within the MHC

Class I and II alleles

The majority of studies examining HLA class I and II frequencies in SLE across many ethnic groups is summarized in Table 21.4. Amongst Caucasoid patients two predominant associations exist. The first is with HLA-DR3, as part of the extended haplotype HLA-A1–B8–DR3–C4AQ*0. The second association in Caucasoid patients is with HLA-DR2. The prevalence of HLA-DR3 in lupus

TABLE 21.4 Class I and II allele associations in different populations of patients with systemic lupus erythematosus

Reference	Ethnic group	Study number Patients	Study number Controls	Class I Allele	Class I Patients (%)	Class I Controls (%)	Class I RR or p-value*	Class II Allele	Class II Patients (%)	Class II Controls (%)	Class II RR or p-value*
Grumet et al. (1971)[a]	Caucasoid	25	82	A8	36	16	$p < 0.05$	Not tested	—	—	—
				W15	36	10	$p < 0.008$				
Rigby et al. (1978)[a]	Caucasoid	30[c] 14[d]	1037	A2, B7	37	13.5	$p < 0.001$	Not tested	—	—	—
				A1, B8	57	18.4	$p < 0.001$				
Reinertsen et al. (1978)[a]	Caucasoid	28	490	None	—	—	—	DR2	57.1	26.4	3.7
								DR3	46.4	22.2	3.0
Gibofsky et al. (1978)[a]	Caucasoid	24	40	Not tested	—	—	—	DR2	50	25	3.0
								DR3	54	20	4.7
Gladman et al. (1979)[a]	Caucasoid	30	258	None	—	—	—	DR2	53	26	3.25
Celada et al. (1980)[a]	Caucasoid	40	123	None	—	—	—	DR3	42.5	11.3	5.75
Scherak et al. (1980)[a]	Caucasoid	28	125	None	—	—	—	DR3	39.2	16	3.4
Black et al. (1982)[a]	Caucasoid	53	188 (class I) 148 (DR)	A1	43.4	26.1	$p = 0.02$	DR2	37.5	21.6	$p = 0.05$
				A2	30.2	51.6	$p = 0.01$	DR3	43.8	28.4	$p = 0.08$
				A19	37.7	20.6	$p = 0.02$				
				B8	43.4	22.3	$p = 0.005$				
Schur et al. (1982)[a]	Caucasoid	106	1024	B8	39	17.1	3.1	DR3	37	22	2
Fielder et al. (1983)[a]	Caucasoid	29	39	A1, B8, Cw	25.9	3.8	—	DR3	69	18	$p < 0.001$
Hochberg et al. (1985)[a]	Caucasoid	111	626	Not reported	—	—	—	DR2	44.1	24.4	$p = 0.00005$
								DR52	67.6	45.9	$p = 0.00003$
								DQ1	69.4	54.6	$p = 0.005$
Reveille et al. (1985)[a]	Caucasoid	15	626	Not tested	—	—	—	DR2	53	24	—
								DR3	100	25	—
Dunckley et al. (1986)[b]	Caucasoid	46	134	Not tested	—	—	—	DR3	54.3	35.8	1.5
Howard et al. (1986)[a]	Caucasoid	63	63	None	—	—	—	DR2	46	22.2	3.0
								DQ1	71	49	2.58
								DR3	38	25.4	NS
Savi et al. (1988)[a]	Italian	80	633	None	—	—	—	None	—	—	—
Hartmung et al. (1989)[a]	Caucasoid	248	2163	A1	42.3	26.4	1.8	DR2	40.6	29.1	1.7
				B8	41.9	21.7	3.2	DR3	40.6	22.6	2.3
				B13	10.9	18.3	2	DQ2	56.5	32.9	2.6
So et al. (1990)[b]	Caucasoid	49	56	—	—	—	—	DR2	21.4	10.7	2.3
								DR3	29.6	13.4	2.7
Schur et al. (1990)[a]	English/Irish	27[e]	144[e]	None	—	—	—	DR3	41	19	$p < 0.02$
	Other Europeans	62[e]	310[e]	None	—	—	—	None	—	—	—
Reveille et al. (1991)[a, b]	Caucasoid	60	66	Not tested	—	—	—	DR3	52	29	2.6
								DQw2.1[k]	50	29	2.6
Reinharz et al. (1991)[b]	Caucasoid	52	418	Not tested	—	—	—	DR3	44.2	16	$p < 0.001$
Hartung et al. (1991)[a]	Caucasoid	356	1926	Not tested	—	—	—	DR2	25.6	15.8	$p = 0.0001$
								DR3	26.8	12	$p < 0.01$
Hartung et al. (1992a)[a]	Caucasoid	356	975	B8	24	9.6	$p = 0.0000001$	DR2	25.6	15.5	$p < 0.0005$
								DR3	26.7	12.5	$p < 0.0005$

TABLE 21.4 Continued.

Reference	Ethnic group	Study number Patients	Study number Controls	Class I Allele	Class I Patients (%)	Class I Controls (%)	Class I RR or p-value*	Class II Allele	Class II Patients (%)	Class II Controls (%)	Class II RR or p-value*
Ruuska et al. (1992)[a]	Finnish	60	270	A1	28	15	2.27	Dw3	49	12	7.07
				A19	23	11	2.41	DR3	52	16	5365
				B8	45	14	4.85				
Yao et al. (1993a)[b]	German	178	207	Not tested	—	—	—	DRB1*02	44.4[g]	28[g]	2.1
								DRB1*03	44.4[g]	18.8[g]	3.4
								DQB1*0201	65.5[g]	35.4[g]	3
Yao et al. (1993b)[b]	German	178	206	Not tested	—	—	—	DPB1*1501	11.3[g]	3.7[g]	$p < 0.0002$
Barren et al. (1993)[b]	Caucasoid	153[f]	160	Not tested	—	—	—	DR3 (DR17)	43	25	2.3
Bettinotti et al. (1993a)[b]	German DR2-positive	62	192	Not tested	—	—	—	DRB1*0101	81	81.8	NS
								DQA1*0102	81	81.8	NS
								DQB1*0602	81	81.8	NS
Goldstein and Sengar (1993)[a,b]	French–Canadian	86	88	None	—	—	—	DQ6	60	37	2.6
	Non–French–Canadian	86	72	B8	55	26	3.5	DR3	58	20	5.6
								DR52	81	51	4.1
								Dw24	56	26	3.7
								DQ2	63	34	3.2
De Juan et al. (1993)[a,b]	Spanish	22[h]	69	B18	55	19	5.2	DR3	68	29	5.2
								DR17.2	59	17	6.9
								DQα2[1]	82	29	10.2
								DQβ2a[1]	82	29	10.2
Mehra et al. (1993)[a]	Northern Indian	58	106	None	—	—	—	DR4	37.5	17.9	$p < 0.03$
Kong et al. (1994)[a]	Malay	55	91	B7	14.5	3.3	4.99	DR2	56.4	14.3	3.28
				Cw7	52.7	27.5	2.94				
Fouad et al. (1994)[a]	Kuwaiti	25	127	A3	44	11.8	5.89	DR2	44	16.7	3.33
Cowland et al. (1994)[b]	Danish	24	102	Not tested	—	—	—	DRB1*0306	83.3	35.5	9.1
								DQA1*0501	70.8	29.7	5.8
								DRB3*0103	70.8	36.1	4.3
Reveille et al. (1995a)[b]	Mexican	55	112	Not tested	—	—	—	DRB1*0301	27	9	$p < 0.005$
Reveille et al. (1995b)[b]	Greek	53	48	Not tested	—	—	—	None	—	—	—
Davies et al. (1995)[b]	Caucasoid	82	59	Not tested	—	—	—	DR3	60	37	$p = 0.009$
								DQA*0501	63	39	$p = 0.004$
Skarsvag et al. (1995b)[b]	Scandinavian	51	121	Not tested	—	—	—	DRB1*0301	41.4	20.2	3.1[i]
								DQA1*0501	41.4	20.2	3.9[j]
								DQB1*0201	41.4	20.2	—
Marintchev et al. (1995)[b]	Bulgarian	59	50	Not tested	—	—	—	DRB1*0301	32	20	NS
Truedsson et al. (1995)[a,b]	Caucasoid	27	121	A1	63	29.8	$p = 0.0017$	DR3	63	25	$p = 0.0004$
				Cw7	92.6	37.1	$p < 0.0001$				
				B8	66.7	24.8	$p < 0.0001$				
Granados et al. (1996)[a]	Mexican	102	200	None	—	—	—	DR3	16.1	11.9	2.56
								DR7	17.6	17.1	3.08

TABLE 21.4 *Continued.*

Reference	Ethnic group	Study number Patients	Study number Controls	Class I Allele	Class I Patients (%)	Class I Controls (%)	Class I RR or p-value*	Class II Allele	Class II Patients (%)	Class II Controls (%)	Class II RR or p-value*
Granados et al. (1997)[a]	Mexican	104	200	None	—	—	—	DR3	17.3	7	2.78
								DR7	19.2	6.5	3.42
								DQ2	26.9	12.5	2.58
Reveille et al. (1998)[b]	Caucasoid	67	200	Not tested	—	—	—	DRB1*0301	51	25	$p = 0.0003$
								DQA1*0501	64	36	$p < 0.05$
								DQB1*0201	60	36	$p < 0.05$
Reveille et al. (1998)[b]	Hispanic	70	105	Not tested	—	—	—	DRB1*0301	20	9	$p = 0.046$
								DRB1*08	40	16	$p = 0.008$
Gladman et al. (1979)[a]	African–American	25	73	None	—	—	—	None	—	—	—
Alarif et al. (1983)[a]	African–American	31	73	None	—	—	—	DR3	58	27	3.67
Wilson et al. (1984a)[a]	African–American	28	137	Not tested	—	—	—	DR7	50	17	$p = 0.005$
Kachru et al. (1984)[a]	African–American	37	147	Aw31	13.5	3.4	4.44	DR2	40.5	18.3	3.03
								DR3	62.2	20.4	6.41
Hochberg et al. (1985)[a]	African–American	36	168	Not reported	—	—	—	None	—	—	—
Howard et al. (1986)[a]	African–American	35	35	None	—	—	—	None	—	—	—
Olsen et al. (1989)[a,b]	African–American	79	68	B35	26	7	$p = 0.003$	None	—	—	—
Reveille et al. (1989a)[b]	African–American	63	57	Not tested	—	—	—	None	—	—	—
Barron et al. (1993)[b]	African–American	126[j]	130	Not tested	—	—	—	DR2	49	28	2.5
								DQB1*0602	40	22	2.5
Rudwaleit et al. (1995b)[b]	Black South African	49	89	Not tested	—	—	—	DRB1*02	34.7	12.6	$p < 0.005$
								DQB1*06	73.3	55.4	$p < 0.06$
								DQB1*0602	51.1	27	$p < 0.005$
Reveille et al. (1998)[b]	African–American	88	88	Not tested	—	—	—	DRB1*1503	38	22	$p = 0.046$
								DQB1*0602	47	28	$p < 0.05$
Hong et al. (1994)[b]	Korean	60	72	Not tested	—	—	—	DRB1*1501	26.9	12.5	3.3
								DR9	23.9	11.1	3.9
								DRB5*0101	26.9	12.5	3.3
Lu et al. (1997)[a]	Taiwanese	105	115	Not tested	—	—	—	DQB1*0501	11.4	0.9	14.7
Sakurami et al. (1982)[a]	Japanese	18	144	Not tested	—	—	—	None	—	—	—
Kameda et al. (1982)[a]	Japanese	45	36	None	—	—	—	DR4 (decrease)	20	58	$p < 0.001$
Hashimoto et al. (1985)[a]	Japanese	86	75	None	—	—	—	DR2	46.5	28	2.24
Hashimoto et al. (1994)[a,b]	Japanese	58	97	B39	16.7	5.2	9.8	DRB1*1501	29.6	8.2	11.86
								DRB5*0101	29.6	8.2	11.86
								DQB1*0602	29.6	8.2	11.86
Hawkins et al. (1987)[a]	Southern Chinese	100	100	None	—	—	—	DR2	62	38	2.66
Doherty et al. (1992)[b]	Southern Chinese	87	66	Not tested	—	—	—	DR2	51.7	18.2	4.8
								DR15	37.9	10.6	5.2
								DQ1	75	57.1	2.3
Rudwaleit et al. (1995a)[b]	Singapore Chinese	26	77	Not tested	—	—	—	None	—	—	—

[a]Serological typing; [b]DNA typing; [c]mild disease; [d]severe disease; [e]number of haplotypes studied; [f]38 children, 115 adults; [g]allele frequencies; [h]SLE patients with diffuse proliferative nephritis; [i]RR values obtained by two-by-two comparisons of pairs DRB1*0301–C4AQ*0 and DQA1*0501–C4A*Q0; [j]29 children, 97 adults; [k]DQ polymorphism identified by RFLP analysis; [l]fragments determined by RFLP analysis. *p-Value if no relative risk (RR) given; NS: not significant.

patients may be due to its linkage disequilibrium with the disease susceptibility gene, C4AQ*0. (This is discussed below.) However, a large European multicentre trial could not demonstrate an increase in the frequency of C4AQ*0 independent of the HLA-B8–DR3–C4AQ*0 haplotype (Hartung et al., 1992a). Conversely, the raised incidence of HLA-DR2 in patients suggests that this allele, or a gene in close proximity to the DR2 allele also functions as a disease-susceptibility gene. In contrast to HLA-DR3, HLA-DR2 is not in linkage disequilibrium with the C4AQ*0 allele.

The situation becomes very complicated when the prevalence of class I and class II alleles is examined in other ethnic groups (see Table 21.4). In African–American patients many of the studies did not show any significant association with class II alleles (Gladman et al., 1979; Hochberg et al., 1985; Howard et al., 1986; Olsen et al., 1989; Reveille et al., 1989a). However, two studies have found an association with HLA-DR3 (Alarif et al., 1983; Kachru et al., 1984) and three with HLA-DR2 (Kachru et al., 1984; Barron et al., 1993; Reveille et al., 1998). With regard to the HLA-DR2 association, no significant HLA-DR2 subtypes have been found using oligotyping and nucleotide sequencing (Reveille et al., 1991b). The most frequent HLA-DRB1 allele, DRB1*1503, occurred in 87% of patients and 78% of HLA-DR2-positive healthy controls. In Japanese and Chinese studies, significant associations with HLA-DR2 are seen in the majority (Kameda et al., 1982; Hashimoto et al., 1985, 1994; Hawkins et al., 1987; Doherty et al., 1992).

The prevalence of *DQ* alleles in lupus patients has been mainly associated with the presence of particular autoantibodies (discussed below), rather than the presence of the disease per se. For example, one recent study of Caucasoid patients found a raised frequency of the HLA-DQA*0501 allele (Davies et al., 1995). This allele is in linkage disequilibrium with both HLA-DR3 and C4AQ*0. Using empirical logistic analysis, C4AQ*0 and HLA-DQA*0501 appeared not to have independent and additive effects on disease susceptibility. However, when anti-La-negative patients were examined the association with HLA-DQA*0501 was lost, whereas the association for C4AQ*0 persisted. Therefore, the presence of DQA*0501 in this population may be associated with autoantibody formation, rather than SLE per se.

There is no evidence of gene interactions between HLA-DR and -DQ alleles in the susceptibility to SLE. There was no increase in either DR or DQ heterozygous and homozygous phenotypic combinations over expected values in a large study of Caucasoid patients (Hartung et al., 1991). Furthermore, a study examining polymorphisms of the HLA-DQA1 promoter region showed that all HLA-DR2-positive patients carry the QAP1.2 polymorphism and all HLA-DR3-positive patients carried the QAP4.1 polymorphism (Yao et al., 1993a). However, analysis of the relative risks of the various alleles demonstrated that these polymorphisms were not independently linked with disease susceptibility.

Studies have failed to demonstrate an association between SLE and the DP alleles (Reveille et al., 1991a; Davies et al., 1994a). One study of 178 German patients, however, found a significant increase in DPB1*0101, but this was associated with the HLA-B8–DR3 haplotype (Yao et al., 1993b).

Significant class I associations have been demonstrated in Caucasoid patients. However, they are almost always with HLA-A1 or HLA-B8 alleles, reflecting linkage disequilibrium within the common extended haplotype (HLA-A1–B8–DR3–C4AQ*0–C4*B1–C2*C–Bf*S) rather than independent disease-susceptibility factors. Furthermore, in cases where HLA-DR2 was significantly associated with SLE, the majority of studies found no class I association.

Complement genes C4 and C2

Hereditary complement deficiency and systemic lupus erythematosus

Hereditary homozygous deficiencies of C4 and C1q are associated with at least a 75% prevalence of SLE (Walport and Lachmann, 1990). Homozygous deficiency of C2 is associated with SLE in approximately 33% of cases. In view of the link between homozygous complement deficiency and SLE, it was hypothesized that partial deficiencies of C4 or C2 could increase disease susceptibility. The data for C4 null alleles and C2 null alleles are discussed separately below.

C4 null alleles and systemic lupus erythematosus

C4 is encoded by two tandemly duplicated genes, C4A and C4B, within the class III MHC region. Human C4A and C4B are highly polymorphic, with variants including non-expressed or null alleles for which no protein product is identifiable (expressed as C4AQ*0 and C4BQ*0, i.e. quantity zero). A single C4 null allele may be seen in up to 30% of healthy Caucasoid subjects, with approximately 4% having homozygous C4A deficiency and 1% homozygous C4B deficiency. Polymorphic allotypes of C4A and C4B can be detected by differences in electrophoretic mobility (Awdeh and Alper, 1980), especially after the removal of C-terminal basic amino acids with carboxypeptidase B (Sim and Cross, 1986). C4 typing, by analysing band intensities on agarose gels using laser densitometry, has been used in many of the studies investigating C4 null allele frequency in SLE. However, this technique has the drawback that patients with heterozygous null alleles at both the C4A and C4B loci are difficult to detect and therefore potentially underestimated. More importantly, phenotyping is difficult in patients with active disease in whom C4 levels are usually very low because of complement activation and consumption *in vivo*. Molecular cloning of the C4 genes has enabled detailed restriction mapping of this region to be performed (Carroll and Porter, 1983). RFLP allows more accurate determination of C4 heterozygous and homozygous deficiency states than is possible using differential staining intensities on agarose gel electrophoresis. Using these techniques the prevalence of C4A alleles is usually higher than that obtained from phenotypic analysis.

There is a strong association between C4AQ*0 null alleles and SLE in Caucasoid patients (Table 21.5). Moreover, in these patients the association shows a gene dose-dependent effect. For example, one study of Caucasoid patients demonstrated a RR in heterozygotes of 3.23, rising markedly in homozygotes to 16.86 (Howard et al., 1986). However, the majority of studies in Caucasoid patients shows that C4AQ*0 null alleles are in strong linkage disequilibrium with HLA-DR3. This prevents separation of the relative contribution of the two genes to disease susceptibility.

Two approaches aimed at resolving this issue have been performed. The first was to analyse the HLA associations of SLE in different ethnic groups where different HLA allotypes are seen. The second was to analyse C4AQ*0 null allele frequency in SLE patients who are HLA-DR3 negative. It is noteworthy that in one Caucasoid study, which demonstrated a marked C4AQ*0 null allele association, an association with HLA-DR2 but not HLA-DR3 was seen (Howard et al., 1986). Using multivariate analysis, HLA-DR2 and C4AQ*0 appeared to contribute independently to disease susceptibility in this study, with a RR of approximately 3 for each allele. When both were present the RR rose to 24.9.

Studies of patients from many other ethnic groups demonstrate that association between C4A deficiency and SLE remains but the association with HLA-DR3 is inconsistent (Table 21.5). For example, in African–American patients the presence of a C4AQ*0 allele confers a RR of 4.5 despite no association with HLA-DR3 (Olsen et al., 1989). Furthermore, in Japanese populations where HLA-DR3 is extremely rare, the association with C4AQ*0 persisted (Yamada et al., 1990). However, interethnic differences for the association between SLE and C4AQ*0 do exist. In a study of French–Canadian and non-French–Canadian lupus patients a significant association with C4AQ*0 could be demonstrated only in the non-French–Candian group (Goldstein and Sengar, 1993). In addition, a Spanish study (De Juan et al., 1993) and a Greek study (Reveille et al., 1995b) failed to detect an association with C4AQ*0 null alleles.

In HLA-DR3-negative Caucasoid lupus patients a significant association was demonstrated for both C4AQ*0 and C4BQ*0 alleles in combination but not for C4AQ*0 alone (Batchelor et al., 1987). The same study analysed the haplotypes of 14 non-Caucasoid, DR3-negative patients and found that 71% had C4 null alleles. This approach is difficult in Caucasoid patients because the frequency of HLA-DR3 negative patients is low.

The association between C4BQ*0 alleles and SLE is much less strong than for C4AQ*0 alleles (Table 21.6). Many studies have found no increase in C4BQ*0 gene frequency in Caucasoid populations and in other ethnic groups. It is therefore

TABLE 21.5 C4AQ*0 allele frequencies in different populations of patients with systemic lupus erythematosus (SLE)

Reference	Ethnic group	Study number Patients	Study number Controls	Homozygous C4AQ*0 deficiency (%) Patients	Homozygous C4AQ*0 deficiency (%) Controls	Homozygous C4AQ*0 deficiency (%) p-Value or RR	Frequency of C4AQ*0 allele (%) Patients	Frequency of C4AQ*0 allele (%) Controls	Frequency of C4AQ*0 allele (%) p-Value or RR	DR3 association with SLE
Fielder et al. (1983)	Caucasoid	29	42	14	0	—	38	8.3	$p < 0.01$	Yes
Christiansen et al. (1983)	Caucasoid[a]	41	176	12	0	—	32[b]	20[b]	—	Not tested
Reveille et al. (1985)	Caucasoid	15	63	13	0	—	60	19	—	Yes
Howard et al. (1986)	Caucasoid	63	63	11	0	$p < 0.006$ 16.86	25.4	9.5	$p = 0.003$ 3.23	No, but DR2 increased
Dunckley et al. (1987)	Caucasoid	63	197	7.9	1.5	$p < 0.05$	31.7	16.9	$p < 0.01$	Not tested
Batchelor et al. (1987)	Caucasoid DR3 negative	30	60	0	0	—	8.3	8.3	NS	All DR3 negative
Kemp et al. (1987)	Caucasoid	88	236	10.2	1.7	$p < 0.002$	35	6	$p < 0.05$	Yes
		38	32	4	0	—	26	12	$p < 0.005$	Yes
So et al. (1990)	Caucasoid	56	62	7	5	—	41	22	$p = 0.03$	Yes
Schur et al. (1990)	Caucasoid	27[c]	144[c]	6	9	—	39.1	—	—	Yes
Kumar et al. (1991)	Caucasoid	32	—	9	—	—	31.8	38.9[d]	—	All DR3 negative
	Mexican	11	9[d]	13	0	—	30	15	—	No
Christiansen et al. (1991)	Australian Aborigines	62	133	0	—	—	11	—	—	No
		9	—	6	—	—	30	—	—	Yes
Reveille et al. (1991a)	Caucasoid	48	66	7.8[e]	0	—	29	12	$p < 0.00001$	Yes
Hartung et al. (1992a)	Caucasoid	396	204	0	0	—	12	6	NS	No, but DQ6 increased
Golesten and Sengar (1993)	French–Canadians	86	88	7	0	—	31	10	$p = 0.001$ 4.3	Yes
	Non-French–Canadians	86	72							
Cornillet et al. (1993)	French	74	130	2.7	0	—	14.9	3.1	4.87	Not tested
De Juan et al. (1993)	Spanish	21[f]	53	—	—	—	23	27	NS	Yes
Steinsson et al. (1995)	Icelandic	28[g]	—	11	—	—	68	—	—	No
Reveille et al. (1995a)	Mexican	55	112	—	—	—	14	7	$p < 0.005$	Yes
Reveille et al. (1995b)	Greek	36	33	0	0	—	11	30	NS	No
Davies et al. (1995)	Caucasoid	82	59	—	—	—	63[h]	41[h]	$p = 0.008$	Yes
Skarsvag (1995a)	Scandinavian	51	121	13.7	1.6	$p = 0.0028$ 9.7	45.1	26.6	$p = 0.0172$ 2.3	Yes
Reveille et al. (1998)	Caucasoid	69	186	—	—	—	30	20	NS	Yes
	Hispanic	68	119	—	—	—	19	13	NS	Yes and DR8
Howard et al. (1986)	African–American	35	35	1	0	—	20	7.1	$p = 0.046$ 3.25	No
Wilson et al. (1988a)	African–American	59	59	1.7	0	—	32.3	15.3	$p = 0.05$	Not tested
Olsen et al. (1989)	African–American	79	68	5	0	—	14.5	3.7	$p = 0.01$ 4.5	No
Reveille et al. (1998)	African–American	88	73	—	—	—	20	20	NS	No, but DR2 increased
Hong et al. (1994)	Korean	60	72	0	0	—	41.7	25	$p < 0.05$	No
Dunckley et al. (1987)	Chinese	75	76	2.7	0	NS	30.4	18.8	$p < 0.05$	Not tested
Hawkins et al. (1987)	Southern Chinese	72	61	—	—	—	30.7	15.5	$p < 0.05$	No, but DR2 increased
Denckley et al. (1987)	Japanese	51	50	11.8	0	$p < 0.05$	34.7	12.1	$p < 0.01$	Not tested
Yamada et al. (1990)	Japanese	59	166	0	0	—	44.1	13.3	$p < 0.005$	No

[a]Two female Burmese and three Australian aborigines included in patient group; [b]minimal estimated C4A null allele frequency; [c]number of haplotypes studied; [d]relatives, [e]$n = 174$ patients ("family study"); [f]SLE patients with diffuse proliferative nephritis; [g]family study; [h]population frequency. RR: Relative risk; NS: not significant.

TABLE 21.6 C4BQ*0 allele frequencies in different populations of patients with systemic lupus erythematosus (SLE)

Reference	Ethnic group	Study number Patients	Study number Controls	Homozygous C4AQ*0 deficiency (%) Patients	Homozygous C4AQ*0 deficiency (%) Controls	Homozygous C4AQ*0 deficiency (%) p-Value or RR	Frequency of C4AQ*0 allele (%) Patients	Frequency of C4AQ*0 allele (%) Controls	Frequency of C4AQ*0 allele (%) p-Value or RR	DR3 association with SLE
Fielder et al. (1983)	Caucasoid	29	42	0	0	–	8.6	14.6	–	Yes
Christiansen et al. (1983)	Caucasoid[a]	41	176	5	4	–	15[b]	10[b]	–	Not tested
Reveille et al. (1985)	Caucasoid	15	63	7	3	–	40	21	–	Yes and DR2
Howard et al. (1986)	Caucasoid	63	63	–	–	–	8.7	11.1	NS	No, but DR2 increased
Dunckley et al. (1987)	Caucasoid	63	197	3.2	4	NS	23.1	19.5	NS	Not tested
Batchelor et al. (1987)	Caucasoid DR3 negative	30	60	0	0	–	25	26.6	NS	All DR3 negative
Schur et al. (1990)	Caucasoid	27[c]	144[c]	0	2	NS	–	–	–	Yes
Christiansen et al. (1991)	Australian	62	133	3.1	–	–	12	17	NS	No
	Aborigines	9	–	2	–	–	33	–	–	No
Hartung et al. (1992a)	Caucasoid	396	204	–	–	–	10	14	NS	Yes
De Juan et al. (1993)	Spanish	21[d]	53	–	–	–	50	36	NS	Yes
Reveille et al. (1995a)	Mexican	55	112	–	–	–	9	5	NS	Yes
Reveille et al. (1995b)	Greek	36	33	–	–	–	11	21	NS	No
Reveille et al. (1998)	Caucasoid	69	186	–	–	–	23	18	NS	Yes
	Hispanic	68	119	–	–	–	21	8	$p = 0.03$	Yes and DR8
Howard et al. (1986)	African–American	35	35	–	–	–	17.1	7.1	NS	No
Wilson et al. (1988a)	African–American	59	59	5.1	5.1	NS	15.6	12.7	NS	Not tested
Reveille et al. (1998)	African–American	88	73	–	–	–	19	19	NS	DR2
Hong et al. (1994)	Korean	60	72	–	–	–	0	4.2	NS	No
Dunckley et al. (1987)	Chinese	75	76	1.3	1.3	NS	14.3	12.6	NS	Not tested
Hawkins et al. (1987)	Southern Chinese	72	61	–	–	–	16.8	14.7	NS	No, but DR2 increased
Dunckley et al. (1987)	Japanese	51	50	0	2	NS	14.2	19.3	NS	Not tested

[a]Two female Burmese and three Australian aborigines included in patient group; [b]minimal estimated C4B null allele frequency; [c]number of haplotypes studied; [d]SLE patients with diffuse proliferative nephritis; RR: Relative risk; NS: not significant.

necessary to explain why partial deficiency of C4A and not C4B predisposes to SLE. The functional differences between C4A and C4B could account for this. The C4A isotype shows preferential binding to amino groups, forming amide bonds, and binds particularly to proteins, for example in immune complexes (Schifferli et al., 1986). C4B is haemolytically more active than C4A. Complement activation by immune complexes interferes with lattice formation, thereby maintaining complexes in solution (Heidelberger, 1941). Hence, deficiency of C4A may lead to less effective processing of immune complexes with deposition in tissues and resultant damage. The role of complement in the processing of immune complexes has been reviewed recently (Walport and Davies, 1996)

Another possibility is that a second disease-susceptibility gene on the HLA-B8–DR3 extended haplotype, in linkage disequilibrium with C4AQ*0 alleles, could be responsible. Recent work has demonstrated that B cells and resting T cells from healthy HLA-B8–DR3-positive individuals express markedly reduced levels of Fas (CD95/APO-1) compared with cells from HLA-B8–DR3-negative matched controls (Stassi et al., 1997). This is of direct relevance to SLE, where inefficient clearance of apoptotic cells has been implicated in the disease pathogenesis (Carroll, 1998).

The molecular basis of a number of the more common C4Q*0 alleles has been characterized. In Caucasoid lupus patients and controls who are HLA-A1–B8–DR3 positive, C4AQ*0 is almost invariably associated with a large deletion involving C4A and the adjacent 21-hydroxylase A pseudogene locus (Carroll et al., 1985; Goldstein et al., 1988; So et al., 1990; Kumar et al., 1991). Amongst individuals who possess C4AQ*0 alleles but not the HLA-A1–B8–DR3 haplotype, no gene deletion is seen (Goldstein et al., 1988; Kumar et al., 1991). In a study of African–American lupus patients, 24% were found to have the deleted C4AQ*0 allele (Olsen et al., 1989). Although all patients with the deletion were either HLA-DR2 or HLA-DR3 positive, no HLA-DR antigen was significantly associated with SLE. Amongst Japanese (Yamada et al., 1990) and Korean (Hong et al., 1994) patients, C4AQ*0 was not associated with deletion of the C4A gene, which is in keeping with the rarity of HLA-DR3 in these populations. The genetic basis for C4AQ*0 null alleles in lupus patients is not known in these populations. In the majority of cases with C4BQ*0 alleles no gene deletion is seen (Carroll et al., 1985).

C2 null alleles and systemic lupus erythematosus

Hereditary C2 deficiency is the most common inherited complement deficiency. The C2 null allele (C2Q*0) occurs, in the majority of cases, in association with the haplotype HLA-A25–B18–Cw–DR2–C4A*4–C4B*2–Bf*S (Agnello, 1978; Awdeh et al., 1981; Hauptmann et al., 1982). The molecular basis of C2 deficiency associated with this extended haplotype consists of a 28 base pair genomic deletion that causes skipping of exon 6 during RNA splicing, resulting in the generation of a premature stop codon (type I C2 deficiency) (Johnson et al., 1992).

The extended haplotype was not found in raised numbers in a study of 248 central European lupus patients (Hartung et al., 1989) and none of the HLA-B18-positive lupus patients in another Caucasoid study possessed the "comploype", C4A*4–C4B*2–Bf*S (Christiansen et al., 1983), suggesting that a single null C2 allele does not confer increased susceptibility to disease. However, one study reported a significant association between SLE and heterozygous C2 deficiency (5.9% vs 1.2%, $p = 0.0009$) (Glass et al., 1976b). Recently, a study of 86 Swedish lupus patients investigated the frequency of the 28 base pair deletion associated with type I C2 deficiency (Truedsson et al., 1993). No homozygous cases were found in either patients or controls. The frequency of C2 null alleles was 5.8%, compared with a frequency in controls of 1% (not significant). A further study showed a heterozygous frequency in patients of 1.6% compared with a frequency in controls of 1.3% (Christiansen et al., 1991). Therefore, the majority of studies has failed to demonstrate either a significant increase in heterozygous C2 deficiency or an increase in its associated haplotype. Hence, partial C2 deficiency does not appear to be a disease-susceptibility factor for the development of SLE.

Tumour necrosis factor polymorphisms

The genes encoding TNFα and TNFβ are located between HLA-B and the C4A genes within the class III region (Carroll et al., 1987). TNFα has been implicated in the pathogenesis of SLE. HLA-DR2 and HLA-DQ1 positivity is associated with low TNFα production, whereas HLA-DR3 and HLA-DR4 positivity is associated with high levels of TNFα production (Jacob et al., 1990). Following work on TNF expression and murine lupus, it has been hypothesized that the strong association between HLA-DR2–DQ1 in lupus patients and nephritis (Fronek et al., 1990) may be related to TNF polymorphisms. Such patients have low levels of TNFα production compared with HLA-DR3-positive lupus patients who have high levels of TNFα production.

The (NZB × NZW)F_1 mouse hybrid, which spontaneously develops a lupus-like illness, expresses a TNFα polymorphism associated with low levels of TNFα production and development of nephritis (Jacob and McDevitt, 1988). Treatment with recombinant TNFα delayed the onset of nephritis and increased survival. Associations between the known polymorphisms of the genes encoding TNFα and TNFβ and SLE are discussed below.

Tumour necrosis factor-α

RFLP of the TNFα loci show two bands of 5.5 and 10.5 kb using the restriction enzyme *Nco*I. The frequency of the 10.5 kb band was reduced in patients with SLE, whilst the frequency of the 5.5 kb band was increased (Fugger et al., 1989). A later, small study of 20 lupus patients also found the frequency of the 5.5 kb NcoI fragment to be significantly higher than in healthy controls (Tomita et al., 1993).

Duff and colleagues have described a polymorphism within the human gene encoding for the TNFα gene (Wilson et al., 1992), which is situated in the promoter region and consists of a single nucleotide substitution (guanine to adenine). The uncommon allele containing adenine at position –308 (known as TNF2 or TNFα-308A) is associated with increased levels of transcription (Wilson et al., 1997). This may explain the association between HLA-DR3 and high production of TNFα. The TNF2 allele was increased compared with controls in a study of 81 Caucasoid lupus patients (Wilson et al., 1994). In northern Europeans, the TNF2 allele is strongly associated with the haplotype, HLA-A1–B8–DR3–DQ2 (Wilson et al., 1993) and it was not possible to demonstrate an independent association between these alleles. A later study confirmed an increase in TNF2 allele frequency in lupus patients, again strongly associated with HLA-DR3 (Danis et al., 1995). Conversely, no association was seen in a study of 123 Italian patients (D'Alfonso et al., 1996).

In order to test the hypothesis that the TNF2 allele is directly involved in the pathogenesis of SLE and not secondary to linkage disequilibrium on the extended haplotype, a similar approach to that used for C4AQ*0 alleles has been utilized, i.e. the frequency of this allele in different ethnic groups has been studied.

In Chinese lupus patients, where no HLA-B8–DR3 association exists, the frequency of the TNF2 allele was not increased in a cohort of 67 patients (Fong et al., 1996) and a cohort of 100 patients (Chen et al., 1997) compared with healthy controls. In a study of two ethnically different populations of patients with SLE (49 Caucasoid patients from the UK and 49 Black South African patients), the TNF2 allele was increased in the Caucasoid patients, where it was strongly associated with HLA-DR3 (Rudwaleit et al., 1996). However, in the South African patients, where HLA-DR2 but not DR3 allele frequency was increased, the TNF2 allele frequency was reduced. These studies indicate that the TNF2 allele does not independently confer susceptibility to lupus.

In contrast, in a recent study of 88 African–Americans with SLE the TNF2 allele was significantly associated with lupus despite no association with HLA-DR3 (Sullivan et al., 1997). The only HLA-DR association was HLA-DRB1*15. Of 53 patients who had undergone C4 allotyping, 37% of the C4A*Q0 group possessed the TNF2 allele, compared with a frequency of 39% in the C4A*Q1 group. Therefore, these two alleles were not in linkage disequilibrium. In African–Americans TNF2 may be an independent genetic disease-susceptibility gene, but this finding needs confirmation.

Tumour necrosis factor-β

TNFβ (also known as lymphotoxin, LT) is the product of the LTA locus in the class III region. A Nco1 polymorphic restriction site is present in the first intron of the LT-encoding gene (Messer et al., 1991). The allele LTA*1 (TNFβ1) possesses this restriction site, whilst LTA*2 (TNFβ2) does not. Individuals homozygous for the *LTA**1 (TNFβ*1) have higher LT (TNFβ) production than those homozygous for LTA*2 (TNFβ*2) (Messer et al., 1991). The frequency of the LTA*1 (TNFβ*1) allele was increased in 173 German lupus patients compared with 192 healthy unrelated controls ($p < 0.002$, RR 1.96) (Bettinotti et al., 1993b). However, this increase could be explained by the increase in the prevalence of the haplotype A1–B8–DR3–DQ2–C4AQ*0, suggesting that the association could be secondary to linkage disequilibrium. LTA*2 (TNFβ*2) homozygosity has been significantly associated with nephritis in Korean lupus patients (Kim et al., 1996). No significant differences were found between the 41 patients without nephritis and controls, consistent with the hypothesis that LTA*2 (TNFβ*2) homozygosity is a susceptibility gene for lupus nephritis.

Tumour necrosis factor microsatellite polymorphisms

In addition to the polymorphisms described above, five microsatellite markers have been identified in the TNF region (Udalova et al., 1993). TNF a and b are upstream of the LTA gene, TNF c is in the first intron of the LTA gene and TNF d and e are downstream of the TNFα encoding gene. In Caucasoid lupus patients, the TNF a2, b3 and d2 allele frequencies were all significantly increased but were part of the extended haplotype, DRB1*0301–DQA1*0501–DQB1*0201–C4A*Q0–TNF(a2,b3,d2) (Hajeer et al., 1997). The TNF alleles were also increased in patients with photosensitivity and Raynaud's phenomenon but not nephritis and were also significantly increased in patients with Ro and La antibodies. In an analysis of 10 Swedish families, the TNF(a2,b3,c1) haplotype was more common in SLE but again was associated with the extended haplotype (Sturfelt et al., 1996). Amongst a cohort of 46 Greek lupus patients, linkage disequilibrium was found between DRB1*1501 and TNF a11 and between DR3 and TNF(a2,b3,d2) (Tarassi et al., 1998). There was a trend suggesting that DRB*1501–TNF a 11 allele frequency was higher in patients with renal disease, but this did not reach statistical significance. Clearly, these microsatellite polymorphisms in Caucasoid lupus patients are not independent of the known disease-susceptibility haplotype and further studies are needed among different ethnic groups.

With the exception of the recent study discussed above (Sullivan et al., 1997) the current evidence suggests that TNFα and TNFβ polymorphisms are not independent susceptibility factors for SLE, certainly within Caucasoid populations. However, the presence of the TNF2 allele or TNFB*1 homozygosity, both associated with increased production of TNFα and LT (TNFβ), respectively, may influence the development of lupus nephritis.

Transporters associated with antigen-processing genes

The TAP1 and TAP2 (transporters associated with antigen processing 1 and 2) genes are located in the class II region between the DPA1 and DQB2 loci (Spies et al., 1990; Trowsdale et al., 1990; Bahram et al., 1991). They belong to the family of ATP-binding membrane transporters and are important in the transport of cytoplasmic peptides into the endoplasmic reticulum, where they combine with MHC class I molecules prior to presentation to cytotoxic T cells. A complex TAP polymorphism was demonstrated to alter the array of class I bound peptides in rodents (Powis et al., 1992). However, TAP polymorphisms in mice and humans did not influence peptide selection (Obst et al., 1995). In humans, four TAP1 alleles, resulting from two single amino acid substitutions, and eight TAP2 alleles, resulting from four single amino acid substitutions, have been demonstrated (Colonna et al., 1992; Powis et al., 1992; Carrington et al., 1993).

TAP polymorphisms have been studied in SLE as potential disease-susceptibility loci. However, no significant differences were seen in either TAP1 or TAP2 allele frequencies in Chinese (Savage et al., 1995) or Japanese (Takeuchi et al., 1996) patients with SLE compared with normal controls. Moreover, in a study of 92 Caucasoid patients, no TAP2 polymorphism was significantly associated with SLE or any clinical or immunological subset of the disease (Davies et al., 1994b). Finally, in a large study of 151 white European and 35 Afro-Caribbean SLE patients, no difference in TAP2 polymorphism between cases and controls within the ethnic groups was found (Ocal et al., 1996). Therefore, no current evidence supports a role for TAP1 and TAP2 as disease-susceptibility genes in SLE.

Other candidate genes within the major histocompatibility complex

Recently discovered genes within the MHC class III region include those encoding heat shock proteins (HSP70-1, HSP70-2, HSP70-hom) (Sargent et al., 1989) and the RD(D6S45) gene (Levi-Strauss et al., 1988) (see Chapter 2). HSP70-2 and RD(D6545) have been studied as potential novel susceptibility factors in SLE.

Heat shock protein 70-2

The gene for HSP70-2 encodes an inducible stress protein important in protein transport and unfolding (Beckmann et al., 1990). An 8.5 PstI polymorphism of HSP70-2 has been significantly associated with Graves' disease (Ratanachaiyavong et al., 1991) and insulin-dependent diabetes mellitus (IDDM) (Caplen et al., 1990). However, in IDDM, the association is not independent of HLA-DR3 and may therefore simply reflect linkage disequilibrium (Pugliese et al., 1992). Similarly, in a study of 90 Spanish SLE patients, an increased incidence of this polymorphism was detected (allele frequency 56% vs 42%, $p = 0.007$), but the association was not independent of HLA-DR3 (Pablos et al., 1995). In addition, in a study involving 46 Italian lupus patients, there was no difference in the frequency of the HSP70-1 promoter region polymorphisms compared with controls (Cascino et al., 1994). Despite these negative findings a recent study detected a significant increase in the frequency of the 8.5 kb PstI polymorphism in a series of 46 American Blacks (allele frequency 80% vs 61%, $p = 0.0044$) (Jarjour et al., 1996). The results of this study did not demonstrate a significant association between SLE and either C4AQ*0 or HLA-DR3 in this group of subjects and the authors suggested that the HSP70-2 polymorphism was an independent disease-susceptibility factor.

RD(D6S45)

The RD(D6S45) gene comprises 10 exons encoding a protein of unknown function (Speiser and White, 1989). The predicated protein sequence contains an unusual core of 24 dipeptide (arginine/aspartic acid) repeats which are encoded by one exon. Sequence homology exists between the RD protein and the U1 small nuclear ribonucleoprotein (snRNP). The Ul snRNP is an autoantigen in SLE and hence polymorphisms of the RD(D6S45) gene were analysed (White et al., 1992). However, no differences were found between 43 lupus patients and controls.

Systemic lupus erythematosus disease subsets

Antiphospholipid antibody syndrome

The antiphospholipid antibody syndrome (APS) is characterized by recurrent miscarriages, arterial and venous thromboses, and thrombocytopenia in the presence of antiphospholipid antibodies (Boey et al., 1983). It occurs predominantly in the context of SLE but can also occur as a separate disease entity (primary APS) (Asherson and Cervera, 1994).

In primary APS, associations in Caucasoid patients have included HLA-DR4 (Asherson et al., 1992; Camps et al., 1995) and HLA-DR53 (Asherson et al., 1992). These studies also detected associations with HLA-DQ7 (Camps et al., 1995) and a non-significant increase in HLA-DQB1*0301 (Asherson et al., 1992). In addition, a study of 20 patients with the lupus anticoagulant showed that HLA-DQ (DQB1*0301), associated with both HLA-DR4 and HLA-DR5 haplotypes, was present in 70% of the patients ($p = 0.002$) (Arnett et al., 1991).

Several studies have examined the HLA associations in patients with SLE and APS to test the hypothesis that these are similar to the associations described for the primary APS. Some studies have looked for HLA associations with anticardiolipin antibody-positive lupus patients, whilst others have examined associations in patients with the antiphospholipid antibody syndrome. This is an important distinction as the presence of anticardiolipin antibodies does not always imply a thrombotic tendency.

The presence of anticardiolipin antibodies in lupus patients has been associated with HLA-DR4 in Caucasoid patients (McHugh and Maddison, 1989) and with HLA-DR7 in an Italian study (Savi et al., 1988), whilst two further studies have found no association (Sebastiani et al., 1991; Gulko et al., 1993). In a large series of 314 Caucasoid lupus patients, anticardiolipin antibodies were associated with an increased frequency of HLA-DR4 and HLA-DR7 and significantly associated with HLA-DR53 but not with any HLA-DQ alleles (Hartung, 1992b). However, recently an increased frequency of HLA-DQB1*06 in patients with lupus anticoagulants, especially in the presence of antiplatelet antibodies, was found (Panzer et al., 1997). In lupus patients with APS, an increased frequency, although not statistically significant, of HLA-DR7, HLA-DR53 and HLA-DQ7 has been reported (Goldstein et al., 1996). When these patients were grouped with patients with primary APS, a highly significant increase for HLA-DR53 (RR 5.1), HLA-DR7 (RR 5.6) and HLA-DQ7 (RR 3.6) was demonstrated compared with lupus patients without APS. In Mexican lupus patients with APS significant increases in HLA-DR7 have also been demonstrated (Granados et al., 1997).

These studies provide evidence that the genetic basis of SLE with APS is similar to that of the primary APS and different to uncomplicated SLE. However, it should be noted that many of these studies involve only small numbers of patients and that associations, when demonstrated, show low relative risk values, implying that other genetic factors, including non-HLA genes, are important. Despite this, Arnett et al. (1991) showed that, in patients with the lupus anticoagulant, the most common associated allele is HLA-DQB1*0301 (DQ7). All of the patients without this allele possessed either HLA-DQB1*0302 (DQ8) or HLA-DQB1*0602–*0603 (DQ6) alleles. Common to all of these DQB1 alleles is a seven amino acid region in the third hypervariable region of the HLA-DQ molecule (TRAELDT) (Arnett et al., 1991). This provides suggestive evidence of a "shared epitope" hypothesis to link the HLA associations described above.

C4B and C4A null alleles have been found in increased frequency on haplotypes in a family study of patients with the primary APS (Wilson et al., 1995), but this finding was not confirmed in another study (Asherson et al., 1992). In studies of patients with SLE and anticardiolipin antibodies, no association was found in Caucasoid patients (Hartung et al., 1992b; Goldstein et al., 1996) and in Mexican patients (Granados et al., 1997). However, one study involving African–Americans, found a significant association for both C4A and C4B null alleles and anticardiolipin antibodies in patients with SLE (Wilson et al., 1988b). Clearly, further studies are needed, but these conflicting reports suggest that if any association exists, it is likely to be weak.

Drug-induced lupus

Drug-induced lupus has been most commonly described for hydralazine, procainamide and isoniazid. The condition has similarities with idiopathic SLE but is particularly associated with antihistone antibodies (Fritzler and Tan, 1978) and resolves after the offending drug is stopped. Moreover, renal disease, hypocomplementaemia and antibodies to double-stranded DNA are rare. Lupus induced by hydralazine, isoniazid and procainamide occurs predominantly in patients who possess the slow acetylator phenotype. This implies that the native drugs rather than their acetylated metabolites are causal. In keeping with this was the failure of acetylprocainamide to induce lupus in a series of 11 patients with a history of procainamide-induced SLE (Kluger et al., 1981).

Studies of HLA associations and hydralazine-induced lupus have demonstrated significant increases in HLA-DR4 compared with both healthy individuals and those with idiopathic SLE (Batchelor et al., 1980). In addition, the frequency of HLA-DR4 remained significantly raised compared with patients with the slow-acetylator phenotype who had been treated with hydralazine for 1 year and who had not developed lupus. HLA-DR4 therefore identifies patients who are at high risk of developing hydralazine-induced lupus. A further study of 21 Caucasoid patients confirmed this association (Speirs et al., 1989). However, no association between hydralazine-induced SLE and HLA-DR4 was found in a study of 15 Australian patients (Brand et al., 1984).

Therefore, the HLA associations for drug-induced SLE are different from those of idiopathic SLE, providing evidence that these conditions are genetically distinct. In addition, it has been demonstrated that the slow acetylator phenotype is not increased in lupus patients (Reidenberg et al., 1993).

In addition to the MHC class II findings it was found that 16 out of 21 (76%) of patients with hydralazine-induced SLE possessed C4 null alleles compared with 35 out of 82 normal controls (43%) (Speirs et al., 1989) which, as discussed above, are strongly associated with idiopathic SLE. Furthermore, family studies have shown that individuals with hydralazine-induced SLE have significantly reduced levels of erythrocyte complement receptor 1 (CR1) (Mitchell et al., 1987), a finding that has been reported in idiopathic SLE. Hydralazine and isoniazid, but not their acetylated metabolites, inhibit the covalent binding reaction of C4 *in vitro* (Sim et al., 1984). These findings imply that the development of hydralazine-induced SLE is related both to genetic factors, such as the presence of C4 null alleles and HLA-DR4, and to a direct effect of the native drug on C4.

Finally, it has recently been shown that sulfasalazine-induced SLE occurs in individuals with MHC haplotypes associated with idiopathic SLE, together with the slow acetylator phenotype (Gunnarsson et al., 1997). Nephritis, persistent disease despite stopping the drug and antibodies to double-stranded DNA were seen in many of the patients. These findings would support the hypothesis that the native drug acts as a trigger that precipitates a disease similar to idiopathic SLE, which contrasts with the transient disease caused by hydralazine or procainamide.

Subacute cutaneous lupus erythematosus

Subacute cutaneous lupus erythematosus (SCLE) is characterized by erythematous, non-scarring annular or psoriasiform lesions that occur particularly on sun-exposed parts of the body (Sontheimer et al., 1979). The HLA-A1–B8–DR3 haplotype is significantly associated with both the annular and papulosquamous subtypes of SCLE (Sontheimer et al., 1981). The association with HLA-DR3 was much stronger in the annular group than in the papulosquamous group (RR 67.1 and 10.8, respectively). Subsequently, it was shown that, among patients with the annular lesions, there was marked concordance of anti-Ro antibodies and HLA-DR3 (Sontheimer et al., 1982). A later study confirmed the association with HLA-DR3 but also noted an increased frequency of HLA-DR2 (Watson et al., 1991). The HLA associations of Ro antibodies are discussed further below.

Neonatal lupus

Neonatal lupus erythematosus is characterized by congenital complete heart block and/or transient non-scarring skin lesions. It is almost invariably associated with the presence of maternal anti-Ro antibodies. The immunogenetics of this condition are those of the maternal disease and there is no evidence that MHC genes in the neonate influence the presence or expression of neonatal lupus. In an early study of six families, all but one of the mothers was HLA-DR3 positive and all had anti-Ro antibodies (Lee et al., 1983). Further studies have consistently found an association with HLA-B8–DR3–DQ2 (Watson et al., 1984; Brucato et al., 1995; Julkunen et al., 1995). Clearly, these associations are identical to those described for the annular subtype of SCLE and in both conditions anti-Ro antibodies are typical. Furthermore, patients with Sjögren's syndrome and anti-Ro antibodies may present with cutaneous manifestations of lupus such as SCLE. Common to these "overlap" patients and mothers of neonatal lupus children is the extended haplotype, HLA-B8–DR3–DQ2–DR52 (Alexander et al., 1989). A recent study of Japanese neonatal lupus using DNA typing has demonstrated significant maternal HLA-DQ associations with alleles containing specific amino acid residues known to be associated with anti-Ro and anti-La antibodies (discussed below) (Miyagawa et al., 1997). The incidence of C4A*QA is increased in mothers of children with neonatal lupus (Watson et al., 1992). However, this is likely to be secondary to its linkage disequilibrium with the HLA-A1–B8–DR3 haplotype.

Systemic lupus erythematosus autoantibody subsets

The relationship between HLA antigens and autoantibody production in SLE is complex. Many of the reported associations are inconsistent and based on small numbers of patients. In addition, the prevalence of particular autoantibodies differs among ethnic groups. However, when demonstrated, these associations are often stronger for the autoantibodies than for SLE itself. Recent data indicate that the HLA associations related to autoantibody subsets are primarily within the DQ region.

Anti-Ro and anti-La antibodies

Bell and Maddison first demonstrated an association between HLA-B8 and HLA-DR3 in anti-Ro antibody-positive lupus patients, the frequency of both alleles being significantly increased compared with anti-Ro antibody-negative lupus patients (Bell and Maddison, 1980). Several subsequent studies have confirmed the association between HLA-DR3 and anti-Ro antibodies in Caucasoid lupus patients (Ahearn et al., 1982; Alvarellos et al., 1983; Arnett et al., 1989a; Hartung et al., 1992c; Skarsvag et al., 1995b). Furthermore, anti-Ro antibodies are present in nearly all patients with Sjögren's syndrome (Alspaugh et al., 1976), which is strongly associated with HLA-DR3 (Chused et al., 1977). The level of anti-Ro antibodies is higher in patients who are HLA-DR3 positive, a finding that also applies to patients with Sjögren's syndrome (Manthorpe et al., 1982). However, not all lupus patients with these antibodies are HLA-DR3 positive.

In addition, in studies of African–Americans, anti-Ro antibodies were strongly associated with HLA-DR7 (Wilson et al., 1984a) and HLA-DR3 (Reveille et al., 1989a), whereas one study found no association (Hochberg et al., 1985).

To complicate matters further, some studies have demonstrated that the HLA associations in Caucasoid patients with anti-Ro antibodies alone are different to those found in patients with both anti-Ro and anti-La antibodies. Patients with both anti-Ro and anti-La antibodies appear to have the strongest association with HLA-DR3 (Hochberg et al., 1985; Hamilton et al., 1988; Hartung et al., 1992c), whilst those with the anti-Ro antibodies alone associate with HLA-DR2 (Hochberg et al., 1985; Hamilton et al., 1988). A difference in DQ alleles has also been shown. Anti-Ro antibody-positive Caucasoid patients showed an increased prevalence of HLA-DR2–HLA-DQ1, whilst those with both antibodies had an increased frequency of HLA-B8–DR3–DR52–DQ2 (Hamilton et al., 1988).

Further analysis of patients with Sjögren's syndrome showed that the entire association between HLA-DQ alleles and anti-Ro antibodies could be attributed solely to the effect of DQ1–DQ2 heterozygotes (Harley et al., 1986). It was therefore hypothesized that hybrid DQ molecules (created by *trans* association of the A and B chains of different HLA-DQ alleles, e.g. DQA1–DQB2 and DQB1–DQA2) may play a role in anti-Ro antibody production. Consistent with these results, it was found also that HLA-DR2–DR3 heterozygotes and HLA-DQ1–DQ2 heterozygotes have much higher levels of anti-Ro antibodies than their respective non-heterozygotes (Hamilton et al., 1988; Harley et al., 1989; Fujisaku et al., 1990).

These associations have been refined by studies of both Caucasoid and African–American patients with SLE or primary Sjögren's syndrome, amongst whom it was found that the strongest association with anti-Ro antibodies (with or without anti-La antibodies) was with heterozygosity for HLA-DQ2.1 and HLA-DQ6 (a subtype of HLA-DQ1) (Arnett et al., 1989b; Reveille et al., 1991c). The most frequent HLA-DQ alleles were HLA-DQA1*0501 and HLA-DQB1*0201. Further evidence supporting this gene complementation effect was demonstrated by analysing RFLP of DQA and DQB alleles in anti-Ro-positive lupus patients (Fujisaku et al., 1990). Ro antibodies were related to the simultaneous presence of the α-chain associated with DQ1 and the β-chain associated with DQ2.

A possible molecular explanation for this association of anti-Ro antibody production with heterozygosity at the DQ locus is starting to emerge. All DQ alleles associated with anti-Ro and anti-La antibodies, in both African–American and Caucasoid patients, were associated with glutamine at position 34 of the α_1 domain of DQα and leucine at position 26 of the β_1 domain of the DQβ chain (Reveille et al., 1991c). Patients with anti-Ro and anti-La antibodies were most likely to have these specific amino acid residues in all four DQ chains, providing evidence of gene dosage effect.

In contrast to these studies showing an association between DQ heterozygosity and anti-Ro antibody production, a study involving 376 Caucasoid patients with lupus showed no significant association of heterozygous DQ combinations with either anti-Ro or anti-La antibodies (Hartung et al., 1992c). Furthermore, a recent study of Norwegian lupus patients found that the haplotype HLA-DRB1*03–DRB3*0101–DQA1*0501–DQB1*0201 had a stronger correlation than DQ heterozygosity alone (Skarsvag et al., 1995b).

In summary, these studies support the hypothesis that the Ro and La autoantibody response in SLE is influenced by particular class II alleles, especially HLA-DQ heterozygosity, but some of the studies are conflicting. In addition, anti-Ro antibodies are especially prominent in hereditary homozygous deficiency of complement C4, which shows no consistent MHC association (Meyer et al., 1985).

These studies raise the possibility that the association between HLA-DR3 and SLE could be attributed solely to the association of this allele with anti-Ro and anti-La antibodies. Consistent with this idea, several studies reported that the frequency of HLA-DR3 in anti-Ro and anti-La negative patients was not significantly different to that in healthy controls (Hochberg et al., 1985; Smolen et al., 1987; Hamilton et al., 1988). However, in the largest study, the HLA-DR3 allele frequency among anti-Ro and anti-La-negative lupus patients remained significantly elevated compared with healthy controls (Ehrfeld et al., 1992). Therefore, the association between HLA-DR3 and anti-Ro antibodies appeared to be independent of the association between HLA-DR3 and susceptibility to SLE.

Anti-Sm and U1-RNP antibodies

Anti-Sm antibodies react with epitopes on several uridine-rich small ribonucleoproteins. The frequency of anti-Sm antibodies in lupus patients is approximately 25% in African–American populations compared with only 10% in Caucasoid populations (Arnett et al., 1988). No HLA association was reported in one study of Caucasoid patients of whom 11 had anti-Sm antibodies (Ahearn et al., 1982) and in a smaller study of Norwegian patients (Skarsvag et al., 1995b). However, another study, again of Caucasoid patients, found an association between HLA-DR7 and anti-Sm antibodies (Schur et al., 1982). In a small study of Japanese patients, the presence of HLA-DR9 was negatively associated with anti-Sm antibodies (Kawai et al., 1994). More recently, a study involving 27 African–American patients found an increased frequency of HLA-DR2 and HLA-DQ1 (DQ6) compared with both healthy controls and anti-Sm-negative African–American lupus patients (Olsen et al., 1993). Only four of the 27 Caucasoid patients had anti-Sm antibodies and two of these were HLA-DR2–DQ1 positive.

Anti-RNP antibodies react with epitopes on U1-RNP and again are more common in Black populations. These antibodies are associated with HLA-DR4 in Caucasoid patients with mixed connective tissue disease (Genth et al., 1987; Hoffman et al., 1990) and in some studies of Caucasoid patients with SLE (Smolen et al., 1987; Harley et al., 1989; Olsen et al., 1993). However, no HLA association was seen in a large study of 178 Caucasoid patients (Yao et al., 1994) and two smaller studies (Ahearn et al., 1982; Skarsvag et al., 1995b). In Japanese patients no association was reported in one study (Kawai et al., 1994), whilst another showed an association with HLA-DQ3 (Nishikai and Sekiguchi, 1985). Recently, an increased frequency of the HLA-DRQ5-associated alleles, DQA1*0101 and DQB1*0501, was seen in both Caucasoid and African–American lupus patients (Olsen et al., 1993). Caucasoid patients in this study also had an increased frequency of the DQ8-associated allele, DQB1*0302. No specific HLA-DR subtypes were associated with these antibodies.

Anti-DNA antibodies

An association between HLA-DR3 and anti-DNA antibodies has been demonstrated in 37 Caucasoid patients with anti-DNA antibodies, 30 of whom met the American Rheumatism Association criteria for the diagnosis of SLE (Griffing et al., 1980). The association remained significant when the lupus patient group was compared with anti-DNA antibody-negative lupus controls. However, this was not confirmed in later studies which demonstrated an association with HLA-DR2 (Alvarellos et al., 1983), with HLA-DR7 (Schur et al., 1982) and no association at all (Ahearn et al., 1982). A study of 36 African–American lupus patients demonstrated a significant increase in HLA-DR2 compared with anti-DNA antibody-negative controls (54% vs 12%, $p = 0.02$) (Hochberg et al., 1985). Conversely, no significant associations were seen in the 111 Caucasoid patients studied. The associations, if any, between particular class II alleles and anti-DNA antibody production appear to be very weak.

Conclusion

There is unequivocal evidence that MHC genes play an important role in determining both disease susceptibility and phenotype in SLE and its subsets. However, there is still uncertainty as to which gene products are involved and the phenomenon of linkage disequilibrium remains a serious obstacle to dissection of the relevant genes. What are the most serious candidates at this stage? There is strong evidence that C4A null alleles are important in SLE. This is firstly because of the evidence that homozygous deficiency of non-MHC linked complement genes, especially C1q, are causal in the induction of SLE. Secondly, an association of C4A null alleles has been confirmed in many different populations, independent of other MHC associations.

The evidence that certain class II alleleic variants may also influence susceptibility to the production of particular autoantibodies is strong, presumably acting through either the T cell receptor repertoire or peptide selection. The DQ associations with anti-Ro antibody production illustrate this most strongly.

PRIMARY SJÖGREN'S SYNDROME

Primary Sjögren's syndrome is an autoimmune exocrinopathy characterized by lymphocytic infiltration of exocrine glands such as the lacrimal and major and minor salivary glands. Important extraglandular features include non-erosive arthritis, Raynaud's phenomenon, vasculitis, peripheral neuropathy, lymphadenopathy and lymphoma. The syndrome can occur in association with other autoimmune diseases, most commonly rheumatoid arthritis and SLE (secondary Sjögren's syndrome).

Early studies in Caucasoid patients with primary Sjögren's syndrome demonstrated an association with HLA-B8 and HLA-DR3 (Table 21.7). However, these associations were not seen in patients with secondary Sjögren's syndrome, where the HLA associations reflected those known to be linked with the primary condition, e.g. HLA-DR4 in patients with rheumatoid arthritis and secondary Sjögren's syndrome (Moutsopoulos et al., 1979). Further analysis demonstrated that the association was strongest for HLA-DR3, suggesting that the association with HLA-B8 was a consequence of its linkage disequilibrium with HLA-DR3 (Chused et al., 1977). An association with HLA-DR52 has also been demonstrated in both Caucasoid and African–American patients (Wilson et al. 1984b). In a study of three families, all patients with primary Sjögren's syndrome carried the HLA-DR52 allele on at least one haplotype (Mann and Moutsopoulos, 1983). In contrast to HLA-DR3, HLA-DR52 was significantly increased in frequency in patients with secondary Sjögren's syndrome in association with SLE or rheumatoid arthritis (Wilson et al., 1984b). No association has been established between primary Sjögren's syndrome and HLA-DPB1 alleles among different ethnic groups (Reveille et al., 1992; Kang et al., 1993).

Studies among patients from different ethnic groups have failed to identify a common diseasesusceptibility allele. The association with HLA-DR3 is increased in prevalence in both Caucasoid and African–American patients (Wilson et al., 1984b). However, an association with HLA-DR11 has been reported in Greek and Jewish patients (Roitburg-Tambur et al., 1990), whilst in Japanese patients HLA-DR53 has been significantly associated with both primary Sjögren's syndrome and Sjögren's syndrome associated with rheumatoid arthritis (Moriuchi et al., 1986). Two other studies in Japanese patients with primary Sjögren's syndrome have reported associations with HLA-DRB1*0401–0407 (Kang et al., 1993) and HLA-DR8 and HLA-DR52 (Miyagawa et al., 1992).

Analysis of deduced common haplotype frequencies among Chinese (HLA-DRB1*0803–DQB1*0601, $p<0.05$), Caucasoid (HLA-DRB1*0301–DQB1*0201, $p<0.0002$) and Japanese (HLA-DRB1*0405–DQB1*0401, $p<0.003$) patients with primary Sjögren's syndrome clearly shows that no single haplotype is common

TABLE 21.7 Class I and II allele associations in different populations of patients with primary Sjögren's syndrome

Reference	Ethnic group	Study number Patients	Study number Controls	Class I Allele	Class I RR or p-value*	Class II Allele	Class II RR or p-value*
Gershwin et al. (1975)[a]	Caucasoid	24[c]	1205	B8	5.3	Not tested	–
Fye et al. (1976)[a]	Caucasoid	31[c]	1205	B8	3.7	Not tested	–
Hinzova et al. (1977)[a]	Caucasoid	29	58	B8	3.96	DR3	19.2
Chused et al. (1977)[a]	Caucasoid	19	91	B8	NS	DR3	$p < 0.00001$
Moutsopoulos et al. (1979)[a]	Caucasoid	22	184	B8	$p = 0.02$	DR3	$p = 0.03$
Manthorpe et al. (1981)[a]	Caucasoid	32	345	–	–	DR2	3.7
		32	334	–	–	DR3	2.8
		35	3301	B8	3.6	–	–
Wilson et al. (1984b)[a]	Caucasoid/Black	50/5	626/168	Not tested	–	DR3	3.3
						DR52	7.8
Vitali et al. (1986)[a]	Italian	28	322	None	–	DR3	$p < 0.002$
						DR52	$p < 0.0001$
Moriuchi et al. (1986)[a]	Japanese	23	114	B54	5.4	DR53	5.0
Papasteriades et al. (1988)[a]	Greeks	46	172	None	–	DR5	$p < 0.007$
Pease et al. (1989)[a]	Caucasoid[f]	22	100	B8	$p < 0.007$	DR3	$p < 0.0035$
						DR52	$p < 0.02$
Miyagawa et al. (1992)[a]	Japanese	40	472	None	–	DR8	5.1
						DR52	3.6
Kang et al. (1993)[b]	Caucasoid	150[d]	270	Not tested	–	DRB1*0301,-*0302	$p < 0.001$
		150[d]	270			DQA1*0401, *0501, *0601	$p < 0.05$
		112[d]	270			DQB1*0201	$p < 0.01$
	Chinese	90	84	Not tested	–	None	–
	Japanese	66	98	Not tested	–	DRB1*0401-0407	$p < 0.05$
		66	98			DQA1*0301,*0302	$p < 0.01$
		64	92			DQB1*0401, 0402	$p < 0.01$
Roitberg-Tambur et al. (1993)[b]	Israeli Jews	17	258	Not tested	–	DR11	3.6
						DQA1*0501	3.6
						DQB1*0301	–
Greeks		22	54	Not tested	–	DR11	–
						DQA1*0501	4.5
						DQB1*0301	2.0
Guggenbuhl et al. (1998)[b]	Caucasoid	42	200	Not tested	–	DRB1*1501-*0301[e]	$p < 0.002$
						DQB1*0201-*0602[e]	$p < 0.006$

[a]Serological typing; [b]DNA typing; [c]included some patients with secondary Sjögren's syndrome; [d]number of chromosomes tested; [e]heterozygous genotype frequency; [f]primary Sjögren's syndrome with anti-Ro and La antibodies. *p-Value if no relative risk (RR) given; NS: not significant.

to these different populations (Kang et al., 1993). However, a region of DQβ from amino acids 59 to 69 was identical among the disease-associated DQ alleles (HLA-DQB1*0201, *0401 and *0601). Therefore, it is possible that the different haplotypes can be explained by the presence of a shared DQβ epitope. However, this epitope is not present on the β-chain encoded by HLA-DQB1*0301 allele, which is associated with the disease in Israeli Jews and non-Jewish Greek patients. In addition, a unique sequence in positions 47 to 56 of the HLA-DQA1 alleles, DQA1*0501, *0401 and *0601, has been reported (Roitberg-Tambur, 1993). However, although HLA-DQA1*0501 is increased in some studies of Caucasoid patients it has not been reported in Japanese or Chinese patients (see Table 21.7).

Antibodies directed against Ro are seen in the majority of patients with primary Sjögren's syndrome (Alspaugh et al., 1976) and the level of these antibodies is higher in patients who are HLA-DR3 or HLA-DR2 positive (Manthorpe et al., 1982; Arnett et al., 1989b). The presence of anti-Ro antibodies is particularly associated with DQ heterozygosity and has been discussed in detail earlier in this chapter (see *Anti-Ro and anti-La antibodies*, p. 351). Consistent with this association, a recent study has shown a significant increase in both HLA-DRB1*1501–*0301 and HLA-DQB1*0201–*0602 heterozygous genotype frequency among Caucasoid patients with primary Sjögren's syndrome (Guggenbuhl et al., 1998). Notably, the frequency of the individual alleles was not significantly different to healthy controls. A strong association between the HLA-DRB1*1501–*0301 heterozygous genotype and the presence of autoantibodies (rheumatoid factor, antinuclear antibodies, anti-Ro and La antibodies), Raynaud's phenomenon and haematological abnormalities such as leucopenias was also demonstrated.

In summary, although many Caucasoid studies have reported an association between primary Sjögren's syndrome and HLA-DR3, this has not been a consistent finding, particularly in studies among other ethnic groups. DQ heterozygosity appears to influence the autoantibody response, being especially associated with the presence of anti-Ro antibodies typical of primary Sjögren's syndrome.

REFERENCES

Abe, J., Kohsaka, T., Tanaka, M. et al. (1993). Genetic study on HLA class II and class III region in the disease associated with IgA nephropathy. *Nephron* **65**: 17–22.

Agnello, V. (1978). Complement deficiency states. *Medicine* **57**: 1–23.

Agnello, V., Chung, R.T. and Kaplan, L.M. (1992). A role for Hepatitis C infection in type II cryoglobulinaemia. *N. Engl. J. Med.* **327**: 1490–1495.

Ahearn, J.M., Provost, T.T., Dorsch, C.A. et al. (1982). Interrelationships of HLA-DR, MB, and MT phenotypes, autoantibody expression, and clinical features in systemic lupus erythematosus. *Arthritis Rheum.* **25**: 1031–1040.

Alarif, L.I., Ruppert, G.B., Wilson, R., Jr et al. (1983). HLA-DR antigens in Blacks with rheumatoid arthritis and systemic lupus erythematosus. *J. Rheumatol.* **10**: 297–300.

Alexander, E.L., McNicholl, J., Watson, R.M. et al. (1989). The immunogenetic relationship between anti-Ro (SS-A)/La(SS-B) antibody positive Sjögren's/lupus erythematosus overlap syndrome and the neonatal lupus syndrome. *J. Invest. Dermatol.* **93**: 751–756.

Alspaugh, M.A., Talal, N. and Tan, E.M. (1976). Differentiation and characterization of autoantibodies and their antigens in Sjögren's syndrome. *Arthritis Rheum.* **19**: 216–222.

Alvarellos, A., Ahearn, J.M., Provost, T.T. et al. (1983). Relationships of HLA-DR and MT antigens to autoantibody expression in systemic lupus erythematosus. *Arthritis Rheum.* **26**: 1533–1535.

Anonymous (1990). Criteria for diagnosis of Behçet's disease, International Study Group for Behçet's Disease. *Lancet* **335**: 1078–1080.

Arend, W.P., Michel, B.A. and Bloch, D.A. (1990). The American College of Rheumatology 1990 criteria for the classification of Takayasu's arteritis. *Arthritis Rheum.* **33**: 1129–1134.

Armstrong, R.D., Behn, A., Myles, A. et al. (1983). Histocompatibility antigens in polymyalgia rheumatica and giant cell arteritis. *J. Rheumatol.* **10**: 659–661.

Armstrong, R.D., Panayi, G.S. and Welsh, K.I. (1984). Polymyalgia rheumatica and rheumatoid arthritis: similarity of HLA antigen frequencies. *Arthritis Rheum.* **27**: 1438–1439

Arnett, F.C., Hamilton, R.G., Roebber, M.G. et al. (1988). Increased frequencies of Sm and nRNP autoantibodies in American blacks compared to whites with systemic lupus erythematosus. *J. Rheumatol.* **15**: 1773–1776.

Arnett, F.C., Hamilton, R.G., Reveille, J.D. et al. (1989a). Genetic studies of Ro (SS-A) and La (SS-B) autoantibodies in families with systemic lupus erythematosus and primary Sjögren's syndrome. *Arthritis Rheum.* **32**: 413–419.

Arnett, F.C., Bias, W.B. and Reveille, J.D. (1989b). Genetic studies in Sjögren's syndrome and systemic lupus erythematosus. *J. Autoimmun.* **2**: 403–413.

Arnett, F.C., Olsen, M.L., Anderson, K.L. et al. (1991). Molecular analysis of major histocompatibility complex alleles associated with the lupus anticoagulant. *J. Clin. Invest.* **87**: 1490–1495.

Asherson, R.A. and Cervera, R. (1994). "Primary", "secondary" and other variants of the antiphospholipid syndrome. *Lupus* **3**: 293–298.

Asherson, R.A., Doherty, D.G., Vergani, D. et al. (1992). Major histocompatibility complex associations with primary antiphospholipid syndrome. *Arthritis Rheum.* **35**: 124–125.

Awdeh, Z.L. and Alper, C.A. (1980). Inherited structural polymorphism of the fourth component of human complement. *Proc. Natl Acad. Sci. USA* **77**: 3576–3580.

Awdeh, Z.L., Raum, D.D., Glass, D. et al. (1981). Complement-human histocompatibility antigen haplotypes in C2 deficiency. *J. Clin. Invest.* **67**: 581–583.

Bahram, S., Arnold, D., Bresnahan, M. et al. (1991). Two putative subunits of a peptide pump encoded in the human major histocompatibility complex class II region. *Proc. Natl Acad. Sci. USA* **88**: 10 094–10 098.

Barrier, J., Bignon, J.D., Soulillou, J.P. et al. (1981). Increased prevalence of HLA-DR4 in giant-cell arteritis. *N. Engl. J. Med.* **305**: 104–105.

Barron, K., Silverman, E., Gonzales, J. et al. (1992). Major histocompatibility complex class II alleles in Kawasaki syndrome – lack of consistent correlation with disease or cardiac involvement. *J. Rheumatol.* **19**: 1790–1793.

Barron, K.S., Silverman, E.D., Gonzales, J. et al. (1993). Clinical, serologic, and immunogenetic studies in childhood-onset systemic lupus erythematosus. *Arthritis Rheum.* **36**: 348–354.

Batchelor, J.R., Welsh, K.I., Tinoco, R.M. et al. (1980). Hydralazine-induced systemic lupus erythematosus: influence of HLA-DR and sex on susceptibility. *Lancet* **i**: 1107–1109.

Batchelor, J.R., Fielder, A.H., Walport, M.J. et al. (1987). Family study of the major histocompatibility complex in HLA DR3 negative patients with systemic lupus erythematosus. *Clin. Exp. Immunol.* **70**: 364–371.

Beckmann, R.P., Mizzen, L.E. and Welch, W.J. (1990). Interaction of Hsp 70 with newly synthesized proteins: implications for protein folding and assembly. *Science* **248**: 850–854.

Bell, D.A. and Maddison, P.J. (1980). Serologic subsets in systemic lupus erythematosus: an examination of auto antibodies in relationship to clinical features of disease and HLA antigens. *Arthritis Rheum.* **23**: 1268–1273.

Bettinotti, M.P., Hartung, K., Deicher, H.R. et al. (1993a). DR2 haplotypes (DRB1, DQA1, DQB1) associated with systemic lupus erythematosus. *Immunogenetics* **38**: 74–77.

Bettinotti, M.P., Hartung, K., Deicher, H. et al. (1993b). Polymorphism of the tumor necrosis factor β gene in systemic lupus erythematosus: TNFB-MHC haplotypes. *Immunogenetics* **37**: 449–454.

Bignon, J.D., Barrier, J., Soulillou, J.P. et al. (1984). HLA DR4 and giant cell arteritis. *Tissue Antigens* **24**: 60–62.

Bignon, J.D., Ferec, C., Barrier, J. et al. (1988). HLA class II genes polymorphism in DR4 giant cell arteritis patients. *Tissue Antigens* **32**: 254–258.

Black, C.M., Welsh, K.I., Fielder, A. et al. (1982). HLA antigens and Bf allotypes in SLE: evidence for the association being with specific haplotypes. *Tissue Antigens* **119**: 115–120.

Boey, M.L., Colaco, C.B., Gharavi, A.E. et al. (1983). Thrombosis in systemic lupus erythematosus: striking association with the presence of circulating lupus anticoagulant. *BMJ* **287**: 1021–1023.

Brand, C., Davidson, A., Littlejohn, G. et al. (1984). Hydralazine-induced lupus: no association with HLA-DR4. *Lancet* **i**: 462.

Bridgeford, P.H., Lowenstein, M., Bocanegra, T.S. et al. (1980). Polymyalgia rheumatica and giant cell arteritis: histocompatibility typing and hepatitis-B infection studies. *Arthritis Rheum.* **23**: 516–518.

Brouet, J.C., Clauvel, J.P., Danon, F. et al. (1974). Biologic and clinical significance of Cryoglobulins. *Am. J. Med.* **57**: 775–788.

Brucato, A., Gasparini, M., Vignati, G. et al. (1995). Isolated congenital complete heart block: longterm outcome of children and immunogenetic study. *J. Rheumatol.* **22**: 541–543.

Calamia, K.T., Moore, S.B., Elveback, L.R. et al. (1981). HLA-DR locus antigens in polymyalgia rheumatica and giant cell arteritis. *J. Rheumatol.* **8**: 993–996.

Camps, M.T., Cuadrado, M.J., Ocon, P. et al. (1995). Association between HLA class II antigens and primary antiphospholipid syndrome from the south of Spain. *Lupus* **4**: 51–55.

Caplen, N.J., Patel, A., Millward, A. et al. (1990). Complement C4 and heat shock protein 70 (HSP70) genotypes and type I diabetes mellitus. *Immunogenetics* **32**: 427–430.

Carrington, M., Colonna, M., Spies, T. et al. (1993). Haplotypic variation of the transporter associated with antigen processing (TAP) genes and their extension of HLA class II region haplotypes. *Immunogenetics* **37**: 266–273.

Carroll, M.C. (1998). The Lupus paradox. *Nat. Genet.* **19**: 3–4.

Carroll, M.C. and Porter, R.R. (1983). Cloning of a human complement component C4 gene. *Proc. Natl Acad. Sci. USA* **80**: 264–267.

Carroll, M.C., Palsdottir, A., Belt, K.T. et al. (1985). Deletion of complement C4 and steroid 21-hydroxylase genes in the HLA class III region. *EMBO J.* **4**: 2547–2552.

Carroll, M.C., Katzman, P., Alicot, E.M. et al. (1987). Linkage map of the human major histocompatibility complex including the tumor necrosis factor genes. *Proc. Natl Acad. Sci. USA* **84**: 8535–8539.

Cascino, I., Galeazzi, M., Salvetti, M. et al. (1994). HSP70-1 promoter region polymorphism tested in three autoimmune diseases. *Immunogenetics* **39**: 291–293.

Celada, A., Barras, C., Benzonana, G. et al. (1980). Increased frequency of HLA-DRw3 in systemic lupus erythematosus. *Tissue Antigens* **15**: 283–288.

Chen, C.J., Yen, J.H., Tsai, W.C. et al. (1997). The TNF2 allele does not contribute towards susceptibility to systemic lupus erythematosus. *Immunol. Lett.* **55**: 1–3.

Christiansen, F.T., Dawkins, R.L., Uko, G. et al. (1983). Complement allotyping in SLE: association with C4A null. *Aust. N. Z. J. Med.* **13**: 483–488.

Christiansen, F.T., Zhang, W.J., Griffiths, M. et al. (1991). Major histocompatibility complex (MHC) complement deficiency, ancestral haplotypes and systemic lupus erythematosus (SLE): C4 deficiency explains some but not all of the influence of the MHC. *J. Rheumatol.* **18**: 1350–1358.

Chused, T.M., Kassan, S.S., Opelz, G. et al. (1977). Sjögren's syndrome association with HLA-Dw3. *N. Engl. J. Med.* **296**: 895–897.

Cid, M.C., Ercilla, G., Vilaseca, J. et al. (1988). Polymyalgia rheumatica: a syndrome associated with HLA-DR4 antigen. *Arthritis Rheum.* **31**: 678–682.

Colonna, M., Bresnahan, M., Bahram, S. et al. (1992). Allelic variants of the human putative peptide transporter involved in antigen processing. *Proc. Natl Acad. Sci. USA* **89**: 3932–3936.

Congia, M., Clemente, M.G., Dessi, C. et al. (1996). HLA class II genes in chronic hepatitis C virus-infection and associated immunological disorders. *Hepatology* **24**: 1338–1341.

Cornillet, P., Pennaforte, J.L., Philbert, F. et al. (1993). Complement C4A gene deletion in patients with systemic lupus erythematosus in France. *J. Rheumatol.* **20**: 1633–1634.

Cotch, M.F., Fauci, A.S. and Hoffman, G.S. (1995). HLA typing in patients with Wegener granulomatosis. *Ann. Intern. Med.* **122**: 635.

Cowland, J.B., Andersen, V., Halberg, P. et al. (1994). DNA polymorphism of HLA class II genes in systemic lupus erythematosus. *Tissue Antigens* **43**: 34–37.

D'Alfonso, S., Colombo, G., Della Bella, S. et al. (1996). Association between polymorphisms in the TNF region and systemic lupus erythematosus in the Italian population. *Tissue Antigens* **47**: 551–555.

Danis, V.A., Millington, M., Hyland, V. et al. (1995). Increased frequency of the uncommon allele of a tumour necrosis factor α gene polymorphism in rheumatoid arthritis and systemic lupus erythematosus. *Dis. Markers* **12**: 127–133.

Davies, E.J., Hutchings, C.J., Hillarby, M.C. et al. (1994a). HLA-DP does not contribute towards susceptibility to systemic lupus erythematosus. *Ann. Rheum. Dis.* **53**: 188–190.

Davies, E.J., Donn, R.P., Hillarby, M.C. et al. (1994b). Polymorphisms of the TAP2 transporter gene in systemic lupus erythematosus. *Ann. Rheum. Dis.* **53**: 61–63.

Davies, E.J., Steers, G., Oilier, W.E. et al. (1995). Relative contributions of HLA-DQA and complement C4A loci in determining susceptibility to systemic lupus erythematosus. *Br. J. Rheumatol.* **34**: 221–225.

D'Cruz, D., Taylor, J., Ahmed, T. et al. (1992). Complement factor 2 deficiency: a clinical and serological family study. *Ann. Rheum. Dis.* **51**: 1254–1256.

De Juan, D., Martin-Villa, J.M., Gomez-Reino, J.J. et al. (1993). Differential contribution of C4 and HLA-DQ genes to systemic lupus erythematosus susceptibility. *Hum. Genet.* **91**: 579–584.

Doherty, D.G., Ireland, R., Demaine, A.G. et al. (1992). Major histocompatibility complex genes and susceptibility to systemic lupus erythematosus in southern Chinese. *Arthritis Rheum.* **35**: 641–646.

Dong, R.P., Kimura, A., Numano, F. et al. (1992). HLA-DP antigen and Takayasu arteritis. *Tissue Antigens* **39**: 106–110.

Dunckley, H., Gatenby, P.A. and Serjeantson, S.W. (1986). DNA typing of HLA-DR antigens in systemic lupus erythematosus. *Immunogenetics* **24**: 158–162.

Dunckley, H., Gatenby, P.A., Hawkins, B. et al. (1987). Deficiency of C4A is a genetic determinant of systemic lupus erythematosus in three ethnic groups. *J. Immunogenet.* **14**: 209–218.

Ehrfeld, H., Hartung, K., Renz, M. et al. (1992). MHC associations of autoantibodies against recombinant Ro and La proteins in systemic lupus erythematosus. Results of a multicenter study. SLE Study Group. *Rheumatol. Int.* **12**: 169–173.

Elkon, K.B., Sutherland, D.C., Rees, A.J. et al. (1983). HLA antigen frequencies in systemic vasculitis: increase in HLA-DR2 in Wegener's granulomatosis. *Arthritis Rheum.* **26**: 102–105.

Fielder, A.H., Walport, M.J., Batchelor, J.R. et al. (1983). Family study of the major histocompatibility complex in patients with systemic lupus erythematosus: importance of null alleles of C4A and C4B in determining disease susceptibility. *Br. Med. J. Clin. Res. Ed.* **286**: 425–428.

Fildes, N., Burns, J., Newburger, J. et al. (1992). The HLA class II region and susceptibility to Kawasaki disease. *Tissue Antigens* **39**: 99–101.

Fong, K.Y., Howe, H.S., Tin, S.K. et al. (1996). Polymorphism of the regulatory region of tumour necrosis factor α gene in patients with systemic lupus erythematosus. *Ann. Acad. Med. Singapore* **25**: 90–93.

Fouad, F., Johny, K., Kaaba, S. et al. (1994). MHC in systemic lupus erythematosus: a study on a Kuwaiti population. *Eur. J. Immunogenet.* **21**: 11–14.

Fritzler, M.J. and Tan, E.M. (1978). Antibodies to histones in drug-induced and idiopathic lupus erythematosus. *J. Clin. Invest.* **62**: 560–567.

Fronek, Z., Timmerman, L.A., Alper, C.A. et al. (1990). Major histocompatibility complex genes and susceptibility to systemic lupus erythematosus. *Arthritis Rheum.* **33**: 1542–1553.

Fugger, L., Morling, N., Ryder, L.P. et al. (1989). NcoI restriction fragment length polymorphism (RFLP) of the tumor necrosis factor (TNF α) region in four autoimmune diseases. *Tissue Antigens* **34**: 17–22.

Fujisaku, A., Frank, M.B. and Neas, B. (1990). HLA-DQ gene complementation and other histocompatibility relationships in man with the anti-Ro/SSA autoantibody response of systemic lupus erythematosus. *J. Clin. Invest.* **86**: 606–611.

Fye, K.H., Terasaki, P.I., Moutsopoulos, H. et al. (1976). Association of Sjögren's syndrome with HLA-B8. *Arthritis Rheum.* **19**: 883–886.

Gelfand, E.W., Clarkson, J.E. and Minta, J.O. (1975). Selective deficiency of the second component of complement in a patient with anaphylactoid purpura. *Clin. Immunol. Immunopathol.* **4**: 269–276.

Genth, E., Zarnowski, H., Mierau, R. et al. (1987). HLA-DR4 and Gm(1,3; 5,21) are associated with U1-nRNP antibody positive connective tissue disease. *Ann. Rheum. Dis.* **46**: 189–196.

Gershwin, M.E., Terasaki, I., Graw, R. et al. (1975). Increased frequency of HL-A8 in Sjögren's syndrome. *Tissue Antigens* **6**: 342–346.

Gibofsky, A., Winchester, R.J., Patarroyo, M. et al. (1978). Disease associations of the Ia-like human alloantigens. Contrasting patterns in rheumatoid arthritis and systemic lupus erythematosus. *J. Exp. Med.* 148: 1728–1732.

Girona, E., Yamamoto-Furusho, J.K., Cutino, T. et al. (1996). HLA-DR6 (possibly DRB1*1301) is associated with susceptibility to Takayasu arteritis in Mexicans. *Heart Vessels* 11: 277–280.

Gladman, D.D., Terasaki, P.I., Park, M.S. et al. (1979). Increased frequency of HLA-DRW2 in SLE. *Lancet* ii: 902.

Glass, D. (1976a). Association between HLA and cutaneous necrotizing venulitis. *Arthritis Rheum.* 19: 945–949.

Glass, D., Raum, D., Gibson, D. et al. (1976b). Inherited deficiency of the second component of complement. Rheumatic disease associations. *J. Clin. Invest.* 58: 853–861.

Goldstein, R. and Sengar, D.P. (1993). Comparative studies of the major histocompatibility complex in French Canadian and non-French Canadian Caucasoids with systemic lupus erythematosus. *Arthritis Rheum.* 36: 1121–1127.

Goldstein, R., Arnett, F.C., McLean, R.H. et al. (1988). Molecular heterogeneity of complement component C4-null and 21-hydroxylase genes in systemic lupus erythematosus. *Arthritis Rheum.* 31: 736–744.

Goldstein, R., Moulds, J.M., Smith, C.D. et al. (1996). MHC studies of the primary antiphospholipid antibody syndrome and of antiphospholipid antibodies in systemic lupus erythematosus. *J. Rheumatol.* 23: 1173–1179.

Gouet, D., Alcalay, D., Azais, I. et al. (1985). HLA-DR antigens in polymyalgia rheumatica and giant cell arteritis. *J. Rheumatol.* 12: 627–628.

Granados, J., Vargas-Alarcon, G., Andrade, F. et al. (1996). The role of HLA-DR alleles and complotypes through the ethnic barrier in systemic lupus erythematosus in Mexicans. *Lupus* 5: 184–189.

Granados, J., Vargas-Alarcon, G., Drenkard, C. et al. (1997). Relationship of anticardiolipin antibodies and antiphospholipid syndrome to HLA-DR7 in Mexican patients with systemic lupus erythematosus (SLE). *Lupus* 6: 57–62.

Granato, J.E., Abben, R.P. and May, W.S. (1980). Familial association of giant cell arteritis. A case report and brief review. *Arch. Intern. Med.* 141: 115–117.

Griffing, W.L., Moore, S.B., Luthra, H.S. et al. (1980). Associations of antibodies to native DNA with HLA-DRw3. A possible major histocompatibility complex-linked human immune response gene. *J. Exp. Med.* 152: 319s–325s.

Grumet, F.C., Coukell, A., Bodmer, J.G. et al. (1971). Histocompatibility (HL-A) antigens associated with systemic lupus erythematosus. A possible genetic predisposition to disease. *N. Engl. J. Med.* 285: 193–196.

Guerne, P.A., Salvi, M., Seitz, M. et al. (1997). Molecular analysis of HLA-DR polymorphism in polymyalgia rheumatica. Swiss Group for Research on HLA in Polymyalgia Rheumatica. *J. Rheumatol.* 24: 671–676.

Guggenbuhl, P., Jean, S., Jego, P. et al. (1998). Primary Sjögren's syndrome: role of the HLA-DRB1*0301-*1501 heterozygotes. *J. Rheumatol.* 25: 900–905.

Gulko, P.S., Reveille, J.D., Koopman, W.J. et al. (1993). Anticardiolipin antibodies in systemic lupus erythematosus: clinical correlates, HLA associations, and impact on survival. *J. Rheumatol.* 20: 1684–1693.

Gunnarsson, I., Kanerud, L., Pettersson, E. et al. (1997). Predisposing factors in suiphasalazine-induced systemic lupus erythematosus. *Br. J. Rheumatol.* 36: 1089–1094.

Hagen, E.C., Stegeman, C.A., D'Amaro, J. et al. (1995). Decreased frequency of HLA-DR13DR6 in Wegener's granulomatosis. *Kidney Int.* 48: 801–805.

Hajeer, A.H., Worthington, J., Davies, E.J. et al. (1997). TNF microsatellite a2, b3 and d2 alleles are associated with systemic lupus erythematosus. *Tissue Antigens* 49: 222–227.

Hamilton, R.G., Harley, J.B., Bias, W.B. et al. (1988). Two Ro (SS-A) autoantibody responses in systemic lupus erythematosus. Correlation of HLA-DR/DQ specificities with quantitative expression of Ro (SS-A) autoantibody. *Arthritis Rheum.* 31: 496–505.

Hansen, J.A., Healey, L.A. and Wilske, K.R. (1985). Association between giant cell (temporal) arteritis and HLA-Cw3. *Hum. Immunol.* 13: 193–198.

Harley, J.B., Reichlin, M., Arnett, F.C. et al. (1986). Gene interaction at HLA-DQ enhances autoantibody production in primary Sjögren's syndrome. *Science* 232: 1145–1147.

Harley, J.B., Sestak, A.L., Willis, L.G. et al. (1989). A model for disease heterogeneity in systemic lupus erythematosus. Relationships between histocompatibility antigens, autoantibodies, and lymphopenia or renal disease. *Arthritis Rheum.* 32: 826–836.

Harris, R. and Jones, H.P. (1970). Polyarteritis nodosa in identical twins. *Ann. Phys. Med.* 10: 241–24.

Hartung, K., Fontana, A., Klar, M. et al. (1989). Association of class I, II, and III MHC gene products with systemic lupus erythematosus. Results of a Central European multicenter study. *Rheumatol. Int.* 9: 13–18.

Hartung, K., Coldewey, R., Krapf, F. et al. (1991). Hetero- and homozygosity of MHC class II gene products in systemic lupus erythematosus. The Members of the Deutsche Multizentrische SLE-Studie. *Tissue Antigens* 38: 165–168.

Hartung, K., Baur, M.P., Coldewey, R. et al. (1992a). Major histocompatibility complex haplotypes and complement C4 alleles in systemic lupus erythematosus. Results of a multicenter study. *J. Clin. Invest.* 90: 1346–1351.

Hartung, K., Coldewey, R., Corvetta, A. et al. (1992b). MHC gene products and anticardiolipin antibodies in systemic lupus erythematosus results of a multicenter study. SLE Study Group. *Autoimmunity* 13: 95–99.

Hartung, K., Ehrfeld, H., Lakomek, H.J. et al. (1992c). The genetic basis of Ro and La antibody formation in systemic lupus erythematosus. Results of a multicenter study. The SLE Study Group. *Rheumatol. Int.* 11: 243–249.

Hashimoto, H., Tsuda, H., Matsumoto, T. et al. (1985). HLA antigens associated with systemic lupus erythematosus in Japan. *J. Rheumatol.* 12: 919–923.

Hashimoto, H., Nishimura, Y., Dong, R.P. et al. (1994). HLA antigens in Japanese patients with systemic lupus erythematosus. *Scand. J. Rheum.* 23: 191–196.

Hata, A. and Numano, F. (1995). Magnetic resonance imaging of vascular changes in Takayasu arteritis. *Int. J. Cardiol.* 52: 45–52.

Hauptmann, G., Tongio, MM., Goetz, J. et al. (1982). Association of the C2-deficiency gene (C2*QO) with the C4A*4, C4B*2 genes. *J. Immunogen.* **9**: 127–132.

Hawkins, B.R., Wong, K.L., Wong, R.W. et al. (1987). Strong association between the major histocompatibility complex and systemic lupus erythematosus in southern Chinese. *J. Rheumatol.* **14**: 1128–1131.

Hazleman, B., Goldstone, A. and Voak, D. (1977). Association of polymyalgia rheumatica and giant-cell arteritis with HLA-B8. *BMJ* **ii**: 989–991.

Heidelberger, M. (1941). Quantitative chemical studies on complement or alexin. *J. Exp. Med.* **73**: 691–694.

Hinzova, E., Ivanyi, D., Sula, K. et al. (1977). HLA–Dw3 in Sjögren's syndrome. *Tissue Antigens* **9**: 8–10.

Hochberg, M.C., Boyd, R.E., Ahearn, J.M. et al. (1985). Systemic lupus erythematosus: a review of clinico-laboratory features and immunogenetic markers in 150 patients with emphasis on demographic subsets. *Medicine* **64**: 285–295.

Hoffman, R.W., Rettenmaier, L.J., Takeda, Y. et al. (1990). Human autoantibodies against the 70-kd polypeptide of U1 small nuclear RNP are associated with HLA-DR4 among connective tissue disease patients. *Arthritis Rheum.* **33**: 666–673.

Hong, G.H., Kim, H.Y., Takeuchi, F. et al. (1994). Association of complement C4 and HLA-DR alleles with systemic lupus erythematosus in Koreans. *J. Rheumatol.* **21**: 442–447.

Howard, P.F., Hochberg, M.C., Bias, W.B. et al. (1986). Relationship between C4 null genes, HLA-D region antigens, and genetic susceptibility to systemic lupus erythematosus in Caucasoid and black Americans. *Am. J. Med.* **81**: 187–193.

Hunder, G.G., Taswell, H.F., Pineda, A.A. et al. (1977). HLA antigens in patients with giant cell arteritis and polymyalgia rheumatica. *J. Rheumatol.* **4**: 321–323.

Hunder, G.G., Arend, W.P., Bloch, D.A. et al. (1990). The American College of Rheumatology 1990 criteria for the classification of vasculitis. Introduction. *Arthritis Rheum.* **33**: 1065–1067.

Ishikawa, K. (1988). Diagnostic approach and proposed criteria for the clinical diagnosis of Takayasu's arteriopathy. *J. Am. Coll. Cardiol.* **12**: 964–972.

Isohisa, I., Numano, F., Maezawa, H. et al. (1978). HLA-Bw52 in Takayasu disease. *Tissue Antigens* **12**: 246–248.

Jacob, C.O. and McDevitt, H.O. (1988). Tumour necrosis factor-alpha in murine autoimmune "lupus" nephritis. *Nature* **331**: 356–358.

Jacob, C.O., Fronek, Z., Lewis, G.D. et al. (1990). Heritable major histocompatibility complex class II-associated differences in production of tumor necrosis factor alpha: relevance to genetic predisposition to systemic lupus erythematosus. *Proc. Natl Acad. Sci. USA* **87**: 1233–1237.

Jarjour, W., Reed, A.M., Gauthier, J. et al. (1996). The 8.5-kb PstI allele of the stress protein gene, Hsp70-2: an independent risk factor for systemic lupus erythematosus in African Americans? *Hum. Immunol.* **45**: 59–63.

Jennette, J.C., Falk, R.J., Andrassy, K. et al. (1994). Nomenclature of systemic vasculitides. Proposal of an international consensus conference. *Arthritis Rheum.* **37**: 187–192.

Johnson, C.A., Densen, P., Hurford, R.K., Jr et al. (1992). Type I human complement C2 deficiency. A 28-base pair gene deletion causes skipping of exon 6 during RNA splicing. *J. Biol. Chem.* **267**: 9347–9353.

Julkunen, H., Siren, M.K., Kaaja, R. et al. (1995). Maternal HLA antigens and antibodies to SS-A/Ro and SS-B/La. Comparison with systemic lupus erythematosus and primary Sjögren's syndrome. *Br. J. Rheumatol.* **34**: 901–907.

Kachru, R.B., Sequeira, W., Mittal, K.K. et al. (1984). A significant increase of HLADR3 and DR2 in systemic lupus erythematosus among blacks. *J. Rheumatol.* **11**: 471–474.

Kameda, S., Naito, S., Tanaka, K. et al. (1982). HLA antigens of patients with systemic lupus erythematosus in Japan. *Tissue Antigens* **20**: 221–222.

Kang, H.I., Fei, H.M., Saito, I. et al. (1993). Comparison of HLA class II genes in Caucasoid, Chinese, and Japanese patients with primary Sjögren's syndrome. *J. Immunol.* **150**: 3615–3623.

Kaslow, R., Bailowitz, A., Lin, F. et al. (1985). Association of epidemic Kawasaki syndrome with the HLA-A2, B44, Cw5 antigen combination. *Arthritis Rheum.* **28**: 938–940.

Kasuya, K., Hashimoto, Y. and Numano, F. (1992). Left ventricular dysfunction and HLA-BwS2 antigen in Takayasu arteritis. *Heart Vessels* **7**(Suppl.): 116–119.

Kato, H., Koike, S., Yamamoto, M. et al. (1975). Coronary artery aneurysms in infants and young children with acute febrile mucocutaneous lymph node syndrome. *J. Pediatr.* **86**: 892–898.

Kato, S., Kimura, M., Tsuji, K. et al. (1978). HLA antigens in Kawasaki disease. *Pediatrics* **61**: 252–255.

Katz, P., Ailing, D.W., Haynes, B.F. et al. (1979). Association of Wegener's granulomatosis with HLA-B8. *Clin. Immunol. Immunopathol.* **14**: 268–270.

Kawai, T., Tani, K., Okubo, T. et al. (1994). HLA antigens in Japanese patients with high titre antiribonucleoprotein antibodies. *Ann. Rheum. Dis.* **53**: 426–427.

Kawasaki, T., Kosaki, F., Okawa, S. et al. (1974). A new infantile acute febrile mucocutaneous lymph node syndrome (MLNS) prevailing in Japan. *Pediatrics* **54**: 271–276.

Kemp, M.E., Atkinson, J.P., Skanes, V.M. et al. (1987). Deletion of C4A genes in patients with systemic lupus erythematosus. *Arthritis Rheum.* **30**: 1015–1022.

Keren, G., Danon, Y., Orgad, S. et al. (1982). HLA Bw51 is increased in mucocutaneous lymph node syndrome in Israeli patients. *Tissue Antigens* **20**: 144–146.

Khraishi, M.M., Gladman, D.D., Dagenais, P. et al. (1992). HLA antigens in North American patients with Takayasu arteritis. *Arthritis Rheum.* **35**: 573–575.

Kim, T.G., Kim, H.Y., Lee, S.H. et al. (1996). Systemic lupus erythematosus with nephritis is strongly associated with the TNFB*2 homozygote in the Korean population. *Hum. Immunol.* **46**: 10–17.

Kimura, A., Kitamura, H., Date, Y. et al. (1996). Comprehensive analysis of HLA genes in Takayasu arteritis in Japan. *Int. J. Card.* **54**(Suppl.): 561–69.

Kluger, J., Drayer, D.E., Reidenberg, M.M. et al. (1981). Acetyl procainamide therapy in patients with previous procainamide-induced lupus syndrome. *Ann. Intern. Med.* **95**: 18–23.

Kong, N.C., Nasruruddin, B.A., Murad, S. et al. (1994). HLA antigens in Malay patients with systemic lupus erythematosus. *Lupus* **3**: 393–395.

Krensky, A., Berenberg, W., Shanley, K. et al. (1981). HLA antigens in mucocutaneous lymph node syndrome in New England. *Pediatrics* **67**: 741–743.

Krensky, A., Grady, S., Shanley, K. et al. (1983). Epidemic and endemic HLA-B and DR associations in mucocutaneous lymph node syndrome. *Hum. Immunol.* **6**: 75–77.

Kumar, A., Kumar, P. and Schur, P.H. (1991). DR3 and non DR3 associated complement component C4A deficiency in systemic lupus erythematosus. *Clin. Immunol. Immunopathol.* **60**: 55–64.

Lee, L.A., Bias, W.B., Arnett, F.C., Jr et al. (1983). Immunogenetics of the neonatal lupus syndrome. *Ann. Intern. Med.* **99**: 592–596.

Leff, R., Harrer, W.V., Baylis, J.C. et al. (1971). Polyarteritis nodosa in two siblings. *Am. J. Dis. Child.* **121**: 67–70.

Lenzi, M., Frisoni, M., Mantovani, V. et al. (1998). Haplotype HLA-B8-DR3 confers susceptibility to hepatitis C virus-related mixed cryoglobulinemia. *Blood* **91**: 2062–2066.

Levi-Strauss, M., Carroll, M.C., Steinmetz, M. et al. (1988). A previously undetected MHC gene with an unusual periodic structure. *Science* **240**: 201–204.

Lhotta, K., Konig, P., Hintner, H. et al. (1990). Renal disease in a patient with hereditary complete deficiency of the fourth component of complement. *Nephron* **56**: 206–211.

Liang, G.C., Simkin, P.A., Hunder, G.G. et al. (1974). Familial aggregation of polymyalgia rheumatica and giant cell arteritis. *Arthritis Rheum.* **17**: 19–24.

Lie, J.T. (1992). Vascular involvement in Behçet's disease: arterial and venous and vessels of all sizes. *J. Rheumatol.* **19**: 341–343.

Love, D.C., Rapkin, J., Lesser, G.R. et al. (1986). Temporal arteritis in blacks. *Ann. Intern. Med.* **105**: 387–389.

Lowenstein, M.B., Bridgeford, P.H., Vasey, F.B. et al. (1983). Increased frequency of IWA-DR3 and DR4 in polymyalgia rheumatica-giant cell arteritis. *Arthritis Rheum.* **26**: 925–927.

Lu, L.Y., Ding, W.Z., Fici, D. et al. (1997). Molecular analysis of major histocompatibility complex allelic associations with systemic lupus erythematosus in Taiwan. *Arthritis Rheum.* **40**: 1138–1145.

McHugh, N.J. and Maddison, P.J. (1989). HLA-DR antigens and anticardiolipin antibodies in patients with systemic lupus erythematosus. *Arthritis Rheum.* **32**: 1623–1624.

McLean, R.H., Wyatt, R.J. and Julian, B.A. (1984). Complement phenotypes in glomerulonephritis: increased frequency of homozygous null C4 phenotypes in IgA nephropathy and Henoch–Schonlein purpura. *Kidney Int.* **26**: 855–860.

Malmvall, B.E., Bengtsson, B.A. and Rydberg, L. (1980). HLA antigens in patients with giant cell arteritis, compared with two control groups of different ages. *Scand. J. Rheum.* **9**: 65–68.

Mann, D.L. and Moutsopoulos, H.M. (1983). HLA DR alloantigens in different subsets of patients with Sjögren's syndrome and in family members. *Ann. Rheum. Dis.* **42**: 533–536.

Manthorpe, R., Morling, N., Platz, P. et al. (1981). HLA-D antigen frequencies in Sjögren's syndrome. Differences between the primary and secondary form. *Scand. J. Rheum.* **10**: 124–128.

Manthorpe, R., Teppo, A.M., Bendixen, G. et al. (1982). Antibodies to SS-B in chronic inflammatory connective tissue diseases. Relationship with HLA-Dw2 and HLA-Dw3 antigens in primary Sjögren's syndrome. *Arthritis Rheum.* **25**: 662–667.

Marintchev, L.M., Naumova, E.J., Rashkov, R.K. et al. (1995). HLA class II alleles and autoantibodies in Bulgarians with systemic lupus erythematosus. *Tissue Antigens* **46**: 422–425.

Mason, J.C., Cowie, M.R., Davies, K.A. et al. (1994). Familial polyarteritis nodosa. *Arthritis Rheum.* **37**: 1249–1253.

Matsuda, I., Hattori, S., Nagata, N. et al. (1977). HLA antigens in mucocutaneous lymph node syndrome. *Am. J. Dis. Child.* **131**: 1417–1418.

Matsumura, N. and Mizushima, Y. (1975). Leucocyte movement and coichicine treatment in Behcet's disease. *Lancet* **ii**: 813.

Mattingly, P.C., Mowat, A.G., Gunson, H.H. et al. (1978). HLA antigens in polymyalgia rheumatica. *BMJ* **i**: 989–990.

Mehra, N.K., Pande, I., Taneja, V. et al. (1993). Major histocompatibility complex genes and susceptibility to systemic lupus erythematosus in northern India. *Lupus* **2**: 313–314.

Mehra, N.K., Rajalingam, R., Sagar, S. et al. (1996). Direct role of HLA-B5 in influencing susceptibility to Takayasu arteritis. *Int. J. Cardiol.* **54** (Suppl.): S71–79.

Messer, G., Spengler, U., Jung, M.C. et al. (1991). Polymorphic structure of the tumor necrosis factor (TNF) locus: an NcoI polymorphism in the first intron of the human TNF-beta gene correlates with a variant amino acid in position 26 and a reduced level of TNF-beta production. *J. Exp. Med.* **173**: 209–219.

Meyer, O., Hauptmann, G., Tappeiner, G. et al. (1985). Genetic deficiency of C4, C2 or C1q and lupus syndromes. Association with anti-Ro (SS-A) antibodies. *Clin. Exp. Immunol.* **62**: 678–684.

Migliorini, P., Bombardieri, S., Castellani, A. et al. (1981). HLA antigens in essential mixed cryoglobulinemia. *Arthritis Rheum.* **24**: 932–936.

Misiani, R., Bellavita, P., Fenili, D. et al. (1994). Interferon alfa-2a therapy in cryoglobulinemia associated with hepatitis C virus. *N. Engl. J. Med.* **330**: 751–756.

Mitchell, J.A., Batchelor, J.R., Chapel, H. et al. (1987). Erythrocyte complement receptor type 1 (CR1) expression and circulating immune complex (CIC) levels in hydralazine-induced SLE. *Clin. Exp. Immunol.* **68**: 446–456.

Miyagawa, S., Dohi, K., Shima, H. et al. (1992). Absence of HLA-B8 and HLA-DR3 in Japanese patients with Sjögren's syndrome positive for antiSSA(Ro). *J. Rheumatol.* **19**: 1922–1924.

Miyagawa, S., Shinohara, K., Fujita, T. et al. (1997). Neonatal lupus erythematosus: analysis of HLA class II alleles in mothers and siblings from seven Japanese families. *J. Am. Acad. Dermatol.* **36**: 186–190.

Mizuki, N., Inoko, H., Tanaka, H. et al. (1992a). Human leukocyte antigen serologic and DNA typing of Behçet's disease and its primary association with B51. *Invest. Ophthalmol. Vis. Sci.* **33**: 3332–3340.

Mizuki, N., Inoko, H., Sugimura, K. et al. (1992b). RFLP analysis in the TNF-beta gene and the susceptibility to alloreactive NK cells in Behçet's disease. *Invest. Ophthalmol. Vis. Sci.* **33**: 3084–3090.

Mizuki, N., Inoko, H., Ando, H. et al. (1993). Behçet's disease associated with one of the HLA-B51 subantigens, HLAB*5101. *Am. J. Ophthalmol.* **116**: 406–409.

Mizuki, N., Olmo, S., Sato, T. et al. (1995). Microsatellite polymorphism between the tumor necrosis factor and HLA-B genes in Behçet's disease. *Hum. Immunol.* **43**: 129–135.

Mizuki, N., Inoko, H. and Ohno, S. (1997a). Pathogenic gene responsible for the predisposition of Behçet's disease. *Int. Rev. Immunol.* **14**: 33–48.

Mizuki, N., Ota, M., Kimura, M. et al. (1997b). Triplet repeat polymorphism in the transmembrane region of the MICA gene: a strong association of six GCT repetitions with Behçet's disease. *Proc. Natl Acad. Sci. USA* **94**: 1298–1303.

Morens, D., Anderson, L. and Hurwitz, E. (1980). National surveillance of Kawasaki disease. *Pediatrics* **65**: 21–25.

Moriuchi, J., Wakisaka, A., Aizawa, M. et al. (1982). HLA-linked susceptibility gene of Takayasu disease. *Hum. Immunol.* **4**: 87–91.

Moriuchi, J., Ichikawa, Y., Takaya, M. et al. (1986). Association between HLA and Sjögren's syndrome in Japanese patients. *Arthritis Rheum.* **29**: 1518–1521.

Moriwaki, R. and Numano, F. (1992). Takayasu arteritis: follow-up studies for 20 years. *Heart Vessels* **7**(Suppl.): 138–145.

Moutsopoulos, H.M., Mann, D.L., Johnson, A.H. et al. (1979). Genetic differences between primary and secondary sicca syndrome. *N. Engl. J. Med.* **301**: 761–763.

Murty, G.E., Mains, B.T., Middleton, D. et al. (1991). HLA antigen frequencies and Wegener's granulomatosis. *Clin. Otolaryngol.* **16**: 448–451.

Naito, S., Arakawa, K., Saito, S. et al. (1978). Takayasu's disease: association with HLA-B5. *Tissue Antigens* **12**: 143–145.

Ninet, J., Gebuhrer, L., Bonvoisin, B. et al. (1987). Distribution of HLA-DR antigens in unrelated giant cell arteritis. *Presse Med.* **16**: 1725–1728.

Nishikai, M. and Sekiguchi, S. (1985). Relationship of autoantibody expression and HLA phenotype in Japanese patients with connective tissue diseases. *Arthritis Rheum.* **28**: 579–581.

Numano, F., Isohisa, I., Maezawa, H. et al. (1979). HLA antigens in Takayasu's disease. *Am. Heart J.* **98**: 153–159.

Obst, R., Armandola, E.A., Nijenhuis, M. et al. (1995). TAP polymorphism does not influence transport of peptide variants in mice and humans. *Eur. J. Immunol.* **25**: 2170–2176.

Ocal, L., Russell, K., Beynon, H. et al. (1996). Genetic analysis of TAP2 in systemic lupus erythematosus patients from two ethnic groups. *Br. J. Rheumatol.* **35**: 529–533.

Ohno, S., Asanuma, T., Sugiura, S. et al. (1978). HLA-Bw51 and Behçet's disease. *JAMA* **240**: 529.

Ohtawa, T. (1976). HLA antigens in arterial occlusive diseases in Japan. *Jpn. J. Surg.* **6**: 1–8.

Olsen, M.L., Goldstein, R., Arnett, F.C. et al. (1989). C4A gene deletion and HLA associations in black Americans with systemic lupus erythematosus. *Immunogenetics.* **30**: 27–33.

Olsen, M.L., Arnett, F.C. and Reveille, J.D. (1993). Contrasting molecular patterns of MHC class II alleles associated with the anti-Sm and anti-RNP precipitin autoantibodies in systemic lupus erythematosus. *Arthritis Rheum.* **36**: 94–104.

Ossi, E., Bordin, M.C., Businaro, M.A. et al. (1995). HLA expression in type II mixed cryoglobulinemia and chronic hepatitis C virus. *Clin. Exp. Rheumatol.* **13**: 591–593.

Pablos, J.L., Carreira, P.E., Martin-Villa, J.M. et al. (1995). Polymorphism of the heat-shock protein gene HSP70-2 in systemic lupus erythematosus. *Br. J. Rheumatol.* **34**: 721–723.

Panzer, S., Pabinger, I., Gschwandtner, M.E. et al. (1997). Lupus anticoagulants: strong association with the major histocompatibility complex class II and platelet antibodies. *Br. J. Haematol.* **98**: 342–345.

Papasteriades, C.A., Skopouli, F.N., Drosos, A.A. et al. (1988). HLA-alloantigen associations in Greek patients with Sjögren's syndrome. *J. Autoimmun.* **1**: 85–90.

Papiha, S.S., Murty, G.E., Ad'Hia, A. et al. (1992). Association of Wegener's granulomatosis with HLA antigens and other genetic markers. *Ann. Rheum. Dis.* **51**: 246–248.

Park, M.H. and Park, Y.B. (1992). HLA typing of Takayasu arteritis in Korea. *Heart Vessels* **7**(Suppl.): 81–84.

Pease, C.T., Shattles, W., Charles, P.J. et al. (1989). Clinical, serological, and HLA phenotype subsets in Sjögren's syndrome. *Clin. Exp. Rheumatol.* **7**: 185–190.

Powis, S.J., Deverson, E.V., Coadwell, W.J. et al. (1992). Effect of polymorphism of an MHC-linked transporter on the peptides assembled in a class I molecule. *Nature* **357**: 211–215.

Pugliese, A., Awdeh, Z.L., Galluzzo, A. et al. (1992). No independent association between HSP70 gene polymorphism and IDDM. *Diabetes* **41**: 788–791.

Ratanachaiyavong, S., Demaine, A.G., Campbell, R.D. et al. (1991). Heat shock protein 70 (HSP70) and complement C4 genotypes in patients with hyperthyroid Graves' disease. *Clin. Exp. Immunol.* **84**: 48–52.

Reidenberg, M.M., Drayer, D.E., Lorenzo, B. et al. (1993). Acetylation phenotypes and environmental chemical exposure of people with idiopathic systemic lupus erythematosus. *Arthritis Rheum.* **36**: 971–973.

Reinertsen, J.L., Klippel, J.H., Johnson, A.H. et al. (1978). B-Lymphocyte alloantigens associated with systemic lupus erythematosus. *N. Engl. J. Med.* **299**: 515–518.

Reinharz, D., Tiercy, J.M., Mach, B. et al. (1991). Absence of DRw15/3 and of DRw15/7 heterozygotes in Caucasoid patients with systemic lupus erythematosus. *Tissue Antigens* **37**: 10–15.

Reveille, J.D., Arnett, F.C., Wilson, R.W. et al. (1985). Null alleles of the fourth component of complement and HLA haplotypes in familial systemic lupus erythematosus. *Immunogenetics* **21**: 299–311.

Reveille, J.D., Schrohenloher, R.E., Acton, R.T. et al. (1989a). DNA analysis of HLA-DR and DQ genes in American blacks with systemic lupus erythematosus. *Arthritis Rheum.* **32**: 1243–1251.

Reveille, J.D., Goodman, R.E., Barger, B.O. et al. (1989b). Familial polyarteritis nodosa: a serologic and immunogenetic analysis. *J Rheumatol.* **16**: 181–185.

Reveille, J.D., Anderson, K.L. and Schrohenloher, R.E. (1991a). Restriction fragment length polymorphism analysis of HLA-DR, DQ, DP and C4 alleles in Caucasoids with systemic lupus erythematosus. *J. Rheumatol.* **18**: 14–18.

Reveille, J.D., Barger, B.O. and Hodge, T.W. (1991b). HLA-DR2-DRB1 allele frequencies in DR2-positive black Americans with and without systemic lupus erythematosus. *Tissue Antigens* **38**: 178–180.

Reveille, J.D., Macleod, M.J., Whittington, K. et al. (1991c). Specific amino acid residues in the second hypervariable region of HLA-DQA1 and DQB1 chain genes promote the Ro (SS-A)/La (55-B) autoantibody responses. *J. Immunol.* **146**: 3871–3876.

Reveille, J.D., Brady, J., MacLeod-St. Clair, M. et al. (1992). HLA-DPB1 alleles and autoantibody subsets in systemic lupus erythematosus, Sjögren's syndrome and progressive systemic sclerosis: a question of disease relevance. *Tissue Antigens* **40**: 45–48.

Reveille, J.D., Moulds, J.M. and Arnett, F.C. (1995a). Major histocompatibility complex class II and C4 alleles in Mexican Americans with systemic lupus erythematosus. *Tissue Antigens* **45**: 91–97.

Reveille, J.D., Arnett, F.C., Olsen, M.L. et al. (1995b). HLA-class II alleles and C4 null genes in Greeks with systemic lupus erythematosus. *Tissue Antigens* **46**: 417–421.

Reveille, J.D., Moulds, J.M., Ahn, C. et al. (1998). Systemic lupus erythematosus in three ethnic groups: I. The effects of HLA class II, C4, and CR1 alleles, socioeconomic factors, and ethnicity at disease onset. LUMINA Study Group. Lupus in minority populations, nature versus nurture. *Arthritis Rheum.* **41**: 1161–1172.

Richardson, J.E., Gladman, D.D., Fam, A. et al. (1987). HLA-DR4 in giant cell arteritis: association with polymyalgia rheumatica syndrome. *Arthritis Rheum.* **30**: 1293–1297.

Rigby, R.J., Dawkins, R.L., Wetherall, J.D. et al. (1978). HLA in systemic lupus erythematosus: influence on severity. *Tissue Antigens* **12**: 25–31.

Roitberg-Tambur, A., Brautbar, C., Markitziu, A. et al. (1990). Immunogenetics of HLA class II genes in primary Sjögren's syndrome in Israeli Jewish patients. *Isr. J. Med. Sci.* **26**: 677–681.

Roitberg-Tambur, A., Friedmann, A., Safirman, C. et al. (1993). Molecular analysis of HLA class II genes in primary Sjögren's syndrome. A study of Israeli Jewish and Greek non-Jewish patients. *Hum. Immunol.* **36**: 235–242.

Rosenthal, M., Muller, W., Albert, E.D. et al. (1975). HL-A antigens in polymyalgia rheumatica. *N. Engl. J. Med.* **292**: 595.

Rudwaleit, M., Gibson, K., Wordsworth, P., et al. (1995a). HLA associations of systemic lupus erythematosus in Chinese from Singapore. *Ann. Rheum. Dis.* **54**: 686–687.

Rudwaleit, M., Tikly, M., Gibson, K. et al. (1995b). HLA class II antigens associated with systemic lupus erythematosus in black South Africans. *Ann. Rheum. Dis.* **54**: 678–680.

Rudwaleit, M., Tikly, M., Khamashta, M. et al. (1996). Interethnic differences in the association of tumor necrosis factor promoter polymorphisms with systemic lupus erythematosus. *J. Rheumatol.* **23**: 1725–1728.

Ruuska, P., Hameenkorpi, R., Forsberg, S. et al. (1992). Differences in HLA antigens between patients with mixed connective tissue disease and systemic lupus erythematosus. *Ann. Rheum. Dis.* **51**: 52–55.

Sakkas, L.I., Loqueman, N., Panayi, G.S. et al. (1990). Immunogenetics of polymyalgia rheumatica. *Br. J. Rheumatol.* **29**: 331–334.

Sakurami, T., Ueno, Y., Iwaki, Y. et al. (1982). HLA-DR specificities among Japanese with several autoimmune diseases. *Tissue Antigens* **19**: 129–133.

Salvarani, C., Macchioni, P., Zizzi, F. et al. (1991). Epidemiologic and immunogenetic aspects of polymyalgia rheumatica and giant cell arteritis in northern Italy. *Arthritis Rheum.* **34**: 351–356.

Sargent, C.A., Dunham, I., Trowsdale, J. et al. (1989). Human major histocompatibility complex contains genes for the major heat shock protein HSP70. *Proc. Natl Acad. Sci. USA* **86**: 1968–1972.

Savage, D.A., Ng, S.C., Howe, H.S. et al. (1995). HLA and TAP associations in Chinese systemic lupus erythematosus patients. *Tissue Antigens* **46**: 213–216.

Savi, M., Ferraccioli, G.F., Neri, T.M. et al. (1988). HLA-DR antigens and anticardiolipin antibodies in northern Italian systemic lupus erythematosus patients. *Arthritis Rheum.* **31**: 1568–1570.

Scherak, O., Smolen, J.S. and Mayr, W.R. (1980). HLA-DRw3 and systemic lupus erythematosus. *Arthritis Rheum.* **23**: 954–957.

Schifferli, J.A., Steiger, G., Paccaud, J.P. et al. (1986). Difference in the biological properties of the two forms of the fourth component of human complement (C4). *Clin. Exp. Immunol.* **63**: 473–477.

Schur, P.H., Meyer, I., Garovoy, M. et al. (1982). Associations between systemic lupus erythematosus and the major histocompatibility complex: clinical and immunological considerations. *Clin. Immunol. Immunopathol.* **24**: 263–275.

Schur, P.H., Marcus-Bagley, D., Awdeh, Z. et al. (1990). The effect of ethnicity on major histocompatibility complex complement allotypes and extended haplotypes in patients with systemic lupus erythematosus. *Arthritis Rheum.* **33**: 985–992.

Sebastiani, G.D., Lulli, P., Passiu, G. et al. (1991). Antic ardiolipin antibodies: their relationship with HLA-DR antigens in systemic lupus erythematosus. *Br. J. Rheumatol.* **30**: 156–157.

Seignalet, J., Janbon, C., Sany, J. et al. (1977). HLA in temporal arteritis. *Tissue Antigens* **9**: 69.

Sharma, B.K., Iliskovic, N.S. and Singal, P.K. (1995). Takayasu arteritis may be underdiagnosed in North America. *Can. J. Cardiol.* **11**: 311–316.

Sim, E. and Cross, S.J. (1986). Phenotyping of human complement component C4, a class-III HLA antigen. *Biochem. J.* **239**: 763–767.

Sim, E., Gill, E.W. and Sim, R.B. (1984). Drugs that induce systemic lupus erythematosus inhibit complement component C4. *Lancet* **ii**: 422–424.

Skarsvag, S. (1995a). The importance of C4A null genes in Norwegian patients with systemic lupus erythematosus. *Scand. J. Immunol.* **42**: 572–576.

Skarsvag, S., Hansen, K.E., Moen, T. et al. (1995b). Distributions of HLA class II alleles in autoantibody subsets among Norwegian patients with systemic lupus erythematosus. *Scand. J. Immunol.* **42**: 564–571.

Smolen, J.S., Klippel, J.H., Penner, E. et al. (1987). HLA-DR antigens in systemic lupus erythematosus: association with specificity of autoantibody responses to nuclear antigens. *Ann. Rheum. Dis.* **46**: 457–462.

So, A.K., Fielder, A.H., Warner, C.A. et al. (1990). DNA polymorphism of major histocompatibility complex class II and class III genes in systemic lupus erythematosus. *Tissue Antigens* **35**: 144–147.

Sontheimer, R.D., Thomas, J.R. and Gilliam, J.N. (1979). Subacute cutaneous lupus erythematosus: a cutaneous marker for a distinct lupus erythematosus subset. *Arch. Dermatol.* **115**: 1409–1415.

Sontheimer, R.D., Stastny, P. and Gilliam, J.N. (1981). Human histocompatibility antigen associations in subacute cutaneous lupus erythematosus. *J. Clin. Invest.* **67**: 312–316.

Sontheimer, R.D., Maddison, P.J., Reichim, M. et al. (1982). Serologic and HLA associations in subacute cutaneous lupus erythematosus, a clinical subset of lupus erythematosus. *Ann. Intern. Med.* **97**: 664–671.

Speirs, C., Fielder, A.H., Chapel, H. et al. (1989). Complement system protein C4 and susceptibility to hydralazine-induced systemic lupus erythematosus. *Lancet* **i**: 922–924.

Speiser, P.W. and White, P.C. (1989). Structure of the human RD gene: a highly conserved gene in the class III region of the major histocompatibility complex. *DNA* **8**: 745–751.

Spencer, S.J., Burns, A., Gaskin, G. et al. (1992). HLA class II specificities in vasculitis with antibodies to neutrophil cytoplasmic antigens. *Kidney Int.* **41**: 1059–1063.

Spies, T., Bresnahan, M., Bahram, S. et al. (1990). A gene in the human major histocompatibility complex class II region controlling the class I antigen presentation pathway. *Nature* **348**: 744–747.

Stassi, G., Todaro, M., De Maria, R. et al. (1997). Defective expression of CD9S (FAS/APO-1) molecule suggests apoptosis impairment of T and B cells in HLA-B8, DR3-positive individuals. *Hum. Immunol.* **55**: 39–45.

Steinsson, K., Arnason, A., Erlendsson, K. et al. (1995). A study of the major histocompatibility complex in a Caucasoid family with multiple cases of systemic lupus erythematosus: association with the C4AQ0 phenotype. *J. Rheumatol.* **22**: 1862–1866.

Strimlan, C.V., Taswell, H.F., Kueppers, F. et al. (1978). HLA-A antigens of patients with Wegener's granulomatosis. *Tissue Antigens* **11**: 129–131.

Sturfelt, G., Helimer, G. and Truedsson, L. (1996). TNF microsatellites in systemic lupus erythematosus – a high frequency of the TNFabc 2-3-1 haplotype in multicase SLE families. *Lupus* **5**: 618–622.

Sullivan, K.E., Wooten, C., Schmeckpeper, B.J. et al. (1997). A promoter polymorphism of tumor necrosis factor alpha associated with systemic lupus erythematosus in African–Americans. *Arthritis Rheum.* **40**: 2207–2211.

Sussman, M., Jones, J.H., Almeida, J.D. et al. (1973). Deficiency of the second component of complement associated with anaphylactoid purpura and presence of mycoplasma in the serum. *Clin. Exp. Immunol.* **14**: 531–539.

Takayasu, M. (1908). A case with peculiar changes of the retinal central vessels. *Acta Soc. Ophthal. Jpn* **2**: 554–555.

Takeno, M., Kariyone, A., Yamashita, N. et al. (1995). Excessive function of peripheral blood neutrophils from patients with Behçet's disease and from HLA-B51 transgenic mice. *Arthritis Rheum.* **38**: 426–433.

Takeuchi, F., Nakano, K., Nabeta, H. et al. (1996). Polymorphisms of the TAP 1 and TAP2 transporter genes in Japanese SLE. *Ann. Rheum. Dis.* **55**: 924–926.

Tarassi, K., Carthy, D., Papasteriades, C. et al. (1998). HLA-TNF haplotype heterogeneity in Greek SLE patients. *Clin. Exp. Rheumatol.* **16**: 66–68.

Terasaki, P.I., Healey, L.A. and Wilske, K.R. (1976). Distribution of HLA haplotypes in polymyalgia rheumatica. *N. Engl. J. Med.* **295**: 905.

Tomita, Y., Hashimoto, S. and Yamagami, K. (1993). Restriction fragment length polymorphism (RFLP) analysis in the TNF genes of patients with systemic lupus erythematosus (SLE). *Clin. Exp. Rheumatol.* **11**: 533–536.

Trowsdale, J., Hanson, I., Mockridge, I. et al. (1990). Sequences encoded in the class II region of the MHC related to the "ABC" superfamily of transporters. *Nature* **348**: 741–744.

Truedsson, L., Sturfelt, G. and Nived, O. (1993). Prevalence of the type I complement C2 deficiency gene in Swedish systemic lupus erythematosus patients. *Lupus* **2**: 325–327.

Truedsson, L., Sturfelt, G., Johansen, P. et al. (1995). Sharing of MHC haplotypes among patients with systemic lupus erythematosus from unrelated Caucasoid multicase families: disease association with the extended haplotype [HLA-B 8, SC01, DR17]. *J. Rheumatol.* **22**: 1852–1861.

Udalova, I.A., Nedospasov, S.A., Webb, G.C. et al. (1993). Highly informative typing of the human TNF locus using six adjacent polymorphic markers. *Genomics* **16**: 180–186.

Vitali, C., Tavoni, A., Rizzo, G. et al. (1986). HLA antigens in Italian patients with primary Sjögren's syndrome. *Ann. Rheum. Dis.* **45**: 412–416.

Volkman, D.J., Mann, D.L. and Fauci, A.S. (1982). Association between Takayasu's arteritis and a B-cell alloantigen in North Americans. *N. Engl. J. Med.* **306**: 464–465.

Walport, M.J. and Davies, K.A. (1996). Complement and immune complexes. *Res. Immunol.* **147**: 103–109.

Walport, M.J. and Lachmann, P.J. (1990). Complement deficiencies and abnormalities of the complement system in systemic lupus erythematosus and related disorders. *Curr. Opin. Rheumatol.* **2**: 661–663.

Watson, R.M., Lane, A.T., Barneff, N.K. et al. (1984). Neonatal lupus erythematosus. A clinical, serological and immunogenetic study with review of the literature. *Medicine* **63**: 362–378.

Watson, R.M., Talwar, P., Alexander, E. et al. (1991). Subacute cutaneous lupus erythematosus-immunogenetic associations. *J. Autoimmun.* **4**: 73–85.

Watson, R.M., Scheel, J.N., Petri, M. et al. (1992). Neonatal lupus erythematosus syndrome: analysis of C4 allotypes and C4 genes in 18 families. *Medicine* **71**: 84–95.

Weyand, C.M., Hicok, K.C., Hunder, G.G. et al. (1992). The HLA-DRB1 locus as a genetic component in giant cell arteritis. Mapping of a disease-linked sequence motif to the antigen binding site of the HLA-DR molecule. *J. Clin. Invest.* **90**: 2355–2361.

Weyand, C.M., Hunder, N.N., Hicok, K.C. et al. (1994). HLA-DRB1 alleles in polymyalgia rheumatica, giant cell arteritis, and rheumatoid arthritis. *Arthritis Rheum.* **37**: 514–520.

White, P.C., Vitek, J., Lahita, R.G. et al. (1992). Polymorphism in the RD (D6545) gene. *Hum. Gen.* **89**: 243–244.

Wilson, A.G., Giovine, F.S. di, Blakemore, A.I. et al. (1992). Single base polymorphism in the human tumour necrosis factor alpha (TNF alpha) gene detectable by NcoI restriction of PCR product. *Hum. Mol. Gen.* **1**: 353.

Wilson, A.G., Vries, N. de, Pociot, F. et al. (1993). An allelic polymorphism within the human tumor necrosis factor alpha promoter region is strongly associated with HLA A1, B8, and DR3 alleles. *J. Exp. Med.* **177**: 557–560.

Wilson, A.G., Gordon, C., Giovine, F.S. di et al. (1994). A genetic association between systemic lupus erythematosus and tumor necrosis factor alpha. *Eur. J. Immunol.* **24**: 191–195.

Wilson, A.G., Symons, J.A., McDowell, T.L. et al. (1997). Effects of a polymorphism in the human tumor necrosis factor alpha promoter on transcriptional activation. *Proc. Natl Acad. Sci. USA* **94**: 3195–3199.

Wilson, W.A., Scopelitis, E. and Michaiski, J.P. (1984a). Association of HLA-DR7 with both antibody to SSA(Ro) and disease susceptibility in blacks with systemic lupus erythematosus. *J. Rheumatol.* **11**: 653–657.

Wilson, R.W., Provost, T.T., Bias, W.B. et al. (1984b). Sjögren's syndrome. Influence of multiple HLA-D region alloantigens on clinical and serologic expression. *Arthritis Rheum.* **27**: 1245–1253.

Wilson, W.A., Perez, M.C. and Armatis, P.E. (1988a). Partial C4A deficiency is associated with susceptibility to systemic lupus erythematosus in black Americans. *Arthritis Rheum.* **31**: 1171–1175.

Wilson, W.A., Perez, M.C., Michalski, J.P. et al. (1988b). Cardiolipin antibodies and null alleles of C4 in black Americans with systemic lupus erythematosus. *J. Rheumatol.* **15**: 1768–1772.

Wilson, W.A., Scopelitis, E., Michaiski, J.P. et al. (1995). Familial anticardiolipin antibodies and C4 deficiency genotypes that coexist with MHC DQB1 risk factors. *J. Rheumatol.* **22**: 227–235.

Wong, V.S., Egner, W., Elsey, T. et al. (1996). Incidence, character and clinical relevance of mixed cryoglobulinaemia in patients with chronic hepatitis C virus infection. *Clin. Exp. Immunol.* **104**: 25–31.

Yamada, H., Watanabe, A., Mimori, A. et al. (1990). Lack of gene deletion for complement C4A deficiency in Japanese patients with systemic lupus erythematosus. *J. Rheumatol.* **17**: 1054–1057.

Yanagawa, H., Nakamura, Y., Yashiro, M. et al. (1988). A nationwide incidence survey of Kawasaki disease in 1985–1986 in Japan. *J. Infect. Dis.* **158**: 1296–1301.

Yao, Z., Kimura, A., Hartung, K. et al. (1993a). Polymorphism of the DQA 1 promoter region (QAP) and DRB1, QAP, DQA1, DQB1 haplotypes in systemic lupus erythematosus. SLE Study Group 3 members. *Immunogenetics* **38**: 421–429.

Yao, Z., Hartung, K., Deicher, H.G. et al. (1993b). DNA typing for HLADPB 1-alleles in German patients with systemic lupus erythematosus using the polymerase chain reaction and DIG-ddUTP-labelled oligonucleotide probes. Members of SLE Study Group. *Eur. J. Immunogen.* **20**: 259–266.

Yao, Z., Seelig, H.P., Ehrfeld, H. et al. (1994). HLA class II genes and antibodies against recombinant U1-nRNP proteins in patients with systemic lupus erythematosus. SLE Study Group. *Rheumatol. Int.* **14**: 63–69.

Yoshida, M., Kimura, A., Katsuragi, K. et al. (1993). DNA typing of HLA-B gene in Takayasu arteritis. *Tissue Antigens* **42**: 87–90.

Yurdakul, S., Gunaydin, I., Tuzun, Y. et al. (1988). The prevalence of Behçet's syndrome in a rural area in northern Turkey. *J. Rheumatol.* **15**: 820–822.

Zhang, L., Jayne, D.R., Zhao, M.H. et al. (1995). Distribution of MHC class II alleles in primary systemic vasculitis. *Kidney Int.* **47**: 294–298.

CHAPTER 22

HLA and Dermatological Disease

Adam Friedmann

INTRODUCTION

The propensity of many diseases, among them autoimmune conditions, to manifest themselves in the skin may result from a faulty recognition by the body's immune system, during immunological maturation in the thymus or peripheral organs, of what constitute the skin's "self-antigens" (Morhenn, 1997). The association of the human major histocompatibilty complex (MHC) with many dermatological diseases supports this hypothesis.

The human leucocyte antigen (HLA) system is associated with a variety of skin diseases. Autoimmune diseases account for most of these HLA associations, but several skin cancers, drug-induced dermatological conditions and sensitivity of skin and mucous membranes to different drugs, chemicals and allergens also show a degree of HLA association. These associations indicate that immune response genes are involved in the predisposition of the skin and mucous membranes to many pathological conditions (Tiwari and Terasaki, 1985; Pleyer et al., 1996).

The amount of data available on the HLA association the different dermatological diseases varies greatly. Some have been intensively investigated over many years and have yielded very informative results but, in others, the HLA associations are still unclear or have not been well studied. Thus, the information available for different diseases varies in depth and detail.

The advent of DNA-based typing techniques has allowed researchers to narrow the range of HLA molecules involved in disease susceptibility (Szafer et al., 1987; Scharf et al., 1988). In addition, the crystallographic studies that revealed the three dimensional structure of the HLA class I and class II molecules, the discovery of the antigen-processing machinery and transgene technology have greatly improved our understanding of the relationship between structure and function, and brought us closer to understanding the mechanism by which HLA molecules function in normal and pathological conditions of the skin (Bjorkman et al., 1987; Jardetzky, 1994; Wucherpfennig, 1995).

THE SKIN IMMUNE SYSTEM

A fully operational immune system is essential in higher organisms in order to combat invading microorganisms and repress malignant transformation of the body cells. The first barrier to invading infectious agents is the skin. The skin plays an active role and is alive with a myriad of immune-system activities that are important for the defence against intruding microorganisms and other agents. The skin-associated lymphoid tissue (SALT), a term coined by Streilin (1983), is now considered to be part of a larger unit known as the skin immune system (SIS). The components of the SIS may be considered at each anatomical layer. In the epidermis there are keratinocytes, Langerhans' cells (LC) and T lymphocytes. In the dermis, dermal antigen-presenting cells (APC; dendritic cells, macrophages), T lymphocytes, the dermal microvascular unit (endothelial cells, pericytes), afferent lymphatics, eosinophils, basophils, neutrophils, mast cells and natural killer (NK) cells are considered part of the SIS. In subcutaneous tissues there are afferent lymphatics and the skin draining lymph nodes. In the epidermis, LC express MHC class I and class II molecules and function as APC. LC process antigens and can thus leave the skin and trigger resting T cells in an antigen-specific class I and class II-restricted fashion in lymphoid organs. Keratinocytes, which comprise over 90% of all epidermal cells, can secrete a plethora of cytokines and chemokines which set in motion the skin immune response (Jahn et al., 1997; Simon, 1997; Stingl, 1997). The genetic background of individuals and ethnic groups, including their HLA genes, combines with environmental factors to determine the pathogenic responses of the skin, leading to a potential myriad of pathological conditions. The most important skin conditions associated with HLA polymorphisms will now be discussed individually.

PEMPHIGUS VULGARIS

Pemphigus vulgaris (PV), a life-threatening autoimmune blistering disease of the mucous membrane and skin, is thought to be mediated by autoantibodies (Ahmed, 1968). Autoantibodies directed against the desmoglein-3 glycoprotein (DG3), also termed the PV antigen (PVA), are thought to mediate the main pathological damage. DG3 is a transmembrane molecule belonging to the cadherin family of Ca^{2+}-dependent adhesion molecules, and it participates in the desmosome structure of keratinocytes (Beutner and Jordan, 1964; Karpati et al., 1994). Anti-DG3 autoantibodies of the immunoglobulin G_1 (IgG_1) and IgG_4 subtypes are elicited in PV patients carrying a rare HLA-DR4 allele, DRB1*0402, or the rare HLA-DQB1 allele, DQB1*0503. High titres of both IgG_1 and IgG_4 autoantibodies seem to be pathogenic and attack DG3, causing a loss of cell adhesion and leading to blister formation (Ahmed, 1968; Callot and Bagot, 1994; Karpati et al., 1994). PV is a disease in which the HLA-associated alleles have been studied thoroughly at the serological, genomic and structural levels, laying the basis for understanding HLA-disease association and the possible underlying mechanism of action. The discovery of the PVA has an important impact on the understanding of HLA association in autoimmunity, enabling the investigation of the interaction of MHC, self-peptide and T cell receptor. This may contribute to the understanding of how and why certain individuals bearing a specific HLA haplotype develop autoimmune disease. PV occurs in the Ashkenazi Jewish population (where the incidence of PV is relatively high), non-Ashkenazi Jews, Caucasians and other races (Ahmed, 1968; Tiwari and Terasaki, 1985; Friedmann and Brautbar, 1993). The onset of PV is usually between the ages of 20 and 30 years, but onset of disease in younger and in older people has also been reported (Friedmann and Brautbar, 1993).

HLA associations with pemphigus vulgaris

An association of PV with HLA was first demonstrated by Park et al. (1979), who showed an association of PV with HLA-DR4. Brautbar et al. (1978) and Amar et al. (1984) narrowed this to an association with HLA-DR4–DW10 in Ashkenazi Jews [relative risk (RR) of 31.9]. Szafer et al. (1987) used restriction fragment length polymorphism (RFLP) to show for the first time that in the Israeli population (including Ashkenazi Jews, non-Ashkenazi Jews and Arabs) PV was associated not only with HLA-DR4–Dw10 but also with DR6. This RFLP study revealed that in patients carrying the DR6 allele, an HLA-DQB RFLP fragment was strongly associated with PV. This 2.2 kb fragment was present in all DR6 PV patients, but was rare (less than 1%) in healthy DR6 controls. In further molecular studies, polymerase chain reaction (PCR) and DNA sequence analysis were used by Scharf et al. (1988, 1989) and Sinha et al. (1988) to study the HLA association at the DNA sequence level. The results of these studies indicated clearly that PV is associated with two different HLA genes. The first is an HLA-DR4 gene allele, DRB1*0402 (1 of 22 known DR4 alleles). Among Ashkenazi Jews, > 93% of PV patients carry DRB1*0402, which is relatively rare in the general population (Scharf et al., 1989). The second haplotype associated with PV, found mainly in non-Ashkenazi Jews,

Caucasians and Japanese, is the rare DQB1 allele, DQB1*0503, in linkage disequilibrium with DRB1*1401 (Scharf et al., 1988, 1989; Sinha et al., 1988; Niizeki et al., 1994). It is noteworthy that PV patients usually carry an extended HLA haplotype: Ashkenazi Jewish patients usually carry HLA-A26–B38–SC21–DR4–DQ3 and/or HLA-B35–SC31–DR4–DQ3, while Caucasians, Japanese and non-Ashkenazi Jewish patients carry HLA-B35–SC31–DR4–DQ3 or HLA-B55–SB45–DR6(DR14)–DQ1. These extended haplotypes may have a diagnostic value (Ahmed et al., 1990; Alper et al., 1990; Bohl et al., 1995).

Significance of the HLA associations in pemphigus vulgaris to the pathogenic process

In the HLA-DR4 β-chain, the polymorphic residues of the P4 pocket are important in determining susceptibility to PV. The P4 pocket is extremely polymorphic and is located in a central position of the HLA-DR peptide binding site. The DRB1*0402 allele carried by a high percentage of PV patients has a glutamic acid at position DRβ71 and an aspartic acid at position DRβ70. Therefore, the P4 pocket of this DRβ-chain is negatively charged at positions DRβ70 and DRβ71. The importance of these residues at positions DRβ70 and DRβ71 in PV for the presentation of the DG3 self-antigen was demonstrated by Wucherpfennig et al. (1995), who used T cell clones isolated from DRB1*0402 PV patients that responded *in vitro* against two human DG3-derived peptides ($DG3_{190-194}$ and $DG3_{206-220}$). Specific T cells responded *in vitro* to the DG3-derived peptides presented by APC expressing the DRB1*0402-coded β-chain. When the amino acid at position DRβ71 was altered (by *in situ* mutagenesis) from glutamic acid to positively charged amino acids such as lysine or arginine (carried by other DR4 subtypes), or the aspartic acid at DRβ70 was mutated to glutamine, specific T cells failed to recognize the $DG3_{190-194}$ and $DG3_{206-220}$ peptides. The results also indicated that position DRβ67 played a lesser role in DG3 antigen presentation. It is predicted that the negative charge at the DRβ71 and DRβ70 positions in the PV-linked DRB1*0402 molecule confers selective binding of self-peptides with a positive charge at P4 (e.g. $DG3_{190-194}$ containing lysine and $DG3_{206-220}$ containing arginine). It was concluded that two negatively charged residues at positions DRβ70 and DRβ71 are critical for presentation of DG3 peptides, providing the basis for understanding the DRB1*0402-linked autoimmunity in PV. It is important to note in this context that another HLA-DR4 allele, DRB1*0404, predisposes to rheumatoid arthritis (RA). Comparison of the DRB1*0402 allele, predisposing to PV, with the RA-related DRB1*0404 is very informative. The two alleles differ only in the amino acids at positions DRβ67, DRβ70 and DRβ71. DRB1*0402 has isoleucine at DRβ67, aspartic acid at DRβ70 and glutamic acid at DRβ71, while DRB1*0404 has leucine at DRβ67, glutamine at DRβ70 and arginine at DRβ71. Therefore, the DRβ67–DRβ71 region of the DRβ-chain of DR4 is important in predisposition to both diseases.

The second HLA allele associated with PV is the rare DQ1 allele consisting of DQB1*0503, which is in linkage disequilibrium with HLA-DR6 (DRB1*01401), and DQA1*101 alleles (Levinson et al., 1997). The DRB1*0503 allele differs from the common DRB1*0501 allele by only one amino acid at position DQβ57, having an aspartic acid instead of valine (Scharf et al., 1988; Levinson et al., 1997): again, a negatively charged amino acid on the α-helix at a position important for peptide binding. The position DQβ57 is also important in diabetes-associated DQB1*0302 and DQB1*0201 alleles (Horn et al., 1988). However, the diabetes-associated HLA alleles do not carry a negative charge at DQβ57 (Tisch and McDevitt, 1996). In PV, results indicate that position *DQβ57* alone is not sufficient to account for PV susceptibility. In the context of the DQB1*0503 allele, the segment around this residue also seems to be important (Scharf et al., 1989).

Correlation of pemphigus vulgaris-linked HLA alleles with pathogenic autoantibodies and cytokine production

PV patients with active disease produce high titres of IgG_1 and IgG_4 antibodies, directed against DG3. *In vitro* experiments show that these antibodies react with two DG3-derived peptides termed Bos1 ($DG3_{50-79}$) and Bos6 ($DG3_{200-229}$),

(Bohl et al., 1995). IgG_1 titres against Bos1 remain elevated in PV patients with remission but IgG_1 and IgG_4 autoantibodies against Bos6 disappear. The anti-Bos1 IgG_1 autoantibody titres of PV patients in remission did not differ from titres of Bos1-specific IgG_1 antibodies in healthy relatives bearing the same HLA haplotypes, but IgG_4 against Bos1 and Bos6 could not be detected. PV patients carrying DRB1*0402 or DRB1*1401–DQB1*503 alleles show no selective synthesis of either IgG_1 or IgG_4 against Bos1 or Bos6 peptides, indicating that both HLA alleles predispose similarly to PV. In addition, no difference in disease severity or anti-DG3 antibodies titres were found in PV patients homozygous or heterozygous for, or carrying only one of the predisposing PV HLA alleles (Levinson et al., 1997). The induction of high titres of DG3-specific autoantibodies depends on T cell production of cytokines that trigger the proliferation of the autoreactive B cells. T cell clones, restricted by DRB1*0402 alleles and specific for the DG3190–204 peptide, are of the Th2 subtype which secrete interleukin-4 (IL-4) and IL-10 cytokines and promote B cell differentiation and antibody production (Paul and Seder, 1994; Wucherpfennig et al., 1995).

HLA association with pemphigus vulgaris in different ethnic groups

Studies of Israeli PV patients (Ashkenazi Jews, non-Ashkenazi Jews, Arabs), Austrian (Szafer et al., 1987; Scharf et al., 1988; Ahmed et al., 1993), Italian (Carcassi et al., 1996; Lombardi et al., 1996), Indian (Wilson et al., 1994; Delgado et al., 1996), English (Wilson et al., 1994) and Japanese patients (Niizeki et al., 1994) all show an association between PV and the two HLA alleles DRB1*0402 and DQB1*0503. In Ashkenazi Jews, DRB1*0402 is more prevalent and, in other ethnic groups, the DRQ1*0503 allele is more prevalent.

HLA alleles and anti-DG3 autoantibodies in pemphigus vulgaris families and relatives

Several publications have described the distribution of HLA alleles within families of PV patients (Brenner et al., 1985; Feinstein et al., 1991; Reohr et al., 1992; Bohl et al., 1996; Levinson et al., 1997).

These studies show that all patients carried one or both of the HLA alleles associated with PV (HLA-DR4DRB1*0402 and DR6DRB1*1401–DQB1*503). Some parents, brothers, sisters and distant relatives of PV patients, who carried PV-predisposing HLA alleles also developed IgG_1 autoantibodies specific for DG3-derived peptides or basal membrane antigens (Bohl et al., 1995; Levinson et al., 1997). These results indicate that PV predisposing HLA alleles by themselves are not sufficient to cause the onset of PV. Therefore, PV is a multifactorial disease with a strong genetic background, in which specific HLA alleles play a crucial role in disease induction. The RR for the two predisposing HLA alleles involved in PV was calculated to be over 80 (Friedmann and Brautbar, 1993).

HLA in drug-triggered pemphigus vulgaris

The induction of PV and PV-like diseases by certain drugs has been reported many times (Degos et al., 1969; Brenner et al., 1990; Mobini and Ahmed, 1993; Matzner et al., 1995). Drug-induced pemphigus occurred mostly in RA patients treated with penicillamine and with other drugs (Zone et al., 1982; Matzner et al., 1995). Inspection of the chemical composition of the drugs thought to induce pemphigus shows that several of them contain sulfur groups (Mobini and Ahmed, 1993). In cases where drug-induced pemphigus developed into full-blown PV and persisted for the rest of patients' life, the HLA association studies demonstrated that all patients carried the DRB1*0402 or DQB1*0503 alleles or both. Drug-induced PV studies again indicate the crucial role that HLA alleles play in the onset of the disease and lend support to the claim that environmental factors are also important in PV induction.

Protective HLA alleles in pemphigus vulgaris

The question of certain HLA alleles exerting a preventive or protective function is not well studied in PV. In insulin-dependent diabetes mellitus (IDDM), protective HLA alleles were described (see Chapter 15). However, HLA haplotype analysis of several hundreds of PV patients have failed to show the presence of HLA-DR3. These data may indicate that DR3 exerts a protective influence against PV in people who carry DR3 and in individuals who carry the PV-associated alleles as well as DR3 (Friedmann and Brautbar, 1993; Wilson et al., 1994). In addition, DR2 was shown to be reduced in Indian PV patients, and thus may have a protecting role in PV in that ethnic group (Wilson et al., 1994).

Conclusion

In summary, PV is associated with the DR4 allele, DRB1*0402, in the Ashkenazi Jewish population and the DQB1 allele, DQB1*0503, in other populations. The association with DRB1*0402 has been mapped to pocket 4 of the DRβ-chain and that with DQB1*0503 is at least in part explained by aspartic acid at position DQβ57. The RR of these two predisposing alleles is over 80.

PEMFIGUS FOLIACEUS (FOGO SELVAGEM)

Endemic pemphigus foliaceus (PF), also called fogo selvagem (FS), is an autoimmune blistering skin disease, characterized by IgG autoantibodies (predominantly of the IgG_4 subclass) to desmoglein 1 glycoprotein (a desmosomal protein embedded in the basal membrane). The histopathological picture of FS is of separation of epidermal cells, leading to blister formation in the mucosal and intraepidermal layers (Eyre and Stanley, 1987; Diaz et al., 1989; Rock et al., 1989; Kunte et al., 1997). FS occurs in certain regions of Brazil and an unknown infectious agent, transmitted probably by a fly (*Simulium pruinosum*) is predicted to be the environmental agent responsible for inducing the disease (Lombardi et al., 1992). However, not all individuals resident in endemic regions develop FS, suggesting that host factors play a part in determining who will develop the disease. A strikingly strong HLA association in three different ethnic groups in Brazil – Brazilian Mestizos, Xavante Indians and Terena Indians suffering from FS – has been reported. DRB1*0404, DRB1*1402 and DRB1*1406 are associated with FS in Xavante and Terena Indians and DRB1*01 in Mestizos (Moraes et al., 1991, 1997; Cerena et al., 1993).

HLA association with fogo selvagem

Xavante and Terena patients with FE show a close to absolute association with the DRB1*0404, DRB1*1402 and DRB1*1406 alleles (RR = 18 for Xavante; 14.1 for Terena). In Mestizos, a different HLA allele, namely DRB1*01 (especially the DRB1*0102 allele), is mainly associated (RR = 5.4). Genomic analysis of the different HLA alleles involved in FS revealed the striking fact that all alleles share a common epitope in the third hypervariable region of the DRB1 gene product. A stretch of amino acids at positions 67–74 of the DRβ-chain has the amino acid motif, LLEQRRAA (Cerena et al., 1993; Moraes et al., 1997). It has, therefore, been postulated that susceptibility to FS is inherited by the LLEQRRAA amino acid epitope at positions 67–74 of the DRβ-chain and is not linked to a particular HLA allele or ethnic group. The different relative risks for FS observed in the three populations can be explained by arguing that other segments of the DRβ molecules, in which the alleles differ in composition, can influence the susceptibility conferred by the shared epitope.

It has also been proposed that the absence of certain alleles carrying the shared epitope in the populations studied may be responsible for their lack of association (Moraes et al., 1997). As in PV, penicillamine has been shown to induce FS in individuals carrying FS-linked HLA alleles (Korman et al., 1991). Individuals living in endemic regions and expressing the HLA-DQ2 gene did not develop the disease, thus implicating this allele in FS resistance (Moraes et al., 1991). The story of HLA association in FS is an important one. It seems that some HLA alleles are able to bind self-peptides and elicit T cell responses to self-antigens because of the amino acid motifs in crucial regions of their HLA molecules. The conclusion of the FS-HLA studies is that different HLA alleles in different ethnic groups predispose to the same disease by virtue of sharing an identical epitope.

Pemphigoid gestationis (herpes gestationis)

Pemphigoid gestationis (PG) is a rare autoimmune disease of the skin which appears in pregnancy or the postpartum period (incidence 1:50 000 pregnancies). PG is characterized by pruritic, urticarial and vesiculobullous eruptions on the body. Genetic predisposition is thought to involve HLA genes (Shornick et al., 1981, 1983, 1995).

HLA associations with pemphigoid gestationis

The major association of PG is with the HLA-DRB1*0301 allele (Shornick et al., 1981, 1995). In addition, the DRB1*0401 and DRB1*040x alleles appear in 85% of the DRB1*0301-positive patients and hence are also thought to be involved in PG susceptibility (Shornick et al., 1995). An interesting association with the polymorphic allele of the complement system was found in 90% of PG patients who carried a polymorphic variant of a C4 null allele, C4*Q0. Whether this allele has a primary genetic association or is related to its linkage disequilibrium with DR3 and DR4 genes has yet to be determined (Shornick et al., 1993). The involvement of a complement factor in skin autoimmune disease deserves special attention as the C4 null gene is linked to other autoimmune skin diseases and may indicate another unknown PG susceptibility gene in the vicinity of this non-expressed complement gene.

PSORIASIS VULGARIS

Psoriasis vulgaris is a chronic inflammatory skin disease, clinically characterized as hyperproliferation of epidermal cells and inflammation (dermatosis), resulting from infiltration of activated T helper cells and mononuclear cells and the release of proinflammatory cytokines (Burch and Rowell, 1965; Breathnach, 1993; Christophers, 1996; McKusick, 1997). Psoriasis vulgaris affects up to 2% of the population in Europe and North America but is relatively rare among Blacks, Japanese and South Americans, and is absent in Eskimos (Farber, 1991). Psoriasis has a definite genetic background and HLA class I and class II genes seem to play a role in disease susceptibility (Henseler, 1997), among several other gene markers on other chromosomes (Trembath, 1996; Henseler, 1997). Two types of psoriasis vulgaris exist: type I, with early onset, associated with HLA-Cw6, B57 and DR7 alleles, and type II, with late onset, where HLA-Cw2 is overrepresented (Russel et al., 1972; Brenner et al., 1978; Tiilikainen et al., 1980; Henseler, 1997).

Genetics of psoriasis

Psoriasis vulgaris has been intensively investigated around the globe, in thousands of patients and matched controls, in hundreds of families of psoriasis patients and in hundreds of twins (Lomholt, 1963; Henseler, 1997; McKusick, 1997). However, despite this enormous effort, the genes linked to psoriasis are not yet known. The disease is probably polygenic, with the involved genes located on several chromosomes; however, the association with the HLA genes on 6p21 shows the strongest genetic association (Nair, 1997; Trembath, 1997). Several family studies and two wide genome scan analyses using several DNA probes and different genetic approaches have indicated the involvement of the following chromosomal loci as candidates for psoriasis susceptibility: 6p21, 16q, a region distal to 17q and 20p. Previous linkages on chromosome 4 or at 17q were dismissed (Kimberling and Dobson, 1973; Pietrzyk et al., 1982; Nair et al., 1997; Trembath et al., 1997). It is thought that psoriasis has a dominant mode of inheritance with partial penetrance and disease heterogeneity. Environmental factors such as bacterial infection (group A, β-haemolytic streptococci) and herpes viruses probably take part in disease onset or aggravation (McKusick, 1997).

HLA associations with psoriasis

Russell et al. (1972), the first to report an HLA association with psoriasis, found that HLA-B13 and HLA-B17 (B57) alleles were present in 27.3% and 22.7% of patients (compared with 3.4% and 9.0% in controls). Later, Brenner et al. (1978) reported that HLA-Cw6 is strongly associated with psoriasis, Tiilikainen et al. (1980) showed linkage with HLA-DR7 and Schmitt Egenolf et al. (1993) reported an association with the haplotype

HLA-Cw6–B57–DRB1*0701/02–DQA1*0201–DQB1*0303 (*DQ9*). Nakagawa et al. (1993) and Jenisch et al. (1995) reported associations with DRB1*0701/02–DQA1*0201–DQB1*0303 and Ikaheimo et al. (1996) reported the association in a Finnish population with A1/2–B13–B57–Cw6–DR7–DQA1*0201. Several other groups (Ishibashi and Juji, 1991; Henseler, 1997), confirmed most or part of these observations. In calculating the RR for the several HLA alleles indicated here, the following alleles seem to be mostly associated with psoriasis: HLA-Cw6, HLA-B13, HLA-B57, DRB1*0701/02, DQA1*0201 and DQB1*0303 (Henseler and Christophers, 1985; Schmitt-Egenoff et al., 1993, 1996; Henseler, 1997). These studies may indicate that linkage disequilibrium of the HLA system, together with recombination events, may have changed the haplotypes in which the HLA-Cw6 is present. For example, the association of the Cw6 allele with HLA-B13 or with HLA-B57 reported several times may indicate that these haplotypes are also involved to some degree with disease susceptibility, or that another gene associated with psoriasis is located near these genes. A hypothesis that the HLA-C molecule may by itself have a role in self-peptide presentation was proposed by Roitberg-Tambur et al. (1995), who claimed that, in Jewish psoriasis patients, the Cw6 molecule has a unique peptide-binding pocket, having the amino acid alanine at position 73 of the HLA-C α-chain and a negatively charged aspartic acid at position 9. Additional support for the role of alanine 73 comes from the study by Asahina et al. (1996), who also found aspartic acid at position 9 to be associated with psoriasis. It is thought that the aspartic acid located on the β-sheet of the α_1-domain influences peptide binding in the C pocket of the groove, as many with the alanine at position 73. This could contribute to the disease susceptibility to psoriasis vulgaris in the immune responses. Different conclusions concerning position 73 were reached by Mallon et al. (1997), who found the HLA-Cw*06 allele C*0602 to be the major HLA allele associated with psoriasis type I. These authors also found alanine at position 73 to be increased in male patients compared with controls. However, since alanine at position 73 appears in other HLA-C alleles, the C*0602 allele additionally exerts its influence through other parts of the molecule.

HLA and psoriasis types I and II

Psoriasis appears, much like diabetes, in two forms: type I with disease onset between the ages of 15 and 25 years, and psoriasis type II with a mean onset age of 60 (females) and 57 years (males) (Henseler and Christophers, 1985). It is important to note that, while over 50% of patients suffering from psoriasis type I had first-degree relatives with psoriasis, patients with late-onset disease very rarely had first-degree psoriatic relatives (Muto et al., 1995). The HLA association of type 1 psoriasis is frequently with the previously reported HLA alleles, HLA-Cw*060x(*0602), HLA-B13/HLA-B57, DRB1*0701/02, DQA1*0201 and DQB1*0303. In contrast, in patients with type II psoriasis, HLA-Cw2 and HLA-B27 were involved, HLA-Cw6 and HLA-DR7 were only slightly elevated and HLA-DQB1*0303 was not observed (Muto et al., 1995; Enerback et al., 1997; Henseler, 1997). Psoriasis type I is a severe disease which tends to relapse, while the type II disease is milder with rare relapses. These observations indicate that the two types of disease are clearly different and have different HLA associations.

HLA and psoriatic families

The study of large numbers of psoriatic families for the psoriasis gene(s) has yielded confusion and indicated that finding these genes may be a complicated task. For example, in a whole-genome scan of several psoriatic families by positional cloning using polymorphic DNA microsatellites to detect markers in linkage with the disease, one such microsatellite marker, in chromosomal locus D17S784, mapped to the distal region of chromosome 17q and segregated in eight families affected by type I disease with susceptibility to psoriasis (Tomfohrde et al., 1994; Bowcock, 1995). Within the largest family of this study, where the marker showed the strongest linkage with the disease, 21 members had the susceptibility haplotype, 20 of whom had psoriasis. In this family and in several others showing the D17S784 marker linkage, no association between HLA-Cw6 and psoriasis was detected. In four other families showing the psoriasis–D17784 linkage, the presence of the HLA-Cw6 allele was explained as contributing to the disease development (Tomfohrde et al., 1994; Bowcock, 1995). It is important to note that other studies which tested the D17S784 locus in other families failed to show the linkage of this marker with psoriasis (Nair et al., 1995). It may be hypothesized that in families in which HLA-Cw6 (alone or in the context of its extended haplotypes) or the D17S784 locus fails to show linkage with the disease, genetic events such as recombination or occurrence of mutations may have abolished the genetic associations. Yet another explanation for these results may lie in the difficulty of typing HLA-C by serological means. The typing may have failed to identify the HLA-Cw6 antigen accurately in all individuals tested (Bunce and Welsh, 1994). In this context, it is appropriate to note the model for psoriasis proposed by Theeuwes and Morhenn (1995), which is based on the assumption that the disease is caused by expansion of short repeats of DNA.

BEHÇET'S DISEASE

Behçet's disease or syndrome (BD) is a multisystem disease characterized by the recurrent ulceration of the genitals, aphthous lesions of the mouth, uveitis or iridocyclitis followed by hypopyon and, in addition, lesions in articular, vascular, neurological pulmonary and gastrointestinal systems. The disease is endemic in the Mediterranean region and Japan and may have more than one form and degree of severity. Morbidity is also increasing in central and north Europe (Grabski and Grabski, 1990; Sakane, 1997; Chen et al., 1997; Roman et al., 1997; Taylor et al., 1997; Hadfield et al., 1997). Hypersensitivity to herpes simplex virus or streptococcal antigens is postulated to be a key trigger of BD. It is thought that the microbial peptide mimics a peptide motif in a human heat shock protein (HSP) present in the skin and mucous membranes and thus elicits an immune response to self-antigens (Mizuki et al., 1991; Sakane, 1997; Sugi et al., 1997). The association of BD with HLA is very strong: HLA-B51 has been reported as a disease susceptibility allele in most populations, but in Chinese studies class II alleles have been implicated (Sakane, 1997). A gene located between the HLA-B and the tumour necrosis factor (TNF) loci is thought to be the true BD susceptibility gene (Mizuki et al., 1997a).

HLA association with Behçet's disease

Several HLA genes have been associated with BD.

HLA-B51 (HLA-B*5101) and HLA-Cw15

Ohno et al. (1973) were the first to report an association of BD with the HLA-B5 antigen (B5 is now split serologically into B51 and B52). BD is associated in the majority of the world with the HLA-B51 allele (Mizuki et al., 1997a; Yazici, 1997). The reports supporting this association came from several ethnic groups: from Israel (Brautbar et al., 1978; Arber et al., 1990), Japan (Mizuki et al., 1992a, b), Spain (Villanueva et al., 1993), Russia (Golosenko, 1990), Turkey (Azizleri et al., 1994), Portugal (de Souza-Ramalho et al., 1991), Brazil (Barra et al., 1991), Italy (Balboni et al., 1992) and northern China (Mineshita et al., 1993). DNA-based typing has confirmed the association with HLA-B5101, an allelic variant of B57. Reports claiming that HLA-Cw14 and Cw15 were associated with BD proved to result from linkage disequilibrium with the HLA-B*5101 allele. Indeed, when the HLA-C allele frequencies were compared for the B51-positive and -negative patients and controls, no significant difference in HLA-C allele frequency was detected between patients and control groups. These facts suggest that the gene associated with BD is not the HLA-C gene, but the HLA-B*5101 allele or a non-HLA gene residing in the centromeric side of the HLA-B gene (Mizuki et al., 1996). Polymorphic analysis of the Tau-a microsatellite located between the HLA-B and TNF genes has indicated that the BD gene is located near the HLA-B gene but is not the HLA-B51 gene itself. Many important genes exist in the region located between the TAP1 and HLA-B genes, including the MIC, PERB and NOB genes (Mizuki and Ohno, 1996).

The MICA gene

The MICA gene, encoding a class I-like chain and having a specialized function in antigen presentation, is located close to the HLA-B gene. A triplet repeat microsatellite polymorphism of $(GCT/AGC)_n$ in the transmembrane region of the MICA gene was found to possess, five distinct alleles in the population. One of them contained an additional one-base insertion, which created a frame-shift mutation, resulting in a premature termination codon in the TM region. Investigating the association of this microsatellite in Japanese patients with BD, it appeared that this microsatellite allele, (GCT/AGC)$_6$, was present at a significantly higher frequency in all HLA-B51-positive patients and in an additional 13 B51-negative patients. These results suggest the possibility of a primary association of BD with an allele of the MICA gene rather than the HLA-B gene (Mizuki et al., 1997b).

TAP1

A linkage disequilibrium between HLA-DQB1*0501 and TAP2B was observed in BD patients, while TAP1C was absent in patients but was found in 12.1% controls. The complete absence of TAP1C alleles in BD patients may indicate that TAP1 has some influence on in the development of BD. Furthermore, the existence of a linkage disequilibrium between HLA-DQB1*0501 and TAP2B in BD patients suggests that the gene conferring susceptibility for BD is inherited as an extended haplotype in the population studied (Gonzalez-Escribano et al., 1995) and that the region to study further in order to identify the BD gene lies between HLA-B*5101 and TAP1, or even beyond. There was found to be no association of BD with LMP7 and TAP2 loci, which are centromeric to TAP1 (Gonzalez-Escribano et al., 1995; Ishihara et al., 1996). These findings can also explain the association of BD with class II genes which has been reported by several groups.

Other HLA alleles in association with Behçet's disease

Golosenko (1990) found that HLA-Cw2 was elevated in the group of BD patients that also typed as HLA-B51. Ando et al. (1997) observed a novel allele, HLA-B*0402, to be significantly elevated in a Saudi Arabian family with BD. Hamza et al. (1988) reported that in a group of Tunisian BD patients (several suffering from uveitis) there was a significant excess of HLA-DQ3. A substantial increase in DR2, DR4 and DR7 was also observed. Another study where HLA class II genes were found in a significant association with BD was reported by Mizuki et al. (1992b) and Castillo et al. (1996), who found that, aside from the presence of HLA-B*0501 in their group of BD patients, PCR genotyping revealed a significant increase in the DRB1*0802, DQA1*0301 and DQB1*0303 alleles. Sun et al. (1993) found a marginally significant increase in DR6 and DR8 antigens in BD patients. Jankowski et al. (1992) reported that a group of Celtic Caucasian BD patients who went on to develop gastrointestinal involvement had the DR4 or the DR7 antigen. Affected males were younger than affected females.

HLA alleles significantly decreased in Behçet's disease

The frequencies of the following haplotypes were significantly reduced in Japanese BD patients compared with controls: HLA-A11, -A33, -B35, -B44, -DQ1, DRB1*1502, DQA1*0103, DQA1*0101, DQB1*0601, DQB1*0501 and the alleles HLA-Cw*0304 and -Cw*01 (Mizuki et al., 1991, 1992b, 1996). Hamza et al. (1990) reported a deficiency of HLA-DR6 and DQ1 in BD patients. DQ5 was also found to be negatively associated with BD, particularly in B51-positive patients (Castillo et al., 1996). The positive and negative associations of class II alleles with the disease can be explained by linkage disequilibrium with HLA-B*5101 or another gene located close to this gene locus. It is also possible that BD is a symptom complex associated with some independent disease.

Association of the HLA-B*5101 allele with different manifestations of Behçet's disease

There are many distinct differences between the clinical manifestations of BD in different ethnic groups and geographical locations. Genetic and environmental factors and severity of the disease are among the parameters involved. The severity and degree of systemic involvement in BD depends on the presence of HLA-B*5101, as a significant correlation was observed between neutrophil hyperfunction and the possession of HLA-B51 phenotype, regardless of the presence of the disease (reviewed by Sakane, 1997). In addition, Azizleri et al. (1994) reported that genital manifestations were significantly higher in BD patients bearing HLA-B*5101 than in those who did not, and that patients with thrombophlebitis expressed HLA-B*5101 less frequently than those without thrombophlebitis. Mizuki et al. (1992a) observed a positive association of HLA-B*5101 with ocular involvement. A gender-related influence of the HLA-B51 antigen was reported by Villanueva et al. (1993), who observed that in a family with BD, in which all siblings were B*5101 positive, only the female siblings, with a positive identical HLA phenotype, A2, A11, B51 (B*0501), B44, Cw6, Cw5, DR4, DR13, DR53, DRw52, DQ7, DQ6, developed the disease symptoms, whereas none of the male siblings was affected. These results suggest that the HLA-B*5101 allele affects BD development as well as clinical symptoms.

Search for the Behçet's disease antigen

Investigation of the aetiology of Behçet's disease has focused predominantly on herpes simplex virus and the streptococci. Aetiological factors might have a common denominator in microbial HSP which shows significant homology with the human mitochondrial HSP. Indeed, the uncommon serotypes of *Streptococcus sanguis* found in BD cross-react with the 65 000 mol. wt HSP which also shares antigenicity with an oral mucosal antigen. T cell epitope mapping has identified four peptides derived from the sequence of the 65 000 mol. wt HSP which stimulate specifically T $\gamma\delta^+$ lymphocytes from patients with BD (reviewed in Lehner, 1997). An earlier report (Hirohata, 1992) describes the stimulation by streptococcal antigens of IL-6 and interferon-γ (IFNγ) production by T cells from BD patients. These data suggest that T cell hypersensitivity to viral or bacterial antigens may play a central role in the pathogenesis of Behçet's disease.

T cells and HLA-B51 antigens in Behçet's disease

HLA-B51 molecules themselves may be responsible, at least in part, for the neutrophil hyperfunction in Behçet's disease (Sakane, 1997). It has been shown that HSP is involved in T cell activation in patients with BD, but not in healthy subjects. T cells expressing certain TCR Vβ subfamilies were increased in the circulation and responded to the HSP peptide. The oligoclonal nature of these TCR suggests that T cells in patients with BD have already been expanded oligoclonally *in vivo* (Kaneko et al., 1997). T cells bearing $\gamma\delta$ TCR chains also play an important role in mucocutaneous immunity. A phenotypically distinct and minor subset of such T cells, CD45RA$^+$ CD45RO$^-$ Vγ9$^+$ Vδ2$^+$, in healthy individuals, was found to increase in number in BD, irrespective of disease activity. This T cell subset produced IL-2Rβ and TNF and contained perforin granules during active disease. Analysis of the cytokines expressed by active T cells in BD reveals that they belong to the Th1 subfamily. Therefore, these T cells, carrying $\alpha\beta$ or $\gamma\delta$ TCR, are thought to contribute to immunological abnormalities which may lead to the complexity of pathophysiology in BD (Sugi et al., 1997).

Diagnosis of Behçet's disease

The study of the HLA antigens in BD can provide useful diagnostic and prognostic information and aid in analysing the different clinical forms of the disease. Certain peptides that specifically stimulate TCR $\gamma\delta^+$ lymphocytes (Sugi et al., 1997) and molecular analysis of polymorphism of the MICA gene-associated microsatellite, Tau-a microsatellite and TAP1 polymorphism may have diagnostic value, especially in BD families.

DERMATITIS HERPETIFORMIS

Dermatitis herpetiformis (DH) is a pruritic, blistering skin disease, characterized in part by the presence of pathognomonic patterns of granular deposits of IgA at the dermal–epidermal junction, an associated gluten-sensitive enteropathy and a strong association (95–100%) with the HLA-DQ2 antigen (Hall and Otley, 1991; Smith and Zone, 1993). Although the precise role of this antigen in the pathogenesis of DH is unclear, one theory proposes that patients with DH possess a unique molecular subtype of DQ2 antigen, which causes immune abnormalities leading to the clinical manifestations of DH (Hall and Otley, 1991; Demoulins-Giacco et al., 1996). Familial DH occurs much less frequently than CD, but cases have been reported in the literature with females affected more than males (Wilson et al., 1995; Reunala et al., 1996). T cell receptor Vβ expression is restricted in DH skin (Garioch et al., 1997). The gliadin peptide and a 97 000 mol. wt protein present in the skin are implicated in DH, and enhanced expression of interstitial collagenase, stromelysin-1 and urokinase plasminogen activator was observed in DH lesions and implicated in the pathogenic process (Airola et al., 1995). It is important to note that DH has a very similar, although not identical, HLA association profile to that of coeliac disease, and displays other common disease characteristics (Garioch et al., 1997). (HLA associations with coeliac disease and DH are discussed in greater detail in Chapter 22.)

HLA associations with dermatitis herpetiformis

DH displays an extremely strong association with HLA-DQ alleles. The majority of the patients in many ethnic groups shows an association with the HLA-DQ2 molecule composed of the product of DQA1*0501 and DQB1*02 alleles. Other DH patients carry another combination: the heterodimer formed by the products of DQA1*03 and DQB1*0302 (Balas et al., 1997; Spurkland et al., 1997). DH patients usually carry the DQA1*0501 and DQB1*02 alleles in *cis* or *trans*. Comparison of observed and expected DQA1–DQB1 genotype distributions in DH patients showed an excess of patients with DQB1*02 on both chromosomes in *cis* or *trans*. This is the consequence of this allele being able to complement DQA1*0501 (Balas et al., 1997). Sequence polymorphism in the DQ2 β-chain at position 135 was shown to distinguish the DR3 and DR7 haplotypes. This polymorphism was present in CD, DH patients, or normal controls of the appropriate DR and DQ genotypes. This DQ2 second-domain polymorphism may explain increased *trans*-associated risk in DH and CD (Hall et al., 1993). Other associations of DH with HLA and other genes were also found: association-specific polymorphisms of HLA-DRB1*0301 and DPB1*0101 have been reported (Hall et al., 1996). In addition, association of DH with extended HLA haplotypes indicates that HLA genes alone may not be the only predisposing genes, but other genes in the HLA region (e.g. TAP2 gene, located between DP and DQ, and the TNF gene, located between HLA-B and DRB), may also be responsible for DH susceptibility. For example, the extended haplotype HLA-A1–B8–DR3–DQ2 was frequently associated with DH, although the association with DQ2 was the strongest (Wilson et al., 1995). It is noteworthy that homozygosity for DQ2 is significantly increased in the CD compared with DH and so differences in dosage of class II genotypes between DH and CD may account for the milder gastrointestinal symptoms characteristic of DH (Hall et al., 1996).

LICHEN PLANUS

Oral, erosive oral and cutaneous lichen planus

Lichen planus (LP), a mucocutaneous syndrome (dermatosis) of uncertain pathogenesis, is heterogeneous and may represent a variety of skin conditions. Oral (OLP), erosive oral (ELP) and cutaneous lichen planus (CLP) are better defined LP conditions that are considered premalignant by some (Zhang et al., 1997). Hepatitis C virus infection, common in LP, is assumed to induce host factors and enhances or triggers the disease (Nagao et al., 1997). The disease is believed to develop as a result of the stimulation by a foreign antigen of LC and keratinocytes followed by the secretion of cytokines, and the induction or upregulation of expression of MHC class II (DR) and adhesion molecule expression. This is associated with the activation of T helper cells that inflict cellular damage (Hedberg and Hunter, 1987; Andre et al., 1990; McCartan and Lamey, 1997). The disease in most clinical forms has a definite HLA association. However, these associations are with different HLA alleles in the different ethnic groups studied to date. The data in some reports are several years old and based only on serological typing. The associations reported here are, in all cases, statistically significant, with RR values of 3–21.

HLA association with lichen planus

Arab patients with CLP show an increase in HLA-DR1 and DR10, the latter being more significant.

A significant decrease in DR5 was also reported (White and Rostom, 1994). In British OLP patients, HLA-B57 was significantly increased and DQ1 decreased (Porter et al., 1993). Another study of European ELP and CLP patients showed increased frequency of DR1 and DQ1 alleles and decreased DQ3. Among the DR1-CLP patients, 78.5% carried the DRB1*0101 allele and 21.4% DRB1*0102, compared with 35.7 and 67.8%, respectively, of controls (Kuwata et al., 1995). In Chinese OLP and ELP patients, a highly significant increase in DR9 was noted (Lin and Sun, 1990). In Israeli Jewish ELP patients, a significant association with DR2 and a decrease in DR4 were noted. Genotyping of the HLA-DR2 alleles showed the presence of all three common variants (DRB1*1501, DRB1*1502 and DRB1*1601) in these patients (Roitberg-Tambur et al., 1994). In Italian OLP patients, serological typing revealed a highly significant increase in DR1 antigen (Valsecchi et al., 1988). In Japanese OLP patients, a significant association with DR9 (carried on B61–Cw3–DR9 haplotypes) and a decreasing trend in the frequency of HLA-B52 was noted (Watanabe et al., 1986). In a Sardinian CLP group, the allele DRB1*0101 was again found to be significantly associated with CLP. DRB1*0101 and DRB1*0102 are associated with the same DQA1 and DQB1 alleles and differ only at two amino acids in positions 85 and 86 of the DRβ-chain. In the Sardinian CLP patients, predisposition to CLP seemed to be correlated with a valine in position DRβ85 and a glycine in position DRβ86 (Carcassi et al., 1994). In Swedish ELP and OLP patients, HLA-DR3 was very significantly increased, and in CLP the A1–B8–DR3 haplotype was found more frequently (Jontell et al., 1987). The heterogeneous HLA association with LP may indicate that other candidate genes within the HLA region between HLA-B and DQ genes are responsible. Further analysis of the alleles involved in LP is needed to understand better its HLA associations.

VITILIGO

Vitiligo is a common skin disease, manifested by patches of depigmentation, resulting from the destruction and disappearance of melanocytes. The disease occurs in all races and is often familial. It has been proposed that three epistatically interacting autosomal diallelic loci are involved in the pathogenesis of vitiligo and that, hence, affected individuals are recessive homozygotes at each of the three loci (Lerner, 1959; Majumder et al., 1993). Several reports indicating an HLA association with vitiligo and the observation that patients with vitiligo have circulating antibodies directed in part to pigment cell antigens (Cui et al., 1995) suggest that autoimmune mechanisms are involved and specific HLA phenotypes may influence the expression of vitiligo. However, the HLA association data are not consistent and, in some cases, only marginally significant. Future studies are needed to decide which HLA alleles or other genes located within the HLA region are involved (Tiwari and Terasaki, 1985; Schallreuter et al., 1993; Buc et al., 1996).

HLA associations with vtiligo

Report HLA associations with vitiligo are listed in Table 22.1 according to chronological date of the reports. Although large numbers of patients and families with vitiligo have been studied, there is very little consistency in the HLA associations observed. It is difficult to find a common denominator for all HLA molecules reported here. Most probably, one of the genes predisposing to vitiligo is located between the HLA-A gene and the DQ locus. It is tempting to predict a location near the C4AQ0 complement locus.

ATOPIC DERMATITIS

The subject of the allergic causation of atopic dermatitis (AD) has been a source of controversy since the 1930s. Atopic dermatitis is characterized by increased production of IgE and IL-4, increased expression of CD23 on mononuclear cells, elevated cyclic AMP-phosphodiesterase activity and decreased production of IFNγ. The most consistent causative agents of atopic conditions are irritants (Hanifin, 1997; Moore et al., 1997).

TABLE 22.1 HLA associations with vitiligo

HLA association	Reference
B13 in young Jewish Moroccan patients; B35 in young Jewish Yemenite patients, who were all B13 negative	Metzker et al. (1980)
DR4 in White American patients	Foley et al. (1983)
DR4, DQ3 in American Blacks; both alleles segregated with positive family history and early onset. DR6 in patients with negative family history and late onset	Dunston and Halder (1990)
DR1 and DR3 in Hungarian patients	Poloy et al. (1991)
BfS, C4A3, C4B1, DR5 (DR11), DQ3 characteristic of the paediatric form of vitiligo and -BfS, C4A3, C4B1, DR7, DQ2 in the adult form of the disease with a significant decrease in DR1 and DR3	Finco et al. (1991)
A30, B27, Cw6 in Italian patients; significant decrease in C4AQ0	Orecchia et al. (1992)
A3 and Gm (3, 25, 5, 10, 11, 13, 14) in Italian patients with marked decrease in C4AQ0	Lorini et al. (1992); Venneker et al. (1993)
A2 (marginally) in German patients and B60 only in adult patients	Schallreuter et al. (1993)
B46 in Japanese familial vitiligo patients and A31 (part of A19 complex), Cw4 in non-familial patients	Ando et al. (1993) Venneker et al. (1993)
Cw6 and DR6 in Duch vitiligo patients	
B21, Cw6 and DR53 in Kuwaiti patients; A19, DR52 significantly decreased	al-Fouzan et al. (1995)
Bw6, DR7 in native Omani patients; DR7 increased significantly in patients with acrofacial disease. DR4 segregated 100% with patients with a positive family history	Venkataram et al. (1995)
A2 and Dw7 in Slovak patients	Buc et al. (1996)

HLA associations with atopic dermatitis

In the past, several class I and II alleles have been associated with AD, but later found not to be significant (Svejgaard et al., 1985). However, a relatively recent study found more striking associations. In AD patients with high serum IgE levels, frequencies of DRB1*1302 and DQB1*0604 were increased in a Japanese population, whereas the frequency of DQB1*0302 was decreased. Analysis of the amino acid epitopes on the HLA-DRβ1 and DQβ1 domains revealed a significant association of AD with glutamic acid at position DRβ71 (RR = 5.71), histidine at DQβ30 (RR = 3.25) and valine at DQβ57 (RR = 3.13). HLA-DM and TAP were not found to be associated with AD (Juji and Shibata, 1995; Saeki et al., 1995; Kuwata et al., 1996).

OTHER DERMATOLOGICAL DISEASES ASSOCIATED WITH THE HLA SYSTEM

Many other diseases which are purely dermatological or have a skin manifestation have been associated with the HLA system. However, the information regarding their associations with HLA is not significant enough or not well investigated. Other diseases that have been associated in the past with HLA have subsequently been shown to have no positive correlation. For completeness, several dermatological diseases which are not described in detail and their putative HLA associations are listed in Table 22.2.

TABLE 22.2 Summary of main HLA allele associations with dermatological diseases

Disease	HLA class I	HLA class II	Other	Significance
Acne conglobata	B7 Creg group			LS
Actinic prurigo		DR4 (DRB1*0407)		HS
	A28			HS
	B39			S
Alopecia aerata		DQB1*0301, *0303		HS
		P at DQβ55		HS
		DRB1*1104		LS
		DQB1*06 (D)		LS
		DR52a (D)		LS
Anti-Ro(SS-A)+ patients		DR52a		HS
		DR3E at DQα34		S
		L at DQβ26		
Atopic dermatitis		DRB1*1302		S
		E at DRβ71		
		DQB1*0604		S
		H at DQβ30, V at DQβ57		S
Atopy allergy		DRB1*07 bee		S
		DRB1*11 wasp		S
		DR4, DR7 asthmatic patients		S
	B16 lab. animal	DR4 lab. animals		
Behçet's disease	B*5101			HS
	B*0402 (Arabs)			HS
	MICA			HS
		DRB1*0802,		S
		DQA1*0303, DQB1*0303		S
			Tau-a	S
			TAP2B	S
			TAP1C (D)	HS
	A11,33(D)			LS
	B35,44(D)			LS
	Cw*01,*0304(D)			S
		DQB1*0501, *0601		LS
		DR6, DRB1*1502(D)		S
		DQA1*0103,*0101(D)		S
Circatrical pemphigoid		DQB1*0301		S
Dermatitis herpetiformis		DQw2		HS
		DQA1*0501–DQB1*02 dimer		
		DQA1*03–DQB1*0302 dimer		
Dermatographism	A2, B16,			LS
familial urticarias	A1, B5			LS
Epidermal necrolysis	A29, B12	DR7		LS
Epidermolysis bullosa		DR2 (DRB1*1501), DR5		S
		DRB1*13		S
Felty's syndrome		Haplotype: B44–Bf*S–C4A*3–C4B*Q0–DR4–DQ7		S
			TAP2D	S
Geographic tongue		DR5, DRW6, DR2(D)		S
Granuloma anulare	A31, 29, B35			LS
Hailey–Hailey's disease	B16			S
Kawasaki syndrome		HLA-DRB3*0301		S
Keloid	B21			S
		DR5		S
		DQ3		HS
Lichen planus		DR1 (DRB1*0101, *0102)		HS
		V at DRβ85, G at DRβ86		
		DR2		S

TABLE 22.2 Continued.

Disease	HLA class I	HLA class II	Other	Significance
		DR2		S
		DQ1		S
		DQ3(D)		S
Lichen sclerosus	A29, B44	DQ7P at DQβ55		S
Pemphigus foliacceus		DRB1*0404		HS
		DRB1*1402, DRB1*1406		HS
		DRB1*0102		
		Alleles share identical epitope: LLEQRRAA at positions DRβ67–74		
		DQ2 (D)		HS
Pemphigus gestationis		DRB1*301		HS
		DRB1*0401, *040x		S
			C4*Q0	HS
Pemphigus vulgaris		DRB1*0402		HS
		D at DRβ70, E at DRβ71		
		DQB1*0503		HS
		D at DQβ57		
		DR3(D)		HS
Psoriasis vulgaris	Cw*0602			HS
Type I	D at Cα9			
	A at Cα73			
	B13, B57			S
		DRB1*0701/02		S
		DQA1*0201, DQB*0303		S
Type II	Cw2			S
		DQB1*0303(D)		S
Psoriasis arthritis		HLA-A*0207		HS
Stevens–Johnson (SJS)	B12			S
Sweet's syndrome	A19			S
	A40(D)			S
Urticaria		DR4		LS
Vitiligo	Many	Many		LS
			C4Q0(D)	HS

S: Significant [risk ratio (RR) > 2]; HS: highly significant (RR > 10); LS: low significance (RR 1–2); (D): decreased alleles (RR < 0.3).

References for diseases described only in Table 22.2: *acne conglobata*: Vasey et al. (1984), Kurc et al. (1987), Rosner et al. (1993); *actinic prurigo*: Menage et al. (1996), Dawe et al. (1997), Hojyo-Tomoka et al. (1997); *alopecia areata*: Duvic et al. (1991, 1995), Welsh et al. (1994), McDonagh and Messenger (1996); *anti-Ro (SS-A)*: Provost and Watson (1993), Miyagawa et al. (1996); *atopic dermatitis*: Aron et al. (1996), Holloway et al. (1996), Oxelius et al. (1996), Sjostedt et al. (1996), Coleman et al. (1997), Faux et al. (1997); *cicatricial pemphigoid*: Chan et al. (1994, 1997); *epidemolysisi bullosa*: Gammon et al. (1988, 1992), Lee et al. (1996); *epidermal necrolysis*: Roujeau et al. (1986, 1987), Mockenhapt (1997); *Felty's syndrome*: Thomson et al. (1988), Clarkson et al. (1990), Hillarby et al. (1996); *geographic tongue*: Fenerli et al. (1993); *granuloma annulare*: Middleton and Allen (1984), Friedman-Birnbaum et al. (1986); *Hailey–Hailey's disease*: Karvonen and Tiilikainen (1976), Malchus et al. (1986), Richard et al. (1990); *Kawasaki syndrome*: Barron et al. (1992), Fildes et al. (1992); *keloid*: Laurentaci and Dioguardi (1977), Rossi and Bozzi (1989); *lichen sclerosus*: Purcell et al. (1990), Marren et al. (1994); *psoriatic arthritis*: Muto et al. (1996); *Stevens–Johnson syndrome*: Roujeau et al. (1986, 1987); *Sweet's syndrome*: Mizoguchi (1996), von den Driesch et al. (1997); *urticaria*: Haas et al. (1996), Lipski et al. (1996), Sabroe and Greaves (1997).

ACKNOWLEDGEMENTS

The author thanks Dr Carmelit Richler and Ms Rivaka Golan for their help and support in gathering the thousands of reprints and in the preparation of the manuscript.

REFERENCES

Ahmed, A.R. (1968). Pemphigus vulgaris. *Dermatology* **1**: 1–17.

Ahmed, A.R., Yunis, E.J., Khatari, K. et al. (1990). *Proc. Natl Acad. Sci. USA* **87**: 7658–7662.

Ahmed, A.R., Mohimen, A., Yunis, E.J. et al. (1993). Linkage of pemphigus vulgaris antibody to major histocompatibility complex in healthy relatives of patients. *J. Exp. Med.* **177**: 419–424.

Airola, K., Vaalamo, M., Reunala, T. and Saarialho-Ker, U.K. (1995). Enhanced expression of interstitial collagenase, stromelysin-1, and urokinase plasminogen activator in lesions of dermatitis herpetiformis. *J. Invest. Dermatol.* **105**: 184–189.

Alper, C.A. (1990). Major histocompatibility complex haplotype studies in Ashkenazi Jewish patients with pemphigus vulgaris. *Proc. Natl Acad. Sci. USA* **87**: 7658–7662.

Amar, A., Rubinstein, S., Hacham-Zadeh, S. et al. (1984). Is predisposition to pemphigus vulgaris in Jewish patients mediated by HLA-Dw10 and DR4? *Tissue Antigens* **23**: 17–22.

Ando, I., Chi, H.I., Nakagawa, H., and Otsuka, F. (1993). Difference in clinical features and HLA antigens between familial and non-familial vitiligo of non-segmental type. *Br. J. Dermatol.* **129**: 408–410.

Ando, H., Mizuki, N., Ohno, S. et al. (1997). Identification of a novel HLA-B allele (B*4202) in a Saudi Arabian family with Behcet's disease. *Tissue Antigens* **49**: 526–528.

Andre, J., Laporte, M. and Delavault, P. (1990). Lichen planus: etiopathogenesis. *Acta Stomatol. Belg.* **87**: 229–231.

Arber, N., Klein, T., Meiner, Z. et al. (1990). Close association of HLA-B51 and B52 in Israeli Behcet patients. *Vestn. Dermatol. Venerol.* **5**: 55–57.

Aron, Y., Desmazes-Dufeu, N., Matran, R. et al. (1996). Evidence of a strong, positive association between atopy and the HLA class II alleles DR4 and DR7. *Clin. Exp. Allergy* **26**: 821–828.

Asahina, A., Kuwata, S., Tokunaga, K. et al. (1996). Study of aspartate at residue 9 of HLA-C molecules in Japanese patients with psoriasis vulgaris. *J. Dermatol. Sci.* **13**: 125–133.

Azizleri, G., Aksungur, V.L., Sarica, R. et al. (1994). The association of HLA-B5 antigen with specific manifestations of Behçet's disease. *Dermatology* **188**: 293–295.

Balas, A., Vicario, J.L., Zambrano, A. et al. (1997). Absolute linkage of celiac disease and dermatitis herpetiformis to HLA-DQ. *Tissue Antigens* **50**: 52–56.

Balboni, A., Pivetti-Pezzi, P., Orlando, P. et al. (1992). Serological and molecular HLA typing in Italian Behçet's patients: significant association to B51 DR5-DQw3 haplotype. *Tissue Antigens* **39**: 141–143.

Barra, C., Belfort Junior, R., Abreu, M.T. et al. (1991). Behcet's disease in Brazil – a review of 49 cases with emphasis on ophthalmic manifestations. *Jpn J. Ophthalmol.* **35**: 339–346.

Barron, K.S., Silverman, E.D., Gonzales, J.C. et al. (1992). Major histocompatibility complex class II alleles in Kawasaki syndrome lack of consistent correlation with disease or cardiac involvement. *J. Rheumatol.* **19**: 1790–1793.

Beutner, E.H. and Jordan, R.E. (1964). Demonstartion of skin antibodies in sera of pemphigus vulgaris patients by indirect immunofluorescent staining. *Proc. Soc. Exp. Biol. Med.* **117**: 505–510.

Bjorkman, P.J., Saper, M.A., Samraoui, B. et al. (1987). Structure of the human class I histocompatibility antigen, HLA-A2. *Nature* **329**: 506–512.

Bohl, K., Natarajan, K., Nagarwalla, N. et al. (1995). Correlation of peptide specificity and IgG subclass with pathogenic and nonpathogenic autoantibodies in pemphigus vulgaris: A model for autoimmunity. *Proc. Natl Acad. Sci. USA* **92**: 5239–5243.

Bohl, K., Yunis, J. and Ahmed, A.R. (1996). Pemphigus vulgaris in distant relatives of two families: association with major histocompatibility complex class II genes. *Clin. Exp. Dermatol.* **21**: 100–103.

Bowcock, A.M. (1995). Genetic locus for psoriasis identified. *Ann. Med.* **27**: 183–186.

Brautbar, C., Chajek, T., Ben-Tuvia, S. et al. (1978). A genetic study of Behcet's disease in Israel. *Tissue Antigens* **11**: 113–120.

Breathnach, S.M. (1993). The skin immune system and psoriasis. *Clin. Exp. Immunol.* **91**: 343–345.

Brenner, S., Dorfman, B. and Himelfarb, M. (1985). Familial pemphigus vulgaris. *Dermatologica* **171**: 929–932.

Brenner, W., Gschnait, F. and Mayr, W.R. (1978). HLA B13, B17, B37, and Cw6 in psoriasis vulgaris. *Arch. Dermatol. Res.* **262**: 337–339.

Brenner, S., Hodak, E., Dascalu, D. et al. (1990). A possible case of drug-induced pemphigus. *Acta Derm. Venereol.* **70**: 357–358.

Buc, M., Busova, B., Hegyi, E. and Kolibasova, K. (1996). Vitiligo is associated with HLA-A2 and HLA-Dw7 in the Slovak populations. *Folia Biol.* **42**: 23–25.

Bunce, M. and Welsh, K.I. (1994). Rapid DNA typing for HLA-C using sequence-specific primers (PCR-SSP): identification of serological and non-serologically defined HLA-C alleles including several new alleles. *Tissue Antigens* **43**: 7–17.

Burch, P.R. and Rowell, N.R. (1965). Psoriasis: aetiological aspects. *Acta Derm. Venerol.* **45**: 366–380.

Burns, A.P., Fisher, M., Li, P. et al. (1995). Molecular analysis of HLA class II genes in Goodpasture's disease. *QJM* **88**: 93–100.

Callot, V., and Bagot, M. (1994). *Pemphigus. Rev. Prat.* **44**: 81–88.

Carcassi, C., Cottoni, F., Floris, L. et al. (1994). The HLA-DRB1*0101 allele is responsible for HLA susceptibility to lichen ruber planus. *Eur. J. Immunogenet.* **21**: 425–429.

Carcassi, C., Cottoni, F., Floris, L. et al. (1996). HLA haplotypes and class II molecular alleles in Sardinian and Italian patients with pemphigus vulgaris. *Tissue Antigens* **48**: 662–667.

Castillo, M.J., Palma, M.J., Sanchez-Roman, J. et al. (1996). Serologic and molecular HLA typing in patients from Andalucia with Behcet's disease. Genetic and clinical correlations. *Med. Clin.* **106**: 121–125.

Cerena, M., Fernandez-Vina, M., Friedman H.C. et al. (1993). Genetic markers for susceptibly to endemic Brazilian pemphigus foliaceus (Fogo Salvagem) in Xavante Indians. *Tissue Antigens* **42**: 138-140.

Chan, L.S., Wang, T., Wang, X.S. et al. (1994). High frequency of HLA-DQB1*0301 allele in patients with pure ocular cicatricial pemphigoid. *Dermatology* **189**(Suppl. 1): 99–101.

Chan, L.S., Hammerberg, C. and Cooper, K.D. (1997). Significantly increased occurrence of HLA-DQB1*0301 allele in patients with ocular cicatricial pemphigoid. *J. Invest. Dermatol.* **108**: 129–132.

Chen, K.R., Kawahara, Y., Miyakawa, S. and Nishikawa, T. (1997). Cutaneous vasculitis in Behçet's disease: a clinical and histopathologic study of 20 patients. *J. Am. Acad. Dermatol.* **36**: 689–696.

Christophers, E. (1996). The immunopathology of psoriasis. *Int. Arch. Allergy* **110**: 199–206.

Clarkson, R., Bate, A.S., Grennan, D.M. et al. (1990). DQw7 and the C4B null allele in rheumatoid arthritis and Felty's syndrome. *Ann. Rheum. Dis.* **49**: 976–979.

Coleman, R., Trembath, R.C. and Harper, J.I. (1997). Genetic studies of atopy and atopic dermatitis. *Br. J. Dermatol.* **136**: 1–5.

Cui, J., Arita, Y. and Bystryn, J.C. (1995). Characterization of vitiligo antigens. *Pigment Cell Res.* **8**: 53–59.

Dawe, R.S., Collins, P., Ferguson, J. and O'Sullivan, A. (1997). Actinic prurigo and HLA-DR4. *Invest. Dermatol.* **108**: 233–234.

Degos, T., Touraine, R. and Blaich, S. (1969). Pemphigus chez un malade traite par penicillamine pour maladie de Wilson. *Bull. Soc. Fr. Dermatol. Syphilogr.* **76**: 751–753.

Delgado, J.C., Yunis, D.E., Bozon, M.V. et al. (1996). MHC class II alleles and haplotypes in patients with pemphigus vulgaris from India. *Tissue Antigens* **48**: 668–672.

Demoulins-Giacco, N., Gagey, V., Teillac-Hamel, D. et al. (1996). Dermatitis herpetiformis occurring in patients with celiac disease in childhood. *Arch. Pediatr.* **3**: 541–548.

Diaz, L.A., Sampaio, S.A., Rivitti, E.A. et al. (1989). Endemic pemphigus foliaceus (fogo selvagem). I. Clinical features and immunopathology. *J. Am. Acad. Dermatol.* **20**: 657–669.

Driesch, P. von den, Simon, M., Jr, Djawari, D. and Wassmuth, R. (1997). Analysis of HLA antigens in Caucasian patients with acute febrile neutrophilic dermatosis (Sweet's syndrome). *J. Am. Acad. Dermatol.* **37**: 276–278.

Dunckley, H., Chapman, J.R., Burke, J. et al. (1991). HLA-DR and -DQ genotyping in anti-GBM disease. *Dis. Markers* **9**: 249–256.

Dunston, G.M. and Halder, R.M. (1990). Vitiligo is associated with HLA-DR4 in black patients. A preliminary report. *Hum. Arch. Dermatol.* **126**: 56–60.

Duvic, M., Hordinsky, M.K., Fiedler, V.C. et al. (1991). HLA-D locus associations in alopecia areata. DRw52a may confer disease resistance. *Arch. Dermatol.* **127**: 64–68.

Duvic, M., Welsh, E.A., Jackow, C., Papadopoulos, E. et al. (1995). Analysis of HLA-D locus alleles in alopecia areata patients and families. *J. Invest. Dermatol.* **104**: 5S–6S.

Enerback, C., Martinsson, T., Inerot, A. et al. (1997). Evidence that HLA-Cw6 determines early onset of psoriasis, obtained using sequence-specific primers (PCR-SSP). *Acta Derm. Venerol.* **77**: 273–276.

Eyre, R.W. and Stanley, J.R. (1987). Human autoantibodies against a desmosomal protein complex with a calcium-sensitive epitope are characteristic of pemphigus foliaceus patients. *J. Exp. Med.* **165**: 1719–1724.

Farber, E.M. and Nall, L. (1991). Epidemiology: natural history and genetics. In *Psoriasis*, 2nd edn (eds H.H. Roenigk and H.I. Maibach), pp. 209–258, Marcel Dekker, New York.

Faux, J.A., Moffatt, M.F., Lalvani, A. et al. (1997). Sensitivity to bee and wasp venoms: association with specific IgE responses to the bee and wasp venom and HLA DRB1 and DPB1. *Clin. Exp. Allergy* **27**: 578–583.

Feinstein, A., Yorav, S., Movshovitz, M. and Schewach-Millet, M. (1991). Pemphigus in families. *Int. J. Dermatol.* **30**: 347–351.

Fenerli, A., Papanicolaou, S., Papanicolaou, M. and Laskaris, G. (1993). Histocompatibility antigens and geographic tongue. *Oral. Surg. Oral. Med. Oral. Pathol.* **76**: 476–479.

Fildes, N., Burns, J.C., Newburger, J.W. et al. (1992). The HLA class II region and susceptibility to Kawasaki disease. *Tissue Antigens* **39**: 99–101.

Finco, O., Cuccia, M., Martinetti, M. et al. (1991). Age of onset in vitiligo: relationship with HLA supratypes. *Clin. Genet.* **39**: 48–54.

Foley, L.M., Lowe, N.J., Misheloff, E. and Tiwari, J.L. (1983). Association of HLA-DR4 with vitiligo. *J. Am. Acad. Dermatol.* **8**: 39–40.

al-Fouzan, A., al-Arbash, M., Fouad, F. et al. (1995). Study of HLA class I/IL and T lymphocyte subsets in Kuwaiti vitiligo patients. *Eur. J. Immunogenet.* **22**: 209–213.

Friedman-Birnbaum, R., Gideoni, O., Bergman, R. and Pollack, S. (1986). A study of HLA antigen association in localized and generalized granuloma annulare. *Br. J. Dermatol.* **115**: 329–333.

Friedmann, A. and Brautbar, C. (1993). Immune response genes and autoimmune diseases. *Israel J. Med. Sci.* **29**: 869–874.

Gammon, W.R. and Woodley, D.T. (1992). HLA association in patients with EBA: evidence that autoimmunity to type VII collagen may be associated with independent risk effects in separate gene pools. *J. Invest. Dermatol.* **93**: 589A.

Gammon, W.R., Heise, E.R., Burke, W.A. et al. (1988). Increased frequency of HLA-DR2 in patients with autoantibodies to epidermolysis bullosa acquisita antigen: evidence that the expression of autoimmunity to type VII collagen is HLA class II allele associated. *J. Invest. Dermatol.* **91**: 228–232.

Garioch, J.J., Baker, B.S., Leonard, J.N. and Fry, L. (1997). T-cell receptor V beta expression is restricted in dermatitis herpetiformis skin. *Acta Derm. Venereol.* **77**: 184–186.

Golosenko, I. (1990). The relationship of the HLA antigen system to Behcet's disease. *Vestn. Dermatol. Venerol.* **5**: 55–57.

Gonzalez-Escribano, M.F., Morales, J., Garcia-Lozano, J.R. et al. (1995). TAP polymorphism in patients with Behçet's disease. *Ann. Rheum. Dis.* **54**: 386–388.

Grabski, J. and Grabski, J., Jr (1990). Clinical aspects of Behçet's syndrome. *Pol. Tyg. Lek.* **45**: 520–522.

Haas, N., Iwen, W., Grabbe, J. et al. (1996). MHC class II antigen expression is increased in different forms of urticaria. *Int. Arch. Allergy Immunol.* **109**: 177–182.

Hadfield, M.G., Aydin, F., Lippman, H.R. and Sanders, K.M. (1997). Neuro-Behçet's disease. *Clin. Neuropathol.* **16**: 55–60.

Hall, R.P. III and Otley, C. (1991). Immunogenetics of dermatitis herpetiformis. *Semin. Dermatol.* **10**: 240–245.

Hall, M.A., Lanchbury, J.S., Lee, J.S. et al. (1993). HLA-DQ2 second-domain polymorphisms may explain increased trans-associated risk in celiac disease and dermatitis herpetiformis. *Hum. Immunol.* **38**: 284–292.

Hall, M.A., Lanchbury, J.S. and Ciclitira, P.J. (1996). HLA class II region genes and susceptibility to dermatitis herpetiformis: DPB1 and TAP2 associations are secondary to those of the DQ subregion. *Eur. J. Immunogenet.* **23**: 285–296.

Hama, M., Ayed, K., Hamzaoui, K. and Bardi, R. (1988). Behçet's disease and major histocompatibility complex class II antigens in Tunisians. *Dis. Markers* **6**: 263–267.

Hamza, M., Ayed, K., Bardi, R. et al. (1990). Behçet's disease and class II and III antigens of the major histocompatibility complex. *Rev. Rheum. Mal. Osteoartic.* **57**: 59–61.

Hanifin, J.M. (1997). Atopic dermatitis: critical evaluation of food and mite allergy in the management of atopic dermatitis. *J. Dermatol.* **24**: 495–503.

Hedberg, N.M. and Hunter, N. (1987). The expression of HLA-DR on keratinocytes in oral lichen planus. *J. Oral. Pathol.* **16**: 31–35.

Henseler, T. and Christophers, E. (1985). Psoriasis of early and late onset: characterization of two types of psoriasis vulgaris. *J. Am. Acad. Dermatol.* **13**: 450–456.

Henseler, T. (1997). The genetics of psoriasis. *J. Am. Acad. Dermatol.* **37**: S1–S11.

Hillarby, M.C., Davies, E.J., Donn, R.P. et al. (1996). TAP2D is associated with HLA-B44 and DR4 and may contribute to rheumatoid arthritis and Felty's syndrome susceptibility. *Clin. Exp. Rheumatol.* **14**: 67–70.

Hirohata, S., Oka, H. and Mizushima, Y. (1992). Streptococcal-related antigens stimulate production of IL6 and interferon gamma by T cells from patients with Behçet's disease. *Cell Immunol.* **140**: 410–419.

Hojyo-Tomoka, T., Granados, J., Vargas-Alarcon, G. et al. (1997). Further evidence of the role of HLA-DR4 in the genetic susceptibility to actinic prurigo. *J. Am. Acad. Dermatol.* **36**: 935–937.

Holloway, J.W., Doull, I., Begishvili, B. et al. (1996). Lack of evidence of a significant association between HLA-DR, DQ and DP genotypes and atopy in families with HDM allergy. *Clin. Exp. Allergy* **26**: 1142–1149.

Horn, G.T., Bugawan, T.L., Long, C.M. and Erlich, H.A. (1988). Allelic sequence variation of the HLA-DQ loci: relationship to serology and to insulin-dependent diabetes susceptibility. *Proc. Natl Acad. Sci. USA* **85**: 6012–6016.

Ikaheimo, I., Tiilikainen, A., Karvonen, J. and Silvennoinen-Kassinen, S. (1996). HLA risk haplotype Cw6,DR7, DQA1*0201 and HLA-Cw6 with reference to the clinical picture of psoriasis vulgaris. *Arch. Dermatol. Res.* **288**: 363–365.

Ishibashi, Y. and Juji, T. (1991). Study of HLA class I, class II and complement genes (C2, C4A, C4B and BF) in Japanese psoriatics and analysis of a newly-found high-risk haplotype by pulsed field gel electrophoresis. *Arch. Dermatol. Res.* **283**: 281–284.

Ishihara, M., Ohno, S., Mizuki, N. et al. (1996). LMP7 polymorphism in Japanese patients with sarcoidosis and Behçet's disease. *Hum. Immunol.* **51**: 103–105.

Jahn, S., Trefzer, U. and Sterry, W. (1997). SALT-skin associated lymphoid tissues: functional aspects. In *Oculodermal Diseases* (eds U. Pleyer, Chr. Hartmann and W. Sterry), pp. 21-31, Aeolus Press, The Netherlands.

Jankowski, J., Crombie, I. and Jankowski, R. (1992). Behçet's syndrome in Scotland. *Postgrad. Med. J.* **68**: 566–570.

Jardetzky, T.S., Brown, J.H., Gorga, J.C. et al. (1994). Three-dimensional structure of a human class II histocompatibility molecule complexed with superantigen. *Nature* **368**: 711–718.

Jenisch, S., Henseler, T., Westphal, E. et al. (1995). Analysis of HLA-class II-extended haplotypes in type I psoriasis: association with HLA-DRB1*0701-DQA1*0201-DQB1*0303. *Skin Pharmakol.* **8**: 79–83.

Jontell, M., Stahlblad, P.A., Rosdahl, I. and Lindblom, B. (1987). HLA-DR3 antigens in erosive oralive oral lichen planus, cutaneous lichen planus, and lichenoid reactions. *Acta Odontol. Scand.* **45**: 309–312.

Juji, T. and Shibata, Y.J. (1995). Lack of primary association between transporter associated with antigen processing genes and atopic dermatitis. *Allergy Clin. Immunol.* **96**: 1051–1060.

Kaneko, S., Suzuki, N., Yamashita, N. et al. (1997). Characterization of T cells specific for an epitope of human 60-kD heat shock protein (hsp) in patients with Behçet's disease (BD) in Japan. *Clin. Exp. Immunol.* **108**: 204–212.

Karpati, S., Amagai, M., Prussick, R. and Stanley, J.R. (1994). Pemphigus vulgaris antigen is a desmosomal desmoglein. *Dermatology* **189**(Suppl. I): 24–26.

Karvonen, J. and Tiilikainen, A. (1976). HLA antigens in Hailey–Hailey's disease. *Tissue Antigens* **8**: 277–278.

Kimberling, W. J. and Dobson, R. L. (1973). The inheritance of psoriasis. *J. Invest. Dermatol.* **60**: 538–540.

Korman, N.J., Eyre, R.W., Zone, J. and Stanley, J.R. (1991). Drug-induced pemphigus: autoantibodies directed against the pemphigus antigen complexes are present in penicillamine and captopril-induced pemphigus. *J. Invest. Dermatol.* **96**: 273–276.

Kunte, C., Barbosa, J.M., Wolff, H. and Meurer, M. (1997). Brazilian pemphigus foliaceus. *Hautarzt* **48**: 228–233.

Kurc, D., De Saint-Pere, R., Madoule, P. et al. (1987). Chronic osteitis and arthritis of palmoplantar pustulosis. A familial form of B-27 negative spondylarthropathy. *Rev. Med. Interne.* **8**: 79–84.

Kuwata, S., Yanagisawa, M., Saeki, H. et al. (1995). HLA antigen distribution in different clinical subgroups demonstrates genetic heterogeneity in lichen planus. *Br. J. Dermatol.* **132**: 897–900.

Kuwata, S., Yanagisawa, M., Nakagawa, H. et al. (1996). HLA-DM gene polymorphisms in atopic dermatitis. *J. Allergy Clin. Immunol.* **98**: S192–S200.

Laurentaci, G. and Dioguardi, D. (1977). HLA antigens in keloids and hypertrophic scars. *Arch. Dermatol.* **113**: 1726.

Lee, C.W., Kim, S.C. and Han, H. (1996). Distribution of class II alleles in Korean patients with epidemolysis bullosa acquisita. *Dermatology* **193**: 328–329.

Lehner, T. (1997). The role of heat shock protein, microbial and autoimmune agents in the aetiology of Behçet's disease. *Int. Rev. Immunol.* **14**: 21–32.

Lerner, A.B. (1959). Vitiligo. *J. Invest. Derm.* **32**: 285–310.

Levinson, D., Brautbar, C., Aharonson, S. et al. (1997). Immune response genes in pemphigus vulgaris. In *Oculodermal Diseases* (eds U. Pleyer, Chr. Hartmann and W. Sterry), pp. 273–2285, Aeolus Press, The Netherlands.

Lin, S.C. and Sun, A. A. (1990). HLA-DR and DQ antigens in Chinese patients with oral lichen planus. *Oral. Pathol. Med.* **19**: 298–300.

Lipski, S., Grabbe, J. and Henz, B.M. (1996). Absence of MHC class II antigen on mast cells at sites of inflammation in human skin. *Exp. Dermatol.* **5**: 120–124.

Lombardi, C., Borges, P.C. and Chaul, A. (1992). Environmental risk factors in endemic pemphigus foliaceus (Fogo selvagem). *J. Invest. Dermatol.* **98**: 847–850.

Lombardi, M.L., Mercuro, O., Tecame, G. et al. (1996). Molecular analysis of HLA DRB1 and DQB1 in Italian patients with pemphigus vulgaris. *Tissue Antigens* **47**: 228–230.

Lomholt, G. (1963). Psoriasis prevalence, spontaneous course and genetics: a census study on the prevalence of skin disease in the Faroe Islands. C.E.C. Gad, Copenhagen.

Lorini, R., Orecchia, G., Martinetti, M. et al. (1992). Autoimmunity in vitiligo: relationship with HLA, Gm and Km polymorphisms. *Autoimmunity* **11**: 255–260.

McCartan, B.E. and Lamey, P.J. (1997). Expression of CD1 and HLA-DR by Langerhans cells (LC) in oral lichenoid drug eruptions (LDE) and idiopathic oral lichen planus (LP). *Oral Pathol. Med.* **26**: 176–180.

McDonagh, A.J. and Messenger, A.G. (1996). The pathogenesis of alopecia areata. *Dermatol. Clin.* **14**: 661–670.

McKusick, V.A. (1997). Psoriasis, susceptibility to no. 177900. OMIM, Online Mendelian Inheritance in Man. Internet. http://www.ncbi.nlm.nih.gov/htbin-post/Omim/dispmim?142840

Majumder, P.P., Nordlund, J.J. and Nath, S.K. (1993). Pattern of familial aggregation of vitiligo. *Arch. Dermatol.* **129**: 994–998.

Malchus, R., Marsch, W.C. and Ehlers, G. (1986). HLA-B 16 in Hailey–Hailey's disease. *Acta Derm. Venereol.* **66**: 264–266.

Mallon, E., Bunce, M., Wojnarowska, F. and Welsh, K. (1997). HLA-Cw*0602 is a suceptibility factor in type I psoriasis, and evidence ala-73 is increased in male type I psoriatics. *J. Invest. Dermatol.* **199**: 183–186.

Marren, P., Millard, P., Chia, Y. and Wojnarowska, F. (1994). Mucosal lichen sclerosus/lichen planus overlap syndromes. *Br. J. Dermatol.* **131**: 118–123.

Matzner, Y., Erlich, H.A., Brautbar, C. et al. (1995). Identical HLA class II alleles predispose to "drug triggered" and idiopathic pemphigus vulgaris. *Acta Derm. Venereol.* **75**: 12–14.

Menage, H. duP., Vaughan, R.W., Baker, C.S. et al. (1996). HLA-DR4 may determine expression of actinic prurigo in British patients. *Invest. Dermatol.* **106**: 362–367.

Merkel, F., Netzer, K. and Weber M. (1997). Immunology of diseases affecting the basement membrane. In *Oculodermal Diseases* (eds U. Pleyer, Chr. Hartmann and W. Sterry), pp. 53–63, Aeolus Press, The Netherlands.

Metzker, A., Zamir, R., Gazit, E. et al. (1980). Vitiligo and the HLA system. *Dermatologica* **160**: 100–105.

Middleton, D. and Allen, G.E. (1984). HLA-A29 HLA antigen frequency in granuloma annulare. *Br. J. Dermatol.* **110**: 57–59.

Mineshita, S., Tian, D., Wang, L.M. et al. (1993). Behçet's disease associated with one of the HLA-B51 subantigens, HLA-B*5101. *Am. J. Ophthalmol.* **116**: 406–409.

Miyagawa, S., Dohi, K., Shima, H. and Shirai, T. (1992). Absence of HLA-B8 and HLA-DR3 in Japanese patients with Sjogren's syndrome positive for anti-SSA(Ro). *J. Rheumatol.* **19**: 1922–1924.

Miyagawa, S., Dohi, K., Shima, H. and Shirai, T. (1993). HLA antigens in anti-Ro(SS-A)-positive patients with recurrent annular erythema. *J. Am. Acad. Dermatol.* **28**: 185–188.

Mizoguchi, M. (1996). Sweet's syndrome. *Nihon. Rinsho. Meneki. Gakkai. Kaishi* **19**: 169–178.

Mizuki, N. and Ohno S. (1996). Immunogenetic studies of Behçet's disease. *Rev. Rhum. Engl. Ed.* **63**: 520–527.

Mizuki, N., Ohno, S., Kamata, K. et al. (1991). Immunogenetic mechanism of Behçet's disease. *Nippon Ganka. Gakkai. Zasshi.* **95**: 783–789.

Mizuki, N., Inoko, H., Mizuki, N. et al. (1992a). Human leukocyte antigen serologic and DNA typing of Behçet's disease and its primary association with B51. *Invest. Ophthalmol. Vis. Sci.* **33**: 3332–3340.

Mizuki, N., Ohno, S., Tanaka, H. et al. (1992b). Association of HLA-B51 and lack of association of class II alleles with Behçet's disease. *Tissue Antigens* **40**: 22–30.

Mizuki, N., Ohno, S., Ando, H. et al. (1996). HLA-C genotyping of patient with Behçet's disease in the Japanese population. *Hum. Immunol.* **50**: 47–53.

Mizuki, N., Inoko, H. and Ohno, S. (1997a). Pathogenic gene responsible for the predisposition of Behçet's disease. *Int. Rev. Immunol.* **14**: 33–48.

Mizuki, N., Ota, M., Kimura, M. et al. (1997b). Triplet repeat polymorphism in the transmembrane region of the MICA gene: a strong association of six GCT repetitions with Behçet disease. *Proc. Natl Acad. Sci. USA* **94**: 1298–1303.

Mobini, N., and Ahmed, A.R. (1993). Immunogenetics of drug-induced bullous diseases. *Clin. Dermatol.* **11**: 449–460.

Mochizuki, M., Morita, E., Yamamoto, S. and Yamana, S. (1997). Self-reactive CD4[+] T cells Th1 and Th2, (but to a minor extent also CD8[+]) play some role in the pathogenesis of Behçet's disease. Characteristics of T cell lines established from skin lesions of Behçet's disease. *J. Dermatol. Sci.* **15**: 9–13.

Mockenhapt, M. (1997). Drug induced acute oculo-muco-cutaneus diseases: erythema exsuativum multiformie majus (EEMM), Stevens Johnson syndrome (SJS) and toxic epidermal necrolysis (TEN). In *Oculodermal Diseases* (eds U. Pleyer, Chr. Hartmann and W. Sterry), pp. 153–168, Aeolus Press, The Netherlands.

Moore, C., Ehlayel, M., Inostroza, J. et al. (1997). Elevated levels of soluble HLA class I (sHLA-I) in children with severe atopic dermatitis. *Ann. Allergy Asthma Immunol.* **79**: 113–118.

Moraes, J.R., Moraes, M.E., Fernandez-Vina, M. et al. (1991). HLA antigens and risk for development of pemphigus foliaceus (fogo selvagem) in endemic areas of Brazil. *Immunogenetics* **33**: 388–391.

Moraes, M.E., Fernandez-Vina, M., Lazaro, A. et al. (1997). An epitope in the third hypervariable region of the DRB1 gene is involved in the susceptibility tendemic pemphigus foliaceus (fogo selvagem) in three different Brazilian populations. *Tissue Antigens* **49**: 35–40.

Morhenn, V.B. (1997). Cell-mediated autoimmune diseases of the skin: some hypotheses. *Med. Hypotheses* **49**: 241–245.

Muto, M., Nagai, K., Mogami S. et al. (1995). HLA antigens in Japanese patients with psoriatic arthritis. *Tissue Antigens* **45**: 362–364.

Muto, M., Date, Y., Ichimiya, M. et al. (1996). Significance of antibodies to streptococcal M protein in psoriatic arthritis and their association with HLA-A*0207. *Tissue Antigens* **48**: 645–650.

Nagao, Y., Sata, M., Abe, K. et al. (1997). Immunological evaluation in oral lichen planus with chronic hepatitis C. *Gastroenterology* **32**: 324–329.

Nair, R.P., Guo, S.W., Jenisch, S. et al. (1995). Scanning chromosome 17 for psoriasis susceptibility: lack of evidence for a distal 17q locus. *Hum. Hered.* **45**: 219–230.

Nair, R.P., Henseler, T., Jenisch, S. et al. (1997). Evidence for two psoriasis susceptibility loci (HLA and 17q) and two novel candidate regions (16q and 20p) by genome-wide scan. *Hum. Mol. Genet.* **6**: 1349–1356.

Nakagawa, H., Akazaki, S., Asahina, A. et al. (1993). Oligonucleotide typing reveals association of type I psoriasis with the HLA-DRB1*0701/2, -DQA1*0201-DQB1*0303 extended haplotype. *J. Invest. Dermatol.* **100**: 749–752.

Niizeki, H., Inoko, H., Narimatsu, H. et al. (1994). HLA class II antigens are associated with Japanese pemphigus patients. *Hum. Immunol.* **31**: 246–250.

Ohno, S., Aoki, K., Sugiura, S. et al. (1973). HL-A5 and Behçet's patients with Behçet's syndrome. *Ann. Rheum. Dis.* **50**: 351–353.

Orecchia, G., Perfetti, L., Malagoli, P. et al. (1992). Vitiligo is associated with a significant increase in HLA-A30, Cw6 and DQw3 and a decrease in C4AQ0 in northern Italian patients. *Dermatology* **185**: 123–127.

Oxelius, V.A., Sjostedt, L., Willers, S. and Low, B. (1996). Development of allergy to laboratory animals is associated with particular Gm and HLA genes. *Int. Arch. Allergy Immunol.* **110**: 73–78.

Park, M.S., Terasaki, P.I., Ahmed, R.A. and Tiwari, J.l. (1979). HLA-DRw4 in 91% of Jewish pemphigus vulgaris patients. *Lancet* **ii**: 441–442.

Paul, W.E. and Seder, R.A. (1994). Lymphocyte responses and cytokines. *Cell* **76**: 241–251.

Pietrzyk, J.J., Turowsk, G. and Kapinska-Mrowka, M. (1982). Family studies in psoriasis. II. Inheritance of HLA genotypes. *Arch. Dermatol. Res.* **273**: 295–300.

Pleyer, U., Bruckner-Tuderman, L., Friedmann, A. et al. (1996). Immunology of bullous-oculo-cutaneous disorders. *Immunol. Today* **17**: 111–113.

Poloy, A., Tibor, L., Kramer, J. et al. (1991). HLA-DR1 is associated with vitiligo. *Immunol. Lett.* **27**: 59–62.

Porter, K., Klouda, P., Scully, C. et al. (1993). Class I and II HLA antigens in British patients with oral lichen planus. *Oral Surg. Oral. Med. Oral. Pathol.* **75**: 176–180.

Provost, T.T. and Watson, R. (1993). Anti-Ro (SS-A) HLA-DR3-positive women: the interrelationship between some ANA negative, SS, SCLE, and NLE mothers and SS/LE overlap female patients. *J. Invest. Dermatol.* **100**: 14S–20S.

Purcell, K.G., Spencer, L.V., Simpson, P.M. et al. (1990). HLA antigens in lichen sclerosus et atrophicus. *Arch. Dermatol.* **126**: 1043–1045.

Reohr, P.B., Mangklabruks, A., Janiga, A.M. et al. (1992). Pemphigus vulgaris in siblings: HLA-DR4 and HLA-DQw3 and susceptability to pemphigus. *J. Am. Acad. Dermatol.* **27**: 189–193.

Reunala, T. (1996). Incidence of familial dermatitis herpetiformis. *Br. J. Dermatol.* **134**: 394–398.

Richard, G., Linse, R., Hadlich, J. and Schubert, H. (1990). Genetics of Hailey–Hailey familial chronic benign pemphigus. *Dermatol. Monatsschr.* **176**: 673–681.

Rock, B., Martins, C.R., Theofilopoulos, A.N. et al. (1989). The pathogenic effect of IgG_4 autoantibodies in endemic pemphigus foliaceus (fogo selvagem). *N. Engl. J. Med.* **320**: 1463–1469.

Roitberg-Tambur, A., Friedmann, A., Korn, S. et al. (1994). Serologic and molecular analysis of the HLA systemin Israeli Jewish patients with oral erosive lichen planus. *Tissue Antigens* **43**: 219–223.

Roitberg-Tambur, A., Friedmann, A., Zfoni, E.E. et al. (1995). Do specific pockets of HLA C molecules involved in predisposition of psoriasis vulgaris in Jews. *J. Am. Acad. Dermatol.* **31**: 964–968.

Roman, J.A., Puig, N., Tornero, C. and Alcaniz, C. (1997). Serological and molecular HLA typing in Andalusian patients with Behçet's disease. Genetic–clinical correlations. *Med. Clin.* **108**: 397–398.

Rosner, I.A., Burg, C.G., Wisnieski, J.J. et al. (1993).The clinical spectrum of the arthropathy associated with hidradenitis suppurativa and acne conglobata. *J. Rheumatol.* **20**: 684–687.

Rossi, A. and Bozzi, M. (1989). HLA and keloids: antigenic frequency and therapeutic response. *G. Ital. Dermatol. Venereol.* **124**: 341–344.

Roujeau, J.C., Bracq, C., Huyn, N.T. et al. (1986). HLA phenotypes and bullous cutaneous reactions to drugs. *Tissue Antigens* **28**: 251–254.

Roujeau, J.C., Huynh, T.N., Bracq, C. et al. (1987). Genetic susceptibility to toxic epidermal necrolysis. *Arch. Dermatol.* **123**: 1171–1173.

Russel, T.J., Schultes, L.M. and Kuban, D.J. (1972). Histocompatibility (HLA-A) antigens associated with psoriasis. *N. Engl. J. Med.* **287**: 738–743.

Sabroe, R.A. and Greaves, M.W. (1997). The pathogenesis of chronic idiopathic urticaria. *Arch. Dermatol.* **133**: 1003–1008.

Saeki, H., Kuwata, S., Nakagawa, H. et al. (1995). Analysis of disease-associated amino acid epitopes on HLA class II molecules in atopic dermatitis. *J. Allergy Clin. Immunol.* **1996**: 1061–1068.

Sakane, T. (1997). New perspective on Behçet's disease. *Int. Rev. Immunol.* **14**: 89–96.

Schallreuter, K.U., Levenig, C., Kuhnl, P. et al. (1993). Histocompatibility antigens in vitiligo: Hamburg study on 102 patients from northern Germany. *Dermatology* **18**: 186–192.

Scharf, S.J., Friedmann, A., Brautbar, C. et al. (1988). HLA class II allelic variation and susceptibility in pemphigus vulgaris. *Proc. Natl Acad. Sci. USA* **85**: 3504–3508.

Scharf, S.J., Friedmann, A., Steinman, L. et al. (1989). Specific HLA-DQβ and DRβl alleles confer susceptibility to pemphigus vulgaris. *Proc. Natl Acad. Sci. USA* **86**: 6215–6219.

Schmitt-Egenolf, M., Boehncke, W.H., Stander, M. et al. (1993). Psoriasis and HLA-Cw6. *Br. J. Dermatol.* **102**: 179–184.

Schmitt-Egenolf, M., Eiermann, T.H., Boehncke, W.H. et al. (1996). Familial juvenile onset psoriasis is associated with the human leukocyte antigen (HLA) class I side of the extended haplotype Cw6–B57–DRB1*0701–DQA1*0201–DQB1*0303: a population- and family-based study. *J. Invest. Dermatol.* **106**: 711–714.

Shornick, J.K., Stasny, J.K. and Gilliam, J.N. (1981). High frequency of histocompatibility antigen HLA-DR3 and DR4 in herpes gestationis. *J. Clin. Invest.* **68**: 553–555.

Shornick, J.K., Bangert, J.L., Freeman, R.G. and Gilliam, R.G. (1983). Herpes gestationis: clinical and histologic features of twenty-eight cases. *J. Am. Acad. Dermatol.* **8**: 214–224.

Shornick, J.K., Artlett, C.M., Jenkins, R.E. et al. (1993). Complement polymorphism in herpes gestationis: association with C4 null allele. *J. Am. Acad. Dermatol.* **29**: 545–549.

Shornick, J.K., Jenkins, R.E., Artlett, C.M. et al. (1995). Class II MHC typing in pemphigoid gestationis. *J. Clin. Exp. Dermatol.* **20**: 123–126.

Simon, J.C. (1997). Immunophysiology of the skin. In *Oculodermal Diseases* (eds U. Pleyer, Chr. Hartmann and W. Sterry), pp. 13–20, Aeolus Press, The Netherlands.

Sinha, A.A., Brautbar, C., Szafer, F. et al. (1988). A novel DQB allele associated with pemphigus vulgaris. *Science* **239**: 1026–1029.

Sjostedt, L., Willers, S. and Orbaek, P. (1996). Human leukocyte antigens in occupational allergy: a possible protective effect of HLA-B16 in laboratory animal allergy. *Am. J. Ind. Med.* **30**: 415–420.

Smith, E.P. and Zone, J.J. (1993). Dermatitis herpetiformis and linear IgA bullous dermatosis. *Dermatol. Clin.* **11**: 511–526.

Souza-Ramalho, P. de, D'Almeida, M.F., Freitas, J.P. and Pinto, J. (1991). Behçet's disease in Portugal. *Acta Med. Port.* **4**: 79–82.

Spurkland, A., Ingvarsson, G., Falk, E.S. et al. (1997). Dermatitis herpetiformis and celiac disease are both primarily associated with the HLA-DQ (alpha 1*0501, beta 1*02) or the HLA-DQ (alpha 1*03, beta030* 1 1 heterodimers). *Tissue Antigens* **49**: 29–34.

Stingl, G. (1997). The skin immune system. In *Oculodermal Diseases* (eds U. Pleyer, Chr. Hartmann and W. Sterry), pp. 1–12, Aeolus Press, The Netherlands.

Streilin, J.W. (1983). Skin-associated lymphoid tissues (SALT): origin and functions. *J. Invest. Dermatol.* **80**: 12–16.

Sugi, N., Nakazawa, M., Nakamura, S. et al. (1997). Analysis of the profile of $CD4^+$ cells in Behçet's disease. *Nippon Ganka. Gakkai. Zasshi.* **101**: 335–340.

Sun, A., Lin, S.C., Chu, C.T. and Chiang, C.P. (1993). HLA-DR and DQ antigens in Chinese patients with Behçet's disease. *J. Oral Pathol. Med.* **22**: 60–63.

Svejgaard, E., Jakobsen, B. and Svejgaard, A. (1985). Studies of HLA-ABC and DR antigens in pure atopic dermatitis and atopic dermatitis combined with allergic respiratory disease. *Acta Derm. Venereol.* **114**(Suppl.): 72–76.

Szafer, F., Brautbar, C., Tzfoni, E. et al. (1987). Detection of disease specific restriction fragment length polymorphism in pemphigus vulgaris linked to the DQW1 and DQW3 alleles of the HLA-D region. *Proc. Natl Acad. Sci. USA* **84**: 6542–6545.

Taylor, C.B., Low, N., Raj, S. et al. (1997). Behçet's syndrome progressing to gastrointestinal perforation in a West African male. *Br. J. Rheumatol.* **36**: 498–501.

Theeuwes, M. and Morhenn, V. (1995). Allelic instability in mitosis model and the inheritance of psoriasis. *J. Am. Acad. Dermatol.* **32**: 44–52.

Thomson, W., Sanders, P.A., Davis, M. et al. (1988). Complement C4B-null alleles in Felty's syndrome. *Arthritis Rheum.* **31**: 984–989.

Tiilikainen, A., Lassus, A., Karvonen, J. et al. (1980). Psoriasis and HLA-Cw6. *Br. J. Dermatol.* **102**: 179–184.

Tisch, R. and McDevitt, H. (1996). Insulin dependent diabetes mellitus. *Cell* **85**: 291–297.

Tiwari, J.L. and Terasaki, P.I. (1985). Pemphigus vulgaris. In *HLA and Disease Association* (eds J.L. Tiwari and P.I. Terasaki), pp. 130–132, Springer, New York.

Tomfohrde, J., Silverman, A., Barnes, R. et al. (1994). Gene for familial psoriasis susceptibility mapped to the distal end of human chromosome 17q. *Science* **264**: 1141–1145.

Trembath, R.C., Clough, R.L., Frodsham, A. et al. (1996). A complete genomic search for susceptibility loci in psoriasis. *J. Invest. Dermatol.* **106**: 901–907.

Trembath, R.C., Clough, R.L., Rosbotham, J.L. et al. (1997). Identification of a major susceptibility locus on chromosome 6p and evidence for further disease loci revealed by a two stage genome-wide search in psoriasis. *Hum. Mol. Genet.* **6**: 813–820.

Valsecchi, R., Bontempelli, M., Rossi, A. et al. (1988). HLA-DR and DQ antigens in lichen planus. *Acta Derm. Venereol.* **68**: 77–80.

Vasey, F.B., Fenske, N.A., Clement, G.B. et al. (1984). Immunological studies of the arthritis of acne conglobata and hidradenitis suppurativa. *Clin. Exp. Rheumatol.* **2**: 309–311.

Venkataram, M.N., White, A.G., Leeny, W.A. et al. (1995). HLA antigens in Omani patients with vitiligo. *Clin. Exp. Dermatol.* **20**: 35–37.

Venneker, G.T., Westerhof, W., Vries, I.J. de et al. (1992). Molecular heterogeneity of the fourth component of complement (C4) and its genes in vitiligo. *J. Invest. Dermatol.* **99**: 853–858.

Villanueva, J.L., Gonzalez-Dominguez, J., Gonzalez-Fernandez, R. et al. (1993). HLA antigen familial study in complete Behcet's syndrome affecting three sisters. *Ann. Rheum. Dis.* **52**: 155–157.

Watanabe, T., Ohishi, M., Tanaka, K. and Sato, H. (1986). Analysis of HLA antigens in Japanese with oral lichen planus. *J. Oral Pathol.* **15**: 529–533.

Welsh, E.A., Clark, H.H., Epstein, S.Z. et al. (1994). Human leukocyte antigen-DQB1*03 alleles are associated with alopecia areata. *J. Invest. Dermatol.* **103**: 758–763.

White, A.G. and Rostom, A.I. (1994). HLA antigens in Arabs with lichen planus. *Clin. Exp. Dermatol.* **19**: 236–237.

Wilson, C., Wojnarowska, F., Mehra, N.K. and Pasricha, J.S. (1994). Pemphigus in Oxford, UK, and New Delhi, India: A comparative study of disease characteristics and HLA antigens. *Dermatology* **189**(Suppl. 1): 108–110.

Wilson, A.G., Clay, F.E., Crane, A.M. et al. (1995). Comparative genetic association of human leukocyte antigen class II and tumor necrosis factor-alpha with dermatitis herpetiformis. *J. Invest. Dermatol.* **104**: 856–858.

Wucherpfennig, K.W. (1995). Selective binding of self peptides to disease-associated major histocompatibility complex (MHC) molecules: a mechanism for MHC-linked susceptibility to human autoimmune diseases. *J. Exp. Med.* **181**: 1597–1601.

Wucherpfennig, K.W., Yu, B., Bohl, K. et al. (1995). Structural basis for major histocompatibility complex (MHC)-linked susceptibility to autoimmunity: charged residues of single MHC binding pocket confer selective presentation of self-peptide in Pemphigus vulgaris. *Proc. Natl Acad. Sci. USA* **92**: 11 935–11 939.

Wucherpfennig, K.W., Catz, I., Hausmann, S. et al. (1997). Recognition of the immunodominant myelin basic protein peptide by autoantibodies and HLA-DR2-restricted T cell clones from multiple sclerosis patients. Identity of key contact residues in the B-cell and T-cell epitopes. *J. Clin. Invest.* **100**: 1114–1111.

Yamashita, N., Kaneoka, H., Kaneko, S. et al. (1997). Role of gammadelta T lymphocytes in the development of Behçet's disease. *Clin. Exp. Immunol.* **108**: 204–212.

Yazici, H. (1997). The place of Behçet's syndrome among the autoimmune diseases. *Int. Rev. Immunol.* **14**: 1–10.

Zhang, L., Michelsen, C., Cheng, X. et al. (1997). Molecular analysis of oral lichen planus. A premalignant lesion? *Am. J. Pathol.* **151**: 323–327.

Zone, J., Ward, J. and Boyce, E. (1982). Penicillamine-induced pemphigus. *JAMA* **247**: 2705–2707.

CHAPTER 23

HLA and Transplantation I: Allorecognition of HLA Molecules in Transplantation

Richard J. Baker

INTRODUCTION

Despite a lack of knowledge about human alloantigens, transplantation of kidneys was able to proceed in the early 1960s. As the T lymphocyte was deemed to be the key mediator of rejection, transplantation progressed with some success by employing the antilymphocyte therapies available at that time. Initially, irradiation and corticosteroids were used but the introduction of azathioprine, derived from the antineoplastic agent 6-mercaptopurine, afforded results that had previously been unimaginable (Calne, 1960). Renal transplantation was then able to leave the experimental arena and emerge as a clinical modality for end-stage renal failure.

The emerging HLA scientific community had discovered the antigens which defined the "foreignness" of tissues and it now remained to show that these were indeed the targets of the infiltrating lymphocytes known to be present in renal allografts undergoing rejection. This certainly seemed likely as Medawar's group had long since shown that incompatibility at the H-2 locus, in mice, determined the fate of skin grafts (Billingham et al., 1954). Supportive evidence came from the *in vitro* mixed lymphocyte culture, which showed that lymphocyte proliferation was broadly proportionate to the degree of HLA incompatibility, particularly at the class II loci (Bach and Hirschhorn, 1964).

It was important to show that the allografts that were undergoing rejection were the same grafts that serologically expressed the most foreign, or mismatched, antigens. If this were indeed the case then a device would be available to predict graft rejection. In turn, this would permit the avoidance of deleterious combinations with a consequent reduction in the incidence of rejection and an increase in graft survival. It seemed logical that this would prove to be the case, as conversely it was already known that transplants between genetically identical individuals were possible without any immunosuppression (Merrill et al., 1956). To this end, the tissue typing fraternity emerged.

The first study was performed by Dausset who, working with Felix Rapaport from New York, transplanted skin grafts between healthy volunteers (Dausset et al., 1965). He concluded that incompatibility for the human leucocyte antigens (HLA) tested appeared to affect graft survival, but their findings were severely limited by incomplete knowledge of the complexity of the HLA antigens. Paul Terasaki and an emerging transplant surgeon from Denver, Thomas Starzl, retrospectively studied 64 recipients of renal transplants. They found that there was an imperfect correlation with the results of serological testing, although similar reactivity patterns were associated with better graft outcome (Starzl et al., 1965). Further studies then emerged confirming the benefit of tissue compatibility with respect to graft outcome. This conclusion applied either to living related grafts where grafts were shown to function better if the donor and recipient shared tissue types (Singal et al., 1969), or to cadaveric grafts (Batchelor and Joysey, 1969). At the third international congress of the Transplantation Society, in The Hague in 1970, many papers were presented supporting the use of tissue typing in transplantation. However, one presentation by Terasaki's group, which honestly reported some negative findings, spawned a huge amount of controversy and was omitted from the *Proceedings* (Mickey et al., 1971). In many ways this was the beginning of a dispute that has continued unabated between a burgeoning tissue typing industry and an equally vociferous group of dissenters.

The exact role of tissue typing in clinical transplantation remains to be defined, although in certain areas, e.g. bone marrow transplantation, it is essential. To evaluate its role in various transplantation settings large databases have been established. The Collaborative Transplant Study (CTS) was established in Europe, and the United Network of Organ Sharing (UNOS) registry in the USA. The retrospective statistical analysis of huge numbers of transplanted organs has enabled an objective evaluation of the effects of matching for HLA antigens. This, in turn, has influenced ongoing clinical practice. The results of some of these studies are presented in the next section on tissue typing.

HLA MOLECULES: ANTIGENS, IMMUNOGENS AND TOLEROGENS?

Allorecognition has been discussed in an earlier chapter, where the unique nature of alloantigens as vigorous stimulators of immune responses was emphasized. Allogeneic HLA molecules, whether presented directly or indirectly, have been considered as uniquely immunogenic. In fact, if allogeneic major histocompatibility complex (MHC) molecules are encountered in the neonatal period then persistent tolerance can be established. This was first demonstrated by Ray Owen, at the University of Wisconsin, in bovine dizygotic twins who invariably share a placental blood supply and thus represent haemopoietic chimaeras. As a consequence, both twins become tolerant to each other's MHC antigens (Owen, 1945). Experiments by Medawar's group in Oxford established the neonatal period as unique in its ability to establish persistent tolerance to foreign MHC molecules (Billingham et al., 1953). This tolerance was assumed to be accomplished by a central mechanism, that is thymic deletion of alloreactive T cells. Subsequent studies have shown that it is certainly more complex than simply thymic deletion of alloreactive clones, but such a putative mechanism has spawned a number of attempts to produce central tolerance to allogeneic MHC molecules. Indeed, the intrathymic injection of rat glomeruli has led to renal allograft acceptance (Remuzzi et al., 1991). Similarly, injections of MHC-derived peptides have also led to graft acceptance, but in this case the T cell tolerance must be induced via indirect mechanisms (Sayegh et al., 1993, 1994; Chowdhury et al., 1996; Shirwan et al., 1997). These developments are clearly limited in their applications to human transplantation and efforts have been made to look at peripheral tolerance induction, a process that is clearly more amenable to therapeutic intervention.

During the course of his experiments on the immunogenicity of thyroid epithelial cells, Kevin Lafferty, working in Australia, suggested that two signals were required to achieve full activation. Influenced by earlier suggestions of a two-signal model by Bretscher and Cohn, he developed the idea that the first signal was provided by the T cell receptor (TCR) interaction with MHC, but that a further, and at that time undefined, second signal was required (Bretscher and Cohn, 1970; Lafferty et al., 1983). In the late 1980s, Ron Schwartz's group, in Bethesda, discovered that this second signal is brought about by non-cognate interactions between pairs of molecules on T cells and antigen-presenting cells (APC), as discussed in an earlier chapter (Quill and Schwartz, 1987; Jenkins et al., 1988). Paramount amongst these is the interaction between CD28 on T cells and B7 molecules (B7.1 and B7.2) on APC (Jenkins et al., 1991; Linsley et al., 1991; Sayegh and Turka, 1998). CD40 Ligand (CD154) on T cells and CD40 on APC form another important receptor–ligand pair

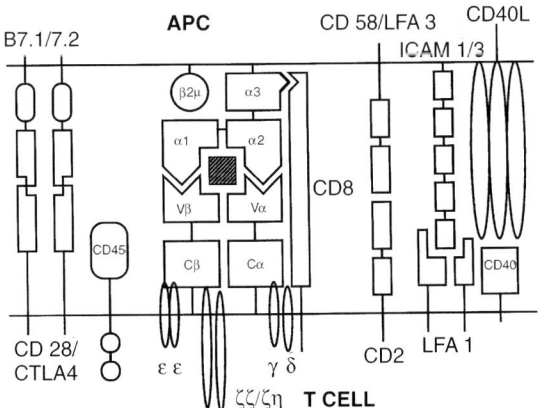

FIGURE 23.1 Besides αβ TCR engagement, by HLA class I loaded with peptide in this case, full activation requires the interaction between non-cognate ligand pairs such as CD28/CTLA4 and B7.1/2.

(Grewal and Flavell, 1998; Niimi et al., 1998) (Figure 23.1). If the TCR is ligated without costimulation then the T cells become anergic, a process that is active and inhibits interleukin-2 (IL-2) production and cellular proliferation. Such anergic cells are rendered unresponsive to further stimulation by their cognate antigen, irrespective of the state of costimulation, on the next occasion that it is encountered. This durability is important because if the host's T cells can be rendered anergic at the time of transplantation, then further non-specific immunosuppression might not be necessary.

So are all allogeneic cells equally immunogenic? Common sense would suggest not as various cell types express widely different levels of costimulatory molecules. The cells with the highest levels of costimulatory molecules are specialized APC, in particular dendritic cells, and these would therefore be expected to be the most immunogenic in an alloresponse. In fact, as early as 1957, Snell had noticed that allogeneic murine tumours that had been treated with corticosteroids were not so immunogenic, taking longer to be rejected (Snell, 1957). He also knew that corticosteroids destroyed lymphoid tissue and he hypothesized that this passenger lymphoid tissue contributed in some way to the immunogenicity of the graft. Lafferty developed this idea, when he took thyroid tissue and maintained it in culture for 12 days before transplanting it into an allogeneic host. As a result the rejection process was markedly delayed. He hypothesized that this was due to the emigration of immunogenic allogeneic cells out of the graft during the culture stage (Lafferty et al., 1975). In a group of experiments with rodents, Robert Lechler and Richard Batchelor, working in London, took kidney grafts from (ASXAUG) F_1 donor rats and "parked" them in parental AS rats under the cover of immunosuppression, which prevented them from being rejected (Lechler and Batchelor, 1982). When these grafts were retransplanted 1 month later into a second AS rat they were accepted without immunosuppression (Figure 23.2). Furthermore, the graft was rejected if the second rat was subsequently injected with dendritic cells from the original donor. Thus, the donor bone marrow-derived APC was established as the main determinant of immunogenicity in an MHC-mismatched graft. This work has stimulated attempts to deplete such leucocytes prior to clinical transplantation (Goldberg et al., 1995).

By implication, the epithelial and endothelial cells that bear class II molecules in the graft are not intrinsically immunogenic, owing to a lack of expression of costimulatory molecules. There is now a body of evidence that this is indeed the case. *In vitro* culture with allogeneic renal and thyroid epithelial cells does not induce an immune response and in certain circumstances may

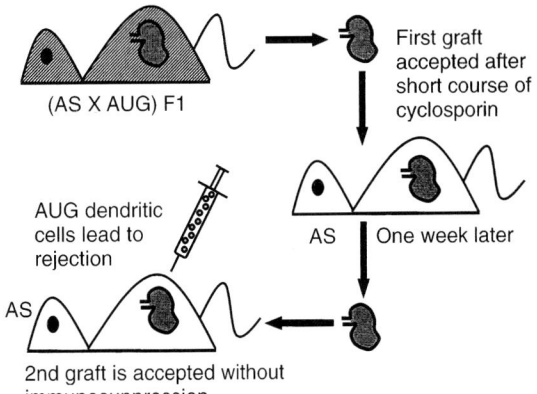

FIGURE 23.2 Transplant of a kidney from an ASXAUG F_1 rat into an AS will lead to rejection. If this graft is covered by a 1-week course of cyclosporin to prevent rejection and then retransplanted into a second AS rat without immunosuppression it is accepted. The injection of AUG dendritic cells leads to graft rejection.

actually induce anergy (Marelli Berg et al., 1997; Frasca et al., 1998). Some groups have also found that human renal endothelial cells, which in contrast to rodent's cells do express class II MHC antigens (Hayry et al., 1980), do not incite alloresponses owing to a lack of costimulatory molecules (Marelli Berg et al., 1996). This latter point is controversial and some groups believe that allogeneic human endothelial cells will stimulate an alloresponse (i.e. are immunogenic) (Savage et al., 1993; Briscoe et al., 1998; Rose, 1998). Interestingly, they suggest that the costimulation is mediated via LFA3 on vascular endothelial cells and CD2 on T cells.

So is it true that the parenchymal cells in the graft, despite expressing copious allogeneic HLA molecules, are not immunogenic? In fact, the data described above would suggest that once the passenger leucocytes have migrated out of the graft, then the recipient may actually become hyporesponsive to the direct presentation of donor HLA molecules owing to the tolerogenic influence of graft parenchymal cells. One way to address this issue is to quantify the T cell response *in vitro*. This can be accomplished by limiting dilution analysis (LDA) which allows an estimate to be made of the frequency of recipient T cells that respond to any given allogeneic target. Inevitably, such assays are subject to a number of limitations. Foremost amongst these is the sampling error inherent in measuring frequencies from peripheral blood mononuclear cells, which may not be representative of the T cells in the immediate vicinity of the graft, i.e. in the local lymph nodes. Certainly rodent data suggest that frequencies of allogeneic T cells are higher in local lymphoid tissues for both direct and indirect pathways (Benichou et al., 1999). Another important consideration is the variation in frequencies over time, whereas the assays only measure responses at a fixed point. In addition, it is far harder to measure indirect frequencies, as they are substantially lower and approach the limit of sensitivity for the assay. Notwithstanding these criticisms, the limited data available suggest that a few months after transplantation the frequencies of alloreactive T cells against donor alloantigens, via the direct pathway, are specifically diminished in renal, hepatic and cardiac transplant recipients (DeBruyne et al., 1995; Mason et al., 1996; de Haan et al., 1998; Hornick et al., 1998). Indeed, this accords with the data above that once passenger leucocytes emigrate out of the graft then the HLA molecules on the graft parenchymal cells are intrinsically tolerogenic, presumably owing to a lack of costimulatory molecules. Although the aetiology of chronic rejection is multifactorial, this work has led a number of authors to suggest that indirect presentation of alloantigens may be a contributory factor. As stated below, there is now some circumstantial evidence to suggest that this may indeed be the case (Vella et al., 1997a; Ciubotariu et al., 1998).

If two signals were required to activate the alloreactive T cell response to mismatched donor HLA antigens then interruption of the second signal at the time of transplantation would be an attractive strategy. This has been made possible by the chimaeric fusion protein CTLA4–Ig, which blocks the CD28–B7.1/2 interaction (Koulova et al., 1991). In rodent models CTLA4–Ig can prevent rejection of renal allografts and allow a state of tolerance to be achieved (Akalin et al., 1996, 1997; Azuma et al., 1996; Judge et al., 1999). This could even be achieved if the animals had been presensitized (Onodera et al., 1997) and perhaps most remarkably it was possible to reverse chronic rejection that had already been established (Chandraker et al., 1998). CTLA4–Ig has also been used with antibodies which block another important costimulatory pathway, that between CD40Ligand (CD154) and CD40. Simultaneous administration of both reagents to mice permitted the acceptance of fully allogeneic skin and heart grafts (Larsen et al., 1996). Curiously, in this case true tolerance to the allogeneic tissue was not achieved as later challenge with a second graft, without any further antibodies, was met with rejection. It is encouraging that this regime has now been tested in primates where, in some cases, it permitted long-term acceptance of renal allografts without any chronic immunosuppression (Kirk et al., 1997). This area of research is clearly very exciting and clinical trials of these reagents are eagerly awaited.

It is obviously an attractive concept to use allogeneic HLA molecules to induce hyporesponsiveness or even tolerance to grafts. This approach would be more applicable in the cases of live-related organ transplants or bone marrow transplantation where the procedure can be planned electively. Experiments in murine models have highlighted the interesting phenomenon of "linked suppression", whereby induction of tolerance to a single allogeneic MHC molecule can prolong the survival of allografts expressing that same antigen together with others to which the recipient has not previously been exposed (Madsen et al., 1988; Davies et al., 1996; Wong et al., 1997). Such a mechanism has also been shown to operate through the indirect pathway (Wise et al., 1998). This phenomenon of "linked suppression" may be mediated by cell surface interactions, secreted "immunosuppressive" cytokines or even steric competition for the APC surface. Whatever the mechanism it offers hope to those engaged in trying to induce tolerance to alloantigens.

Some clinical data suggest that prior exposure to alloantigens may produce hyporesponsiveness to later allografts. There is the evidence from studies of preoperative blood transfusions in renal transplant recipients which suggests that previous transfusion leads to improved renal allograft survival (Solheim et al., 1980; Sanfilippo et al., 1990). More recently, during the cyclosporin era, this effect has been questioned but a recent prospective multicentre study showed that transfused patients had a significantly better graft survival (Opelz et al., 1997). In this study the blood was not tissue typed so it is unclear whether the preoperative transfusions matched the donor allograft in any way. Previous evidence has suggested that sharing one HLA-DR antigen is beneficial (Lagaaij et al., 1989; Jackson et al., 1997; van der Mast and Balk, 1997). It should be emphasized that the mechanism for this effect is far from clear. A second piece of evidence comes from an intriguing recent study looking at renal allograft survival in grafts from live-related sibling donors. This showed that graft survival was significantly better if the recipient received a graft expressing the non-inherited maternal haplotype than a graft expressing the non-inherited paternal haplotype, implying that foetal exposure to maternal HLA antigens had somehow induced hyporesponsiveness (Burlingham et al., 1998a). Curiously, the former group actually experienced more episodes of early rejection despite their increased survival.

THE CLINICAL CONSEQUENCES OF ALLOREACTIVITY

Central role of the $CD4^+$ cell

The individual cellular and molecular interactions that comprise the alloresponse can be dissected *in vitro*, but inevitably the whole is more complex than simply the sum of the parts. Understanding the co-ordination and regulation of the different immunopathological mechanisms remains an elusive goal. However, it is certain that $CD4^+$ T cells play a central role in orchestrating the immune response to HLA molecules. They are able to initiate a number of effector mechanisms that in turn are responsible for the clinical manifestations of allograft rejection. Recent evidence shows that the $CD4^+$ cell, once activated, facilitates the priming of $CD8^+$ cytotoxic T cells by activating APC. This is mediated by interactions through CD40L (CD154) on the $CD4^+$ T cell and CD40 on the APC (Ridge et al., 1998; Schoenberger et al., 1998). In addition, $CD4^+$ T cells are able to interact with B cells in the germinal centre of lymphoid organs, permitting isotype switching and affinity maturation of alloreactive antibodies (Clark and Ledbetter, 1994). Finally, $CD4^+$ cells can circulate to the graft, activating macrophages with a potent mixture of proinflammatory cytokines. This process forms part of a delayed-type hypersensitivity reaction (DTH). Only natural killer (NK) cells may act independently of $CD4^+$ T cells, but their significance in the rejection of solid organ allografts is unclear (Manilay and Sykes, 1998).

After activation by HLA molecules, $CD4^+$ T cells may differentiate into either Th1 or Th2 type effector cells and this process affects the nature of the alloresponse (Mosmann et al., 1986; Del Prete et al., 1991). These phenotypes are defined by their cytokine secretion patterns upon activation (Abbas et al., 1996; O'Garra, 1998). Th1 cells, secreting the cytokines interferon-γ (IFNγ), tumour necrosis factor-β (TNFβ) and interleukin-2 (IL-2), typically support DTH-type responses, the development of cytotoxic $CD8^+$ cells and the production of some IgG antibody subclasses. Th2 cells, in contrast, support the production of other classes of antibody and secrete the cytokines IL-4, IL-5 and IL-10. It has been suggested that skewing the $CD4^+$ T cells towards the Th2 phenotype, away from a Th1-inducing response, may facilitate tolerance induction to transplanted tissues (Lowry et al., 1993). While it would appear to be oversimplistic to equate Th1 responses to rejection and Th2 responses to graft acceptance, it is true that in animal models of tolerance there are several features of Th2-type responses (Nickerson et al., 1997).

Hyperacute rejection

Hyperacute rejection (HAR) is the term applied to very early graft destruction, usually within the first 48 h. It occurs when preformed antibodies are present in the recipient's serum, specific for donor antigens expressed on graft vascular endothelial cells. Such antibodies constitute two main groups. Low-affinity IgM antibodies, which are specific for the ABO blood group antigens, form the first group and they mandate ABO blood group matching in solid organ transplantation. Similar antibodies exist in humans against the galactose–1-3-galactose epitope present in all other mammals and constitute one of the major impediments to successful xenotransplantation (Parker et al., 1994). The second group of antibodies consists of high-affinity IgG antibodies directed against HLA class I antigens. These usually occur as a result of previous immunization, by blood transfusions, pregnancies or failed allografts. They also occur in 1% of the population for no obvious reason (Scornik et al., 1988). As both blood group and class I HLA antigens are present on the vascular endothelium, it follows that as soon as blood flows into the allograft, then the antibodies come into contact with their respective antigens. The binding of these antibodies triggers activation of clotting, complement and kinin cascades, leading to intravascular thrombosis, ischaemia and subsequent necrosis.

The problem of blood group compatibility was appreciated early on in the history of renal transplantation and was rectified by ABO matching (Hume et al., 1955). Despite this, HAR continued to occur sporadically, and in 1966 the Danish scientist Kissmeyer described the phenomenon in two multiparous patients who had been heavily transfused (Kissmeyer Nielsen et al., 1966). These patients had high-affinity IgG antibodies to endothelial HLA class I antigens, a specificity that was confirmed in later series (Williams et al., 1968). To avoid this adverse outcome, the pretransplant cross-match is carried out, whereby donor lymphocytes are mixed with the recipient's serum to detect the presence of cytotoxic antibodies. This process will only detect antibodies to antigens present on the lymphocytes themselves and the specimens are usually predominantly composed of T lymphocytes and express HLA class I only. As the most problematic antibodies are those against class I HLA molecules, they will be detected by this approach. However, rarely HAR has been described as a result of other preformed antibodies with specificities for other endothelial antigens (Jordan et al., 1988). Antibodies have also been described against donor class II HLA molecules which can be detected with the B cell cross-match. A positive B cell cross-match does not cause HAR but there is some evidence that it predicts early rejection episodes and a poor short-term outcome (Lazda, 1994). The pretransplant cross-match has already proved to be successful at virtually eliminating HAR from clinical practice, but there are now other more sensitive methods of detecting antibodies such as flow cytometry cross-matching and antigen-specific enzyme-linked immunosorbent assays (ELISA). These techniques are certainly more sensitive and yield more positive results; however, these positives usually do not correlate with subsequent HAR. One study looked at patients who had had a negative conventional T cell cross-match. They showed that a positive flow cytometric cross-match correlated with early episodes of acute rejection but there were no cases of HAR (Kimball et al., 1998). Similarly, positive cross-matches by ELISA-based methods have been shown to predict patients with early graft loss (Monteiro et al., 1997). In contrast, another study looking at flow cytometric positive cross-matches showed that a positive test had no effect on 1 year graft survival and, perhaps more importantly, if these positive tests had been acted upon, 28% of patients in the series would have been denied a transplant (Christiaans et al., 1996). Although positive cross-matches by these more sensitive techniques may portend a worse prognosis, their place in clinical transplant practice remains to be defined.

It is important to point out that there is a considerable difference between different solid organs with respect to outcome associated with a positive cross-match. In hepatic transplantation HAR usually only occurs with ABO incompatibility and a positive conventional cross-match rarely results in HAR, although it is associated with decreased graft survival (Donaldson and Williams, 1997). Similarly, in cardiac transplantation a positive cross-match does not necessarily lead to HAR, but once again it is associated with a worse outcome (Smith et al., 1993).

Patients awaiting renal transplantation who have high levels of broadly reactive anti-HLA antibodies must therefore wait until a transplant becomes available, with a negative cross-match, or alternatively the antibodies can be temporarily removed by either plasma exchange or immunoadsorption (Alexandre et al., 1987; Palmer et al., 1989; Ross et al., 1993). The early return of antibodies after transplantation does not always result in HAR and graft loss (Reding et al., 1987). For some reason the binding of the anti-HLA antibodies at this later stage does not bring about the same disastrous conclusion. This phenomenon, called accommodation, remains incompletely understood.

Acute rejection

If preformed antibodies are not present, solid organ allografts succumb to rejection in the absence of immunosuppression. In the clinical situation, with current immunosuppressive protocols, this form of rejection is seen in 15–50% of patients, occurring between 5 days and 3 months after transplantation. The immune effector mechanisms involved in this type of rejection fuelled much controversy in the 1950s, as scientists argued in favour of humoral or cellular mechanisms. It was Peter Gorer who had demonstrated allospecific antibodies to tumour cells which could retard the growth of leukaemic cells *in vitro* (Gorer, 1942), but Peter Medawar's group, working in Oxford, had demonstrated that immunity to allogeneic tumours in mice was transferable by cells but not serum (Mitchison, 1953). Similarly, accelerated second set rejection could only be adoptively transferred by lymphocytes (Billingham et al., 1954). In another elegant experiment, allogeneic tissue was placed in a diffusion chamber which permitted the passage of small molecules, including antibodies, but not cells. These tissues were not rejected (Algire, 1957). Thus, the lymphocyte was established as the protagonist in the rejection process.

During the alloresponse it is probable that the main afferent mechanism involves the emigration of donor leucocytes, in particular dendritic cells, out of the graft and into the regional lymph nodes where they can prime donor lymphocytes by the mechanisms alluded to above. The $CD4^+$ T cell would be central in this process and principal in orchestrating the effector mechanisms. What effector mechanisms are central to acute cellular rejection? Looking specifically at renal transplantation, the histological findings reveal a diffuse interstitial cellular infiltrate composed of both $CD4^+$ and $CD8^+$ T cells. This picture is dominated by $CD8^+$ T cells with an activated or memory, $CD45RO^+$, phenotype (Ibrahim et al., 1995). Macrophages are also present to a lesser extent. Initially it was thought that the $CD8^+$ T cell was the foremost effector cell in the acute rejection process, as it was demonstrated that they were directly cytotoxic to allogeneic tissue *in vitro* (Hayry and Defendi, 1970). There is evidence that donor-specific CTL frequencies rise at the time of acute rejection and drop again when it resolves (Ouwehand et al., 1993; Mestre et al., 1996). $CD8^+$ cytotoxicity is mediated by perforin, granzyme B and Fas-mediated pathways (Sharma et al., 1996; Vasconcellos et al., 1998; Wever et al., 1998). In some animal models it is notable that both $CD4^+$ T cell and $CD8^+$ T cell populations can reject solid organ allografts independently, while in others there is some evidence that $CD4^+$ cells are an absolute requirement (Krieger et al., 1996, 1997). In the clinical setting it is likely that both cell subtypes are involved in the rejection process. One property shared by all effector mechanisms is the binding of cognate antigen receptors to allogeneic HLA molecules on cell surfaces. This binding may involve B cells producing anti-HLA antibodies, $CD8^+$ T cells lysing cells bearing allogeneic class I molecules or $CD4^+$ T cells recognizing allogeneic class II molecules and releasing a potent cocktail of proinflammatory cytokines. One notable feature of this response is its fine specificity for allogeneic cellular targets. If syngeneic skin grafts are placed side by side with allogeneic grafts, on the same tissue bed, only the allogeneic graft is destroyed, with a very precise line of demarcation between the two (Rosenberg et al., 1989). Furthermore, skin grafts from allophenic donors, that is donors derived from four parents by artificially fusing two embryonic F_1 cells, were initially overwhelmed by a vigorous immune response against skin cells that represent a mosaic of the two genotypes. After this process had subsided some cells were observed to survive and these were all cells that were syngeneic to the host (Mintz and Silvers, 1967; Rosenberg and Singer, 1988).

The role of the indirect pathway in acute rejection is unclear, although recent experiments in mice show that it is already active, although comparatively weak when compared with the direct response (Benichou et al., 1999). Indeed, the vigour of the direct response at this stage has tended to obscure the role of indirect allorecognition. Despite this, human studies have been able to identify evidence of indirect allorecognition in patients undergoing acute rejection of cardiac, hepatic and renal grafts (Liu et al., 1996a, b; Molajoni et al., 1997).

Notwithstanding the well-documented preeminence of cellular elements in acute rejection there is some evidence for the involvement of antibodies, directed against allogeneic HLA molecules. Data in rodent models suggest that alloantibodies can effect the acute rejection process and indeed can come to prominence when the T cell compartment is suppressed (Brandle et al., 1998). In renal transplantation, about 10% of biopsies for acute rejection show evidence of intravascular fibrinoid necrosis, typical of antibody-mediated damage, with variable degrees of cellular infiltration. These cases sometimes show evidence of antibody and complement deposition by immunofluorescence. Monitoring of patients after transplantation shows that donor-specific alloantibodies develop in more than 50% of recipients of first transplants (Feucht and Opelz, 1996). A recent study showed that antibodies directed against HLA molecules, in particular class II, detected by ELISA, predicted rejection episodes and poor graft outcome (Feucht and Opelz, 1996; Schonemann et al., 1998). The significance of these findings is uncertain, although in general it seems that patients with antibodies may fare worse than those without (Lobo et al., 1995).

Chronic rejection

Most of the research into chronic rejection has been carried out in renal transplantation and this is reflected in the discussion below. Chronic rejection is a severe problem in renal transplantation and constitutes the main cause of late graft loss. The aetiology involves a number of immunological and non-immunological factors and there is evidence to substantiate both kinds of mechanism (Massy et al., 1996; Kasiske, 1997) (Table 23.1). Evidence that immunological factors are important stems from a number of sources. First, it has been shown that episodes of acute rejection strongly predict the development of chronic rejection, implying that the two processes may share common aetiologies (Almond et al., 1993; Tesi et al., 1993). In addition, chronic rejection is less common in grafts that are better matched, whether they are from live or cadaveric donors (Cecka, 1998). This is reflected in the longer half-lives of better-matched grafts, whether they are from live or cadaveric donors. Finally, in animal models, the induction of tolerance can prevent the development of chronic rejection and indeed established rejection can be reversed with therapy that interferes with T cell activation (Azuma et al., 1996; Subbotin et al., 1997; Chandraker et al., 1998).

Histological analysis reveals evidence of interstitial fibrosis with infiltrates of lymphocytes, plasma cells and mast cells associated with tubular atrophy. Characteristically, arterioles are thickened with fibrosis and deposition of extracellular matrix material with a variable cellular infiltrate. Immunofluorescence often reveals deposits of

TABLE 23.1 Non-immunological factors implicated in chronic rejection

Calcineurin inhibitor-induced nephrotoxicity
Hypertension
Hyperfiltration by reduced nephron mass
Dyslipidaemia
CMV infection

both antibodies and complement. These findings suggest that antibody mediated effects may be important. Certainly, if antidonor antibodies are transferred into mice that have a functioning cardiac allograft and no native antibodies, then graft atherosclerosis can be brought about (Russell et al., 1994). Indeed, it has been shown in human recipients of renal, cardiac and lung allografts that the development of anti-HLA antibodies is linked to the development of chronic rejection (Suciu Foca et al., 1991a, b; Sundaresan et al., 1998). In some of these studies it was demonstrated that the pathogenicity of these antibodies could be modulated by soluble HLA antigens and anti-idiotypic antibodies, both of which could potentially "mop up" such pathogenic antibodies. Subsequent work has suggested that chronic rejection in renal allografts may be specifically caused by antibodies to class II HLA antigens (Abe et al., 1997). Recent rodent experiments suggest that the endothelial expression of MHC class II molecules may contribute to the development of transplant arteriosclerosis, but that class I expression may be protective (Shi et al., 1999). These observations need to be confirmed.

Research into the cellular response in chronic rejection has often yielded evidence of hyporesponsiveness to direct presentation of alloantigens in both $CD4^+$ and $CD8^+$ T cell populations (DeBruyne et al., 1995; Mason et al., 1996; Mestre et al., 1996; de Haan et al., 1998; Hornick et al., 1998). Interest has therefore focused on the indirect pathway as a means of continuing T cell activation in the chronic phase. There is evidence from rodent models that indirect allorecognition can contribute to the rejection of allografts. First, the injection of allogeneic donor peptides can cause rejection, the implication being that these peptides can only be recognized by the indirect pathway (Dalchau et al., 1992; Fangmann et al., 1992; Parker et al., 1992; Benham et al., 1995). In fact, T cell clones can be generated from such animals that are restricted by self-MHC and are specific for allogeneic MHC peptides (Waaga et al., 1998). Comparative studies between the two pathways have not been carried out in humans but such studies in rodents would suggest that the direct pathway is dominant in the early stages (Benichou et al., 1999). Thereafter, the indirect pathway may come to the fore, as the direct pathway seems to be downregulated. Rodent models have yielded evidence that the indirect pathway is active in models of chronic rejection (Vella et al., 1997b). Two separate studies, looking at cardiac and renal transplant recipients with confirmed chronic rejection have showed evidence, by LDA, of sensitization to indirectly processed allogeneic donor-derived HLA peptides (Vella et al., 1997a; Ciubotariu et al., 1998), while a further study has documented evidence of indirect recognition of peptides derived from donor splenocyte preparations in patients with chronic cardiac rejection (Hornick et al., 1999). As stated previously, this phenomenon could be secondary to tissue damage, but the recent demonstration that lesions resembling chronic rejection could be transferred by a self-restricted T cell line specific for an allogeneic MHC-derived peptide suggests that it may indeed be causal (Valujskikh et al., 1998).

Graft-versus-host disease

Although graft-versus-host disease (GVHD) may occur after blood transfusion and vascularized organ transplants, it is predominantly a problem encountered in bone marrow transplantation (BMT). BMT has become an accepted form of therapy for haematological malignancies and it is also employed for other cancers, hereditary blood disorders, immunodeficiencies and certain metabolic disorders (Armitage, 1994). For certain diseases such as chronic myeloid leukaemia (CML) it offers the only chance of a complete cure (Clift et al., 1993). Not surprisingly, grafts from monozygotic twins fare the best, followed by those from HLA-matched siblings (Madrigal et al., 1997). However, these will only provide transplants for about 30% of patients on the waiting list and has prompted a number of groups to explore the use of volunteer unrelated donors (VUD) (Hansen et al., 1998). A major cause of morbidity and mortality following allogeneic BMT is the development of GVHD. GVHD is caused by the T cells in the donor marrow which recognize foreign determinants on the new host and initiate an immune response against them (Theobald, 1995; Nash and Storb, 1996). This principle is well illustrated by bone-marrow transplantation (BMT) in the treatment of children with severe combined immunodeficiency. These children accept BMT without immunosuppression and by T cell depletion of donor marrow the incidence of GVHD is markedly reduced, even in one-haplotype mismatched sibling combinations (Buckley et al., 1999). When GVHD did occur it was usually associated with placental transmission of maternal T cells, so the disease was actually graft-versus-graft as opposed to GVH. These observations underpin the role of T cells in GVHD. However, in BMT for severe combined immunodeficiency states associated with normal NK cell function, the incidence of GVHD in one-haplotype mismatched grafts rises, which suggests a possible role for NK cells becoming activated by a lack of self-HLA molecules, as detailed above (Haddad et al., 1998).

Accordingly, the incidence of GVHD in BMT for haematological malignancies can be reduced by T cell depleting the donor marrow prior to infusion but unfortunately this abrogates a beneficial effect of the T cells, that of the graft-versus-leukaemia (GVL) effect. GVL represents the destruction of residual leukaemic cells in the body of the recipient by the alloreactive donor T cells in the graft. This effect is beneficial in reducing the risk of relapse. An elusive goal in BMT is to dissociate these two effects so that GVHD can be abrogated without losing the GVL effect. The use of T cell clones directed at disease-specific peptides may represent one forward step in this dilemma.

In animal models GVHD is mediated by both $CD4^+$ and $CD8^+$ T cells expressing $\alpha\beta$ TCRs. In some models there may also be a role for $\gamma\delta$ T cells (Blazar et al., 1996). Examination of human biopsy material reveals an epithelium infiltrated by $\alpha\beta$ TCR^+, $CD3^+$ and $CD45RO^+$ cells (Diamond et al., 1995). The role of $\gamma\delta$ T cells is uncertain in humans. In HLA-identical sibling BMT, the antigens recognized by these infiltrating T cells are minor histocompatibility antigens (i.e. non-HLA) presented by autologous HLA molecules in a conventional self-restricted manner. These antigens have been extremely difficult to identify but there is now convincing clinical evidence that mismatches at these loci are associated with the development of GVHD (Goulmy et al., 1996). However, with the expansion of the donor pool to include VUD, allogeneic HLA molecules have become one of the principal targets of alloreactive T cells. Accordingly, transplants from VUD are associated with higher incidences of GVHD (Ash et al., 1990). This arises despite efforts to match completely the donor and recipient at the four major loci (HLA-A, -B, -DR and -DQ). Some HLA mismatching still inevitably occurs and this has a number of causes. First, retrospective analysis, performed by molecular genotyping, reveals "cryptic" mismatches for HLA-A, -B, -DR and -DQ (see below) (Petersdorf et al., 1998; Sasazuki et al., 1998). Some donors are mismatched, often only for a small number of amino acid residues, but such small differences have been shown *in vitro* and in animal models to be sufficient to generate an alloresponse (Ewenstein et al., 1980; Hunt et al., 1990; Pullen et al., 1994). When sibling transplants are carried out, if a two-haplotype match is identified then it is almost certain that the whole MHC regions on both chromosomes are identical, including HLA-DP and HLA-C. With VUD there is no guarantee that the intervening untyped alleles will be matched. Typing for the HLA-C locus has proved problematic, with a dearth of good-quality serological reagents, although this will improve with new molecular techniques (Madrigal et al., 1997). Mismatching for HLA-C has been associated with graft failure and the generation of high frequencies of CTL precursors in otherwise well-matched pairs when measured by LDA (Barnardo et al., 1996; Petersdorf et al., 1997). Similarly, HLA-DP has been hard to characterize serologically because of its low expression on the cell surface. HLA-DP inheritance is relatively weakly linked to HLA-DR and HLA-DQ and so it is regularly mismatched in VUD transplantation (Moreau and Cesbron, 1994). HLA-DP incompatibility alone is certainly sufficient to stimulate an *in vitro* mixed lymphocyte culture and mismatch, particularly at position 69 of HLA-DPB1, is sufficient to generate high frequencies of helper T cell frequencies by LDA (Olerup et al., 1990; Potolicchio et al., 1996). However, the role of HLA-DP mismatch in the development of GVHD remains controversial (Petersdorf et al., 1993).

Summary

The pathological processes above represent the clinical correlates of the alloresponse. Mechanistically, they all involve cognate recognition of allogeneic HLA molecules by antigen receptors of the immune system (TCR, Ig or killer cell inhibiting receptors). Important alloresponses also occur against non-HLA antigens (minor histocompatibility or tissue-specific antigens), particularly in BMT between siblings, but in the majority of clinical settings the major problem is the immune response against allogeneic HLA molecules. The challenge is therefore to derive strategies to avoid exposure to alloantigens. This is the *raison d'être* of the tissue-typing fraternity. To minimize the exposure to allogeneic HLA molecules on donor tissue (or host tissue in the case of GVHD), HLA typing is carried out to minimize the number of alloantigens to which the recipient is exposed. In addition the recipient is screened for preformed antibodies against donor HLA antigens on the graft tissue by the pretransplant cross-match as detailed above. The impact of these tests on clinical transplantation is discussed in the next section.

SOLUBLE HLA MOLECULES IN TRANSPLANTATION

During the early trials of liver transplantation in pigs, Calne's group in Cambridge described one notable long-term survivor in the absence of immunosuppression (Calne et al., 1967). They speculated that the liver might secrete tolerogenic donor antigens. This proposal was taken up by Van Rood's group in Leiden who proposed that these antigens might be soluble HLA molecules, a hypothesis that they were able to prove in 1970 (Van Rood et al., 1970). It transpires that soluble HLA antigens derived from both class I and II HLA molecules are present in the serum and other body fluids of virtually all normal individuals (McDonald and Adamashvili, 1998). Detection of these soluble HLA (sHLA) molecules has been aided by the development of accurate ELISA assays (Ferreira et al., 1983; Sakaguchi et al., 1988; Comuzio et al., 1991; McDonald et al., 1992). Biochemical studies have revealed that the same sHLA is the consequence of cellular shedding whilst other molecules are actively secreted. This means that sHLA circulates in different forms with variable molecular weights. Total levels of sHLA class I were reported to be elevated in acute GVHD (Tsuji et al., 1985) and then also in acute rejection of renal allografts (Zavazava et al., 1990). Similarly, increased total levels of sHLA class I have been associated with rejection of heart, liver and kidney–pancreas allografts (Rhynes et al., 1993; Zavazava et al., 1993; Puppo et al., 1994; Abendroth et al., 1995; Renna Molajoni et al., 1995). The conclusion to be drawn from the current evidence is that total sHLA class I levels have been shown to rise in all forms of allograft rejection but bear no relationship to allograft survival. It may be that sHLA class I levels reflect generalized immune activity. The total level of class II sHLA does not correlate with rejection, infection or HLA class I levels and further studies need to be performed to define its role.

More specifically, investigators have studied donor sHLA class I levels. This effort has been facilitated by the development of allospecific monoclonal antibodies. A study of HLA-A2-negative recipients of A2-positive kidney or kidney–pancreas transplants revealed that sHLA-A2 levels rose during acute rejection (DeVito-Haynes et al., 1994). Another group studied a mixture of 12 liver, 18 kidney and eight heart transplant recipients. They found that all the liver recipients had donor sHLA in their sera, compared with 72% of kidney and 50% of heart recipients. High and stable sHLA levels characterized stable graft function (McDonald et al., 1997). Suciu-Foca's group in New York, studying renal transplant recipients, showed that sHLA antigens interact with anti-HLA antibodies and anti-idiotypic antibodies, in a way that may obscure the detection of anti-HLA antibodies owing to immune complex formation. They suggested a role for sHLA in self-tolerance (King et al., 1989). It would seem that sHLA of donor origin is associated with most organ transplants but especially hepatic grafts. This sHLA may be complexed to anti-HLA antibodies and the concentrations may vary with time. The periods of variability may be during times of antibody excess or near equivalence. Periods of sHLA excess are associated with good graft function. In experimental systems this donor sHLA need not necessarily be derived from cells within the graft. There is abundant experimental evidence that sHLA can downregulate immune responses (Buelow et al., 1995). Its possible tolerogenic role has been speculated upon, particularly in hepatic transplantation where postoperative donor sHLA is almost universally found. There has also been consideration that anti-idiotypic antibodies might circulate in the serum, which could mimic the structure of soluble HLA molecules and thus potentially have an immunoregulatory role. Indeed, such antibodies, mimicking HLA-A201, have been demonstrated, but whether such a mechanism operates *in vivo* is uncertain (Burlingham et al., 1998b).

OTHER FUNCTIONS OF HLA MOLECULES IN TRANSPLANTATION

In 1987 two papers revealed that peptides derived from the α_1 and α_2-helices of HLA-A2 could inhibit cell-mediated cytotoxicity (Clayberger et al., 1987; Parham et al., 1987). These were allele-specific peptides and were of little practical benefit. However, further studies soon unveiled the immunosuppressive properties of other non-polymorphic regions of the α_1-helix of class I HLA molecules (Clayberger et al., 1993). Two such peptides, HLA-B7 (75–84) and HLA-B27 (75–84), have commenced clinical trials under the brand name ALLOTRAP® (Sangstat). Initially, these human-derived peptides were shown not only to be immunosuppressive but also to induce donor-specific tolerance in animal models (Nisco et al., 1994; Murphy et al., 1997). A small randomized trial of ALLOTRAP® in first renal transplants has been reported to show reduced levels of *in vitro* NK cytotoxicity, the consequence of which is unclear (Giral et al., 1997). The mechanism of action may involve binding to heat shock proteins in the T cell cytoplasm, but exactly what the function of the resulting complex is remains unknown, although there has been speculation that it might involve interference with TCR signal transduction (Nossner et al., 1996). It is also possible that it interferes with NK function by binding to the NK inhibitory receptor (Krensky and Clayberger, 1997) (Figure 23.2). Non-polymorphic peptides from class II MHC molecules have also been shown to be effective in animal models (Shen et al., 1996). The first such human peptides, derived from the HLA-DQ α-chain, have been shown to have immunosuppressive properties *in vitro* (Magee and Sayegh, 1997).

The immunomodulatory properties of these peptides, which are present in normal individuals, have prompted some authors to question whether this is a further, as yet undefined, physiological function for HLA molecules.

CONCLUSION

This chapter has attempted to follow the historical discovery of the HLA antigens and their role in transplantation. The crucial role of HLA antigens in allorecognition has been outlined and the variable outcome of the interaction between T cell receptor and allogeneic HLA molecule has been discussed. The integration of these pathways into the clinical syndromes that are seen in patients undergoing transplantation has been described. Clearly, despite the fact that HLA molecules evolved for an entirely different physiological function, they are now intimately related to the field of transplantation. The study of transplantation has progressed from two principal directions: pioneering surgeons performing organ allografts and scientists studying the MHC gene complex. The frontier between these two areas is occupied by the science of tissue typing, which has attempted to reconcile clinical observations with scientific theory in a logical manner. In solid organ transplantation the exact role of tissue typing remains to be defined. There is unquestionable benefit to receiving a well-matched allograft but equally a number of other factors significantly affects outcome (Table 23.2). The debate relates to whether these other factors should take precedence over the considerable efforts taken to

TABLE 23.2 Factors affecting renal transplant outcome

Degree of histocompatibility
Donor source
Number of previous transplants
Drug regime
Donor age
Recipient age
Cold ischaemia time
Presensitization
Recipient race
Recipient medical status
Recipient body mass index
Centre effect

obtain a well-matched graft (Chang, 1996). The establishment of such databases as the UNOS and the CTS in order to analyse the influence of HLA matching on clinical transplantation should enable such decisions to be made on a rational basis.

ACKNOWLEDGEMENT

R.J.B. is a Medical Research Council Clinical Training Fellow.

REFERENCES

Abbas, A.K., Murphy, K.M. and Sher, A. (1996). Functional diversity of helper T lymphocytes. *Nature* **383**: 787–793.

Abe, M., Kawai, T., Futatsuyama, K. et al. (1997). Postoperative production of anti-donor antibody and chronic rejection in renal transplantation. *Transplantation* **63**: 1616–1619.

Abendroth, D., Storck, M., Techt, B. and Zavazava, N. (1995). Analysis of the rejection markers tumor necrosis factor, ICAM-1, neopterin, interleukin-10, and soluble HLA in simultaneous pancreas and kidney transplantation with bladder drainage. *Transplant Proc.* **27**: 3114–3115.

Akalin, E., Chandraker, A., Russell, M.E. et al. (1996). CD28-B7 T cell costimulatory blockade by CTLA4Ig in the rat renal allograft model: inhibition of cell-mediated and humoral immune responses in vivo. *Transplantation* **62**: 1942–1945.

Akalin, E., Chandraker, A., Sayegh, M. and Turka, L.A. (1997). Role of the CD28:B7 costimulatory interaction in alloimmune responses. *Kidney Int. Suppl.* **58**: S8–10.

Alexandre, G.P., Squifflet, J.P., De Bruyere, M. et al. (1987). Present experiences in a series of 26 ABO-incompatible living donor renal allografts. *Transplant Proc.* **19**: 4538–4542.

Algire, G. (1957). Studies on tissue homotransplantation in mice, using diffusion-chamber methods. *Ann. N.Y. Acad. Sci.* **64**: 1009–1013.

Almond, P.S., Matas, A., Gillingham, K. et al. (1993). Risk factors for chronic rejection in renal allograft recipients. *Transplantation* **55**: 752–756.

Armitage, J.O. (1994). Bone marrow transplantation. *N. Engl. J. Med.* **330**: 827–838.

Ash, R.C., Casper, J.T., Chitambar, C.R. et al. (1990). Successful allogeneic transplantation of T-cell-depleted bone marrow from closely HLA-matched unrelated donors. *N. Engl. J. Med.* **322**: 485–494.

Azuma, H., Chandraker, A., Nadeau, K. et al. (1996). Blockade of T-cell costimulation prevents development of experimental chronic renal allograft rejection. *Proc. Natl Acad. Sci. USA* **93**: 12439–12444.

Bach, F. and Hirschhorn, K. (1964). Lymphocyte interaction; a potential histocompatibility test *in vitro*. *Science* **143**: 813–814.

Barnardo, M.C., Davey, N.J., Bunce, M. et al. (1996). A correlation between HLA-C matching and donor antirecipient CTL precursor frequency in bone marrow transplantation. *Transplantation* **61**: 1420–1423.

Batchelor, J.R. and Joysey, V.C. (1969). Influence of HL-A incompatibility on cadaveric renal transplantation. *Lancet* **i**: 790–792.

Benham, A., Sawyer, G. and Fabre, J. (1995). Indirect T cell allorecognition of donor antigens contributes to the rejection of vascularised kidney allografts. *Transplantation* **59**: 1028–1032.

Benichou, G., Valujskikh, A., Heeger, P.S. et al. 1999. Contributions of direct and indirect T cell alloreactivity during allograft rejection in mice. Induction of allograft nonresponsiveness after intrathymic inoculation with donor class I allopeptides. II. Evidence for persistent chronic rejection despite high levels of donor microchimerism. *J. Immunol.* **162**: 352–358.

Billingham, R., Brent, L. and Medawar, P. (1953). 'Actively acquired tolerance' of foreign cells. *Nature* **172**: 603–606.

Billingham, R., Brent, L., Medawar, P. and Sparrow, E. (1954). Quantitative studies of transplantation immunity. I. Survival times of skin homografts exchanged between members of different inbred strains of mice. *Proc. R. Soc. Biol.* **143**: 43–58.

Blazar, B.R., Taylor, P.A., Panoskaltsis-Mortari, A. et al. (1996). Lethal murine graft-versus-host disease induced by donor gamma/delta expressing T cells with specificity for host nonclassical major histocompatibility complex class Ib antigens. *Blood* **87**: 827–837.

Brandle, D., Joergensen, J., Zenke, G. et al. (1998). Contribution of donor-specific antibodies to acute allograft rejection: evidence from B cell-deficient mice. *Transplantation* **65**: 1489–1493.

Bretscher, P. and Cohn, M. (1970). A theory of self-nonself discrimination. *Science* **169**: 1042–1049.

Briscoe, D.M., Alexander, S.I. and Lichtman, A.H. (1998). Interactions between T lymphocytes and endothelial cells in allograft rejection. *Curr. Opin. Immunol.* **10**: 525–531.

Buckley, R., Schiff, S., Schiff, R. et al. (1999). Hematopoietic stem-cell transplantation for the treatment of severe combined immunodeficiency. *N. Engl. J. Med.* **340**: 508–516.

Buelow, R., Burlingham, W.J. and Clayberger, C. (1995). Immunomodulation by soluble HLA class I. *Transplantation* **59**: 649–654.

Burlingham, W.J., Grailer, A.P., Heisey, D.M. et al. (1998a). The effect of tolerance to noninherited maternal HLA antigens on the survival of renal transplants from sibling donors. *N. Engl. J. Med.* **339**: 1657–1664.

Burlingham, W.J., Jankowska-Gan, E., DeVito-Haynes, L. et al. (1998b). HLA (A*0201) mimicry by anti-idiotypic monoclonal antibodies. *J. Immunol.* **161**: 6705–6714.

Calne, R. (1960). The rejection of renal homografts. Inhibition in dogs by 6-mercaptopurine. *Lancet* **i**: 417–418.

Calne, R.Y., White, H.J., Yoffa, D.E. et al. (1967). Observations of orthotopic liver transplantation in the pig. *BMJ* **ii**: 478–480.

Cecka, M. (1998). Clinical outcome of renal transplantation. Factors influencing patient and graft survival. *Surg. Clin. North Am.* **78**: 133–148.

Chandraker, A., Azuma, H., Nadeau, K. et al. (1998). Late blockade of T cell costimulation interrupts progression of experimental chronic allograft rejection. *J. Clin. Invest.* **101**: 2309–2318.

Chang, R.W. (1996). How should cadaver kidneys be allocated? *Lancet* **348**: 453–454.

Chowdhury, N.C., Murphy, B., Sayegh, M.H. et al. (1996). Acquired systemic tolerance to rat cardiac allografts induced by intrathymic inoculation of synthetic polymorphic MHC class I allopeptides. *Transplantation* **62**: 1878–1882.

Christiaans, M.H., Overhof, R., ten Haaft, A. et al. (1996). No advantage of flow cytometry crossmatch over complement-dependent cytotoxicity in immunologically well-documented renal allograft recipients. *Transplantation* **62**: 1341–1347.

Ciubotariu, R., Liu, Z., Colovai, A.I. et al. (1998). Persistent allopeptide reactivity and epitope spreading in chronic rejection of organ allografts. *J. Clin. Invest.* **101**: 398–405.

Clark, E.A. and Ledbetter, J.A. (1994). How B and T cells talk to each other. *Nature* **367**: 425–428.

Clayberger, C., Parham, P., Rothbard, J. et al. (1987). HLA-A2 peptides can regulate cytolysis by human allogeneic T lymphocytes. *Nature* **330**: 763–765.

Clayberger, C., Lyu, S.C., Pouletty, P. et al. (1993). Peptides corresponding to T-cell receptor-HLA contact regions inhibit class I-restricted immune responses. Inhibition of alloreactive cytotoxic T lymphocytes by peptides from the alpha 2 domain of HLA-A2. *Transplant Proc.* **25**: 477–478.

Clift, R.A., Appelbaum, F.R. and Thomas, E.D. (1993). Treatment of chronic myeloid leukemia by marrow transplantation [Editorial; comment]. *Blood* **82**: 1954–1956.

Comuzio, S., Puppo, F., Ruzzenenti, R. et al. (1991). Simple ELISA method for the evaluation of soluble HLA class I antigens in human serum. *J. Clin. Lab. Anal.* **5**: 278–283.

Dalchau, R., Fangmann, J. and Fabre, J.W. (1992). Allorecognition of isolated, denatured chains of class I and class II major histocompatibility complex molecules. Evidence for an important role for indirect allorecognition in transplantation. *Eur. J. Immunol.* **22**: 669–677.

Dausset, J., Rapaport, F.T., Colombani, J. and Feingold, N. (1965). A leucocyte group and its relationship to tissue histocompatibility in man. *Transplantation* **3**: 701–705.

Davies, J.D., Leong, L.Y., Mellor, A. et al. (1996). T cell suppression in transplantation tolerance through linked recognition. *J. Immunol.* **156**: 3602–3607.

DeBruyne, L.A., Renlund, D.G. and Bishop, D.K. (1995). Evidence that human cardiac allograft acceptance is associated with a decrease in donor-reactive helper T lymphocytes. *Transplantation* **59**: 778–783.

Del Prete, G.F., De Carli, M., Mastromauro, C. et al. (1991). Purified protein derivative of Mycobacterium tuberculosis and excretory-secretory antigen(s) of *Toxocara canis* expand *in vitro* human T cells with stable and opposite (type 1 T helper or type 2 T helper) profile of cytokine production. *J. Clin. Invest.* **88**: 346–350.

DeVito-Haynes, L.D., Jankowska-Gan, E., Sollinger, H.W. et al. (1994). Monitoring of kidney and simultaneous pancreas–kidney transplantation rejection by release of donor-specific, soluble HLA class I. *Hum. Immunol.* **40**: 191–201.

Diamond, D.J., Chang, K.L., Jenkins, K.A. and Forman, S.J. (1995). Immunohistochemical analysis of T cell phenotypes in patients with graft-versus-host disease following allogeneic bone marrow transplantation. *Transplantation* **59**: 1436–1444.

Donaldson, P.T. and Williams, R. (1997). Cross-matching in liver transplantation. *Transplantation* **63**: 789–794.

Ewenstein, B.M., Uehara, H., Nisizawa, T. et al. (1980). Biochemical studies on the H-2K antigens of the MHC mutants bm3 and bm11. *Immunogenetics* **11**: 383–395.

Fangmann, J., Dalchau, R. and Fabre, J. (1992). Rejection of skin allografts by indirect allorecognition of donor class I major histocompatibility peptides. *J. Exp. Med.* **175**: 1521–1529.

Ferreira, A., Revilla, Y., Bootello, A. and Gonzalez-Porque, P. (1983). Use of the ELISA to detect and quantify histocompatibility antigens and their subunits. *J. Immunol. Methods* **65**: 373–381.

Feucht, H.E. and Opelz, G. (1996). The humoral immune response towards HLA class II determinants in renal transplantation [Editorial]. *Kidney Int.* **50**: 1464–1475.

Frasca, L., Marelli Berg, F., Imami, N. et al. (1998). Interferon-gamma-treated renal tubular epithelial cells induce allospecific tolerance. *Kidney Int.* **53**: 679–689.

Giral, M., Cuturi, M.C., Nguyen, J.M. et al. (1997). Decreased cytotoxic activity of natural killer cells in kidney allograft recipients treated with human HLA-derived peptide. *Transplantation* **63**: 1004–1011.

Goldberg, L.C., Bradley, J.A., Connolly, J. et al. (1995). Anti-CD45 monoclonal antibody perfusion of human renal allografts prior to transplantation. A safety and immunohistological study. CD45 Study Group. *Transplantation* **59**: 1285–1293.

Gorer, P. (1942). Role of antibodies in immunity to transplanted leukaemia in mice. *J. Pathol. Bacteriol.* **54**: 51–65.

Goulmy, E., Schipper, R., Pool, J. et al. (1996). Mismatches of minor histocompatibility antigens between HLA-identical donors and recipients and the development of graft-versus-host disease after bone marrow transplantation. *N. Engl. J. Med.* **334**: 281–285.

Grewal, I.S. and Flavell, R.A. (1998). CD40 and CD154 in cell-mediated immunity. *Annu. Rev. Immunol.* **16**: 111–135.

Haan, A. de, Berg, A.P. van den, Hepkema, B.G. et al. (1998). Donor-specific hyporeactivity after liver transplantation: prominent decreases in donor-specific cytotoxic T lymphocyte precursor frequencies independent of changes in helper T lymphocyte precursor frequencies or suppressor cell activity. *Transplantation* **66**: 516–522.

Haddad, E., Landais, P., Friedrich, W. et al. (1998). Long-term immune reconstitution and outcome after HLA-nonidentical T-cell-depleted bone marrow transplantation for severe combined immunodeficiency: a European retrospective study of 116 patients. *Blood* **91**: 3646–3653.

Hansen, J.A., Gooley, T.A., Martin, P.J. et al. (1998). Bone marrow transplants from unrelated donors for patients with chronic myeloid leukemia. *N. Engl. J. Med.* **338**: 962–968.

Hayry, P. and Defendi, U. (1970). Mixed lymphocyte cultures produce effector cells: *in vitro* model for allograft rejection. *Science* **168**: 133–135.

Hayry, P., Willebrand, E. von and Andersson, L.C. (1980). Expression of HLA-ABC and -DR locus antigens on human kidney, endothelial, tubular and glomerular cells. *Scand. J. Immunol.* **11**: 303–310.

Hornick, P.I., Mason, P.D., Yacoub, M.H. et al. (1998). Assessment of the contribution that direct allorecognition makes to the progression of chronic cardiac transplant rejection in humans. *Circulation* **97**: 1257–1263.

Hornick, P., Baker, R.J., Hernandez-Fuentes, M. et al. (1999). Significant frequencies of T cells with indirect anti-donor specificity in patients with chronic cardiac allograft rejection. *Circulation* (in press).

Hume, D., Merrill, J., Miller, B. and Thorn, G. (1955). Experiences with renal homotransplantation in the human: report of nine cases. *J. Clin. Invest.* **34**: 327–382.

Hunt, H.D., Pullen, J.K., Dick, R.F. et al. (1990). Structural basis of Kbm8 alloreactivity. Amino acid substitutions on the beta-pleated floor of the antigen recognition site. *J. Immunol.* **145**: 1456–1462.

Ibrahim, S., Dawson, D.V. and Sanfilippo, F. (1995). Predominant infiltration of rejecting human renal allografts with T cells expressing CD8 and CD45RO. *Transplantation* **59**: 724–728.

Jackson, A., McSherry, C., Butters, K. et al. (1997). Pretransplant exposure to donor HLA-DR antigen in random transfusion units and the development of donor antigen-specific hyporeactivity. *Hum. Immunol.* **55**: 148–153.

Jenkins, M.K., Ashwell, J.D. and Schwartz, R.H. (1988). Allogeneic non-T spleen cells restore the responsiveness of normal T cell clones stimulated with antigen and chemically modified antigen-presenting cells. *J. Immunol.* **140**: 3324–3330.

Jenkins, M.K., Taylor, P.S., Norton, S.D. and Urdahl, K.B. (1991). CD28 delivers a costimulatory signal involved in antigen-specific IL-2 production by human T cells. *J. Immunol.* **147**: 2461–2466.

Jordan, S.C., Yap, H.K., Sakai, R.S. et al. (1988). Hyperacute allograft rejection mediated by anti-vascular endothelial cell antibodies with a negative monocyte crossmatch. *Transplantation* **46**: 585–587.

Judge, T.A., Wu, Z., Zheng, X.G. et al. (1999). The Role of CD80, CD86, and CTLA4 in alloimmune responses and the induction of long-term allograft survival. *J. Immunol.* **162**: 1947–1951.

Kasiske, B.L. (1997). Clinical correlates to chronic renal allograft rejection. *Kidney Int. Suppl.*: 0098–6577.

Kimball, P., Rhodes, C., King, A. et al. (1998). Flow cross-matching identifies patients at risk for postoperative elaboration of cytotoxic antibodies. *Transplantation* **65**: 444–446.

King, D.W., Reed, E. and Suciu-Foca, N. (1989). Complexes of soluble HLA antigens and anti-HLA autoantibodies in human sera: possible role in maintenance of self-tolerance. *Immunol. Res.* **8**: 249–262.

Kirk, A.D., Harlan, D.M., Armstrong, N.N. et al. (1997). CTLA4-Ig and anti-CD40 ligand prevent renal allograft rejection in primates. *Proc. Natl Acad. Sci. USA* **94**: 8789–8794.

Kissmeyer Nielsen, F., Olsen, S., Petersen, V.P. and Fjeldborg, O. (1966). Hyperacute rejection of kidney allografts, associated with pre-existing humoral antibodies against donor cells. *Lancet* **ii**: 662–665.

Koulova, L., Clark, E.A., Shu, G. and Dupont, B. (1991). The CD28 ligand B7/BB1 provides costimulatory signal for alloactivation of $CD4^+$ T cells. *J. Exp. Med.* **173**: 759–62.

Krensky, A.M. and Clayberger, C. (1997). HLA-derived peptides as novel immunosuppressives [Editorial]. *Nephrol. Dial. Transplant.* **12**: 865–868.

Krieger, N.R., Yin, D.P. and Garrison Fathman, C. (1996). $CD4^+$ but not $CD8^+$ cells are essential for allorejection. *J. Exp. Med.* **184**: 2013–2018.

Krieger, N.R., Ito, H. and Fathman, C.G. (1997). Rat pancreatic islet and skin xenograft survival in CD4 and CD8 knockout mice. *J. Autoimmun.* **10**: 309–315.

Lafferty, K.J., Cooley, M.A., Woolnough, J. and Walker, K.Z. (1975). Thyroid allograft immunogenicity is reduced after a period in organ culture. *Science* **188**: 259–261.

Lafferty, K.J., Prowse, S.J., Simeonovic, C.J. and Warren, H.S. (1983). Immunobiology of tissue transplantation: a return to the passenger leukocyte concept. *Annu. Rev. Immunol.* **1**: 143–173.

Lagaaij, E.L., Hennemann, I.P., Ruigrok, M. et al. (1989). Effect of one-HLA-DR-antigen-matched and completely HLA-DR-mismatched blood transfusions on survival of heart and kidney allografts. *N. Engl. J. Med.* **321**: 701–705.

Larsen, C.P., Elwood, E.T., Alexander, D.Z. et al. (1996). Long-term acceptance of skin and cardiac allografts after blocking CD40 and CD28 pathways. *Nature* **381**: 434–438.

Lazda, V.A. (1994). Identification of patients at risk for inferior renal allograft outcome by a strongly positive B cell flow cytometry crossmatch. *Transplantation* **57**: 964–969.

Lechler, R.I. and Batchelor, J.R. (1982). Restoration of immunogenicity to passenger cell-depleted kidney allografts by the addition of donor strain dendritic cells. *J. Exp. Med.* **155**: 31–41.

Linsley, P.S., Brady, W., Grosmaire, L. et al. (1991). Binding of the B cell activation antigen B7 to CD28 costimulates T cell proliferation and interleukin 2 mRNA accumulation. *J. Exp. Med.* **173**: 721–730.

Liu, Z., Colovai, A.I., Tugulea, S. et al. (1996a). Indirect recognition of donor HLA-DR peptides in organ allograft rejection. *J. Clin. Invest.* **98**: 1150–1157.

Liu, Z., Harris, P.E., Colovai, A.I. et al. (1996b). Indirect recognition of donor MHC class II antigens in human transplantation. *Clin. Immunol. Immunopathol.* **78**: 228–235.

Lobo, P.I., Spencer, C.E., Stevenson, W.C. and Pruett, T.L. (1995). Evidence demonstrating poor kidney graft survival when acute rejections are associated with IgG donor-specific lymphocytotoxin. *Transplantation* **59**: 357–360.

Lowry, R.P., Takeuchi, T., Cremisi, H. and Konieczy, B. (1993). Th2-like effectors may function as antigen-specific suppressor cells in states of transplantation tolerance. *Transplant Proc.* **25**: 324–326.

McDonald, J.C. and Adamashvili, I. (1998). Soluble HLA: a review of the literature. *Hum. Immunol.* **59**: 387–403.

McDonald, J.C., Gelder, F.B., Aultman, D.F. et al. (1992). HLA in human serum—quantitation of class I by enzyme immunoassay. *Transplantation* **53**: 445–449.

McDonald, J.C., Adamashvili, I., Zibari, G.B. et al. (1997). Serologic allogeneic chimerism. *Transplantation* **64**: 865–871.

Madrigal, J.A., Arguello, R., Scott, I. and Avakian, H. (1997). Molecular histocompatibility typing in unrelated donor bone marrow transplantation. *Blood Rev.* **11**: 105–117.

Madsen, J.C., Superina, R.A., Wood, K.J. and Morris, P.J. (1988). Immunological unresponsiveness induced by recipient cells transfected with donor MHC genes. *Nature* **332**: 161–164.

Magee, C.C. and Sayegh, M.H. (1997). Peptide-mediated immunosuppression. *Curr. Opin. Immunol.* **9**: 669–675.

Manilay, J. and Sykes, M. (1998). Natural killer cells and their role in graft rejection. *Curr. Opin. Immunol.* **10**: 532–538.

Marelli Berg, F.M., Hargreaves, R.E., Carmichael, P. et al. (1996). Major histocompatibility complex class II-expressing endothelial cells induce allospecific nonresponsiveness in naive T cells. *J. Exp. Med.* **183**: 1603–1612.

Marelli Berg, F.M., Weetman, A., Frasca, L. et al. (1997). Antigen presentation by epithelial cells induces anergic immunoregulatory $CD45RO^+$ T cells and deletion of $CD45RA^+$ T cells. *J. Immunol.* **159**: 5853–5861.

Mason, P.D., Robinson, C.M. and Lechler, R.I. (1996). Detection of donor-specific hyporesponsiveness following late failure of human renal allografts. *Kidney Int.* **50**: 1019–1025.

Massy, Z.A., Guijarro, C., Wiederkehr, M.R. et al. (1996). Chronic renal allograft rejection: immunologic and non-immunologic risk factors. *Kidney Int.* **49**: 518–524.

Mast, B.J. van der and Balk, A.H. (1997). Effect of HLA-DR-shared blood transfusion on the clinical outcome of heart transplantation. *Transplantation* **63**: 1514–1519.

Merrill, J., Murray, J., Harrison, J. and Guild, W. (1956). Successful homotransplantation of the human kidney between identical twins. *JAMA* **160**: 277–282.

Mestre, M., Massip, E., Bas, J. et al. (1996). Longitudinal study of the frequency of cytotoxic T cell precursors in kidney allograft recipients. *Clin. Exp. Immunol.* **104**: 108–114.

Mickey, M.R., Kreisler, M., Albert, E.D. et al. (1971). Analysis of HL-A incompatibility in human renal transplants. *Tissue Antigens* **1**: 57–67.

Mintz, B. and Silvers, W.K. (1967). "Intrinsic" immunological tolerance in allophenic mice. *Science* **158**: 1484–1486.

Mitchison, N. (1953). Passive transfer of transplantation immunity. *Nature* **157**: 267–268.

Molajoni, E.R., Cinti, P., Orlandini, A. et al. (1997). Mechanism of liver allograft rejection: the indirect recognition pathway. *Hum. Immunol.* **53**: 57–63.

Monteiro, F., Buelow, R., Mineiro, C. et al. (1997). Identification of patients at high risk of graft loss by pre- and posttransplant monitoring of anti-HLA class I IgG antibodies by enzyme-linked immunosorbent assay. *Transplantation* **63**: 542–546.

Moreau, P. and Cesbron, A. (1994). HLA-DP and allogeneic bone marrow transplantation. *Bone Marrow Transplant.* **13**: 675–681.

Mosmann, T.R., Cherwinski, H., Bond, M.W. et al. (1986). Two types of murine helper T cell clone. I. Definition according to profiles of lymphokine activities and secreted proteins. *J. Immunol.* **136**: 2348–2357.

Murphy, B., Kim, K.S., Buelow, R. et al. (1997). Synthetic MHC class I peptide prolongs cardiac survival and attenuates transplant arteriosclerosis in the Lewis–Fischer 344 model of chronic allograft rejection. *Transplantation* **64**: 14–19.

Nash, R.A. and Storb, R. (1996). Graft-versus-host effect after allogeneic hematopoietic stem cell transplantation: GVHD and GVL. *Curr. Opin. Immunol.* **8**: 674–680.

Nickerson, P., Steiger, J., Zheng, X.X. et al. (1997). Manipulation of cytokine networks in transplantation: false hope or realistic opportunity for tolerance? *Transplantation* **63**: 489–494.

Niimi, M., Pearson, T.C., Larsen, C.P. et al. (1998). The role of the CD40 pathway in alloantigen-induced hyporesponsiveness *in vivo. J. Immunol.* **161**: 5331–5337.

Nisco, S., Vriens, P., Hoyt, G. et al. (1994). Induction of allograft tolerance in rats by an HLA class-I-derived peptide and cyclosporine A. *J. Immunol.* **152**: 3786–3792.

Nossner, E., Goldberg, J.E., Naftzger, C. et al. (1996). HLA-derived peptides which inhibit T cell function bind to members of the heat-shock protein 70 family. *J. Exp. Med.* **183**: 339–348.

O'Garra, A. (1998). Cytokines induce the development of functionally heterogenous T helper cell subsets. *Immunity* **8**: 275–283.

Olerup, O., Moller, E. and Persson, U. (1990). HLA-DP incompatibilities induce proliferation in primary mixed lymphocyte culture in HLA-A, -B, -C, -DR, and -DQ-identical individuals. *Transplant. Proc.* **22**: 141–142.

Onodera, K., Chandraker, A., Schaub, M. et al. (1997). CD28-B7 T cell costimulatory blockade by CTLA4Ig in sensitized rat recipients: induction of transplantation tolerance in association with depressed cell-mediated and humoral immune responses. *J. Immunol.* **159**: 1711–1717.

Opelz, G., Vanrenterghem, Y., Kirste, G. et al. (1997). Prospective evaluation of pretransplant blood transfusions in cadaver kidney recipients. *Transplantation* **63**: 964–967.

Ouwehand, A.J., Baan, C.C., Roelen, D.L. et al. (1993). The detection of cytotoxic T cells with high-affinity receptors for donor antigens in the transplanted heart as a prognostic factor for graft rejection. *Transplantation* **56**: 1223–1229.

Owen, R. (1945). Immunogenetic consequences of vascular anastomoses between bovine twins. *Science* **102**: 400–401.

Palmer, A., Taube, D., Welsh, K. et al. (1989). Removal of anti-HLA antibodies by extracorporeal immunoadsorption to enable renal transplantation. *Lancet* **i**: 10–12.

Parham, P., Clayberger, C., Zorn, S.L. et al. (1987). Inhibition of alloreactive cytotoxic T lymphocytes by peptides from the alpha 2 domain of HLA-A2. *Nature* **325**: 625–628.

Parker, K.E., Dalchau, R., Fowler, V.J. et al. (1992). Stimulation of $CD4^+$ T lymphocytes by allogeneic MHC peptides presented on autologous antigen-presenting cells. Evidence of the indirect pathway of allorecognition in some strain combinations. *Transplantation* **53**: 918–924.

Parker, W., Bruno, D. Holzknecht, Z.E. and Platt, J.L. (1994). Characterization and affinity isolation of xenoreactive human natural antibodies. *J. Immunol.* **153**: 3791–3803.

Petersdorf, E.W., Smith, A.G., Mickelson, E.M. et al. (1993). The role of HLA-DPB1 disparity in the development of acute graft-versus-host disease following unrelated donor marrow transplantation. *Blood* **81**: 1923–1932.

Petersdorf, E.W., Longton, G.M., Anasetti, C. et al. (1997). Association of HLA-C disparity with graft failure after marrow transplantation from unrelated donors. *Blood* **89**: 1818–1823.

Petersdorf, E.W., Gooley, T.A., Anasetti, C. et al. (1998). Optimizing outcome after unrelated marrow transplantation by comprehensive matching of HLA class I and II alleles in the donor and recipient. *Blood* **92**: 3515–3520.

Potolicchio, I., Brookes, P.A., Madrigal, A. et al. (1996). HLA-DPB1 mismatch at position 69 is associated with high helper T lymphocyte precursor frequencies in unrelated bone marrow transplant pairs. *Transplantation* **62**: 1347–1352.

Pullen, J.K., Tallquist, M.D., Melvold, R.W. and Pease, L.R. (1994). Recognition of a single amino acid change on the surface of a major transplantation antigen is in the context of self peptide. *J. Immunol.* **152**: 3445–3452.

Puppo, F., Pellicci, R., Brenci, S. et al. (1994). HLA class-I-soluble antigen serum levels in liver transplantation. A predictor marker of acute rejection. *Hum. Immunol.* **40**: 166–170.

Quill, H. and Schwartz, R.H. (1987). Stimulation of normal inducer T cell clones with antigen presented by purified Ia molecules in planar lipid membranes: specific induction of a long-lived state of proliferative nonresponsiveness. *J. Immunol.* **138**: 3704–3712.

Reding, R., Squifflet, J.P., Latinne, D. et al. (1987). Early postoperative monitoring of natural anti-A and anti-B isoantibodies in ABO-incompatible living donor renal allografts. *Transplant Proc.* **19**: 1989–1990.

Remuzzi, G., Rossini, M., Imberti, O. and Perico, N. (1991). Kidney graft survival in rats without immunosuppressants after intrathymic glomerular transplantation. *Lancet* **337**: 750–752.

Renna Molajoni, E., Puppo, F., Brenci, S. et al. (1995). Serum HLA class I soluble antigens: a marker of acute rejection following liver transplantation. *Transplant Proc.* **27**: 1155–1156.

Rhynes, V.K., McDonald, J.C., Gelder, F.B. et al. (1993). Soluble HLA class I in the serum of transplant recipients. *Ann. Surg.* **217**: 485–491.

Ridge, J.P., Di Rosa, F. and Matzinger, P. (1998). A conditioned dendritic cell can be a temporal bridge between a CD4+ T-helper and a T-killer cell. *Nature* **393**: 474–478.

Rose, M.L. (1998). Endothelial cells as antigen-presenting cells: role in human transplant rejection. *Cell Mol. Life Sci.* **54**: 965–978.

Rosenberg, A.S. and Singer, A. (1988). Evidence that the effector mechanism of skin allograft rejection is antigen-specific. *Proc. Natl Acad. Sci. USA* **85**: 7739–7742.

Rosenberg, A.S., Katz, S.I. and Singer, A. (1989). Rejection of skin allografts by CD4+ T cells is antigen-specific and requires expression of target alloantigen on Ia-epidermal cells. *J. Immunol.* **143**: 2452–2456.

Ross, C.N., Gaskin, G., Gregor Macgregor, S. et al. (1993). Renal transplantation following immunoadsorption in highly sensitized recipients. *Transplantation* **55**: 785–789.

Russell, P.S., Chase, C.M., Winn, H.J. and Colvin, R.B. (1994). Coronary atherosclerosis in transplanted mouse hearts. II. Importance of humoral immunity. *J. Immunol.* **152**: 5135–5141.

Sakaguchi, K., Ono, R., Tsujisaki, M. et al. (1988). Anti-HLA-B7,B27,Bw42,Bw54,Bw55,Bw56, Bw67,Bw73 monoclonal antibodies: specificity, idiotypes, and application for a double determinant immunoassay. *Hum. Immunol.* **21**: 193–207.

Sanfilippo, F., Thacker, L. and Vaughn, W.K. (1990). Living-donor renal transplantation in SEOPF. The impact of histocompatibility, transfusions, and cyclosporine on outcome. *Transplantation* **49**: 25–29.

Sasazuki, T., Juji, T., Morishima, Y. et al. (1998). Effect of matching of class I HLA alleles on clinical outcome after transplantation of hematopoietic stem cells from an unrelated donor. Japan Marrow Donor Program. *N. Engl. J. Med.* **339**: 1177–1185.

Savage, C.O., Hughes, C.C., McIntyre, B.W. et al. (1993). Human CD4+ T cells proliferate to HLA-DR+ allogeneic vascular endothelium. Identification of accessory interactions. *Transplantation* **56**: 128–134.

Sayegh, M.H., Perico, N., Imberti, O. et al. (1993). Thymic recognition of class II major histocompatibility complex allopeptides induces donor-specific unresponsiveness to renal allografts. *Transplantation* **56**: 461–465.

Sayegh, M.H., Perico, N., Gallon, L. et al. (1994). Mechanisms of acquired thymic unresponsiveness to renal allografts. Thymic recognition of immunodominant allo-MHC peptides induces peripheral T cell anergy. *Transplantation* **58**: 125–132.

Sayegh, M.H. and Turka, L.A. (1998). The role of T-cell costimulatory activation pathways in transplant rejection. *N. Engl. J. Med.* **338**: 1813–1821.

Schoenberger, S.P., Toes, R.E., Voort, E.I. van der et al. (1998). T-cell help for cytotoxic T lymphocytes is mediated by CD40–CD40L interactions. *Nature* **393**: 480–483.

Schonemann, C., Groth, J., Leverenz, S. and May, G. (1998). HLA class I and class II antibodies: monitoring before and after kidney transplantation and their clinical relevance. *Transplantation* **65**: 1519–1523.

Scornik, J.C., Salomon, D.R., Howard, R.J. and Pfaff, W.W. (1988). Evaluation of antibody synthesis in broadly sensitized patients. *Transplantation* **45**: 95–100.

Sharma, V.K., Bologa, R.M., Li, B. et al. (1996). Molecular executors of cell death – differential intrarenal expression of Fas ligand, Fas, granzyme B, and perforin during acute and/or chronic rejection of human renal allografts. *Transplantation* **62**: 1860–1866.

Shen, X., Hu, B., McPhie, P. et al. (1996). Peptides corresponding to CD4-interacting regions of murine MHC class II molecules modulate immune responses of CD4+ T lymphocytes *in vitro* and *in vivo*. *J. Immunol.* **157**: 87–100.

Shi, C., Feinberg, M.W., Zhang, D. et al. (1999). Donor MHC and adhesion molecules in transplant arteriosclerosis. *J. Clin. Invest.* **103**: 469–474.

Shirwan, H., Wu, G.D., Barwari, L. et al. (1997). Induction of allograft nonresponsiveness after intrathymic inoculation with donor class I allopeptides. II. Evidence for persistent chronic rejection despite high levels of donor microchimerism. *Transplantation* **64**: 1671–1676.

Singal, D.P., Mickey, M.R. and Terasaki, P.I. (1969). Serotyping for homotransplantation. 23. Analysis of kidney transplants from parental versus sibling donors. *Transplantation* **7**: 246–258.

Smith, J., Danskine, A., Laylor, R. et al. (1993). The effect of panel reactive antibodies and donor-specific cross-match on graft survival after heart and heart–lung transplantation. *Transplant Immunol.* **1**: 60–65.

Snell, G. (1957). The homograft reaction. *Annu. Rev. Microbiol.* **11**: 439–458.

Solheim, B.G., Flatmark, A., Halvorsen, S. et al. (1980). The effect of blood transfusions on renal transplantation. Studies of 395 patients registered for transplantation with a first cadaveric kidney. *Tissue Antigens* **16**: 377–386.

Starzl, T.E., Marchioro, T.L., Terasaki, P.I. et al. (1965). Chronic survival after human renal homotransplantation. Lymphocyte-antigen matching, pathology and influence of thymectomy. *Ann. Surg.* **162**: 749–787.

Subbotin, V., Sun, H., Aitouche, A. et al. (1997). Abrogation of chronic rejection in a murine model of aortic allotransplantation by prior induction of donor-specific tolerance. *Transplantation* **64**: 690–695.

Suciu Foca, N., Reed, E., Marboe, C. et al. (1991a). The role of anti-HLA antibodies in heart transplantation. *Transplantation* **51**: 716–724.

Suciu Foca, N., Reed, E., B'Agali, V.D. et al. (1991b). Soluble HLA antigens, anti-HLA antibodies, and antiidiotypic antibodies in the circulation of renal transplant recipients. *Transplantation* **51**: 593–601.

Sundaresan, S., Mohanakumar, T., Smith, M.A. et al. (1998). HLA-A locus mismatches and development of antibodies to HLA after lung transplantation correlate with the development of bronchiolitis obliterans syndrome. *Transplantation* **65**: 648–653.

Tesi, R.J., Elkhammas, E.A., Henry, M.L. et al. (1993). Acute rejection episodes: best predictor of long-term primary cadaveric renal transplant survival. *Transplant Proc.* **25**: 901–902.

Theobald, M. (1995). Allorecognition and graft-versus-host disease. *Bone Marrow Transplant* **15**: 489–498.

Tsuji, K., Kusama, M., Nakatsuji, T. and Kato, S. (1985). Biological significance of Ss (serum soluble) HLA-class I antigens in bone marrow transplantation. *Tokai J. Exp. Clin. Med.* **10**: 169–174.

Valujskikh, A., Matesic, D., Gilliam, A. et al. (1998). T cells reactive to a single immunodominant self-restricted allopeptide induce skin graft rejection in mice. *J. Clin. Invest.* **101**: 1398–1407.

Van Rood, J., Van Leeuwan, A. and Van Santen, M. (1970). Anti-HLA-A2 inhibitor in normal sera. *Nature* **226**: 366.

Vasconcellos, L.M., Asher, F., Schachter, D. et al. (1998). Cytotoxic lymphocyte gene expression in peripheral blood leukocytes correlates with rejecting renal allografts. *Transplantation* **66**: 562–566.

Vella, J.P., Spadafora, F.M., Murphy, B. et al. (1997a). Indirect allorecognition of major histocompatibility complex allopeptides in human renal transplant recipients with chronic graft dysfunction. *Transplantation* **64**: 795–800.

Vella, J.P., Vos, L., Carpenter, C.B. and Sayegh, M.H. (1997b). Role of indirect allorecognition in experimental late acute rejection. *Transplantation* **64**: 1823–1828.

Waaga, A.M., Chandraker, A., Spadafora Ferreira, M. et al. (1998). Mechanisms of indirect allorecognition: characterization of MHC class II allopeptide-specific T helper cell clones from animals undergoing acute allograft rejection. *Transplantation* **65**: 876–883.

Wever, P.C., Boonstra, J.G., Laterveer, J.C. et al. (1998). Mechanisms of lymphocyte-mediated cytotoxicity in acute renal allograft rejection. *Transplantation* **66**: 259–264.

Williams, G.M., Hume, D.M., Hudson, R.P. Jr et al. (1968). "Hyperacute" renal-homograft rejection in man. *N. Engl. J. Med.* **279**: 611–618.

Wise, M.P., Bemelman, F., Cobbold, S.P. and Waldmann, H.T. (1998). Linked suppression of skin graft rejection can operate through indirect recognition. *J. Immunol.* **161**: 5813–5816.

Wong, W., Morris, P.J. and Wood, K.J. (1997). Pretransplant administration of a single donor class I major histocompatibility complex molecule is sufficient for the indefinite survival of fully allogeneic cardiac allografts: evidence for linked epitope suppression. *Transplantation* **63**: 1490–1494.

Zavazava, N., Leimenstoll, G. and Muller-Ruchholtz, W. (1990). Measurement of soluble MHC class I molecules in renal graft patients: a noninvasive allograft monitor. *J. Clin. Lab. Anal.* **4**: 426–429.

Zavazava, N., Bottcher, H. and Ruchholtz, W.M. (1993). Soluble MHC class I antigens (sHLA) and anti-HLA antibodies in heart and kidney allograft recipients. *Tissue Antigens* **42**: 20–26.

CHAPTER 24

HLA and Transplantation II: The Role of HLA Matching in Clinical Transplantation

Gerhard Opelz

INTRODUCTION

It has been described above how the first studies of kidney grafts from related donors yielded high success rates (Singal et al., 1969). This observation was made, even though only a limited number of human leucocyte antigen (HLA)-A and HLA-B specificities was known at the time (the HLA-DR locus had not yet been discovered). The available serological typing technique, which was primitive by current standards, allowed the identification of siblings who possessed identical HLA chromosomes. Had unrelated donors and recipients been typed rather than siblings, it would have been quite impossible to define the quantitative degree of HLA incompatibility given the incomplete knowledge of the HLA system at that time. This very point explains the controversy that has surrounded the role of HLA matching in cadaver kidney transplantation since the late 1960s. Significant associations of matching with graft outcome were published and contradicted numerous times. This is not surprising, when the simplistic matching schemes of the early years are compared with modern-day analyses based on current knowledge of HLA polymorphisms. The data below are predominantly extracted from the Collaborative Transplant Study (CTS), which represents the largest organ transplant database for transplants performed during the years 1986–1996. The study has been conducted since 1982 with 5-yearly reporting cycles. Centres that failed to respond to more than three consecutive mailings were excluded. Currently, transplant centres in 47 countries are actively participating: 298 kidney transplant centres, 101 cardiac transplant centres and 63 liver transplant centres. With each mailing, all data that a centre reported to the study are returned for verification and correction.

Graft and patient survival rates for the present report were computed according to the method of Kaplan and Meier (1958). No exclusions were made. Patients dying with functioning grafts were counted as graft failures. Log rank test and weighted regression analysis were used to determine statistical significance. Transplants performed from 1986 to 1996 were analysed.

Data from the USA have been collected by the United Network for Organ Sharing (UNOS) (Registry, 1997). They have collected data from over 62 000 kidney transplants since 1 January 1988, performed in 249 centres. In the following section the CTS data are discussed unless stated otherwise.

HLA CHROMOSOME COMPATIBILITY AND KIDNEY GRAFT OUTCOME

The well-established fact that compatibility for the HLA chromosomes has a major impact on kidney

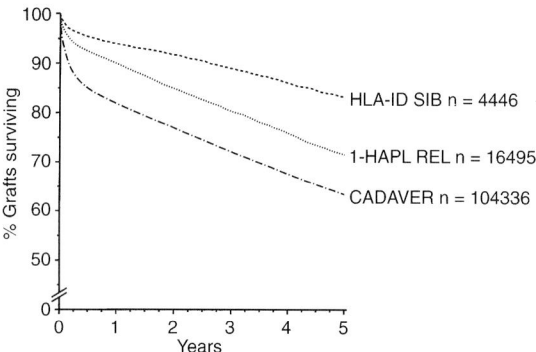

FIGURE 24.1 Kidney transplant survival according to compatibility of recipients and donors for the HLA chromosomes. Grafts from HLA-identical sibling donors (both HLA chromosomes matched) have the best outcome. Numbers of transplants studied are indicated at the end of each curve.

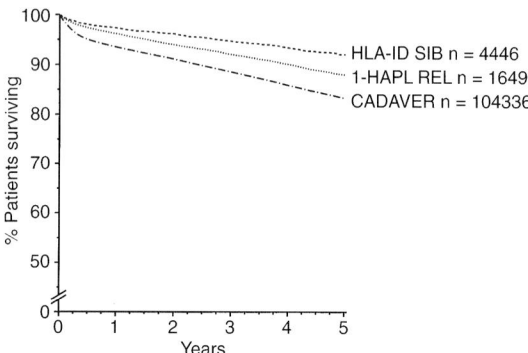

FIGURE 24.2 Patient survival rates in relation to compatibility for HLA haplotypes. The association of chromosome matching with patient survival is statistically highly significant (regression $p < 0.001$).

graft outcome is confirmed based on the transplant data obtained during the 1990s. As shown in Figures 24.1 and 24.2, both graft survival and patient survival are highest in transplants matched for both HLA chromosomes (HLA-identical siblings), followed by the results of transplants in which one HLA chromosome was shared between recipient and donor (HLA one-haplotype matched related transplants), whereas transplants from cadaver donors (two foreign HLA chromosomes) had the lowest success rate.

Figure 24.3 illustrates that this overriding HLA chromosome effect can also be demonstrated in second and third transplants. Importantly, whereas a "sensitization effect" in terms of a progressively lower success rate of second and third transplants is apparent for grafts with one or two mismatched HLA chromosomes, grafts from HLA-identical siblings have an exceptionally good outcome even if they are repeat transplants (after previous failures of transplants from HLA mismatched donors).

HLA MATCHING ANALYSIS OF SIBLING DONOR TRANSPLANTS

Among siblings, the constellation of HLA identity occurs in 25% of sibling pairs, whereas 50% of siblings are matched for one HLA chromosome (haplotype) and another 25% are mismatches for 2 HLA chromosomes (haplotypes). An analysis of

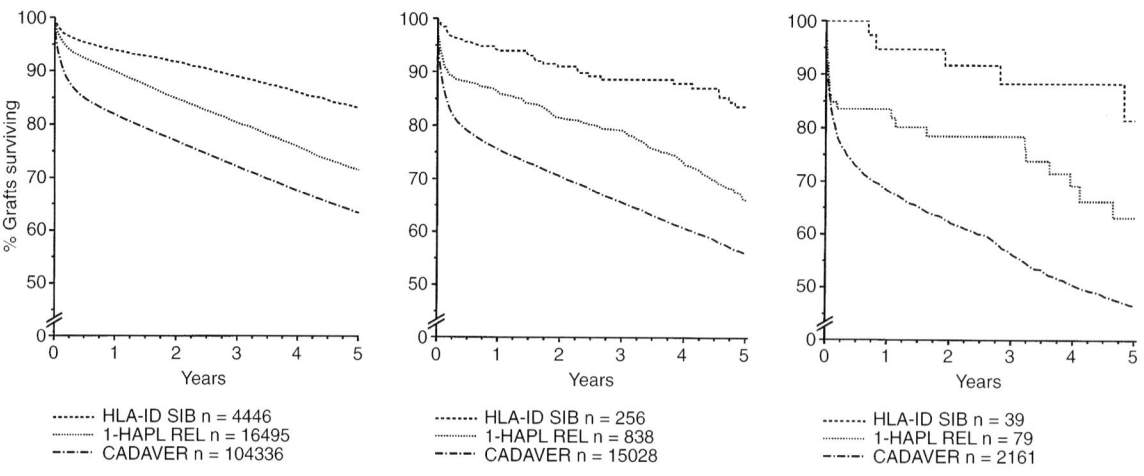

FIGURE 24.3 Effect of matching for HLA chromosomes on graft survival of first, second and third kidney transplants.

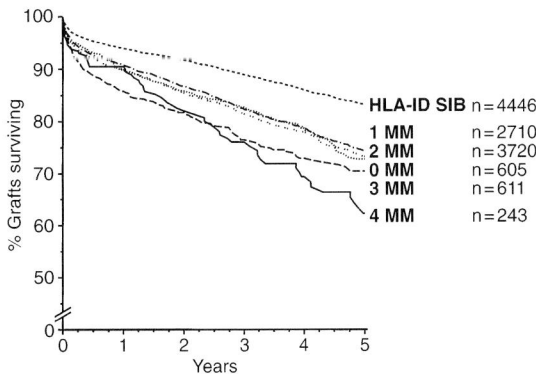

FIGURE 24.4 Analysis of kidney transplants from sibling donors. The numbers of HLA-A + B mismatches and numbers of patients studied are indicated. First grafts were analysed.

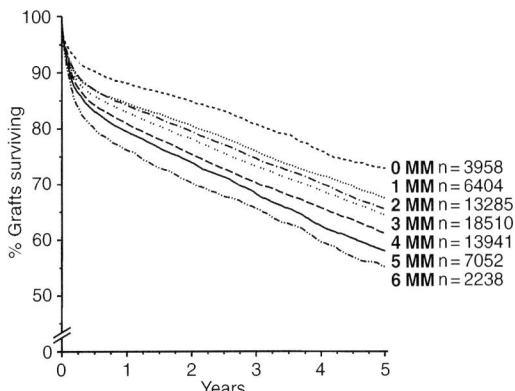

FIGURE 24.6 Combined effect of mismatches for HLA-A, HLA-B and HLA-DR antigens on survival of first cadaver kidney transplants. The association of matching with graft outcome was statistically significant (regression $p < 0.0001$).

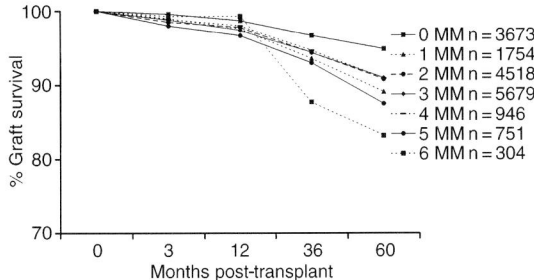

FIGURE 24.5 Graft survival of live-related renal allografts from the UNOS registry, according to number of mismatches for HLA-A, HLA-B and HLA-DR.

sibling donor transplants according to the number of HLA-A and -B locus mismatches is illustrated in Figure 24.4. The survival curves separate into three main groups: HLA-identical siblings, one-haplotype matched siblings (zero, one or two HLA-A + B mismatches), and two-haplotype mismatched siblings (three or four HLA-A + B mismatches). Similar data are available from the UNOS registry and are shown in Figure 24.5.

HLA COMPATIBILITY AND OUTCOME OF CADAVERIC KIDNEY TRANSPLANTS

The influence of mismatches at the HLA-A, -B and -DR loci is illustrated in Figure 24.6. Graft survival decreases stepwise from zero to six A-, B-, DR mismatches (regression $p < 0.0001$). Because our analyses have consistently shown that the consideration of broad specificities at the HLA-A and -B loci (e.g. A9 or B12) does not result in as good a correlation with outcome as the analysis of split specificities (e.g. A23, A24 or A44, A45), the current analysis was restricted to transplants for which split specificities were reported to the study centre for both the recipient and the donor. Again, similar results are available from the UNOS registry and are shown in Figure 24.7.

In addition to a significant impact on graft survival, HLA mismatches also have a statistically highly significant influence on patient survival (Figure 24.8, regression $p < 0.0001$). In all likelihood, the association of matching with patient

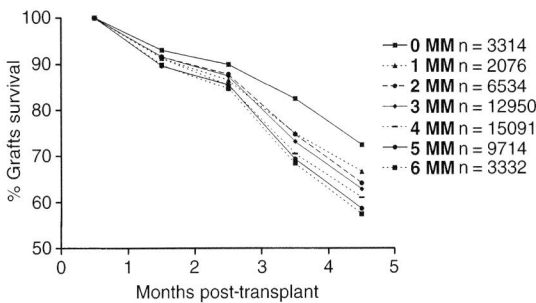

FIGURE 24.7 Graft survival of cadaveric renal allografts from the UNOS registry, according to number of mismatches for HLA-A, HLA-B and HLA-DR.

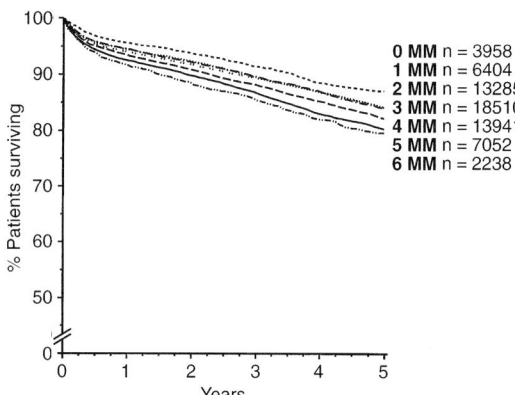

FIGURE 24.8 Effect of HLA matching on patient survival in recipients of first cadaver kidney grafts (regression $p < 0.0001$).

survival is due to clinical complications (e.g. infections) resulting from the necessity of intensified immunosuppressive treatment in patients with immunological rejections.

When the effect on graft outcome of the HLA-A locus, HLA-B locus and HLA-DR locus is analysed separately, it is apparent that all three loci have an influence on graft outcome. During the first year after transplantation, the influence of HLA-DR mismatches outweighs that of HLA-A or HLA-B mismatches (Figure 24.9). During subsequent years of follow-up, however, all three loci can be seen to have a similar impact (Figure 24.10). The influence of the three loci is additive, so that the combination of HLA-A and HLA-B or the combination of HLA-B and HLA-DR has a stronger impact than any individual HLA locus. The strongest influence, however, is seen in a combined analysis of the three loci, as shown for the period following the first year in Figure 24.11 and for the overall analysis in Figure 24.6.

Even though the DR locus has a stronger influence than the A or B locus during the early post-transplant period, if long-term graft outcome is considered, the simple addition of the number of mismatches at the three loci gives a nearly optimal result which, for practical purposes and reasons of transparency, is the analysis method preferred by the CTS.

As zero mismatched transplants are only accomplished in a limited number of cases, a great deal of interest has developed in the concept of permissible mismatches. If certain mismatches are acceptable then the demands placed on any matching system are less stringent and a higher proportion of patients could potentially receive "functionally matched" grafts. Genotypically distinct HLA antigens can be grouped into public or cross-reactive groups (CREG) which consist of sets of allelic types with similar reactivities to antibodies with broad

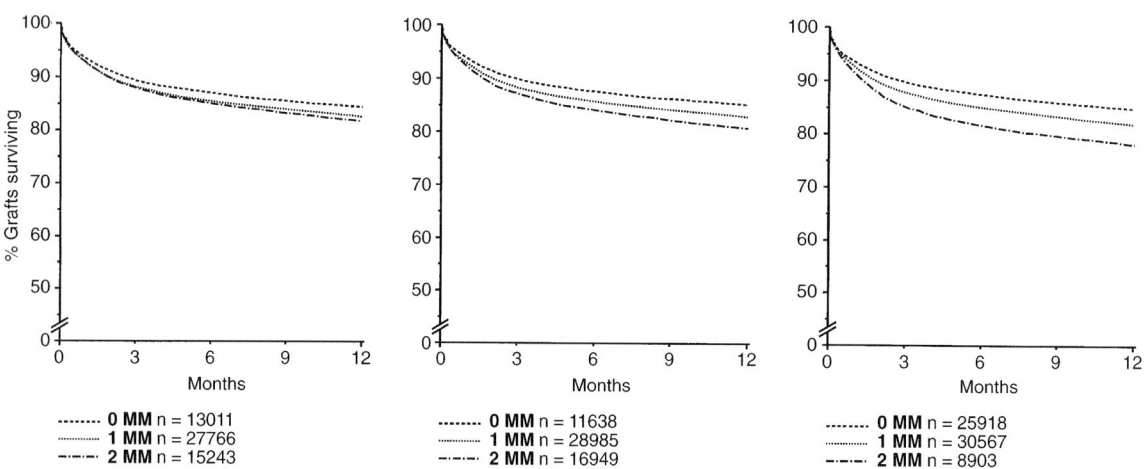

FIGURE 24.9 Separate analysis of HLA-A, HLA-B or HLA-DR locus mismatches on graft outcome during the first post-transplant year. The influence was smallest for the A-locus, intermediate for the B-locus and strongest for the DR-locus.

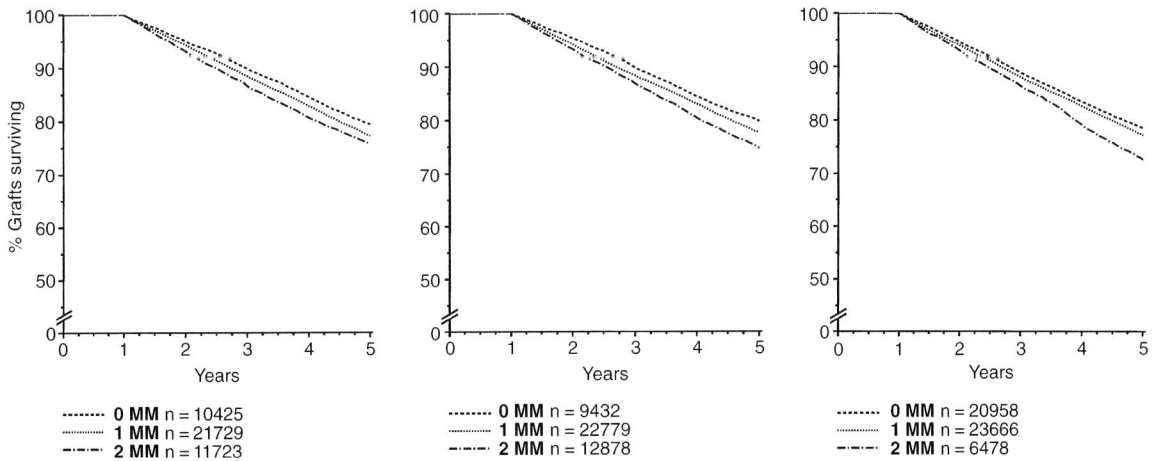

FIGURE 24.10 Separate analysis of the three HLA loci during the period following the first year. The three loci had a similar impact.

specificity. Antibodies to CREG are more common than those against private epitopes, suggesting that these specificities are perhaps more important (Rodey et al., 1994). One analysis of UNOS data predicted that CREG matched transplants with no DR mismatches could be anticipated in 43% of all transplants (Takemoto et al., 1994). Moreover, in these cases graft survival could be expected to be intermediate between that of zero mismatch and one mismatch transplants. Such a system would also partially redress some of the racial imbalance inherent in the American system, where Blacks are underrepresented in the donor pool and therefore undergo a disproportionately low number of transplants owing to HLA incompatibility (Cecka, 1997). Similarly, small studies in Europe and the USA, usually studying patients with one mismatch, suggest that certain mismatches are more acceptable than others (Doxiadis et al., 1996; Starzl et al., 1997).

However, the CTS has not been able to confirm these reports in the literature on the reliable identification of "permissible" or "taboo" mismatches, or claims concerning the superiority of CREG matching.

INFLUENCE OF SHORT PRESERVATION TIME

It has been suggested that the deleterious effect of HLA mismatches could be eliminated by keeping the cold ischaemic preservation time very short. Such a strategy would favour local transplantation and argue against the sharing of kidneys over long distances. The CTS data demonstrate that kidneys with short ischaemic preservation of < 6 h still show a very strong effect of HLA matching (Figure 24.12). Moreover, shared, well-matched kidneys have better success rates than locally transplanted, poorly matched ones, regardless of the length of ischaemia (Figure 24.13).

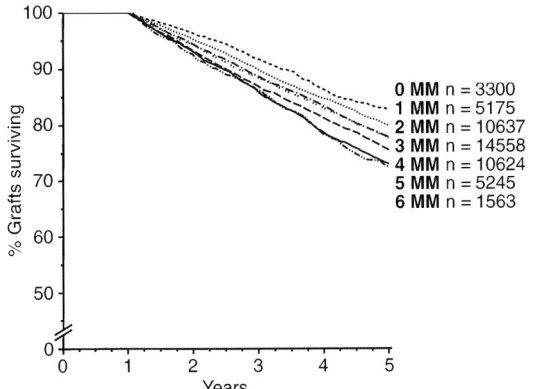

FIGURE 24.11 Combined influence of HLA-A + B + DR-locus mismatches during the period following the first post-transplant year.

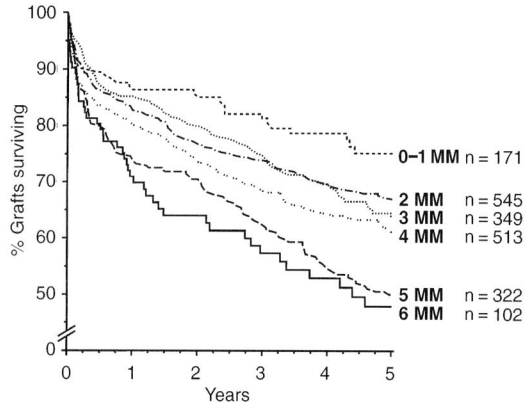

FIGURE 24.12 Effect of HLA-A + B + DR matching on transplants from first cadaver donors with < 6 h of ischaemic preservation time (regression $p < 0.0001$).

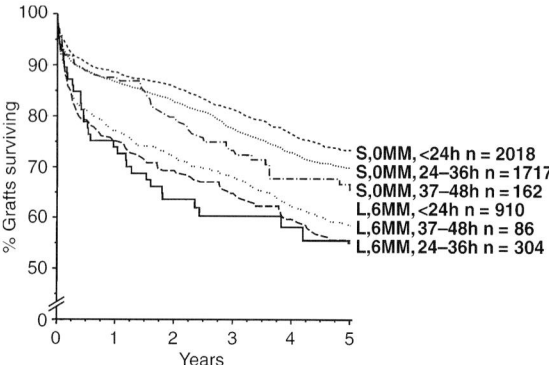

FIGURE 24.13 Comparison of well-matched (0 HLA-A + B + DR mismatch) shipped (S) kidneys with ischaemia times of < 2 h, 24–36 and 37–48 h, with poorly matched (six HLA-A + B + DR mismatches) locally transplanted (L) kidneys. Well-matched (S) kidneys outperformed poorly matched (L) kidneys, regardless of ischaemia time.

TRANSPLANTS FROM LIVE UNRELATED DONORS

Recently, there has been a lively debate regarding the use of live unrelated donors. This issue has gained prominence with the growing pressure to augment the donor pool to cope with increasing demands for transplantation. A study based on the UNOS registry showed that HLA identical grafts from siblings had the best survival, but that one-haplotype matched grafts from child to parent, or vice versa, had the same 3-year survival as unrelated grafts between spouses or friends, which in

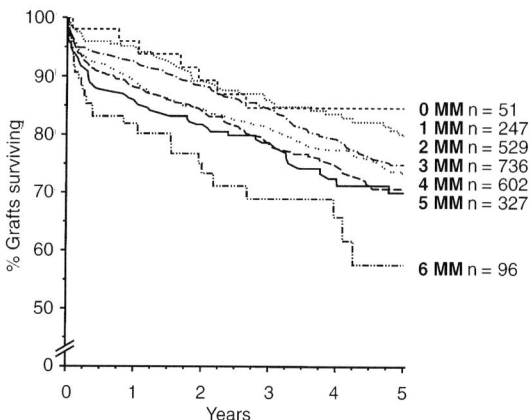

FIGURE 24.14 Influence of HLA-A + B + DR mismatches on outcome of kidney grafts from live unrelated donors (regression $p < 0.001$).

turn had a mean of 4.1 mismatches (Terasaki et al., 1995). These data have subsequently been confirmed in 1396 unrelated grafts, with an overall graft survival at 3 years of 83%, compared with 83% in parent-to-child grafts and 84% in child-to-parent grafts (Suzuki et al., 1997). The authors therefore argued that if completely healthy kidneys could be obtained from cadavers, in particular without *pre mortem* damage, then these kidneys should yield the same survival as parental donor grafts. However, claims that the good overall success rate of kidney grafts from live unrelated donors argues against an effect of HLA matching under "ideal conditions" are contradicted by the results from the CTS (Opelz, 1997). Figure 24.14 demonstrates a highly significant impact of HLA matching on the outcome of kidney transplants from live unrelated donors.

INFLUENCE OF PREFORMED LYMPHOCYTOTOXIC ANTIBODIES

The effect of HLA matching is stronger in patients with preformed antibodies than in patients without evidence of presensitization. This is evident both in first cadaveric kidney transplants (Figure 24.15) and in retransplants (Figure 24.16). Thus, sensitization of the recipients against HLA antigens enhances the matching effect, even though direct antidonor sensitization is routinely excluded in the pretransplant cross-match tests.

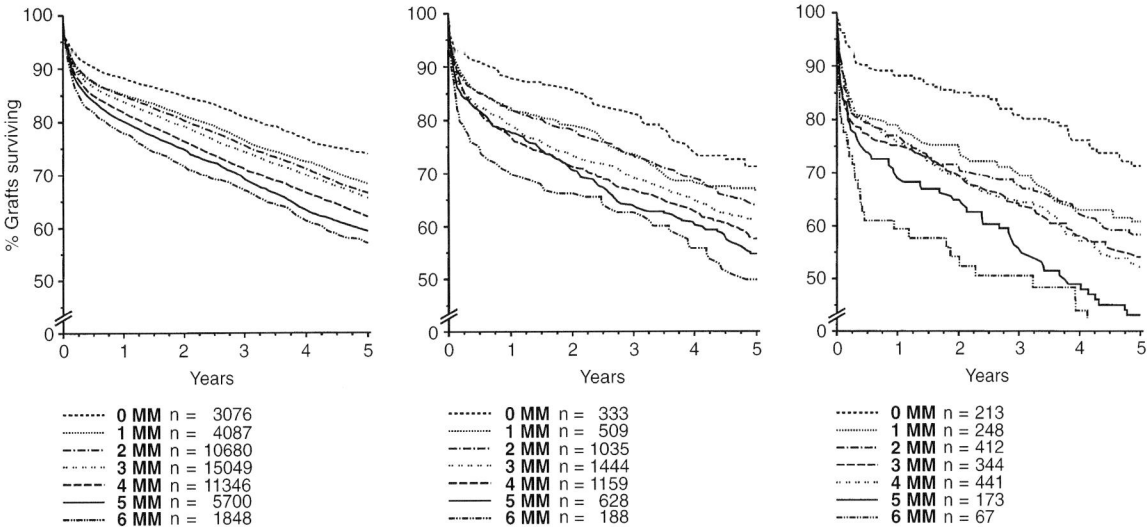

FIGURE 24.15 Analysis of HLA-A + B + DR mismatches on outcome of first cadaver kidney transplants in recipients with < 10%, 10–50% or > 50% preformed lymphocytotoxic antibodies. The association of matching with graft outcome was statistically significant in all three analyses (regression $p < 0.00011$).

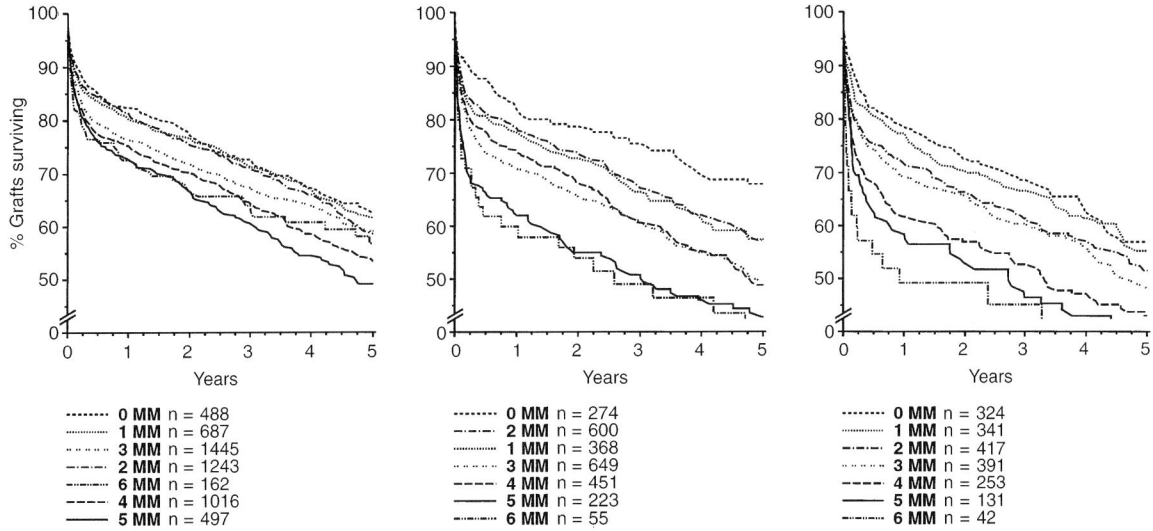

FIGURE 24.16 Effect of HLA-A + B + DR matching on outcome of cadaver donor retransplants in patients with < 10%, 10–50% or > 50% antibody reactivity (regression $p < 0.00011$).

HLA COMPATIBILITY AND HEART TRANSPLANTATION

The outcome of cardiac transplants is statistically significantly influenced by the number of HLA-A, -B, -DR mismatches. Figure 24.17 illustrates the results for transplants performed from 1986 to 1996, confirming an earlier CTS report (Opelz et al., 1994). This conclusion is supported by data from the UNOS registry, for both heart and lung transplantation (Hosenpud et al., 1996).

These results of HLA matching in cardiac transplantation are of special interest because they demonstrate an HLA matching effect in a situation where no effort of prospective matching was made. In cadaver kidney transplantation, the argument has been put forward that donor kidneys with good HLA matches might be assigned

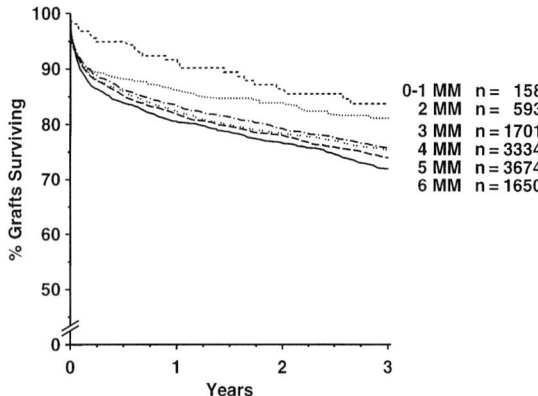

FIGURE 24.17 Association of HLA-A + B + DR matching with outcome of first heart transplants (regression $p < 0.0001$.)

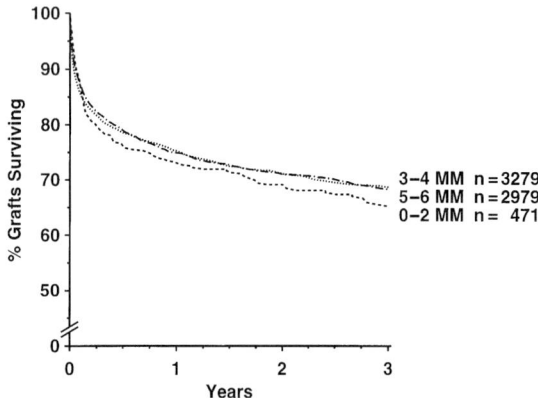

FIGURE 24.18 Analysis of HLA-A + B + DR matching in first liver transplants. There was no significant impact of matching on graft survival.

preferentially to good transplant candidates who are clinically stable, whereas clinically unfavourable patients might receive poorly matched grafts. It was reasoned that this type of selection might produce an artificial HLA matching effect. The heart transplant results, however, strongly argue against such a hypothesis. Based simply on a random assignment of donor hearts without any prospective consideration of the HLA match, the retrospective analysis shows a significant impact of matching in spite of the fact that heart transplant recipients receive stronger immunosuppressive treatment than kidney recipients.

HLA MATCHING AND LIVER TRANSPLANTS

The CTS data do not show any effect of HLA matching in liver transplantation. The results for HLA-A, -B, -DR are shown in Figure 24.18. Separate analysis of the three loci did not show any significant effect. It is intriguing that some studies have shown a dualistic effect of HLA on liver transplantation. In these studies fewer episodes of rejection occur in well-matched transplants but these same grafts have a significantly lower survival, especially if matched for HLA-DR (Markus et al., 1988). The putative mechanism for this survival disadvantage involves the possibility that if the HLA antigens are matched then the disease-producing epitopes, in either autoimmune or chronic viral disease, are recreated on the graft, permitting the reactivation of the original pathogenic T cell clones. This hypothesis remains unproven but a recent study suggests that a partial class I match is related to the development of acute rejection in patients with preoperative viral infection (Ontanon et al., 1998). Another study also suggested that full compatibility for HLA-A and HLA-B might be deleterious (Donaldson et al., 1993). However, some studies have shown benefits of HLA matching, especially for HLA-DR (Nikaein et al., 1994).

IMPACT OF MOLECULAR TYPING TECHNIQUES

With the collaboration of more than 100 tissue typing laboratories and transplant centres, the CTS is conducting a large DNA typing project in order to compare the HLA typing results obtained with serological or DNA typing methods, and to evaluate any potential advantage of improved HLA typing using DNA techniques. An update of previously published results is provided (Opelz et al., 1991, 1997; Mytilineos et al., 1997a, b).

MATCHING FOR HLA-DRB

The key result comparing cadaver kidney transplants that were reported as HLA-A, -B, -DR matched based on serological typing is shown in Figure 24.19. If the absence of an HLA-DR

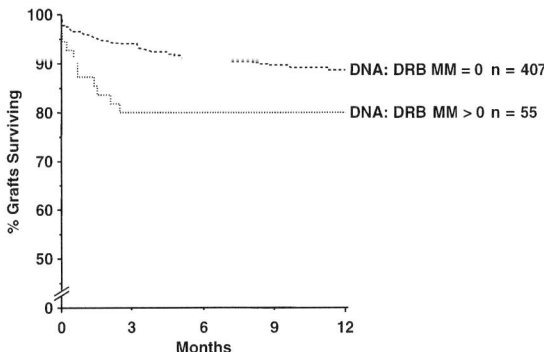

FIGURE 24.19 Impact of DNA typing on association of HLA matching with graft survival. All transplants were reported to the Collaborative Transplant Study as having no mismatches for HLA-A + B + DR. Retrospective DNA typing revealed an HLA-DR mismatch in 55 cases. The survival rate of "truly matched" transplants was significantly better than that of the 55 grafts which turned out to be mismatched (log rank $p < 0.04$).

mismatch was confirmed by DNA typing, graft survival was significantly higher than when retrospective DNA typing revealed a mismatch that had not been recognized by serological typing.

HLA-A AND HLA-B

A similar result was obtained in a recent extension of the project to DNA typing for HLA-A and HLA-

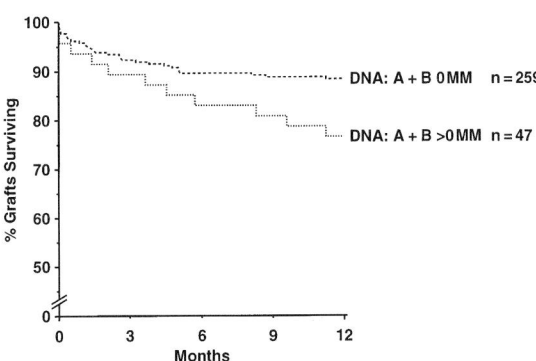

FIGURE 24.20 Cadaveric transplant survival rates comparing patients in whom the absence of HLA-A and HLA-B mismatches was confirmed by DNA typing with patients in whom DNA typing revealed mismatches that had not been detected by serological typing (log rank $p = 0.02$).

B. Among transplants reported to the CTS as having zero mismatches, those confirmed as zero mismatches by DNA typing had a significantly better graft outcome than those shown to be mismatched (Figure 24.20).

HLA-DR SPLIT SPECIFICITIES

Among first cadaveric kidney grafts, the analysis of HLA-DR matching showed no significant difference between an analysis of the broad specificities DR1–DR10 or the additional split specificities DR11–DR18 (Figure 24.21). Among retransplants, however, extended matching including the DR11–DR18 splits resulted in an improved correlation of matching with graft survival (Figure 24.22).

HLA-DP

Typing for HLA-DP is not routinely done and the HLA-DP antigens are not considered for clinical histocompatibility matching. Based on retrospective HLA-DP typing based on DNA methodology, an analysis of a possible impact of DP mismatches on kidney graft survival was performed. Whereas HLA-DP had no influence on the outcome of

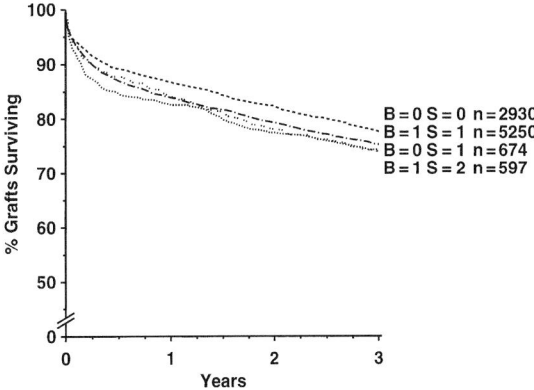

FIGURE 24.21 Analysis of matching for broad or split specificities at the HLA-DR locus. First cadaver kidney grafts were analysed. Whether the consideration of broad (B) or split (S) specificities resulted in the same number of mismatches (B = 0, S = 0 or B = 1, S = 1) or whether the number of mismatches increased when splits were considered (B = 0, S = 1 or B = 1, S = 2) did not make much difference.

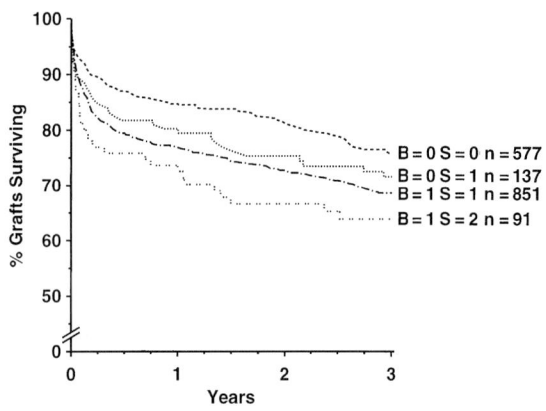

FIGURE 24.22 Analysis of broad or split mismatches for HLA-DR in cadaver donor retransplants. If consideration of splits resulted in a higher number of mismatches (B = 0, S = 1 or B = 1, S = 2), the graft survival rate decreased.

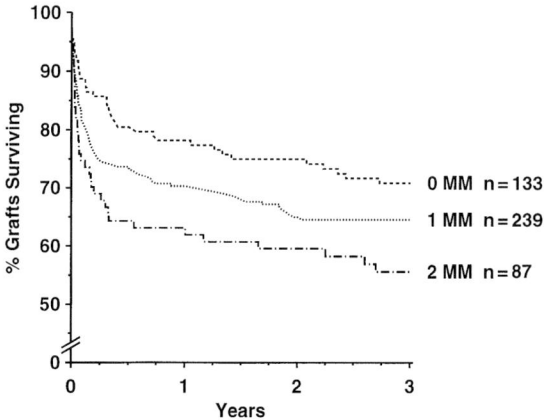

FIGURE 24.24 Impact of HLA-DP locus mismatches on survival rate of cadaver donor retransplants in patients with >50% preformed lymphocytotoxic antibodies (regression $p = 0.02$).

primary kidney grafts, there was a significant influence in retransplants (Figure 24.23). The DP matching effect was pronounced when retransplants in patients with preformed lymphocytotoxic antibodies were analysed (Figure 24.24).

BONE-MARROW TRANSPLANTATION

Most bone-marrow transplants have previously been carried out between siblings who are HLA identical and thus attention has been focused on the immune responses to minor antigens. However an increasing number of transplants have carried out between matched unrelated pairs and one haplotype-matched siblings. These results clearly show that the degree of HLA mismatch is significantly related to both the graft failure and the incidence of graft-versus-host disease (GVHD) (Anasetti et al., 1989, 1990; Hansen et al., 1998). The degree of HLA mismatch has also been shown to be a significant predictor of outcome after placental blood transplantation from unrelated donors (Rubinstein et al., 1998).

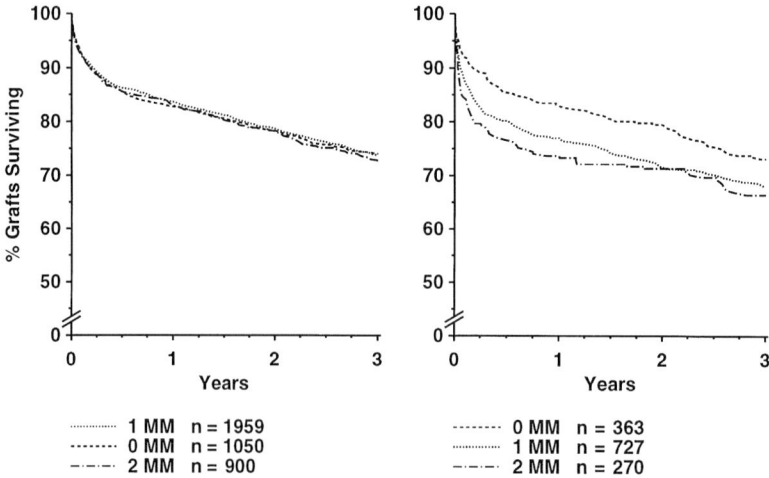

FIGURE 24.23 Analysis of HLA-DP mismatches in primary or retransplants from cadaver donors. There was no significant influence in first grafts, whereas there was a significant influence in regrafts (regression $p = 0.04$).

When serotypically matched volunteer unrelated donors (VUD) were reanalysed retrospectively according to HLA-DRB1 genotype it was found that a number of patients was mismatched at the molecular level (Petersdorf et al., 1995). For patients who were truly genotypically matched at this locus, overall survival was increased and the incidence of GVHD was significantly decreased. Two major recent studies have looked back at large series of VUD, reanalysing their patients according to genotype. They found that indeed a number of patients is actually mismatched at the molecular level. These mismatched patients had significantly more GVHD and worse survival compared with the truly matched patients (Petersdorf et al., 1998; Sasazuki et al., 1998). Herein lies one of the great problems with tissue typing. The greater the degree of matching at a molecular level, then the greater the benefit in terms of outcome, but as the molecular resolution becomes finer then a diminishing number of patients will have a suitable matched donor (van Rood and Oudshoorn, 1998).

ORGAN ALLOCATION AND HLA MATCHING

Liver transplantation

With respect to liver transplantation, the results of the retrospective analysis shown herein provide no justification for prospective HLA matching. This applies to the HLA-A, -B, and -DR loci; it leaves open the possibility that other HLA loci may have an impact on graft outcome. It is not known why the HLA-A, -B and -DR loci do not influence liver graft survival. There has been speculation that the large lymphatic mass of the transplanted liver may induce immunological unresponsiveness or alternatively that the release of soluble HLA molecules may be contributory (see below).

Heart transplantation

In contrast, heart transplants are clearly influenced by the degree of compatibility for HLA-A, -B and -DR. Initial attempts at prospective matching are currently made in Great Britain. The accepted limit for cold ischaemic preservation of donor hearts is approximately 5 h. This restriction makes large-scale organ sharing problematic, although regional sharing over shorter distances is possible, especially if HLA typing is performed prior to donor organ removal. It remains to be seen whether prospective HLA matching will gain acceptance in the clinical heart transplant community. The advent of rapid molecular typing has clearly been beneficial towards making such efforts possible (Bunce et al., 1997).

Kidney transplantation

For cadaver kidney grafts, HLA matching has been practised since the 1970s, although the allocation criteria have varied greatly among different geographical regions. Probably the most stringent matching policy has been followed by the Eurotransplant organization. Much has been learned during the many years of experience. Accordingly, the matching policies have been adjusted over time. Because of the extensive polymorphism of the HLA system, large pools of potential transplant recipients have to be formed in order to increase the likelihood of obtaining good matches. For that reason, national and international waiting pools containing several thousand patients have been formed. The Eurotransplant waiting list currently contains more than 10 000 patients, while the waiting list in the USA contains more than 30 000 patients.

The main problem with respect to donor kidney allocation according to the HLA match is that patients with rare HLA phenotypes are at a disadvantage. They are usually bypassed in the allocation process and become "long waiters" who sometimes spend 10 years or more waiting for a donor kidney. Since rare HLA phenotypes by definition are rare, cadaveric donors usually have phenotypes that match better with patients who have a common HLA phenotype. As the years passed, patients on the Eurotransplant waiting list no longer presented a normal distribution of HLA phenotypes. Instead, rare phenotypes were overrepresented and the fraction of long-waiting patients continuously increased. In the USA, a similar problem was recognized, with the added difficulty that Black patients received transplants at an inadequate rate because they possessed rare HLA specificities. It was therefore suggested that HLA matching should be disregarded altogether and that donor organs should be allocated based on other criteria, such as the length of time for which a patient was listed on the waiting list.

The author has taken a different approach, with the idea that patients with rare HLA phenotypes should not have to wait unreasonably long for a transplant, but that the benefits of HLA matching should be taken advantage of as well as possible. With this philosophy in mind, an allocation model was developed that considered, among other factors, the likelihood of a patient receiving a well-matched kidney (Wujciak and Opelz, 1993a, b). For a patient with a rare HLA phenotype, the "most likely achievable match" was calculated and an appropriate number of points was assigned to a weighted point algorithm. Because points are also given for waiting time, a patient with a rare phenotype accumulates waiting-time points, which in turn lower the limit of acceptable HLA compatibility. As a result, patients with rare phenotypes on average still do not receive as many well-matched grafts as patients with common phenotypes. However, the best possible overall distribution of good HLA matches is achieved and excessively long waiting times are avoided. Computer simulations show that with this type of donor kidney allocation, approximately 8000 years of graft function can be gained for every 10 000 transplanted patients who are followed over a 10-year period, in comparison to a random allocation of organs or allocation simply based on the waiting time in the pool.

This kidney allocation model was implemented at Eurotransplant in the spring of 1996. Several modifications were made; for example, priority points for paediatric recipients were added. A first analysis 18 months after implementation showed that the new allocation algorithm works exceedingly well (De Meester et al., 1998). More long-waiting patients have been transplanted, while the fraction of excellent HLA matches (0 HLA-A + B + DR mismatches) was a high 23%. This dynamic and self-adjusting system is thought to be the best currently available system for the allocation of cadaver donor kidneys. It allows the utilization of HLA matching with its well-established benefits, without discriminating against patients with rare HLA phenotypes.

REFERENCES

Anasetti, C.D., Amos, P.G., Beatty, F.R. et al. (1989). Effect of HLA compatibility on engraftment of bone marrow transplants in patients with leukemia or lymphoma. *N. Engl. J. Med.* **320**: 197.

Anasetti, C.P.G., Beatty, R., Storb, P.J. et al. (1990). Effect of HLA incompatibility on graft-versus-host disease, relapse, and survival after marrow transplantation for patients with leukemia or lymphoma. *Hum. Immunol.* **29**: 79.

Bunce, M., Young, N.T. and Welsh, K.I. (1997). Molecular HLA typing – the brave new world. *Transplantation* **64**: 1505.

Cecka, J.M. (1997). The role of HLA in renal transplantation. *Hum. Immunol.* **56**: 6.

De Meester, J., Persijn, G.G., Wujciak, T. et al. (1998). The new Eurotransplant Kidney Allocation System: report one year after implementation. Eurotransplant International Foundation. *Transplantation* **66**: 1154.

Donaldson, P., Underhill, J., Doherty, D. et al. (1993). Influence of human leukocyte antigen matching on liver allograft survival and rejection: "the dualistic effect". *Hepatology* **17**: 1008.

Doxiadis, II, Smits, J.M., Schreuder, G.M. et al. (1996). Association between specific HLA combinations and probability of kidney allograft loss: the taboo concept. *Lancet* **348**: 850.

Hansen, J.A., Gooley, T.A., Martin, P.J. (1998). Bone marrow transplants from unrelated donors for patients with chronic myeloid leukemia. *N. Engl. J. Med.* **338**: 962.

Hosenpud, J.D., Edwards, E.B., Lin, H.M. et al. (1996). Influence of HLA matching on thoracic transplant outcomes. An analysis from the UNOS/ISHLT Thoracic Registry. *Circulation* **94**: 170.

Kaplan, E. and Meier, P. (1958). Nonparametric estimation from incomplete observations. *J. Am. Statist. Assoc.* **53**: 457.

Markus, B.H., Duquesnoy, R.J. and Gordon, R.D. (1988). Histocompatibility and liver transplant outcome. Does HLA exert a dualistic effect? *Transplantation* **46**: 372.

Mytilineos, J., Deufel, A. and Opelz, G. (1997a). Clinical relevance of HLA-DPB locus matching for cadaver kidney retransplants: a report of the Collaborative Transplant Study. *Transplantation* **63**: 1351.

Mytilineos, J.M., Lempert, D. Middleton, F. et al. (1997b). HLA class I DNA typing of 215 "HLA-A, -B, -DR zero mismatched" kidney transplants. Clinical relevance of HLA-DPB locus matching for cadaver kidney retransplants: a report of the Collaborative Transplant Study. *Tissue Antigens* **50**: 355.

Nikaein, A., Backman, L., Jennings, L. et al. (1994). HLA compatibility and liver transplant outcome. Improved patient survival by HLA and cross-matching. *Transplantation* **58**: 786.

Ontanon, J., Muro, M., Garcia-Alonso, A.M. et al. (1998). Effect of partial HLA class I match on acute rejection in viral pre-infected human liver allograft recipients. *Transplantation* **65**: 1047.

Opelz, G. (1997). Impact of HLA compatibility on survival of kidney transplants from unrelated live donors. *Transplantation* **64**: 1473.

Opelz, G., Mytilineos, J., Scherer, S. et al. (1991). Survival of DNA HLA-DR typed and matched cadaver kidney transplants. The Collaborative Transplant Study. *Lancet* **338**: 461.

Opelz, G., Wujciak, T., Wujciak, T. et al. (1994). The influence of HLA compatibility on graft survival after heart transplantation. *N. Engl. J. Med.* **330**: 816.

Opelz, G., Scherer, S. and Mytilineos, J. (1997). Analysis of HLA-DR split-specificity matching in cadaver kidney transplantation: a report of the Collaborative Transplant Study. *Transplantation* **63**: 57.

Petersdorf, E.W., Longton, G.M., Anasetti, C. et al. (1995). The significance of HLA-DRB1 matching on clinical outcome after HLA-A, B, DR identical unrelated donor marrow transplantation. *Blood* **86**: 1606.

Petersdorf, E.W., Gooley, T.A., Anasetti, C. et al. (1998). Optimizing outcome after unrelated marrow transplantation by comprehensive matching of HLA class I and II alleles in the donor and recipient. *Blood* **92**: 3515.

Registry, U.D. (1997). 1997 UNOS Annual Data Report.

Rodey, G.E., Neylan, J.F., Whelchel, J.D. et al. (1994). Epitope specificity of HLA class I alloantibodies. I. Frequency analysis of antibodies to private versus public specificities in potential transplant recipients. *Hum. Immunol.* **39**: 272.

Rood, J.J. van and Oudshoorn, M. (1998). The quest for a bone marrow donor – optimal or maximal HLA matching? [Editorial; comment]. *N. Engl. J. Med.* **339**: 1238.

Rubinstein, P., Carrier, C., Scaradavou, A. et al. (1998). Outcomes among 562 recipients of placental-blood transplants from unrelated donors. *N. Engl. J. Med.* **339**: 1565.

Sasazuki, T., Juji, T., Morishima, Y. et al. (1998). Effect of matching of class I HLA alleles on clinical outcome after transplantation of hematopoietic stem cells from an unrelated donor. Japan Marrow Donor Program. *N. Engl. J. Med.* **339**: 1177.

Singal, D.P., Mickey, M.R. and Terasaki, P.I. (1969). Serotyping for homotransplantation. 23. Analysis of kidney transplants from parental versus sibling donors. *Transplantation* **7**: 246.

Starzl, T. E., Eliasziw, M., Gjertson, D. et al. (1997). HLA and cross-reactive antigen group matching for cadaver kidney allocation. *Transplantation* **64**: 983.

Suzuki, M., Cecka, M.J.M. and Terasaki, P.I. (1997). Unrelated living donor kidney transplants. *Br. Med. Bull.* **53**: 854.

Takemoto, S., Terasaki, P.I. and Gjertson, D.W. (1994). Equitable allocation of HLA-compatible kidneys for local pools and minorities. *N. Engl. J. Med.* **331**: 760.

Terasaki, P.I. Cecka, J.M. and Gjertson, D.W. (1995). High survival rates of kidney transplants from spousal and living unrelated donors. *N. Engl. J. Med.* **333**: 333.

Wujciak, T. and Opelz, G. (1993a). Computer analysis of cadaver kidney allocation procedures. *Transplantation* **55**: 516.

Wujciak, T. and Opelz, G. (1993b). A proposal for improved cadaver kidney allocation. *Transplantation* **56**: 1513.

CHAPTER 25

HLA and Psychiatric Disease

Padraig Wright, Vishwajit Nimgaonkar, Rohan Ganguli and Robin M. Murray

INTRODUCTION

So far, it has not been possible to determine the cause of a single major psychiatric disease. Research is thwarted by heterogeneous clinical presentation, the uncertain validity of subjective diagnosis and the probability that any individual psychiatric disorder represents a stereotyped response by the brain to an array of aetiological agents. Furthermore, most psychiatric diseases probably result from interactions between multiple predisposing genes and one or more environmental influences. Therefore, any single aetiological agent, whether genetic or environmental, is likely to explain only a small proportion of the variance. The search for such aetiological agents has included detailed investigations of putative aetiological factors associated with non-psychiatric diseases. One such disease, narcolepsy, both exhibits the most marked human leucocyte antigen (HLA) association known and is an archetypal neuropsychiatric disease. It is not surprising, therefore, that HLA association studies have been undertaken in a variety of psychiatric diseases. This chapter will review HLA association studies in schizophrenia, affective disorders, anxiety disorders, eating disorders, autism and Tourette's syndrome. Potential HLA associations with the efficacy and safety of psychotropic medication will also be addressed.

SCHIZOPHRENIC DISORDERS

Having a schizophrenic relative is the single most powerful risk factor for schizophrenia (Gottesman, 1991) but no major susceptibility genes have been identified to date. At least 33 class I HLA association studies of schizophrenia have been undertaken since 1974 and a number of these has reported a weak but significant association with HLA-A9 (Eberhard et al., 1975; Julien et al., 1977; Asaka et al., 1981; Wright et al., 1995), -A10 (Smeraldi et al., 1976a; Crowe et al., 1979; Asaka et al., 1981) and -A28 (Ivanyi et al., 1976; Ivanyi et al., 1978; Rosler et al., 1980). These results are difficult to interpret because strict diagnostic criteria were not always applied and control subjects were sometimes inappropriate. McGuffin and Sturt (1986) have provided a meta-analysis of class I schizophrenia association studies and have concluded that, while schizophrenia as a whole is probably not an HLA-associated disorder, paranoid schizophrenia may be weakly associated with HLA-A9. Most investigations of the HLA-A23 and -A24 subspecificities of -A9 have not found any significant associations (Crowe et al., 1979; Alexander et al., 1990; Gibson et al., 1997), although the excess of -A9 in the present authors' sample of Caucasian schizophrenic patients was almost wholly accounted for by -A24 (Wright et al., 1995). Turner (1979) provided inconclusive evidence (recombination fraction = 0.15, lod score = 2.57) for linkage of an HLA association with schizophrenia spectrum disorders (referred to as "schizotaxia" by the author), but McGuffin et al. (1983) were unable to replicate this finding in another sample.

Class II HLA studies in schizophrenia have been less numerous but more rigorous than class I investigations [Miyanaga et al., 1984; Ruddock et al., 1984a, b; Rabin et al., 1987 (quoted in Nimgaonkar et al., 1992); Metzer et al., 1988; Nimgaonkar et al., 1992, 1995, 1997; Sasaki et al., 1994; Zamani et al., 1994; Blackwood et al., 1996; Gibson et al., 1997; Wright et al., 1997a, b]. A number of serotyping studies has reported associations with HLA-DR8 in Japanese (Miyanaga et al., 1984), -DR6 in African–American [Rabin et al., 1987 (quoted in Nimgaonkar et al., 1992)] and -B35 in Scottish patients (Blackwood et al., 1996). Other investigators have been unable to replicate the association of either HLA-DR8 or -DR6 with schizophrenia, while HLA-B35 awaits further studies.

All class II HLA genotyping studies reported to date have used operationalized diagnostic criteria. One study has studied the HLA-DPB1 locus (Zamani et al., 1994), five at the -DQB1 locus (Nimgaonkar et al., 1992, 1995, 1997; Wright et al., 1996, 1997a) and six at the -DRB1 locus (Sasaki et al., 1994; Zamani et al., 1994; Wright et al., 1996, 1997a, b; Gibson et al., 1997), and patients of African–American (Nimgaonkar et al., 1992, 1997), Caucasian (Nimgaonkar et al., 1992; Zamani et al., 1994; Wright et al., 1996, 1997a, b), Japanese (Sasaki et al., 1994) and Chinese (Nimgaonkar et al., 1995) ethnicity have been studied. Insulin-dependent diabetes mellitus (IDDM) may be inversely associated with schizophrenia (Finney, 1989) and the presence of DQB1*0602 appears to protect against it (Pugliese et al., 1995). The HLA-DQB1 locus has therefore been investigated and a negative association of DQB1*0602 with schizophrenia has been reported in African–American (Nimgaonkar et al., 1992, 1997) and Chinese (Nimgaonkar et al., 1995) patients, but not in Caucasians (Nimgaonkar et al., 1992; Wright et al., 1996).

Rheumatoid arthritis may also be inversely associated with schizophrenia (Eaton et al., 1992) and it is associated with HLA-DRB1*04 alleles. A negative genetic association of DRB1*04 with schizophrenia has been reported in one study (Wright et al., 1996), although three other investigations found no associations (Sasaki et al., 1994; Zamani et al., 1994; Gibson et al., 1997). However, the serotyping investigation of Blackwood et al. (1996) reported a reduced frequency of HLA-DR4 in Scottish schizophrenic patients, and a trend for the preferential non-transmission of DRB1*04 alleles from heterozygous parents to schizophrenic offspring (16 of 23 alleles not transmitted, $\chi^2 = 3.5$, $p = 0.06$) was found in another study (Wright et al., 1997b). A significantly reduced frequency of DQA1*0301/2 alleles, with which DRB1*04 exhibits tight linkage disequilibrium, has also been reported in schizophrenic patients (Wright et al., 1997a).

Zamani et al. (1994) reported a negative association of the DPB1*0101 allele with schizophrenia in Belgian patients but in the present authors' most recent sample, 17 (18.1%) of 94 White UK schizophrenic patients and 17 (16.2%) of 105 controls had DPB1*0101. Finally, Schwab et al. (1995) described linkage of CAR, a marker in strong linkage disequilibrium with DQB1, with schizophrenia.

MOOD (AFFECTIVE) DISORDERS

Mood or affective disorders (bipolar disorder or manic depressive psychosis, and unipolar disorder or major depression), like schizophrenia, are thought to have a major genetic aetiological component (Mynett-Johnson et al., 1997). Thus, almost as many association studies of affective disorders have been undertaken since the first was reported by Shapiro et al. in 1976 (as have just been described in schizophrenia). All reported studies have utilized serotyping techniques.

Of the published studies, nine reported no differences in antigen frequencies between patients and controls (Targum et al., 1979; James et al., 1980; Johnson et al., 1981; Mendlewicz et al., 1981; Propert et al., 1981; Goldin et al., 1982; Cambell et al., 1984; Fieve et al., 1984; Small et al., 1989). However, both positive and negative associations have been described at the HLA-A (Shapiro et al., 1976; Beckman et al., 1978; Smeraldi et al., 1978; Temple et al., 1979; Ventura et al., 1990), HLA-B (Shapiro et al., 1976; Shapiro et al., 1977; Beckman et al., 1978; Smeraldi et al., 1978; Temple et al., 1979; Sorokina et al., 1987; Ventura et al., 1990) and HLA-C (Temple et al., 1979) loci, although the only findings replicated even once are of an increased frequency of HLA-B7 and -B16 and a reduced frequency of -B8. Shapiro et al. (1976) first reported an excess of HLA-B7 and -B16 when 47 Danish patients with manic-depressive disorders were compared with controls, and replicated this finding in an expanded sample of 107 patients (1977). Temple et al. (1979) subsequently investigated 38 Eastern European Jewish Caucasians with the affective disorder and reported an increased frequency of HLA-B16, but only in patients without a family history of psychiatric disease. Shapiro et al. also reported a reduced frequency of HLA-B8 in their preliminary (1976) and expanded (1977) samples, but neither this nor the positive associations already described with HLA-B7 and -B16 have been further replicated.

An increase in HLA haplotype sharing over random expectation was found in the Ontario follow-up of 117 pedigrees ($p < 0.01$) and was more marked ($p < 0.002$) if pedigrees with three or more affected siblings were excluded from the analysis (Stancer et al., 1988). The authors concluded that HLA genes were one of a number of aetiological factors for affective disorders (Weitkamp and Stancer, 1989). One linkage study has reported a conventionally significant lod score of 3.901 ($\theta = 0.10$) for HLA with depression-susceptibility genes (Kruger et al., 1982). However, these results were derived from only two pedigrees. Four studies found no linkage of affective disorders to HLA (Saurez et al., 1982; Waters et al., 1988; Payami et al., 1989; Price et al., 1989). The study by Price et al. (1989) included 136 pedigrees, utilized both sib-pair and linkage (with several models) analysis and concluded that linkage of affective disorder to HLA had been excluded to a distance of 20–25 centiMorgans (cM).

Three further reports of HLA studies in subgroups of patients with affective disorder warrant mention. Whalley et al. (1982) found that HLA (and other) genetic markers did not distinguish between puerperal and non-puerperal subgroups of women with affective psychosis, while Schiffer et al. (1988) investigated risk factors for affective disorder in patients with multiple sclerosis and concluded that female gender and DR antigens increased susceptibility. Finally, Staner et al. (1991) noted that sleep-onset rapid eye movements (SOREM) occur in both narcolepsy and major depression and investigated the relationship between SOREM and HLA in patients with major depression. No significant differences were observed in the frequencies of HLA-DR or -DQ antigens between patients and controls.

ANXIETY DISORDERS

In contrast to the major psychotic disorders, few HLA investigations of anxiety disorders have been undertaken. Panic disorder was investigated by Cottraux et al. (1989) and by Vorob'eva et al. (1992), who reported an excess of HLA-Cw6 when 30 patients were compared with controls. Ayuso Gutierrez et al. (1991) investigated three generations of a multiply affected pedigree at the HLA-A, -B and -C loci, and reported that the HLA-A3–B18 haplotype was present in all affected members and absent in all others. This finding is in keeping with the earlier report by Surman et al. (1983), who described two HLA-identical sibling pairs with panic disorder.

Perhaps not surprisingly, given its possible viral origin, a number of HLA investigations of chronic fatigue syndrome (CFS) has been performed. Preliminary studies suggested an overrepresentation of HLA-DR4 in patients with CFS but a larger Northern Ireland genotyping study found only a non-significant increase in HLA-DR14 alleles (Middleton et al., 1991). A further study of 57 UK patients with CFS found no significant associations with any HLA-A, -B, -DR or -DQ antigen or allele.

EATING DISORDERS

Biederman et al. (1984) reported an increased frequency of HLA-B16 and of the HLA-A26–B38 haplotype in 37 patients with anorexia nervosa compared with controls. Kiss et al. (1988) were unable to confirm these findings and found no other association in their own sample of 37 patients with anorexia nervosa. Kiss et al. (1989) also found no associations between HLA and bulimia nervosa.

AUTISM

Autism has a major genetic component (Rutter, 1997). The initial HLA association study of autism investigated 20 autistic children and their parents and found statistically insignificant increases in the frequency of HLA-A2 in the patients and -A10 in their fathers (Stubbs and Magenis, 1980). Subsequent investigation reported an increased frequency of the extended haplotype HLA-B44–SC30–DR4 in 22 autistic children and confirmed this finding in a further sample of 23 subjects (Daniels et al., 1995). The extended haplotype was present in 40% of the autistic children or their mothers, in comparison to 2% of unrelated control subjects. This Utah group also demonstrated an increased frequency of the C4B null allele and of two alleles of the DRB1 gene in autism (Warren et al., 1996a) and subsequently reported that the third hypervariable region (HVR-3) of several DRB1*04 alleles was present on extended haplotypes in 46% of autistic children and 7.5% of controls, while the HVR-3 of DRB1*0701 was carried on the extended haplotypes of 32% of autistic subjects and 10.1% of controls (Warren et al., 1996b).

TOURETTE'S SYNDROME

Early studies of HLA and Tourette's syndrome found no evidence of linkage at the HLA-A, -B, -C or -DR loci (Caine et al., 1985). However, studies of the Canadian Old Order Mennonites, a genetic isolate with the world's largest reported aggregation of IDDM, suggest an association of Tourette's syndrome with HLA-DR4 (Jaworski et al., 1989), while Min et al. (1991) found an increased frequency of HLA-A11 and -A26, and a reduced frequency of HLA-A24 and -B13, in Korean patients, with HLA-A24 being particularly uncommon in patients with affected relatives.

HLA AND PSYCHOTROPIC MEDICATION

A small number of investigations has been undertaken to test the hypothesis that HLA may modify therapeutic and adverse responses to antipsychotic medications and to lithium.

HLA and treatment efficacy

Smeraldi et al. (1976b) reported that, of 33 schizophrenic patients, those with HLA-A1 responded best to antipsychotic treatment with chlorpromazine, while patients with HLA-A2 responded poorly. These authors also found that clinical improvement was more marked in HLA-A1-positive than -A1-negative schizophrenic patients from a second group of 17 patients whose symptoms were evaluated quantitatively before and after initiation of chlorpromazine treatment. In marked contrast, Bersani et al. (1989) found that HLA-A1-negative and -A2-positive patients with paranoid schizophrenia experienced a better response to antipsychotic medication.

As with the clinical response to antipsychotic treatment, the first HLA study of response to lithium prophylaxis was reported by Smeraldi et al. (1978). Ninety-one patients were studied and the frequency of HLA-B5 was found to be increased in non-relapsed patients compared with relapsed patients. Perris et al. (1979) investigated 82 patients who had received lithium prophylaxis for at least 6 months and compared the 34 patients who had relapsed with the 48 who had not. They found an increased frequency of HLA-A3 in the former and a total absence of HLA-B18 in the latter. Some years later, Maj et al. (1985) studied two samples of patients receiving lithium prophylaxis and concluded that the presence of HLA-A3 was associated with an unfavourable response to treatment. The most recent investigation of association between HLA and response to lithium found an increased frequency of HLA-A2 in patients who did not respond to therapy (Ventura et al., 1990)

HLA and treatment tolerance

HLA studies of parkinsonism and tardive dyskinesia (TD) (extrapyramidal movement disorders associated with antipsychotic treatment), and of clozapine-induced agranulocytosis, have been undertaken. Metzer et al. (1989) found that HLA-B44 was present in 55.2% of patients with parkinsonism and in 13.0% of patients without parkinsonism (relative risk = 7.16), and concluded that possession of HLA-B44 may confer an immunogenetic susceptibility to this side-effect. The present authors were unable to replicate this finding and found no other differences at the HLA A, B, DRB, DQA, DQB or -DPB loci in a study of 95 schizophrenic patients, 48 of whom had parkinsonism (Wright, 1995).

Antipsychotic drugs induce autoantibodies such as antinuclear antibody (Canosa et al., 1977). Canosa et al. (1982) found that such autoantibodies were more likely to develop in antipsychotic-treated patients with HLA-B44 and because autoantibodies against neural antigens have been described in chorea (Husby et al., 1976) they suggested that neuroleptic-induced autoimmunity might be responsible for TD. They assessed 66 patients and found that the presence of both autoantibodies and HLA-B44 was associated with TD (Canosa et al., 1986). In two further investigations, Metzer et al. (1990) studied 53 white, male, schizophrenic patients and found that HLA-DR4 was more common in those with TD (relative risk = 3.04), while Brown and White (1991) found no association between TD and any HLA antigen in a study of 40 schizophrenic patients, half of whom had TD. The present authors were unable to replicate any of the above findings, and found no other associations with TD at the HLA-A, -B, -DRB, -DQA, -DQB or -DPB loci in a study of 95 schizophrenic patients (Wright, 1995).

An HLA association has been demonstrated in Jewish patients between what is arguably the most serious adverse effect of any neuroleptic, clozapine-induced agranulocytosis (CIA), and the extended haplotype HLA-B38–DR4–DQ3 (Lieberman et al., 1990). In keeping with this finding, Pfister et al. (1992) reported a native American with the HLA-B16 (of which B38 is a subspecificity)–DR4–DQ3 haplotype who developed CIA. However, the largest investigation of HLA and CIA was performed by Claas et al. (1992), who compared 103 patients with CIA with 95 matched controls and found no significant associations at the HLA-A, -B, -C, -DR or -DQ loci.

CONCLUSIONS

Many HLA association studies of psychiatric diseases, especially schizophrenia and bipolar disorders, have been undertaken. Taken together, the results of these are conflicting. However, the few findings that have been replicated at least once provide possible evidence that schizophrenia may exhibit a weak positive association with HLA-A9 and a negative association with HLA-DQB1*0602 and DRB1*04 alleles, that affective disorders may be positively associated with HLA-B16 and that the HVR-3 of certain DRB1*04 alleles may be present in an excess of autistic children. However, it must be concluded that there is no conclusive evidence at present that any psychiatric disease exhibits HLA association. Only further studies, utilizing improved diagnostic criteria, larger and ethnically more homogeneous samples of patients, and HLA genotyping, will resolve this issue.

REFERENCES

Alexander, R.C., Coggiano, M., Daniel, D.G. et al. (1990). HLA antigens in schizophrenia. *Psychiatr. Res.* **31**: 221–233.

Asaka, A., Okazaki, Y., Namura, I. et al. (1981). Study of HLA antigens among Japanese schizophrenics. *Br. J. Psychiatr.* **138**: 498–500.

Ayuso Gutierrez, J.L., Llorente Perez, L.J., Ponce de Leon, C. et al. (1991). Genetic studies of panic disorder. *Arch. Neurobiol.* **54**: 104–110.

Beckman, L., Perris, C., Strandman, E. et al. (1978) HLA antigens and affective disorders. *Hum. Hered.* **28**: 96–99.

Bersani, G., Valeri, M., Bersani, I. et al. (1989). HLA antigens and neuroleptic response in clinical subtypes of schizophrenia. *J. Psychiatr. Res.* **23**: 213–220.

Biederman, J., Rivinus, T.M., Herzog, D.B. et al. (1984). High frequency of HLA-Bw16 in patients with anorexia nervosa. *Am. J. Psychiatry* **141**: 1109–1110.

Blackwood, D.H., Muir, W.J., Stephenson, A. et al. (1996). Reduced expression of HLA B35 in schizophrenia. *Psychiatr. Genet.* **6**: 51–59.

Brown, K.W. and White, T. (1991). Human leucocyte antigens and tardive dyskinesia. *Br. J. Psych.* **158**: 270–272.

Caine, E.D., Weitkamp, L.R., Chiverton, P. et al. (1985). Tourette syndrome and HLA. *J. Neurol. Sci.* **69**: 201–206.

Campbell, J., Crowe, R.R., Goeken, N. et al. (1984). Affective disorder not linked to HLA in a large bipolar kindred. *J. Affect. Disord.* **7**: 45–51.

Canoso, R.T. and Hutton, R.A. (1977). A chlorpromazine induced inhibitor of blood coagulation. *Am. J. Haem.* **2**: 183–191.

Canoso, R.T., Lewis, M.E. and Yunis, E.J. (1982). Association of HLA Bw44 with chlorpromazine induced autoantibodies. *Clin. Immunol. Immunopathol.* **25**: 278–282.

Canoso, R.T., Romero, J.A. and Yunis, E.J. (1986). Immunogenetic makers in chlorpromazine induced tardive dyskinesia. *J. Neuroimmunol.* **12**: 247–252.

Claas, F.H., Abbott, P.A., Witvliet, et al. (1992). No direct clinical relevance of the human leucocyte antigen (HLA) system in clozapine induced agranulocytosis. *Drug Safety* 7(Suppl. 1): 3–6.

Cottraux, J., Gebuhrer, L., Bardi, R. et al. (1989). HLA system and panic attacks. *Biol. Psych.* 25: 505–508.

Crowe, R., Thompson, J., Flink, R. et al. (1979). HLA antigens and schizophrenia. *Arch. Gen. Psychiatry* 30: 231–233.

Daniels, W.W., Warren, R.P., Odell, J.D. et al. (1995). Increased frequency of the extended or ancestral haplotype B44–Sc30–DR4 in autism. *Neuropsychobiology* 32: 120–123.

Eaton, W.W., Hayward, C. and Ram, R. (1992). Schizophrenia and rheumatoid arthritis: a review. *Schizophr. Res.* 6: 181–192.

Eberhard, G., Franzen, G. and Low, B. (1975). Schizophrenia susceptibility and HLA antigens. *Neuropsychobiology* 1: 211–217.

Fieve R.R., Go, R., Dunner, D.L. and Elston, R. (1984). Search for biological/genetic markers in a long-term epidemiological and morbid risk study of affective disorders. *J. Psychiatr. Res.* 18: 425–445.

Finney, G.O. (1989). Juvenile onset diabetes and schizophrenia. *Lancet* ii: 1214–1215.

Gibson, S., Hawi, Z., Straub, R.E., Walsh, D., Kendler, K.S., Gill, M. (1991). HLA and schizophrenia: reputation of reported associations with A9 (A23/A24), DR4, and DQ beta 1*0602. *Am. J. Med. Genet.* 88: 416–421.

Goldin, L.R., Clerget-Darpoux, F. and Gershon, E.S. (1982). Relationship of HLA to major affective disorder not supported. *Psychiatry Res.* 7: 29–45.

Gottesman, I.I. (1991). *Schizophrenia Genesis: The Origins of Madness*, pp. 104–132, W.H. Freeman, New York.

Husby, G., Van de Rijn, I. and Zabriskie, J.B. (1976). Antibodies reacting with cytoplasm of subthalamic caudate nuclei neurons in chorea and acute rheumatic fever. *J. Exp. Med.* 144: 1094–1110.

Ivanyi, D., Zemek, P. and Ivanyi, P. (1976). HLA antigens in schizophrenia. *Tissue Antigens* 8: 217–220.

Ivanyi, D., Zemek, P. and Ivanyi, P. (1978). HLA antigens as possible markers of heterogeneity in schizophrenia. *J. Immunogenet.* 5: 165–172.

James, N.M., Smouse, P.E., Carroll, B.J. and Haines, R.F. (1980). Affective illness and HLA frequencies: no compelling association. *Neuropsychobiology* 6: 208–216.

Jaworski, M.A., Severini, A., Mansour, G. et al. (1989). Inherited diseases in North American Mennonites: focus on Old Colony (Chortitza) Mennonites. *Am. J. Med. Genet.* 32: 158–168.

Johnson, G.F., Hunt, G.E., Robertson, S. and Doran, T.J. (1981). A linkage study of manic-depressive disorder with HLA antigens, blood groups, serum proteins and red cell enzymes. *J. Affect. Disord.* 3: 43–58.

Julien, R.A., Mercier, P. and Chouraqui, P. (1977). Schizophreniques et systeme d'histocompatibilitie tissulaire. Mise en evidence de l'augmentation des antigenes A9 et Cw4 dans la schizophrenie paranoide. *Ann. Med. Psychologique* 35: 939–944.

Kiss, A., Hajek-Rosenmayr, A. and Haubenstock, A. (1988). Lack of association between HLA antigens and anorexia nervosa. *Am. J. Psychiatry* 145: 876–877.

Kiss, A., Hajek-Rosenmayr, A., Wiesnagrotzki, S. et al. (1989). Lack of association between HLA antigens and bulimia. *Biol. Psychiatry* 25: 803–806.

Kruger, S.D., Turner, W.J. and Kidd, K.K. (1982). The effects of requisite assumptions on linkage analyses of manic-depressive illness with HLA. *Biol. Psychiatry* 17: 1081–1099.

Lieberman, J.A., Yunis, J., Egea, E. et al. (1990). HLA B38, DR4, DQw3 and clozapine induced agranulocytosis in Jewish patients with schizophrenia. *Arch. Gen. Psychiatry* 47: 945–948.

McGuffin, P., Festenstein, H. and Murray, R.M. (1983). A family study of HLA antigens and other genetic markers in schizophrenia. *Psychol. Med.* 13: 31–43.

McGuffin, P. and Sturt, E. (1986). Genetic markers in schizophrenia. *Hum. Hered.* 36: 65–88.

Maj, M., Arena, F., Lovero, N. et al. (1985). Factors associated with response to lithium prophylaxrophylaxis in DSM III major depression and bipolar disorder. *Pharmacopsychiatry* 18: 309–313.

Mendlewicz, J., Verbanck, P., Linkowski, P. and Govaerts, A. (1981). HLA antigens in affective disorders and schizophrenia. *J. Affect. Disord.* 3: 17–24.

Metzer, W.S., Newton, J.E. and Paige, S.R. (1988). HLA-A1 and schizophrenia. *Biol. Psychiatry* 24: 364–365.

Metzer, W.S., Newton, J.E.O., Steele, R.W. et al. (1989). HLA antigens in drug induced parkinsonism. *Movement Disord.* 4: 121–128.

Metzer, W.S., Newton, J.E.O., Steele, R.W. et al. (1990). HLA antigens and tardive dyskinesia. *J. Neuroimmunol.* 26: 179–181.

Middleton, D., Savage, D.A. and Smith, D.G. (1991). No association of HLA class II antigens in chronic fatigue syndrome. *Dis Markers* 9: 47–49.

Min, S.K., Lee, H., Park, K.I. et al. (1991). Tourette disorder and HLA typing. *Yonsei Med. J.* 32: 315–318.

Miyanaga, K., Machiyama, Y. and Juji, T. (1984). Schizophrenic disorders and HLA-DR antigens. *Biol. Psychiatry* 19: 121–129.

Mynett-Johnson, L.A., Murphy, V.E., Manley, P. et al. (1997). Lack of evidence for a major locus for bipolar disorder in the pericentromeric region of chromosome 18 in Irish pedigrees. *Biol. Psychiatry* 42: 486–494.

Nimgaonkar, V.L., Ganguli, R., Rudert, W.A. et al. (1992). A negative association of schizophrenia with an allele of the HLA DQB1 gene among African–Americans. *Schizophr. Res.* 8: 199–209.

Nimgaonkar, V.L., Rudert, W.A., Zhang, X.R. et al. (1995). Further evidence for an association between schizophrenia and the the HLA DQB1 gene locus. *Schizophr. Res.* 18: 43–49.

Nimgaonkar, V.L., Rudert, W.A., Zhang, X.R. et al. (1997). Negative association of schizophrenia with HLA DQB1*0602: evidence from a second African–American cohort. *Schizophr. Res.* 23: 81–86.

Payami, H., Dubay, C. and Valenzuela, R.C. (1989). HLA may be involved in resistance and susceptibility to affective disorders. *Genet. Epidemiol.* 6: 293–298.

Perris, C., Strandman, E. and Wahlby, L. (1979). HL-A antigens and the response to prophylactic lithium. *Neuropsychobiology* **5**: 114–118.

Pfister, G.M., Hanson, D.R., Roerig, J.L., et al. (1992). Clozapine induced agranulocytosis in a native American: HLA typing and further support for an immune mediated mechanism. *J. Clin. Psychiatry* **53**: 242–244.

Price, R.A. (1989). Affective disorder not linked to HLA. *Genet. Epidemiol.* **6**: 299–304.

Propert, D.N., Tait, B.D. and Davies, B. (1981). HLA antigens and affective illness. *Tissue Antigens* **18**: 335–340.

Pugliese, A., Gianani, R., Moromisato, R., et al. (1995). HLA DQB1*0602 is associated with dominant protection from diabetes even among islet-cell antibody positive relatives of patients with IDDM. *Diabetes* **44**: 608–613.

Rosler, M., Bellaire, W., Gressnich, N. et al. (1980). HLA antigens in schizophrenia, major depressive disorder and schizoaffective disorder. *Med. Microbiol. Immunol.* **172**: 57–65.

Rudduck, C., Franzen, G., Low, B. and Rorsman, B. (1984a). HLA antigens and clinical subgroups of schizophrenia. *Hum. Hered.* **34**: 18–26.

Rudduck, C., Franzen, G., Low, B. and Rorsman, B. (1984b). HLA antigens in patients with and without a family history of schizophrenia. *Hum. Hered.* **34**: 291–296.

Rutter, M. (1997). Implications of genetic research for child psychiatry. *Can. J. Psychiatry* **42**: 569–576.

Sasaki, T., Kuwata, S., Dai, X.Y. et al. (1994). HLA DR types in Japanese schizophrenica: analysis by group-specific PCR amplification. *Schizophr. Res.* **14**: 9–14.

Schiffer, R.B., Weitkamp, L.R., Wineman, N.M. and Guttormsen, S. (1988). Multiple sclerosis and affective disorder. Family history, sex, and HLA-DR antigens. *Arch. Neurol.* **45**: 1345–1348.

Schwab, S.G., Albus, M., Hallmayer, J. et al. (1995). Evaluation of a susceptibility gene for schizophrenia on chromosome 6p by multipoint sib-pair linkage analysis. *Nature Genet.* **11**: 325–327.

Shapiro, R.W., Bock, E., Rafaelsen, O.J. et al. (1976). Histocompatibility antigens and manic-depressive disorders. *Arch. Gen. Psychiatry* **33**: 823–825.

Shapiro, R.W., Ryder, L.P., Svejgaard, A. and Rafaelsen, O.J. (1977). HLA antigens and manic-depressive disorders: further evidence of an association. *Psychol. Med.* **7**: 387–396.

Small, G.W., Matsuyama, S.S., Komanduri, R. et al. (1989). HLA antigens in depressed, demented, and nondemented elderly. *J. Geriatr. Psychiatry Neurol.* **2**: 70–75.

Smeraldi, E, Bellodi, L. and Cazzullo, C.L. (1976a). Further studies on the major histocompatibility complex as a genetic marker for schizophrenia. *Biol. Psychiatry* **11**: 655–671.

Smeraldi, E., Bellodi, L., Sacchetti, E. and Cazzullo, C.L. (1976b). The HLA system and the clinical response to treatment with chlorpromazine. *Br. J. Psychiatry* **129**: 486–489.

Smeraldi, E., Negri, F., Melica, A.M. et al. (1978). HLA typing and affective disorders: a study in the Italian population. *Neuropsychobiology* **4**: 344–352.

Sorokina, T.T., Evsegneev, R.A., Levin, V.I. and Semenov, G.V. (1987). Features of the distribution of HLA-antigens among patients with endogens psychoses. *Zh. Nevropatol. Psikhiatr.* **87**: 885–888.

Stancer, H.C., Weitkamp, L.R., Persad, E. et al. (1988). Confirmation of the relationship of HLA (chromosome 6) genes to depression and manic depression. II. The Ontario follow-up and analysis of 117 kindreds. *Ann. Hum. Genet.* **52**: 279–298.

Staner, L., Bouillon, E., Andrien, M. et al. (1991). Lack of association between HLA-DR antigens and sleep-onset REM periods in major depression. *Biol. Psychiatry* **15**: 1199–1204.

Stubbs, E.G. and Magenis, R.E. (1980). HLA and autism. *J. Autism Dev. Disord.* **10**: 15–19.

Suarez, B.K. and Croughan, J. (1982). Is the major histocompatibility complex linked to genes that increase susceptibility to affective disorder? A critical appraisal. *Psychiatry Res.* **7**: 19–27.

Surman, O.S., Sheehan, D.V., Fuller, T.C. and Gallo, J. (1983). Panic disorder in genotypic HLA identical sibling pairs. *Am. J. Psych.* 1983; **140**: 237–238.

Targum, S.D., Gershon, E.S., Van Eerdewegh, M. and Rogentine, N. (1979). Human leukocyte antigen system not closely linked to or associated with bipolar manic-depressive illness. *Biol. Psychiatry* **14**: 615–636.

Temple, H., Dupont, B. and Shopsin, B. (1979). HLA antigen and affective disorders: a report and critical assessement of histocompatibility studies. *Neuropsychobiology* **5**: 50–58.

Turner, W.D. (1979). Genetic markers for schizotaxia. *Biol. Psychiatry* **14**: 177–205.

Ventura, T., Lobo, A. and Marco, J.C. (1990). HLA antigens in bipolar patients. *Actas Luso Esp. Neurol. Psiquiatr. Cienc. Afines* **18**: 339–343.

Vorob'eva, O.V., Tanonov, A.T. and Kurmyshkin, A.A. (1992). The genetic aspects in the genesis of autonimic crises. *Zh. Nevropatol. Psikhiatr. Im. S.S. Korsakova* **92**: 57–58.

Warren, R.P., Odell, J.D., Warren, W.L. et al. (1996a). Strong association of the third hypervariable region of HLA-DR beta 1 with autism. *J. Neuroimmunol.* **67**: 97–102.

Warren, R.P., Singh, V.K., Averett, R.E. et al. (1996b). Immunogenetic studies in autism and related disorders. *Mol. Chem. Neuropathol.* **28**: 77–81.

Waters, B., Sengar, D., Marchenko, I., et al. (1988). A linkage study of primary affective disorder. *Br. J. Psychiatry* **152**: 560–562.

Weitkamp, L.R. and Stancer, H.C. (1989). Analysis of the Toronto–Rochester Depression Study follow-up data confirms an HLA-region gene contribution to susceptibility to affective disorder. *Genet. Epidemiol.* **6**: 305–310.

Whalley, L.J., Roberts, D.F., Wentzel, J. and Wright, A.F. (1982). Genetic factors in puerperal affective psychoses. *Acta Psychiatr. Scand.* **65**: 180–193.

Wright, P. (1995). Immunogenetic studies of schizophrenia and neuroleptic induced movement disorder. Doctorate of Medicine thesis, University College, Dublin.

Wright, P., Donaldson, P.T., Curtis, V.A. et al. (1995). Immunogenetic markers in schizophrenia: HLA A9 revisited. *Schizophr. Res.* **15**: 1, 2, 50.

Wright, P., Donaldson, P.T., Underhill, J.A. et al. (1996). Genetic association of the HLA DRB1 gene locus on chromosome 6p21.3 with schizophrenia. *Am. J. Psych.* **153**: 1530–1533.

Wright, P., Donaldson, P.T., Underhill, J.A. et al. (1997a). Evidence for a resistance locus for schizophrenia close to the HLA DQA1 and DRB1 gene loci on chromosome 6p21.3. *Schizophr. Res.* **24**: 1,2, 50–51.

Wright, P., Dawson, E., Donaldson, P.T. et al. (1997b). Evidence from a transmission/disequilibrium study that alleles of the DRB1*04 gene on chromosome 6p21.3 may protect against schizophrenia. *Schizophr. Res.* **24**: 1, 2, 51.

Zamani, M.G., De Hert, M., Spaepen, M. et al. (1994). Study of the possible association of HLA Class II, CD4 and CD3 polymorphisms with schizophrenia. *Am. J. Med. Genet.* **54**: 372–377.

SECTION III

Definition of HLA Polymorphism

CHAPTER 26

Serological Methods in HLA Typing

Philip A. Dyer, Susan Martin and Rachel E. Stanford

INTRODUCTION

Following its introduction in the early 1960s (Terasaki and McClelland, 1964) the complement-dependent cytotoxicity (CDC) microassay became the universally accepted method for the serological definition of human leucocyte antigen (HLA) specificities. Although the technique has been modified and refined over time, its simplicity and robustness ensured its place as the preferred tissue-typing technique for over 25 years. Peripheral blood lymphocytes (PBL) are used as targets and tested against HLA specific alloantisera for the definition of HLA-A, -B and -C specificities. HLA-DR specificities were first officially recognized in 1977 (Bodmer et al., 1978) following analysis of the reaction patterns of B lymphocyte reactive sera. The CDC test was then modified and cell separation techniques were refined so that B cells could be used for the definition of HLA-DR and subsequently -DQ (Albert et al., 1984) specificities. Although universally used for HLA-DR and -DQ typing over many years the CDC test was never as robust for -DR and -DQ as for -A, -B and -Cw typing. This arose from both the paucity of alloantisera monospecific for -DR and DQ antigens and difficulties in B cell isolation.

The term "serology" applies to the use of antisera in the test. Although in the field of histocompatibility and immunogenetics serology has been used synonymously with CDC it could equally be applied to techniques where antibody binding, after incubation with alloantisera, is detected not by complement activation but by the use of fluorochromes and flow cytometry. The use of flow cytometry is inappropriate for complete HLA phenotyping owing to the number of tests needed to cover all specificities, but the methodology has been applied to testing for the presence of HLA-B27 (Reynolds et al., 1994) and HLA-DR4 and also for -DP typing.

The development of immunoprecipitation/isoelectric-focusing techniques in some laboratories (Yang, 1989) provided higher resolution and hence enabled the definition of HLA specificities not identified by CDC, but the methods were too cumbersome and time consuming to be applicable to routine patient and organ donor phenotyping.

As molecular biology typing techniques for DNA sequences developed they were initially applied to the identification of HLA-DR, -DQ and -DP alleles (Middleton and Williams, 1997). Methods involving DNA amplification using the polymerase chain reaction (PCR) and sequence-specific oligonucleotide probes (SSOP) enabled precise definition of HLA gene polymorphisms using small amounts of DNA and were ideally suited to disease association studies (see Chapter 7). Further advances using sequence-specific primers (SSP) have made the techniques more rapid and they are now increasingly applied to HLA-A, -B, -Cw as well as -DR and -DQ typing.

NOMENCLATURE

Since 1965, International Histocompatibility Workshops (IHW) have been organized approximately every 4 years with the aim of standardizing and improving the detection and definition of genes of the HLA genetic system and their polymorphisms. After each IHW a World Health Organization (WHO) Nomenclature Committee has met to officially recognize newly defined genes, specificities and alleles. The progress in identifying the number of HLA genes and the number of specificities encoded by these genes is shown in Table 26.1.

In the 1987 report, HLA alleles were first included and with the subsequent advent of rapid sequencing methods interim WHO Committee reports have been issued more frequently. In the same year the number of recognized HLA-Cw specificities was reduced by one, with the discovery that the previously designated Cw11 specificity was identical in sequence to Cw8. The recognition of new specificities is now limited to submissions which include sequencing data. The most recent WHO Committee report followed the 12th IHW held in Paris in 1995 (Bodmer et al., 1997; Charron, 1997); a complete list of currently recognized serological HLA specificities is shown in Table 26.2. The association of HLA-B specificities with the "supertypic" HLA-Bw4 and -Bw6 epitopes is shown in Table 26.3.

By convention, HLA-A and -B specificities are numbered exclusively. Thus, there is an HLA-A2 but no -B2 and a -B5 but no -A5. Note that there is no HLA-A20 or -B20 since this specificity was reassigned to the C locus. In earlier WHO reports some specificities were prefixed by a letter "w" to indicate that definition was provisional or "workshop"; when acceptably defined, full status was indicated by omission of the "w". This practice ceased in 1989 when allele nomenclature and the requirement for gene sequencing was introduced; the "w" was retained for the HLA-C locus to avoid confusion with components of the complement system, some of which (C2, C4 and Bf) have genes located within the major histocompatibility complex (MHC). The "w" is retained for HLA-DP specificities since they were originally defined by methods using cellular reagents and not by serology. The same is true for the "Dw" specificities. An informative review of the importance of using a common HLA nomenclature is recommended to all working in the field (Bodmer, 1997).

SOME DEFINITIONS

The successful use of a common nomenclature has depended on agreed understanding of some commonly used terms.

HLA genes

The region of DNA sequence which encodes polymorphic histocompatibility genes has been termed the MHC, reflecting its biological function (Klein, 1979). The extent of the MHC is accepted to be limited centromerically by the HLA-DPB2 gene and telomerically by the HLA-F gene. The WHO Committee has included genes which are transcribed and also pseudogenes in its nomenclature. Most of the functional genes (HLA) code

TABLE 26.1 The number of HLA genes and specificities officially recognized by World Health Organization HLA Nomenclature Committees

Year	Number of genes	Number of specificities					
		HLA-A	HLA-B	HLA-C	HLA-DR	HLA-DQ	HLA-DP
1965	?	12					
1970	1	17					
1972	1	21					
1977	4	18	28	3	7		
1980	4	20	42	8	10		
1984	6	23	49	8	16	3	6
1987	19	24	52	11	20	9	6
1991	35	27	59	10	24	9	6
1994	37	27	59	10	24	9	6
1996	37	31	61	10	24	9	6

TABLE 26.2 Complete listing of HLA specificities recognized by the World Health Organization (1996)

HLA-A	HLA-B	HLA-Cw	HLA-DR	HLA-DQ	HLA DP
$n = 28$	$n = 61$	$n = 10$	$n = 24$	$n = 9$	$n = 6$
A1	B5	Cw1	DR1	DQ1	DPw1
A2	B7	Cw2	DR103	DQ2	DPw2
A203	B703	Cw3	DR2	DQ3	DPw3
A210	B8	Cw4	DR3	DQ4	DPw4
A3	B12	Cw5	DR4	DQ5(1)	DPw5
A11	B13	Cw6	DR5	DQ6(1)	DPw6
A9	B14	Cw7	DR6	DQ7(3)	
A23(9)	B15	Cw8	DR7	DQ8(3)	
A24(9)	B16	Cw9(3)	DR8	DQ9(3)	
A2403	B17	Cw10(3)	DR9		
A10	B18		DR10		
A25(10)	B21		DR11(5)		
A26(10)	B22		DR12(5)		
A19	B27		DR13(6)		
A29(19)	B2708		DR14(6)		
A30(19)	B35		DR1403		
A31(19)	B37		DR1404		
A32(19)	B38(16)		DR15(2)		
A33(19)	B39(16)		DR16(2)		
A34(19)	B3901		DR17(3)		
A36	B3902		DR18(3)		
A43	B40		DR51		
A66(19)	B4005		DR52		
A28	B41		DR53		
A68(28)	B42				
A69(28)	B44(12)				
A74(19)	B45(12)				
A80	B46				
	B47				
	B48				
	B49(21)				
	B50(21)				
	B51(5)				
	B5102				
	B5103				
	B52(5)				
	B53				
	B54(22)				
	B55(22)				
	B56(22)				
	B57(17)				
	B58(17)				
	B59				
	B60(40)				
	B61(40)				
	B62(15)				
	B63(15)				
	B64(14)				
	B65(14)				
	B67				
	B70				
	B71(70)				
	B72(70)				
	B73				
	B75(15)				
	B76(15)				
	B77(15)				
	B78				
	B81				
	Bw4				
	Bw6				

TABLE 26.3 HLA-Bw4 and HLA-Bw6 associated specificities

Bw4	B5, B5102, B5103, B13, B17, B27, B37, B38(16), B44(12), B47, B49(21), B51(5), B52(5), B53, B57(17), B58(17), B59, B63(15), B77(15)
	and
	A23(9), A24(9), A2403, A25(10), A32(19)
Bw6	B7, B703, B8, B2708, B35, B39(16), B3901, B3902, B4005, B41, B42, B45(12), B46, B48, B50(21), B54(22), B55(22), B56(22), B60(40), B61(40), B62(15), B64(14), B65(14), B67, B71(70), B72(70), B73, B75(15), B76(15), B78, B81

HLA polymorphisms

The HLA genes are the most polymorphic in the human genome. Until 1984 the polymorphism was detected entirely by serological methods, and the alloantigenic epitopes of mature HLA molecules determined by the amino acid sequence are termed specificities. The polymorphisms detected at the DNA sequence level are termed alleles and they are greater than the number of specificities since DNA base changes do not always alter amino acid sequences. Conventionally, alleles must occur at a frequency of more than 1% in a population and less frequent polymorphisms are termed variants, but for HLA, where there is such a large number of polymorphisms, it is common practice to term even an HLA DNA sequence change from a single individual as an allele.

The frequency of HLA specificities varies within and between populations and the 11th IHW reports on frequencies world-wide (Tsuji et al., 1992). The frequencies of HLA specificities from a large number of individuals resident in England are set out in Figures 26.1–26.5. These represent the HLA specificities typical of a north-western European Caucasoid population, although some specificities may vary significantly

for cell surface molecules which present peptide to T cell receptors. However, the region also includes genes which have a role in antigen processing (e.g. TAP and LMP). The HLA molecules have usefully been termed histoglobulins, reflecting their homology with immunoglobulin structure (Bodmer, 1987), but this term is not in common use and clumsy phraseology such as "HLA-A molecules" is frequently used. Based on structural and functional similarities, Klein grouped HLA-A, -B and -C gene protein products as "MHC class I" molecules and HLA-DR, -DQ and -DP as "MHC class II".

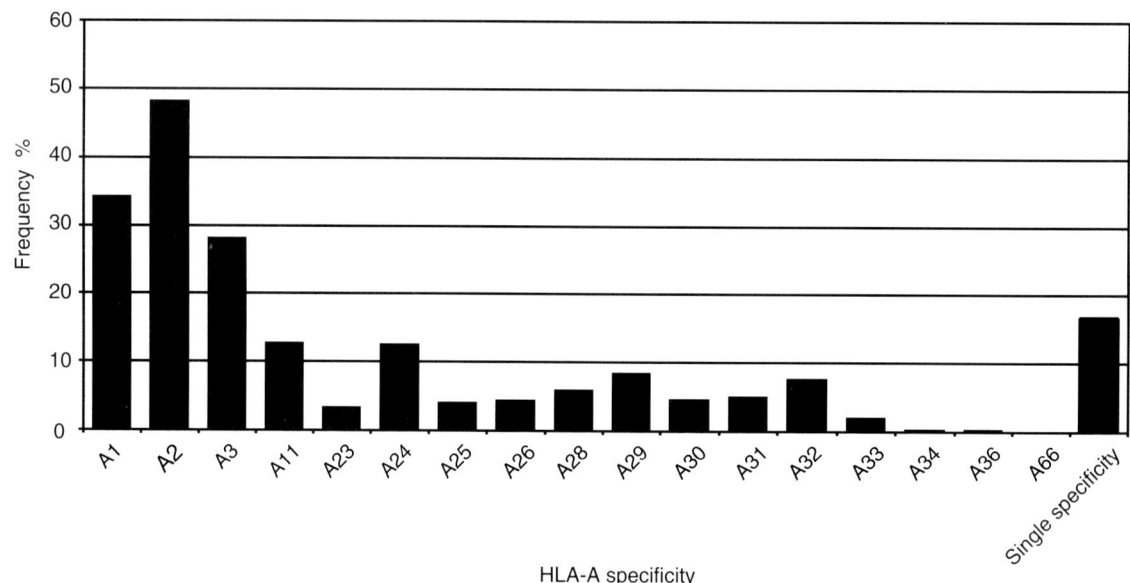

FIGURE 26.1 Frequencies of HLA-A specificities in 1563 kidney donors.

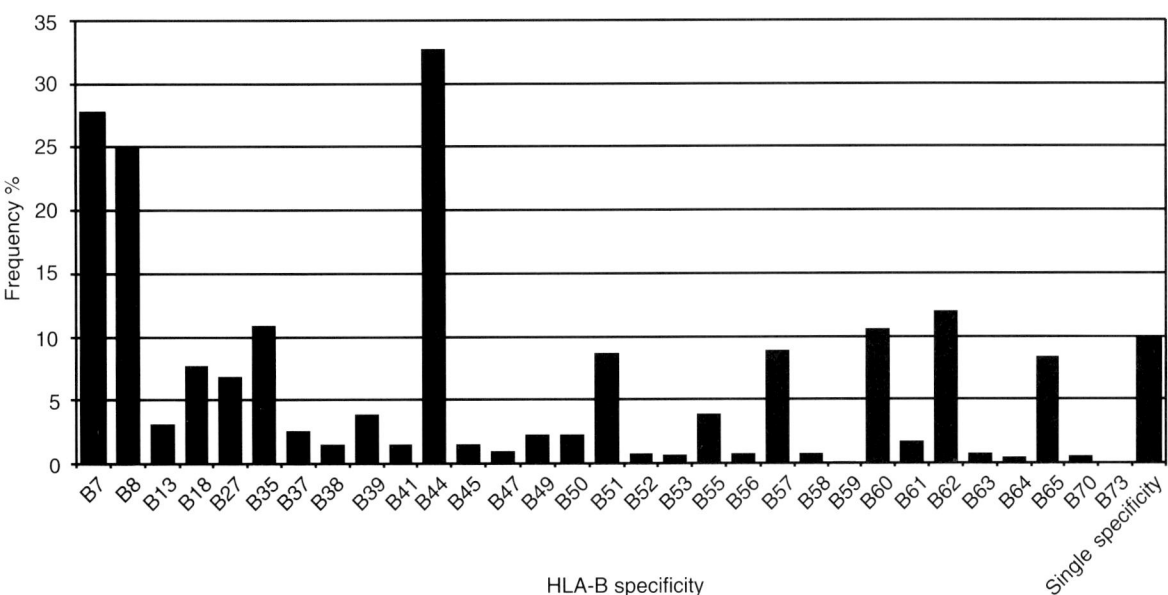

FIGURE 26.2 Frequencies of HLA-B specificities in 1563 kidney donors.

in frequency on a narrow geographical basis. Any study of HLA associations in a disease population must use carefully collected local healthy control HLA frequencies. A useful listing of HLA allele frequencies in healthy and disease populations is also available (Terasaki and Gjertson, 1997).

MICROLYMPHOCYTOTOXICITY

The lymphocytotoxicity assay is a three-stage, complement-dependent reaction in which viable lymphocytes are incubated with HLA specific antisera, rabbit serum is added as a source of complement and then the percentage of viable

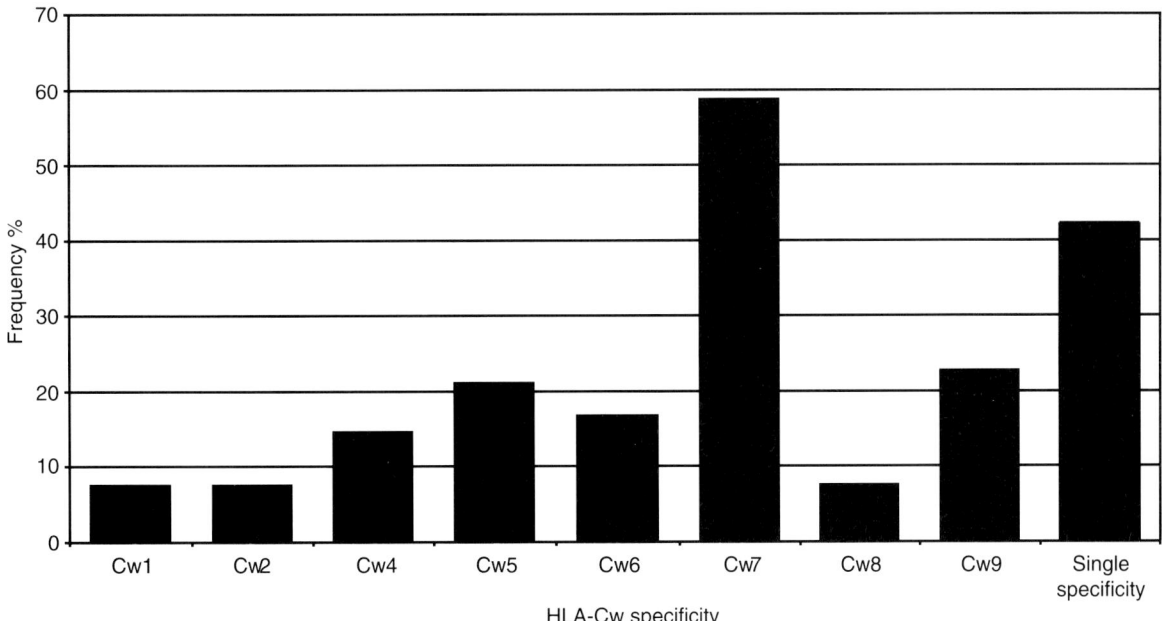

FIGURE 26.3 Frequencies of HLA-Cw specificities in 721 kidney donors.

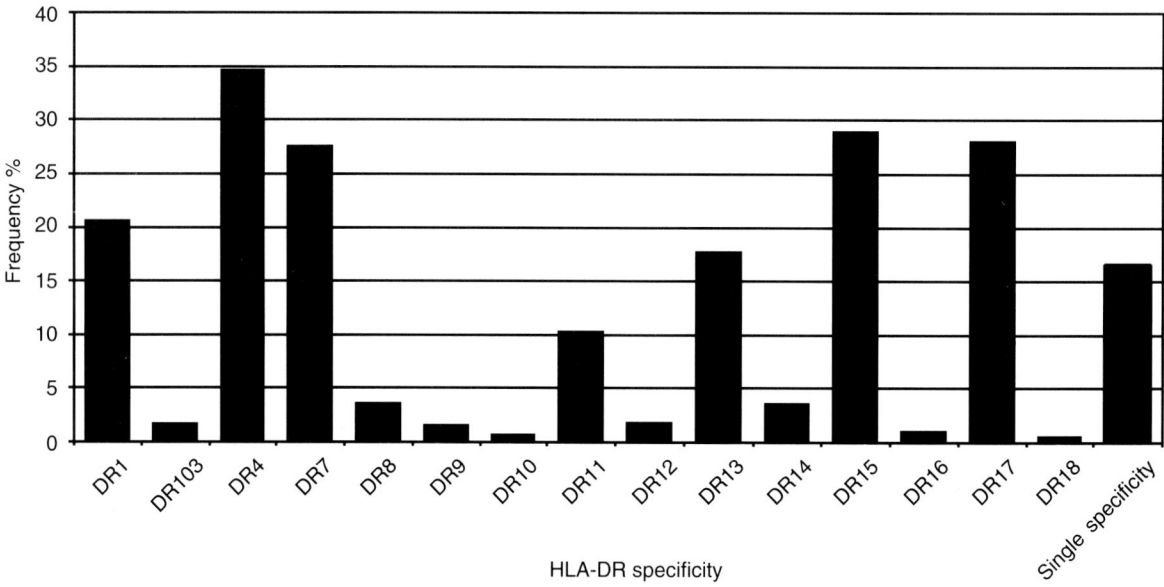

FIGURE 26.4 Frequencies of HLA-DR specificities in 1554 kidney donors.

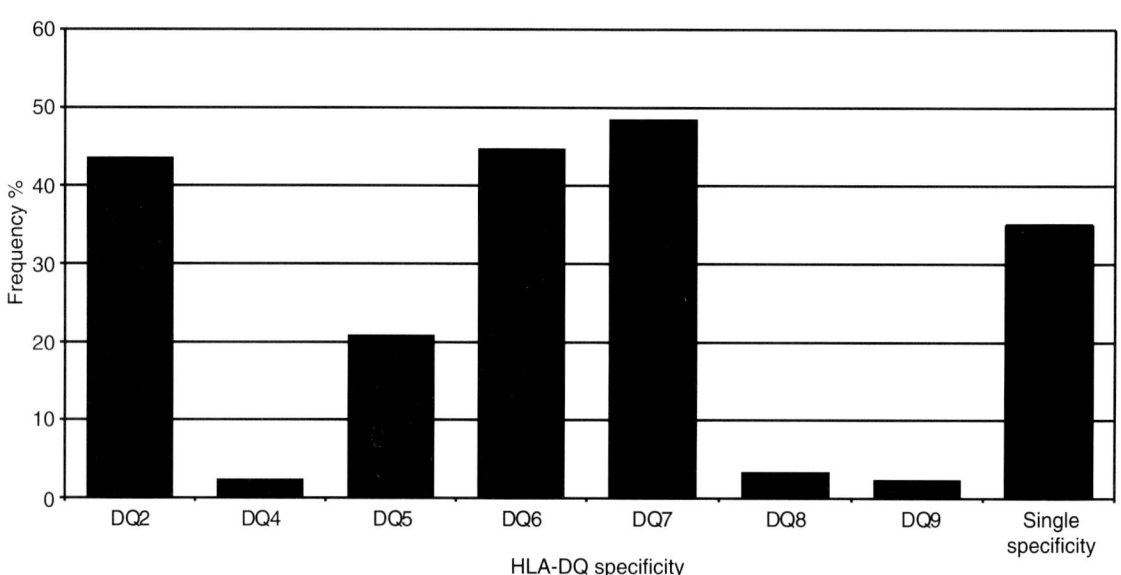

FIGURE 26.5 Frequencies of HLA-DQ specificities in 729 kidney donor.

cells is determined, usually by microscopic evaluation following staining. Effective standardization of this technique is essential, with participation in external quality-assurance schemes being strongly advisable. HLA typing reagents fall into two categories: alloantisera and monoclonal antibodies.

Alloantisera

HLA specific alloantibodies can be produced *in vivo* in response to pregnancy, blood transfusion and transplantation of allogeneic organs. Owing to the high degree of polymorphism of HLA specificities, alloimmunization is a potentially frequent event moderated only by differential antigenicity between specificities. Alloantisera are polyclonal in nature and can contain mixtures of antibodies directed either against different epitopes on the same molecule or against antigenic determinants of different molecules. Both complement-binding cytotoxic and non-cytotoxic antibodies can be produced.

Pregnancy

HLA specific antibodies can be produced during pregnancy, most commonly during the second or subsequent pregnancies, in response to paternal HLA antigens expressed by the foetus. This response can give rise to high-titre, high-affinity antibodies of restricted HLA specificity and so have been the main source of HLA typing reagents.

Blood transfusion and transplantation

HLA specific antibodies can be produced following blood transfusion owing to the presence of donor leucocytes and in response to HLA-mismatched organ and tissue transplants, but the response is only rarely of restricted monospecificity. More usually, antibodies are directed against a common epitope(s) shared by cross-reactive groups of antigens and are consequently not useful typing reagents. In addition, a serum may contain a mixture of antibodies conferring multiple specificity.

Screening for alloantibodies

The procurement of HLA typing reagents is an expensive and a time-consuming process. Potential antibody-containing sera are screened, using the CDC assay, against panels of T and B lymphocytes from HLA-A, -B, -Cw, -DR, -DQ-typed individuals to determine the presence of antibodies. The panel must include, whenever possible, all known HLA antigens in suitable combinations. As the quality of the screening will only be as good as the quality of the panel cell typing, panel cells must be typed to the highest possible degree of resolution. Statistical analyses of the serum reactions are based on 2×2 contingency tables and give chi-square (χ^2) and correlation coefficient (R) values (Table 26.4). The quality of a serum as a typing reagent improves as R approaches unity and for common specificities an R-value of at least 0.9 is desirable. Operationally monospecific antibodies to less common specificities may, however, only be found in conjunction with others, so that a lower R-value has to be accepted until a better serum is found.

Although CDC assays are microassays, using only 1 μl of serum per test, when a potentially useful antibody has been identified, a large donation of serum must be requested to provide adequate reagent volume for stock and for exchange with other laboratories. The serum may need manipulating to provide a reagent of the required standard. This could involve dilution to remove unwanted antibodies that give "extra" reactions, purification and concentration of weak antibodies and platelet absorption to remove -A, -B and -Cw but leave -DR and -DQ activity.

Laboratories in blood transfusion centres that have a reagent procurement programme screen regular female blood donors and can expect to find approximately 3% with HLA-specific antibodies. Other laboratories have programmes for screening sera from women attending antenatal clinics, although in the region of only 1% of these will produce useful typing reagents. Over the last

TABLE 26.4 The 2×2 table for identification of serological typing reagents

Serum specificity	Concordant positive reactions serum+/cell+	Extra reactions +/−	Missed reactions −/+	Concordant negative reactions −/−	χ^2	R
HLA-B8	26	2	1	71	85.6	0.93
HLA-Cw4	15	4	2	79	63.8	0.80
HLA-DR7	26	6	4	64	58.8	0.77

Definition of specificity is determined by the χ^2 correlation coefficient (R) values, calculated using the concordant positive and negative reactions from serum and all samples. R approaches unity as the quality of the serum used as a typing reagent improves.

25 years there has been international co-operation and serum exchange, with sera also being made available through national transplant and organ exchange organizations in an attempt to improve and standardize the quality of serological typing. However, as laboratories move towards the use of monoclonal antibodies and/or DNA-based typing techniques and also suffer increasing staffing and financial pressures, reagent procurement programmes are being dropped or moved to the commercial sector and a wide range of alloantisera is now available commercially, either individually or as test kits.

All laboratories collecting HLA typing serum reagents should ensure that each serum donation is tested for potentially infective viruses. Serum donors should be counselled so that their informed consent is obtained.

The CDC assay is carried out using microlitre volumes of reagents under tissue culture oil to prevent evaporation, in Terasaki microtitre trays. Trays containing alloantisera can be prepared in bulk and stored at –40°C until required. The required volume of serum (usually 0.5 or 1.0 µl) is dispensed into each well using a six-barrel microsyringe or preferably an automatic serum dispenser.

Monoclonal antibodies

An increasing number of cytotoxic monoclonal antibodies is becoming available commercially. They are valuable typing reagents which are gradually replacing alloantisera. The majority of monoclonal antibodies used as typing reagents have been generated by fusion of a mouse myeloma cell line with spleen or lymph node cells from a mouse previously immunized with the chosen immunogen. The hybrid cell lines secrete the antibody produced by the immunized cells. Monoclonal antibodies have the great advantage as a reagent that, once stable antibody-secreting lines are established, an unlimited supply of standard reagent can be made available long term. In contrast to alloantisera, they are homogeneous and their specificity can be precisely defined at the epitope level. Many monoclonal antibodies react with more than one HLA specificity by recognizing a shared epitope.

Some groups have invested considerable effort in the production of HLA-specific monoclonal antibodies which have predominantly been applied to the definition of epitopes. However, some have been made available to supplement alloantisera for HLA typing. The extremely high titre of the monoclonal antibodies necessitates considerable care in handling in order to optimize their working concentration and avoid carryover from one reaction well to the next. The use of monoclonal antibodies for HLA typing has now been facilitated by the availability of commercially prepared and standardized monoclonal antibody typing trays which have the complement as well as the antibodies added to the trays. These were initially introduced for HLA-DR and -DQ typing but are now also available for HLA-A and -B. They are valuable tools for the HLA serologist when the supplies of alloantisera are diminishing.

Another use of monoclonal antibodies is to elucidate the expression of HLA molecules in tissue sections (Connor and Stern, 1990).

Commercial availability of serological HLA typing reagents

HLA typing antisera are conventionally obtained through a local screening programme and subsequently by exchange of reagents with other laboratories. With the decline in screening programmes, the advent of *in vitro* production of monoclonal antibodies and the move to molecular biological techniques, commercial companies have entered the HLA serological reagent market. The benefits of buying HLA typing reagents include:

- reliability of reagents backed by company liability
- ease of supply in small and bulk quantities
- standardization
- absence of need for long-term storage
- accompanying supply of high-quality complement.

Having chosen to purchase HLA serological typing reagents it is advisable to liaise with the commercial supplier to test batches of reagents with target cells of known HLA phenotypes. This establishes the quality of the reagents and the efficiency of the recommended methodology which accompanies the reagents. Some suppliers provide target cell isolation kits based on antibody-coated magnetizable microspheres to standardize the assay further.

Target cells

A prerequisite for the CDC assay is a pure population of lymphocytes with over 90% viability. They should be free of contamination by red blood cells, which do not express HLA molecules, and platelets, granulocytes and monocytes which do, and so may reduce antibody activity. Density gradient separation of leucocytes from anticoagulated blood was first described by Boyum, and the procedure has become standard for lymphocyte isolation with detailed methods widely published (Dyer and Darke, 1993) and separation media commercially available. T and B cells have historically been isolated from lymphocyte populations by T cell-enriching rosetting techniques or nylon wool column B cell depletion, although the use of these methods has now largely been superseded by the use of commercially available monoclonal antibody-coated magnetizable microspheres (Dyer, 1992). For the purpose of HLA-A, -B and -Cw typing, a whole-lymphocyte suspension will usually suffice since >85% of cells will be T cells. T cell isolation using monoclonal antibody coated magnetizable microspheres is highly advantageous if there is high granulocyte contamination or poor cell viability or if starting from whole blood. In contrast, owing to the limited cell expression of HLA-DR and -DQ specificities, it is essential to have a pure B cell suspension for -DR and -DQ typing. This is best achieved by the use of monoclonal antibody-coated magnetizable microspheres specific for HLA-DR and -DQ monomorphic epitopes or a B cell determinant.

It has been established that cytotoxicity assays are not blocked by the presence of monoclonal antibody-coated magnetizable microspheres bound to cells; indeed, the sensitivity of assays is increased and incubation times can be shortened as a result.

Complement

Pooled rabbit serum is the most effective source of complement for use in the CDC assay and it can be obtained commercially. It is a recognized variable in the CDC assay and each batch must be standardized carefully before use. In the UK the National Institute for Biological Standards and Controls provides a standard reagent against which fresh batches of rabbit serum can be compared. Since the rabbit serum may contain rabbit antihuman antibodies that can reduce the viability of the lymphocytes it must be checked for toxicity. It must also be checked for complement efficiency: this is usually achieved by testing it in dilution in a "chequerboard" against HLA typing sera, also in dilution.

Incubations

Target cells are incubated at 22°C with antiserum in Terasaki trays for 30 min to allow specific antibody to bind to antigens which may be expressed on the cell surface. After this period, complement is added and incubation proceeds for a further 60 min to allow cytotoxicity to occur. These incubation times can be varied to alter the sensitivity of the assay, but practice has shown the times quoted to be optimal (Zachary et al., 1995).

Staining and reading reactions

The test is halted by the addition of ethylene diamine tetra-acetic acid (EDTA), which prevents further cytotoxicity. A cocktail of ethidium bromide, to stain dead target cells (positive reaction), and acridine orange, to stain viable cells (negative reaction), is added and, under blue light illumination using an inverted microscope, the observer estimates the positivity of each reaction by the percentage of target cells killed. For HLA typing, strong reactions of more than 60% target cells killed are required for assignment of positivity.

Interpretation of serological typing

Over time, as antibody screening programmes expanded and newly informative antisera were discovered, it became possible to divide some specificities into subtypes. The original specificities are termed "broad" while the subtypes are termed "splits". The WHO Nomenclature provides for identification of split specificities by inclusion of the broad specificity bracketed after the split specificity; thus, for HLA-A24(9) the broad specificity A9 is represented in this phenotype by the split specificity A24. In practice, the antibodies used to define specificities can either be mixtures of antibodies or react with antigenic epitopes which occur on several HLA molecules. In both cases, comprehensive analysis will reveal multireactivity leading to definition of subtypes. All HLA-B molecules possess a bi-allelic dimorphism assigned as either HLA-Bw4 or -Bw6. The HLA-A9 specificity also carries HLA-Bw4. Thus, HLA-B specificities can be grouped according to Bw4 and Bw6 reactivity (see Table 26.3). The reactivity of a target cell with Bw4 and Bw6 typing sera is a useful aid to assigning some HLA-B specificities and splits; for example, the B44 split of B12 is Bw4 positive, while the B45 split of B12 is Bw6 positive, thus allowing assignment of B12 split specificities in the absence of B44 and B45 specific antisera but by using Bw4 and Bw6 typing reagents.

AUTOMATION OF HLA SEROLOGICAL TESTING

Owing to the relatively low demand for HLA typing, in part restricted by the limited availability of serological reagents, widespread automation has not occurred and any equipment commercially available is costly. Paradoxically, commercially available equipment has been marketed only in recent times, when many laboratories are changing to molecular biology techniques. The value of automation is twofold, both by increasing the number of tests possible and by enhancing the accuracy of tests through standardized processing. The following steps in CDC testing have been automated:

- dispensing tissue culture oil into Terasaki trays
- dispensing serological reagents into Terasaki trays
- dispensing target cells into Terasaki trays
- semiautomated microscope-based reading of tests and recording of test results on computer
- fully automated microscope-based reading of tests using fluorochrome stains
- data analysis.

HLA TYPING FOR HLA-A, -B AND -C BY IMMUNOPRECIPITATION AND ISOELECTRIC FOCUSING

In the attempt to improve definition of HLA polymorphisms the technique of immunoprecipitation of HLA molecules followed by isoelectric focusing of the specific immunoprecipitate was developed (Yang, 1989). The technique is most successful for HLA-A, -B and -C molecules when locus-specific monoclonal antibodies are available. HLA-DR and -DQ typing by this method is problematic since monoclonal antibodies often coprecipitate other molecules, leading to difficulty in interpreting the final gels or needing a second dimension of electrophoresis to separate products on a molecular size basis.

The basic principles of the technique are:

- metabolic radiolabelling of target cells with [^{35}S]methionine
- lysis of labelled cells with detergent
- precipitation of ^{35}S-labelled HLA antigens with monoclonal antibody
- separation of precipitated ^{35}S-labelled HLA antigens according to their isoelectric points in a pH gradient gel
- fixing and staining of gel to reveal banding patterns
- interpretation of HLA phenotype using common β_2-microglobulin band as reference.

A major drawback is the time-frame of several days needed to obtain a result. However, it is possible to identify some HLA-A, -B and -C alleles as well as most specificities using this technique. With the development of molecular biological techniques, immunoprecipitation and one-dimensional isoelectric focusing have ceased to be used but they would have advantages for studies of HLA antigen expression.

HLA SPECIFICITY FREQUENCY DATA

The frequency of HLA specificities in a population should be determined for the purposes of control data to compare with frequencies in a disease population. A large sample is essential to avoid unintentional sampling bias. In Figures

TABLE 26.5 Some examples of allelic associations

Specificities	Number of individuals positive or negative for each specificity.					
	+/+	+/−	−/+	−/−	% +/+	Δ × 10⁻³
HLA-A1, -B8	302	232	89	940	19.3	70
HLA-B8, -Cw7	138	35	286	262	19.1	40
HLA-B8, -DR3	128	261	311	715	8.2	40
HLA-DR3, -DQ2	183	13	133	400	25.1	100

Data derived from Figures 26.1–26.5.

26.1–26.5, the frequencies of HLA specificities determined by CDC in a population of 1563 cadaver organ donors tested between 1979 and 1997 in the Manchester, UK transplant centre are shown. These data can be considered to represent HLA specificity frequencies in a north-western European Caucasoid population, all of whom were apparently healthy within hours of their sudden death caused by either trauma or intracerebral haemorrhage.

Using frequency data of HLA specificities from a population it is possible to demonstrate genetic phenomena such as allelic association (sometimes called linkage disequilibrium), which is the occurrence together of two or more genetic alleles (or specificities) at a frequency which exceeds that predicted by random assortment. Many HLA specificities exhibit allelic association and some examples identified from the data presented in Figures 26.1–26.5 are set out in Table 26.5. The strength of allelic association is estimated by Δ (delta; Mattiuz et al., 1970), with a positive value indicating an association. The association of HLA specificities reflects favourable selection mechanisms and/or close genetic linkage of coding genes, which results in conserved stretches of DNA sequences termed haplotypes.

APPLICATION OF HLA SEROLOGICAL TYPING TO HLA AND DISEASE STUDIES

Nearly all significant associations of HLA polymorphisms with specified diseases are based on original studies which used HLA serological typing. This is largely because molecular biological techniques were not available when such studies were initiated, but subsequent studies using molecular biology have largely confirmed the findings of serology. It is unlikely that future studies would use serology since molecular biology techniques are cheaper, more efficient and more accurate.

The significant advantage of HLA typing by molecular biology is comprehensive definition of HLA polymorphisms at the allele level, thus avoiding the potential pitfall of false-negative observations. This was always a problem when using less efficient serological typing, a good example being the linkage of HLA-B47 with 21-OH deficiency as few, if any, B47-specific typing sera have ever been available.

It is debatable whether DNA-based HLA allele typing has added significantly to our understanding of HLA and disease associations beyond the hypotheses established from data obtained by serological typing (Thorsby, 1997). This may be because relevant mechanisms involve HLA molecules and their varying structures rather than DNA sequence changes. However, it is more likely that the efficiency of HLA allele typing by molecular biology has facilitated more comprehensive typing and more robust data for analysis. Studies of rheumatoid arthritis have been extended with the ability to subdivide the HLA-DR4 specificity into its 26 alleles (MacGregor et al., 1995) and, because of the particular inefficiency of HLA-DQ serological typing, studies of susceptibility to coeliac disease as influenced by DQ2 have revealed interesting findings (Spurkland et al., 1997). Studies of diseases in remote populations have become possible as a result of the robust nature of DNA molecules contrasted with the need to transport viable target cells for serological typing. Thus, multigene screening of patients with infectious diseases with the inclusion of HLA typing has been performed, generating significant results (Hill et al., 1991).

THE FUTURE ROLE OF SEROLOGICAL HLA TYPING IN DISEASE STUDIES

Although DNA-based techniques will almost certainly replace serology for the definition of HLA-A, -B and -Cw as well as -DR and -DQ specificities, there remains a role for serology in -A, -B, -Cw typing, particularly using monoclonal antibodies. In the longer term, serology will remain important for the investigation of the cell-surface expression of HLA alleles and variants defined at the DNA level. The use of serology may help to elucidate whether the expression of an HLA antigen as well as the presence of the allele is associated with a particular disease.

REFERENCES

Albert, E.D., Baur, M.P. and Mayr, W.R. (1984). *Histocompatibility Testing 1984*, Springer, Berlin.

Bodmer, J.G., Marsh, S.G.E. and Albert, E.D. (1997). Nomenclature for factors of the HLA system 1996. *Tissue Antigens* **49**: 297–321.

Bodmer, W.F. (1987). The HLA system: structure and function. *J. Clin. Pathol.* **40**: 948–958.

Bodmer, W.F. (1997). HLA: what's in a name? A commentary on HLA nomenclature development over the years. *Tissue Antigens* **46**: 293–296.

Bodmer, W.F., Batchelor, J.R., Bodmer, J.G. et al. (1978). *Histocompatibility Testing 1977*, Munksgaard, Copenhagen.

Charron, D. (1997). *Proceedings of the Twelfth International Histocompatibility Workshop and Conference* (ed. D. Charron), EDK, Paris.

Connor, M.E. and Stern, P.L. (1990). Loss of MHC class I expression in cervical carcinomas. *Int. J. Cancer* **46**: 1029–1034.

Dyer, P.A. (1992). Immune mediated cell separation methods. *Biochem. Soc. Trans.* **20**: 230–233.

Dyer, P.A. and Darke, C. (1993). Clinical HLA typing by cytotoxicity. *Histocompatibility Testing: A Practical Approach* (eds P.A. Dyer and D.C. Middleton), pp. 51–80, Oxford University Press, Oxford.

Hill, A.V.S., Allsopp, C.E.M., Kwiatkowski, D. et al. (1991). Common West African HLA antigens are associated with protection from severe malaria. *Nature* **352**: 595.

Klein, J. (1979). The major histocompatibility complex of the mouse. *Science* **203**: 516–521.

MacGregor, A., Ollier, W.E., Thomson, W. et al. (1995). HLA DRB1*0401/0404 genotype and rheumatoid arthritis: increased association in men, young age at onset and disease severity. *J. Rheumatol.* **22**: 1032–1036.

Mattiuz, P.L., Inde, D., Piazza, A. et al. (1970). New approaches to the population genetics and segregation analysis of the HLA system. In *Histocompatibility Testing 1970*, p. 193, Munksgaard, Copenhagen.

Middleton, D.C. and Williams, F. (1997). A history of DNA typing for HLA. *HLA 1997* (eds P.I. Terasaki and D.W. Gjertson), pp. 61–84, UCLA Tissue Typing Laboratory, Los Angeles, CA.

Reynolds, W.M., Evans, P.R. and Lane, A.C. (1994). Automated HLA-B27 testing using the FACSPrep/FACScan system. *Cytometry* **18**: 109–115.

Spurkland, A., Ingvarsson, G., Falk, E. et al. (1997). Dermatitis herpetiformis and celiac disease are both primarily associated to the HLA-DQ($\alpha1*0501$, $\beta1*02$) or the HLA-DQ($\alpha1*03$, $\beta1*0302$) heterodimers. *Tissue Antigens* **49**: 29–33.

Terasaki, P.I. and Gjertson, D.W. (1997). *HLA 1997*. UCLA Tissue Typing Laboratory, Los Angeles, CA.

Terasaki, P.I. and McClelland, J.D. (1964). Microdroplet assay of human serum cytotoxins. *Nature* **204**: 998–1000.

Thorsby, E. (1997). HLA associated diseases. *Hum. Immunol.* **53**: 1–11.

Tsuji, K., Aizawa, M. and Sasazuki, T. (1992). *HLA 1991*. Oxford University Press, Oxford.

Yang, S.Y. (1989). A standardised method for detection of HLA-A and HLA-B alleles by one-dimensional isoelectric focusing (IEF) gel electrophoresis. In *Immunobiology of HLA*, Vol. 1 (ed. B. Dupont), pp. 332–335, Springer, New York.

Zachary, A.A., Klingman, L., Thorne, N. et al. (1995). Variations of the lymphocytotoxicity test. *Transplantation* **60**: 498–503.

CHAPTER 27

Cellular Methods in Testing Histocompatibility

Philip D. Mason and Robert I. Lechler

INTRODUCTION

The necessity and degree of human leucocyte antigen (HLA) matching in transplantation depend on a number of factors, including the recipient's past history and degree of sensitization and the organ transplanted. Although rapid genetic tissue typing is becoming increasingly available (at least for HLA class II), matching has generally not been feasible for heart or liver transplantation because of the necessary short interval between retrieval and implantation and the limited number of available donors. Improved matching in renal transplantation and the avoidance of antigens to which the recipient is known to be sensitized result in significantly better short-term and long-term graft survival. In bone-marrow transplantation (BMT) even better matching is necessary, since severe graft-versus-host disease (GVHD) commonly accompanies HLA mismatches between donor and recipient. However, of particular interest is the observation that some BMT recipients with known HLA mismatches do not develop acute GVHD, while in others mismatches result in severe and sometimes lethal GVHD.

Functional *in vitro* assays of recipient versus donor (or in BMT donor versus recipient) immunoreactivity were initially developed in the hope that they would predict the *in vivo* clinical outcome following transplantation and be of value in selecting donors, and possibly also provide a means of monitoring alloreactivity post-transplant. As discussed in this chapter, some techniques may have clinical relevance in BMT, but in solid organ transplantation, although studied extensively, they remain research tools.

The major techniques that will be discussed are the mixed lymphocyte culture or reaction (MLC or MLR) and the measurement of alloreactive T (helper or cytotoxic) cell precursor frequencies by limiting dilution analysis (LDA).

MIXED LYMPHOCYTE REACTION

Even before the HLA complex was well defined, it was known that mixing peripheral blood mononuclear cells from non-identical individuals resulted in proliferation of lymphocytes, and this became known as the MLR or MLC (Bach and Hirschhorn, 1964; Bain et al., 1964). Indeed, studies of MLC between family members led to the conclusion that a genetic complex, which became known as the HLA region, controlled reactivity. By preventing the proliferation of lymphocytes from one party (usually by X- or γ-irradiation) the MLR can be rendered "one-way". It is quantitated by comparing the proliferation (e.g. by the uptake of [^{3}H]thymidine) of the MLR with a culture of responding lymphocytes alone (i.e. without irradiated stimulator cells). The ratio of these values is referred to as the stimulation index. Alternatively, the measured proliferation may be expressed as a percentage of maximum response using pooled random alloreactive lymphocytes as stimulator cells.

The magnitude of the MLR is predominantly (but not exclusively) determined by the degree of HLA class II mismatch and is therefore dependent on dendritic cells, monocytes and B cells from the stimulator cells and responder T cells. The response is most discriminative during the logarithmic phase of cell proliferation (usually at 5 days but customarily assessed at days 3, 5 and 7 days). Consequently, prospective MLR are only suitable for live transplantation. The technique was widely used in the preparatory work-up to live donor renal transplantation in the early days of serological HLA typing, but following the development of more precise serological and genetic typing, class II differences have proved more reliable than MLR for predicting clinical outcome (Jeffrey et al., 1984).

There is no clear relationship between pre-transplant or post-transplant MLR and early rejection episodes after renal transplantation. However, a number of studies has reported low antidonor, but not anti-third-party, MLR reactivity in long-term transplant recipients with good graft function (Starzl et al., 1993). This undoubtedly reflects the development of donor-specific hyporesponsiveness in such recipients, but is not helpful prospectively.

The MLR has been used extensively in BMT. Although early reports indicated that the stimulation index of the MLR correlated with the risk of GVHD, most subsequent reports have found no such correlation (Lim et al., 1988) and high-resolution, allelic HLA typing has proved a better, but by no means perfect, predictor of GVHD. There have been several variations on the MLR test including the use of donor epidermal cells (Bagot et al., 1988) in BMT recipients, but the tests are difficult to perform and have not been adequately substantiated.

LIMITING DILUTION ANALYSIS

LDA provides a sensitive and precise method of quantitating the frequencies of cells with some functionally measurable parameter. The requirements for the method are that:

1. only the cells to be measured are present in limiting numbers; any other nutrients, growth factors, stimulating or accessory cells and antigen must all be present in excess

2. a sensitive and reproducible readout must be available to measure the reactivity of cells of interest (e.g. antibody or cytokine production, proliferation or cytotoxicity)
3. statistical analysis of the experimental results indicate that the data are consistent with single hit kinetics (see below: this should follow from points 1 and 2).

Technical aspects and theoretical background

The mathematical basis of LDA is complex; a full discussion of this is beyond the scope of this chapter and the reader is directed to other sources for a more detailed description (Lefkovits and Waldmann, 1979). Immune responses depend on the stimulation, activation and often proliferation of a usually small set of antigen-specific lymphocytes. Theoretically, if a single cell is stimulated by a specific antigen in the well of a microtitre plate in an appropriate environment with the necessary cofactors, nutrients and growth factors, clonal proliferation will occur, leading eventually to numbers of cells that can be detected. The culture conditions and method of detection depend on which population of antigen-specific cells is of interest (e.g. cytotoxicity for $CD8^+$ cytotoxic T cells, cytokine production for helper T cells or antibody production for B cells).

The assay requires that only antigen-specific cells are activated and proliferate. In the case of a mixture of antigen-specific and other cells at a ratio of 1:9, if exactly 10 cells were aliquoted into 100 wells, each well would contain, on average, one antigen-specific cell per well. In reality, some wells would contain no specific cells while others would contain one and a smaller number of wells would contain two or more specific cells. Detection methods identify wells with one or more specific responding cells (i.e. the well either contains responders or it does not). The distribution of specific cells is described by the Poisson distribution, which predicts the proportion of wells, F_0, likely *not* to have a responding cell for a given number of input cells, u:

$$F_0 = e^{-u}$$

So, for $u = 1$ (i.e. average of one cell per well)

$$F_0 = e^{-1}$$
$$F_0 = 0.37.$$

This means that when the fraction of wells that are negative for the measured parameter is 0.37 they are likely to contain, on average, one responding cell; if the input cell number was 10 000, then the precursor frequency would be 1 in 10^4.

The formula can be rearranged as:

$$u = -\ln F_0.$$

There should therefore be a linear relationship between the input cell number and the natural logarithm of the fraction of negative wells. The determination of whether an individual well is positive requires the measured parameter [e.g. interleukin-2 (IL-2) or other cytokine production or ^{51}Cr release in a cytotoxicity assay] to be compared with wells containing no responder cells. Any well in which the measured parameter exceeds the mean + 3 standard deviations of control wells is treated as a positive. Data from a hypothetical experiment are presented graphically in Figure 27.1. Increasing the number of dilutions (≥ 6) of responder cell numbers and the number of replicate wells (≥ 24) for each of these improves the precision of the result. However, this has to be balanced against the amount of clinical material required and the size and expense of the resulting assays, especially since it is desirable to test simultaneously each responder cell population against several target (stimulator) cells and/or to test responder cells taken from patients at different times.

A variety of statistical methods has been developed to calculate frequencies, with the maximum

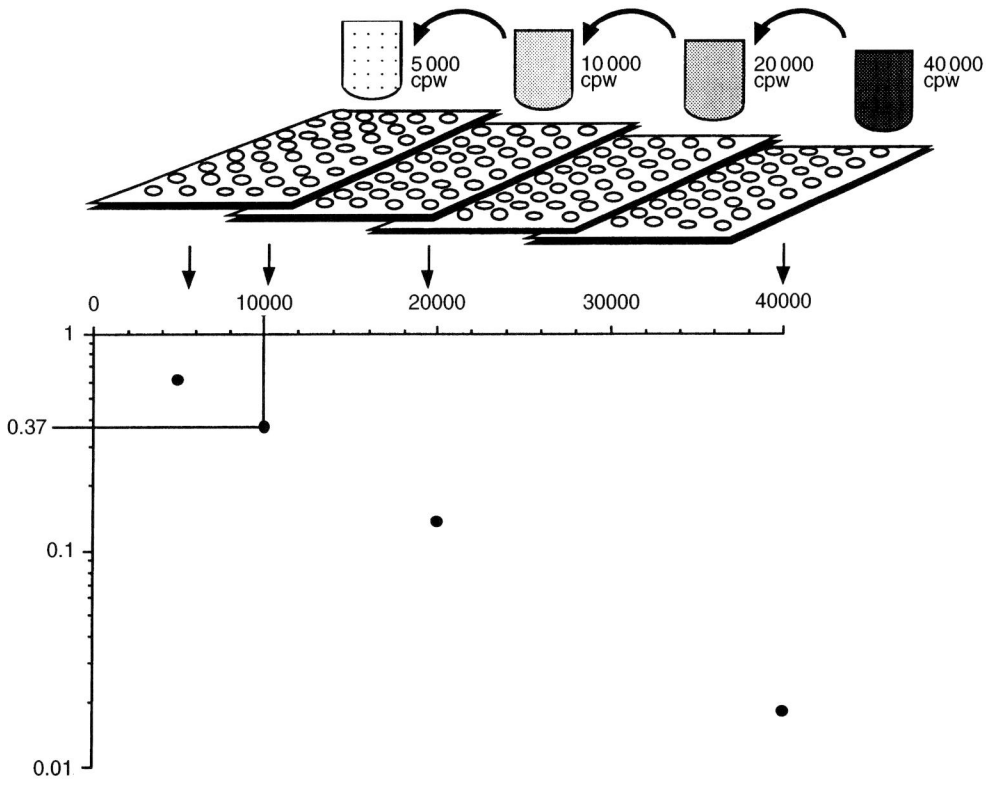

FIGURE 27.1 Schematic outline of a limiting dilution assay. Replicate dilutions of study cells are added to wells containing irradiated stimulator cells in supplemented medium. After an appropriate culture period each well is tested for the presence of responding cells and the fraction of negative wells (F_0) is plotted against the original input cell number. One responding cell per well corresponds to $F_0 = 0.37$; in the example this occurs at an original input cell number of 10 000 cells per well and the frequency of responders is therefore 1 in 10 000.

likelihood method being widely used. Computer programs that calculate frequencies from experimental data are available. As stated above, a meaningful frequency is only generated from the data if the experiment has conformed to single-hit kinetics. This means that only one cell type (the one being measured) is limiting the generation of the parameter being detected (e.g. proliferation, IL-2 production, cytotoxicity); other necessary cells (stimulator or antigen-presenting cells) and antigen and nutrients, etc. must be present in excess. If this is not the case then there will not be a linear relationship between input cell number and the logarithm of F_0. The likelihood of the measured data conforming to single-hit kinetics must be confirmed. This is usually done using the chi-square test, which translates into the probability that the data conform ($p > 0.05$), and this is usually calculated by one of the available computer programs. It is also usual to calculate 95% confidence intervals, which enable comparison between frequencies and definition of those that are significantly different (i.e. confidence intervals do not overlap). Figure 27.2 illustrates the use of a LDA to demonstrate a significant change (reduction) in donor-specific alloreactive frequency in a patient following renal transplantation.

Clearly, only antigen-specific cells capable of being activated and proliferating under the culture conditions can be detected, and this must be borne in mind when interpreting results. Even if all antigen-specific cells were able to undergo proliferation and produce the requisite cytokine or cytotoxicity, the sensitivity of these assays depends partly on the culture conditions (including the care with which they are set up) and partly on the sensitivity of the readout. When optimized, LDA assays are potentially capable of detecting single antigen-specific cells (Table 27.1) (Mason et al., 1996). However, in reality since every antigen-specific cell may not be in a state suitable to be stimulated to proliferate and the readout may have suboptimal sensitivity, the assay may only be able to detect a proportion of the antigen-specific cells (Table 27.1; clones G8, G11 and G12). Thus, LDA may be better at comparing relative frequencies rather than measuring absolute values. Despite the many assay-specific factors that can potentially affect the result, the frequencies are generally reproducible when repeated using cells cryopreserved at the same time and using standard antigens in healthy individuals over time (Mason et at, 1996). However, it is safest to compare frequencies generated using

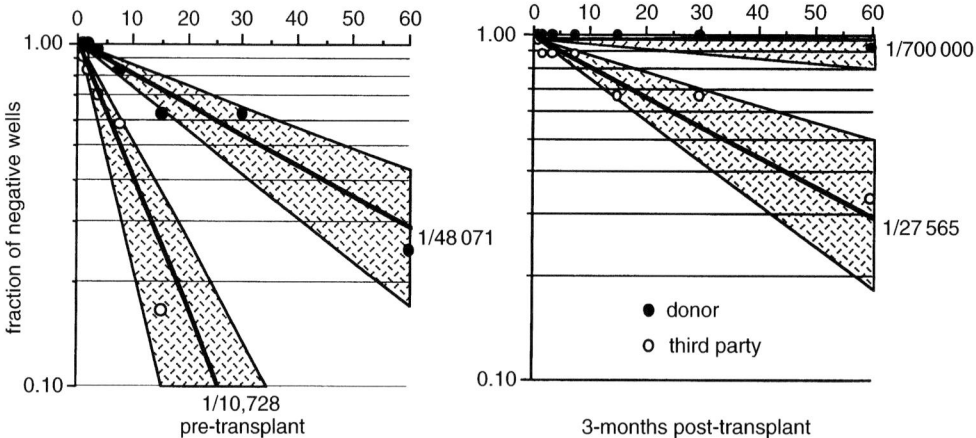

FIGURE 27.2 Data from a LDA experiment to measure the precursor frequency of alloreactive T cells specific for donor (closed symbols) and third party (open symbols) in the peripheral blood of a renal transplant recipient taken before (left panel) and 3 months after (right panel) transplantation. The shaded areas represent the 95% confidence interval. The anti-third-party frequency falls slightly after transplantation, while the antidonor frequency falls by more than an order of magnitude.

TABLE 27.1 Measured precursor frequencies of anti-DR1 clones

Clone	1/Frequency	95% CI	χ^2	p-Value
G3	1.5	1.1–2.0	4.6	> 0.1
G3	1.8	1.4–2.4	2.9	> 0.1
G8	2.4	1.8–3.1	1.4	> 0.1
G11	4.2	3.2–7.1	5.6	> 0.1
G12	2.4	1.9–3.2	5.9	> 0.1

A range of anti-DR1 specific human T cell clones was used as responder cells in limiting dilution assays to measure frequencies. The frequencies against no stimulator cells and DRw17-expressing cells were undetectable. CI: Confidence interval.

exactly the same reagents at the same time. To standardize assays, an internal standard such as measuring the precursor frequency of T cells against a reference antigen (e.g. tetanus toxoid) alongside the antigen(s) of interest has been suggested, but there is no way of knowing whether the frequency against this reference antigen has changed.

HELPER AND CYTOTOXIC CELL FREQUENCIES

Assays to measure helper T lymphocyte (HTL) frequencies measure cytokine production, usually IL-2, but others including interferon-γ and IL-4 have also been used. IL-2 can be detected by enzyme-linked immunosorbent assay (ELISA) but the most sensitive method that is currently available is a bioassay involving the use of CTLL cells (an IL-2-dependent murine T cell line) (Bishop and Orosz, 1989). The optimum duration of the assay depends on several factors, including the responder population (naive or memory cells) and the sensitivity of the detection system, but assays are usually incubated for 2–5 days.

Cells of interest, e.g. alloreactive responding T cells, are mixed with an excess of stimulator cells that have been irradiated (generally around 20–30 Gy) to prevent proliferation and reduce IL-2 production. However, unless heavily irradiated (which significantly reduces the immunogenicity of stimulator cells), T cells in the stimulator population are capable of producing IL-2 in response to class II-expressing cells in the responder cell population, a phenomenon known as back-stimulation. Back-stimulation can be prevented or reduced by depleting T lymphocytes from the stimulator cells. This can be done using monoclonal antibodies and magnetic beads, but it is important to choose monoclonal antibodies that will not stimulate T cells (such as anti-CD3). A combination of anti-CD4 and anti-CD8 is often used. Major histocompatibility complex (MHC) class II-bearing cells can also be depleted from the responder cell population to reduce further the possibility of back-stimulation. Some authors believe that these measures abrogate the need to irradiate stimulator cells.

Responder cells are added to excess stimulator cells in titrated numbers, usually in doubling dilutions to ensure a range of 0.5–5 responsive cells per well of input cell numbers. Statistical considerations require a minimum of 24 replicates for each input cell number and at least six dilutions of responder cells.

Cytotoxic T lymphocyte (CTL) LD assays are set up in a similar way but require sufficient responding T cells to lyse target cells in a cytotoxicity assay. Therefore, the assays usually need to be incubated for slightly longer and often need to be fed with fresh medium containing essential growth factors (especially IL-2) during the culture period. The readout cytotoxicity assay (release of radioactive chromium or non-radioactive vital dye from dead cells) identifies wells in which lysis above background levels has occurred. It should be remembered that if the effector:target ratio in the test wells is too low a negative well will be recorded despite the presence of specific cytotoxic cells limiting the sensitivity. It is important that every assay is carefully designed with the particular cells being studied in mind; for instance, cytotoxic cells killing by inducing apoptosis rather than by membrane damage may need longer cytotoxicity assays.

OTHER METHODS OF ESTIMATING ALLOREACTIVE FREQUENCIES

The use of LDA to estimate precursor frequencies has been criticized. The method is dependent not only on the presence of specific T cells, but also on the ability of those cells to be stimulated under the culture conditions and for effector cells to survive and expand sufficiently during culture *in vitro*. These limitations have been addressed by the recent development of two alternative methods of detecting antigen-specific T cells. The ELISPOT technique detects cytokine release by individual cells using an ELISA plate coated with a cytokine-specific monoclonal antibody. After incubation with antigen-stimulated cells released cytokine is detected using a second, labelled, monoclonal antibody. Specific T cells can also be directly stained with MHC–peptide tetrameric complexes and measured by flow cytometry. These techniques detect much higher frequencies than LDA, at least when studying antiviral T cells. For instance, one study reported anti-Epstein–Barr virus frequencies detected by the ELISPOT method to be four to five times higher than those measured by LDA. Direct staining of virus-specific T cells using a combination of MHC-virus-specific peptide tetramers revealed frequencies five to six times higher than measured by ELISPOT (Tan et al., 1999). In a mouse model of lymphocytic choriomeningitis virus infection as many as 50–70% of the activated CD8 T cells (by tetramer staining) appeared to be virus peptide-specific at the peak of infection (Murali-Krishna et al., 1998). The interpretation of these data is controversial. It is possible that the latter techniques are detecting T cells of low avidity or reactivity, which may not be clinically relevant. Furthermore, these techniques have not yet been successfully applied to the "allo" setting. ELISPOT assays using cells as allostimulators have high backgrounds. Furthermore, the tetramer technique requires the peptide(s) to be known and production of class II MHC tetramers is considerably more difficult than for class I. None the less, in contexts in which significant numbers of effector cells are anticipated, such as during an acute immune response, these alternative techniques may well be important in estimating frequencies.

CLINICAL STUDIES WITH LIMITING DILUTION ANALYSIS

Bone-marrow transplantation

Modern genetic methods of HLA typing allow identification and matching of almost all class II alleles but similar techniques for class I typing are not yet refined sufficiently for routine use. Cytotoxic T cells are able to detect allelic class I differences that are not differentiated by serological typing methods (Kaminski et al., 1991) and these almost certainly contribute to acute GVHD (aGVHD) following allogeneic BMT from an unrelated donor (Keever et al., 1994). With these data in mind and following the demonstration that MLC is not sensitive enough to predict aGVHD (and in any case is predominantly dependent on class II differences; see above) several groups have studied the value of measuring alloreactive cytotoxic T lymphocyte precursor frequencies (CTLp). Higher frequencies are detected with HLA mismatches, whether or not they are detected by conventional serology (Breur-Vriesendorp et al., 1991) or a combination of serology, one-dimensional isoelectric focusing for class I and low-resolution DNA-based class II typing (O'Shea et al., 1997). Several series have been reported in which aGVHD is associated with a high CTLp frequency. In 48 unrelated BMT patients those with a CTLp frequency $> 1/10^5$ had a much increased risk (relative risk of 9.0) of severe (grade III or IV) GVHD compared with recipients with a CTLp of $< 1/10^5$ (Spencer et al., 1995). In another study of 77 patients receiving unrelated BMT, CTLp frequencies were significantly associated with severe grades of GVHD, although this effect was lost when adjusted for other significant covariables including known HLA mismatches (Keever-Taylor et al., 1997). In the presence of known mismatches the CTLp frequency is not always high and so it has been suggested that these assays will still be valuable even if high-resolution class I HLA matching becomes routine. This important observation raises the possibility that functional assays such as LDA may allow the definition of "permissible" mismatches.

Assays to measure HTLp frequencies were developed because they are quicker and simpler to perform and seem to have greater sensitivity than CTLp between well-matched donor–recipient pairs (Figure 27.3). Indeed, frequencies are sometimes measured between HLA-identical sibling pairs, leading to the suggestion that other minor antigens may be detectable. High HTLp frequencies also correlate with severe grades of aGVHD (Schwarer et al., 1993; Lachance et al., 1997; Weston et al., 1997). CTLp and/or HTLp fre-

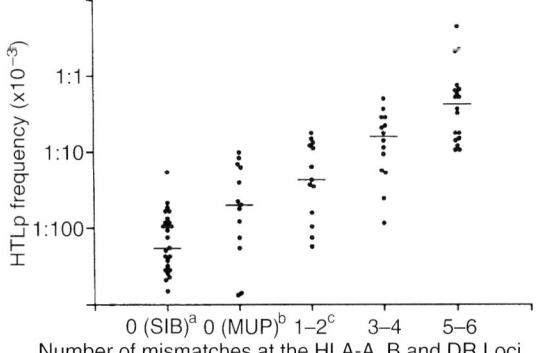

FIGURE 27.3 HTLp frequencies of 85 responder/stimulator pairs by degree of HLA-A, -B and -DR loci. The horizontal lines indicate median frequencies in each group. [a]Genotypically HLA-identical sibling pairs; [b]phenotypically HLA-matched unrelated pairs; [c]1–2, 3–4, or 5–6 HLA-mismatched pairs.

quencies are currently used by some BMT units for selecting matched unrelated bone-marrow donors.

Solid organ transplantation

LDA in solid organ transplantation is used largely for monitoring antidonor reactivity after transplantation and is currently an experimental tool. CTLp and HTLp frequencies have been studied in a variety of clinical settings using a variety of assay systems and in peripheral blood as well as in cells obtained from the transplanted organs by biopsy. The latter need careful interpretation since clonal expansion may well occur locally.

Although high CTLp frequencies in peripheral blood against donor cells have been found by some to be related to acute rejection episodes in renal transplants (Jooss et al., 1989), this has not been reproduced by others (Steinmann et al., 1994). In heart transplant recipients, a correlation between an increase in CTLp frequency and rejection has been found more consistently (van Emmerik et al. 1996, 1997). In most of these studies (except for DeBruyne et al., 1995), the frequencies were determined from T cells obtained from endomyocardial biopsies following expansion by *in vitro* culture with IL-2 and it is difficult to relate the results of these measurements to the frequencies of CTL in the starting population.

Several groups have reported low donor-specific, but not anti-third party, CTLp in a proportion of patients with stable long-term transplants (as has also been described for MLC; see above). Some reports have suggested that detection of high avidity cells (as indicated by insensitivity to anti-CD8 antibody or cyclosporin) after transplantation is a better predictor of outcome, but these were mainly amongst graft infiltrating cells in heart transplant recipients (Vaessen et al., 1994; van Emmerick et al., 1996).

Given the central role played by T helper cells in orchestrating the immune response to allografts, development of donor-specific hyporesponsiveness or tolerance in the helper T cell repertoire is of considerable interest. There have been a few reports of a fall in HTLp frequencies following transplantation in a proportion of patients. Recent evidence suggests that the fall in frequency affects particularly high avidity donor-specific alloreactive helper T cells (Mason and Lechler, unpublished observations).

The use of LDA assays in solid organ transplantation is currently experimental, but they may prove to be valuable tools to improve our understanding of the changing immunological recipient–donor relationship following transplantation. It is an exciting possibility that such assays may support the reduction of non-specific immunosuppression in selected patients, thus reducing the significant long-term drug-related morbidity. However, despite the development of donor-specific hyporesponsiveness, perhaps in the majority of patients, chronic rejection is still a common cause of solid organ graft failure. It follows that direct alloreactivity, as measured by the assays described in the literature discussed, does not drive this process. Although there are several possible explanations, including non-immunological factors, experimental animal data suggest that indirect allorecognition may play an important role in this process (Sayegh, 1996). Limiting dilution assays have been developed to measure the precursor frequencies of potential indirect alloreactive T cells in clinical transplantation (Lui et al., 1996; Hornick and Lechler, unpublished observations).

SUMMARY

Cellular methods of defining histocompatibility do not currently have wide clinical applications. Mixed lymphocyte cultures are crude measures of alloreactivity and generally less predictive than the rapidly improving methods of genetic HLA typing. LDA assays probably play a role in choosing matched unrelated bone marrow donors and they also are providing valuable and exciting data in solid organ transplantation as a means of monitoring the emergence of donor-specific hyporesponsiveness.

REFERENCES

Bach, F. and Hirschhorn, K. (1964). Lymphocyte interaction: a potential histocompatibility test *in vitro*. *Science* **143**: 813–814.

Bagot, M., Mary, J.Y., Heslan, M. et al. (1988). The mixed epidermal cell lymphocyte reaction is the most predictive factor of acute graft versus host disease in bone marrow graft recipients. *Br. J. Haematol.* **70**: 403–409.

Bain, B., Vas, M.R. and Lowenstein, L. (1964). The development of large immature mononuclear cells in mixed leukocyte cultures. *Blood* **23**: 108–116.

Bishop, K.D. and Orosz, C.G. (1989) Limiting dilution analysis for alloreactive, TCGF-secretory T cells. *Transplantation* **47**: 671–677.

Breur-Vriesendorp, B.S., Vingerhoed, J., van-Twuyver, E. et al. (1991). Frequency analysis of HLA-specific cytotoxic T lymphocyte precursors in humans. *Transplantation* **51**: 1096–1103.

DeBruyne, L.A., Renlund, D.G. and Bishop, D.K. (1995). Evidence that human cardiac allograft acceptance is associated with a decrease in donor-reactive helper T lymphocytes. *Transplantation* **59**: 778–783.

Emmerick, N.E. van, Loonen, E.H., Vaessen, L.M. et al. (1997). The avidity, not the mere presence, of cytotoxic T-lymphocytes for donor human leucocyte class II antigens determines their clinical relevance after heart transplantation. *J. Heart Lung Transplant.* **16**: 240–249.

Emmerick, N.E. van, Vaessen, L.M., Balk, A.H. et al. (1996). Progressive accumulation of CTL with high avidity for donor antigens during the development of acute cardiac rejection. *Transplantation* **62**: 529–536.

Jeffery, J.R., Cheung, K., Masniuk, J. and Taylor, D. (1984). Mixed lymphocyte culture responses. Lack of correlation with cadaveric renal allograft survival and blood transfusions. *Transplantation* **38**: 42–45.

Jooss, J., Eiermann, T.H., Wagner, H. and Kabelitz, D. (1989). Interleukin 2 production by alloantigen-stimulated $CD4^+$ and human T cell subsets: frequency of HLA class I or class II-reactive precursor cells and clonal specificity of activated T cells. *Immunobiology* **179**: 366–381.

Kaminski, E.R., Hows, J.M., Bridge, J. et al. (1991). Cytotoxic T lymphocyte precursor (CTL-p) frequency analysis in unrelated donor bone marrow transplantation: two case studies. *Bone Marrow Transplant.* **8**: 47–50.

Keever, C.A., Leong, N., Cunningham, I. et al. (1994). HLA-B44-directed cytotoxic T cells associated with acute graft-versus-host disease following unrelated bone marrow transplantation. *Bone Marrow Transplant.* **14**: 137–145.

Keever-Taylor, C.A., Passweg, J., Kawanishi, Y. et al. (1997). Association of donor-derived host-reactive cytolytic and helper T cells with outcome following alternative donor T cell-depleted bone marrow transplantation. *Bone Marrow Transplant.* **19**: 1001–1009.

Lachance, S., Le Gouvello, S., Roudot, F. et al. (1997). Predictive value of host-specific donor helper T-cell precursor frequency for acutegraft-versus-host disease and relapse in HLA-identical siblings receiving allogeneicbone marrow transplantation for hematological malignancies. *Transplantation* **64**: 1147–1152.

Lefkovits, I. and Waldmann, H. (1979). *Limiting Dilution Analysis of Cells in the Immune System*, Cambridge University Press, Cambridge.

Lim, S.H., Patton, W.N., Jobson, S. et al. (1988). Mixed lymphocyte reactions do not predict severity of graft versus host disease (GVHD) in HLA DR compatible, sibling bone marrow transplants. *J. Clin. Pathol.* **41**: 1155–1157.

Lui, Z., Colovai, A.I., Tgulea, S. et al. (1996). Indirect recognition of donr HLA-DR peptides in organ allograft rejection. *J. Clin. Invest.* **98**: 1150–1157.

Murali-Krishna, K., Altman, J.D., Suresh, M. et al. (1998). Counting antigen-specific CD8 T cells: a reevaluation of bystander activation during viral infection. *Immunity* **8**: 177–187.

O'Shea, J., Madrigal, A., Davey, N. et al. (1997) Measurement of cytotoxic T lymphocyte precursor frequencies reveals cryptic HLA class I mismatches in the context of unrelated donor bone marrow transplantation. *Transplantation* **64**: 1353–1356.

Sayegh, M.H. and Carpenter, C.B. (1996). Role of indirect allorecognition in allograft rejection. *Int. Rev. Immunol.* **13**: 221–229.

Schwarer, A.P., Jiang, Y.Z., Brookes, P.A. et al. (1993). Frequency of anti recipient alloreactive helper T cell precursors in donor blood and graft versus host disease after HLA identical sibling bone marrow transplantation. *Lancet* **341**: 203–205.

Spencer, A., Brookes, P.A., Kaminski, E. et al. (1995). Cytotoxic T lymphocyte precursor frequency analyses in bone marrow transplantation with volunteer unrelated donors. Value in donor selection. *Transplantation* **59**: 1302–1308.

Starzl, T.E., Demetris, A.J., Trucco, M. et al. (1993). Chimerism and donor specific nonreactivity 27 to 29 years after kidney allotransplantation. *Transplantation* **55**: 1272–1277.

Steinmann, J., Kaden, May-G., Schroder, K. et al. (1994). Failure of *in vitro* T-cell assays to predict clinical outcome after human kidney transplantation. *J. Clin. Lab. Anal.* **8**: 157–162.

Tan, L.C., Gudgeon, N., Annels, N.E. et al. (1999). A re-evaluation of the frequency of CD8+ T cells specific for Epstein–Barr virus in healthy virus carriers. *J. Immunol.* (in press).

Vaessen, L.M., Baan, C.C., Ouwehand, A.J. et al. (1994). Differential avidity and cyclosporine sensitivity of committed donor-specific graft-infiltrating cytotoxic T cells and their precursors. Relevance for clinical cardiac graft rejection. *Transplantation* **57**: 1051–1059.

CHAPTER 28

Polymerase Chain Reaction-based Methods of HLA Typing

Henry A. Erlich

INTRODUCTION

The initial reports of human leucocyte antigen (HLA) disease associations all relied on the serological definition of class I and class II antigens and, to a lesser extent, the cellular definition of HLA class II antigens by the mixed lymphocyte reaction was also used. During the 1980s and 1990s, molecular genetic techniques have been used to isolate the genes encoding the HLA class I and class II molecules and to characterize their genomic organization as well as their allelic sequence diversity. The sequence polymorphism at the HLA class I and II loci, revealed by extensive sequencing studies on various human populations, is far greater than the antigenic variation detected by conventional serological typing (Bodmer et al., 1997). For example, there are about 200 DRB1 alleles defined by the second exon sequence but only about 15 different DR serological specificities or serotypes. For the DPA1 and DPB1 loci, which encode the DP molecule, no serological typing system has been available and, consequently, most functional investigations of DP polymorphism, including disease association studies, have been possible only since the development of DNA-based HLA typing methods (Bugawan et al., 1988). For all of the HLA class I and class II loci, the study of allelic sequence diversity as well as the development of simple and rapid DNA-based typing methods has been greatly facilitated by the development of polymerase chain reaction (PCR) amplification in the mid-1980s (Saiki et al., 1985, 1988b; Mullis and Faloona, 1987).

In general, DNA-based analysis of HLA polymorphism allows not only a much more accurate (fewer errors) and precise (more discriminating) method of typing than serology but also the use of synthetic, standardized reagents. In addition, DNA methods permit the typing of a much wider variety of samples because, unlike serological methods, the viability of the cells or the expression of the relevant antigen on the cell surface is not required. Thus, HLA typing can, with PCR-based methods, be carried out on buccal swabs, hairs and dried blood spots, as well as archival material, such as paraffin-embedded tissue biopsy sections. This capability has made possible valuable retrospective epidemiological (Apple et al., 1994) and population genetics studies (Bugawan et al., 1994b) as well as forensics applications (Blake et al., 1990). In addition, PCR-based typing data from which the nucleotide sequence of the sample can be inferred can reveal how and where alleles differ, allowing the analysis of the role of specific polymorphic amino acid residues in peptide binding and presentation, as well as in disease association and clinical transplantation. For example, the DR4 specificity can be encoded by any of 22 DRB1 alleles; some DR4-associated diseases, e.g. insulin-dependent diabetes mellitus (IDDM) and rheumatoid arthritis (RA), are positively associated with specific DR4 alleles, and negatively associated with others (Caillat-Zucman et al., 1992; Cucca et al., 1993; Erlich et al., 1993). The negatively and positively associated alleles can differ by as little as a single codon. These critical polymorphic sequence motifs can be now be readily distinguished by PCR-based HLA typing methods. Although space constraints do not permit an exhaustive survey of the wide variety of PCR-based HLA typing methods that have been reported, the principal PCR methods that have been developed since the late 1980s are reviewed in this chapter.

HLA ALLELIC SEQUENCE DIVERSITY

The allelic sequence diversity of the HLA class I and class II loci is the highest among mammalian coding sequence polymorphisms, with the number of alleles at some loci approaching 200 as of 1997 (Bodmer et al., 1997). The functional significance of this extensive polymorphism as well as the genetic mechanisms and the evolutionary forces that have generated and maintained this sequence diversity continue to be the subject of many immunological, genetic and evolutionary investigations (Klein and Figueroa, 1986; Bergstrom et al., 1998; Erlich and Gyllensten, 1991). For the HLA class II genes, the loci encoding the α- and β-chains of the DR, the DQ and the DP antigens, virtually all of the polymorphism is localized to the second exon. This exon encodes the α-helical "walls" and the β-pleated sheet "floor" of the peptide-binding groove formed by the α–β heterodimer. Among the class II α chain loci, only the DQA1 locus, with over 15 alleles, shows extensive polymorphism. The β-chain loci, however, are highly polymorphic. Population surveys in a variety of human populations had identified over 77 alleles at the DPB1 locus and over 184 at the DRB1 locus as of 1997 (Bodmer et al., 1997). A small number of these alleles is identical in amino acid sequence and differs only at the nucleotide level through silent substitutions. The other DRB loci (i.e. DRB3) show a relatively modest number of alleles, although some of the alleles at DRB3 differ at many different sites.

In contrast, the HLA class I HLA-A, -B and -C antigens contain polymorphic α_1 and α_2 protein domains that comprise the peptide-binding groove encoded within a single heavy-chain locus. The α_1 and α_2 domains are encoded in the second and third exons, respectively. Like the class II loci, the class I loci are highly polymorphic: population studies have revealed 80 alleles at the HLA-A locus, 175 alleles at the HLA-B locus and 38 alleles at the HLA-C locus (Bodmer et al., 1997). In general, a given population contains only a subset of the class I or class II alleles that have been identified world-wide and different populations often have significantly different allele frequency distributions. Different populations can also have varying patterns of linkage disequilibrium, that is, the non-random association of alleles at the linked HLA loci.

In general, the pattern of allelic sequence diversity at all of the polymorphic HLA loci is a patchwork of discrete sequence motifs. This patchwork pattern of polymorphism is thought to reflect the operation of gene conversion-like events that have generated the extensive sequence diversity observed in human populations by recombining these short sequence motifs (Erlich and Gyllensten, 1991). One consequence of this pattern is that, in PCR-based HLA typing, a large number of different alleles can be distinguished by using a relatively modest number of oligonucleotide primers or probes (see below) that recognize these sequence motifs. However, sometimes a given pattern of sequence motifs, detected either with probes or primers or by sequencing, may be consistent with more than a single genotype because the observed sequence motifs can be combined into more than a unique pair of alleles. This issue of ambiguity in DNA-based typing is discussed below.

CORRELATING SEROLOGICAL SPECIFICITIES WITH DNA-DEFINED ALLELES

In general, the allele groups, designated by the first two digits (i.e. DRB1*04 or *08), correspond to serological specificities (i.e. DR4 or DR8). There are, however, exceptions. For the HLA-B locus, all alleles within a single allele group designation (e.g. HLA-B*15) do not necessarily encode the same serological type. For example, several different serological specificities are associated with different B15 alleles (e.g. B*1522 with B35 and B*1501, B*1504, B*1505-1508, B*1515,

B*1520, B*1524 with B62). Recently, a "dictionary" relating alleles and serological specificities for DR, A, B and C has been reported (Hurley et al., 1997).

HLA TYPING: THE DEVELOPMENT OF DNA-BASED TECHNIQUES

Introduction

Historically, HLA typing has been performed by using a combination of the serological microcytotoxicity test (Terasaki et al., 1964) and, in special cases, mixed lymphocyte reaction (MLR) assays (Bach and Voynow, 1966) for class II antigens. The microcytotoxicity assay requires viable cells and uses antisera obtained primarily from individuals who have been sensitized to HLA differences such as multiparous women (women who have had multiple pregnancies) or individuals who have received multiple transfusions. Consequently, these reagents are limiting in quantity and difficult to standardize. Although, as noted, over 150 DRB1 alleles (i.e. with different amino acid sequences) have been identified, serological reagents can distinguish only 15 different groups of DR molecules encoded by these alleles. This illustrates the limitations of the serological methods of HLA typing.

The initial approach to HLA typing at the DNA level involved the restriction fragment length polymorphism (RFLP) method (Wake et al., 1982; Erlich et al., 1983). Radioactively labeled cDNA clones or genomic fragments were used as hybridization probes for genomic DNA that had been digested with a restriction enzyme, size separated on an agarose gel and then transferred to a membrane. Different HLA serotypes exhibited different banding patterns. In some cases, the probes cross-hybridized to related genomic sequences, yielding a complex pattern of bands (restriction fragments). The use of small locus-specific probes, often derived from the 3'UT region, typically simplified the banding patterns and facilitated typing. Following this initial phase, non-radioactive probes were introduced (Erlich et al., 1986). RFLP typing required neither cell-surface expression nor cell viability and, provided that several restriction enzymes were used, could often subdivide HLA class II serotypes. Although an informative approach, RFLP typing required a large amount of high molecular weight genomic DNA and the Southern blotting procedure was somewhat cumbersome. This approach, consequently, was not well suited to routine clinical typing. In addition, since many of the polymorphic restriction sites used for RFLP typing were not in exon-2, this method relied heavily on linkage disequilibrium with the exon-2 polymorphism. Most importantly, RFLP typing failed to distinguish much of the HLA class II sequence polymorphism.

The development of the PCR greatly facilitated the analysis of sequence polymorphism at both the class I and II loci and the HLA region was a primary model system for developing PCR genetic typing methods. The second region, after β-globin, to be amplified from genomic DNA using this newly developed amplification method was the HLA-DQA1 second exon (Saiki et al., 1986; Scharf et al., 1986). By generating billions of copies of the specific target sequence from a complex template like genomic DNA, this technique enabled the use of simple, non-radioactive methods to analyse sequence information. Using PCR, genetic typing can be carried out on samples containing minute amounts of DNA; samples containing degraded DNA can be readily typed as well. Based on the available database of class I and II allelic sequence diversity, a variety of relatively simple and rapid PCR-based methods has been developed to carry out HLA typing at the DNA level.

Most typing methods involve the design of primer pairs that are capable of amplifying all alleles at the target locus with the polymorphic sequence motifs localized between the primer sites. The sequences between the primers are subsequently characterized by a variety of approaches, including hybridization probes, restriction enzymes, chain-termination sequencing reactions, or inferred from the conformation-based mobility of the PCR products using gel electrophoresis. The other main approach to HLA typing uses the PCR itself as a method of distinguishing polymorphisms by exploiting the specificity of oligonucleotide primer extension and places the 3' end of the primer at the polymorphic site.

Sequence-specific oligonucleotide probes

The first PCR-based approaches to HLA typing utilized labeled sequence-specific oligonucleotide (SSO) probes to hybridize to PCR products amplified using the primers specific for non-polymorphic regions and immobilized on a Nylon or nitrocellulose filter, the "dot-blot" method (Saiki et al., 1986) (Figure 28.1). SSO probes had been previously applied to restriction enzyme-digested genomic DNA in gels (Conner et al., 1983; Angelini et al., 1986) but the probes often cross-hybridized to non-target genomic fragments and the procedure was very complex and cumbersome. In SSO dot-blot typing, under appropriate hybridization and wash conditions, the labeled SSO probes would bind only to the complementary sequence in the amplified DNA and were able to distinguish single nucleotide differences. Using a panel of probes specific for informative sequence motifs, the HLA alleles in the sample could be inferred from the pattern of probe reactivity. The initial methods utilized ^{32}P-labeled probes for typing of the HLA-DQA1 locus but, shortly thereafter, non-radioactively labeled probes were introduced (Saiki et al., 1988a) and are now commonly used. The probes can be labeled either directly with an enzyme, e.g. horseradish peroxidase (HRP) or with biotin. The biotin-labeled probes can be detected with streptavidin conjugated to HRP or alkaline phosphatase (AP) and can be used with chromogenic or chemoluminescent substrates for sensitive, -simple and robust detection. More recently, digoxigenin-labeled probes and antibody to digoxigenin detection methods have been introduced (Nevinny-Stickel et al., 1991; Shaffer et al., 1992). Currently, computer programs are routinely used to infer the genotype from the probe reactivity pattern. Given enough primers and probes, the PCR/SSO method is, in principle, capable of distinguishing all of the alleles at a given HLA locus. To achieve allele level typing, however, it is sometimes necessary to amplify separately the alleles of a heterozygote (see below) and analyse the probe reactivity pattern for each individual allele.

In the SSO probe typing approach, the sequence of the primers and the amplification reaction conditions (thermal profile, Mg^{2+} concentration, primer concentrations, etc.) are designed so that all alleles at the target locus are amplified and, whenever possible, the amplification is locus specific. In some cases, such as generic DRB1 typing, the same primer pairs amplify the DRB1, DRB3, DRB4 and DRB5 second exons. Amplification of more than one locus can complicate the interpretation of probe reactivity if the complementary sequence motif is found in more than one locus. In the design of SSO probes, the position of the mismatched base pair should

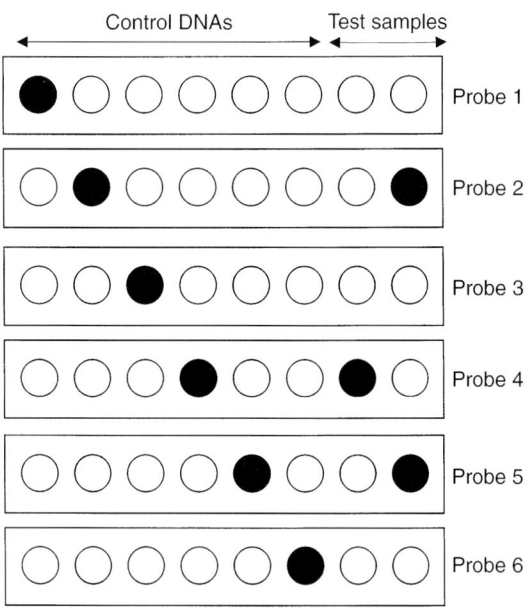

FIGURE 28.1 SSO Filter Probing. Control and test PCR products are immobilized on to filters and each filter is then hybridized to a locus-specific oligonucleotide. The conditions are chosen such that an oligonucleotide probe will remain bound to the filter only if it can form a perfect annealed duplex with correct Watson–Crick base pairing of probe to template PCR DNA product. The probes are locus specific and detect the respective control locus DNA target, and the same allele if it is present in the test DNA samples. In the example shown, the first test DNA sample would be scored homozygous for the probe 4 allele, whereas the second sample is heterozygous for alleles 2 and 5. Detection of the probe hybridization can utilize radioactivity or, more recently, enzymatic labels (e.g. HRP) and chromogenic substrates.

generally be placed toward the middle of the probe to minimize the stability of the mismatched probe–target duplex. Not all base-pair mismatches are equally destabilizing; the G-T mismatch is the least destabilizing and, if possible, should be avoided to minimize cross-hybridization to the non-target allele. To simplify the hybridization conditions for multiple probes, buffer systems, such as those including tetramethylammonium chloride (TMAC), have been developed to minimize the effect of GC% differences between probes. However, owing to its toxicity and cost, many laboratories have designed probes and conditions to simplify hybridization analysis without the addition of TMAC.

Following the initial SSO probe typing system for DQA1, PCR/SSO dot-blot typing systems for DRB1, DQB1, DPB1 and DPA1 have been developed (Bugawan et al., 1990; Bugawan and Erlich, 1991; Erlich et al., 1991; Scharf et al., 1991), as well as, more recently, for the HLA-A, -B and -C loci (Yoshida et al., 1992; Oh et al., 1993). Reverse hybridization approaches using immobilized SSO probe arrays, developed to simplify SSO typing, are described below.

Sequence specific priming/allele-specific amplification/amplification refractory mutation system

Another PCR-based approach, based on the specificity of primer extension rather than that of probe hybridization, has also been applied to HLA typing. This method is known variously as allele-specific amplification (ASA) (Wu et al., 1989), sequence specific priming (SSP) (Olerup and Zetterquist, 1992) and the amplification refractory mutation system (ARMS) (Newton et al., 1989; Browning et al., 1993). Here, a specific primer pair is designed for each polymorphic sequence motif or pair of motifs and the presence of the targeted polymorphic sequence in a sample is detected as a positive PCR, typically identified as a band on a gel (Figure 28.2). In SSP typing, if the PCR is negative and no product is detected, the sample is assumed to lack one or both of the specific motifs. Since inhibition of the PCR or absence of template also yields a negative result, each reaction should include a positive control, that is, PCR primers for an unrelated

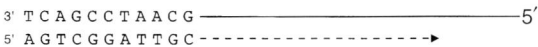

FIGURE 28.2 Principle of the ARMS and ASA techniques. An allele-specific oligonucleotide primer is hybridized to the target locus sequence on the template DNA. A perfect match between the primer and template enables the DNA polymerase enzyme to extend the primer and yield an amplification DNA product (top). However, if the allele is different (bottom) then the primer will anneal yielding a mismatched primer/template combination. The mismatch at the 3′ end prevents the DNA polymerase from extending the primer and so no amplification is possible.

monomorphic target sequence that produces a fragment that can be resolved from the HLA PCR product. SSP/ARMS typing has also been applied to class I typing (Bunce and Welsh, 1994, 1995a; Krausa and Browning, 1996) and robotics have, in some cases, simplified this procedure (Bunce et al., 1995b). Recently developed detection methods that are not based on visualizing a band in a gel and can, therefore, eliminate the gel electrophoresis step, are just starting to be applied (see below). Although informative and relatively fast for small numbers of samples, the SSP approach requires many separate PCR to achieve intermediate- or high-level typing and, in its current format, is not suited to rapid throughput of large sample numbers. As noted above, allele-specific amplification can be used in conjunction with SSO probe typing for high-resolution typing by allowing the separate amplification of the two alleles in a heterozygote.

Polymerase chain reaction–restriction fragment length polymorphism, single-strand conformation polymorphism, directed heteroduplex analysis and double-stranded conformation analysis

Other PCR-based methods involve the use of multiple restriction endonucleases and gel electrophoresis (PCR-RFLP) to characterize the polymorphisms in the PCR product; these approaches have been developed for class I and class II typing (Maeda et al., 1989; Olerup, 1990; Urya et al., 1990; Ota et al., 1991; Yunis et al., 1991). Conformation-based gel mobility analyses, such as PCR-single-strand conformation polymorphism (SSCP), have also been applied (Carrington et al., 1992; Hoshino et al., 1992). Another conformation and mobility-based approach to HLA typing, PCR-directed heteroduplex analysis (DHA), has also been developed for class II typing (Zimmerman et al., 1993) but these approaches, unlike SSO or SSP methods, are not widely used in clinical settings. A recent modification of the directed heteroduplex approach, termed double-stranded conformation analysis (DSCA), utilizes a fluorescent single-stranded probe specific for a particular class I allele (e.g. A*0101) to hybridize to the class I PCR products (e.g. the two HLA-A alleles) amplified from the sample, thereby generating labeled heteroduplex molecules. The mobility of these molecules can be systematically compared to a standard set of markers for each allele (Argüello et al., 1998) (Figure 28.3).

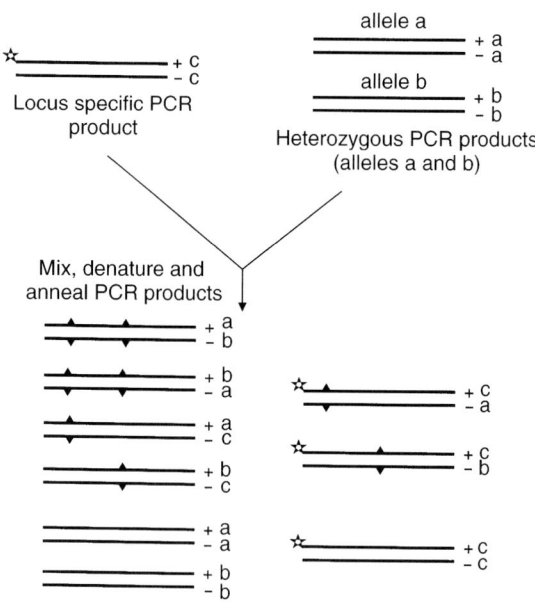

FIGURE 28.3 Principle of double-stranded conformational analysis (DSCA). A locus-specific PCR DNA product is labeled at the 5' end of only one DNA strand with a fluorescent dye label (DNA strands, c). This double-stranded DNA is mixed with unlabelled PCR targets from a test patient, containing two alleles, a and b. The DNA mixture is denatured and the DNA strands are allowed to hybridize back together. As all of the DNA comes from the same locus and is essentially the same DNA sequence, the strands can re-anneal back in any combination. There are nine possible re-annealed products, as shown. The three original DNA duplexes are perfect matches (aa, bb and cc). The other six duplexes contain DNA strands originating from different alleles (a and b) or the locus specific product (c). In the example shown the black triangles represent mismatches in the DNA duplexes due to sequence differences in alleles a and b relative to the test allele c. Allele a has one difference relative to c, and so has allele b. However, the positions of the polymorphisms are such that alleles a and b have mismatches at two positions. Although the nine possible DNA fragments are all the same size, the DNA composition varies owing to the polymorphisms at the allelic sites. On non-denaturing gels the fragments will thus have different mobilities relative to each other. Of the nine possible duplexes formed only three will be detected by analysis on an automated DNA sequencer, as they are the only duplexes to contain the fluorescent dye (represented by the star symbol). These are the top dye-labeled c+ DNA strands base paired with the bottom c-, a- or b-strands.

Reverse hybridization with immobilized sequence-specific oligonucleotide probe arrays

A recently developed approach to SSO probe analysis of HLA polymorphism has greatly facilitated PCR-based typing with multiple probes (Saiki et al., 1989). The conventional dot-blot involved an immobilized PCR product that is hybridized to each of many labeled SSO probes. The reverse-blot (or immobilized probe) method is based on the hybridization of PCR product, labeled with biotinylated primers during the amplification, to an array of immobilized probes on a membrane. Although initially the probes were immobilized by ultraviolet-cross-linking polythymidine "tails" that had been added to the 3' end of the probe with terminal transferase, a variety of immobilization methods can now be used. The presence of the PCR product bound to a specific probe is detected using streptavidin-HRP and a chromogenic substrate. This procedure, illustrated in Figure 28.4, requires only a single PCR and a single hybridization

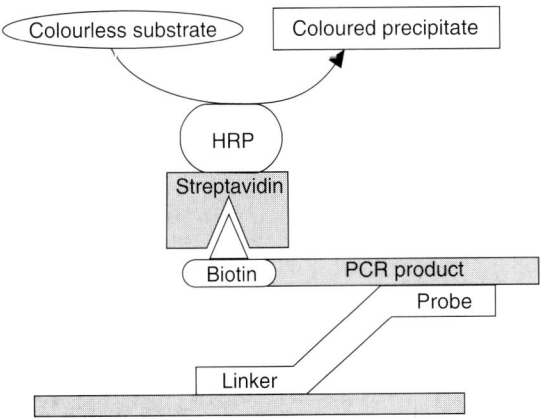

FIGURE 28.4 Schematic of the reverse hybridization assay utilizing immobilized probe arrays.

reaction to obtain information from the entire SSO probe panel; all of the probe reactivity information is contained on a single membrane, making it amenable to automated data interpretation. The critical challenge in this approach is to design a large number of SSO probes that will hybridize specifically under a single set of conditions. In addition, because the PCR product is denatured and hybridized in solution to the immobilized complementary probe, secondary structure in the labeled target single strand that would prevent binding to the probe must be minimized.

This approach has been applied to a variety of genetic typing systems, including direct mutation detection for β-thalassemia and cystic fibrosis (Saiki et al., 1989; Erlich et al., 1991b) and was the basis for the first commercial PCR test, the Amplitype HLA-DQ-alpha Forensic test, introduced in 1990 and used, since that time, in thousands of forensic cases. In order to accommodate more probes on the membrane, the probes are now immobilized as lines rather than as dots (Begovich and Erlich, 1995). A commercially available HLA-DRB test, the ReLi HLA-DRB test, uses 36 probes and provides intermediate-level typing for alleles at the DRB1, DRB3, DRB4 and DRB5 loci. Immobilized probe tests using the same PCR thermal profile and hybridization conditions have been developed for DQB1 as well as for the HLA-A, -B and -C loci (Bugawan et al., 1994; Trachtenberg et al., 1994; Blair et al., 1997). The present author's current B-locus typing sys-

tem uses 82 probes for exons 2 and 3, which are coamplified with two primer pairs. An instrument to automate the hybridization, wash and colour development of these immobilized probe strips is commercially available (SLT Profiblot), as are a scanner and software to read and interpret the probe reactivity patterns. The immobilized probe approach is well suited to the rapid typing of a few samples as well as to high-throughput typing of large sample numbers. Reverse hybridization approaches have also utilized the microtiter format (Cross et al., 1992), and other microtiter formats using biotinylated PCR product capture with streptavidin and hybridization with enzyme-labeled SSO probes have been reported (Lazaro et al., 1993).

Sequence-based typing

The application of semiautomated chain-termination sequencing using either fluorescent primers or fluorescent dideoxy terminators, with the development of appropriate sequence analysis software, has become a powerful approach to allele-level HLA typing (McGinnis et al., 1995; Vooter et al., 1997) (Figure 28.5). Several commercial approaches to automating the gel electrophoresis step are now available and, recently, the introduction of modified thermostable DNA polymerases capable of efficiently incorporating fluorescent dideoxy nucleotide monophosphates has made sequencing a more robust typing method. However, in its current format, sequence-based typing (SBT) is still a somewhat expensive procedure and not ideally suited to large-scale typing or routine clinical typing.

HLA TYPING REQUIREMENTS

In general, the requirements for HLA typing and the desired performance characteristics of PCR-based typing methods will depend on the application. Bone-marrow donor registry screening and recruitment requires a method that provides intermediate-level resolution typing and that is low cost and high throughput. The nature of the typing data and the level of resolution should be such as to facilitate the storage and search of the donor typings in the registry for possible

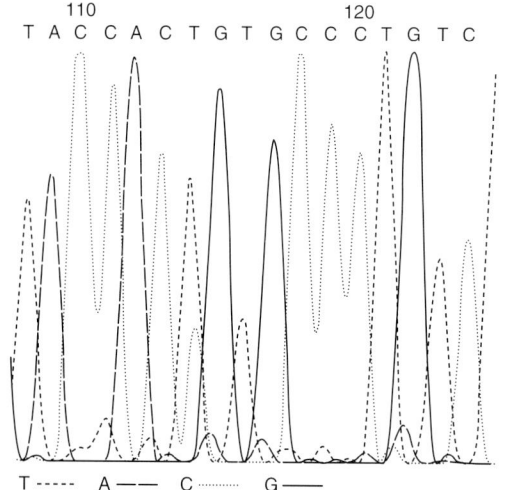

FIGURE 28.5 DNA sequence analysis. Automated DNA sequence analysis relies on labeling DNA fragments with a different fluorescent dye specific for each of the four bases in DNA. The DNA bands are resolved on a denaturing polyacrylamide gel and the individual DNA strands are detected as they pass a scanning laser. The figure shows a section of a computer trace of the fluorescence detected, with each base represented by a different line (T: ----; A: ——; C: ......; and G: ——). The computer program interprets the data and calls the respective base above the corresponding peak in the electropherogram. All of the bases called are clearly unambiguous, apart from position 114, where the laser has detected a signal for both C and T bases, and the resulting base called is Y (meaning a mixture of the two pyrimidine bases C and T). The DNA sample is thus heterozygous, containing a mixture of two alleles, one with C at position 114 and the other with T at 114.

matches with the patient/recipient. Although HLA typings for most donor registries now involve serological HLA-A and -B and PCR-based or serological typing for HLA-DR, it is likely that future donor typings will be PCR based for all three loci. The final HLA typing for matching bone marrow donors and recipients should be high resolution because preliminary data suggest that the clinical outcomes (both survival and graft-versus-host disease) can be influenced by allelic differences (e.g. DRB1*1101 vs *1104) in subtypes within a given serotype (Petersdorf et al., 1997). In this setting, cost and throughput are less critical issues. Solid organ transplantation, in contrast, demands a fairly rapid typing method; here, speed, rather than cost and throughput is the critical issue and an intermediate-level typing is probably sufficient. Research applications of HLA typing, such as disease associations or population genetics studies, should be carried out at a high level of resolution, when possible, because often a particular allele or haplotype, defined at the DNA sequence level, is much more strongly associated with a disease than is the serotype. Similarly, anthropological genetic analyses of populations are most informative when performed at the allele or haplotype level.

THE PROBLEMS OF NEW ALLELES AND AMBIGUITY

As more and more PCR-based HLA typing is being carried out in more and more populations, new (i.e. previously unreported) alleles at both class I and class II loci are being detected, leading to a slow but steady increase in the number of alleles. The vast majority of these involves new combinations of previously known sequence motifs and, thus, they can be identified without adding additional probes. None the less, they can create problems for typing strategies. The identification of new alleles requires frequent updating of the files in the genotyping software that relate sequence motifs to alleles. The addition of these new alleles can lead to increased ambiguities in the genotype interpretation of the primary typing data, such as the SSO probe reactivity or SSP patterns. An additional consequence of these newly discovered alleles is to modify the interpretation of typing results obtained prior to their discovery. For example, a pattern of reactive SSO probes that was consistent with a given genotype at one time might, following the identification of the new allele, be consistent with additional genotypes. This ambiguity can, in principle, be avoided by complete sequencing of the two separated alleles in a heterozygous sample but this is not, at the moment, a practical solution for most clinical diagnostic settings.

Most HLA typing systems, with the exception of sequencing separated alleles, can occasionally generate an ambiguous result; here, ambiguity is defined as HLA typing data (e.g. a probe reactivity pattern) consistent with more than one pair of alleles. The need to resolve these ambiguities is a function of the typing application. In some cases, if a given probe reactivity pattern (or other typing data) is consistent with either genotype X or Y, consideration of the genotype frequencies in the relevant populations and the likelihood that the sample is X or is Y may be appropriate in interpreting the typing data. For bone-marrow transplantation, however, any ambiguity in the patient typing results should be resolved by additional typing. A proposal for storing intermediate-level typing data in bone marrow donor registries by entering the probe reactivity patterns rather than the interpreted genotypes is one approach to dealing with ambiguity (Hurley, 1997). This proposal solves some of the problems associated with storing typing data but the issue of interpreting these patterns as genotypes remains.

HLA DNA TYPING AND DISEASE ASSOCIATIONS

The ability to analyse the polymorphism in the HLA class I and class II loci, typed by a variety of PCR-based methods, has clearly revolutionized the field of HLA disease associations. Comparing a sequence that is disease associated with closely related sequences that are not can point to specific polymorphic sequence motifs and, given the known structure, to specific pockets in the peptide-binding groove as being functionally important in disease susceptibility or resistance. For example, comparing DRB1*0402 to some of the other DR4 alleles or DRB1*0103 to the other DR1 alleles implicates the motif "I-DE" in the third hypervariable region. The role of position 57 of DQB1 in IDDM susceptibility, based on the negative association of Asp^{57} with disease, is well known (Todd et al., 1987; Horn et al., 1988), but the susceptibility of a given haplotype cannot be predicted by the presence of an individual amino acid residue. Clearly, there are some haplotypes that contain DQB1 Asp^{57} (e.g. DRB1*0801–DQB1*0402 in Caucasians or DRB1*0405–DQB1*0402 in the Japanese) that are positively associated with IDDM (Nepom and Erlich, 1991; Noble et al., 1997).

In general, HLA DNA typing has emphasized the potential role of multiple MHC loci in HLA-associated diseases and, using familial material and the ability to analyse DNA-defined extended haplotypes, has made it possible to separate the effects of specific combinations of alleles from associations due solely to linkage disequilibrium with high-risk DR-DQ alleles. The ability of DNA typing to implicate specific peptide-binding pockets and, based on the binding motifs of the disease-associated allele, to infer putative disease-associated peptides, can lead to new approaches to therapeutic intervention. The inference of possible *Plasmodium falciparum* peptides from a protective association of HLA-B53 and severe cerebral malaria in the Gambia represents an classic example of this approach (Hill et al., 1991). HLA DNA typing methods are likely to evolve rapidly; whatever their format, they will continue to make critical contributions to our understanding of HLA-associated diseases as well as our efforts to prevent them.

ACKNOWLEDGEMENTS

The author is grateful to Ray Apple for reviewing this manuscript and to Ann Begovich, Dory Bugawan, Ray Apple, Jim Novotny, Alan Blair, Sean Boyle and Priscilla Moonsamy for their contributions to the development of PCR-based HLA typing systems.

REFERENCES

Angelini, G., Preval, C.D., Gorski, J. and Mach, B. (1986). High-resolution analysis of the human HLA-DR polymorphism by hybridization with sequence-specific oligonucleotide probes. *Proc. Natl Acad. Sci. USA* **83**: 4489.

Apple, R.J., Erlich, H.A., Klitz, W. et al. (1994). HLA DR-DQ disease associations with cervical carcinoma show papillomavirus-type specificity. *Nature Genet.* **6**: 157.

Argüello, J.R., Little, A., Pay, A.L. et al. (1998). Mutation detection and typing of polymorphic loci through double-strand conformation analysis. *Nature Genet.* **18**: 192–194.

Bach, F.H. and Noynow, N.K. (1966). One-way stimulation in mixed leukocyte cultures. *Science* **153**: 545–547.

Begovich, A. and Erlich, H.A. (1995). HLA typing and bone marrow transplantation: the clinical impact of new PCR-based methods for matching donors and recipients. *JAMA* **273**: 586.

Blair, A., Bugawan, T.L. and Erlich, H.A. (1997). PCR based DNA typing for the HLA-C locus using an immobilized oligonucleotide probe array in the line blot format. *Hum. Immunol.* **55**: 144 (Abstract).

Bodmer, J.G., Marsh, S.G.E., Albert, E.D. et al. (1997). Nomenclature for factors of the HLA system, 1996. *Tissue Antigens* **49**: 297–321.

Browning, M.J., Krausa, P., Rowan, A. et al. (1993). Tissue typing the HLA-A locus from genomic DNA by sequence-specific PCR: comparison of HLA genotypes and surface expression on colorectal tumor cell lines. *Proc. Natl Acad. Sci. USA* **90**: 2842.

Bugawan, T.L. and Erlich, H.A. (1991). Rapid typing of HLA-DQB1 DNA polymorphism using non-radioactive oligonucleotide probes and amplified DNA. *Immunogenetics* **33**: 163–170.

Bugawan, T.L., Begovich, A.B. and Erlich, H.A. (1990). Rapid HLA-DPA typing using enzymatically amplified DNA and non-radioactive sequence specific oligonucleotide probes: application to tissue typing for transplantation. *Immunogenetics* **32**: 231–241.

Bugawan, T.L., Apple, R. and Erlich, H.A. (1994a). A method for typing polymorphism at the HLA-A locus using PCR amplification and immobilized oligonucleotide probes. *Tissue Antigens* **44**: 137–147.

Bugawan, T.L., Chang, J.D., Klitz, W. and Erlich, H.A. (1994b). PCR/oligonucleotide probe typing of HLA class II alleles in a Filipino population reveals an unusual distribution. *Am. J. Hum. Genet.* **54**: 331–340.

Bunce, M. and Welsh, K.I. (1994). Rapid DNA typing for HLA-C using sequence-specific primers (PCR-SSP): identification of serological and non-serologically defined HLA-C alleles including several new alleles. *Tissue Antigens* **43**: 7.

Bunce, M., Fanning, G.C. and Welsh, K.I. (1995a). Comprehensive, serologically equivalent DNA typing for HLA-B by PCR using sequence-specific primers (PCR-SSP). *Tissue Antigens* **45**: 81.

Bunce, M., O'Neill, C.M. and Barnardo, M.C.N.M. (1995b). Phototyping: comprehensive DNA typing for HLA-A, B, C, DRB1, DRB2, DRB4, DRB5 & DQB1 by PCR with 144 primer mixes utilizing sequence-specific primers. *Tissue Antigens* **46**: 355.

Caillat-Zucman, S., Garchon, H.J., Timsit, J. et al. (1992). Age dependent HLA genetic heterogeneity of type I insulin-dependent diabetes mellitus. *J. Clin. Invest.* **90**: 2242–2250.

Carrington, M., Miller, T. and White, M. (1992). Typing of HLA-DQA1 and DQB1 using DNA single-strand conformation polymorphism. *Hum. Immunol.* **33**: 208–212.

Conner, B.J., Reyes, A.A., Morin, C. et al. (1983). Detection of sickle cell beta S-globin allele by hybridization with synthetic oligonucleotides. *Proc. Natl Acad. Sci. USA* **80**: 278–282.

Cros, P., Allibert, P., Mandrand, B. et al. (1992). Oligonucleotide genotyping of HLA polymorphism on microtiter plates. *Lancet* **340**: 870.

Cucca, F., Muntoni, F. and Lampis, R. (1993). Combinations of specific DRB1, DQA1, DQB1 haplotypes are associated with insulin dependent diabetes mellitus in Sardinia. *Hum. Immunol.* **37**: 85–94.

Erlich, H.A. and Gyllensten, U. (1991). Shared epitopes among HLA class II alleles: gene conversion, common ancestry, and balancing selection. *Immunol. Today* **11**: 411–414.

Erlich, H.A., Stetler, D., Sheng-Dong, R. et al. (1983). Segregation and mapping analysis of polymorphic class I restriction fragments: detection of a novel fragment. *Science* **222**: 72.

Erlich, H.A., Sheldon, E.K. and Horn, G. (1986). HLA typing using DNA probes. *Bio/Technology* **4**: 975–981.

Erlich, H.A., Bugawan, T., Begovich, A. et al. (1991a). HLA-DR, DQ, and DP typing using PCR amplification and immobilized probes. *Eur. J. Immunogenet.* **18**: 33–55.

Erlich, H.A., Gelfand, D. and Sninsky, J.J. (1991b). Recent advances in the Polymerase Chain Reaction. *Science* **252**: 1643–1651.

Erlich, H.A., Zeidler, A., Chang, J. et al. (1993). HLA class II alleles and susceptibility and resistance to insulin dependent diabetes mellitus in Mexican–American families. *Nature Genet.* **3**: 358–364.

Hill, A.V.S., Allsopp, C.E.M. and Kwiatkowski, D. (1991). Common West African HLA antigens are associated with protection from severe malaria. *Nature* **352**: 595–600.

Horn, G.T., Bugawan, T.L., Long, C. and Erlich, H.A. (1988). Allelic sequence variation of the HLA-DQ loci: relationship to serology and insulin-dependent diabetes susceptibility. *Proc. Natl Acad. Sci. USA* **85**: 6012–6016.

Hoshino, S., Kimura, A., Fukuda, A. et al. (1992). Polymerase chain reaction-single-strand conformation polymorphism analysis of polymorphism in DPA1 and DPB1 genes: a simple, economical and rapid method for histocompatibility testing. *Hum. Immunol.* **33**: 98–107.

Hurley, C.K., (1997). Acquisition and use of DNA-based HLA typing data in bone marrow registries. *Tissue Antigens* **49**: 323–328.

Hughes, A.L. and Nei, M. (1988). Pattern of nucleotide substitution at major histocompatibility complex class I loci reveals overdominant selection. *Nature* **335**: 167–170.

Hurley, C.K., Schreuder, M.T., Marsh, S.G.E. et al. (1997). The search for HLA-matched donors: a summary of HLA-A*, -B*, -DRB1/3/4/5* alleles and their association with serologically defined HLA-A, -B, -DR antigens. *Tissue Antigens* **50**: 401–418.

Krausa, P. and Browning, M.J. (1996). A comprehensive PCR-SSP typing system for identification of HLA-A locus alleles. *Tissue Antigens* **47**: 237.

Lazaro, A.M., Fernandez-Vina, M.A., Liu, Z. and Stastny, P. (1993). Enzyme-linked DNA oligotyping. A practical method for clinical HLA-DNA typing. *Hum. Immunol.* **36**: 243.

McGinnis, M.D., Conrad, M.P., Bouwens, A.G.M. et al. (1995). Automated, solid-phase sequencing of DRB region genes using T7 sequencing chemistry and dye-labeled primers. *Tissue Antigens* **46**: 173.

Maeda, M., Murayama, N. and Ishi, H. (1989). A simple and rapid method for HLA-DQA1 genotyping by digestion of PCR-amplified DNA with allele specific restriction endonucleases. *Tissue Antigens* **34**: 290–298.

Mason, P.D., Robinson, C.M. and Lechler, R.I. (1996). Detection of donor-specific hyporesponsive following late failure of human renal allografts. *Kidney Int.* **50**: 1019–1025.

Mullis, K.B. and Faloona, F. (1987). Specific synthesis of DNA in vitro via polymerase catalyzed chain reaction. *Meth. Enzymol.* **155**: 335–350.

Nepom, G. and Erlich, H.A. (1991). MHC class II molecules and autoimmunity. *Annu. Rev. of Immun.* **9**: 493–525.

Nevinny-Stickel, C., Bettinotti, M. and Andreas, A. (1991). Non-radioactive HLA class II typing using polymerase chain reaction and digoxigenen-11-2'-3'-dideoxyuridinetriphosphate-labeled oligonucleotide probes. *Hum. Immunol.* **31**: 7.

Newton, C.R., Graham, A. and Heptinstall, L.E. (1989). Analysis of any point mutation in DNA. The amplification refractory mutation system (ARMS). *Nucl. Acids Res.* **17**: 2503–2516.

Noble, J.A., Valdes, A.M., Cook, M. et al. (1996). The role of HLA class II genes in insulin-dependent diabetes mellitus: molecular analysis of 180 Caucasian, mulitplex families. *Am. J. Hum. Genet.* **59**: 1134–1148.

Oh, S.H., Fleischhauer, K. and Yang, S.Y. (1993). Isoelectric focusing subtypes of HLA-A can be defined by oligonucleotide typing. *Tissue Antigens* **41**: 135–142.

Olerup, O. (1990). HLA class II typing by digestion of PCR-amplified DNA with allele-restriction endonucleases will fail to unequivocally identify the genotypes of many homozygous and heterozygous individuals. *Tissue Antigens* **36**: 83.

Olerup, O. and Zetterquist, H. (1992). HLA-DR typing by PCR amplification with sequence-specific primers (PCR-SSP) in 2 hours: an alternative to serological DR typing in clinical practice including donor-recipient matching in cadaveric transplantation. *Tissue Antigens* **39**: 225–235.

Ota, M., Seki, T. and Nomura, N. (1991). Modified PCR-RFLP method for HLA-DPB1 and -DQA1 genotyping. *Tissue Antigens* **38**: 60.

Petersdorf, E.W., Kollman, C., Hurley, C. et al. (1997). Analysis of 390 marrow transplants for the treatment of chronic myeloid leukemia from unrelated donors facilitated by the U.S. Nation Marrow Donor Program (NMDP): effect of HLA-DRB1 allele disparity on clinical outcome. *Hum. Immunol.* **55**: 30 (Abstract).

Saiki, R.K., Scharf, S. and Faloona, F. et al. (1985). Enzymatic amplification of b-globin genomic sequences and restriction site analysis for diagnosis of sickle cell anemia. *Science* **230**: 1350.

Saiki, R.K., Bugawan, T.L., Horn, G.T. et al. (1986). Analysis of enzymatically amplified β-globin and HLA-DQβ DNA with allele-specific oligonucleotide probes. *Nature* **324**: 163–166.

Saiki, R.K., Chang, C., Levenson, C.H. et al. (1988a). Diagnosis of sickle cell anemia and β-thalassemia with enzymatically amplified DNA and non-radioactive allele-specific oligonucleotide probes. *N. Engl. J. Med.* **319**: 537.

Saiki, R.K., Gelfand, D.H., Stoffel, S. et al. (1988b). Primer-directed enzymatic amplification of DNA with a thermostable DNA polymerase. *Science* **239**: 487–491.

Saiki, R.K., Walsh, P.S., Levenson, C.H. and Erlich, H.A. (1989). Genetic analysis of amplified DNA with immobilized sequence-specific oligonucleotide probes. *Proc. Natl Acad. Sci. USA* **86**: 6230–6234.

Scharf, S.J., Horn, G.T. and Erlich, H.A. (1986). Direct cloning and sequence analysis of enzymatically amplified genomic sequences. *Science* **223**: 1076.

Scharf, S.J., Griffith, R.L. and Erlich, H.A. (1991). Rapid typing of DNA sequence polymorphism at the HLA-DRB1 locus using the polymerase chain reaction and non-radioactive oligonucleotide probes. *Hum. Immun.* **30**: 190–201.

Shaffer, A.L., Falk-Wade, J.A. and Tortorelli, V. (1992). HLA-DRw52-associated DRB1 alleles: Identification using polymerase chain reaction-amplified DNA, sequence-specific oligonucleotide probes, and a chemiluminescent detection system. *Tissue Antigens* **39**: 84.

Tan, L.C., Gudgeon, N., Annels, N.E. et al. (1999). A re-evaluation of the frequency of CD8+ T cells specific for EBV in healthy virus carriers. *J. Immunol.* **162**: 1827–1835.

Terasaki, P.I., Mandell, M., Van de Water, J. and Edginton, T.E. (1964). Human blood lymphocyte cytotoxicity reactions with allogeneic antisera. *Ann. N.Y. Acad. Sci.* **120**: 322–334.

Todd, J.A., Bell, J.I. and McDevitt, H.O. (1987). HLA-DQ beta gene contributes to susceptibility and resistance to insulin-dependent diabetes mellitus. *Nature* **329**: 599–604.

Trachtenberg, E.A., Bugawan, T.L., Apple, R. et al. (1994). Efficient typing of the HLA-B locus using the PCR and immobilized oligonucleotide probes. *Hum. Immunol.* **40**: 43.

Urya, N., Maeda, M., Ota, M. et al. (1990). A simple and rapid method for HLA-DRB and -DQB typing by digestion of PCR-amplified DNA with allele specific restriction endonucleases. *Tissue Antigens* **35**: 20.

Vooter, C.E.M., Rozemuller, E.H., Bruyn-Geraets, D. de et al. (1997). Comparison of DRB sequence-based typing using different strategies. *Tissue Antigens* **1994**: 44–93.

Wake, C.T., Long, E.O. and Mach, B. (1982). Allelic polymorphism and complexity of the genes for HLA-DRB chains – direct analysis by DNA-DNA hybridization. *Nature* **300**: 372.

Weston, L.E., Geczy, A.F. and Farrell, C. (1997). Donor helper T-cell frequencies as predictors of acute graft-versus-host disease in bone marrow transplantation between HLA-identical siblings. *Transplantation* **64**: 836–841.

Wu, D.Y., Ugozzoli, L., Pal, B.K. and Wallace, R.B. (1989). Allele-specific enzymatic amplification of beta-globin genomic DNA for diagnosis of sickle cell anemia. *Proc. Natl Acad. Sci. USA* **86**: 2757–2760.

Yoshida, M., Kimura, A., Numano, F. and Sasazuki, T. (1992). Polymerase-chain-reaction-based analysis of polymorphism in the HLA-B gene. *Tissue Antigens* **41**: 135–142.

Yunis, I., Salazar, M. and Yunis, E.J. (1991). HLA-DR generic typing by RFLP. *Tissue Antigens* **38**: 78.

Zimmerman, P.A., Carrington, M.N. and Nutman, T.B. (1993). Exploiting structural differences among heteroduplex molecules to simplify genotyping the DQA1 and DQB1 alleles in human lymphocyte typing. *Nucl. Acids Res.* **21**: 4541–4547.

Index

1C7 gene 42
2 × 2 contingency table 130, 132, 133

accommodation 393
acridine orange 437
Actimomyces thermophilus rectivirgula 266
acute anterior uveitis (AAU) 289
acute inflammatory demyelinating polyneuropathy (AIDP) 224–5
  see also Guillain–Barré syndrome
acute posterior multifocal placoid pigment epitheliopathy (APMPPE) 293–4
acute rejection 393–4
Addison's disease 244–5
affective disorders 420–21
agranulocytosis
  antithyroid drug-induced 241
alanine, position 73 371
alkaline phosphatase (AP) 454
alleles, new 458
allele-sharing methods 110–11
allele-specific amplification 455
allelic association 439
allelic sequence diversity 452
alloantibodies, screening 435–6
alloantigens 388
alloantisera 434–6
allografts 280, 283
  immunity 18
allograft inflammatory factor-1 (AIF-1) 42
allophenic donors 394
allorecognition 283
  HLA molecules 387–99
  indirect 447
ALLOTRAP® 398
Alzheimer's disease 243
amplification refractory mutation system 455
Amplitype HLA-DQ-alpha Forensic test 457
ANAP phenomenon 6
anhydride-induced asthma 270
animal models

congenic mice 114, 119
  speed congenics 119
eye disease 281–4
Fas antigen 117
genetic analysis 116
non-obese diabetic (NOD) strain 117–18
recombinant inbred (RI) strains 118
ankylosing spondylitis (AS) 134, 143, 187, 272, 274
  HLA-B27 187
anorexia nervosa 421
anterior chamber-associated immune deviation (ACAID) 279–80
antibody
  preformed 410
  testing and type I diabetes 235
anticardiolipin antibodies 348–9
anti-DNA antibodies 352–3
antiendomysial antibodies 257
antigen presenting cells
  antigen presentation 73–5
  antigen processing 73–5
  non-professional 58, 77
  professional 58
  transgenic mice models 58–9
antigen processing
  antigen protolysis 78–9
  class II molecules 77–8
  endoproteases 78
  protease inhibitors 78
antigens 388–91
antigen-specific enzyme-linked immunosorbent assays (ELISA) 392
anti-human gamma globulin antibodies 4
anti-La antibodies 351–2
anti-lymphocyte globulin (ALG) 16
anti-lymphocyte therapies 387
anti-neutrophil cytoplasmic antibody (ANCA) 212–14
  associated vasculitis 213
antinuclear antibodies (ANA) 289–90
antiphospholipid antibody syndrome 348–9
antipsychotic treatment 422

anti-Ro antibodies 351–2
anti-Sm antibodies 352
antithyroid drugs 241
anxiety disorders 421
arrestin 281
asbestosis 271–2
aspartic acid 371
association studies
   juvenile myoclonic epilepsy 227
   multiple sclerosis 221
association test 108–9
   chi square test 109
asthma 269–70
astrovirus 255
atopic dermatitis 376–7
autism 422
autoantibody screening 235
autoimmune
   Addison's disease 244–5
   hepatitis (AIH) 252
   hypothyroidism 242–3, 286
      family studies 243
      population studies 242–3
   polyglandular syndrome (APS) 244–5
automation, of serological testing 438
azathioprine 387

B cells 51
   alloantibody production 65
   B cell receptor 74–5
   Fc receptor 76
   receptor mediated antigen uptake 74
   T cell interaction 65
B8, DR3 association 223–4
B71 (CD80) molecule 60
B72 (CD86) molecule 60
back-stimulation 445
bacterial infections 315–18
BAT1 45
Behçet's disease 336–7, 372–4
berylliosis 263–6
beryllium 263–4
   hypersensitivity 263–5
biotin 454
bipolar disorder 420
bird breeder's disease 266, 267
birdshot chorioretinopathy (BSCR) 292
blood group systems, discovery 3
blood transfusion 435
bone marrow transplantation (BMT) 395, 414–15, 441, 442, 446–7
Borrelia burgdorferi 318
bulimia nervosa 421
Burkitt's lymphoma (BL) 312–3

C2 null alleles 345–6
C4 null alleles 342–5
cadaver donors 406
caesin kinase II β-subunit 45
Campylobacter infection 224

Caplan's syndrome 272
cardiac transplantation 411–12
$CD4^+$ T cells
   central role 391–2
   coeliac disease 258
   berylliosis 263
$CD8^+$ T cells 57, 393
CD40 molecule 60
cell separation techniques 429
cell-mediated lympholysis test 13
cervical
   carcinoma 314
   intraepithial neoplasia (CIN) 313
chain-termination sequencing reactions 453
chi-square test 6, 117–18, 129–31, 133
   null hypothesis 130
Chlamydia trachomatis 286
chlorpromazine 422
choroidal disorders 292–4
chronic
   bronchitis 272
   fatigue syndrome 421
   inflammatory demyelinating polyneuropathy (CIDP) 224–5
      see also Guillain–Barré syndrome
   Lyme arthritis 318
   myeloid leukaemia 395
   rejection 394–5
Churg–Strauss syndrome 336
cicatricial pemphigoid 287
clinical studies, using LDA 446–7
clozapine-induced agranulocytosis 422–3
coal worker's pneumoconiosis and silicosis 272–3
cobalt-induced fibrosis 271
coeliac disease (CD) 252–8
Cogan's syndrome 287
Collaborative Transplant Study 405
complement
   C2 40
   C4 40
   factor B 40
   fixing antibodies and graft rejection 280
   systemic lupus erythematosus 341–8
complement-dependent cytotoxicity (CDC) assay 6, 429–39
complementry determining regions (CDRs) 30
complete Freund's adjuvant 282
congenital adrenal hyperplasia 245
conjunctival disorders 286–9
corneal
   disorders 286–9
   transplantation 283–4, 288–9
Creb-rp 43
Crohn's disease 249
cross-match, pretransplant 392
cross-reactive groups (CREG) 408
cryptogenic fibrosing alveolitis (CFA) 273–4
CTLL cells, bioassay 445
cutaneous

leucocytoclastic vasculitis 335
lichen planus 375
CYNAP 6
cytokine production, PV linked HLA alleles 367–8
cytotoxic
   antibodies 392
   cell frequencies 445
   T lymphocyte(CTL)
      LD assays 445
      MHC recognition 14
      response
         to Epstein–Barr virus 311
         to human immunodeficiency virus 307
         to malaria 301–2

degenerate binding 307
delayed-type hypersensitivity and graft rejection 280
dendritic cells 393
   class II synthesis 77
   cornea 279–80
   phagocytosis 76
   professional APC 77
dengue fever virus 314–5
dengue haemorrhagic fever 315
dermatitis herpetiformis (DH) 252, 374–5
dermatological disease 365–79
Dermatophagoides farinae 270
Dermatophagoides pteronissimus 270
desmoglein-3 glycoprotein (DG3) 366–8
digoxigenin detection methods 454
diisocyanate-induced asthma 270
directed heteroduplex analysis 455–6
disease associations 458
   studies 439–40
DNA
   amplification 429
   based techniques 451–8
   markers
      microsatellites 108
      single-nucleotide polymorphisms (SNPs) 108
donor
   lymphocytes 393
   specific tolerance 398
dot-blot method 454
double-stranded conformation analysis 455–6
Down's syndrome 243
drug-induced
   lupus 349–50
   pemphigus vulgaris 368
dust hypersensitivity pneumonitis, organic 266–7
Dw2 halotype
   multiple sclerosis 219–20
dysthyroid ophthalmopathy 285–6

eating disorders 421
ELISPOT technique 446
emphysema 272
endocrine disease 239–46

enzyme linked immunosorbent assay (ELISA) 445
epistasis 119, 232
   genotpye risk ratio (GRR) 119
epitopes, DQ2 binding 257
Epstein–Barr virus (EBV) 310–13
   vaccine strategies 313
erosive oral lichen planus 375
ethidium bromide 437
ethylene diamine tetra-acetic acid (EDTA) 437
Eurotransplant 17, 415–16
experimental allergic encephalomyelitis 285
experimental autoimmune encephalomyelitis 221
experimental autoimmune uveitis (EAU) 282
eye disease 279–94

falciparum malaria 142, 301–2
   HLA-B53 142
family studies 135–6
   infectious diseases 300
   psoriasis vulgaris 371
farmer's lung 266, 267
fibrotic lung disorders 270–4
filariasis 304
Fischer's exact test 131–2
flow cytometry 446
   cross-matching 392
fogo selvagem 369–70
frequency data, HLA specificities 438–9
Fuchs' heterochromic iridocyclitis 290
functional screen 114
   congenic mice 114, 119
   YACs 114–115

G1 gene 42
G7a gene 44–45
G9 gene 41
G9a gene 44
G11 gene 44
G11a gene 44
G15 gene 40
G17 (PBX2) gene 39
gastrointestinal diseases 249–59
gel electrophoresis 453, 455–6
genomic screening techniques 115
giant cell arteritis (GCA) 327–8, 329
gliadin proteins 257
gliadin-specific T cells 257–8
glomerulonephritis 195
   HLA alleles 195–196
glutamic acid 257
gluten 252, 257
gluten-reactive T-cells 255–7
gluten-specific T-cells 255–6
gold hypersensitivity 265
   pneumonitis 266
Goodpasture's disease 139, 142, 195–203
   α2(IV)NC1 201, 203–204
   anti glomerular basement membrane (GMB) antibodies 196

class II molecules 195, 200, 202–203
    affinity for Goodpasture antigens 201
    structure 201
    DRB1 198–201
    DR15 198, 201–202, 203
        antigen presentation 202,
    DR4 198, 199
    DR7 203
    DQ allele 196–197
    HLA class II 199–200
    HLA-DR2 196–197
    pocket 4 200–201
    T cells 201–203
grafts 279–81
graft-versus-host disease (GVHD) 395–6, 414, 441, 442, 446
graft-versus-leukaemia (GVL) 396
granulomatous lung disorders 267–9
Graves' disease 239–41, 285–6
    family studies 240–1
    population studies 239–40
Guillain–Barré syndrome (GBS) 224–5
    see also acute inflammatory demyelinating polyneuropathy

Hantavirus 314
hard-metal lung disease 271
Hashimoto's thyroiditis 242–3, 286
heart transplantation 411–12
    organ allocation 415
heat shock protein 41–2, 70-2, 348, 372
    gp96 41
    HSC70 41
    HSP70 41
helper T lymphocyte, frequencies 445
Henoch–Schönlein purpura 335
hepatitis 252
hepatitis B virus 308–9
    vaccine strategies 309
hepatitis C virus 309–10
hereditary homozygous deficiency 341–6
Hermanski–Pudlack syndrome 274
herpes
    gestationis 369–70
    simplex virus (HSV) 96, 282, 374
herpetic
    corneal disease 282
    retinal disease 282
histocompatibility, cellular testing methods 441–8
histoglobulins 432
HLA antigens
    disease associations 15
    transplantation antigens 16
        skin grafting technique 16
        kidney graft survival 17
HLA class I alleles 23
    antigen binding 25
    eye 280
    hepatitis B infection 308

hepatitis C infection 310
HIV 305
HLA-A 163, 166–7, 172
HLA-B 163, 165, 167, 172
HLA-C 163, 165, 167
HLA-E 163–4, 165
HLA-F 163, 165
HLA-G 163–4, 165
leprosy 315
MHC class I-related gene family (MIC) 164
polymorphism 26–8
systemic lupus erythematosus 337–41
subunit structure 24–5
HLA class II alleles 23
    eye 280
    hepatitis B infection 308–9
    hepatitis C infection 310
    HIV 305
    HLA-DR 164, 165–6, 167, 172–3
        HLA-DRB 166
    HLA-DQ 164, 165–6, 167, 172–3
    HLA-DP 164, 165, 167, 172
    leprosy 315
    systemic lupus erythematosus 337–41
HLA class III alleles 23, 164
    antigen binding 25–6
    hepatitis B infection 309
    hepatitis C infection 310
    HIV 307
    polymorphism 26–8
    subunit structure 25
HLA disease studies 132–4
    ankylosing spondylitis (AS) 134, 143
    Bonferoni correction factor 134
    DR types 129–33
    family studies 135–6
    HLA-B27 145
    receptors for microbes 143–4
    relative risk ratio 132
    serological typing 439
    statistical significance 133
HLA genes, definition 431–2
HLA matching 405–16
    corneal transplantation 288–9
    in graft acceptance 64
HLA mismatches 406–14, 441
HLA molecules
    better fit model 62
    soluble 397
HLA polymorphisms, definition 432–3
HLA specificities, recognized 431
HLA system
    chromosome 6 14, 163
    class I 163
    class II 163
    class III 163
    HLA sequence database 155
    Human Genome Project 160
    lineage 165–7
    linkage disequilibrium 164

nomenclature 147–60
  D-region genes 149, 156–7
polymorphism 167
structure 14
subtypes 437–8
workshop 9, 148–50
World Wide Web 159
  EMBL 159–60
  GenBank 159–60
HLA typing
  requirements 457–8
  serological methods 429–40
HLA-A
  identification of locus antigens 6, 10
  matching and clinical transplantation 413
HLA-B
  identification of locus antigens 6, 10
  matching and clinical transplantation 413
HLA-B27 187–92
  asbestosis 272
  ankylosing enthesopathy (ankent) 190
  arthritogenic peptide hypothesis 188
  class II molecules 189
  cysteine-67 189
  formation of homodimers 189
  functional studies 191
  microbial superantigens 189–90
  molecular mimicry hypothesis 190
  murine β2m 190–1
  receptor hypothesis 190
  spondylarthropathies 187–8
  subtypes 191–2
  thymic expression 188
HLA-B51 antigens
  in Behçet's disease 374
HLA-B53 molecule 54
HLA-C, discovery of 10
HLA-DP
  coeliac disease 254
  matching and clinical transplantation 413–14
  multiple sclerosis 221
  sarcoidosis 268
  structure 264
HLA-DPβGlu^{69} 264–5
  sarcoidosis 268
HLA-DQ
  coeliac disease 254
  diabetes 231–35
  molecular properties 233–5
  multiple sclerosis 221
  narcolepsy 225
  prediction of type I diabetes 235
HLA-DQ3.2 233–4
HLA-DR
  coeliac disease 254
  discovery of 10–11, 12–13, 18
  epidemiological studies 15
  gene frequencies 15
  multiple sclerosis 221

  narcolepsy 225
  split specificities 413
HLA-DR2 219–20
  Dw2 association and narcolepsy 225
HLA-DRB
  matching and clinical transplantation 412–13
HLA-Dw2 219–20
Hodgkin's disease 15, 313
homozygous typing cells (HTC) 11
horseradish peroxidase (HRP) 454
house dust mite allergy 270
human immunodeficiency virus (HIV) 143, 304–7
  vaccine strategies 307
human leucocyte antigen (HLA)
  definition 35
  see also HLA
human papillomavirus (HPV) 313–14
hybridization probes 453
hydralazine 349–50
21-hydroxylase deficiency 245
hyperacute rejection (HAR) 392
hypersensitivity disorders
  lower respiratory tract 263–7
  pneumonitis 266
  vasculitis 335
hyperthyroidism 239

IDDM see insulin dependent diabetes mellitus
idiopathic
  intermediate uveitis 291
  membranous nephropathy (IMN) 203–4
    DQ alleles 205–7
    DR alleles 204-5, 206–7
    Goodpasture's disease 204
    HLA class II molecules 204, 206–7
    HLA-DR3 antigen 204–5, 206
  pulmonary fibrosis (IPF) 274–4
IKBL 42
immobilized probe method 456–7
immune privilege 279–81
immune response (Ir) genes 23
  class II molecule 53
  defects 28
  idiosyncratic effects 52–4
  phenomenon 14
immune system
  adaptive 51
  innate 51
immunogens 388–91
immunologic ignorance 279
immunoprecipitation 429, 438
incubation 437
infectious diseases 299–318
infectious mononucleosis 312
inflammatory bowel disease (IBD) 249
  and uveitis 290
inflammatory demyelinating polyneuropathies 224–5
insulin autoimmune syndrome 241
insulin dependent diabetes mellitus (IDDM) 119–23, 142, 231–5

heat shock protein 70-2 348
insulin gene (IDDM2) 120-1
MHC haplotype (IDDM1) 120, 123
non-obese diabetic mouse 120–3, 141–2
    epistasis 121–2
    H-2A 141–2
    H-2E 141–2
schizophrenia 420
see also type I diabetes
interferon-γ
    gluten-specific T-cells 256
intermediate uveitis 290–1
International Histocompatibility Workshops (IHW) 429
interphotoreceptor binding protein (IRBP) 281, 282
isocyanate-induced asthma 270
iso-electric-focusing techniques 429, 438
isoniazid 349

juvenile myoclonic epilepsy (JME) 226–7
juvenile rheumatoid arthritis (JRA)
    uveitis associated with 289–90

Kawasaki disease 332
kidney graft 405–6
kidney transplants
    cadaveric 407–9, 411
    organ allocation 415–16
Klebsiella 143

Leishmania sp. 303
leishmaniasis 303–4
leprosy 315–7
leucocyte antibodies
    agglutination of 4, 7, 8
    groups 4
    non-haemolytic transfusion reactions (NHTR) 5, 9
leucocyte specific transcript (LST-1) 42
lichen planus 375–6
limiting dilution analysis (LDA) 390, 442–5
linkage analysis 109–10
    autoimmune disease 110
    markers 109
    maximum lod score (MLS) 109
linkage disequilibrium for loci 112–14
    ESTs 113–14
    IDDM2 112
    IDDM7 113
    physical mapping 113
linkage studies
    elliptocytosis 134
    infectious diseases 300
    juvenile myoclonic epilepsy 226
    log odds ratio (lod score) 135
    multiple sclerosis 222
linked suppression 391
lipopolysaccharide S 282
lithium prophylaxis 422
live unrelated donors 410

liver transplants 397, 412
    organ allocation 415
lower respiratory tract, hypersensitivity disorders 263–7
lung
    immune system 263
    T lymphocyte density 263
Lyme disease 318
lymphatic drainage 279
lymphoblastoid cell lines (LCL) 311
lymphocytic hypophysitis 245
lymphocytotoxicity assay 433
lymphotoxin see TNFβ
lysophospatidic acid acetyltransferase (LPAAT) 40

major depression 420
major histocomaptibility complex see MHC
malaria 301–3
    vaccine strategies 303
manic depressive psychosis 420
Marek's disease virus (MDV) 176
massive pulmonary fibrosis 272
menopause 245
mesangial IgA nephropathy (IgAN) 211–12
    HLA alleles 212
    HLA-DR4 213
mesangiocapillary nephritis 214
methimazole 241
MHC 35
    associated diseases 107–8
    antigen presentation to T cells 24
    allorecognition 61
        direct pathway 61
        indirect pathway 61
    disease association 175
    microrecombination 172
    peptide-binding groove and polymorphism 27
    polymorphism 171–7
        molecular arms race 174–5
    point mutation 171
    restriction 14, 52
MHC class I 35
    antigen presentation 83
        exogenous antigens 93–4
        TAP-independent pathway 87–8
    antigen processing 83–9
        cytotoxic T lymphocytes 88–9
    calnexin 90–1
    calreticulin 90–1
    downregulation of class I MHC 94–6
    endoplasmic reticulum 89–1
    gpUS2 97
    gpUS11 97
    heavy chain-β2m complex 89–1
    HLA-A 35, 45
    HLA-B 35, 45
    HLA-C 35, 45
    HLA-E 37
    HLA-F 37
    HLA-G 37

hsp96 94
molecular mapping 35–6
OVA-specific CD8$^+$ T cells 93–4
proteosome 84–5
MHC class II 37–8, 144
    αβ heterodimer 79–81
    ABC (ATP-binding cassette) transporter superfamily
        TAP1 38, 45
        TAP2 38, 45
    antigen processing 77–8
    class II-associated invariant-chain peptide (CLIP) 79–81
    HLA-DM 81
    invariant chain 79–80
    LMP2 38, 45
    LMP7 38, 45
    MIIC (or CIIV) compartment 81, 82
    recycling of class II molecules 81–2
    RING3 38
    uptake of antigen 74
MHC class III 39
MHC-peptide complex (pMHC) 29, 446
MICA 43, 159, 336, 372–3
MICB 43
microlymphocytotoxicity 433–8
microscopic polyarteris/polyangiitis 332–4
microships carrying oligonucleotide probes 116
microtiter formats 457
microtoxicity test 6, 10
minimal change nephrotic syndrome, steroid sensitive (SSNS) 207–8
    class II alleles 210-11
    DQ alleles 208–9, 210
    DQ3 209, 210
    DR associations 209, 210
    HLA-DR7 alleles 208–10, 211
minor histocompatibility antigen
    group five system 9
MISIS 11
mixed essential cryoglobulinaemia 334–5
mixed lymphocyte culture/reaction (MLC/MLR) 9, 10–11, 441–2
    inhibition by HLA antibodies 11
molecular
    biology techniques 439
    mimicry 143
    typing techniques 412
monoclonal
    antibodies 436
    antibody-coated magnetizable microspheres 437
mood disorders 420–1
Mooren's ulcer 287
MS *see* multiple sclerosis
mucocutaneous lymph node syndrome 332
mucosal lesions 252
multiple sclerosis (MS) 219–22
    experimental allergic encephalomyelitis, model of 140

eye disease 284–5
family studies 135–6
pars planitis 291
MHC association 140
myelin basic protein (MBP) 140–1
relapsing/remitting 221
myasthenia gravis (MG) 222–4
mycobacterial infections 315–17
    vaccine strategies 317
Mycobacterium leprae 315
Mycobacterium tuberculosis 76, 315
myelin
    basic protein (MBP) 285
    oligodendrocyte glycoprotein (MOG) 285

nacolepsy 139, 225–6, 419
nasopharyngeal carcinoma 313
natural killer (NK) cells 37, 65–7, 163
    HLA molecules 65–6
    killer activating receptors (KARs) 66
    killer inhibitory receptors (KIRs) 66
    repression of NK cell activity 97
neoantigens 266
neonatal lupus erythematosus 350
nephropathia epidemica 314
neurological diseases 219–27
nickel hypersensitivity 265, 266
nomenclature 429–30
NOTCH-4 43
nylon wool column B cell depletion 437

occupational asthma 270
ocular
    immune privilege 279–81
    infection, models 282
    inflammatory disease 280
Onchocerca volvulus 304
onchocerciasis 304
optic neuritis (ON) 222, 284–5
    treatment trial (ONTT) 284
oral lichen planus 375
organ allocation 415–16

P4 pocket 367
P450c21 enzyme 244, 246
palmitoyl-protein thioesterase (PPT1) 43
panel cells 435
panic disorder 421
paranoid schizophrenia 419
parasitic infections 301–4
parkinsonism 422
pars planitis 291
pathogenic autoantibodies, PV linked HLA alleles 367–8
pemphigoid gestationis 369–70
pemphigus
    foliaceus 369–70
    vulgaris 366–9

penicillamine 368, 369
peptide-binding
  of DQ2 256
  high-affinity binding 256
peripheral
  blood lymphocytes 429
  nerve demyelination 224–5
permissible mismatches 409, 446
phagocytosis
  activated macrophages 76
  dendritic cells 76–7
  macropinocytosis 76
phosducin 282
phospholipids 40
Plasmodium falciparum 301
polyarteritis nodosa (PAN) 330
polymerase chain reaction (PCR) 115–16, 429
  amplification 451–9
    with sequence-specific oligonucleotide probes (PCR-SSO) 221
    with sequence-specific primers (PCR-SSP) 221
polymyalgia rheumatica (PMR) 327–8, 329
postpartum thyroiditis 243
post-streptococcal glomerulonephritis 214
preformed lymphocytotoxic antibodies 410
pregnancy 435
premature ovarian failure 245
preservation time 409
primarily chronic progressive multiple sclerosis 221
primary
  biliary cirrhosis (PBC) 251–2
  myxoedema 242–3
  sclerosing cholangitis (PSC) 251
primers 453
probes 454
procainamide 349
promiscuous recognition 307
prostaglandins 40
  arachidonic acid 40
proteosome 84–5
  class I MHC 85
  immunodominance of peptide determinants 88–9
  LMP2 84
  LMP7 84
  MECL-1 84
psoriasis vulgaris 370–2
psychiatric disease 419–23
psychotropic medication 422–3
pulmonary
  fibrosis 273–4
  tuberculosis 317
PV antigen 366

rabbit serum 437
RAGE 39–40
ragweed allergy 269
RD(D6S45) 348
reactive arthritis (ReA) 187
  HLA-B27 187
recoverin 281

regional immunology 280
ReLi HLA-DRB test 457
renal transplantation 394, 398
respiratory disease 263–74
restriction enzymes 453
restriction fragment length polymorphism (RFLP) 221, 453, 455–6
retinal
  disorders 292–4
  vasculitis 292
reverse hybridization assay 456–7
reverse-blot method 456–7
Rhesus (Rh) antibodies 5
rheumatoid arthritis (RA) 181–4
  HLA-DR4 alleles 182–4
  HLA-DRB1 181–3
  tumour necrosis factor (TNF) locus 181
  conjunctival disorders 287–8
  schizophrenia 420
rheumatoid pneumoconiosis 272
rhodopsin 281
rous sarcoma virus (RSV) 176
rye-grass pollen allergy 269

Sag *see* superantigens
Salmonella 143
sandwich immunofluorescence test 12
Sangstat 398
S-antigen 281, 282
sarcoidosis, pulmonary 267–9
schizophrenic disorders 419–20
sequence motifs 220, 452
  new 458
sequence-based typing (SBT) 221, 457
sequence-specific oligonucleotide probes (SSOP) 429, 454–5
sequence-specific primers (SSP) 429
sequence-specific priming 455
serial analysis of gene expression (SAGE) 116
serological testing 429–40
severe combined immunodeficiency 396
Shigella 143
sibling pairs 110–12
  identity-by-descent (IBD) 110–11
  identity-by-state (IBS) 110–11
    affected-pedigree-number (APM) 111
  polymorphic markers 111–12
sibling donor transplants 406–7
sibling risk 107
silica toxicity 272
silicosis 272–3
single-strand conformation polymorphism 455–6
single-hit kinetics 444
Sjögren's syndrome, primary 353–5
skin
  associated lymphoid tissue 365
  grafts 279
  immune system 365–6
sleep-onset rapid eye movement 421

solid organ
  allografts 393
  transplantation 441, 447, 458
soluble HLA (sHLA) 397–8
Southern blotting procedure 453
spinal muscular atrophy 114
spondylarthropathies 187–8
staining 437
  direct 446
staphylococcal infections 317–18
steroid 21-hydroxylase 43–4
stimulation index 441
streptococcal infections 317–18
Streptococcus sanguis 374
subcute
  cutaneous lupus erythematosus 350
  thyroiditis 243–4
sulfasalazine-induced SLE 350
summer-type HP 266, 267
superantigens 317–18
sympathetic ophthalmia (SO) 293
systemic lupus erythamatosus (SLE) 116, 139, 144, 337–53
  membranous nephropathy 203
systemic vasculitides 327–37

T cells 51
  alloreactivity 63–64
  antigen presenting cells 58
  Behçet's disease 374
  CD4$^+$ 57
  CD28 60
  CD40L 60
  clonal ignorance 59–60
  costimulatory interactions 60
  CTLA4 (CD152) 60
  gliadin-specific 257–8
  gluten-reactive 255–7
  gluten-specific 255–6
  high determinant density theory 61–2
  indirect allorecognition 64
    graft rejection 64
  multiple determinant theory 63
  peripheral tolerance 58
    clonal deletion 59
  self-non-self-discrimination 55
  αβ TCR T cells 56
  V$_\beta$ 6 T cells 56
T cell-enriching resetting technique 437
T cell receptors (TCRs) 54–5
  α chain 30
  αβ dimer 55
  β chain 30
  Fab fragment of antibody 29–30
  γδ dimer 55
  negative selection 57–8
  positive selection 58
  signalling 60
  TCR-transgenic mice 59
  variable domain 30, 55
  X-ray crystal structure 29
Takayasu arteritis 328, 330, 331
TAP 85–7, 89–92, 159, 347–8
  gp33 LCMV epitope 87–8
  gpUS6 96
  heavy chain-β2m complex 89–91
  HLA-A2 87, 88
  interferon (INF)γ 86
  TAP2 85–7, 89
  TAP-associated glycoprotein (tapasin) 89–93
    HLA-B class I molecules 92–3
  TAP-mediated translocation 86
TAP1 85–7, 89, 373
  Tapasin molecule 38
  Cim2 gene 39
  HLA-DM 37
  HLA-DO 37
  HLA-DP 37, 45
  HLA-DQ 37, 45
  HLA-DR 37, 45
tardive dyskinesia 422–3
TCR co-receptors
  CD4 29, 31
  CD8 29, 31
TCR-MHC-peptide interaction 30
Terasaki trays 436, 437
tetramethylammonium chloride 455
thrombophlebitis 374
Thygeson's superficial punctate keratopathy 288
thymoma 223
thymus 56
  negative selection of T-cell receptors 57–8
  positive selection of T-cell receptors 58
thyroid-associated ophthalmopathy (TAO) 239–41
tissue typing 398
  techniques 412
tolerogens 388–91
Tourette's syndrome 422
toxic shock syndrome 317
trachoma 286–7
trans-encoded DQ heterodimers 232
transglutaminase (TGase) 257
transmission disequilibrium test (TDT) 112
transplant arteriosclerosis 395
transplantation 279, 387–416, 435
  clinical 405–16
  immunobiology 279
transporter associated with antigen processing *see* TAP
Trichosporon cutaneum 266
Trypamosoma cruzi 143
tuberculosis (TB) 315, 317
tumor graft 279
tumour necrosis factor (TNF) ligand superfamily 42
  lymphotoxin-α 42

lymphotoxin-β 42
TNFα
　cerebral malaria 302
　systemic lupus erythematosus 346
TNFβ
　systemic lupus erythematosus 346–7
　polymorphisms 245–8
　　microsatellite polymorphisms 347
type I diabetes 231–5
　and the eye 284
　see also insulin dependent diabetes mellitus
typing reagents
　commercial availability 436
　procurement 435

U1-RNP antibodies 352
ulcerative colitis (UC) 249
uncomplicated pneumoconiosis 272
unipolar disorder 420
United Network for Organ Sharing (UNOS) 405
upper respiratory tract allergies 269–70
uveitis 281–2
　acute anterior uveitis (AAU) 289
　inflammatory bowel disease 290
　intermediate 290–1
　juvenile rheumatoid arthritis 289–90

vaccine strategies
　Epstein–Barr virus 313
　hepatitis B virus 309
　human immunodeficiency virus 307
　malaria 303
　mycobacterial infections 317
viral infections 304–15
viruses
　cytomegalovirus (CMV) 94, 96, 97
　downregulation of class I MHC expression 94–6
　evasion of immune surveillance 94
　herpes simplex virus (HSV) 282, 374
　　ICP47 96
　HIV-1 *vpu* 97
　prevention of viral antigen proteolysis 96
vitiligo 376
vitreous cavity disorders 289–91
Vogt–Koyanagi–Harada (VKH) disease 292–3
volunteer unrelated donors (VUD) 395, 415

Wegener's granulomatosis 213, 214, 332–4

XA gene 44
XB gene 44

Yates' continuity correction 130–31